HANDBOOK OF ELECTRONIC SECURITY AND DIGITAL FORENSICS

HANDBOOK OF ELECTRONIC SECURITY AND DIGITAL FORENSICS

edited by

Hamid Jahankhani
University of East London, UK

David Lilburn Watson
Watson Business Solutions Ltd., UK

Gianluigi Me
Università degli Studi di Roma "Tor Vergata", Italy

Frank Leonhardt
Independent Consultant and Commentator

World Scientific

NEW JERSEY · LONDON · SINGAPORE · BEIJING · SHANGHAI · HONG KONG · TAIPEI · CHENNAI

Published by

World Scientific Publishing Co. Pte. Ltd.
5 Toh Tuck Link, Singapore 596224
USA office: 27 Warren Street, Suite 401-402, Hackensack, NJ 07601
UK office: 57 Shelton Street, Covent Garden, London WC2H 9HE

British Library Cataloguing-in-Publication Data
A catalogue record for this book is available from the British Library.

HANDBOOK OF ELECTRONIC SECURITY AND DIGITAL FORENSICS

ISBN-13 978-981-283-703-5
ISBN-10 981-283-703-5

Typeset by Stallion Press
Email: enquiries@stallionpress.com

Printed in Singapore by Mainland Press Pte Ltd.

FOREWORD

The widespread use of information and Communications Technology (ICT) has created a global platform for the exchange of ideas, goods and services; the benefits of which are enormous. However, it has also created boundless opportunities for fraud and deception. Cyber-crime is one of the biggest growth industries around the globe, whether it is in the form of violation of company policies, fraud, hate crime, extremism, child pornography, or terrorism. It is therefore paramount that the security industry raises its game to combat these threats.

Today's top priority is to use computer technology to fight computer crime as our commonwealth is protected by firewalls rather than firepower. This is an issue of global importance as new technology has provided a world of opportunity for criminals.

Advancement in the field of information security and digital forensics are expected to continue increasing during the foreseeable future. The requirements of industry are varied, challenging and continually changing, with cyber security at the forefront of the knowledge explosion.

The handbook of Electronic Security and Digital Forensics is a compilation of collaboration between the researchers and practitioners in the security field and provides a comprehensive literature on current and future e-security needs across application, implementation, testing to Investigative techniques, judicial processes and Criminal Intelligence. The intended audience will include members in academia, public and private sector, students and those who are interested in and will benefit from this handbook.

Dr Hamid Jahankhani
London, England
November 2009

CONTENTS

Chapter 1

OS AND NETWORK SECURITY

ROBERTO DI PIETRO and NINO VINCENZO VERDE
Università di Roma Tre – Dipartimento di Matematica
L.go S. Leonardo Murialdo Roma – Italy

1.1. Introduction

Often, in past years, the computer systems were isolated from other systems; each system had its specific job and worked alone to execute it. Today, computer systems have grown exponentially both in number and in size. They, generally, are interconnected and collaborated to each other exchanging information and sharing resources. The problem with this pervasive presence of computer systems and networks is that malicious attacks can cause loss of data, can deny regular services provided by these systems, can cause disclosure of secret information and moreover can cause loss of money.

In this chapter, we will talk about two main topics: operating systems (OS) security and networks security. Both are needed in order to achieve a robust system able to resist different types of attacks.

A good starting point to achieve security is a secure OS. The OS manages directly the devices (CPU, memory, display, etc.); it is a layer between the user programs and the underlying hardware. A secure OS can protect the system using mechanisms like the reference monitor, host intrusion detection systems, access control, among others. In Sections 1.2 and 1.3, we will talk about this aspect, focusing also in the criteria to evaluate the OS security. In Section 1.4, the concept of hard existing OSs is introduced, since an harden system can assure more security and more reliability than a standard configurations.

Network security is considered in Section 1.5, introducing some of the typical attacks used by malicious users. Moreover, the main concepts and the main devices used to secure a network are introduced. Intrusion detection systems are analysed in Section 1.6 as well as the classification among host intrusion detection systems and network intrusion detection systems. It also provides a short indication about intrusion prevention systems. Section 1.7 concludes this chapter.

1.2. Operating Systems (OS) Security

The job of an OS is to manage devices (one or more processors, the memory, the display, some network interfaces, etc.) and hides the underlying hardware to the programmers. In a simplified architecture, it is a layer between the hardware and the system software like the command interpreter, compilers, editors and other applications. Many times, it is not clear what it is part of an OS and what it is not. Generally, we can say that the OS is the portion of the software that runs in *kernel mode* or *supervisor mode*, protected from user tampering by the hardware. Other applications run in *user mode*. This is not completely true, because there are applications that run in user mode but which help the OS and which perform privileged functions. This is the case of the application that allows the users to change their password. Naturally, this program is not part of the OS and does not run in kernel mode, but executes a sensitive operation and has to be protected by the OS.

At this point, we will analyse the internal structure of an OS that can be *monolithic*, *layered*, *virtual machine*, *exokernel* or *client-server*.

- A *monolithic system* is a system without a well-defined structure; it is a collection of procedures, each of which may use any of the other ones. Each procedure has well-defined input and output interfaces and each one is visible to every other procedure. In these systems, it is possible to have a little structure on three levels: there is a main procedure that invokes the requested service procedure, a set of service procedures and a set of utility procedures, each of which is used by one or more service procedures.
- A *layered system* is a generalisation of the previous approach. In this design, there is a hierarchy of layers, each one build upon the one below it. Generally, the first layer deals with the allocation of the processor, the interrupts and also other hardware operations. Other layers use the functionality of the one below it, the idea is that high layers are more abstract and so more useful for the end user.
- A *virtual machine* is composed building one or more abstract machines upon a real machine. Generally, abstract machines are an exact copy of the underlying system including kernel/user mode, I/O interrupts and everything else that the real machine has. In this way, each virtual machine can use different OSs.
- *Exokernels systems* are designed like virtual machines. Moreover, it is also possible to execute each virtual system in a protected environment so that they cannot

steal data or damage other ones. The exokernel is the program at the bottom layer; it allocates resources to virtual machines and then checks attempts to use them.

- *Client-server model*: It is often used in modern OSs. The main idea is to move the code up into higher layers and move away as much as possible from kernel mode, the final result is a microkernel. Usually, most of the functions of the OS are implemented in user processes. To request a service, a user process sends a request to a server process, the job of the microkernel is to handle the communication between clients and servers.

1.2.1. *Security Principles*

In the past years, researchers have identified some general principles that should be used as a guide when designing a secure system.

- The default should be no access. This means that generally when something is in doubt the system should answer "no";
- The system should be public. A secret system is not more secure than a public system, sooner or later an intruder will make the secret effort fruitless. Moreover, a public system can be tested and improved by any one, indeed it is easy to find and to repair security holes;
- Check for current authority. Many systems check for permissions when a file is opened and not afterwards, so a user that opens a file and keeps it open for a week will continue to have access, even if the owner have changed the file protection. The system should check for access every time that a permission change is executed and, more generally, every time that a critical operation is executed;
- The principle of least privileges: every user or process should operate using the minimum set of privileges necessary to do its job. This principle reduces damages that could result from an illegal or an accidental operation. Usually, this is achieved using a *complete mediation*: every access to any object must be checked. This is the task of the *reference monitor*;
- The protection mechanism should be simple and should be builded into the lowest layers of the system. It is nearly impossible to build a secure system upon an insecure one;
- The scheme chosen must be user-friendly and psychologically acceptable. If the operation of securing a file is too much difficult, probably a user just will avoid it and
- A complicated system cannot be secure. If the system is simple and it has a few guiding principles, then it has a chance to be secure. The main idea is that more code a system has, more bugs there will be.

Operating System security is not a trivial matter. In theory, it is possible to have a completely secure OS, but in practice such a system does not exist. There are many reasons. One is the flexibility of the system: commercial OS needs to be easily used without particular configurations. Moreover, a secure system is often a slow system; there are more checks and more operations to be executed than in a normal system. Often, security policies have to be relaxed in order to allow particularly task required by users.

An ideal design of a secure system must incorporate in an appropriate and formal way:

- The intended use of a system
- A definition of authorisation
- The objects that will be used and
- The kind of use required

Together, these elements form a formal abstract specification of a secure system.

1.2.2. *Reference Monitor*

In the past, not much consideration was granted to systems security, often for performance reasons. In the kernel, for example, structured programming was not used for many years to avoid the overhead associated with that style of programming. Only in the 1970's, the security requirement became important, especially to support-trusted functions and to be sure that desired functionality was implemented correctly. The reference monitor (RM) concept was introduced by the Anderson Report in 1972 [2]. The main idea is that it encapsulates all access mechanisms that support a certain security policy, so every process that wants access to a resource have to ask the permission to it.

Definition 1.1. A RM is an abstract machine that enforces the authorised access relationships between subjects and objects.

A RM must satisfy three properties [11, 9]:

(1) *Completeness: The access mediation mechanism must be always invoked.* This makes sure that any call to system resources is managed by the RM, in this way all the RM mechanisms are applied to all programs including the OS itself.
(2) *Isolation: The access mediation mechanism must be tamperproof.* Naturally, if the RM mechanism can be altered, its integrity cannot be guaranteed.
(3) *Verifiability: It must be small enough to be subject to analysis and tests, the completeness of which can be assured.*

As you can see in Fig. 1.1, subjects must use the RM to access the objects. The RM has access to the security kernel database that lists the access privileges of each subject and the protection attributes of each object; moreover, it will create an audit file to log any detected security violation and authorised changes. In other words, the RM takes charge of verifying whether an access request to an object

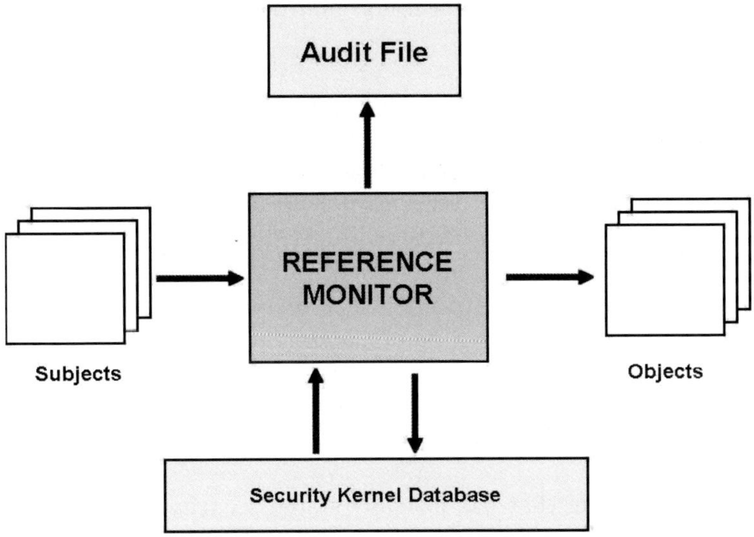

Fig. 1.1. Reference monitor.

(like sending a document to a printer or reading a file) is allowed, or the requesting process has permission to perform an operation on the object.

One drawback using a RM is that it is too complex and requires the developer to start with a totally new OS design. Moreover, early attempts to reproduce it in actual software and hardware are met with only minimal success, since introduces unexpectedly high overhead and a general system degradation [21, 19].

The RM is often part of a Trusted Computing Base (TCB), that is the core of a Trusted OS and will be introduced in Section 1.3.

1.2.3. *Discretionary and Mandatory Access Control*

Access control is the ability to permit or deny access to a specified resource by a particular entity. There are different approaches to build an access control mechanism in a system, two of these are: Discretionary Access Control (DAC) and Mandatory Access Control (MAC).

DAC: Bases access rights on the identity of the subject and identity of the object involved; it is the subject that sets access control mechanisms to allow or deny access to an object. In DAC, the owner of an object grants the access permissions to other subjects, an example of this model is the Access Control List (ACL). This is the simplest framework for describing a protection system: each object has an associated list containing all the entities that may access it. The owner of the object can change its rights to grant access to other users. There are different rights that can be considered, but generally are *read*, *write* and *execute*. Often, ACL is considered a group of users (a set of users) and permissions can be grant to users or to entire groups.

MAC: The subjects do not manage completely access permissions, but there exists a system mechanism that controls them. This mechanism will check information associated with the subjects and the objects to allow access. This means that users are constrained to follow system security policies and they cannot bypass them. Usually, these kind of access mechanisms are used to control information flows in military but also in commercial environment, preventing unauthorised disclosure of information or preventing its integrity. The two most known MACs are Bell La Padula and Biba: the first focuses on confidentiality; the second focuses on data integrity. Both will be introduced in Section 1.3.1.

1.3. Trusted OSs

A trusted OS is a system that has formally stated security requirements. It meets this requirements using a TCB that is the core of such an OS. Following Ref. 22,TCB consists of:

- Software (a portion of the OS kernel and all critical programs that have superuser power);
- Hardware and
- Procedural components that enforce the security policy.

In this way, if an attacker would break the system, it must subvert the TCB. For example, the TCB of a military system may include computer security mechanisms to assure access control and user authentication, a set of policies to restrict access to information in transit across the network, availability mechanisms such as backup procedures, etc. Mainly, the trusted computing base has five tasks: create processes, schedule processes, manage the memory, manage I/O and manage the file system. A crucial aspect on designing and implementing a TCB is its size: it should be kept as small as possible; this reduces security risks and it could be possible to give a proof of security of the system. An important component of the TCB is the RM. As we have seen in Section 1.2, it is the only place where security decisions are taken, with no possibility of bypassing it.

1.3.1. *Multi-Level Security*

Some systems give the control of objects to the hands of the person who creates them. The owner can change the permissions of the object, granting read, write and execute operations to other users or groups. As explained before, this kind of policy is called DAC. In others environments, where higher security is required, there are some rules stated by the organisation that all the users have to follow. This is the case of a MAC. In these systems, every user is assigned a defined level of authorisation. The MAC regulates the flow of information assigning security labels or classifications to system resources and allowing access only to users with the right levels of authorisation. Using MAC, users are not able to change the policy or to override it because it is centralised and controlled only by a policy administrator.

The most known mandatory model was introduced by Bell and La Padula [4]. This model was designed to be applied in a military environment but it is also applicable to other organisations. In this model, active objects are called *subjects* and passive objects are called *objects*; it could be that some objects are also subjects and vice versa. A current access by a subject to an object is represented by a triplet *(subject, object, access-attribute)*, where *access-attributes* are the operations that the subject can execute on the object (read, append, write, etc.). At each subject is assigned a clearance level and at each object is assigned a sensitivity level. Both of these levels are called *access classes.* An *access class* is composed by a security level and a category set. The first is a hierarchical classification like Top Secret, Secret, Unclassified; the second specifies the membership to a specified category, as example Nato, Nuclear or Crypto.

The formal method is based on two properties:

- Simple Security Property (also called SS-property): Informally, it says that a subject can read only objects that are in the same or in lower levels and
- $*$-property: Informally, it says that a subject can write only objects that are in the same or in higher levels.

In addition to the above, there is a third property: Discretionary Security Property. This property uses a simple access matrix to allow DAC. This policy allows an individual to extend access to a document based on his/her own discretion to any individual allowed by non-discretionary policy. This means that the discretionary policy cannot override the non-discretionary policy involved by the $*$ and SS properties.

Using these properties, a subject cannot read information on an higher level but it can read all data on lower levels (*no read-up*), moreover it can write only to same or to higher levels (*no write-down*), and this is shown in Fig. 1.2.

The Bell-La Padula model focuses on data confidentiality instead of a similar model proposed by Biba [5] that focuses on data integrity. In this model, a subject cannot read lower integrity level objects (Simple Integrity condition) and it cannot write on higher integrity level objects (Integrity $*$ property). This is exactly the dual problem of the Bell-LaPadula model [6]. Biba's model uses integrity labels. They are assigned and maintained separately from security labels, because the reasons behind the labels are different. Security labels primarily limit the flow of information; integrity labels primarily inhibit the modification of information. Let S be the set of subjects and O be the set of objects; if we indicate with $i : S \cup O \rightarrow I$, where I is the set of integrity levels, then the rules of this model are

Fig. 1.2. Bell-LaPadula.

as follows:

(1) $s \in S$ can read $o \in O$ if and only if $i(s) \leq i(o)$
(2) $s \in S$ can write to $o \in O$ if and only if $i(o) \leq i(s)$ and
(3) $s_1 \in S$ can execute $s_2 \in S$ if and only if $i(s_2) \leq i(s_1)$

Note that adding the notion of discretionary controls is the full dual problem of Bell-LaPadula model.

1.3.2. *Problem of the Covert Channels*

Bell-LaPadula and Biba's models have a formal proof of security but there exists a method that can be used to transfer information in a manner that violates the system's security policy, namely the *covert channels*. This method is generally used to bypass any multi-level security systems.

It uses a shared resource as the channel, it could be the file system, the memory or any other. For example, we can think to the processor in a multi-user system as this resource. A subject A would transfer an information to a subject B but the policy of the security model do not allow this transfer. A can transfer a bit to B in a simple way: it can use the processor intensively for a certain amount of time. B can check the usage of the processor and read 0 if it is not used intensively, 1 otherwise. In Ref. 16, is proposed a categorisation of the covert channels according to three aspects:

- *Scenarios*: There is a differentiation between storage and timing covert channels. In the first case, it is used either directly or indirectly a finite system resource shared among subjects with different privileges. In the other case, it is used as the modulation of certain resources in order to exchange information, as instance it can be used the CPU computing timing.

- *Noise*: Covert channels can be noisy, like any other communication channel. It is very rare to find a perfect channel where the probability of the receiver receiving exactly what the sender has transmitted is unity. For this reason, correction codes are often used. The drawback with this kind of codes is the narrowing of the bandwidth.
- *Information flows*: As in the conventional lines where are applied different techniques to increase the bandwidth, also in covert channel can be used them. Aggregated channels are channels in which there are many flows between the source and the receiver.

In the last few years, there is a growing concern about the use of covert channel in the TCP/IP protocol [13]. An attacker can use steganography to hide secret information in some fields of the TCP/IP packets, bypassing any security control in the network. He/she could use the packet header to embed some information in the last significant bit of a fields; it could be the packet identification field, the initial sequence number field, the acknowledged sequence number field, etc. Also, the attacker can use a sorting packet algorithm, known to the receiver to send restricted or secret data on the outside. It is a real challenge today to detect and fix this type of covert channel and many researches focus on this.

1.3.3. *Evaluation Criteria*

In this section, we give a sketch about criteria to evaluate the security of a computer system. There are many standards to quantify this parameter such as, Trusted Computer System Evaluation Criteria (TCSEC), European Information Technology Security Evaluation Criteria (ITSEC), Common Criteria (CC) and others.

The TCSEC [7] is commonly known as the *Orange Book*. It was developed by the National Computer Security Center in 1983 for the US Department of Defense and it is the formal implementation of the Bell-LaPadula model. The TCSEC was developed to achieve three objectives: *measurement*, *guidance* and *acquisition*. *Measurement* allows to compare different computer systems using a common metric. *Guidance* identifies standard security requirements that vendors have to build into systems to achieve a given trust level. *Acquisition* provides a standard for specifying acquisition requirements and identifying systems that meet those requirements. The system defines both functional and assurance requirements within the context of the evaluation classes, also TCSEC defines seven different evaluation classes: *D*, *C1*, *C2*, *B1*, *B2*, *B3* and *A1*, where *A1* is the higher security level with verified protection. The requirements considered to evaluate systems and to assign them a class are both functional and assurance requirements. Functional are the requirements that the finished product has and they may be:

DAC: Requirements address propagation of access rights, granularity of control and access control lists.

MAC: Not required until class B1, comprises the simple security condition and the *-property from the Bell-LaPadula Model. These requirements include a description of the hierarchy of labels.

Label Requirements: They are not required until class B1, enable enforcement of mandatory access controls. Both subjects and objects have human-readable labels.

Object Reuse: Memory and disc sector contents should not be transmitted to a new user.

Identification and Authentication (I&A): The system has the ability to identify and to authenticate users or groups, protecting authentication data and associating identity with auditable actions.

Trusted Path: It is not required until class B2, provide a communications path that is guaranteed to be between the user and the TCB.

Audit Requirements: These address the existence of an audit mechanism as well as protection of the audit data.

The assurance requirements refer to the development process of the product; they are listed below:

Configuration Management: Identification, correspondence mapping and documentation of configuration items and code for generating the TCB.

Trusted Distribution: Level $A1$ only. A controlled process from source code to customer delivery that protects the integrity of the product.

TCSEC System Architecture: Begins at $C1$. The aim is to keep the TCB small and simple; these requirements mandate modularity and minimisation of complexity.

Design Specification and Verification: Informal security policy model at $B1$. Top level specification and a formal security policy model at $B2$. System specification must be shown to meet the model at $B3$. Formal top level specification and mapping to the source code at $A1$.

Testing: Also a search for covert channels at higher levels.

Product Documentation: These requirements are divided into a Security Features User's Guide and a Trusted Facility Manual.

Following we will analyse each security class, starting from the less secure, describing for each one functional and assurance requirements.

Class $C1$ is called *discretionary protection.* As functional requirements, it only has I&A and discretionary access controls. The assurance requirements are also minimal, covering testing and documentation only.

Class $C2$ is called *controlled access protection.* In addition to functional requirements of the class $C1$, this class requires object reuse and auditing. Moreover, it contains a more stringent security testing requirements. This was the most commonly used class for commercial products.

Class $B1$ is called *labelled security protection.* It requires mandatory access controls, but these controls can be restricted to a specified set of objects. Labelling supports the MAC implementation. Security testing requirements are more stringent. An informal model of the security policy completes class $B1$. In the past, many OS vendors offered a class $B1$ product in addition to their primary products. Unfortunately, the $B1$ products did not always receive the updates in technology that the main line received, and they often fell behind technically.

Class $B2$ is called *structured protection.* In this class, MAC is required for all objects. Labelling is expanded, and a trusted path for login is introduced. Class $B2$ requires the use of the principle of least privilege to restrict the assignment of privilege to the users least required to perform the specific task. Assurance requirements include more stringent documentation, covert channel analysis, configuration management and a formal model of the security policy that has been proven to be consistent with its axioms.

Class $B3$ is called *security domains.* It implements the full reference validation mechanism, increases the trusted path requirements, constrains how the code is developed in terms of modularity, simplicity and uses of techniques such as layering and data hiding. It has significant assurance requirements that include all the requirements of class $B2$ and other more stringent testing, an administrator's guide and a design documentation.

Table 1.1. ITSEC functionality classes and evaluation levels

F class	E level	Description
NA	E0	Equivalent to TCSEC level D
F-$C1$	E1	Equivalent to TCSEC level $C1$
F-$C2$	E2	Equivalent to TCSEC level $C2$
F-$B1$	E3	Equivalent to TCSEC level $B1$
F-$B2$	E4	Equivalent to TCSEC level $B2$
F-$B3$	E5	Equivalent to TCSEC level $B3$
F-$B3$	E6	Equivalent to TCSEC level $A1$
F-IN	NA	TOE (Target od Evaluation)s with high integrity requirements
F-AV	NA	TOEs with high availability requirements
F-DI	NA	TOEs with high integrity requirements during data communication
F-DC	NA	TOEs with high confidentiality requirements during data communication
F-DX	NA	Networks with high confidentiality and integrity requirements

Class $A1$ is called *verified protection.* It has the same functional requirements as class $B3$. The difference is in the assurance, since this class requires significant use of formal methods in covert channel analysis, design specification and verification. It also requires trusted distribution and increases both test and design documentation requirements.

The European ITSEC, unlike TCSEC, addresses confidentiality, integrity, availability and the chance to evaluate an entire system called TOE instead of a single computing system [8]. The ITSEC classifies system evaluating functionality and assurance separately. There are 7 evaluation levels (E) and 10 functionality classes (F); the table in this page summarises this classification.

Common Criteria (CC) combines the best aspects of different source criteria (ITSEC, TCSEC and others) solving the conceptual and technical differences between them. It is a contribution to the development of an international standard and opens the way to worldwide mutual recognition of evaluation results.

1.4. OS Hardening

Operating system hardening is the process of securely configuring a system to protect it against malicious users and software, but also making the system more reliable. System hardening is necessary since some OSs are designed to be easy to use rather than secure. The OS hardening is not so simple as it seems to be; every system is different from any other one, depending on the role of the system in the network and the applications installed on it, their version, the patch applied, etc. Many researchers think that OS hardening is not what makes a system secure because it is difficult, or impossible to make secure a system that was not designed to be secure. They think that using OS hardening, it is possible to obtain a reasonably secure system where most of possible attacks are blocked.

The main idea of OS hardening is to minimise the exposure to current and future threats by configuring the OS and removing unnecessary applications. Generally, OS hardening is composed of the following steps:

(1) Patch the OS and its components.
(2) Enable higher security options, and disable lower security options.
(3) Turn off or disable unnecessary OS services.
(4) Enable appropriate logging options and
(5) Appropriately set user access permissions.

The first step assures that the OS is up-to-date and new discovered bugs are patched. In the second step, the administrator must analyse the security system option, installing or activating softwares that can increase system security. Third step is useful to restrict possible points of failure, if less services are active then less services could be attacked by a malicious user. Enable appropriate logging options can be extremely useful to detect intrusion activities. The latter step provides an higher degree of security.

In literature, there are many papers about OS hardening, each one of them analyses different aspects trying to increment security systems. In Ref. 23 the authors have addressed the problem to build a security enhancement architecture for Commercial Off-The-Shelf (COTS) OSs in practice a hardening architecture. It was designed to meet three goals: *security*, *compatibility* and *portability*. *Security* because it integrates into a COTS OS a MAC, an authentication system, an event auditing and other features required by the TCSEC class $B1$. Moreover it also integrates encryption algorithms and encrypting devices. *Compatibility* because neither source code nor binary level modifications are required and the added security mechanisms should be transparent to both users and applications. *Portability* because the architecture is OS independent to the most extent.

Prabhakaran *et al.* [18] address the problem of hardening disc failures. File systems and storage systems designers have assumed in past years, that discs operate in a fail stop manner; this means that they work perfectly until they are stopped to work, then they fail definitively. Modern disc drives are more complex because they could have only a block or a set of blocks inaccessible, or sometimes they could have silently corrupted blocks, other times they exhibit transient performance problems. Using IRON file systems, the authors show that redundancy techniques greatly increase the robustness of the file system under partial failures while incurring modest time and space overheads.

1.5. Network Security

Originally, the aim of the Internet was to provide connection among research and educational communities. This environment was considered trusted and security was not considered a priority. With the exponential growth of Internet and other private networks, security has become essential. All the original network protocols were not focused on security. The Internet Protocol (IP), for instance, do not has strong mechanisms to verify the contents of IP packets and the packet's source, so many attacks can be done using those weaknesses [17].

Considering network security, there are two main actors: the attacker and the victim. Usually, the victim is an organisation's network, a company or a private user. According to McHugh *et al.* [15], the attacker's point of view focuses in different questions:

- What is the target?
- What vulnerabilities exist in the target system?
- What damage or other consequences are likely?
- What exploit scripts or other attack tools are available?, and
- What is my risk of exposure?

On the other hand, the victim point of view focuses on mitigating these attacks, so he/she must answer to:

- What happened?
- Who is affected and how?
- Who is the intruder?
- Where and when did the intrusion originate?, and
- How and why did the intrusion happen?

If the victim answers all these questions, he/she will be able to solve his/her problems and to build a secure network. Note that any network is truly secure; this is because in many cases, there is the need to relax some policies to allow specific tasks, or simply because new attacks may be discovered and executed.

1.5.1. *Network Traffic and Network Intrusion*

Network traffic consists of million of packets exchanged among hosts on LAN or in the Internet. Packets introduced to consume resource uselessly, to interfere with some system function or to gain system knowledge (and then use this information) constitute network intrusions. Usually, the goals of a network intrusion is compromise integrity, confidentiality or availability of a specified resource.

A simple and historic example of network intrusion is the *land* attack. It is based on a weakness of early IP implementation protocol: a packet with identical source and destination IP addresses in some older OSs can cause a crash. The *smurf* attack is also simple as the land attack: the attacker spoofs the source addresses and sets it equal to the target system address, then it broadcasts an echo request to hundreds of machines that will answer to the target IP addresses.

Probably, the most known attack in network security is the Denial of Service (DoS). The target of this attack is to disrupt the service provided by a server or network. We can classify DoS attacks in two classes. In the first class, there are all attacks that use a combination of carefully crafted packets that exploit a software vulnerability, causing the reboot, the crash or the freeze of the target system. An example is the "ping of death": a large International Control Message Protocol (ICMP) ping packet is sent to the target system segmented into multiple fragment, in some system this causes many security problems. In the second class, there are all the attacks that use massive volume of useless traffic occupying all the resources of a target system. The attacked resource could be the CPU, the stack space in network protocol software, the internet link, etc. If the attacker is able to exhaust these critical resources, then it can prevent legitimate users from accessing the service.

Naturally, the volume of the traffic generated towards the target must be large enough to consume all the target's resources, often this is achieved using Distributed Denial of Service (DDoS). In this case, the attacker controls more than one computer to generate the traffic; usually, he/she, uses others compromised systems called *zombies* that after an attack command launch the attack against the target victim.

It is very difficult to prevent these attacks for many reasons: they can generate an amount of traffic that can exceed 10 Gb/s even if the traffic from each attack source is not conspicuous to constitute a powerful attack, the source field in the packets could be tampered, the attack could be geographically distributed making IP source trace-back extremely difficult. Generally, an attacker must compromise a large number of system to lunch an effective DDoS, but this is not a difficult task today using many different tools freely (but often not legally) downloadable.

1.5.2. *Networks Hardening*

Network hardening and networks designing can be easily confused. Designing concerns which services the network must provide and how the infrastructure to provide those services will be. A good network design is the starting point of a secure network. *Network hardening* is about making sure a network, protecting machines that cannot protect themselves. These two concepts are very closed because hardening can cause a network redesigning and vice versa. Network hardening is helpful when there is the necessity to relax security in some environments. A secure network would never interact with an untrusted system, but often that is not practical. In many networks, there is the need to access external untrusted resources from within the network, access internal resources from untrusted external systems, or exchange efficiently information within the network. To allow these information flows, network hardening could be used. Mainly, it can provide a central point where administrators can control internal or external access, it can keep data secure travelling across untrusted networks, it can prioritise the use of shared resources and it can also restrict access based on per-user authentication.

Typical hardening machines are the firewalls [1, 12]. The main goal of these machines is to filter packets travelling between an outside network and a private internal system, that could be a network or a computer. The firewall either accepts or rejects packets using the information in the header: if it does not meet certain fixed criteria, then it is rejected.

Also Network Address Translator (NAT) and Virtual Private Network (VPN) can be used to harden a network. The first allows a single device to act as an agent between the inside and the outside of a network, using only a single IP address to group the entire network. The NAT can be used to create a one-way trapdoor, where the traffic from the outside to the inside is allowed only if it was initiated by the inside. On the other hand, VPNs are used to allow secure connection from different end-point using an untrusted network like Internet. The VPN is often used by companies to connect all branches and exchange information securely.

1.5.3. *Network Security Analysis and Penetration Testing*

The best way to test the network security is to try to attack the network itself. A penetration test is a method used by many companies to find and correct security problems: it is also called *ethical hacking*. An ethical hacker would employ the same tools and techniques as the intruders, but his/her goal is to evaluate the target systems' security and report back to the owner the vulnerabilities found and possible solutions.

The first phase in this method is an active analysis of the system, mining any vulnerabilities that may result from poor or improper system configuration. According to the information available to the tester, we can categorise penetration tests in two classes: *black box* and *white box* testing. In the first case, the tester does not know anything about the infrastructure to be tested, he/she must gather all information to find possible weaknesses. This simulates an attack from someone who is unfamiliar with the target system. On the other hand, the white box testing assumes that the attacker has a complete knowledge of the infrastructure (network diagram, source code, IP address, etc.); this could be the case of an internal attacker that can easily have access to this information.

In the second phase, the tester exploits weakness found, using different techniques and tools. All data acquired in this phase is probably secret and must be protected by the tester. It will give back the entire documentation about exploit and weaknesses found to the customer.

Identifying the vulnerabilities that exist in a network helps the organisation to find an information security solution. Having this information, the target organisation may redesign the network, may solve security problems on the services provided, may add firewalls, Intrusion Detection Systems (IDSs), etc.

1.6. Intrusion Detection Systems (IDSs)

Definition 1.2. Intrusion detection is the process of monitoring the events occurring in a computer system or in a network, analysing them for signs of intrusions, reporting any unauthorised activities and helping to block them.

1.6.1. *Approach to IDS*

The intrusion detection problem is similar to the signal-detection problem. We can consider the intrusion manifestations like the signal to be detected and normal operations like the noise. In the classical problem, both signal and noise distributions are known, and a decision process must determine if the signal observed belong to noise distribution or not. Otherwise, in the intrusion detection approach, the decision is taken based on only one distribution because the other one is unknown: the signal or the noise. If it is known as the signal (the intrusion manifestation), the IDS process will be indicated as *signature-based*, otherwise if it is known as the noise distribution, the IDS process will be indicated as *anomaly-based* [14, 20]. Each approach has strengths and weaknesses, but both suffer from the difficulty of characterising the two distributions.

A *signature-based* IDS detects attacks using a possessed attack description that can be matched to sensed attack manifestations. It is possible to use different types of matching, from the simplest one that matches a portion of a network packets, to the most complicated one like a state machine or a neural network that can map the attacks to an abstract representation of them. If this representation is enough appropriate to describe an attack, then it could be used to identify previously unseen attacks that are abstractly equivalent to known patterns. The drawback is that they are unable to detect truly novel attacks and suffer from false alarms.

An *anomaly-based* IDS uses the known of the noise distribution to discover unusual or abnormal operations. It recognises as an intrusion any observation that does not appear to be noise alone, in this way the IDS can easily find novel attacks and then alert the administrator. The drawback is the necessity of training the system on noise and successively natural noise modifications: if a service or a system in the network changes, then it can cause false alarms, while intrusion that appears to be normal activities can cause missed detections.

From a system architecture perspective, the components of intrusion detection systems are: an audit data pre-processor, a detection engine, a decision engine and a knowledge base. Since system activities are observable, the audit data pre-processor reads the audit record and extracts activity data. The detection engine processes the data using some detection model and sends an alarm to the decision engine if the activity is abnormal. Decision engine, alarmed by the detection engine, uses a decision table to take an action or to send a report.

1.6.2. *Host-Based vs Network-Based*

The IDSs are often classified using the information source [3]. Some IDSs analyse information sources generated by an OS or by a software, this is the case of host-based IDSs (HIDSs). Others analyse network packets captured from network backbones, they are called network-based IDSs (NIDs).

The NIDS capture network packets, analyse them and report (and possibly stop) network level attacks. They listen to a network segment or switch, so one NIDS can monitor the network traffic affecting multiple hosts that are connected to the network segment. Often a NIDS is not a unique machine but it consists of a set of single purpose sensors placed in a network. Each of these monitors network traffic, performs local analysis and then reports attacks to a central management console. Often, sensors run in a stealth mode in order to make it more difficult for an attacker to determine their presence and their location. Moreover, these sensors are limited to running the IDS, so it is easy secure them against attacks. Using NIDS, attacks can be stopped before they reach hosts or applications but the drawback is that encrypted data cannot be read. Moreover, most NIDSs cannot know if an attack was successful; they can only know that an attack was initiated. This last factor is the main difference with host IDS.

The HIDSs analyse activities determining exactly which processes and users are involved in a particular attack. Unlike NIDSs, they can know the outcome of an attempted attack accessing and monitoring directly the data files and system processes targeted by the attack. It is important to know that an HIDS collects information from within an individual computer system, but some HIDSs are designed to support a centralised IDS management and a reporting infrastructure that can allow a single management console to track many hosts. Two types of data sources are used in HIDS: OS audit trails and system logs. Usually, the first ones are generated at the kernel level, are more detailed and better protected than system logs; system logs are much smaller and easily readable. HIDS can be categorised into three classes:

Filesystem Monitoring HIDS: This type of HIDS checks the filesystem using a trusted database. It can check the timestamp of a file, its size and other parameters. An example of filesystem monitoring HIDS is *TripWire* that will be introduced in Section 1.6.2.2;

Log Monitoring HIDS: This type of HIDS analyses system logs to discover illegal activities. An example of this HIDSs is *Swatch*;

OS Monitoring: This type of HIDS checks main activities of the OS using kernel modules. An example is LIDS or *grsecurity*.

The advantages of this type of IDSs are that encrypted data can be read, each host contributes to the detection process and it is possible to detect some types of attacks that cannot be seen by a NIDS.

1.6.2.1. *An Example of NIDS: SNORT*

In the open-source world, there are many network intrusion detection systems, securely the most known is SNORT. This software can efficiently analyse in real-time network packets and it can log information useful to a future debug. It can detect different attacks and probes, buffer overflow, stealth port scanning, SMB probes, etc. It can work in different ways:

Packet Sniffer: It is able to intercept packets on the net decoding the application level and cataloguing the traffic.

Packet Logger: It is able to log intercepted packets and then it can send an alert. It logs packets in a readable way, usually in a directory where files are categorised based on the source address.

Intrusion Detection: It is a powerful but lightweight IDS, which combines signatures and anomaly-based inspection methods. It is non-intrusive and easily configurable. Even more, it utilises familiar methods for rules development.

The SNORT uses a simple language to describe what kind of traffic must be captured or ignored and to detect what packets are suspicious. The core of SNORT is composed by rules. Each rule has a header and, sometimes, an option's field. The header specifies the actions to do, the protocol, the source and destination IP address and the ports. The options are the alert messages and the information about what packets section has to be analysed. The general form of a rule in SNORT is like this:

```
func protocol source_IP/mask source_port ->
  dest_IP/mask dest_port (option1;option2;...)
```

- `func` is the action: it can be alert (send an alert and log the packet), log (log the packet), pass (ignore the packet), activate (send an alert and then activate a dynamic action) or dynamic (is called by an activate action);
- `protocol` is the packet protocol: it can be TCP, UDP or ICMP;
- `source_IP/mask dest_IP/mask` are respectively the source IP address and the destination IP address with their mask. As instance `192.168.1.0/24` is the block of addresses between `192.168.1.1` and `192.168.1.255`;
- `source_port and dest_port` are respectively the source port and the destination port. It can be a single number or an interval and
- `->` is the traffic direction. In this case, it indicates the traffic from the source to the destination. It can be also `<-` (from the destination to the source) or `<>` (in both directions).

The main keywords that can be in the `options` field are:

- `content`: to find a pattern in the packet payload;
- `msg`: to print an alert message;
- `logto`: to redirect the output in a specified file;
- `ttl, tos, id, ipoption, seq, ack, flags`: to check the respective field in the IP packet;
- `dsize`: to check the payload size to prevent the buffer overflow;
- `ack`: to check the acknowledgement TCP field. It is used to detect NMAP TCP ping;
- `content-list`: to find a set of pattern in the packet payload and
- `resp`: to send an active response.

There are many other options that can be combined together to detect and classify interesting packets, in this way all options present in the rule must be true to generate the corresponding action. Using SNORT, it is possible to download rules from the Internet, containing the lasts signatures for the new attacks, but naturally every one can write his/her own rules.

1.6.2.2. *An Example of HIDS: Trip Wire*

TripWire is a filesystem monitoring HIDS developed in 1992 by Gene H. Kim and Eugene H. SpaKord. The main idea is that the administrator can choose what files the system must check, then TripWire executes a copy of some information about these files in a trusted database and then, when needed, it checks differences between the file system and the information stored in the database. Initially, many HIDS copied only the last modification date of the files to check its integrity, but TripWire uses cryptographic algorithms like MD5, SHA, DES and others to generate a digest of the file, then it uses this digest to find incongruences. To increment the security of the database, it uses digital signatures, in this way any manumission can be discovered when TripWire executes the check.

When TripWire is installed, the user must specify what files must be protected, TripWire will execute a snapshot of the system. Naturally, it is assumed that the system is in a secure state and there are not manumissions in that moment. When the user asks TripWire to check the system, it will execute a new snapshot. Then it will compare this snapshot with that one stored in the trusted database; there are two possible cases: there are some differences between them or there is any difference. The latter case is the simplest, the system is yet secure. If some differences are discovered, TripWire will produce a report, containing all differences found. Successively, the administrator can check this report and can decide if the modifications are legal or not. If these are legal, then the new snapshot will replace the one stored in the DB; if these are not legal, then the administrator can restore the system using the data stored by TripWire.

1.6.3. *Intrusion Prevention Systems (IPSs)*

The IPSs can be considered as an extension of IDSs. The IDSs are typically used to monitor potential intrusions passively, but IPSs are focused on identifying and *blocking* attack traffic. They can react in real-time to block and prevent malicious activities before it does any damage rather than simply raising an alert as the malicious payload has been delivered.

Beside features of any IDS, an IPS requires the following ones:

In-line Operation: Only working in-line, an IPS device can discard all suspect packets immediately, before that they reach the destination.

Reliability and Availability: An IPS failure can close a vital network path and thus causes a DoS condition. It must has an extremely low failure rate in order to maximise up-time.

Resilience: Cooperating with other in-line sensors in a fail-group will ensure that the IPS is not a single point of failure.

Low Latency: Packets should be processed quickly, in order not to slow down the entire network.

There are two main categories of IPS: Host IPS and Network IPS. Host IPS, as HIDS, relies on agents installed directly on the system being protected. It works together the OS kernel monitoring and intercepting system calls in order to prevent attacks as well as log them. It can also check file locations and registry settings.

The drawback with these mechanisms is the integration needed with the OS: upgrade could cause problems. Moreover, on host IPS agent must be very reliable, must not negatively impact performance and must not block legitimate traffic. Network IPS (NIPS) is sometimes known as an *In-line IDS* or *Gateway IDS* (GIDS). It combines features of a standard IDS and a firewall. The NIPS has at least two interfaces, one is internal and the other is external. When packets appear at the either interface is checked by the detection engine, if it is a malicious packet, the NIPS, in addition to raising an alert, will discard the packet and mark that flow as bad. Any other packet that makes up that particular TCP session will be discarded immediately. Legal packets are passed through to the second interface and on to their intended destination.

1.7. Summary

In this chapter, we have analysed different aspects of OS and network security. Both are needed to build a robust system able to detect and block attacks.

As we have seen, network security is able to block malicious packets before they arrive at the target host but some type of activities at this level are undetectable. To detect and try to stop attacks, often are used mechanisms like IDSs and IPSs. In both cases, the approach could be either *host-based* or *network-based*, distinguishing if the IDS (or IPS) is installed on a specified host or is a separated hardware relying on the network. Today, it is also used a kind of ethical hacking to test and try to penetrate in secure networks, helping designers and architects to find security hole. The OS security is needed to build a secure system, with mechanisms to discipline the access of the subjects on the objects. This is achieved using, for instance, a RM together with an access control mechanisms like a MAC or a DAC. Sometimes, especially in commercial OS, the easiest of use is preferred to the security, so many software and hardware are dedicated to hardening such systems.

Surely, the better way to reach a good degree of security is to combine OS and Network security, also applying hardening in both them. Recently, a lot of value is given to the Autonomous Decentralised System (ADS). The main idea is to build an open and common mechanism for communicating between different devices that are used to secure the networks [10]. Actually, many companies are already implementing these concepts, but in a proprietary and independent fashion. Using an open standard, these proprietary programs could inter-operate and live together in a shared network environment. Many researchers think that ADSs are the next step in the security network environment.

1.8. Glossary

Anomaly-based IDS: It uses the known of the noise distribution to discover unusual or abnormal operations.

Common Criteria (CC): It is a contribution to the development of an international standard of evaluation results of OSs

COTS: Commercial Off-The-Shelf OS.

Covert channels: Method used to bypass multi-level security systems using a shared resource.

Dicretionary access control (DAC): With this poilcy, it is the subject that sets access control mechanisms to allow or deny access to an object.

Ethical hacking: It is used by many companies to find and correct security problems using the hackers' works.

Host-Based IDSs: It is an IDS installed on an host that, accessing and monitoring directly the data files and system processes, finds and reports attacks.

Intrusion Detection Systems (IDSs): It is the process of monitoring the events occurring in a computer system or in a network, analysing them for signs of intrusions, reporting any unauthorized activities and helping to block them.

Intrusion Prevention Systems (IPSs): They are focused on identifying and locking attack traffic

Land attack: The simplest attack used in network intrusion, is based on a weakneass of the early IP implementation protocol.

Mandatory access control (MAC): The subjects do not manage completely access permissions, but it exists as a system mechanism that controls them.

NAT: Network Address Translator, often used to hard a network.

Network-Based IDS (NIDS): Listening on a network segment, a NIDS capture network packets, analyse them and report (and possibly stop) network level attacks.

OS Hardening: It is the process of securely configuring a system to protect it against malicious user and software, but also making the system more reliable.

Reference Monitor (RM): It is an abstract machine that enforces the authorised access relationships between subjects and objects.

Signature based IDS: It detects attacks using a possessed attack description that can be matched to sensed attack manifestations.

TOE: Target of evaluation, used in evaluation of OSs, spacially in ITSEC.

Trusted Computing Base (TCB): It is composed by software, hardware and procedural components that enforce the security policy.

References

1. J. Albanese and W. Sonnenreich, *Network Security Illustrated* (McGraw-Hill, 2003), ISBN 0321247442.
2. J. P. Anderson, Computer security technology planning study, *Tech. Rep.* Volume II (1972) Electronic Systems Division, Air Force Systems Command, Hanscom Field, Bedford, http://csrc.nist.gov/publications/history/ande72.pdf.
3. R. Bace and P. Mell, Intrusion detection systems nist special publication on intrusion detection systems, *National Institute of Standards and Technology* (2001).
4. D. E. Bell, and L. J. La Padula, Secure computer systems: Mathematical foundations and model, *Tech. Rep.* M74-244 (1973) The MITRE Corporation, The MITRE Corporation, Bedford, MA, the basis of multi level security.
5. K. J. Biba, Integrity considerations for secure computer systems, *Tech. Rep.* ESD-TR-76-372 (1977) USAF Electronic Systems Division, Hanscom Air Force Base, Bedford, Massachusetts.
6. M. Bishop, *Introduction to Computer Security* (Addison-Wesley Professional, 2004), ISBN 0321247442.
7. Dod, Trusted computer system evaluation criteria (orange book), citeseer.ist.psu.edu/632050.html (1985), department of Defense Standard 5200.28-STD.
8. M. Gehrke, A. Pfitzmann and K. Rannenberg, Information technology security evaluation criteria (itsec) — A contribution to vulnerability? *Proceedings of the IFIP 12th World Computer Congress on Education and Society — Information Processing '92 — Volume 2* (North-Holland Publishing Co., Amsterdam, The Netherlands, 1992), ISBN 0-444-89750-X, pp. 579–587.
9. V. D. Gligor, 20 years of operating systems security, *sp* **00** (1999) p. 0108, doi: http://doi.ieeecomputersociety.org/10.1109/SECPRI.1999.766904.
10. Y. Guo and C. Wang, Autonomous decentralized network security system, *Networking, Sensing and Control* (2005), pp. 279–282, doi: http://doi.ieeecomputersociety.org/10.1109/ICNSC.2005.1461201.
11. C. E. Irvine, The reference monitor concept as a unifying principle in computer security education, (1999). URL citeseer.ist.psu.edu/299300.html.
12. B. Komar, J. Wettern and R. Beekelaar, *Firewalls for Dummies* (John Wiley & Sons, Inc., New York, NY, USA, 2003), ISBN 0764540483.
13. D. Llamas, C. Allison and A. Miller, Covert channels in internet protocols: a survey, *Proceedings of the 6th Annual Postgraduate Symposium about the Convergence of Telecommunications, Networking and Broadcasting (2005), PGNET 2005* .
14. G. A. Marin, Network security basics, *IEEE Security and Privacy* **3**(6) (2005) 68–72, doi: http://dx.doi.org/DE65FB44-50A6-4739-AAAC-64C11E07B30E.
15. J. McHugh, A. Christie and J. Allen, Defending yourself: the role of intrusion detection systems, *IEEE Softw.* **17**(5) (2000) 42–51, doi: http://dx.doi.org/10.1109/52.877859.
16. NCSC, A guide to understanding covert channel analysis of trusted systems, version 1, NCSC-TG-030, Library No. S-240,572, (1993) tcsec Rainbow Series Library.
17. T. Peng, C. Leckie and K. Ramamohanarao, Survey of network-based defense mechanisms countering the dos and ddos problems, *ACM Comput. Surv.* **39**(1) (2007) 3, doi: http://doi.acm.org/10.1145/1216370.1216373.
18. V. Prabhakaran, L. N. Bairavasundaram, N. Agrawal, H. S. Gunawi, A. C. Arpaci-Dusseau and R. H. Arpaci-Dusseau, Iron file systems, *SIGOPS Oper. Syst. Rev.* **39**(5) (2005) 206–220, doi: http://doi.acm.org/10.1145/ 1095809.1095830.
19. T. Rooker, The reference monitor: an idea whose time has come, *NSPW '92-93: Proceedings on the 1992–1993 Workshop on New Security Paradigms*

(ACM, New York, NY, USA, 1993), ISBN 0-8186-5430-9, pp. 192–197, doi: http://doi.acm.org/ 10.1145/283751.283847.

20. R. Sekar, Y. Guang, S. Verma and T. Shanbhag, A high-performance network intrusion detection system, *CCS '99: Proceedings of the 6th ACM Conference on Computer and Communications Security* (ACM, New York, NY, USA, 1999), ISBN 1-58113-148-8, pp. 8–17, doi: http://doi.acm.org/10.1145/319709.319712.
21. V. Skoularidou and D. Spinellis, Securing the network client, *Proceedings of the Third International Network Conference INC '02*, 2002, pp. 389–396, URL http://www.spinellis.gr/pubs/conf/2002-INC-Sec/html/inc2002.html.
22. A. S. Tanenbaum, *Modern Operating Systems* (Prentice Hall PTR, Upper Saddle River, NJ, USA, 2001), ISBN 0130313580.
23. C. Zemao, L. Yi, S. Changxiang, L. Jingchao and Z. Libing, A security enhancement architecture for cots operating system, *isdpe* **0** (2007) 434–438, doi: http://doi.ieeecomputersociety.org/10.1109/ISDPE.2007.28.

Chapter 2

AUTHENTICATION

ROBERTO DI PIETRO and NINO VINCENZO VERDE
Università di Roma Tre-Dipartimento di Matematica
L.go S. Leonardo Murialdo Roma-Italy

2.1. Introduction

In any commercial or private service, if is essential to assure some degree of security. In consequence, there is the need to identify the parties involved in any process and application. This is the first step towards communications and information's security and it is called *authentication*. Basic actors in an authentication process are a system and an entity. The system, using a set of credentials provided by the entity, ascertains that this is really what it claims to be. These processes are largely used to authenticate users, messages, images and also other processes in many different environments. Every day, we use authentication: for instance, it is needed to show the passport to use a credit card, we use a remote control to open a car and we use a PIN to turn on the phone. All these processes are based on the fact that the system (in these cases the merchant, the car or the phone) would confirm the user identity after it provides some additional information (the passport, the remote control or the PIN). Usually, authentication is combined with the authorisation, that is the process to find out if the entity, once identified, has the rights to use the resource.

The rest of the chapter is organised as follows. This section introduces the main concept used in the rest of the chapter. Section 2.2 shows the authentication methods in standard systems, particulary we will refer about password, challenge-response protocols, biometric and location-based authentication. Section 2.3 considers a classification of authentication protocols in standard systems, distinguishing between authentication with symmetric key (using or not a third party) and authentication with public key. In Sections 2.4 and 2.5 respectively, we will deal with authentication in GSM networks and authentication of multimedia contents.

In Section 2.6 we will propose some research directions. Section 2.7 summaries this chapter.

2.1.1. *Principles of authentication*

Authentication is the process with which a system identifies an entity for certain. The entity could be a user, a message, an image or other subjects. An authentication scheme provides the way to achieve the binding between the entity and his/her identity in the system, using one or more of the following information.

(1) What the entity knows (a password, an algorithm or a secret information)
(2) What the entity has (a badge or a card)
(3) What the entity is (fingerprints, voice or retinal characteristics) and
(4) Where the entity is (a room, a city or a country)

For example, we can think the entity as a message sent by a user U. The system or other users want to be sure that this message was sent by U and that no one could alter it after the send. Naturally, an attacker should not be able to forge a message which the receiver will accept as authentic. It is implicit that only a fraction out of the total number of possible messages are accepted as legal and that the transmitter only use a sub-set of these; this is an essential feature in all authentication schemes. The conditions that determine what is the sub-set of authentic messages that can be created and successively accepted are determined by the specific authentication scheme. Usually, this involves some form of cryptographic operation that transmitter and receiver can do since they know a secret key unknown to any other subject.

2.1.2. *Authentication, authorisation and accounting*

The term Authentication, Authorisation and Accounting (AAA) denotes a framework for controlling access to resources, enforcing policies and providing the information needed by billing services.[9] These three aspects are fundamental for any IP network because they provide the tools necessary to ensure the proper use and management of resources.

Using a set of criteria, the *authentication process* allows the user to be identified by the system. Usually, the system compares the credentials received by the user with those stored in a database and, if they match, the user is granted access to the network, otherwise authentication fails and network access is denied.

When a user is authenticated, then he/she must gain authorisation for doing certain tasks: this is the aim of the *authorisation process.* After logging into a system, for instance, a user may try to issue some command. Is the authorisation process that determines if the user has the authority to issue such commands. More generally, this process determines what activities, services or resources are permitted to the user.

Accounting measures the quantity of resources that a user consumes. This could be the time of connection, the amount of data transmitted or other parameter useful to billing, trend analysis, capacity planning activities and resource utilisation.

We are interested specially in authentication because from it depends authorisation and also accounting. Naturally, if we know for certain the identity of a user U we can authorise and account his/her operations and we can be sure that no other one can impersonate U.

Often AAA are provided by a system called AAA server that performs all these functions. Today, probably, the most used is Remote Authentication Dial-In User Service (RADIUS).[16]

2.1.3. *Cryptographic pre-requisites*

Cryptography is widely used in authentication protocols. It is the science to maintain secret information that one does not want to publish, in the way that only one person or a group of people can read it. Information are hidden using the *encryption* process; the operation of calculation the plaintext starting from the ciphertext is called *decryption*. It exists in two types of cryptography, one based on symmetric keys and the other one based on asymmetric keys.

In symmetric key cryptography, the encryption key K and relative decryption key K^{-1} are equal and used as an input of a unidirectional invertible function. To achieve the secrecy of the message, this key must be known only by the transmitter T and the receiver R. This is because any entity that knows K can decipher the message and also forge other correct messages. The most known protocol using symmetric key cryptography is called Data Encryption Standard (DES).[13] The main problem with this type of cryptography is the exchange of the key K between T and R.

Asymmetric cryptography solves this problem. In this system, there are two keys: one of this is secret and is used by the owner to cypher messages, the other one is public and used to decipher messages. The most known protocol that uses asymmetric keys is RSA, from the name of its inventors: Rivest, Shamir and Adleman.[17] In RSA, the public key pk is the pair (d, N), the private key sk is equal to (e, N). N is the k-bit product of two big integers p and q (usually k is bigger than 1,024). The integers d and e are in Z_N^* and they must satisfy the following conditions:

$$de \equiv 1 \pmod{\Phi(N)}, \quad \text{where } \Phi(N) = (p-1)(q-1).$$

d is such that:

$$gcd(d, \Phi(N)) = 1.$$

where $gcd(a, b)$ denotes the greatest common divisor of a and b. Given a message m, the relative ciphertext is equal to $[m]_{sk} \equiv m^d \pmod N$. To decipher the message, the key e is used $m \equiv (m^d)^e \pmod N$.

2.1.4. *Classification of authentication schemes*

In designing an authentication schemes, the main point is the security of the system. Often, the designer needs to find the right trade-off between the strength of the security and further important qualities, such as efficiency and practicality. According to the security level, authentication schemes can be categorised into three groups [18]: *computationally secure*, *provably secure* and *unconditionally secure.*

2.1.4.1. *Authentication schemes computationally secure*

Schemes computationally secure are based on the amount of the computational work required by an attacker to break the system using the best currently known methods. The main idea is that these schemes can be broken trying all the possible keys in sequence but the time needed to find the correct key on the fastest today's computer would be months, years or millions of years. Naturally, the security of these schemes is bounded with the development of new cryptanalytic techniques. An example of this kind of scheme is DES: it is possible to break it trying the possible keys in sequence, but in practice such attacks are considered unfeasible.

2.1.4.2. *Authentication schemes provably secure*

Schemes provably secure are based on security proofs that assume the supposed intractability of an underlying hard problem. This is the case of RSA or ElGamal, where the security is based on factorising large numbers or computing discrete logarithms in certain finite groups. Today, these problems are considered hard to solve but nobody has found a proof for this yet. The difference between *provably* and *computationally secure* schemes is that the first must have a proof of security proving that the scheme is secure at least as the underlying problem. This means that it is proved that if an attacker wants to subvert the protocol, it must solve the equivalent hard problem. For easiness, in literature, these schemes are usually combined, speaking only of *computationally secure schemes.*

2.1.4.3. *Authentication scheme unconditionally secure*

Schemes unconditionally secure are based on information theory. They cannot be broken even if all keys could be tried within short time, imposing no limit on the adversary's computational power. Speaking about authentication, often these schemes are called *authentication codes.* They are in a strict sense, a mathematical dual error detecting and correcting codes. In both cases, redundant information is introduced into the sequence of symbols that are transmitted, resulting in only a fraction out of the set of all possible sequences being available for use by the transmitter[18]. Authentication codes introduce redundancy in the message in such a way that for any message the transmitter may send, the altered messages that an attacker can forge are spread as uniformly as possible over the set of all possible messages.

2.2. Standard Techniques

Authentication consists of a binding between an entity and a subject in the system. The authentication process, determinate if data provided by an entity is really associated with that entity. An authentication system consists of five components:

- A: Authentication information. That is the set of information used to prove the identity of an entity;
- C: Complementary information. The set of information stored in the system and used to validate the authentication information;
- F: Complementation functions. That is the set of complementation functions that generate the complementary information from the authentication information. For $f \in F, f\colon A \to C$;
- L: Authentication functions. The set of the functions that verify an identity. For $l \in L, l : A \times C \to true, false$ and
- S: Selection functions. This is the set of functions that provide the entity the way to change or alter the set A and the set C.

2.2.1. *Password systems and attacks*

The simplest way used to authenticate an entity is a password, namely a sequence of characters. Systems using a password are based on what the entity knows: if the password provided by the entity is the same one stored in the system, then the entity is authenticated, otherwise it is not. We will see now two password systems, the first one is the simplest and the second is the one used by UNIX.

In the simplest password system, the set A is the set of all sequences of characters that can be passwords and if the passwords are stored in clear, the set C is identical to A. F is the singleton set of the identity function and L is composed by the element *eq*, namely the equality test. S is the function to change the password. It is easy to understand that the system has to store somewhere the set C of complementary information. If this set contains the passwords in clear, an attacker could get this file and read every password associated to any entity.

In UNIX, the set of complementation functions is composed by 4,096 hashing functions based upon a permutation of the DES. A password is a sequence of up to eight ASCII characters, so the set A is composed by approximately 6.9×10^{16} elements. The set C contains strings of 13 characters: 2 to identify the used hash function f and 11 that contains the output of the function f. Usually, UNIX system stores these strings in the file*/etc/passwd* that is readable to all users. Other UNIX versions use shadow passwords, which implies that only the superuser can read the file containing these strings. The main authentication functions in UNIX are *login* and *su* and the main selection functions are *passwd* and *nispasspwd*.

In order to find a password, the simplest attack is the *dictionary attack*. In this type of attack, a malicious user (the attacker) guesses the password repeating trials and errors. We can classify these attacks into two types, discriminating if the complementary information or the complementation functions are available or not.

- *Off-line*: The attacker takes each guess g and $\forall f \in F$ computes $f(g)$. If $f(g)$ corresponds to the complementary information for entity E, then g authenticates E under f.
- *On-line*: If complementary information or complementation function are unavailable, the attacker uses authentication function $l \in L$. If the guess $l(g)$ returns true, g is correct.

In a password system, the probability that an attacker guesses a password in a specified period of time T is greater than $\frac{TG}{|A|}$, where G is the number of guesses that can be tested in one time unit. Using this formula, due to Anderson, we can evaluate, for example, the minimum length of a password to assure a fixed probability to discover a password in time T. This formula assumes that the time to test a password does not depend on the password itself and also that every password has the same probability to be chosen. If the first assumption is reasonable, the second one is not always assured.

2.2.2. *Probability distribution of passwords and relative problems*

The effectiveness of a *dictionary attack* is related to the size of the set P; naturally, if P is too small, the authentication scheme will be vulnerable to a dictionary attack. If it takes time T to test a guess and S chooses the passwords on an uniform random distribution, then the expected time X for this type of attack is $E(X) = \frac{1}{2}|P|T$.

But a *dictionary attack* could be effective also under other conditions. When the password is selected manually by the user, usually it is easily guessable. Many times, it is a simple word based on his/her name, his/her birth-place, etc. Usually, these passwords are very short and composed by only letters, numbers or a combination of those. In this case, the passwords are not uniformly distributed over the set A and an attacker will have a high probability of guessing a correct password if he/she starts with the most probable passwords. In Fig. 2.1 are shown two possible distributions of passwords. The axes *Prob* reflect the probability of a password being chosen, the

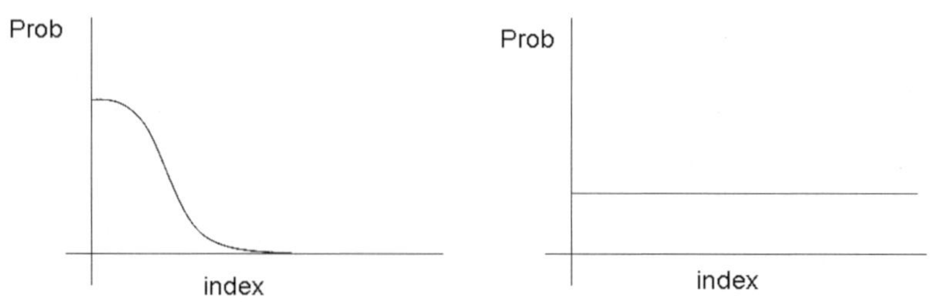

Fig. 2.1. Two possible distributions of passwords.

axes *index* reflect an ordering of the passwords in the set of possible passwords. The area under the two lines is 1, but on the left is distributed only on some passwords, on the right any password has the same probability to be chosen. To avoid this problem, we could select a random password for the user. Following this way, there are other problems: naturally, the password will not be so easy to remember and probably too long, we could leave out this but this is not the only case when a uniform distribution in not assured[3]. If the password is chosen at random, we have to pay attention to the period of the pseudo-random number generator, indeed if it long enough, many of the possible passwords $p \in A$ will not ever be selected. Otherwise, if the period of the generation function is the same as $|A|$, then the distribution induced by S will be as generation function's one.

2.2.3. *Challenge-response*

Suppose that an attacker sniffs a password, then he/she can replay the password and authenticate itself on the system. The main problem with passwords is that they are reusable. This suggests that the transmitted password must change each time, so if the attacker tries to replay the sniffed password, then the system will reject it. This is the main idea of the challenge-response systems shown in Fig. 2.2. In this authentication scheme, the user U and the system S share a secret function f. When the user sends a request to authenticate, the system answers with a challenge, namely a random message r. Successively, the user sends to the system the value $f(r)$. In this way, the secret is never transmitted and the system can check if the value received from u is really equivalent to $f(r)$, in this case, it authenticates the user U. This is the main concept used in Ref. 19.

In some algorithms, password is valid for only one use; these algorithms are called one-time password[7]. The challenge-response algorithm is actually one-time password considering the challenge as the number of authentications and the response as the password for that number. The main problems with this kind of mechanism are the synchronisation of the user and the system, and the generation of random passwords.

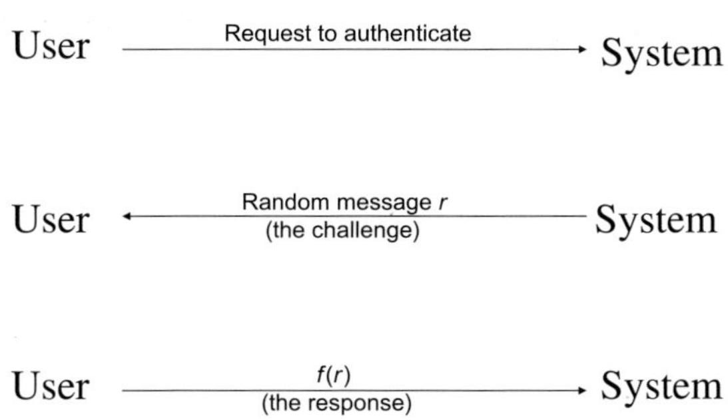

Fig. 2.2. Challenge-response mechanisms.

If the attacker knows the challenge, the response and the function f, challenge-response algorithms are vulnerable to dictionary attacks. On this assumption, the *dictionary attack* works exactly as in the password systems. To avoid dictionary attacks, Bellovin and Merritt[1] proposed a technique called encrypted key exchange. The idea is that random challenges are never sent in clear so that the attacker cannot verify when he/she has correctly deciphered them.

2.2.4. *Biometric authentication*

Biometric is the automated measurement of some biological features that can uniquely identify a person. Usually, these systems are based on approximations of the feature measured, that could be eyes, fingerprints, voices, etc.

2.2.4.1. *Fingerprint authentication*

Probably, fingerprints are the most used biometric authentication methods. They can be scanned optically and converted into a graph. Then a graph, matching algorithm based on isomorphisms, is applied in order to authenticate the user. Because there are imprecisions in measurements, the matching algorithm is an approximation.

2.2.4.2. *Voice authentication*

Often, biometric authentication is achieved by voice authentication. These methods are based on voice characteristics or verbal information verification. The first case uses some statistical techniques to test if the speaker is really the one he/she says to be. In order to authenticate, the user says the pass-phrase or a set of words. In the second case, the system asks a set of questions and checks if the answers of the user match with those in the database. The difference between these two types of voice authentication is that the former is speaker-dependent, but the latter is speaker-independent and relies only on the content of the answers.

2.2.4.3. *Eyes authentication*

Eyes authentication can be achieved using the iris and the retina. The retinal scan is highly intrusive because it uses a laser to check the patterns made by blood cessels at the back of the eye. It is used only in the most secure systems. The iris authentication, instead, is achieved comparing the pattern of the iris that is unique for each person. It is less invasive than the retinal scans and it is more used.

2.2.4.4. *Faces authentication*

This type of authentication relies on the recognition of faces. The differences in lighting, the distortion, the noise can cause problem using this authentication method. If the user places his/her face in a pre-determined position, the problem becomes somewhat easier but also hairs and glasses may make the recognition harder. In some cases, it is checked some facial features such as the distance between eyes, or distance between the nose and the chin or the angle of the line drawn from one to the other.

2.2.4.5. *Keystrokes authentication*

This type of authentication requires a signature based on keystroke intervals, keystroke pressure, keystroke duration and where the key is struck. Keystroke recognition can be both static and dynamic. Static recognition is done during authentication time, usually, typing a fixed string, but then an attacker can capture the connection or take over the terminal without detection. If the recognition is done throughout the session, then it is called dynamic[8]. Naturally, the signature must be chosen so that variation within an individual's session do not cause the authentication to fail. This is not absolutely trivial.

In the last years, many researches have combined different techniques of biometric authentication, achieving an higher degree of accuracy compared with systems that use only one biometric measure.

2.2.5. *Location-based authentication*

This authentication method is based on the position of the entity that wants to be authenticated in the system. Suppose that a user usually logs in from his/her house in Rome, at this point the system can deny any other connection coming from other cities. Naturally, to use this authentication, we need special-purpose hardware, namely a Global Position System (GPS) and a Location Signature Sensor (LSS)[5]. With the help of this hardware, the system is able to know where the user is at any time. The GPS position signal is signed together with the information of the time and it is sent to the system to authenticate the user. This signature is created by a LSS. Note that the physical location of a particular user at any time is unique, moreover the signature and its derived location are virtually impossible to forge. The main advantage with these techniques is that even a user's location signature is stolen, it could never be used elsewhere to gain access to the system. Moreover, the attacker cannot execute a reply attack because if it intercepts a location signature transmitted during login, it cannot use this information from any other place to gain an authorised access to the system.

2.2.6. *Mixed authentication*

All the authentication methods that we have already seen can be combined and used together. Naturally, for an attacker, it is more complicated to break different authentication methods because it should know more information. Mixed authentication methods combine, for example, what a user knows with what he/she has or where the user is with what he/she has and so on.

It is possible to use different authentication methods for different purposes to assure that delicate jobs are more secure that other ones. In this way, we can create a kind of hierarchy where the most important entity, the one with more privileges, must login using many different authentication techniques. On the other hand, entities without any privilege can login using only a method (for example, a password).

2.3. Authentication in Standard Systems

Authentication systems can be categorised according to the cryptographic approach taken[4]. There are authentication systems based on symmetric or public key. Among those based on symmetric key, we can distinguish protocols using or not a trusted third party (TTP).

2.3.1. *Authentication with symmetric key without TTP*

The simplest protocol that does not use a TTP is the ISO One-pass symmetric key Unilateral Authentication Protocol. In this scheme, there is only one single message sent from the user U to the system S.

(1) $U \rightarrow S : \text{Text2}, \{[T_u|N_u], S, \text{Text1}\}_{K_{us}}$

With the notation $\{X\}_{K_{ab}}$, we denote a message X ciphered with the key K_{ab} that is a known key to A and B. In the message 1, the text fields are optional and depend on the implementation of the specific protocol. As we can see, the user U sends a message ciphered with the key K_{us} to the system, then S, knowing k, can check its correctness. If an attacker tries to use a reply attack, S can reject the attack thanks to the presence of T_u or N_u, namely a timestamp or a random number generated by the user U. By receiving the message, the system can deduce that A has recently sent it if the sequence number is appropriate or if the timestamp has a recent value.

Simple challenge response algorithm, introduced in Section. 2.2.3 is the best-known protocol among those that does not use a TTP[12]. This protocol uses two messages: one containing Ns (the random number generated by the system) and the one containing the answer $\{Ns\}_{K_{us}}$. Authenticating in a challenge-response system, usually has three messages but the first is the request of authentication from U to S. The core of the protocol is composed of the two following messages:

(1) $S \rightarrow U : Ns$ and
(2) $U \rightarrow S : \{Ns\}_{K_{us}}$

In literature, there are many variations of the challenge-response algorithm, the main difference among them is about the challenge: sometimes, it is a random number; sometimes, it is predictable but never before used, sometimes is encrypted and sometimes it is not. Other algorithms known in literature are the ISO Two-Pass Unilateral Authentication Protocols and the ISO Three-Pass Mutual Authentication Protocols.

2.3.2. *Authentication with symmetric key with TTP*

Protocols using a TTP are very numerous in literature. Probably, the most known is the Needham–Schroeder Symmetric Key Authentication protocol[14]. It consists of the following five messages:

(1) $A \rightarrow S : A, B, Na;$
(2) $S \rightarrow A : \{Na, B, K_{ab}, \{K_{ab}, A\}_{K_{bs}}\}_{K_{as}};$
(3) $A \rightarrow B : \{K_{ab}, A\}_{K_{bs}};$
(4) $B \rightarrow A : \{Nb\}_{K_{ab}}$ and
(5) $A \rightarrow B : \{Nb - 1\}_{K_{ab}}$

where S is the TTP and A is the user that wants to authenticate itself in B. In the first message, A sends a request to the TTP for a key to communicate with B. The TTP answers with a message ciphered with the key K_{as} containing the number generated by A, B, the key K_{ab} and a message ciphered with the key K_{bs}. Then, A can send this last message to B and B, knowing the key K_{bs}, can decipher K_{ab} and use it to cipher next messages. The last two messages assure that the message 3 is authentic and is not a simple replay of an old message. This protocol has a drawback: suppose that a key K'_{ab} has been compromised by an attacker Z and that the message 3 has been recorded by it. Then, Z can resend the message 3 to B, for B, the message will be regular and it will continue the protocol with the attacker Z. This problem can be fixed with timestamps or using an extra handshake at the beginning of the protocol.

2.3.3. *Authentication with public key*

In literature, authentication protocols using public key cryptography are numerous. Unfortunately, asymmetric protocols are slower than symmetric ones. Often, these protocols are used in the first part of a communication, to exchange a symmetric key and to continue the communication with this key. In the Needham and Schroeder paper[14], it was presented the following algorithm.

(1) $A \rightarrow S : A, B$
(2) $S \rightarrow A : \{K_b, B\}_{K_{s^{-1}}}$
(3) $A \rightarrow B : \{Na, A\}_{K_b}$
(4) $B \rightarrow S : B, A$
(5) $S \rightarrow B : \{K_a, A\}_{K_{s^{-1}}}$
(6) $B \rightarrow A : \{Na, Nb\}_{K_a}$
(7) $A \rightarrow B : \{Nb\}_{K_b}$

S is the trusted server, usually called *certification authority* (CA). A starts the process by sending a request message to S containing its identity and the identity of B. With the message 2 (and 5), CA distributes the public key of the user to the counterpart ciphering the message with its own private key. The third message is used by A to inform B that it wants to start communicating. In this message is included a value Na. Then B asks S the public key of A (message 4 and 5) and sends to A a message ciphered with the A's public key which contains Na and a new value Nb. The last message is used by A to confirm that it is what it says to be, this is because A is the only one that knows the private key K_a used to decipher message 6.

Many other authentication protocols are based on public key cryptography (RSA, DES, etc.) to distribute the keys and to establish secure communication channels. Secure Socket Layer (SSL) and its successor Transport Layer Security (TSL) are the most famous. In these protocols, only one part is authenticated (the server); this means that the users can be sure with whom they are communicating.

2.4. Authentication in GSM Networks

In this part, the aim is to show how the authentication works in *Global Systems for Mobile communication* (GSM)[2]. The entire architecture of such systems is very complicated and out of scope of this chapter, so we will only give a sketch of the main objects involved in the authentication protocol, taking care on the protocol itself. In these systems, a Mobile Station (MS) is the end terminal, property of the final users and is used to communicate with other users. A MS is composed of a Mobile Equipment (ME) and a Subscriber Identity Module (SIM). The ME is the hardware univocally identified by an International Mobile Equipment Identifier (IMEI); SIM is a *kind of memory* where all the information about the user's subscription is stored, particularly:

- A serial number identifying the SIM;
- An identifier for the user called International Mobile Subscriber Identity (IMSI);
- The key K_i used in the authentication process;
- The ciphering key K_C that changes within the time;
- Two algorithms: one for the authentication process called $A3$ and another one called $A8$ used to calculate the key K_C and
- A temporary identifier called Temporary Mobile Subscriber Identity (TMSI) assigned by the network and that replaces IMSI in radio communication.

The MSs speak directly with the Base Station Sub-system (BSS) that is the sub-system that covers all radio sides for communications. All other aspects are controlled by the Network Switching Sub-system (NSS), that is mainly composed of Mobile-services Switching Centre (MSC), Visitor Location Register (VLR), Home Location Register (HLR), Authentication Centre (AuC). The MSC is a commutation centre that controls services provided by the network, the inter-networking and also that manages the user's mobility. The VLR controls MSs that are in the area covered by the associate MSC; HLR is a database that manages users information like their position, the VLR associate at this position, etc. The main subject involved in the authentication process is naturally the AuC. It is used to store information related with GSM security features. Later, we will see in detail how it is shaped.

2.4.1. *Security procedures in GSM networks*

In GSM networks, there have been introduced some different processes to secure the system and to anticipate any illegal access. We can categorise them in four different groups: authentication, ciphering , re-allocation of TMSI and identification of the ME. We are interested in authentication process and partially in ciphering. With the authentication process, the system verifies if a MS has the rights to access to the network using the following information:

- The user's key K_i is composed of 128 bit. This key is stored in the SIM and in the AuC.
- A 128-bit random number (RAND), created by the AuC and sent to the MS after his/her request to access.
- Authentication algorithm, called $A3$. It is stored both in the SIM and in the AuC. This algorithm, starting from K_i and RAND, generates the answer (Signed RESponse–SRES) that the MS must send to the AuC to authenticate and
- Algorithm $A8$, stored both in the SIM and in the AuC, that starting from RAND and K_i generates the ciphering key K_C.

Another information used in GSM system is a triplet (RAND, SRES, K_C) that is associated with an IMSI. These triplets are generated by the AuC, stored in the HLR and provided to VLR when requested. When a MS wants to be authenticated, it sends to the system an access request which contains TMSI or IMSI and successively receives a random number RAND, the rest of the process is represented in Fig. 2.3. From Fig. 2.3, we can note that, using the algorithm $A3$ with input RAND and the key k_i, the MS generates the answer SRES and sends it to the network. The network checks if the received SRES and the one generated by the AuC are equal, in this case, it authenticates the MS, otherwise there are two alternatives: first, if the MS has used TMSI to request the access, the network asks the IMSI and the procedure will be repeated, on the other hand if the MS has used the IMSI, the access is denied and the MS can only generate emergency calls. Triplets are very important because they contain all the information that the network should know to authenticate the user. They can be pre-calculated by the AuC and distributed to VLR, increasing the speed of the authentication process. Ciphering process is almost the same, but in this case networks and MS use the algorithm $A8$ with K_i and RAND as input to calculate the key K_C. Then any communication is ciphered

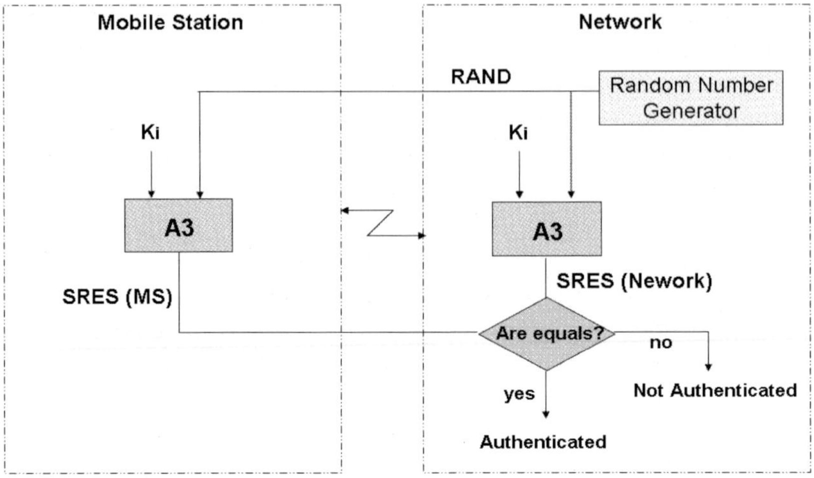

Fig. 2.3. The GSM authentication system.

with the key K_C. This protocol is a challenge-response authentication, where the random number is the challenge and the key K_i, shared between the MS and the network, is the secret used to generate the response.

2.5. Authentication in Watermarking

With the growth of Internet and Peer-to-peer p2p networks, sharing digital contents become more easy than in the past. Furthermore, tools to modify this information have begun communally used. In many practical applications, there is a need for verification or authentication of the integrity of the digital content.

Definition 2.1. Watermarking is the process to protect the integrity of digital contents from unauthorised tampering and usage.

We can use watermarking to introduce a kind of sign in multimedia documents and successively verify their authenticity. Using this approach, we can protect the duplication, as well as we can check and eventually track modified data. Generally, watermarking technique involves inserting imperceptible and sub liminal information in a multimedia content. With the embedded message, we can detect if the document was modified by one or more non-destructive attacks, sometimes we can detect the regions tampered and then try to recover them.

Watermarking can be categorised into two groups: fragile and semi-fragile techniques. Fragile techniques are developed to identify any change to image pixels, localising tampering, detecting geometric transformation (cropping, scaling, etc.) or the addition of foreign objects, signalling the removal of original objects and alerting other image processing operations such compression and filtering.

The drawback with fragile watermarking schemes is that some operations like conversion of images in different formats are recognised as not allowed and successively people cannot authenticate the content. In some cases, the owner would allow this operations: semi-fragile watermarking is the solution to this problem.

The choice to use either fragile or semi-fragile watermarking techniques depends on the type of image and its use. If we compress an artistic image with JPG protocol, probably the perception of the image is not affected, so in this case is preferable to use a semi-fragile protocol. Otherwise, if we think at medical or military use of images, the compression can lose meaningful information and the image will be totally useless: in this case is preferable use fragile watermarking. Moreover, in this field, the embedded watermarking information could invalidate the image. Two different methods exist so as to face this drawback: invertible watermarking schemes and schemes based on separation of zones. In the first one, if the image is deemed authentic, the distortion due to the watermarking process can be removed to obtain the original image. In the second one as, the image is divided into region of interest (ROI) and region of non-interest; ROI not affected by watermarking process and all embedded information is stored in the region of non-interest.

2.5.1. *Generic image authentication system: requirements*

According to Rey and Dugelay[15], to be effective, a generic image authentication system must satisfy the following criteria:

- Sensitivity: The system must be sensitive to malicious manipulations such as cropping or altering the image in specific areas. In this way, the authenticator can be sure that the content of the image is valid and that no one can paste and cut any portion of the authenticated image.
- Tolerance: The system must tolerate some loss of information and more generally non-malicious manipulations. Some operations like compression and conversion into different formats are not illegal and therefore are allowed; the drawback is that applying this operation over an authenticated image, we could lose the marking information. If an image is processed to the degree that the watermark cannot be recognised, later, we cannot authenticate the image in any way.
- Localisation of altered regions: The system should be able to locate precisely any malicious alteration made to the image and verify other areas as authentic. Some protocols use a block watermarking techniques; they divide the original image in different blocks of the same size and then apply a cryptographic function on this area. Like this, we can detect altered zones and at the same time authenticate the rest of the image.
- Reconstruction of altered regions: The system may need the ability to restore, even partially, altered or destroyed regions in order to allow the user to know what was the original content of the manipulated areas. In other words, if the original image has been altered, watermarking must detect altered parts and try

to reconstruct original data. Ideally, this reconstruction could be achieved using information embedded in the original data image, even if the original data has been altered.

There are some other technical features that must be taken into account:

(1) Storage: One way to achieve authentication is using digital signatures, but in this way, the signature will be an information external to the image, often another file. The outcome is that it will be more difficult and less convenient to share and transfer two files rather than one. The deduction is that it is better that authentication data will be embedded in the image rather than in a separate file.
(2) Mode of extraction: During the extraction process, it could be indispensable, or not, to know the original media. According to this distinction, watermarking systems can be classified as blind, semi-blind and non-blind. Blind methods require neither the original media nor the watermark, whereas semi-blind methods require not the original media but the watermark or other kind of information. Non-blind methods need the original media and so do not make sense for an authentication service.
(3) Asymmetrical algorithm: An authentication service requires an asymmetrical watermarking, in this way any user can recognise the authenticity of an image but only the author can authenticate it.
(4) Visibility: The watermark should be invisible under normal observation. This means that the approximation due to the introduction of the watermark into the image must be as weak as possible. Visibility can be assured also using invertible algorithms that remove embedded information due to watermarking to obtain the original data.
(5) Robustness and security: Like any other authentication system, it must not be possible to forge or manipulate authenticated data.
(6) Protocols: This is very important for any image authentication systems. They are necessary to define the set of rules and operations to achieve an high degree of accuracy.

2.5.2. *Fragile watermarks*

When we want to authenticate digital contents like an image and we want to identify any modification that happens on it, we can use fragile watermarking[10,21]. Generally, this method of watermarking use a simple idea that is to insert a specific watermark such that any attempt to modify or alter the content will also alter the watermark itself Fig. 2.4. Authenticating the content involves locating any distortion on the watermark in order to locate the tampered region. The major drawback with these approaches is that we cannot distinguish between legal and illegal alteration.

Fig. 2.4. Watermarking.

An example of legal alteration could be the compression or the conversion to other formats, the filtering, the adjustment of brightness or contrast, etc.

One of the first approaches to verify the authenticity of an image is to embed checksum into the Least Significant Bit (LSB). In Ref. 20, Walton proposes a technique that uses a key-dependent pseudo-random walk on the image. This protocol calculates a checksum, summing the numbers determined by the seven most significant bits and taking a reminder operation with a large integer N. The checksum is successively inserted in the LSB of selected pixels. This process can be repeated for many disjoint random walk or for one random walk over all the existing pixels. To the human eye, the differences between original image and processed image will be undetectable. There are different implementations of this algorithm, some of that work for grey images and others also for RGB (red, green, blue) images. Here, we present the basic version:

ALGORITHM 1 (embedding process)

(1) Let N be a large integer;
(2) Divide the image into 8×8 blocks and
(3) For each block B:
    (a) Define a pseudo-random walk through all 64 pixels, according to the secret key and the block number, and denote the pixels as $(p_1, p_2, \ldots, p_{64})$;
    (b) Generate a pseudo-random sequence of 64 integers $(a_1, a_2, \ldots, a_{64})$ comparable in size to N;
    (c) Calculate $S = \sum_{j=1}^{64} a_j \cdot g(p_j) \mod N$, where $g(p_j)$ is the grey level of the pixel p_j determined using the seven Most Significant Bit (MSBs);
    (d) Encrypt the binary form of S and
    (e) Embed the encrypted sequence into the LSB of the image block.

To check if an image is authentic, for each block, we can calculate the checksum using the MSB and then compare it with the LSB of the image block. The basic idea is very simple, and furthermore, is not possible for an attacker to swap or duplicate entire blocks because the random walk of the pixels p_j and the coefficient a_j are block-dependent. But, if an attacker merges two authenticated images protected with the same key swapping homologous blocks, it could forge a legal image. A simple solution is to make the watermark dependent on the image content.

The algorithm proposed by Fridrich and Goljan[6] tries to self-embed an image into itself. After this process, it is possible to recover portions of the authenticated image that have been tampered. This technique can be used to retrieve the original content rather than just indicate modified blocks.

2.5.3. *Semi-fragile watermarks*

Sometimes, we want to distinguish between legal and illegal operations executed over the images. Semi-fragile watermarks are a type of authentication more robust and less sensitive to user modification allowed. This method focuses on discriminate global operations preserving the semantic content of the image from malicious operations. These techniques are very useful when the author of an image wants to preserve the authenticity but he/she does not want to limit some types of modification such as compression and filtering. One of the most famous semi-fragile schemes that support JPEG compression is proposed by Lin and Chang[11]. The algorithm is found on these two invariance properties of discrete cosine transform (DCT).

(1) If we modify a DCT coefficient to an integral multiple of a quantisation step Q'_m, which is larger than the steps used in later JPEG compressions, then this coefficient can be reconstructed after JPEG compression and
(2) The relationship between two DCT coefficients of the same coordinates position from two blocks will not be changed after the quantisation process. (I can insert the motivation).

The first of these properties is used to embed the authentication data, whereas the second to generate the authentication bits. The algorithm divides the original image into 8×8 blocks and then, using a pre-determinate secret mapping function, forms pairs of blocks. For each block pair (p, q), select a set B_p of n DCT coefficients and generate the binary signature ϕ such that:

$$\phi_P(v) = \begin{cases} 1, & F_p(v) - F_q(v) \geq 0, \\ 0, & F_p(v) - F_q(v) \leq 0, \end{cases}$$

where $v \in B_p$, $F(v)$ is the value of v. The binary signature is then partly embedded in the two block p and q.

When we want to authenticate the image, we first extract the authentication bits from the watermarked areas of the image, and then we use these bits to verify that the DCT coefficient relationship in the signature matchs the predicted criteria. In this case, the image is considered authentic, otherwise the image has been tampered and we know what are the possible blocks manipulated.

Another kind of semi-fragile watermarking protocols are called block-based. These approaches are based on probability. They consist of dividing the image into blocks of 64×64 pixels and applying a "robust" mark into each blocks. When we want to test the authenticity of a such a watermarked image, we have to test the presence or absence of the mark in each block. If it is present with high probability in each block, the image can be considered as authentic. Generally, protocols, such as this, use one or more user-defined thresholds, in this way, we can measure several levels of authenticity.

Other protocols are based on features of the image (feature-based watermark). Behind these kinds of protocols, there is a simple idea that we can extract some features of the image and then embed them within a robust and an invisible watermark. When we want to check if an image has been altered, we compare his/her own features with those of the original image recovered from the watermark. Usually, the features are selected considering the use that we will be of the image and the type of image alteration that we wish to detect.

2.5.4. *Attacks and countermeasures*

The target of an attacker is to modify a protected image forging a falsified authenticated image. To achieve this target, the attacker can try to modify the image without modifying the embedded watermark. A second opportunity for the attacker is to try to create a new watermark, like the original one, so that the final user will consider as authentic. This last chance is the most used attacking a watermarking system.

Let an authenticated image using a fragile watermark, where the watermark is independent of the image content and embedded in the LSB of its pixel. Generally, if the attacker tampers the image without attention to what bits he/she changes, we can detect easily not authentic images. But if the attacker do not alter the LSB of the image, the embedded watermark will remain intact and for us the altered image will be authentic.

This is the drawback in any protocol where the watermarks do not depend on the image's contents. In this case, it is possible to forge an attack that could generate a falsified image starting from a valid image. So, the image generated becomes protected and authenticated such as the starting image. Another kind of attack is the "*Collage Attack*". The basic idea is the same for the previous attacks, but in this case the attacker forges a falsified image using parts of a group of images protected by the same authenticator using the same mark and the same key. The attacker creates a collage combining parts from different images while keeping their relative positions within the image. The simplest solution to avoid both these attacks would be to make the fragile watermarks depend on the image content.

A third type of attack is the most known in any security system: Brute Force attack. Naturally, when the key has been found, the attacker can easily falsify a watermark of an image that has been protected by this key. Using a long and complicated key, we can avoid this type of attack since the attacker would need high computing time.

2.6. Research Directions

As we have seen, authentication is used in many systems. The subject to authenticate can be a user, a message or some other thing. Often, when we speak about images, video or in general other multimedia support we think the way to be sure that the content has not been tampered. Authentication can also be used to assure copyrights on multimedia contents, denying illegal copy of supports and files.

If on one side we have copyrights, on the other side, we have p2p networks where often authentication is required but at the same time anonymity. This is not so simple because often legal rules impose that users must be known to the system, so they cannot be completely anonymous.

In wireless networks, this is a real challenge because these kind of networks are more weak than wired ones. A malicious user could easily impersonate a legal user or simply use an open wireless network to execute some kind of illegal matter: a strong authentication system could help or resolve these aspects.

2.7. Summary

In this section, we have analysed different aspects of the authentication in computer systems: this is the first step towards security. Standard authentication techniques are largely used every day, using the phone or the car, accessing the bank or showing the passport. We can distinguish authentication schemes in to two groups: authentication with symmetric key and with public key. The first group uses a shared key, on the other hand, the second group uses public cryptography to authenticate the subject.

The most-known authentication technique is surely the password, or the pass phrase. It is largely used in the Internet to access in websites or in mail accounts, but also in many other software applications. Over the last few years, other than passwords, the biometric authentication is often used, namely the measurement of some biological features that can uniquely identify a person. This is achieved using eyes, faces, keystrokes, fingerprints, voices authentication and many times, a mix of them. In this chapter, also the location-based authentication has been introduced, namely the authentication restricted to a specific location or a region. We have presented the authentication process in GSM networks, giving a sketch of the entire architecture of such systems, and also the authentication of multimedia content using watermarking techniques. In Section 2.5, we have presented fragile and semi-fragile watermarking, giving some examples of existing algorithms and also of attacks and relative countermeasures.

2.8. Glossary

AAA: Authorisation, authentication and accounting; It denotes a framework for controlling access to resources, enforcing policies and providing the information needed by billing services.

Challenge-response: It is used to authenticate a user using challenge mechanisms.

Collage attack: A kind of attack used in watermarking.

DCT: Discrete cosine transform.

Eyes authentication: Authentication using the retina or the iris.

Faces authentication: Authentication using faces recognition.

Fingerprint authentication: Authentication using the fingerprints.

Keystrokes authentication: Authentication based on keystroke intervals, keystroke pressure, keystroke duration, etc.

Location-based authentication: Authentication based on the position of the entity that wants to be authenticated.

LSB: Least Significant Bit.

LSS: Location Signature Sensor: hardware used in location-based authentication.

ME: The hardware of a MS in GSM networks.

MS: Mobile Station, used in GSM networks.

MSB: The Most Significant Bit.

Password: A sequence of characters used to authenticate a user.

ROI: Region of interest, used in watermarking to specify a particular region.

Scaling: Geometric transformation of a digital content.

SIM: Subscriber module identity, a component of the MS.

SSL: Secure socket layer.

TTP: Trusted third party.

Voice authentication: Authentication using the voice.

Watermarking: It is the process to protect the integrity of digital contents from unauthorised tampering and usage.

References

1. S. M. Bellovin and M. Merritt, Encrypted key exchange: password-based protocols secureagainst dictionary attacks, *SP '92: Proceedings of the 1992 IEEE Symposium on Security and Privacy* (IEEE Computer Society, Washington, DC, USA), ISBN 0-8186-2825-1, 1992, p. 72.
2. O. Bertazioli and L. Favalli, *GSM-GPRS* (Hoepli: Milano 2 ed., 2002).
3. M. Bishop, Comparing authentication techniques, 1991. URL `citeseer.ist.psu.edu/bishop91comparing.html`.
4. J. A. Clark and J. L. Jacob, A survey of authentication protocol literature, 1997, URL `citeseer.ist.psu.edu/clark97survey.html`.
5. D. E. Denning and P. F. MacDoran, *Location-based Authentication: Grounding Cyberspace for Better Security* (ACM Press/Addison-Wesley Publishing Co., New York, NY, USA, 1998), ISBN 0-201-30820-7.

6. J. Fridrich and M. Goljan, Protection of digital images using self embedding, 1999, URL `citeseer.ist.psu.edu/fridrich99protection.html`.
7. N. M. Haller, The S/KEY one-time password system, *Proceedings of the Symposium on Network and Distributed System Security*, 1994, pp. 151–157, URL `citeseer.ist.psu.edu/haller94skey.html`.
8. S. Hocquet, J.-Y. Ramel and H. Cardot, Fusion of methods for keystroke dynamic authentication, *AUTOID '05: Proceedings of the Fourth IEEE Workshop on Automatic Identification Advanced Technologies* (IEEE Computer Society, Washington, DC, USA), ISBN 0-7695-2475-3, 2005, pp. 224–229, doi: http://dx.doi.org/10.1109/AUTOID.2005.30.
9. C. D. Laat, G. Gross, L. Gommans, J. Vollbrecht and D. Spence, Generic AAA Architecture, RFC 2903 (Experimental), 2000, URL `http://www.ietf.org/rfc/rfc2903.txt`.
10. C. Li, Digital fragile watermarking scheme for authentication of jpeg images, *VISP* **151**(6), 2004, pp. 460–466.
11. C. Lin and S. Chang, Semi-fragile watermarking for authenticating JPEG visual content, *SPIE Security and Watermarking of Multimedia Content II, San Jose*, 2000, pp. 140–151, URL `citeseer.ist.psu.edu/lin00semifragile.html`.
12. C. Mitchell, Limitations of challenge-response entity authentication, *Electronics Letters* **25**(17), 1989, pp. 1195–1196.
13. National Institute of Standards and Technology, *FIPŞ PUB 46-3: Data Encryption Standard (DES)* (pub-NIST, pub-NIST:adr), 1999, URL `http://www.itl.nist.gov/fipspubs/fip186-2.pdf`.
14. R. M. Needham and M. D. Schroeder, Using encryption for authentication in large networks of computers, *Commun. ACM* **21**(12), 1978, pp. 993–999, doi: http://doi.acm.org/10.1145/359657.359659.
15. C. Rey and J. L. Dugelay, A survey of watermarking algorithms for image authentication, *EURASIP J. Appl. Signal Process.* **2002**(1), (2002), 613–621.
16. C. Rigney, S. Willens, A. Rubens and W. Simpson, Remote authentication dial in user service (RADIUS), RFC 2865 (Draft Standard), 2000, URL `http://www.ietf.org/rfc/rfc2865.txt`, updated by RFCs 2868, 3575, 5080.
17. R. L. Rivest, A. Shamir and L. Adleman, A method for obtaining digital signatures and public-key cryptosystems, *Commun. ACM* **21**(2), (1978), 120–126, doi: http://doi.acm.org/10.1145/359340.359342.
18. G. J. Simmons, A survey of information authentication, *Contemporary Cryptology, The Science of Information Integrity*, (1998), 603–620.
19. W. Simpson, PPP Challenge Handshake Authentication Protocol (CHAP), RFC 1994 (Draft Standard), 1996, URL `http://www.ietf.org/rfc/rfc1994.txt`, updated by RFC 2484.
20. S. Walton, Information authentification for a slippery new age, *DR. Dobbs Journal* **20**(4), (1995), 18–26.
21. F. Yeh and G. Lee, Toral fragile watermarking for localizing and recovering tampered image, *Intelligent Signal Processing and Communication Systems*, (2005), 321–324.

Chapter 3

CLASSIFICATION OF ATTACKS ON CRYPTOGRAPHIC PROTOCOLS

LUCA SPALAZZI
Università Politecnica delle Marche
Dipartimento di Ingegneria Informatica
Gestionale e dell'Automazione

SIMONE TACCONI
Ministero dell'Interno
Dipartimento di Pubblica Sicurezza
Servizio Polizia Postale e delle Comunicazioni

3.1. Introduction

In recent years, computer networks have received enormous development. These systems provide users with more and more different services, each of them requiring peculiar security considerations. In order to satisfy such requirements, various cryptographic protocols have been proposed and implemented, but, unfortunately, many of them have been shown to have flaws, sometimes a long time after their publication.[a] This is a consequence of the fact that the definition of a cryptographic protocol involves many subtleties that often lead designers into error. This chapter aims to help who deals with cryptographic protocols. Indeed, in protocol design, this work can make designers aware of the most common kinds of attacks affecting existing protocols. Moreover, it can also help in protocol verification, since our classification can be useful to provide case-studies for tuning and testing formal verification methods [71, 58, 57]. For this purpose, this chapter presents a systematic description of protocol attacks. This is accomplished thanks to a classification of attacks from three different points

[a]A noteworthy example is the Needham–Schroeder shared key protocol. Such a protocol was proposed in 1978 [64]. In 1981, it was shown to have a flaw and it was amended [28]. In 1997, 16 years later, the amended version has been shown to be insecure as well [47].

of view:

- *Protocol flaws.* We describe attacks from the point of view of flaws that make protocols vulnerable. It is possible to classify flaws into design flaws (i.e. incorrect or incomplete specifications), implementation flaws (protocol implementation does not satisfy specifications) and configuration flaws (protocol installation and configuration does not satisfy specifications). In this chapter, we focus only on design flaws.
- *Attack strategies.* We describe how an adversary can exploit different protocols, sessions and messages in order to compromise the security of a given protocol.
- *Attack consequences.* Finally, we describe attacks in terms of their consequences, i.e. violation of security requirements. In this classification, we take into account only requirements that are known to be compromised by attacks on cryptographic protocols, i.e. authentication, privacy and secret freshness.

Usually, a given protocol attack may fall into one or more categories of the above classifications. In other words, an attack may be due to one or more flaws, involve one or more strategies and have one or more consequences. Figure 3.1 depicts the result of these classifications. For each category, Fig. 3.1 reports vulnerable protocols and their corresponding attacks.

Information about such protocols is summarised in Tables 3.1 and 3.2. Table 3.1 reports, for each protocol, the reference where the protocol was published, how

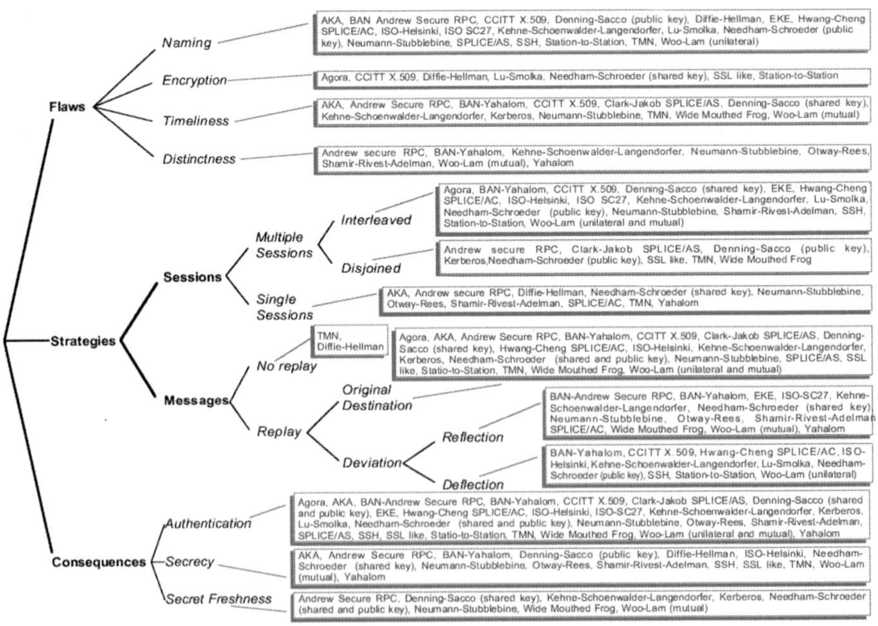

Fig. 3.1. A taxonomy of protocol attacks.

Table 3.1. Some cryptographic protocols and their bibliographic references.

Protocol	References	Attacks	Attack references
Agora	[35]	# 1	[46]
AKA	[72]	# 1	[1]
Andrew Secure RPC	[73]	# 1	[19, 47]
		# 2	[23]
AuthA	[12]	# 1	[76]
BAN-Andrew Secure RPC	[19]	# 1	[50]
BAN-Yahalom	[19]	# 1	[79]
		# 2	[79]
CCITT X.509	[87]	# 1	[19]
Clarck–Jacob SPLICE/AS	[26]	# 1	[47]
Denning–Sacco (secret key)	[28]	# 1	[47]
Denning–Sacco (public key)	[28]	# 1	[7, 3]
Diffie–Hellman	[30]	# 1	[13]
EKE	[14]	# 1	[24]
HB^+	[44]	# 1	[36]
Hwang–Cheng SPLICE/AS	[40]	# 1	[26]
ISO-Helsinki	[63]	# 1	[39]
ISO-SC27	[43]	# 1	[16]
Kehne–Schoenwalder–Langendorfer	[45]	# 1	[50]
		# 2	[50]
Kerberos 5.0	[67]	# 1	[15]
Lu-Smolka variant of SET	[52]	# 1	[69]
Needham–Schroeder (secret key)	[64]	# 1	[28]
		# 2	[18, 25]
Needham–Schroeder (public key)	[64]	# 1	[49]
Neumann–Stubblebine	[66]	# 1	[22, 81]
		# 2	[41, 24]
Otway–Rees	[68]	# 1	[18, 21]
		# 2	[23]
Shamir–Rivest–Adelman three pass	[75]	# 1	[21]
		# 2	[21, 24]
SPLICE/AS	[88]	# 1	[40]
		# 2	[40]
SSH (June 1996 version)	[89]	# 1	[1]
SSL-like	[4]	# 1	[4]
Station-to-station	[31]	# 1	[50]
TMN	[83]	# 1	[51, 77]
		# 2	[51]
		# 3	[51]
Wired Equivalent Privacy (WEP)	[42]	# 1	[17, 10, 20]
		# 2	[33, 20]
Wide-mouthed frog	[19]	# 1	[23, 7]
		# 2	[23, 24]
		# 3	[47]
Woo–Lam (unilateral)	[85]	# 1	[3, 85]
Woo–Lam (mutual)	[85]	# 1	[50]
		# 2	[24, 25]
Yahalom	[19]	# 1	[24]

Table 3.2. Classification of known attacks against cryptographic protocols.

Attack	Flaws	Strategies		Violations
		Sessions	Messages	
Agora	Encryption	Interleaved	Original	Authentication
AKA	Naming, timeliness	Single	Original	Authentication, privacy
Andrew Secure RPC (# 1)	Timeliness	Disjoined	Original	Freshness
Andrew Secure RPC (# 2)	Distinctness	Single	Original	Privacy
AuthA	Distinctness	Interleaved	Original	Authentication
BAN–Andrew Secure RPC	Naming	Interleaved	Reflection	Authentication
BAN-Yahalom (# 1)	Distinctness	Interleaved	Original Deflection, Reflection	Authentication, privacy
BAN-Yahalom (# 2)	Timeliness	Interleaved	Original, Deflection, Reflection	Authentication
CCITT X.509	Timeliness, naming Encryption	Interleaved	Original, Deflection	Authentication
Clark–Jacob SPLICE/AS	Timeliness	Disjoined	Original	Authentication
Denning–Sacco (secret key)	Timeliness	Interleaved	Original	Authentication, freshness
Denning–Sacco (public key)	Naming	Disjoined	Deviation	Authentication, privacy
Diffie–Hellman	Encryption, naming	Single	No-replay	Privacy
EKE	Naming	Interleaved	Reflection	Authentication
HB^+	Encryption	Interleaved	Original	Privacy
Hwang–Cheng SPLICE/AS	Naming	Interleaved	Original, Deflection	Authentication
ISO-Helsinki	Naming	Interleaved	Original, Deflection	Authentication, privacy
ISO-SC27	Naming	Interleaved	Reflection	Authentication
Kehne–Schoenwalder–Langendorfer (# 1)	Naming, timeliness, Distinctness	Interleaved	Original, reflection	Authentication, freshness
Kehne–Schoenwalder–Langendorfer (# 2)	Naming, Distinctness	Interleaved	Original, Deflection, Reflection	Authentication, freshness
Kerberos 5.0	Timeliness	Disjoined	Original	Authentication, freshness
Lu–Smolka variant of SET	Naming, encryption	Interleaved	Deflection	Authentication
Needham–Schroeder (secret key) (# 1)	Timeliness	Disjoined	Original	Authentication, privacy, Freshness
Needham–Schroeder (secret key) (# 2)	Encryption	Single	Reflection	Authentication

(*Continued*)

Table 3.2. (*Continued*)

Attack	Flaws	Strategies: Sessions	Strategies: Messages	Violations
Needham–Schroeder (public key)	Naming	Interleaved	Original, Deflection	Authentication, freshness
Neumann–Stubblebine (# 1)	Distinctness	Single	Reflection	Authentication, privacy, Freshness
Neumann–Stubblebine (# 2)	Naming, timeliness	Interleaved	Original, reflection	Authentication, freshness
Otway–Rees (# 1)	Distinctness	Single	Reflection	Authentication, privacy
Otway–Rees (# 2)	Distinctness	Single	Reflection	Privacy
Shamir–Rivest–Adelman (# 1)	Distinctness	Single	Reflection	Authentication, privacy
Shamir–Rivest–Adelman (# 2)	Distinctness	Interleaved	Reflection	Authentication, privacy
SPLICE/AS (# 1)	Naming	Single	Original	Authentication
SPLICE/AS (# 2)	Naming	Single	Original, reflection	Authentication
SSH	Naming	Interleaved	Deflection	Authentication, privacy
SSL-like	Encryption	Disjoined	Original	Authentication, privacy
Station-to-station	Naming, encryption	Interleaved	Original, Deflection	Authentication
TMN (# 1)	Naming	Single	No-replay	Authentication, privacy
TMN (# 2)	Naming, timeliness	Disjoined	Original	Authentication, privacy
TMN (# 3)	Naming	Single	No-replay	Authentication, privacy
Wide-mouthed frog (# 1)	Timeliness	Disjoined	Reflection	Authentication, freshness
Wide-mouthed frog (# 2)	Timeliness	Disjoined	Original	Authentication, freshness
Wide-mouthed frog (# 3)	Timeliness	Disjoined	Original	Authentication, freshness
Wired equivalent privacy (WEP) (# 1)	Encryption, timeliness	Disjoined	No-replay	Authentication, privacy
Wired equivalent privacy (WEP) (# 2)	Encryption	Disjoined	No-replay	Authentication, privacy
Woo–Lam (unilateral)	Naming	Interleaved	Original, Deflection	Authentication
Woo–Lam (mutual) (# 1)	Timeliness	Interleaved	Original	Freshness
Woo–Lam (mutual) (# 2)	Distinctness	Interleaved	Reflection	Authentication, privacy
Yahalom	Distinctness	Single	Reflection	Authentication, privacy

Table 3.3. Notation adopted.

Notation	Description
$C = \{M\}_K$	Encryption of a plain text M with the key K
$M = \{C\}_{K^{-1}}$	Decryption of a cipher text C with the key K^{-1}
$D = H(M)$	Hash function of a message M
$\text{sign}_A\{m\} = \{H(m)\}_{K_a^{-1}}$	Digital signature of a message m by A

many attacks to this protocol have been discovered and references describing these attacks, while Table 3.2 shows which kinds of attack violate it according to our classification. The most significant examples of such protocols and attacks are also discussed in corresponding sections of this chapter. Moreover, Table 3.3 reports the notation adopted for cryptographic operations.

3.1.1. *Related Work*

Literature about protocols and attacks is vast. Most of it simply proposes a given protocol (see for example, the column "References" in Table 3.1), reports specific attacks on protocols (see for example, the column "Attack references" in Table 3.1) or describes verification techniques (see for example, Refs. 19, 32, 55, 61, 86). For what concerns introductory papers about this matter, relevant publications to point out are as follows: Refs. 74, 60 are books on cryptography; they also introduce protocols and attacks on protocols; Ref. 24 is a vast catalogue of protocols and attacks; Ref. 11 is a survey on e-commerce protocols. Concerning attacks, Refs. 27, 84 contain a long list of attacks on information systems and communication protocols. Our focus is different with respect to all the above papers. Indeed, this chapter aims to survey attacks on cryptographic protocols. It is also an attempt at providing a systematic description by classifying attacks according to three different points of view. Let us consider the work related to each specific perspective.

- *Protocol flaws.* For what concerns this perspective, an important work is Ref. 21. Indeed, the author was the first to introduce this perspective (even if some categories are different from ours). Papers similar to the previous one are Refs. 23, 38. In fact, they propose the same classification described in Ref. 21. In general, design flaws are caused by incorrect (functional) or incomplete (implementation-dependent) specifications. Therefore, a different way of considering this issue consists of taking into account design principles that help a protocol designer in avoiding design flaws. This is what Refs. 3, 8, 80 do (the last two are limited to the design of public key protocols). Implementation-dependent specifications are an important source of flaws; therefore, they have received a particular attention in literature. Remarkable articles that deepen this aspect are Refs. 6, 18, 25, 53, 54, 62.

- *Attack strategies.* An approach similar to ours can be found in Ref. 79. The main difference is that Ref. 79 only takes into account replay attacks, i.e. attacks based on the replay of (a fragment of) a previous message as it is. In our work, we extend this classification in two aspects. First, we take into account non-replay attacks. Second, we consider a replayed message as the result of manipulations upon a fragment of a previous message.
- *Attack consequences.* As far as we know, this is a novel perspective for classifying attacks.

3.2. Attacks and Attackers

In the context of protocols that employ cryptographic primitives, an attack can be directed at cryptographic algorithms adopted in a given protocol or at the protocol itself. In our discussion, we assume that the perfect-encryption assumption holds. In practice, the above assumption is not true, since it is, possible, to attack some algorithms by means of cryptanalytic techniques. Nevertheless, it is, in principle, reasonable because it allows designers to separate protocol analysis/design from cryptographic algorithm analysis/design. For this reason, we focus our attention on attacks exploiting the conceptual structure of protocols.

An attack on a protocol is a sequence of actions performed by an attacker, by means of any hardware or software tool, in order to subvert protocol goals. The attacker (also called "adversary") is a dishonest entity that may be a member of the system, i.e. a legitimate user, or external to the system. In the latter case, the attacker is often referred as "intruder" or "penetrator."

3.2.1. *Attack Notation*

For the sake of readability, here we introduce terms and notation that are usually adopted in literature to formalise protocols and attacks. A protocol is a sequence of n steps; each step is denoted by the following expression:

$$(i) X_i \rightarrow Y_i : m_i$$

where $0 \leq i \leq n$, is the step i of the protocol, X_i is the sender of the step i, Y_i is the receiver and m_i is the message. In the rest of the chapter, we usually denote the honest participants to the protocol as $A, B, \ldots,$ the authentication server (if any) as S. Parties involved in a cryptographic protocol transaction are also designated as "principals". The principal that starts a protocol execution sending the first message is called "initiator", whereas the other principal is called "responder". Finally, the attacker is usually denoted as I. Each message is composed of several components, separated by commas, which can involve cryptographic operators. Concerning the notation of attacks, Fig. 3.2 depicts a generic trace of attack. P, Q are protocols, X, Y, V, W are principals, I is the attacker, χ, ψ are protocol session identifiers, x, y are steps of P and Q, respectively, $m_1, \ldots, m_i, \ldots, m_n$ are messages and $f(\cdot)$

$$\ldots$$

$$(P.\chi.x)\ X \rightarrow I|I, Y|I(Y) : m_i$$

$$\ldots$$

$$(Q.\psi.y)\ I|I(V) \rightarrow W : m = f(m_1, \ldots, m_n)$$

Fig. 3.2. A generic attack trace.

expresses the functional dependency (if any) between message m and previous messages (i.e. $m_1, \ldots, m_i, \ldots, m_n$). Notice that session identifiers are not objects belonging to protocol but, as it is usual in literature, they are used for notational purposes, i.e. to distinguish messages originating from different sessions. The vertical bar denotes alternative actions. I, Y means that I overhears a message for Y $I(Y)$ means that I receives and removes a message for Y. Finally, $I(V)$ describes that I sends a message acting as V.

3.2.2. *Attacker Capabilities*

In the study of protocol attacks, a very important aspect concerns the adversary's capabilities, i.e. the set of actions that the attacker can perform during protocol executions. First of all, any attacker must have at least the same capabilities as honest principals: knowledge about the protocol and related cryptographic primitives. Moreover, the attacker must have as complete control as possible of the communication network. In a *passive attack*, the adversary does not affect the normal execution of protocols, but simply observes the messages to obtain useful information about parties involved in the transaction. This kind of attack, also denominated "eavesdropping", is difficult to detect and thus protocols try to be resistant to *passive attacks* rather than detect them. In an *active attack*, the adversary can alter protocol execution. In this case, the adversary can manipulate and even completely delete outgoing messages. It can generate new messages using its initial knowledge and overheard messages, for delivery to any legitimate entity, thus impersonating any other entity. In a typical scenario, the attacker is able to participate in different simultaneous executions of the (same or a different) protocol, playing different roles and in disguise, to obtain from one transaction information useful in attacking another. Let us emphasise that an attacker does not have to be a complete outsider; it may be a legitimate system user, i.e. the adversary may be one of the parties explicitly involved in the protocol.

3.2.2.1. *Basic actions*

In general, an adversary that wishes to attack a protocol adopts behaviour that may involve two kinds of action, typically interleaved each other: communication acts and internal acts. The former are actions that imply an interaction between the adversary and the communication channel, whereas the latter are actions performed in order to obtain sensible information. The basic actions are listed as follows.

3.2.2.1.1. Communication acts

- *Receive*: The attacker legitimately receives a message m expressly sent to it by X. This action is denoted as follows:

$$X \rightarrow I : m$$

- *Intercept*: The attacker eavesdrops a message m sent from X to Y, without removing it from the communication network. This action is denoted as follows:

$$X \rightarrow I, Y : m$$

- *Remove*: The attacker deletes an intercepted message in transit from X to Y, preventing from its arrival to the designate receiver Y. This action is denoted as follows:

$$X \rightarrow I(Y) : m$$

- *Send*: The attacker sends to Y, impersonating X or itself, a message m received/intercepted in the past or generated from its knowledge. In the first case, the action is denoted as follows:

$$I \rightarrow Y : m$$

  in the latter:

$$I(X) \rightarrow Y : m$$

  Obviously, the case where I impersonates itself is possible only if I is a member of the system.

3.2.2.1.2. Internal acts

- *Wait*: The attacker waits for the transit of an interesting message or the propitious moment to send a message.
- *Store*: The attacker saves in its local storage device a received, intercepted or an elaborated message in order to send it later on, augmenting its knowledge in this way.
- *Elaborate*: The attacker manipulates messages in its memory in order to deduce pieces of information and generate new messages (see Section 3.2.2.2).

3.2.2.2. *Inferential capabilities*

The internal act *elaborate* consists of producing a message, derived from previously stored messages, i.e. computing the function $f(\cdot)$ of Fig. 3.2. In general, a message is produced by combining basic operations that the attacker is able to perform. In the following, we report the most common basic operations performed by a given attacker.

3.2.2.2.1. No elaborations

When an attacker re-uses a previous message with no-manipulation upon it, we have the so-called "simple replay" (see Section 3.4.2).

3.2.2.2.2. String elaboration

Any message can be considered as a bit string, therefore an adversary can use classical string operations as the following:

- *Generation*: Generation of a new message component, i.e. a component that is not derived from the set of messages intercepted by an attacker.
- *Composition*: Concatenation of two message components in order to obtain a compound message.
- *Decomposition*: The reverse of composition, in order to extract the components of a given message.

3.2.2.2.3. Cryptographic elaborations

Execution of a cryptographic protocol involves computation of cryptographic algorithms or mathematical functions involved in these algorithms. As a consequence, an attacker can perform operations as the following:

- *Encryption*: Encryption of a message with a known key.
- *Decryption*: Decryption of a message with a proper key.
- *Hashing*: Hash of a message and
- *Cryptanalysis*: Deduction of an unknown key from a set of messages, some of them encrypted with that key.

3.2.3. *Cryptanalytic Assumptions*

A crucial issue in cryptography is to keep the plain text and the key secret from adversaries. For this reason, an important area is cryptanalysis [74, 78, 60], the science of recovering the plain text of a message without the knowledge of the key. A successful cryptanalytic attack may allow recovery of the message or the key. In this context, fundamental assumptions are Kerchhoffs Principles, a set of requirements for encryption algorithm. The most important of these desiderata [74, 60], may be formulated as follows:

Kerchhoffs principle. *The security of a cryptographic algorithm does not rely on the secrecy of the algorithm but resides only in the secrecy of the key.*

In other words, it is permissible for encryption algorithms to be known. This principle may also be extended to cryptographic protocols as follows:

Kerchhoffs principle for protocols. *The security of a cryptographic protocol does not rely on the secrecy of operations that it requires to its participants, but only in the secrecy of the key.*

In the context of cryptographic protocols, analysis and design is accomplished under the hypothesis that the cryptographic algorithms are perfect. Thus, we can formulate another important assumption:

Perfect Encryption. *The decryption key must be known in order to extract the plain text from the cipher text.*

While the above assumption is obviously not true, it is in practice, reasonable because it allows us to separate protocol analysis/design from cryptographic algorithm analysis/design.

3.3. Protocol Flaws

In this section, we explain the classification of attacks from the point of view of protocol flaws, i.e. the reasons of protocol vulnerability. A remarkable distinction regards the concepts of "flaw" and "attack". A flaw is an undesired property of the protocol and thus it represents an intrinsic feature of the conceptual structure of the protocol itself, while an attack is a sequence of actions that exploit flaws for succeeding. Unfortunately, although it could seem that there is a strict correlation between flaws and attacks, often this connection is unclear or difficult to find. In general, flaws may arise at any stage of the protocol development and they can occur because of incomplete or erroneous specifications. Furthermore, even correct and complete specifications do not assure the correctness of a given implementation. Generally, it is possible to distinguish among three categories of cryptographic protocol flaws [21]:

- *Functional specification flaws.* These are flaws which can be considered as high-level logical or functional deficiencies in the design of a protocol. In other words, they affect the correctness of the specifications.
- *Implementation-dependent flaws.* They are flaws which are due to an incomplete specification and thus they can lead to different implementations, at least one of them correct and at least one wrong. Typically, these errors concern the completeness of specifications.
- *Implementation flaws.* They are flaws that occur when a correct and complete specification is incorrectly implemented.

While the above classification takes into account the abstraction level where flaws arise, the classification that we propose (see Fig. 3.1) is based on conceptual

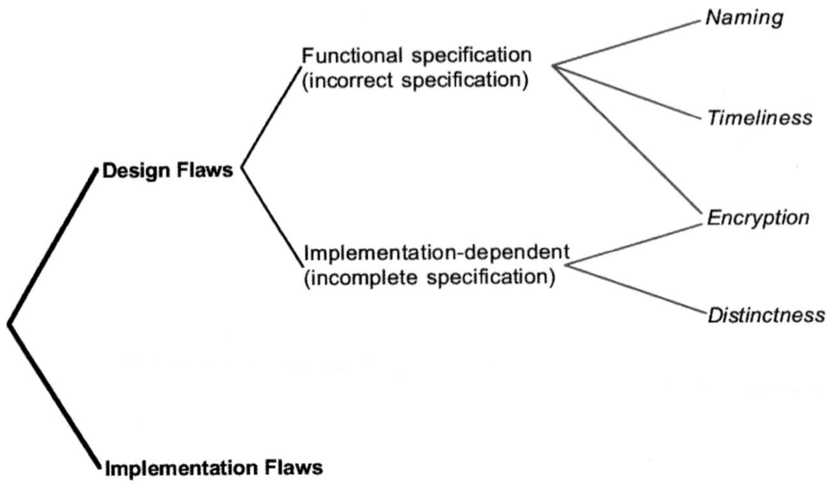

Fig. 3.3. A taxonomy of protocol flaws.

nature of flaws. Indeed, such a classification is inspired by the principles for designing cryptographic protocols enunciated in the valuable work [3]. Therefore, we consider only the first two categories (that deal with design) and neglect implementation flaws. The relationship between the two classifications is shown in Fig. 3.3.

The taxonomy that we propose must not be interpreted in a rigid manner, since in many cases, a protocol may be affected from different kinds of flaws.

3.3.1. *Naming Flaws*

Flaws that belong to this category involve the absence of a principal identifier in a message when the explicit presence of identity is essential to the meaning of the message itself. Such flaws are functional specification flaws. Often, the identity of parties involved in a protocol execution can be derived from other pieces of information or from applied encryption keys. Therefore, designers try to omit principal identities every time they think it is possible, in order to obtain shorter messages and thus to reduce the amount of encryption. Indeed, encryption, typically, requests a considerable computational effort and for this reason it is necessary to find a trade-off between efficiency and security. The general principle to follow, say *Principle of Explicit Communication* in Ref. 3 and *Principle of Full Information* in Ref. 85, states that if the identity of a principal is essential to a correct and univocal interpretation of a message, it is prudent to mention the principal's name explicitly in the message. Unfortunately, in literature, there are several examples of cryptographic protocols where the omission of an essential identity leads to serious consequences for the reliability of protocols themselves. A very famous example is the Needham–Schroeder Public-Key protocol [64]. The original protocol involves seven steps and can be considered as the interleaving of two logically distinct

(1) $A \rightarrow B : \{N_a, A\}_{Kb}$	(α.1) $A \rightarrow I : \{N_a, A\}_{Ki}$
(2) $B \rightarrow A : \{N_a, N_b\}_{Ka}$	(β.1) $I(A) \rightarrow B : \{N_a, A\}_{Kb}$
(3) $A \rightarrow B : \{N_b\}_{Kb}$	(β.2) $B \rightarrow I(A) : \{N_a, N_b\}_{Ka}$
(a)	(α.2) $I \rightarrow A : \{N_a, N_b\}_{Ka}$
	(α.3) $A \rightarrow I : \{N_b\}_{Ki}$
	(β.3) $I(A) \rightarrow B : \{N_b\}_{Kb}$
	(b)

Fig. 3.4. The attack (b) on the Authentication phase of the Needham-Schroeder public key protocol (a).

protocols: one aims to deliver the public-keys to involved parties, the other one, shown in Fig. 3.4, concerns the authentication of principals. We focus our attention to the second one.

In the first step, the initiator A sends a message to responder B containing a nonce N_a and its identity$A$, both encrypted with B's public key. When B receives this message, it decrypts the message and obtains the nonce N_a, then it generates a new nonce N_b and sends to A the concatenation of the two nonces encrypted with A's public key. When A receives this message, it decrypts the message and compares the nonce of step (1) with the first nonce extracted from the current message. If they match, then A should be sure that it is communicating with B. At this point, A extracts the second nonce and uses it to build the third message. When B receives and decrypts this message, if the received nonce matches with the nonce N_b sent in the second message, then it should be sure of the other party's identity. Unfortunately, this is not true because the protocol is vulnerable to the attack reported in Ref. 49 (see Fig. 3.3). In step (α.1), the principal A starts a regular session with the dishonest principal I, sending it a nonce N_a. At this point, I decrypts the above message with its private key K_i^{-1} and encrypts the plain text with B's public key K_b. In step (β.1), I sends the resulting message to B, in order to establish a bogus session masquerading as A. B receives the message and responds in step (β.2) with the proper message. I cannot extract the nonce N_b and thus I forwards the above message to A. This message has the form expected by A for step (α.2) and therefore A responds in step (α.3) to I, completing in this way the session α. I can decrypt this message to obtain the nonce N_b expected by B in the step (β.3) and thus the attacker can complete the session, persuading B that A has established a communication session with it. In other words, the adversary succeeds in impersonating A. As argued by Lowe, the feasibility of the above attack is due to the absence of B's identity in the second message of the protocol. Indeed, if we include the responder's identity:

$$(2)\ B \rightarrow A : \{B, N_a, N_b\}_{Ka}.$$

The intruder cannot successfully replay the message (β.2) in the step (α.2).

3.3.2. *Encryption Flaws*

An elementary kind of encryption flaw is the absence of proper encryption for preserving the secrecy of critical information (e.g. secrets or private keys). This is another case of a functional specification flaw. For example, in the Nesset two-party protocol [65], a session key K_{ab} is encrypted with initiator's private key and sent to B, without any guarantee for the K_{ab}'s secrecy. Another example is a signature affixed to encrypted data:

$$A \rightarrow B : \{\{\text{message}\}_{Kb}\}_{K_a^{-1}}$$

where A first encrypts the message with B's public key for obtaining secrecy and then it signs the cipher text with its own private key. This scheme allows an attacker I to impersonate the author of the message by stripping A's signature and re-signing the message with the K_i^{-1} private key. It is common to use a document's signature for guaranteeing that the signer knew its content. Instead, when a signature is affixed to encrypted data, as in the above example, a third party cannot assume that the signer has any knowledge of the document. For avoiding this problem, the simplest solution is the inversion of the sequence of encryptions, namely it must sign the secret data before it is encrypted, i.e.:

$$A \rightarrow B : \{\{\text{message}\}_{K_a^{-1}}\}_{K_b}.$$

For example, as pointed out in Refs. 9, 19, the first two messages of CCITT X.509 protocol [87] (see Fig. 3.5) are afflicted with this kind of flaw.

Another unhappy situation arrises when a cryptographic algorithm exhibits particular algebraic properties that could compromise the security of a protocol. This can be considered as an implementation-dependent flaw. For example, consider the basic challenge-response scheme [37], reported in Fig. 3.6.

If the encryption algorithm is commutative:

$$\{\{m\}_{K_1}\}_{K_2} = \{\{m\}_{K_2}\}_{K_1}$$

$$(1)\ A \rightarrow B : A, \{T_a, N_a, B, X_a, \{Y_a\}_{Kb}\}_{K_a^{-1}}$$

$$(2)\ B \rightarrow A : B, \{T_b, N_b, A, N_a, X_b, \{Y_b\}_{Ka}\}_{K_b^{-1}}$$

$$(3)\ A \rightarrow B : A, \{N_b\}_{K_b^{-1}}$$

Fig. 3.5. The CCITT X.509 protocol.

$(1)\ A \rightarrow B : \{N_a\}_{K_{ab}}$ $(\alpha.1)\ A \rightarrow\ I(B) : \{N_a\}_{K_{ab}}$

$(2)\ B \rightarrow A : N_a$

(a) $(\beta.1)\ I(B) \rightarrow A : \{\{N_a\}_{K_{ai}}\}_{K_{ab}} = \{\{N_a\}_{K_{ab}}\}_{K_{ai}}$

$(\beta.2)\ A \rightarrow I(B) : \{N_a\}_{K_{ai}}$

$(\alpha.2)\ I(B) \rightarrow A : N_a$

(b)

Fig. 3.6. The attack (b) on the basic challenge-response scheme (a).

$(1)\ A \rightarrow S : A, B, N_a$ $(\alpha.1)\ A \rightarrow S : A, B, N_a$

$(2)\ S \rightarrow A : \{N_a, B, K_{ab}\ ,$ $(\alpha.2)\ S \rightarrow A : \{N_a, B, K'_{ab}, \{K'_{ab}, A\}_{K_{bs}}$

$\{K_{ab}, A\}_{K_{bs}}\}_{K_{as}}$

$(3)\ A \rightarrow B : \{K_{ab}, A\}_{K_{bs}}$ $(\alpha.3)\ A \rightarrow I, B : \{K'_{ab}, A\}_{K_{bs}}$

$(4)\ B \rightarrow A : \{N_b\}_{K_{ab}}$ $(\alpha.4)\ B \rightarrow A : \{N_b\}_{K'_{ab}}$

$(5)\ A \rightarrow B : \{N_b + 1\}_{K_{ab}}.$ $(\alpha.5)\ A \rightarrow B : \{N_b + 1\}_{K'_{ab}}$

(a) $(\beta.3)\ I(A) \rightarrow B : \{K'_{ab}, A\}_{K_{bs}}$

$(\beta.4)\ B \rightarrow\ I(A) : \{N'_b\}_{K'_{ab}}$

$(\beta.5)\ I(A) \rightarrow\ B : \{N'_b + 1\}_{K'_{ab}}$

(b)

Fig. 3.7. The attack (b) on the Needham–Schroeder secret key protocol (a).

then the attack reported in Fig. 3.7 is possible [25]. With this attack, the adversary can impersonate B in both sessions. These situations may occur when protocols are designed without any reference to a particular cryptographic algorithm, since an unfortunate combination of adopted algorithm and protocol is sometimes enough to compromise protocol reliability.

3.3.3. *Timeliness Flaws*

Another kind of functional specification flaw regards temporal information. Usually, it is called "timeliness flaw" or "freshness flaw". A well-designed protocol should guarantee parties that the exchanged messages are fresh and not replays of the old ones. For this reason, it is very important that parties have some mechanisms to correctly deduce message freshness. In order to satisfy this requirement, protocols usually exploit nonces (also called freshness identifiers) as part of a challenge-response scheme. The purpose is to guarantee that the response message is made after the challenge message has been sent. This goal is accomplished by introducing a functional dependency in the response message from the challenge message. Nonces can be implemented in many ways: random numbers, timestamps or counters. In the design of a protocol, this choice is crucial, since any solution guarantees different levels of freshness. Unfortunately, sometimes designers have a poor awareness of the different properties assured by freshness identifiers. This leads to serious mistakes in protocol design (see for example [3]). In other cases, the flaw resides not in an incorrect use of nonces, but in a complete lack of them in a critical message of protocol. A classic example is the Needham–Schroeder secret key protocol, [64, 70] shown in Fig. 3.7.

The goal of this protocol is the authentication of parties and the exchange of a fresh session key generated by an authentication server. The main problem in this protocol is that responder B has no way of deducing that the message (3) is fresh [28]. As a consequence, an attacker can compromise an old key K'_{ab} exchanged in a session α and then replay the appropriate message (3) in a later session β, fooling B into accepting the compromised key as new. This attack is showed in Fig. 3.7. After this attack, I can start a communication with B masquerading as A and understanding messages encrypted with the bogus key K'_{ab}. However, the above scenario requires that the attacker is able to guess the old session key K'_{ab} (by cryptanalysis or exhaustive key-search) or to steal the key by exploiting a weakness of the system implementing the protocol.

3.3.4. *Distinctness Flaws*

Another delicate question involved in protocol design is assuring that principals can recognise messages for what they are and to associate them correctly with the current step of the protocol they are executing. As stated in Ref. 3, the possible forms of confusion (which could happen together) are, first, between the expected message and an homologous message from a previous execution of the protocol; and second, between the expected message and a message belonging to the same protocol or to another protocol. These are implementation-dependent flaws. Indeed, the reason such confusions are possible is that, at an abstract level, a message is a sequence of components each of which have some value but, at a concrete level, a message is represented by a sequence of bits. The above confusion occurs if the recipient of a message accepts that message as valid but it performs an interpretation on the bit sequence that does not correspond to the intended meaning of its creator. In order to avoid this problem [22, 21], all the atomic elements of a message must be distinguishable for a principal, i.e. the principal must be able to recognise their type univocally. This goal may be achieved by means of an explicit typing of components, i.e. adding a type encoding to each component of a given message. This redundancy introduces easily predictable information in

(a)	(b)
(1) $A \to B : \{N_a\}_{K_{ab}}$	(1) $A \to B : \{N_a\}_{K_{ab}}$
(2) $B \to A : \{N_a + 1, N_b\}_{K_{ab}}$	(2) $B \to I, A : \{N_a + 1, N_b\}_{K_{ab}}$
(3) $A \to B : \{N_b + 1\}_{K_{ab}}$	(3) $A \to I(B) : \{N_b + 1\}_{K_{ab}}$
(4) $B \to A : \{K'_{ab}, N'_b\}_{K_{ab}}$	(4) $I(B) \to A : \{K'_{ab} = (N_a + 1), N_b\}_{K_{ab}}$

Fig. 3.8. The attack (b) on the Andrew secure RPC protocol (a).

the encrypted fragments and thus makes the message vulnerable to cryptanalysis. In order to avoid this undesirable consequence, types could be implicitly encoded if principals know characteristic constraints on length, structure and contents of variables and messages. The disadvantage of this solution is that often is difficult, or even impossible, to establish univocal criteria for message rejection. In literature, this kind of error is often denominated "type flaw". A protocol where this problem occurs is the Andrew Secure RPC protocol [73], reported in Fig. 3.8.

The protocol is intended to allow an initiator A to obtain a new session key K'_{ab} from a responder B, given that they already share a key K_{ab}. If nonces and keys are both implemented as bit sequences of the same length, then the protocol is vulnerable [23] to the attack shown in Fig. 3.8. In the above attack, the adversary exploits the similarity between messages (2) and (4). Indeed, both messages have same length and are encrypted with the same key K_{ab}. Therefore, an attacker can replay message (2) as message (4). The principal A is not able to detect the cheat and thus it is fooled into accepting the nonce value $N_a + 1$ as the new session key. Even if the nonce is random and fresh, it may not be generated with the care required by a good session key.

3.4. Attack Strategies

This analysis aims at outlining how an adversary can exploit different protocols, sessions and messages to attack a protocol. In our classification, we take into account both no-replay and replay attacks. Furthermore, from our perspective, a replayed message is the result of manipulations upon previous messages (as sketched in Fig. 3.2). In the following, we present two classifications of attack strategies: strategies about protocols and sessions involved in an attack, strategies about messages used by an attacker.

3.4.1. *Strategies About Sessions*

This classification is based on protocol sessions from which original information is exploited by an attacker, in comparison with a session where the deduced information is used. The classification structure is reported in Fig. 3.1. It envisages a multi-tasking environment which may run simultaneous protocol sessions involving several users. Moreover, the same system may run different protocols and, for practical reasons, the underlying public key infrastructure permits the use of a given key in different protocols. In this situation, we say that two protocols P and Q interact when they use the same keys and when the information derived from P's execution can be used to attack Q's session. The feasibility of interaction could not be accidental, but achieved by means of a plausible tailor-made protocol, also referred to as chosen-protocol. Many examples of this kind of attack, denominated as "multi-protocol attacks" or "chosen-protocol attacks", are reported in Refs. 4, 5, 46.

3.4.1.1. *Multiple session attacks*

This situation occurs when two different protocol sessions (χ and ψ in Fig. 3.2) are involved in an attack. In this situation, we have:

$$\chi \neq \psi.$$

In other words, the attacker derives information from outside the current execution of the protocol. Notice that this is possible only for replay attacks. There are two possible kinds of multiple session attacks: interleaved, when both executions χ and ψ simultaneously run; disjoined, when the execution of ψ starts after the end of χ. In both cases, a replay can involve executions of the same protocol (mono-protocol attack, $P = Q$ in Fig. 3.2) or different protocols (multi-protocol attack, $P \neq Q$ in Fig. 3.2). In Refs. 16, 21, 23, 24, multiple session attacks are often denominated as "external session attacks", "parallel session attacks", or "oracle attacks". The latter expression is used when in session χ, an unaware principal provides to attacker the response to a challenge sent in the other session ψ, acting as an oracle in this way.

3.4.1.1.1. Interleaved session

In the above sections, we have already described some examples of mono-protocol interleaved session attacks. Another noteworthy example is a complex attack [47] on the CCITT X.509 protocol that requires an interleaving of three simultaneous executions. An interesting example of multi-protocol, multiple-session attack is on Agora [35], an e-commerce protocol for pay-per-view web pages (see Fig. 3.9).

In the first step, customer A sends to merchant B a message containing a request for a price quotation R_P, the merchant responds with a signed message having a session number N, the price P and its certificate C_B. When the customer receives this message, it must decide on whether to proceed with the purchase. If so, it verifies the certificate and the signature and thus, if everything is all right, it sends the merchant the signed message including its certificate C_A, the session number N sent in message (2), and the price P. After the verification of this message, the merchant delivers the required pages. Even though the above protocol seems to

$$(1)\ A \rightarrow B : R_P$$

$$(2)\ B \rightarrow A : C_B,\ N, P, \text{sign}_B\{C_B, N, P\}$$

$$(3)\ A \rightarrow B : C_A,\ N, P, \text{sign}_A\{C_A, N, P\}$$

(a)

$$(P.\alpha.1)\ A \rightarrow I(B) : R_V$$

$$(Q.\beta.1)\ I(A) \rightarrow B : R_P$$

$$(Q.\beta.2)\ B \rightarrow I(A) : C_B, N, P, \text{sign}_B\{C_B, N, P\}$$

$$(P.\alpha.2)\ I(B) \rightarrow A : N_1 \equiv (N, P)$$

$$(P.\alpha.3)\ A \rightarrow I(B) : C_A, N_1, \text{sign}_A\{C_A, N_1\}$$

$$(Q.\beta.3)\ I(A) \rightarrow B : C_A, (N, P) \equiv N_1, \text{sign}_A\{C_A, (N, P) = N_1\}$$

(b)

Fig. 3.9. The attack (b) on the Agora protocol (a).

$$(1)\ A \rightarrow B : R_V$$

$$(2)\ B \rightarrow A : N_1$$

$$(3)\ A \rightarrow B : C_A, N_1, \text{sign}_A\{C_A, N_1\}$$

Fig. 3.10. The age-verification protocol.

be secure, it may present unexpected problems when executed in an environment where there are other protocols employed. Consider, for example, the protocol shown in Fig. 3.10. It allows a merchant the verification of the age of a given client.

Client A sends to server B a request R_V containing the pages that it wants to see. At this point, the customer challenges A sending it a nonce N_1. B must answer with a signed message containing the same nonce and its certificate. After the receipt of the last message, if the age reported in the certificate is appropriate, the web pages are sent to the client. If the nonce N_1 has the same size of the concatenation of components N and P involved in the Agora protocol, then it is possible [46] the attack reported in Fig. 3.9. After this attack, B believes to have executed a session of the Agora protocol with A, while A believes to have executed a session of the Age-Verification protocol with B. In other words, an attacker can buy whatever service from B impersonating A and debiting it with every purchase.

3.4.1.1.2. Disjoined session

A very famous example of mono-protocol disjoined replay occurs in the attack on Needham–Schroeder protocol (see Fig. 3.7). Another interesting attack that exploits this kind of replay can be found in an attack on Andrew Secure RPC protocol [19].

$$(1)\ A \rightarrow B : C_a^p$$

$$(2)\ B \rightarrow A : C_b^p, C_b, K_b^t, \mathrm{sign}_B\{K_b^t\}$$

$$(3)\ A \rightarrow B : \{K_{ab}^{\mathrm{pm}}\}_{K_b^t}$$

$$(4)\ B \rightarrow A : \{H(M_{\mathrm{ex}})\}_{K_{ab}^{\mathrm{pm}}}$$

(a)

$$(P.\alpha.1)\ I(A) \rightarrow B : A, B, \{K_i^t\}_{K_b}$$

$$(P.\alpha.2)\ B \rightarrow I(A) : B, A, \mathrm{sign}_B\{K_i^t\}$$

$$(Q.\beta.1)\ A \rightarrow I(B) : C_a^p$$

$$(Q.\beta.2)\ I(B) \rightarrow A : C_b^p, C_b, K_i^t, \mathrm{sign}_B\{K_i^t\}$$

$$(Q.\beta.3)\ A \rightarrow I(B) : \{K_{ab}^{\mathrm{pm}}\}_{K_i^t}$$

$$(Q.\beta.4)\ I(B) \rightarrow A : \{H(M_{\mathrm{ex}})\}_{K_{ab}^{\mathrm{pm}}}$$

(b)

Fig. 3.11. The attack (b) on the SSL-like protocol (a).

A valuable example [4] of multi-protocol disjoined session attack concerns a protocol similar to SSL,[b] reported in Fig. 3.11.

The primary goal of such a protocol is to establish a secret key between two parties, typically a client and a server, in order to obtain privacy of subsequent communication. By sending message (1), client A indicates to server B its cryptographic preferences C_a^p, namely it proposes to the server a set of encryption and signature algorithms that it is able to employ. In step (2), the server indicates its choice, provides a copy of its certificate and supplies a temporary public key K_b^t, signed to preserve its integrity. At this point, A sends in step (3) a pre-master key K_{ab}^{pm} to B, encrypted with K_b^t to preserve its secrecy. Such a pre-master key is used to compute keys that will protect and authenticate the next communication between A and B. Finally, in the message (4), B sends to A the hash of all the exchanged information M_{ex}, encrypted with the recently negotiated key, in order to detect possible alterations of received messages. Now, consider the protocol reported in Fig. 3.12.

In this protocol, principal A sends a secret message to another principal B in order to obtain a signature authenticating the receipt of the message itself. Since the server is forced to sign a hash of a message that it has not built, this protocol can be exploited to violate the SSL-like protocol, as shown in Fig. 3.11. The attacker creates its own temporary public key K_i^t and sends it to B in step

$$(1)\ A \rightarrow B : A, B, \{m\}_{K_b}$$

$$(2)\ B \rightarrow A : B, A, \mathrm{sign}_B\{m\}$$

Fig. 3.12. Signed receipt protocol.

[b]SSL stands for Secure Socket Layer, the protocol defined by Netscape [34] that inspired the Transport Layer Protocol (TLS)[29].

($P.\alpha.1$) masquerading as A. B responds with the signature of the key. At this point, the attacker waits until A begins a session β with B. During this session, it exploits $\text{sign}_B\{K_i^t\}$ in step ($P.\beta.2$), leading A to accept its own temporary public key K_i^t as the public temporary key of server B. In this way, the attacker is able to decrypt message ($P.\beta.3$) and thus it obtains pre-master key K_{ab}^{pm}. As a consequence, the adversary is able to impersonate the server B with the client A and decrypt all the messages encrypted with keys computed from pre-master key K_{ab}^{pm}.

3.4.1.2. *Single session attacks*

A single session attack occurs when an adversary either generates or derives from the current protocol session the messages that it needs. In other words, with respect to Fig. 3.2, we have:

$$\chi = \psi, \quad P = Q.$$

In literature, we can find many protocols that are vulnerable to this kind of attack. A remarkable example is the Otway–Rees protocol, a shared-key protocol [68] intended for authentication and key exchange of a secret session key, with the help of an authentication server. Each party A and B must already share a secret key with the server. The protocol is reported in Fig. 3.13.

There are at least two attacks on this protocol. One of these [18] is a single session attack, as shown in Fig. 3.13. The result of this attack is an adversary deceiving initiator A into believing it is in the presence of B. Moreover, A accepts as a shared-key a fake key K_{ab} constituted by public information (M, A, B). Notice that attack feasibility depends on protocol implementation, since such an attack exploits a type flaw. Unfortunately, the occurrence of this attack it is not a remote possibility, since it is not unlikely that an implementation employs 16 bits for principal identifiers, 32 bits for nonces and 64 bits for keys.

(a)	(b)
(1) $A \to B : M, A, B, \{N_a, M, A, B\}_{K_{as}}$	(1) $A \to I(B) : M, A, B, \{N_a, M, A, B\}_{K_{as}}$
(2) $B \to S : M, A, B, \{N_a, M, A, B\}_{K_{as}},$	(2) omitted
$M, A, B, \{N_b, M, A, B\}_{K_{bs}}$	(3) omitted
(3) $S \to B : M, \{N_a, K_{ab}\}_{K_{as}}, \{N_b, K_{ab}\}_{K_{bs}}$	(4) $I(B) \to A : M, \{N_a, K_{ab} = (M, A, B)\}_K$
(4) $B \to A : M, \{N_a, K_{ab}\}_{K_{as}}$	

Fig. 3.13. The attack (b) on the Otway–Rees protocol (a).

3.4.2. *Strategies About Messages*

Each component of any attacker message may have been generated by the attacker itself or extracted/deduced from a message received/intercepted in the past. In the first case, we have a "no-replay". In the other cases, when an adversary sends a message containing information coming out from the elaboration of intercepted messages, we have a "replay". In the original meaning, a replay is an attack in which a past message is played back to the same recipient. In a broader meaning (see Fig. 3.2), the above expression is used ([37, 79]) to mean the capture of a message m_i that is employed at later time by attacker to send a message m to any recipient. The message m may contain the whole original message m_i, just a piece of it, or, in general, information derived from m_i. When the replayed message m is equal to message m_i, we have the so-called "simple replay", whereas when the attacker performs manipulations upon the past message, we use the expression "derived replay". With respect to the general form of Fig. 3.2, the function $f(.)$ is the identity function in the case of simple replays, a more sophisticated function in the case of derived replays. An adversary in a single attack on a protocol execution may exploit different kinds of messages, for this reason, there are many possible attack scenarios. This classification is depicted in Fig. 3.1.

3.4.2.1. *No-replay attacks*

An example where an adversary can attack a protocol with no-replay is the TMN [83]. The aim of this protocol (reported in Fig. 3.14) is authentication and key-exchange for users of a mobile communication system.

We denote with $V(K_a, K_b)$ the Vernam encryption, namely the bit-wise exclusive, or between the two keys K_a and K_b. When initiator A wants to set up a secure session with responder B, it chooses a key K_a, encrypts K_a with the public key K_s of server, and sends the result to server S in step (1). After the receipt of the above message, the server informs B of A's intention by means of message (2). At this point, the responder B chooses a session key K_b, encrypts K_b with the public key K_s, and sends the resulting message to S. The server computes the Vernam encryption between K_a and K_b, and forwards the result to A in step (4). Since

(a)	(b)
(1) $A \to S : A, S, B, \{K_a\}_{K_s}$	(1) $I(A) \to S : A, S, B, \{K_i\}_{K_s}$
(2) $S \to B : S, A, B$	(2) $S \to B : S, A, B$
(3) $B \to S : B, S, A, \{K_b\}_{K_s}$	(3) $B \to S : B, S, A, \{K_b\}_{K_s}$
(4) $S \to A : S, A, B, V(K_a, K_b)$	(4) $S \to I(A) : S, A, B, V(K_i, K_b)$

Fig. 3.14. An attack (b) on the TMN protocol (a).

$V(K_a, V(K_a, K_b)) = K_b$, the initiator A, knowing K_a is able to deduce the session key K_b. The above protocol is subject to several attacks [7, 77, 51]. Here, we show one of those illustrated in Ref. 51 while another one is discussed in Section 3.5.1. In the attack, the adversary can impersonate initiator A with responder B, acting as shown in Fig. 3.14. At the end of this attack, B erroneously believes that it is communicating with A and the attacker is also able to deduce the session key K_b used by B to encrypt the messages intended for A. Notice that the attacker does not perform message replays.

3.4.2.2. *Replay attacks*

Replay attacks can be classified with respect to the recipient of the replayed message m, in comparison with the intended recipient of the old message m_i (see Fig. 3.1).

3.4.2.2.1. Original destinations

We have an original destination when the message m is replayed to the same recipient of the old message m_i. Therefore, with respect to Fig. 3.2, we have:

$$Y = W.$$

In such a situation, we have that the attacker is able to intercept all the messages, eventually manipulate them, and send them to designated receiver. A famous protocol vulnerable in this way is the CCITT X.509 protocol (see Fig. 3.5). There, we report a complex attack [47] on this protocol that requires an interleaving of three simultaneous executions, as shown in Fig. 3.15:

In session α, A performs a normal session with B, but the attacker intercepts message (α.1) and replays it to original receiver (message (β.1)). In this way, it starts a new session β with B in order to impersonate A. For this purpose, the

$$(\alpha.1) A \to I, B : A, \{T_a, N_a, B, X_a, \{Y_a\}_{K_b}\}_{K_a^{-1}}$$
$$(\alpha.2) B \to A : B, \{T_b, N_b, A, N_a, X_b, \{Y_b\}_{K_a}\}_{K_b^{-1}}$$
$$(\alpha.3) A \to B : A, \{N_b\}_{K_b^{-1}}$$
$$(\beta.1) I(A) \to B : A, \{T_a, N_a, B, X_a, \{Y_a\}_{K_b}\}_{K_a^{-1}}$$
$$(\beta.2) B \to I(A) : B, \{T_b', N_b', A, N_a, X_b', \{Y_b'\}_{K_a}\}_{K_b^{-1}}$$
$$(\gamma.1) A \to I : A, \{T_a', N_a', I, X_a', \{Y_a'\}_{K_i}\}_{K_a^{-1}}$$
$$(\gamma.2) I \to A : I, \{T_i, N_b', A, N_a', X_i, \{Y_i\}_{K_a}\}_{K_i^{-1}}$$
$$(\gamma.3) A \to I : A, \{N_b'\}_{K_a^{-1}}$$
$$(\beta.3) I(A) \to B : A, \{N_b'\}_{K_a^{-1}}$$

Fig. 3.15. The attack on the CCITT X.509 protocol.

attacker needs the A's signature of nonce N'_a for message (3) of the protocol. As a consequence, it engages A to initiate a normal session γ where I is the responder. Notice that no manipulation of intercepted messages is required in such an attack, i.e. we have only simple replay messages.

3.4.2.2.2. Deviations

A deviation occurs when the message m is replayed to other than the intended recipient of the old message m_i. There are two possible kinds of deviations. With respect to Fig. 3.2, if the message m is replayed to the sender of the old message m_i, we have a reflection:

$$Y \neq W, \quad W = X$$

whereas, if the message m is replayed to a third party, we have a deflection:

$$Y \neq W, \quad W \neq X.$$

Reflections There is a reflection attack on the ISO-SC 27 protocol [43]. This protocol is constituted by two challenge-response exchanges between two parties that already share a secret key. The protocol is shown in Fig. 3.16. Inspite of its simplicity, this protocol is subject to an attack [16], as reported in Fig. 3.16.

In such an attack, all the replays performed by the attacker are reflections. Indeed, whenever A sends a message to B, attacker intercepts the message, removes it from the communication channel and replays it to the original sender. As a consequence of this attack, the adversary deceiving A masquerades as B in both sessions.

Deflections An interesting attack of this kind regards the BAN-Yahalom protocol [19], a variant of the Yahalom protocol [19] due to Burrows, Abadi and Needham. The protocol, shown in Fig. 3.17, is intended to provide both authentication and key-exchange between two parties, involving an authentication server.

(1) $A \rightarrow B : N_a$

(2) $B \rightarrow A : \{N_a, N_b\}_{K_{ab}}$

(3) $A \rightarrow B : N_b$

(a)

(α.1) $A \rightarrow I(B) : N_a$

(β.1) $I(B) \rightarrow A : N_a$

(β.2) $A \rightarrow I(B) : \{N_a, N_b\}_{K_{ab}}$

(α.2) $I(B) \rightarrow A : \{N_a, N_b\}_{K_{ab}}$

(α.3) $A \rightarrow I(B) : N_b$

(β.3) $I(B) \rightarrow A : N_b$

(b)

Fig. 3.16. The attack (b) on the ISO-SC 27 protocol (a).

(1) $A \to B : A, N_a$

(2) $B \to S : B, N_b, \{A, N_a\}_{K_{bs}}$

(3) $S \to A : N_b, \{B, K_{ab}, N_a\}_{K_{as}}, \{A, K_{ab}, N_b\}_{K_{bs}}$

(4) $A \to B : \{A, K_{ab}, N_b\}_{K_{bs}}, \{N_b\}_{K_{ab}}$

(a)

(α.1) $A \to I(B) : A, N_a$

(β.1) $I(B) \to A : B, N_a$

(β.2′) $A \to I(S) : A, N_a', \{B, N_a\}_{K_{as}}$

(β.2″) $I(A) \to S : A, N_a, \{B, N_a\}_{K_{as}}$

(β.3) $S \to I(B) : N_a, \{A, K_{ab}, N_a\}_{K_{bs}}, \{B, K_{ab}, N_a\}_{K_{as}}$

(β.4) omitted

(α.2) omitted

(α.3) $I(S) \to A : N_i, \{B, K_{ab}, N_a\}_{K_{as}}, \{A, K_{ab}, N_b\}_{K_{bs}}$

(α.4) $A \to I(B) : \{A, K_{ab}, N_a\}_{K_{bs}}, \{N_i\}_{K_{ab}}$

(b)

Fig. 3.17. The attack (b) on the BAN-Yahalom protocol (a).

The attack [79], reported in Fig. 3.17, involves original destination, reflection and deflection attacks. In the step (β.1), the attacker performs a reflection, replaying to A the message (α.1) sent by A itself to B. After the manipulation of message (β'.2), the attacker sends the altered message to designated receiver S, thus it executes a replay to the original destination. Finally, the attacker performs a deflection, as it captures the message (β.3) intended for B and, after its manipulation, sends the resulting message to A in step (α.3).

3.5. Attacks Consequences

The consequence of any attack is the violation of one or more security requirements. For this reason, in this section, we classify attacks from the point of view of requirements compromised by attacks reported in literature. There are many requirements achievable by means of cryptographic protocols. Because of the importance of this issue, there are many attempts to formalise security requirements [2, 19, 48, 59, 82]. Our focus is different. We are not interested in formalising requirements, but in classifying attacks. Therefore, we simply provide the reader an informal overview of requirements. For what concerns the classification of attack consequences, notice that there are requirements that are not compromised by known attacks on cryptographic protocols. Indeed, as shown in Table 3.2, attacks reported in literature compromises one or more of the following requirements:

authentication, *privacy* and *secret freshness*. There are various reasons that can justify this assertion. First, in the field of cryptographic protocols, authentication protocols (aiming typically to satisfy *authentication*, *privacy* and *secret freshness*) have been the first protocols published and thus the most inspected (also by means of formal methods). Therefore, the more recent protocols types (conceived to satisfy other requirements also) have been designed with more awareness and accuracy; also thanks to formal methods devised in recent years, they are probably more secure. Second, various security requirements are not independent of each other, namely the fulfilment of a requirement can imply satisfaction of other requirements. As a consequence, a violation of one property can imply the violation of others. A formal analysis of how security requirements are tied is a matter of research. Finally, the intrinsic complexity of certain requirements, sometimes, obliges designers to relax them. For example, in the SET specifications [56],[c] it is explicitly stated that *non-repudiation* is not a requirement of SET.

3.5.1. *Violation of Authentication*

One of the most common purposes of a cryptographic protocol is the verification of real identity of principals. If an attacker succeeds in compromising the authentication mechanism of a system, the typical consequence for its members is an incorrect perception of the identity of partners and therefore the opportunity for an attacker to impersonate another entity. The most common aim when compromising an authentication protocol is the impersonation by attacker I of one party X of another Y. X could be active in a communication with another member of the system or even not connected to the system. After a similar attack, a given principal Y has a communication session α apparently open with a principal X, but in fact Y communicates with an adversary. As a consequence, Y believes that it can send to X messages that in fact X will not receive. However, if the attacker does not know the appropriate key, it can neither understand these messages nor generate messages to send to Y acting as X. Nevertheless, it can have unauthorised access (i.e. violation of *access control*) to some resources of system. In the above sections, we have described many examples of attack leading to such consequences concerning authentication. Now, we will show another example, the attack [69] on the Lu–Smolka variant [52] of the SET protocol. Such a protocol is invoked during a web-based commercial transaction, when a given client (the cardholder), after the selection of the goods/services that it wishes to purchase/request, performs an on-line payment. Figure 3.18 reports the Lu and Smolka protocol. C denotes the cardholder, M the merchant and P the payment gateway. TID represents the unique identifier of the transaction; it can be considered as a nonce. PA is the amount that the cardholder is supposed to pay, PY is the amount that the cardholder is willing

[c]SET stands for Secure Electronic Transactions, an electronic-commerce protocol jointly developed by Visa and Mastercard in order to guarantee secure transactions over open networks.

(1) $C \rightarrow M : TID\_request$

(2) $M \rightarrow C : \text{sign}_M\{TID\}$

(3) $C \rightarrow M : \text{sign}_C\{TID\},$

$\{TID, PA\}_{K_m},$

$\text{sign}_C\{\{TID, PY, CA\}_{K_p}\}$

(4) $M \rightarrow P : \text{sign}_C\{\{TID, PY, CA\}_{K_p}\},$

$\text{sign}_M\{TID\}, \{TID, AA, MA\}_{K_p}$

(5) $P \rightarrow M : \text{sign}_P\{TID, Tr\}$

(6) $M \rightarrow C : \text{sign}_P\{TID, Tr\}$

(a)

(α.1) $C \rightarrow I : TID\_request$

(β.1) $I(C) \rightarrow M : TID\_request$

(β.2) $M \rightarrow I(C) : \text{sign}_M\{TID\}$

(α.2) $I \rightarrow C : \text{sign}_I\{TID\}$

(α.3) $C \rightarrow I : \text{sign}_C\{TID\}, \{TID, PA\}_{K_i},$

$\text{sign}_C\{\{TID, PY, CA\}_{K_p}\}$

(β.3) $I(C) \rightarrow M : \text{sign}_C\{TID\}, \{TID, PA\}_{K_m},$

$\text{sign}_C\{\{TID, PY, CA\}_{K_p}\}$

(β.4) $M \rightarrow P : \text{sign}_C\{\{TID, PY, CA\}_{K_p}\},$

$\text{sign}_M\{TID\}, \{TID, AA, MA\}_{K_p}$

(β.5) $P \rightarrow M : \text{sign}_P\{TID, Tr\}$

(β.6) $M \rightarrow I(C) : \text{sign}_P\{TID, Tr\}$

(b)

Fig. 3.18. The attack (b) on the Lu–Smolka variant of the SET protocol (a).

to pay and AA is the charge amount that the merchant requests for authorisation. CA and MA are the identifiers of cardholder's account and merchant's account, respectively. Finally, Tr denotes the response of the gateway indicating either an authorisation or a decline.

Intuitively, the protocol works as follows. In step (1), C sends to M the request for a unique transaction identifier. M assigns TID to the transaction, signs it and returns it to C in step (2). At this point, in step (3), C sends to M the Ordering Information (OI), i.e. TID and PA, and the Payment Instruction (PI), i.e. TID, PY and CA. In step (4), M sends to P the PI, the signed TID, and TID, AA, MA encrypted with the public key of gateway. In step (5), P authorises C's payment card, checks information received, performs appropriate account operations and returns the transaction result Tr to M. Finally, in step (6), M reads the response and forwards the message to C. This protocol can be attacked in several ways; such an attack is shown in Fig. 3.18. The dishonest merchant I waits for a victim to start a session α with it. When this happens, it opens an interleaved session β, acting as the client C towards another merchant M. The TID provided by M in step (β.2) is sent by I to C as transaction identifier of session α. In this way, in step (α.3), I obtains the two components signed by C that can be used to impersonate C in step (β.3). Moreover, the component $\text{sign}_C\{\{TID, PY, CA\}_{K_p}\}$ is involved in the step (β.4), where M requires authorization payment to gateway P. The gateway is unable to detect the cheat and thus authorises the payment for M. For this reason, the session β ends, allowing I to masquerade as C towards M. This means that authentication is violated. From a financial point of view, in the session β, C pays for non-purchased goods. Moreover, if the goods are deliverable by means of the network system, such as the content of a web-page, I obtains those goods. In the session α, if I does not leave the session, C obtains the required goods from I, otherwise C receives nothing. Notice that this is a kind of attack where the dishonest entity is a legitimate participant of the protocol. Another example is the attack [3] on the Denning–Sacco public key protocol [28]. However, as we have seen in the previous sections, many attacks violate authentication without requiring membership of the adversary to the system, namely I must not explicitly take part in a run of protocol.

Recently, a particular family of attacks on authentication has been found [47]. These attacks, denominated as "multiplicity attacks", cause a principal Y to think that another principal X is attempting to set up two or more simultaneous session with it, when in fact X is trying to establish only one session. These attacks may lead to serious consequences since the attacker can eavesdrop messages sent in the genuine session and then replay them in the fake session. However, in this situation, the adversary cannot decrypt these messages. Sometimes, these attacks are very easy to perform, as the attacker must simply replay the last message of the protocol to the designed recipient. Another type of attack compromising authentication involves repeated authentication protocols [26, 50, 81]. These protocols can be attacked both in the initial authentication part and subsequent authentication part. An adversary may impersonate a principal in the initial authentication part. When this situation occurs, an attacker is able to repeat the impersonation also in the subsequent authentications. Sometimes, an adversary can attack subsequent authentication without violating initial authentication. In this case, the attacker succeeds in exploiting tickets overheard from initial authentication to masquerade as principal in the subsequent authentication. In this way, an attacker performs a fake renewal of authentication to a principal.

3.5.2. *Violation of Privacy*

Privacy is the most immediate security objective achievable by means of cryptography. With the concept of perfect encryption, we assume that encryption algorithms are unbreakable and therefore we do not examine violations of privacy through cryptanalytic techniques. Since the privacy of an encrypted message relies only in the secrecy of decryption key, the main issue of privacy concerns reliable distribution of keys to parties. Often, cryptographic protocols do not achieve this goal, because they do not guarantee a secure way to exchange a session key. If an attacker is able to compromise a session key, the most obvious consequence is the impossibility of keeping private sensible information exchanged between parties (say X and Y) during subsequent communication. Unfortunately, when this happens, there are also serious consequences about the *integrity* of messages. Moreover, by knowing a session key K, the adversary is able to create a new message of the form $\{\text{msg}\}_K$ and send it to Y, leading Y to believe that it comes from X. We have already seen attacks that influence privacy. For example, one of these is the attack on the Otway–Rees protocol reported if Fig. 3.12. This attack compromises both authentication and privacy. However, sometimes, privacy may be subverted without violating authentication. For example, in a different attack [23] on the Otway–Rees protocol, parties perform a correct authentication, but key-exchange is compromised by an adversary. Indeed, the parties accept the fragment (M, A, B) as session key K_{ab}, similarly to the attack of Fig. 3.12. In this case, A and B have an effective communication session open and they exchange messages, but the adversary has a complete control over the messages.

In some attacks, the adversary is not able to discover the session key exchanged by means of a protocol execution, but it succeeds in fooling a principal into accepting a faked key. In this case, violation of privacy may occur through a type flaw upon the message where one party sends the session key to the other. For example, in the attack on the Andrew Secure RPC protocol reported in Fig. 3.8, the responder is fooled into accepting the nonce value $N_a + 1$ as the new session key. Even if the nonce is secret, it may not be generated with the care required for a good session key, thus that bogus key could be easily guessed by an attacker.

However, sometimes privacy can be compromised without exploiting a type flaw. There are some attacks where an adversary belonging to a system, in a session α shares a key K_{xi} with principal X and in a session β fools another principal Y into accepting the same key masquerading as X. In this way, in the session β, Y sends encrypted messages to X that X does not receive. Moreover, the attacker is able to understand these messages. This kind of attack may occur on an early version of SSH (the one specified in the June 1996 Internet Draft [89]), a protocol having the goal of authentication and privacy between two entities without the involvement of a third trusted party. In the case where the initiator is authenticated by means of public key cryptography, the protocol requires messages reported in Fig. 3.19.

The symbols K_{bh}, K_{bs} represent two public keys for B. In SSH terminology, the first is a long-term host key, whereas the second is a longer-term server key. The string prev_msgs denotes concatenation of first three messages of a protocol run. The protocol allows establishment of the session key K_{ab} between parties for subsequent communication. In every execution of the protocol, in order to guarantee the privacy of the last message, a secret key K' is employed, derived from K_{ab}, N_a and N_b. The protocol can be attacked by an adversary [1] as reported in Fig. 3.19. Whenever the entity A starts a session α with I, the adversary immediately begins an interleaved session β where it plays the role of an initiator with a third entity B. The attacker, proposing in the session β the same nonce N_a received from A in the session α, is able to spend the signed component $\{H(A, N_a, N_b)\}_{K_a^{-1}}$. In this

(1) $A \to B : N_a$

(2) $B \to A : N_b$

(3) $B \to A : K_{bh}, K_{bs}$

(4) $A \to B : \{H(\text{prev\_msgs}), K_{ab}\}_{K_{bs}}\}_{K_{bh}}$

(5) $A \to B : \{A, K_a, \{H(A, N_a, N_b)\}_{K_a^{-1}}\}_{K'}$

(a)

(α.1) $A \to I : N_a$

(β.1) $I(A) \to B : N_a$

(β.2) $B \to I(A) : N_b$

(α.2) $I \to A : N_b$

(β.3) $B \to I(A) : K_{bh}, K_{bs}$

(α.3) $I \to A : K_{ih}, K_{is}$

(α.4) $A \to I : \{H(\text{prev\_msgs}), K_{ab}\}_{K_{is}}\}_{K_{ih}}$

(β.4) $I(A) \to B : \{H(\text{prev\_msgs}'), K_{ab}\}_{K_{bs}}\}_{K_{bh}}$

(α.5) $A \to I : \{A, K_a, \{H(A, N_a, N_b)\}_{K_a^{-1}}\}_{K'}$

(β.5) $I(A) \to B : \{A, K_a, \{H(A, N_a, N_b)\}_{K_a^{-1}}\}_{K'}$

(b)

Fig. 3.19. The attack (b) on the SSH protocol (a).

way, I successfully terminates the session β and thus impersonates A with B. The above attack implies serious consequences about privacy as well as impersonation. In the message (α.4), the adversary receives a session key K_{ab} that is accepted by B in the session β as session key to share with A.

3.5.3. *Violation of Secret Freshness*

In Section 3.3.3, we have debated about protocol flaws related to temporal issues. In this context, we focus our attention on violation of secret freshness. Timeliness flaw and violation of freshness are closely related, since violation of a secret freshness is often allowed by a timeliness flaw in the protocol. However, as it can be deduced from Table 3.2, there is not a one-to-one relationship between the two issues. For example, the second attack on the KSL protocol [45], reported in Ref. 50, compromises the freshness of the session key, but it is due to the naming and distinctness flaws. Generally, sensible information must be fresh as well as secret, for at least two reasons. First, typically there is a relationship of cause and effect between receiving secret information and accomplishing a critical action by a recipient. When freshness is not guaranteed for a given secret, an attacker is able to obtain again the accomplishment of an action by means of a simple replay of the secret to the recipient. Eventually, this can lead to dramatic consequences in commercial and financial transactions where, for example, a dishonest party can enforce a payment twice (this is a violation of *authorisation* and *non-repudiation* as well). The second question about freshness is related to the maintenance of a secret, especially when the secret is a key. The longer a key is used, the greater is the chance that it will be compromised. Indeed, time is an advantage to those who wish to obtain the secret by means of cryptanalytic techniques or through other means. Obviously, the proper lifetime of a key depends on application. Encryption keys used to encrypt data file for storage, or private keys for public key cryptography application cannot be changed frequently. Conversely, a session key must be used only in the current communication session.

Although freshness requirement is crucial, often cryptographic protocols do not achieve this objective. The most classic example of an attack involving freshness regards the Needham–Schroeder secret-key protocol, illustrated in Fig. 3.7. In this case, the adversary is able to fool the responder into accepting as new, an old compromised session key.

Another valuable example of violation of freshness is given by an attack on the Woo–Lam mutual authentication protocol [85], a protocol that aims at achieving both authentication and secrecy. The protocol is shown in Fig. 3.20.

$$
\begin{aligned}
&(1)\ A \to B : A, N_a\\
&(2)\ B \to A : B, N_b\\
&(3)\ A \to B : \{A, B, N_a, N_b\}_{K_{as}}\\
&(4)\ B \to S : \{A, B, N_a, N_b\}_{K_{as}},\\
&\qquad \{A, B, N_a, N_b\}_{K_{bs}}\\
&(5)\ S \to B : \{B, N_a, N_b, K_{ab}\}_{K_{as}},\\
&\qquad \{A, N_a, N_b, K_{ab}\}_{K_{bs}}\\
&(6)\ B \to A : \{B, N_a, N_b, K_{ab}\}_{K_{as}},\\
&\qquad \{N_a, N_b\}_{K_{ab}}\\
&(7)\ A \to B : \{N_b\}_{K_{ab}}
\end{aligned}
$$

(a)

$$
\begin{aligned}
&(\alpha.1)\ A \to I : A, N_a\\
&\quad(\beta.1)\ I \to A : I, N_a\\
&\quad(\beta.2)\ A \to I : A, N_{a'}\\
&(\alpha.2)\ I \to A : I, N_{a'}\\
&(\alpha.3)\ A \to I : \{A, I, N_a, N_{a'}\}_{K_{as}}\\
&(\alpha.4)\ I \to S : \{A, I, N_a, N_{a'}\}_{K_{as}}, \{A, I, N_a, N_{a'}\}_{K_{is}}\\
&(\alpha.5)\ S \to I : \{I, N_a, N_{a'}, K_{ai}\}_{K_{as}}, \{A, N_a, N_{a'}, K_{ai}\}_{K_{is}}\\
&(\alpha.6)\ I \to A : \{I, N_a, N_{a'}, K_{ai}\}_{K_{as}}, \{N_a, N_{a'}\}_{K_{ai}}\\
&(\alpha.7)\ A \to I : \{N_{a'}\}_{K_{ai}}\\
&\quad(\beta.3)\ I \to A : \{I, A, N_a, N_{a'}\}_{K_{is}}\\
&\quad(\beta.4)\ A \to I(S) : \{I, A, N_a, N_{a'}\}_{K_{is}},\\
&\qquad\qquad \{I, A, N_a, N_{a'}\}_{K_{as}}\\
&\quad(\beta.5)\ I(S) \to A : \{A, N_a, N_{a'}, K_{ai}\}_{K_{is}},\\
&\qquad\qquad \{I, N_a, N_{a'}, K_{ai}\}_{K_{as}}\\
&(\beta.6)\ A \to I : \{A, N_a, N_{a'}, K_{ai}\}_{K_{is}}, \{N_a, N_{a'}\}_{K_{ai}}\\
&\quad(\beta.7)\ I \to A : \{N_{a'}\}_{K_{ai}}
\end{aligned}
$$

(b)

Fig. 3.20. The attack (b) on the Woo–Lam mutual authentication protocol (a).

In the first three messages, parties exchange two nonces and claim their identities. By means of messages (3) and (4), the responder B contacts a trusted party S and obtains from it the session key K_{ab}, whereas the last two messages assure each party that the other party has the proper key. The protocol uses the nonces N_a and N_b in order to guarantee a principal, the freshness of received messages. Unfortunately, this objective is not achieved, as demonstrated by the attack reported in Fig. 3.20 [24, 25]. The above is not a particularly strong attack, but an abnormal situation that should not be permitted by a well-designed protocol. After that session α started, the malicious responder I initiates another session β using in the step (β.1) the same nonce N_a received from A in step (α.1). Similarly, in step (α.2), I uses the same nonce $N_{a'}$ received from A in the step (β.2). After the completion of session α that leads parties A and I to share the session key K_{ai}, I intercepts the message (β.4) and replays to A as message (β.5), the message (α.5) with the order of encrypted components switched. In this way, A is fooled into re-accepting K_{ai}, the key previously issued in session α. A more dangerous attack on this protocol can be found in Ref. [50].

Another protocol vulnerable to attacks having consequences about freshness is the wide-mouthed frog protocol [19]. It allows the transfer of a session key from A to B through S in only two messages, delegating to initiator A the choice of a session key. The above protocol admits different attacks [7, 3, 19, 24, 23]. In one of these [7, 24], an attacker impersonates, in turn, A and B. Moreover, it induces parties to share an old key. The attacker may continue until it wishes to induce A and B to accept the old key K_{ab} again. In other words, the real expiration time of the session key is decided by an adversary.

3.6. Concluding Remarks

In this chapter, we have provided a systematic description and a classification of attacks on cryptographic protocols. The classification is generated according to three different points of view: protocol flaws, attack strategies and attack consequences. We have also illustrated the resulting categories with several examples.

It could seem quite bizarre surveying cryptographic protocols from the perspective of attacks. Our review does not aim to persuade the reader that all the existing protocols are vulnerable. On the contrary, there are several secure protocols. Moreover, cryptographic protocols are perhaps the most effective solution for security problems in computer networks. This work is inspired by the awareness of difficulties in this field. Therefore, we hope that our approach responds to the immediate general need of making designers aware of the most common problems in protocol design. Besides, we hope that our classification also helps those who devises formal methods in testing and improving methodologies of protocol analysis.

Acknowledgements

This chapter is dedicated in memory of Maurizio Panti, who strongly supported us during our work. This work has been supported by a joint project between Polizia di Stato (Italy) and Università Politecnica delle Marche (Ancona, Italy).

References

1. M. Abadi, Explicit communication revisited: Two new attacks on authentication protocols, *IEEE Transactions on Software Engineering* **3**(3) (1997) 185–186.
2. M. Abadi, Security protocols and their properties, *Foundations of Secure Computation, 20th International Summer School* (2000) pp. 39–60.
3. M. Abadi and R. Needham, Prudent engineering practice for cryptographic protocols, *IEEE Transactions on Software Engineering* **22**(1) (1996) 6–15.
4. J. Alves-Foss, Multi-protocol attacks and the public-key infrastructure, *Proceedings of 21st National Information Systems Security Conference* (1998) 566–576.
5. J. Alves-Foss, Provably insecure mutual authentication protocols: the two-party symmetric-encryption case, *Proceedings of 22nd National Information Systems Security Conference*, 1999.
6. R. Anderson, Why cryptosystems fail, *Communications of the ACM* **37**(11) (1994) 32–40.
7. R. Anderson and R. Needham, Programming Satan's computer, *LNCS* (1995) 426–441.
8. R. Anderson and R. Needham, Robustness principles for public key protocols, *Advances in Cryptology–CRYPTO* (1995) 236–247.
9. C. l'Anson and C. Mitchell, Security defects in the CCITT recomendation X.509, *ACM Computer Communications Review* 2 (1990) 30–34.
10. W. Arbaugh, An inductive chosen plaintext attack against WEP/WEP2, *IEEE Document* (2001) 802.11-02/230.
11. N. Asokan, P. A. Janson, M. Steiner and M. Waidner, The state of the art in electronic payment systems, *IEEE Computer* **30**(9) (1997) 28–35.
12. M. Bellare and P. Rogaway, The AuthA protocol for password-based authenticated key exchange, *Contribution to IEEE P1363*, 2000.
13. S. M. Bellovin and M. Merritt, An attack on the interlock protocol when used for authentication, *IEEE Transactions on Information Theory* **40**(1) (1994) 273–275.
14. S. M. Bellovin and M. Merritt, Encrypted key exchange: password based protocols secure against dictionary attacks, *Proceedings of IEEE Symposium on Research in Security and Privacy* (1992) 72–84.
15. S. M. Bellovin and M. Merritt, Limitations of the Kerberos authentication system, *USENIX Conference Proceedings* (1991) 253–267.
16. R. Bird, I. Gopal, A. Herzberg, P. Janson, S. Kutten, R. Molva and M. Yung, Systematic design of a family of attack-resistant authentication protocols, *IEEE Journal on Selected Areas in Communications* **11**(5) (1993) 679–693.
17. N. Borisov, I. Goldberg and D. Wagner, Intercepting mobile communications: the insecurity of 802.11, *Proceedings of the 7th International Conference on Mobile Computing and Networking* (2001) 180–189.
18. C. Boyd, Hidden assumptions in cryptographic protocols, *Proceedings of the IEEE*, number **6** (1990) 433–436.
19. M. Burrows, M. Abadi and R. Needham, A logic of authentication, *ACM Transactions on Computer Systems* 1 (1990) 18–36.

20. N. Cam-Winget, R. Housley, D. Wagner and J. Walker, Security flaws in 802.11 data link protocols, *Communications of the ACM* **46**(5) (2003) 35–39.
21. U. Carlsen, Cryptographic protocol flaws, *Proceedings of the 7th IEEE Computer Security Foundations Workshop* (1994) 192–200.
22. U. Carlsen, Using logics to detect implementation-dependent flaws, *Proceedings 9th Annual Computing Security Applications Conference (ASAC)* (1993) 64–73.
23. J. Clark, Attacking authentication protocols, *High Integrity Systems* **1**(5) (1996) 465–474.
24. J. Clark and J. Jacob, A survey of authentication protocol literature, *Version 1.0. Technical Report*, 1997, University of York.
25. J. Clark and J. Jacob, Implementation dependencies and assumption in authentication protocols, *Technical Report*, 1997, University of York.
26. J. Clark and J. Jacobs, On the security of recent protocols, *Information Processing Letters* **56**(3) (1995) 151–155.
27. F. Cohen, Information system attacks: A preliminary classification scheme, *Computers & Security* **16**(5) (1997) 29–46.
28. D. E. Denning and G. M. Sacco, Timestamps in key distribution protocols, *Communications of the ACM* **24**(8) (1981) 533–536.
29. T. Dierks and C. Allen, The TLS protocol version 1.0, 1999, IETF RFC 2246. 1999.
30. W. Diffie and M. Hellman, New directions in cryptography, *IEEE Transactions on Information Theory* **22**(6) (1976) 644–654.
31. W. Diffie and P. C. van Oorschot and M. J. Wiener, Authentication and authenticated key exchanges, *Designs, Codes and Cryptography* (1992) 107–125.
32. D. Dolev and A. Yao, On the security of public key protocols, *IEEE Transactions on Information Theory* **29**(2) (1983) 198–208.
33. S. R. Fluhrer, I. Mantin and A. Shamir, Weaknesses in the key scheduling algorithm of RC4, *Proceedings of the 8th Workshop on Selected Areas in Cryptography* (2001) 1–24.
34. A. Frier, P. Karlton and P. Kocher, The SSL 3.0 protocol, Netscape Communications Corp., 1996.
35. E. Gabber and A. Silberschatz, Agora: A minimal distributed protocol of electronic commerce, *2nd USENIX Workshop on Electronic Commerce Proceedings* (1996) 223–232.
36. H. Gilbert, M. Robshaw and H. Sibert, Active attack against HB^+ : a provably secure lightweight authentication protocol, *Electronics Letters* **41**(21) (2005) IEE.
37. L. Gong, Variations on the themes of message freshness and replay or, the difficulty of devising formal methods to analyse cryptographic protocols, *Proceedings of the 6th Computer Security Foundations Workshop*, 1993, 131–136.
38. S. Gritzalis and D. Spinellis, Cryptographic protocols over open distributed systems: a taxonomy of flaws and related protocol analysis tools, *Proceedings of the 16th International Conference on Computer Safety, Reliability and Security (SAFECOM)* (1997) 123–137.
39. G. Horng and C. Hsu, Weakness in the Helsinki protocol, *Electronic Letters* **34**(4) (1998) 354–355.
40. T. Hwang and Y. H. Chen, On the security of SPLICE/AS: the authentication system in WIDE internet, *Information Processing Letters* **53** (1995) 97–101.
41. T. Hwang, N. Y. Lee, C. M. Li, M. Y. Ko and Y. H. Chen, Two attacks on Neuman-Stubblebine authentication protocols, *Information Processing Letters* **53** (1995) 103–107.

42. IEEE, Wireless LAN medium access control (MAC) and physical layer (PHY) specifications, *LAN/MAN Standards Committee of the IEEE Computer Society*, 1999, IEEE Standard 802.11.
43. ISO, ISO-IEC JTC1.27.02.2(20.03.1.2) entity authentication using symmetric techniques, International Organization for Standardization (ISO), 1990.
44. A. Juels and S. A. Weis, Authenticating pervasive devices with human protocols, in *Advances in Cryptology (Crypto 05)* (Springer-Verlag, 2005).
45. A. Kehne, J. Schonwalder and H. Landendorfer, A nonce-based protocol for multiple authentications, *ACM Operating Systems Review* **26**(4) (1992) 84–89.
46. J. Kelsey, B. Schneier, D. Wagner, Protocol interactions and the chosen protocol attack, *Proceedings of 5th International Workshop on Security Protocols* (1997) 91–104.
47. G. Lowe, A family of attacks upon authentication protocols, *Technical Report*, 1997, Department of Matematics and Computer Science–University of Leicester.
48. G. Lowe, A hierarchy of authentication specifications, *Proceedings of 10th IEEE Computer Security Foundations Workshop* (1995) 31–43.
49. G. Lowe, An attack on the Needham–Schroeder public-key authentication protocol, *Information Processing Letters* **56**(3) (1995) 131–136.
50. G. Lowe, Some new attacks upon security protocols, *Proceedings of the 8th IEEE Computer Security Foundations Workshop*, 1996.
51. G. Lowe and B. Roscoe, Using CSP to detect errors in the TMN protocol, *IEEE Transactions on Software Engineering* **23**(10) (1997).
52. S. Lu and S. A. Smolka, Model checking the secure electronic transaction (SET) protocol, *Proceedings of 7th International Symposium on Modeling, Analysis and Simulation of Computer and Telecommunication Systems* (1999) 358–365 (IEEE Computer Society).
53. W. Mao and C. Boyd, Classification of cryptographic techniques in authentication protocols, *Workshop on Selected Areas in Cryptography* (1994) 95–104.
54. W. Mao and C. Boyd, Development of authentication protocols: some misconceptions and a new approach, *Proceedings of the 7th Computer Security Foundations Workshop*, 1994 (IEEE Computer Society Press).
55. W. Marrero, E. M. Clarke and S. Jha, Model checking for security protocols, *DIMACS Workshop on Design and Formal Verification of Security Protocols*, 1997.
56. Mastercard and Visa, *SET Secure Electronic Transaction Specification*, Mastercard & Visa, 1997, Available at http://www.setco.org.
57. C. Meadows, Formal methods for cryptographic protocol analysis: Emerging issues and trends, *IEEE Journal on Selected Areas in Communications* **21**(1) (2003) 44–54.
58. C. Meadows, Open issues in formal methods for cryptographic protocol analysis, *Proceedings of DISCEX 2000* (2000) 237–250.
59. C. Meadows and P. Syverson, A formal specification of requirements for payment transactions in the SET protocol, *Proceedings of the 2nd Financial Cryptography*, 1998.
60. A. Menezes, P. van Oorschot and S. Vanstone, *Handbook of Applied Cryptography* (CRC Press, 1997).
61. J. Mitchell, V. Shmatikov and U. Stern, Finite-state analysis of SSL 3.0, *Proceedings of 7th USENIX Security Symposium* (1998) 201–216.
62. J. H. Moore, Protocol failures in cryptosystems. Number 5, 1988, 594–602.
63. ISO/IEC CD 11770-3 (SC27 N974), Key management — Part 3: mechanisms using asymmetric techniques, International Organization for Standardization (ISO), 1995.
64. R. M. Needham and M. D. Schroeder, Using encryption for authentication in large networks of computers, *Communications of the ACM* **21**(12) (1978) 993–999.

65. D. M. Nessett, A critique of the Burrows, Abadi and Needham logic, *ACM Operating Systems Review* **24**(2) (1990) 35–38.
66. B. C. Neuman and S. G. Stubblebine, A note on the use of timestamps as nonce, *ACM Operating Systems Review* **27**(1) (1993) 10–14.
67. B. C. Neuman and T. Ts'o, Kerberos: An authentication service for computer networks, *IEEE Communications* **32**(9) (1994) 33–38.
68. D. Otway and O. Rees, Efficient and timely mutual authentication, *ACM Operating Systems Review* **21**(1) (1987) 8–10.
69. M. Panti, L. Spalazzi and S. Tacconi, Classifications of attacks on security protocols, *Technical Report*, Istituto di Informatica — University of Ancona, 2001.
70. R. M. Needham and M. D. Schroeder, Authentication revisited, *ACM Operating Systems Review* **21**(1) (1987) 7–7.
71. A. D. Rubin and P. Honeyman, Formal methods for the analysis of authentication protocols, *C.I.T.I. Technical Report*, 1993.
72. D. Safford, D. Schales and D. Hess, University anarchistic key authorization (AKA), *Proceedings of the 6th Security Symposium* (1996) 179–185.
73. M. Satyanarayanan, Integrating security in a large distributed systems, *CMU Technical Report*, 1987.
74. B. Schneier, *Applied Cryptography: Protocols, Algorithms and Source Code* (C. John Wiley & Sons, 1996).
75. A. Shamir, R. L. Rivest and L. M. Adleman, Mental poker, *Technical Report*, MIT Laboratory for Computer Science, 1978.
76. K.-A. Shim, Potential weaknesses of AuthA password-authenticated key agreement protocols, *Computer Standards & Interfaces* **29** (2007) 580–583.
77. G. Simmons, Cryptanalysis and protocol failures, *Communications of the ACM* **37**(11) (1994) 56–65.
78. D. R. Stinson, *Cryptography: Theory and Practice* (CRC Press, 1995).
79. P. Syverson, A taxonomy of replay attacks, *Proceedings of the 7th IEEE Computer Security Foundations Workshop* (1994) 131–136.
80. P. Syverson, Limitations on design principles for public key protocols, *Proceedings of the IEEE Symposium on Security and Privacy* (1996) 62–73.
81. P. Syverson, On key distribution protocols for repeated authentication, *ACM Operating Systems Review* **27**(4) (1993) 24–30.
82. P. Syverson and C. Meadows, A formal language for cryptographic protocol requirements, *Designs, Codes, and Cryptography* **1/2** (1996) 27–59.
83. M. Tatebayashi, N. Matsuzaki and D. B. Newman, Key distribution protocol for digital mobile communication systems, *Proceedings of Crypto-Advances in Cryptology* (1989) 324–333.
84. V. L.Voydock and S. T. Kent, Security mechanisms in high-level networks protocolsm, *ACM Computing Surveys* **15**(2) (1983) 135–171.
85. T. Y. C. Woo and S. S. Lam, A lesson on authentication protocol design, *ACM Operating Systems Review* **28**(3) (1984) 24–37.
86. T. Y. C. Woo and S. S. Lam, A semantic model for authentication protocols, *Proceedings of the IEEE Symposium on Research in Security and Privacy*, 1993.
87. C. C. I. T. T. Recommendation X.509, *The Directory-Authentication Framework*, 1988.
88. S. Yamaguchi, K. Okayama and H. Miyahara, Design and implementation of an authentication system in WIDE internet environment, *Proceedings of 10th Regional Conference on Computer and Communication Systems*, 1990.
89. T. Ylonen, SSH Transport layer protocol, Internet draft, 1996.

Chapter 4

WI-FI SECURITY

SUFIAN YOUSEF
Anglia Ruskin University

4.1. Introduction

Many people setting up wireless home networks rush through the job to get their Internet connectivity working as quickly as possible. That is totally understandable. It is also quite risky as numerous security problems can result. Today's *Wi-Fi* networking products do not always help the situation, as configuring their security features can be time-consuming and non-intuitive.

Wireless technology can provide numerous benefits in the business world. By deploying wireless networks, customers, partners and employees are given the freedom of mobility from within and outside of the organisation. This can help businesses to increase productivity and effectiveness, lower costs and increase scalability, improve relationships with business partners and attract new customers. Indeed, there are numerous reasons to deploy wireless technology, but like most technologies, it is not without its risks and downfalls.

There are many different ways to overcome the imperfections native to wireless networking. This chapter is designed to help you understand these flaws and to assist you in making your wireless networks a secure and beneficial asset.

When we are discussing wireless security in this chapter, we are referring to 802.11 networks. The Institute of Electrical and Electronic Engineers (IEEE) 802 committee defines Ethernet, with the series of standards designated 802.11, which is a set of standards for radio communications used in wireless local area networks, or WLANs. The IEEE is an organisation composed of engineers, scientists and students that specialise in creating standards for the computer and electronics industry in order to ensure smooth operability and compatibility. The organisation uses a number system to represent the standards it produces for different technologies. The IEEE uses the number 802 to categorise standards for Ethernet local and wide area networks (WANs), while the number 11 narrows that down to wireless area networks. In our discussions, you will also notice certain letters that appear after the number 11. These letters represent the different versions of the protocol, which specify things such as what frequency they operate in, and bandwidth they employ. These letters can also specify different security methods, as well.

The 802.11 networks are everywhere. The number of shipped 802.11-enabled hardware devices was estimated to exceed 40 million units by the year 2006. Because of the popularity of this communications standard and its prevalence in the world of organisational wireless networking, our focus in this text will be primarily on 802.11 WLANs. By familiarising yourself with the various aspects of the 802.11 standards, you will also be familiarising yourself with the same technologies that are employed within the business world.

4.2. The Benefits of Wireless Networks

Wireless LAN and Wi-Fi offer the following productivity convenience and cost advantages over wired networks:

- Mobility: Wireless LAN systems can provide LAN users with access to real-time information anywhere in their organisation. This mobility supports productivity and service opportunities not possible with wired networks.

  There are now thousands of universities, hotels and public places with public wireless connection. These free you from having to be at home or at work to access the Internet.
- Installation speed and simplicity: Installing a wireless LAN system can be fast and easy and can eliminate the need to pull cable through walls and ceilings.
- Reduced cost-of-ownership: While the initial investment required for wireless LAN hardware can be higher than the cost of wired LAN hardware, overall installation expenses and life-cycle costs can be significantly lower. Long-term cost benefits are greatest in dynamic environments requiring frequent moves and changes.
- Scalability: Wireless LAN systems can be configured in a variety of topologies to meet the needs of specific applications and installations. Configurations are easily changed and range from peer-to-peer networks suitable for a small number of users to full infrastructure networks of thousands of users that enable roaming over a broad area.

4.3. Security of Home Wireless Network

The recommendations below summarise the steps you should take to improve the security of a Wi-Fi home wireless network.

4.3.1. *Change Default Administrator Passwords (and Usernames)*

At the core of most Wi-Fi home networks is an access point or router. To set up these pieces of equipment, manufacturers provide a web-page interface that allows owners to enter their network addresses and account information. These web configuration tools are protected with a login screen (username and password) so that only the rightful owner can do this. However, for any given piece of equipment, the logins provided are simple and very well-known to hackers on the Internet. Change these settings immediately.

4.3.2. *Turn on (compatible) WPA/WEP Encryption*

All Wi-Fi equipment supports some form of *encryption*. Encryption technology scrambles messages sent over wireless networks so that they cannot be easily read by humans. Several encryption technologies exist for Wi-Fi today. Naturally, you will want to pick the strongest form of encryption that works with your wireless network. However, the way these technologies work, all Wi-Fi devices on your network must share the identical encryption settings. Therefore, you may need to find a "lowest common denominator" setting.

4.3.3. *Change the Default SSID*

Access points and routers all use a network name called the *SSID*. Manufacturers, normally, ship their products with the same SSID set. For example, the SSID for Linksys devices is normally "linksys". Whilst knowing the SSID does not, by itself, allow your neighbours to break into your network, it is a start. More importantly, when someone finds a default SSID, they see it as a poorly configured network and are much more likely to attack it. Change the default SSID immediately when configuring wireless security on your network.

4.3.4. *Enable MAC Address Filtering*

Each piece of Wi-Fi gear possesses a unique identifier called the *physical address* or *MAC address*. Access points and routers keep track of the *MAC addresses* of all devices that connect to them. Many such products offer the owner an option to enter the *MAC addresses* of their home equipment and restrict the network to only allow connections from those devices. Do this, but also know that the feature is not as powerful as it may seem. Hackers can use software to fake *MAC addresses* easily.

4.3.5. *Disable SSID Broadcast*

In Wi-Fi networking, the wireless access point or router typically broadcasts the network name (SSID) over the air at regular intervals. This feature was designed for businesses and mobile hotspots where Wi-Fi clients may roam in and out of range. At home, this roaming feature is unnecessary, and it increases the likelihood someone will try to log in to your home network. Fortunately, most Wi-Fi access points allow the SSID broadcast feature to be disabled by the network administrator.

4.3.6. *Do Not Auto-Connect to Open Wi-Fi Networks*

Connecting to an open Wi-Fi network such as a free wireless hotspot or your neighbour's router exposes your computer to security risks. Although not normally enabled, most computers have a setting available allowing these connections to happen automatically without notifying you (the user). This setting should not be enabled unless needed in a particular situation.

4.3.7. *Assign Static IP Addresses to Devices*

Most home networkers gravitate toward using *dynamic IP addresses. DHCP* technology is indeed easy to set up. Unfortunately, this convenience also works to the advantage of network attackers, who can easily obtain valid IP addresses from your network's DHCP pool. Turn off DHCP on the router or access point, set a fixed IP address range instead, then configure each connected device to match. Use a *private IP address range* (like 10.0.0.x) to prevent computers from being directly reached from the Internet.

4.3.8. *Enable Firewalls On Each Computer and the Router*

Modern network routers contain built-in firewall capability, but the option also exists to disable them. Ensure that your router's firewall is turned on. For extra protection, consider installing and running *personal firewall software* on each computer connected to the router.

4.3.9. *Position the Router or Access Point Safely*

Wi-Fi signals normally reach the exterior of a home. A small amount of signal leakage outdoors is not a problem, but the further this signal reaches, the easier it is for others to detect and exploit. Wi-Fi signals often reach through neighbouring homes and into streets, for example. When installing a wireless home network, the position of the access point or router determines its reach. Try to position these devices near the centre of the home rather than near windows to minimise leakage.

4.3.10. *Turn Off the Network During Extended Periods of Non-Use*

The ultimate in wireless security measures, shutting down your network will most certainly prevent outside hackers from breaking in. While impractical to turn off and on the devices frequently, at least consider doing so during travel or extended periods offline. Computer disc drives have been known to suffer from power cycle wear-and-tear, but this is a secondary concern for broadband modems and routers.

4.4. Cracking Wi-Fi Security Protocol

Cracking the Wi-Fi security protocol WEP is a probability game. The number of packets required to successfully decrypt the key depends on various factors, luck included.

When WEP *was compromised* in 2001, the attack needed more than five million packets to succeed. During the summer of 2004, a hacker named KoreK published a new WEP attack (called chopper) that reduced by an order of magnitude the number of packets requested, letting people crack keys with hundreds of thousands of packets, instead of millions.

Three researchers Erik Tews, Andrei Pychkine and Ralf-Philipp Weinmann developed *a faster attack* (based on a cryptanalysis of RC4 by Andreas Klein), that works with ARP packets and just needs 85,000 packets to crack the key with a 95% probability. This means getting the key in less than two minutes.

4.5. Wi-Fi Alliance to Beef up Security

Security remains the key issue deterring enterprise users from making major investments in Wi-Fi, despite all the improvements over the past year. Whether real or perceived, the security risks of wireless LANs are still holding deployments back. Conscious of this, the Wi-Fi Alliance is trying to beef up standard security still further.

It has been already agreed to a dual-layer security approach, with Wi-Fi Proctectd Access (WPA2) (the commercial name for the 802.11i standard) supporting advanced functions including AES encryption, while the more basic WPA — originally an interim standard en route to 802.11i — will be kept for devices that require less stringent security and lower costs, particularly in the consumer space.

Now the group intends WPA2 to enforce a higher level of encryption, reiterating a decision it tentatively made to require 64-character passwords. This move was in response to a spate of rogue access point attacks and new question marks over Wi-Fi's resistance to hackers.

It is even possible that the "light" version of WPA will be dropped later this year, making it compulsory even for consumer devices to support WPA2 in order to be certified as Wi-Fi compatible.

The risk of a very stringent requirement is that vendors at the budget end of the market bypass testing and certification altogether and so undermine the Wi-Fi Alliance's brand. Many consumers have been found to recognise generic terms such as wireless LAN, and individual brands such as Centrino, more than the term Wi-Fi, so it can be argued that Wi-Fi certification is not essential to success in the home market.

Craig Mathias, an analyst at Farpoint Group, believes in keeping the two-tier system. "I do not think everyone will need AES. I also think higher level security of the 802.1x or VPN variety can effectively substitute for AES in many cases", he said.

Throughout this year, the Alliance will be adding various strains of EAP (Extensible Authentication Protocol) to its testbed. As the Wi-Fi community seeks to instil confidence in its technologies, WiMAX now faces the same challenges. Although the 802.16 standards have far greater security functionality built into the base than Wi-Fi did, the perception of their safety will have to be high before they win the trust of enterprise and carrier users. Terabeam, a developer of WiMAX-ready equipment, is one company that believes there are significant security gaps to be filled in 802.16-2004. In particular, it claims WiMAX's authentication facilities are limited and its encryption method, DES 3, is less robust than AES. A combination of standards activity — particularly with an eye to government customers — and third-party enhancements will be essential for commercial WiMAX products to pass the grade. Already, Intel has submitted proposals for incorporating AES into 802.16 too.

Authentication, based on X.509 digital certificates, is included in the media access control layer and gives every 802.16 customer transceiver its own built-in certificate, plus one for the manufacturer, allowing the base station to authorise the end user. Link privacy is implemented as part of another MAC sub-layer, the privacy sub-layer. It is based on the Privacy Key Management protocol that is part of the DOCSIS BPI+ specification.

As in other standards, many advances will come from individual vendors, whether enhancements that differentiate an individual product, or work that may be fed back into the standards' process. One example is Airspan's work with Hifn, a specialist maker of security co-processors. In December, Airspan said that it would use its partner's 7955 co-processor in its base stations as "a suitable encryption solution for IEEE 802.16-2004, that would also be able to support the evolving 802.16e standard".

The Hifn 7955 is designed for networking applications like virtual private networking (VPN) broadband routers, wireless access points, VPN edge router/gateways, firewall/VPN appliances and other network and customer premise equipment. It accelerates a variety of IPsec and SSL/TLS protocols including DES, 3 DES, AES and public key. In addition to IPsec and SSL protocols, it also supports the temporal key integrity protocol (TKIP) and AES countermode encryption.

Although WiMAX may be inherently more secure than its local area cousin, such enhancements will be important if its uptake is not to be delayed, like Wi-Fi's, by the lack of user confidence. In the end, this will be a more important factor in the speed of adoption of 802.16 than the much publicised delays in equipment certification.

4.6. Poor Wi-Fi Security Leaves Users at Serious Risk

A quarter of the UK population is at "serious risk" of identity theft and cyber-fraud because of an inadequate or even non-existent Wi-Fi security.

The claims come after price comparison *website Moneysupermarket.com* commissioned an amateur "hacker" to test the quality of wireless security on the streets of Liverpool, Manchester and Chester.

The hacker was able to tap into 25% of domestic wireless connections using just a laptop.

The study claimed that thousands of Internet users had not enabled the security in their wireless modems, leaving hackers free to access their home PCs.

Hackers would be able to steal bank details and identity information, upload illegal content or pornography or even finance or execute a scam from someone else's connection.

Jason Lloyd, head of broadband at Moneysupermarkct.com, said: "This is a serious problem which leaves Internet users completely exposed".

"Our results found an average one in four of all residential wireless routers unsecured, meaning that anyone can gain access. Inviting identity theft or fraud, an unsecured internet connection can become an open door for criminals".

The repercussions can be severe, according to Lloyd. "It is bad enough if your neighbour can use your internet connection for free, but this becomes far more sinister if someone uses your wireless connection for criminal activity", he said.

"This could be accessing your internet connection to download obscene material, gathering personal information to defraud you or even stealing your identity".

The study was conducted across Liverpool, Manchester and Chester, and Moneysupermarket.com detected that 25% of residential wireless routers were unsecured. This discovery was made using only a standard laptop.

"While we were scanning for unsecured networks we found that the scale of the problem appears to be excessive, and there were many *small business* routers which were unsecured", said Lloyd.

"You would have imagined that, with potentially many people's livelihood at stake, there would be reasonable measures in place".

4.7. Take Control of Your Wi-Fi Security

If you have not set up security properly, someone could be eavesdropping on your e-mail, plucking your passwords out of the air, or sending spam through your Internet connection right now. When you are using a wireless network — whether a Macintosh with AirPort gear, Windows with any Wi-Fi equipment, or a Wi-Fi handheld like the iPhone — you are exposed to risk unless you take steps.

Wireless networking experts Glenn Fleishman and Adam Engst have spent years researching and covering wireless security issues on Glenn's Wi-Fi Networking News blog and in two editions of *The Wireless Networking Starter Kit*. Now they have distilled that experience into this essential guide for anyone using wireless networks, whether at home, at work or on the road. You will learn how to evaluate your real security risks; the best way to restrict access to your network using WPA; how to secure your data in transit with PGP, SSL, SSH and VPNs; and how to protect your computers from viruses and attacks. The book provides extra advice on how to secure small office wireless network, including details on choosing VPN hardware and software and on setting up 802.1X for secure Wi-Fi logins. The final section of the book helps you to determine how successful your security efforts have been by showing you how to perform a detailed security audit on your wireless network using the same freely available tools that crackers might use against you.

4.8. The Five Deadly Dangers of Unsecured Wi-Fi Networks

Once hackers have an access to your Wi-Fi network, they can readily capture personal and business information. There are two types of Wi-Fi attacks. Passive attacks, where the hacker captures your network traffic, are almost impossible to detect because the hacker never joins your network. They can sit silently with their antenna tuned into your network and capture gigabytes of network traffic for off-line analysis at a later time. Active attacks, where the hacker joins the network, can be the most devastating because they can launch active attacks into the network and onto your devices on the network.

There are five attacks that Wi-Fi hackers can very easily and readily perform on your wireless network with very little effort or expense. The first two are passive attacks, and the last three are active attacks. But make no mistake — all of these attacks can be deadly.

Deadly attack #1: Account and password capture. There are several applications that send your account and passwords in clear text over the network. For example, every time a POP3 mail account checks for new e-mail, the account name and password are in the clear as part of the data transfer. Anyone sniffing the network traffic can easily get your e-mail account information. Once they have that information, they can access your e-mail account at their leisure, monitoring for personal information without leaving a trace. From there, any confidential information they can get from your account just escalates their attack.

Deadly attack #2: E-mail, IM and Web Site Traffic Capture — It is very easy to monitor and capture all of the e-mail traffic sent over an unsecured wireless network. Since most e-mails are sent in clear-text, and instant messaging is sent in HTML, it is very simple to capture the traffic and analyse it off-line for any information at a later time. By monitoring your wireless traffic, all of the unencrypted HTML data can be captured and reconstituted as web pages on the hackers' PC to see exactly what web sites and content you are surfing over the wireless network.

Deadly attack #3: Accessing data on your PC. Let us face it, it is pretty easy to turn file sharing on, and then forget to turn it off when you attach to an open Wi-Fi network. Once file sharing has been left on or the personal firewall is misconfigured, a hacker can readily access you PC and hard drive across the wireless network.

Firewalls are also easy to misconfigure or turn off, and forget to turn back on. With older versions of Windows (Pre-Windows XP SP2), if improperly configured, it is an easy prey for a hacker to get in over the network, log-in as a null session and take over your platform.

Deadly attack #4: Access to the corporate network. If your wireless network is connected to a corporate network through a site-to-site VPN, an open wireless network punches a hole through the network, and opens up both sides of the VPN to anyone attaching to the network. Another threat is with improperly configured client VPNs, which can be more easily compromised to provide the hacker access through the VPN.

Deadly attack #5: SPAM and Virus launching over the wireless network. Unsecured networks provide an ideal launch point from which hackers can launch SPAM and virus attacks because it is very difficult to track the source back to them. From a distance, the Spammer can send without repudiation, even from your e-mail account if he or she sniffed your e-mail account info. When the ISP or authorities tracks down the violator, the trail points to *your* network, and possibly your e-mail account. The liabilities to the owner of the unsecured network are still newly contended battlegrounds for the lawyers.

4.9. Best Practices to Secure Your Wireless Network

The good news is that simple tools are available to properly secure your wireless network and avoid the dangers discussed above.

The Wi-Fi Alliance designated WPA as the recommended security practices for consumer and business networks. The WPA comes in two forms: WPA-PSK which offers a lower-level security for consumers, and WPA-Enterprise which offers a higher level of security for enterprises. Solutions like *Witopia* and *WiFi Login Pro* deliver enterprise level security with the consumer-level simplicity that can be easily and quickly deployed in home offices, small offices and medium businesses.

WPA–PSK (Pre-shared key)–WPA–PSK provides a relatively secure solution for consumer networks. If you are technically competent, and feel comfortable configuring the security parameters of your wireless access point or router, you can configure your wireless network to support WPA–PSK. By entering a common 64-digit hexadecimal key or an ASCII pass phrase into every device on the network, you can properly encrypt all network traffic to and from the access point. The *LucidLink WiFi Client* can automatically detect if a network requires WPA–PSK and simplifies the client configuration.

The WPA–PSK has fixed many of the problems associated with pre-shared keys used in WEP. While it is quite awkward to properly enter a 64-digit hexadecimal key into each device on the network, if done carefully, it can provide strong encryption of network traffic and ward off hackers. A random ASCII passphrase (random to avoid a dictionary attack) can be used to avoid the hexadecimal key entry.

One of the common complaints with WPA–PSK, however, is that it uses a common key across all the devices and PCs on the network. If you, an employee, or your child innocently shares this key with anyone, the integrity of the network can be compromised. If any person leaves an organisation or needs to be denied access to the network, every PC on the network needs to be reprogrammed with a new 64-digit pre-shared key. The need to re-key every device on the network if a single user is removed can become a heavy burden to maintain a small business network.

WPA-Enterprise uses the same type of network security used by enterprises and ISP over the last decade to protect access to wired networks. Unlike WPA–PSK, each user accessing the network is given unique credentials. These credentials may be in the form of passwords or electronic certificates.

For a user to access the network, they provide the unique credentials which are verified by a designated PC providing access management using a security protocol called 802.1x. When the server acknowledges the user as having valid credentials, the user is given access to the network and given a new encryption key every time they enter the network. The encryption key is used to encrypt and secure the network traffic between the user's PC and the network access point. Without proper credentials, the user is denied access.

One of the benefits of WPA-Enterprise is that it offers a much higher level of manageability. User access can be controlled on a user-by-user basis. A user can be removed from the network without re-keying every device on the network.

4.10. Further Reading

Wireless networking experts Glenn Fleishman and Adam Engst have spent years researching and covering wireless security issues on Glenn's Wi-Fi Networking News blog and in two editions of *The Wireless Networking Starter Kit*. Now they have distilled that experience into this essential guide for anyone using wireless networks, whether at home, at work or on the road. You will learn how to evaluate your real security risks; the best way to restrict access to your network using WPA; how to secure your data in transit with PGP, SSL, SSH and VPNs; and how to protect your computers from viruses and attacks. The book provides extra advice on how to secure small office wireless network, including details on choosing VPN hardware and software and on setting up 802.1x for secure Wi-Fi logins.

References

1. http://kbserver.netgear.com/kb_Web_files/N100688.asp.
2. http://compnetworking.about.com/od/wirelesssecurity/tp/wifisecurity.htm.
3. http://www.theregister.co.uk/2005/02/14/wi_fi_security/
4. http://www.vnunet.com/vnunet/news/2187063/quarter-uk-surfers-face-serious.
5. http://www.oreilly.com/catalog/9780975950395/
6. http://www.lucidlink.com/2007/04/five-deadly-dangers-of-unsecured-wifi.html.
7. http://www.lucidlink.com/2007/05/best-practices-to-secure-your-wireless.html.

Chapter 5

AUDITING, PENETRATION TESTING AND ETHICAL HACKING

FRANK LEONHARDT
Independent Commentator and Consultant

5.1. Introduction

It is only when you have locked yourself out of your house that you discover how easy it is to get in without a key. There is always an open window, a locksmith who does not seem to need keys or some point of entry that can be forced easily and with minimal damage. Doubtless, you thought you had taken all the good advice about fitting five-lever deadlocks, installing window locks — you may even have an alarm so that the whole neighbourhood would come running at the first sign of trouble, with the flying squad bringing up the rear in a blaze of blue flashing lights and wailing sirens.

I hate being the bearer of bad news, but security on computer systems is no different. Whatever the security products you have purchased (and even installed), and whatever policies you have in place, every system has its weaknesses and the only way you are going to find them is by determined looking.

With computer systems, you are faced with another problem. Unlike a house — basically a sturdy brick building with weak points at the doors and windows — a computer system is a large, complex thing with multiple attack points. It is next to impossible to check everything, so where do you start looking?

The obvious first step is a quick audit — see what you have got. Then a risk assessment is in order. Getting back to our house, an audit reveals that a front garden with a low wall and a gate. How good is the lock on the gate? Is there even a lock fitted? A risk assessment should determine that there is minimal risk from allowing free entry to the front garden — in fact it is necessary if you want anything delivered by mail. Locking the front gate will not help as anyone can step over the wall. Therefore you are wasting your time examining the gate further. So, instead of attempting to audit every aspect of every computer system in sight, it makes sense to target limited resources on what really matters.

Then you can set about checking the security of what is important by pretending you have lost the key. Of course, you may not be that skilled at breaking and entering, so if you really one to know where you are vulnerable, you should find someone who is, do not give them the key and see how far they get.

5.2. Audit

It is way beyond the scope of one chapter in a handbook to cover the subject of auditing in any depth, but you can go a long way with a good technical understanding and a methodical, common-sense approach.

One of the first things to define is what you are actually looking for. Reading the sensationalist news media gives the impression that you are major threat is from gangs of malicious black hats trying to break in and steal your secrets. Whilst this does go on, in practice, the major risk factors come from hardware failure, accidents and rogue employees.

Given that a determined attacker is almost sure to break in if the prize is worth enough — especially if it is an inside job — there is little point in considering security against attack without security against disaster and harmonising procedures and systems to cope with all threats at once.

For example, let us suppose we are looking at the computer holding a small company's accounts. What would be the impact if it was compromised?

Clearly, this is an important asset of the company; the data on the hard disc is vital and must be kept securely. It is also obvious that simply restricting unauthorised access, although important, is not enough (Table 5.1).

Whilst a slight digression from the subject of this chapter, by example, it would be reasonable to preclude Internet access (including e-mail) from the accounting machine, place it behind a firewall on the LAN, keep it under lock-and-key, ensure

Table 5.1. Risk assessment for accounting machine.

Threat	Effect	Impact
Loss of hardware (e.g. fire, hard-disc failure)	No access to accounts — who owes you money?	Inability to trade
Theft of hardware	As above, plus someone else has access to your financial details.	Inability to trade and major vulnerability to fraud.
Unauthorised login or rogue employee	Accounts can be altered to obfuscate fraud. Commercially, sensitive information available to competitors. Possible access to control bank account.	Loss of commercial advantage and vulnerability to fraud.

strong passwords were used and prevent the installation of other software. Also regular off-site snapshots of the data would be prudent should corruption occur, and for financial audit purposes. Hardware should obviously be mirrored — how much down-time is acceptable while a new machine is sourced, software installed and the data restored from a backup?

If you doubt any of this, just threaten to "pretend" an accounts server has been stolen by unplugging it — call it a disaster recover exercise. Only then, will the IT manager discover that all the backups are in some weird tape format, the drives for which have been unavailable for the last three years, and the accounting software discs might be in Jim's draw; we will have to ask him when he gets back from leave.

Now let us consider someone's desktop PC — this probably has a standard software installation, might well need web and e-mail access for its user to function effectively and contains relatively insensitive data, the important elements of which are stored up on a suitably secure central server. The risk here is that the machine may be compromised by malware; this risk can be reduced but by no means can it be eliminated. In the event of a disaster (either caused malware or hardware failure), it can be easily replaced, but what is the impact of this user's connection to the LAN should it be compromised?

Whilst the scenarios presented above are a very long way from exhaustive in their scope, they should give an insight into the methodology used when auditing from the common-sense approach.

A common alternative to conducting an audit and risk assessment is to use some industry "Best Practice" — a generic risk assessment and set of mitigation measures. This has the advantage that unskilled IT managers can offload responsibility for their installation at minimal cost.

5.3. Penetration Testing

Once a list of targets has been defined and prioritised, the task of penetration testing can begin — always bearing in mind that new targets may be discovered in the course of testing. Penetration testing is simply the process of attempting to break in, and to do this properly requires creativity.

It may be advantageous to specify the scenario under which the supposed attacker will be operating. In particular, is any inside knowledge or access assumed at the outset? On one hand, an external attack scenario will expose unexpected weaknesses in perimeter defences not addressed by an internal attack. On the other hand, internal attacks are commonplace, and potentially more dangerous.

5.3.1. *Automated Tools*

There exist many automated penetration testing tools which purport to scan an organisation's network and generate a report highlighting vulnerabilities. Some of these are publically available (such as SARA[a]) and others are licensed commercially (such as Core Impact[b] and Nessus[c]). Microsoft's MBSA covers the latest Microsoft software environments, and specific scanners are available for particular classes of application — for example web sites.

However, automated tools tend to produce long lists of obscure potential vulnerabilities, giving the impression that a network is riddled with problems and leaving the user little closer to identifying real issues amongst the noise. In addition, automated tools are unlikely to identify serious weakness discoverable by an experienced human expert.

This is not to say that penetration test tools do not have their place. With expert interpretation, they provide a valuable source of overall information and can pick up on items too labour-intensive to check for manually.

5.3.2. *Denial of Service*

Keeping data secure is important, but it is of limited use if its rightful users cannot access it. Testing should include denial-of-service (DoS) attacks as far as is possible, to see how critical infrastructure could cope. Of course, not all DoS events are malicious attacks; the effect can be caused by an inappropriate network configuration, use of rogue software on a network (peer-to-peer applications, being an obvious example) or a software/hardware malfunction.

Fortunately, there are many tools available for creating DoS events, and it is trivially easy to create specific new ones using "C" or scripting languages, especially using netcat[d] to handle the low-level communications.

For stress testing web servers, a GPL package called Funkload is popular, and so is "flood" from the Apache Foundation.[e] Super-smack[f] is good for exercising databases. Modern programs like Yersinia[g] are excellent at stressing data networks internally using the latest known vulnerabilities and forms of attack, such as interfering with hot-standby router protocol. Specialist test and DoS tools exist for specific vulnerabilities, such as starving a DHCP server of addresses or interfering with VoIP protocols.

As technology and threats evolve rapidly, it is impossible to make positive recommendations of the latest tools in print. Specialist web sites and mailing lists exist to discuss current developments and it is strongly recommended that these are consulted. Insecure.org, from the publishers of nmap, is a very good place to start.

[a] http://www-arc.com/sara/
[b] www.coresecurity.com.
[c] www.nessus.org.
[d] A simple command-line utility for sending and receiving packets over an IP network, available for most UNIX-like systems. The original 1995 program is not actively maintained, but there are many new and improved variations such as socat or ncat. On open BSD systems, it is known as nc. Driven by a simple shell script, it can be used to simulate most things.
[e] http://httpd.apache.org/test/flood/
[f] http://vegan.net/tony/supersmack/
[g] http://www.yersinia.net.

Simulating some DoS events is surprisingly difficult in spite of the number of tools available, especially attacks perpetrated by criminals using a large number of hijacked computers (a botnet). Whilst a perimeter firewall can easily be configured to drop excessive requests from a particular IP address, it is almost impossible to detect and block spurious requests from multiple random IP addresses. One method of testing is to temporarily suspend such firewall rules and ignore the fact that the traffic comes from a single source in order to gauge the resilience of other aspects of the system. However, this obviously requires the collusion of the network administrator, who may otherwise be unaware that a penetration test is being carried out.

5.3.3. *Human Testing*

The starting point for a human tester is to discover as much information about the target as possible, particularly if an external penetration test is being carried out.

A lot of valuable information can be determined simply from sources such as the telephone directory, company web site or Internet registration databases. Some companies (and some suppliers) are foolish enough to brag about the supply or purchase of some major IT resource — how useful is to know that corporation X has just been supplied with Bluenut model MK14 intrusion detection firewalls when you are trying to probe their defences?

Using information obtained by this passive research, the next stage might be a network scan of appropriate addresses using the UNIX nmap network mapping tool, looking for visible machines running old and potentially vulnerable operating software. If you wish to prove that a vulnerability is more than theoretical, systems such as Metasploit can be employed to take control of an internal machine and from there the LAN can be explored from inside the perimeter defences. By this stage, you are probably going to need the services of the so-called ethical hacker — it is a specialised job, and outside the scope of this handbook.

5.3.4. *Social Engineering*

Much is made of technological attacks and counter measures, but the easiest method of penetrating any organisation's system is though its staff.

Why bother guessing a password when you have a fairly good chance of calling the helpdesk and asking directly? Gartner research suggests that 35% of IT helpdesk time is spent sorting out user's account password problems — after a while anyone calling with a plausible enough story should get results.

To obtain information from a company database, the easy way is to pay a dishonest employee to obtain it, with very little subtly required. This is not normal when penetration testing, but it is an ever-present risk and should not be ignored.

Gaining physical access to an origination's premises during office hours can be very easy. For example, one tried-and-trusted method is to brazenly walk through the main entrance wearing a bright fluorescent jacket. Drawing more attention to yourself by carrying an awkward ladder in both hands and helpful employees will hold doors open for you, bypassing awkward swipe-card access controls. It is also useful to gain access to false ceilings when you want to investigate network cables.

The scenarios presented above might appear to be scenes from a comedic crime caper, but they are very real examples of genuine threats that should be considered. If an organisation's computing infrastructure is important, it is vital to harden its physical security and ensure staff have enough knowledge and motivation to prevent direct physical infiltration. There is no point fitting expensive window locks and leaving the door wide open.

5.4. Standards and Methodologies

As with most industries, there now exist standards and methodologies for penetration testing.

5.4.1. *OSSTMM Methodology*

The Open Source Security Testing Methodology Manual (OSSTMM), focuses on the technical details of which items need testing and how the results should be measured. It is a peer-reviewed document and as such, should be up-to-date with technological and legal advances.

The OSSTMM test cases are divided into five channels, and collectively test:

- Computer and telecommunications networks;
- Fraud and social engineering control levels;
- Information and data controls;
- Military bases;
- Mobile devices;
- Perimeters;
- Personnel security awareness levels;
- Physical locations such as buildings;
- Physical security access controls;
- Security processes and
- Wireless devices

Another feature of OSSTMM is its "Rules of Engagement", used to define how the tester and target organisation are to conduct themselves — ranging from definitions of what is and is not acceptable, to the format of the final report.

5.4.2. *ISSAF Methodology*

The Information Systems Security Assessment Framework (ISSAF) is a new peer-reviewed framework produced by the Open Information Systems Security Group. It currently provides a methodology for assessing an organisation's real-world security requirements.

5.4.3. *NIST Methodology*

The National Institute of Standards and Technology (NIST) produces a document called "Special Publication 800-42, Guideline on Network Security Testing". It is less comprehensive and current than OSSTMM, but has the advantage that it is more likely to be accepted by official regulators, and in order to overcome its shortcomings, it sensibly refers to OSSTMM.

5.4.4. *CHECK Scheme*

This is administered by CESG, formally known as the "Communications and Electronic Security Group" from GCHQ. It was the only form of accreditation in the UK for many years and is mandatory for Central Government testing, although EU rules mean it cannot be made compulsory for Local Government or government agency situations.

Because if its longevity and government acceptance, many large companies also use it.

Organisations subscribed to the scheme follow a code of conduct and certified staff have to be vetted to at least SC security clearance level (check of ID plus criminal records).

5.4.5. *CREST*

The Council of Registered Ethical Security Testers (CREST) is a non-profit organisation whose members it considers are competent penetration testers. It has a mandatory code of practice, and similar requirements to CREST for background checks and ethical standards. It runs its own examinations for membership, the syllabus for which is available on their web site.[h]

5.4.6. *INFOSEC Assessment Methodology*

The (American) National Security Agency (NSA) publishes an Information Security Evaluation Methodology and a corresponding Assessment Methodology. This is sometimes known as Infrastructure Evaluation Methodology and is also abbreviated to IEM, but this leads to confusion with the older Information Engineering Methodology.

It is used as the basis for certifying penetration testers and auditors in the USA; a course and exam can be completed in four days.

5.4.7. *TIGER Scheme*

This is managed by a committee of industry organisations with the intention of setting standards in the penetration testing industry. It has good acceptance in

[h]http://www.crest-approved.org/

some areas because of its close links with customers, its perceived independence and its use of university-based examinations.

Government-backed testing also exists in the US with standards such as the NSA IEM.

5.5. Glossary

Black Box Test

A test, or series of tests, carried out where the tester has no prior knowledge of the system to be tested. This simulates an attack from outside an organisation.

Black Hat

Person exploiting weaknesses in computer systems in pursuit of criminal activity.

Blue Hat

A term apparently originating from and mainly used by, Microsoft to describe third-party vulnerability testers engaged by a company to test its pre-release products.

Cracker

Someone skilled at breaking into computer systems, normally for illicit purposes.

Grey Hat

An individual of questionable ethnics engaged in probing and breaking computer security.

Hacker

Someone familiar enough with the internal workings of a computer's operating environment that they can alter the way it works in order to suit their needs. It comes from the verb "to hack", which in computer terminology refers to a quick-and-dirty change in code to solve a problem. In popular culture, this is often confused with the term "Cracker", which means someone cracking or breaking security, and for this reason the term has acquired negative connotations with the general public.

Script Kiddie

This refers to the phenomena where unskilled people attack computer systems using automated tools (normally written in a scripting language). The impact should not be underestimated as great skill, may have been used to actually produce the tools. However, there is also truth in the idea that dogs might chase cars, but if they ever catch one, they discover they can not drive.

White Box Test

Penetration testing carried out with full or partial technical knowledge of the system under test. Some people describe a partial knowledge test as "grey box". This process simulates an attack within an organisation, or at least with inside help.

White Hat

Person skilled at probing and breaking computer security for legitimate reasons — sometimes called ethical hacking.

Chapter 6

VoIP SECURITY ISSUES

FRANK LEONHARDT
Independent Commentator and Consultant

6.1. Introduction

VoIP (Voice over IP, or routing telephone calls over IP networks) has been with us for many years, with no major security problems being reported. However, existing installations have been deployed using proprietary protocols and equipment over private networks, managed by vendor-trained or supplied specialists.

In the last few years, the open-standard Session Inititation Protocol (SIP) protocol has gained a significant foothold, with a plethora of low-cost handsets and switching servers appearing on the market, together with a number of gateway services competing for early adopters. This, combined with cheap always-on Internet connections, is leading to VoIP being deployed by SMEs and domestic users.

Before SIP came to prominence, the H.232 from the ITU performed a similar function and is still actively used today. However, because SIP-based equipment is currently being deployed rapidly, it is of more interest from a security perspective. Whilst the general issues raised in this paper apply whether SIP, H.232 or any other signalling protocol is in use, specific examples will be limited to SIP.

Equipment using the SIP protocol is designed to inter-operate, so networks are naturally going to be multi-vendor.

As no vendor is responsible for the entire system, no vendor is responsible for the security aspects, and there is little incentive to educate potential customers.

A SIP handset is, fundamentally, a computing device with a network connection, speaker and microphone running appropriate software. It is possible to run such a software on a desktop PC or laptop to achieve the same functionality, and several "soft 'phones" are available for users wishing to turn their workstation into a telephone. These present their own security issues.

As well as the mainstream SIP and H.323 softphones, proprietary systems, notably Skype, are popular with certain types of user. Skype is built on a peer-to-peer file sharing protocol — the technology used for swapping pirate software.

Up until now, VoIP appears to have been spared the attentions of the cyber-criminals. It could be that the proprietary networks use highly secure technology and are operated by professionals well versed in the security aspects of their job. It could be that running on private networks makes them inconvenient to break in to. It could also be that security breaches go unreported by the vendor and users for commercial reasons.

However, this closed environment is changing rapidly with multi-vendor open protocol equipment over the public Internet gaining a significant foothold in the market, especially with domestic users and SMEs.

The year 2007 looks set to see an explosion in the use of VoIP, sometimes rolled out with undue haste by small companies in a land-grab exercise. Criminals have a natural tendency to attack obvious targets, and to a cyber criminal they do not come more obvious that an unprotected mass-market telephone network.

6.2. Background to SIP

SIP — or Session Initiation Protocol — is the control protocol used to set up calls between one SIP handset and another. A handset is properly referred to a "User Agent" in SIP terminology, and need not be a physical telephone handset as such, but the term "handset" will be used in the reminder of this chapter as it is appropriate to the context.

The SIP has nothing to do with transmitting the actual data part of the call; in fact, SIP is not limited in its application to VoIP at all. However, as the common controlling protocol, the term "SIP" has become synonymous with the collection of open-standard protocols common to many VoIP handsets.

The SIP is often compared to HTTP, although the comparison quickly breaks down when the details is examined. To make a SIP call, the caller sends an INVITE packet to the destination's handset. If it is on-line, the destination will reply with a status code of "Trying", followed by one or more "Ringing" codes until the 'phone is picked up, at which point the caller should receive an "OK" status and respond with an ACK.

The call then proceeds until one of the parties sends the other a BYE packet and receives an "OK" in response.

In order for the caller to find the destination handset, it is common to use a registrar server. This is a central server, often located using the DNS. Handsets send a REGISTER packet to the registrar when they boot up, giving details of their location on the network. This gets around the problem of users' handsets changing their IP address; when a caller wishes to contact a user, they locate the handset using the registrar. It also gets around the problem of entering the IP address or

URL of a destination handset when a call is being made, as the server can map this onto something looking like a telephone number, which can be dialled on a numeric keypad.

It is common for registrars to act as gateways to the PSTN, with one- or two-way translation of telephone numbers allowing SIP users to call normal phones or PSTN users to call SIP handsets.

Once SIP has set up a call (in conjunction with the SDP protocol), the actual data is transferred using RTP, and the sound is encoded using one of several CODECs — typically G.711 or G.729.

The RTP is commonly found on proprietary VoIP systems running underneath different call control protocols. It is an IETF standard documented in RFC 3550. A significant point is that RTP packets, normally sent using UDP, have sequence numbers and time stamps.

The G.711 CODEC is simple Pulse Code Modulation (PCM), whereas the low-bandwidth G.729 is ADPCM (Adaptive Differential PCM). This is significant because of the ease with which they can be decoded.

SIP, RTP and the CODECs do not contain encryption, although the RTP sequence numbers start from a random point to make it difficult to reverse engineer the key, should encryption be used.

6.3. Eavesdropping

When security and VoIP are mentioned in the same sentence to potential users, the most common immediate concern is the ability of criminals to eavesdrop on conversations. It is relatively easy to eavesdrop on conventional telephone conversations by means of a hidden microphone or tapping the wires in the street, so the ability to intercept VoIP data should not be considered as a new danger.

6.3.1. *Listening to RTP*

Intercepting a call being carried over RTP, assuming it is not running under an external encryption protocol, is still problematic as the eavesdropper needs physical access to the packets. This can only be achieved by compromising some equipment along the call's route, which is no more practical than intercepting a conventional telephone call. However, if the call is carried on a LAN and a host on the LAN can be compromised, it should be relatively easy to listen in on conversations by re-routing the packets to pass through the compromised host. This can be achieved on switched Ethernet LANs by injecting the compromised host's MAC address into ARP caches in place of the gateways — a common data eavesdropping technique.

In such a scenario, where a compromised host routes the calls, it would also be possible to modify the call data contents, although RTP's timestamps and sequence numbers make this difficult. This would have to be done in real-time and it is hard to imagine any practical purpose, other than to cause annoyance.

Calls could also be intercepted by connecting into the debug ports on smart switches or routers, either physically or by compromising and re-programming the equipment concerned.

6.3.2. *Registrar*

If gaining access to the victims LAN proves difficult, calls could still be intercepted by compromising an external registrar server's integrity. This could be achieved by re-programming a legitimate registrar server to forward calls to the criminals, or the registrar server could be spoofed.

If the registrar trusts the attacker because, for example, the attack comes from the LAN, it is possible to simply register a second handset at a different address on the same account using a SIP REGISTER message and intercept all calls. The attacker can optionally drop the original handset using the same mechanism by sending a REGISTER message for it with an expiry time of zero. This process may be password protected, but it is optional and passwords can be sniffed and cracked, and where the LAN is considered safe this security feature may not have been enabled for the sake of convenience.

While no instances of this occurring on the open Internet appear to have been documented, there is no technical reason why this could not happen in the future. Registrar servers are being set up in increasing numbers by less-experienced operators. These are often running on standard Linux platforms and over time there is no reason to believe that they cannot be compromised in the same way as any other server platform has been in the past.

If a handset's registrar cannot be compromised, then a handset could be still be tricked into registering with a different server controlled by the attacker using DNS poisoning or other methods. One quick and easy way of spoofing a registrar is to log onto a VoIP handset's set-up page, assuming that the password could be broken if set, and edit the legitimate registrar to an address of your choice.

Gaining access to a handset's set-up page may be easier than anticipated. Because NAT routers can cause problems for SIP, it is common for SIP handsets to be available in a DMZ, leading to the web set-up interface being available on the public Internet too. Some have even been indexed by Google, to save the criminals having to scan for them.

6.3.3. *CALEA*

Communications Assistance for Law Enforcement Act (CALEA) is an American law that requires communications providers to give American government agencies the ability to eavesdrop on voice communications. It was passed in 1994 and covered standard telephones, but in 2004 steps were taken to extend this to VoIP, with the new rules coming in to force in May 2007.

The important point about CALEA is that it requires the service provider to intercept and present the unencrypted voice traffic on demand and at the service provider's expense. This effectively means that anyone selling VoIP equipment in America must provide a tapping point — in other words there must be a built-in eavesdropping mechanism. It is unlikely that this will be removed for equipment destined for other markets and therefore presents another theoretical weak point in VoIP security.

In its present form, CALEA does not appear to affect Skype but sources close to eBay have refused to comment on whether it is technically possible to intercept and decode Skype calls. The whole question about CALEA and VoIP remains unresolved, with various legal challenges taking place in the American courts.

6.4. DoS

It is hard to imagine anything more "mission critical" to an organisation than its telephone system. Losing e-mail for a time causes annoyance and inconvenience, but business can be continued using the telephone. However, picking up a handset and not hearing a dial-tone is so unusual these days that a few organisations consider the possibility of it happening.

VoIP systems are vulnerable to DoS attacks in the same way as any other data network. This assumes that the attacker has access to the network the VoIP system is using, which is why separating it from the IT system is a very good idea. Unfortunately, VoIP is often sold on the basis that it can combine voice and data communications, and some customers may choose to actually do this. If the network is connected to the public Internet, then DoS attacks can be launched from anywhere.

The classic DoS attack is to swamp the user's bandwidth, preventing legitimate packets from getting through. VoIP is particularly susceptible to this as degraded bandwidth is useless if it cannot carry real-time call data. If the incoming connection is on the Internet, there are numerous well-documented attack methods for swamping the line.

There is no reason to believe that more subtle techniques such as a SYN flood would be ineffective (where a handset's connection-pending buffer is overfilled with bogus requests for a TCP session), although good routers and firewalls should be able to filter these out before swamping the handsets themselves.

A DoS attack aimed at computers on a LAN would disrupt VoIP handsets on the same LAN too. It is a good practice to keep VoIP on a separate VLAN, but even then the data flow will be combined at some point and the bandwidth could easily be overwhelmed.

There are also numerous VoIP-specific DoS attacks possible. For example, sending periodic INVITE packets to a handset will cause it to ring incessantly. Eventually, the user will give up trying to answer calls (and probably disconnect the ringing 'phone), effectively disabling the service. Such an attack is cheap and easy to launch and hard to prevent.

Flaws have been discovered in the software on several handsets, such that sending them invalid data causes them to re-boot (dropping calls), freeze or behave erratically. Such exploits are being discovered on a regular basis, and although they are curable there are logistical problems in deploying updated firmware. Handset users are not used to having to update the software in their 'phones, and there is no notification mechanism to inform them that it is even needed.

If a host on the same LAN as VoIP handsets becomes infected with specially written malware, it could launch a highly effective DoS attack on them, which may be extremely difficult to track down and could not be mitigated using a perimeter firewall. As with so many VoIP security issues, such a software is currently just a theoretical possibility imagined by the author, but there appears no technical reason to believe it could not happen in the future.

6.4.1. *Sharing a LAN with Computers*

As previously mentioned, VoIP is often sold with the promise of using existing Ethernet infrastructure and WAN connections to carry voice calls in addition to data. Many handsets even have twin Ethernet sockets so that they can be placed in-line with a desktop PC and share the same wall-port. This is a very dangerous idea.

Placing VoIP on the same LAN and WAN as PCs makes handsets susceptible to DoS attacks aimed at PCs or their bandwidth — even resource consumption caused by malware such as a worm would impact telephones as bandwidth is swamped and smart switches or routers isolate problem segments. Not only do users' PCs go down, but so does every telephone in the office (which may be some relief to a beleaguered IT department).

Sharing with PCs also exposes VoIP to attacks from compromised local computers, allowing eavesdropping, spoofing and disruption.

There are more subtle security implications to mixing VoIP and data. For example, in order for SIP to pass through a firewall, special conditions must be met — especially where NAT is involved. This can lead to firewalls' rules being degraded to accommodate SIP, leading to a security weakness for the PCs. This can be mitigated by using properly programmed SIP-aware firewalls but this will take time to deploy.

One solution to the issue of mixing PCs and handsets is to use VLAN technology to logically isolate VoIP from data traffic. However, there is still some risk as VLAN integrity can be compromised, and two networks have to meet at some common vulnerable point.

A further complication are softphones, which run on a PC and must therefore be connected to the same data network as PCs. Network administrators should consider the risks involved very carefully, although softphones are unlikely to replace regular handsets and their unavailability is less likely to cause the same level of inconvenience.

6.5. Spam Over Internet Telephony (SPIT)

The most remarkable thing about spam voice messages being sent over VoIP is that it is not happening yet. Criminals have been sending spam over e-mail for years, and there is no technical reason why voice messages could not be delivered in a similar way.

Most of the discussion on SPIT centres on the possibility of injecting messages into the PSTN and targeting arbitrary handsets (not just VoIP ones). The snag with this approach for the spammers is that at some point someone is paying for a PSTN call, which means gateways are wary of allowing themselves to be abused in this way.

It is also the case that corporate VoIP systems employ a proxy server, which must be bypassed to reach the handsets, preventing SPIT from entering traditional VoIP installations.

However, the widespread adoption of VoIP handsets in the mass market will change this situation. If you can locate a SIP handset on the public Internet, then you can place a call to it. To find a SIP handset simply requires sending INVITES to port 5060/UDP and waiting for a response, although criminals will doubtless trade in lists of such handsets in the way e-mail addresses are currently obtained and circulated.

Handsets can theoretically refuse to accept calls from arbitrary locations, but they all do not support this feature, and it is turned off by default.

Calls made over the PSTN do actually cost money, although this is now very little when calls are relatively local. Nevertheless, access to the PSTN has always been limited and accounted. Where IP to PSTN gateways exist, the operator will control access, even when users are not charged for individual calls. Because of this, it is reasonable to assume that SPIT will not be placed on the PSTN using "open relays" in a similar way to the abuse of SMTP e-mail. Any operator offering a gateway that can be abused is going to lose a lot of money very quickly; commercial reality will sort out the problem.

There is, however, a scenario that would allow SPIT onto the PSTN at no cost to the criminals. Should an intruder break a host on a network with a VoIP to PSTN gateway (such as an IP PABX), it could be used for channelling SPIT up to the capacity of the PSTN lines available. In practice, this could limit abuse to hundreds rather than millions of unwanted messages, but it would still be a significant problem. This is another good reason to keep computers and VoIP handsets on separate networks.

However, calls direct to IP handsets are very cheap to make. The caller will ultimately pay for the bandwidth, but this costs little and will not increase with the distance over which the call is carried — allowing criminals to operate off-shore.

It is also reasonable to assume that criminals engaged in this activity will steal bandwidth from unsuspecting users by compromising and remote controlling unguarded hosts. This practice is widespread for sending spam e-mails, where the obfuscation of the sender's IP address is also important to the criminals.

Guarding against SPIT, if and when it arrives, will be considerably more difficult than dealing with spam. Filtering on content has practical difficulties when dealing with a voice call in real time — the recipient will spot what is happening and hang up long before any voice recognition system has figured out that call is not genuine. Scanning voicemail boxes for SPIT, where the whole message is available and time is not such an issue, will still be technically very difficult and compute-intensive. This is easily demonstrated by attempting to use existing continuous text speech-recognition systems. Although HMM voice recognition techniques show promise, they are far easier to defeat than get working correctly; in practice, they require training to a particular user's voice.

It would be possible for a handset or IP switch to refuse connections from particular sources by IP address or caller-ID using either blacklists or whitelists. To be effective, a remote database of known SPIT sources must be set up. Unfortunately, if the criminals are using remote-controlled PCs, then the IP addresses will keep changing, making this technique far less effective. The ease with which caller-ID can be spoofed also detracts from this strategy's effectiveness — and should it be employed it would be possible to engineer the blacklisting of a particular user by "borrowing" their ID and denying them service as a result.

Handsets do not generally have support for complex access control at present, and although filtering could be provided by a firewall, most of the unprotected handsets are in installations where no suitable firewall is available. Some can be configured to refuse calls not referred by their registrar server, if used. With e-mail spam, criminals set up accounts at no cost using services such as Hotmail and other's eager to expand their user base without regard for security. There are parallels in this with the current "land grab" by VoIP providers.

Other SPIT filtering techniques involve diverting suspect calls to an automated system to determine whether the caller is human, with methods ranging from answering a simple question to determining the level of response to a recorded message played to the caller (the theory being that a real person will stop talking whereas a recording will continue to play). The NEC is launching a product called VoIP SEAL that employs a variety of techniques to give a score to incoming calls in order to determine the likelihood of them being SPIT. This multi-test approach has worked well for e-mail Spam, but there is no way to determine its effectiveness against SPIT until real SPIT starts appearing.

Whatever technological solutions may appear to guard against SPIT at the recipient's end, they will not be readily available to simple handsets in a domestic or home-office environment, and there appears to be no practical defence for a single handset connected to the public Internet.

On the legal front, various laws are being drafted against SPIT. One such approach is to outlaw calls that play the recipient a recorded message (Federal Trade Commission, October 2006). Another bill (H.R. 251–110th Congress (2007): Truth in Caller ID Act of 2007) was introduced in the USA in January, and proposes to make forging caller-ID illegal. As SPIT is unlikely to use genuine caller-IDs, this may be relevant, and as an anti-fraud measure it may be important in securing convictions.

However, given the ineffectiveness of anti-e-mail spam laws at tackling that problem, there is no reason to suppose the pending anti-SPIT laws will be any more successful.

6.6. Fraud

VoIP offers many new possibilities to criminals engaged in fraudulent activity. Often, these are existing criminal activities modified to utilise VoIP technology.

6.6.1. *Session Hijacking*

In Section 6.3, "Eavesdropping", a scenario was described where an attacker placed a host in the path of a call in order to route packets and intercept them at the same time — a classic man-in-the-middle attack. Manipulating the packets passing *en route* is difficult as the RTP sequence numbers and timestamps would have to be maintained. However, it does offer the possibility of taking over a session in its entirety.

The fraudulent application of such a technology, should it be developed, is obvious and extensive. Imagine a call to a bank being intercepted after the caller has passed through the security procedure — the change of a caller's voice could be covered by hijacking the session during a call transfer at the bank. Alternatively, callers to the bank could be diverted and end up talking to the criminals after they have dialled the bank's number.

6.6.2. *CLID Spoof*

Calling Line ID (CLID), or caller-ID, is the system that displays the number of the caller on the recipient's handset. Using conventional telephony, it is relatively reliable. In order to generate a false caller-ID requires the connection of corrupt telephone exchange into the standard network — a logistically difficult and an expensive exercise.

To fake a caller-ID using VoIP technology is very simple. Using SIP, for example, any string in quotes at the start of the "From:" field is presented on the handset as the number of the caller — there is no check on its validity and it can be set to just about anything.

If a call is made using the SIP protocol direct to another IP handset, then a caller can appear to be anyone they wish; if the call enters the PSTN via a gateway, then it is up to the gateway to force a valid ID. There are several companies in America that offer gateway services specifically tailored to spoofing IDs. Examples are Telespoof and SpoofTel at the time of writing this chapter, although such companies do not always stay in business for long periods.

Telephone users tend to trust caller-ID; it is used in some organisation to route calls to particular recipients and has been used by banks to verify that they are talking to valid customers. It is also used by law enforcement authorities to see who is calling whom. As caller-IDs can be spoofed easily, the information they contain should now be considered unreliable, but it will take some time before users are educated enough to distrust it in the same way they suspect the claimed sender of an e-mail.

6.6.3. *VoIP Phishing*

The first documented instance of a VoIP phishing scam (sometimes called Vishing) appeared in January 2007, where a spam e-mail-induced victims to call a false telephone number where they were greeted with an automated attendant asking them to enter their bank account details. This is similar to normal phishing scams, with the fake website replaced by a fake telephone number.

Subsequent scams have appeared, including one targeting Bank of America customers.

6.6.4. *Combined Attacks*

Fraud using VoIP clearly has many possibilities when used on its own, but it becomes even more serious when used in a combined attack. For example, imagine the scenario where criminals send out a spam e-mail to a company's customers telling them that their website has moved, and asking them to log in to the new site. Most users would be wary of such a tactic, but if this was combined with a DoS attack against the real website and their VoIP-based telephone system, and a helpline mentioned in the e-mail backing up the criminals claim, it seems reasonable that more people would be fooled. At present, customers will normally call the company concerned to verify such a claim arriving by e-mail, but this option can be removed. The scenario could be presented as "Our main call centre has been destroyed by fire and we have switched to a back-up call centre and website. Sorry for any inconvenience", or "We have just been taken over by company X — please use the following new contact details".

6.7. Skype and Softphones

A softphone is a VoIP handset implemented entirely in software running on a standard workstation or other computing device. Most of these use open standards, or even open source, but various proprietary systems are in use too — often embedded in instant messaging software.

Because a softphone must, by definition, run on the data side of any network, there is an obvious problem keeping VoIP and data segregated for security reasons.

The most popular softphone is currently Skype (rhymes with "hype"), which was launched in late 2003 as a follow-on project from the Kazaa "file sharing" application. It rapidly gained a following amongst computer users of a particular profile (generally those concerned with swapping files) and has spread to a significant user base since then. It was sold to eBay in 2005.

The business model used by Skype involves charging users for calls made through their PSTN gateway known as SkypeOut. Unlike SIP users, who have a choice of gateway services, Skype has just one.

Skype uses a proprietary protocol, and its writers have gone to considerable lengths to prevent reverse engineering. Apparently based on the Kazaa/Morpheus FastTrack peer-to-peer system, it is designed to work through firewalls and to be hard to detect and block. This was considered necessary because network administrators spend and put a lot of effort into keeping file-sharing applications off their network. As a consequence, the protocol jumps around on different ports makes itself hard to detect.

The protocol itself does appear to be secure, using encryption. However, the presence of Skype on a network is a security risk in itself.

One questionable feature of the software is the ability to transfer files among users. The threat from this is compounded because there are no hooks provided to allow anti-virus filters to scan the content on the PC. As the traffic is both encrypted and hard to identify, it cannot be scanned at the perimeter.

The Skype software is also running a service and is vulnerable to normal buffer-overrun and other exploits associated with services. As Skype is normally user-installed (instead of being under control of a network administrator), the risk is increased. Experience shows that users frequently do not check for and install security updates.

Anyone can register any name on the Skype network, which means users cannot be traced and malicious users could pose as anyone they wish.

It is also possible to connect to the Skype network using the same ID at multiple locations, and hide the presence of the extra logins. If an attacker has obtained a user's password, then calls could be intercepted without their knowledge.

The eBay, Skype's current owner, is taking steps to address some of these issues by making the software less of a risk in a corporate environment but its proprietary and unruly protocol, lack of central control and user anonymity are unlikely to impress those responsible for network security.

6.8. Conclusion and Best Practices

There is a tendency to rely on "Best Practice" where security is concerned. In the case of VoIP, this is probably a mistake. The ground is shifting so fast that by the time a practice becomes accepted, it is unlikely to be relevant, and real risks on new equipment may be overlooked.

Following best practice, removes the need to understand and assess the risks in a new environment and gives those responsible an excuse when things go wrong.

Nonetheless, there are some good practices worth following when deploying VoIP.

(1) Keep the voice and data networks as separate as possible. Separate VLANs are a good start, but separate cables, firewalls, switches, routers and WAN links are better.
(2) Treat the VoIP network like a data network — close unused services; restrict access to everything that is not essential, use encryption and manage passwords in the same way.
(3) Consider very carefully whether softphones should be used at all, and if they must be allowed, isolate them from the network running the hardware handsets.
(4) Develop a survival strategy for a DoS attack appropriate to the risk, bearing in mind that new forms of attack are very likely and prevention may be impossible in the short or even medium term and
(5) Consider security at the start of any VoIP project, not at the end.

VoIP offers many advantages over analogue and ISDN telephony, but with apparently greater risk. Most of these risks have always existed in conventional telephony, but the system still works. However, the PSTN was developed slowly and incrementally by large monopolistic telecos able to exercise complete control. Existing corporate VoIP technology has been similarly managed.

The real danger with VoIP is that the technology is being developed rapidly and deployed by individuals and small companies without the ability to understand it fully. It is not possible to say that any system is secure until it has been proven secure over time. There are many unresolved security issues with poorly designed VoIP deployments and they are probably not getting the attention they require, and this may persist until it is too late for some.

Resources

Information on VoIP Security is available, although it is relatively new topic. One book of particular interest is *Hacking VoIP Exposed* by David Endler and Mark Collier (ISBN-13: 9780072263640).

On the Web there VoIPSA, the VoIP Security Alliance, an organisation formed from companies and individuals with an interest in VoIP security. The accompanying weekly podcast, www.blueboxpodcast.com was required listening for several years, although it is currently (2008) being published less frequently.

Other documents of interest are:

[1] S. A. Baset and H. Schulzrinne, An analysis of the Skype peer-to-peer internet telephony protocol, Department of Computer Science, Columbia University, September 15, 2004.
[2] H. Max and T. Ray, *Skype: The Definitive Guide*, Que, ISBN 032140940X, May 2006.
[3] J. Rosenberg and C. Jennings (Cisco), The Session Initiation Protocol (SIP) and Spam, Internet-Draft draft-ietf-sipping-spam-03, October-2006.

Chapter 7

SECURE BY DESIGN: CONSIDERING SECURITY FROM THE EARLY STAGES OF THE INFORMATION SYSTEMS DEVELOPMENT

HARALAMBOS MOURATIDIS
School of Computing, IT and Engineering
University of East London, UK

This chapter argues for the need to embed security considerations from the early stages of the information systems development. Security comprises of two dimensions. A technical dimension that is related to the available technology and the infrastructure of information systems, and a social dimension that is related to the impact of the human factor on the security of a system. Both dimensions need to be looked at in order to effectively develop secure information systems. Towards such a treatment of security, it is really important that security is considered from the early stages of the development process in order to gain an in-depth understanding of the technical and social issues that might affect the security of an information system. On the other hand, it is equally important to provide the appropriate knowledge to assist information systems developers in dealing with the technical and the social dimensions of security and also educate system users on issues related to the security of information systems. This chapter motivates the establishment of a *secure by design* philosophy of secure information systems development; it discusses its characteristics, vision, principles and practices and it identifies a list of important challenges related to it.

7.1. Introduction

We live in a world where information systems are widely used not only by major corporations and governments but also by individuals. A large amount of critical information is stored in these systems and as a result the need to secure them has been widely argued both in academia and in industry. However, and despite the large pool of research and the efforts of major corporations, surveys indicate that we are far from developing acceptable secure information systems [6, 10].

An important reason for this situation is that security is mostly perceived as a technical problem and the usual approach involves the introduction of a set of standard security mechanisms, such as authentication, after the system's design has been completed. This approach raises two important problems. First of all, this treatment of security results in enforcing security mechanisms into a pre-existing design, which in turn results into conflicts of the system requirements, leading to security vulnerabilities. Second, as it has been argued in the literature, a substantial number of security-related problems raise from the human factor. However, in most of the cases, this social dimension of security is ignored. As a result, a number of systems, although demonstrate strong security mechanisms, are still vulnerable to security attacks.

Moreover, and despite a wide acceptance that security is an important issue in the development of information systems, on one hand researchers and practitioners have not yet understood in depth the security issues involved in the development of information systems; and on the other hand users are not well educated regarding their role in the security of information systems.

It is therefore of paramount importance for the development of secure information systems that security is neither treated only as a technical problem nor it is isolated from the rest of the systems' development process. What is really needed is a *secure by design* philosophy which will form the foundation for understanding in depth the security issues involved in the development of information systems; provide the appropriate knowledge to assist information systems developers in dealing with the technical and the social dimensions of security and also educate system users on issues related to the security of information systems.

This chapter advocates the establishment of such a philosophy that will not treat security in isolation but it will consider it along with its related concepts, such as trust and safety, within the context of the environment of the information system.

Section 2 provides an introduction to the subject of security of information systems, motivating the need to introduce a *secure by design* philosophy. Section 3 discusses the characteristics, vision, principles and practices of secure information systems engineering, and Section 4 describes current problems and limitations. Section 5 discusses some important challenges and Section 6 concludes this chapter.

7.2. Security of Information Systems

Computer security is definitely not a new topic since its history starts in the 1960s [32]. However, the importance and the perception of security have considerably changed since then. In fact, the importance of considering security in information systems has been driven by the characteristics of the different "generations" of information systems. Initially, information systems were stand alone systems without interconnections and connectivity. At that point, the only security considerations were associated to physical security measures to protect the stand alone system against theft and/or destruction. Later, the World Wide Web changed the way information systems operate by introducing characteristics such as connectivity, distribution and communication. Information system developers and users realised that just physical security was not any more adequate, and this led to the development of a large number of computer security solutions such as authentication mechanisms, access control models and so on.

Over the last few years, the technological advances on information systems and the transition towards open and autonomous systems have introduced new requirements for the development of secure information systems. As recent research indicates [9, 26, 38], just the introduction of a set of pre-defined computer security solutions to an information system will not result in solving the security problems. In contrast, "*security concepts must inform every phase of software development, from requirements engineering to design, implementation, testing, and deployment*" [9]. Taking security into account alongside the functional requirements throughout the development stages helps to limit the cases of security/functional requirements conflicts by avoiding them from the very beginning or by isolating them very early in the development process. To adequately consider security issues during the software development life cycle, security should be integrated within software engineering languages, methods, methodologies and processes.

Moreover, we believe that security does not only demonstrate a technical dimension, related to the available technology and the infrastructure of information systems, but also a social dimension. The social dimension is related to the impact of the human factor on the security of a system. Therefore, to effectively develop secure information systems, both dimensions should be considered simultaneously. As Yu *et al.* argue [41] "*All information systems are ultimately embedded in some human social environment, and therefore the effectiveness of the system depends very much on the forces in that environment*". As an example consider a typical social engineering attack on a health information system. Social engineering is a non-technical kind of intrusion that relies on human interaction and it involves tricking users of the information system (doctors and/or nurses in the case of health information systems) to break normal security procedures. A typical scenario involves a private detective calling in a health professional's office or a hospital and introduces himself/herself as a doctor in an emergency unit; he/she then asks for information about the medical record of a particular patient. That way, the "attacker" is able to extract important information from the system without attacking the security technology applied to the system. This example demonstrates that considering only the technical dimension of security will not produce the desirable output.

Therefore, the existence of secure information systems cannot be achieved just by employing formal models, methodologies and security mechanisms (although these are useful) during their development neither by ad hoc approaches to solve the various problems involved in securing information systems. What is really needed is a development philosophy that will form the basis to understand in depth the security issues involved in the development of information systems; provide the appropriate knowledge to assist information systems developers in dealing with the technical and the social dimensions of security and also educate system users on issues related to the security of information systems.

7.3. Secure by Design

The *secure by design* philosophy advocates the consideration of security from the early stages of the information system development process. Its underlying aim is to improve the quality of information systems by reducing the number of security vulnerabilities that these systems demonstrate.

The focus of study or the fundamental research question of such a philosophy can be formulated as "*how to develop secure information systems?*". In answering such a question, many sub-questions need to be formulated and answered. For example, what we mean by "*secure information systems?*", "*what is good security?*", "*how do we define security requirements*?". Usually, different researchers and practitioners will answer differently to such questions. However, it is imperative that common answers are established in such fundamental issues, in order to provide a well-founded base in which we will be able to support further research questions leading us closer to answer the fundamental research question.

Moreover, it is important for a research area to be developed based on the *secure by design* philosophy. However, it is equally important that such a research area does not exist in isolation but it is related to reference areas. Reference research areas are existing bodies of knowledge that help to establish the new research area. Formally, referencing-related areas recognize the contributions of existing knowledge and provide a logical link to the new area. Without this linkage, researchers in existing research areas may question the grounding theories of a new research area and dismiss its importance [20]. A research area based on the *secure by design* principle should therefore build upon the knowledge, theories and methods of several existing research areas such as security requirements engineering (for example [1, 8, 38, 16]) security modelling (for example [23, 18, 14, 28, 29]), secure information system development (for example [11, 27, 29]), and security policies/models/ontology/principles (for example [3, 40, 19]).

We envisage the maturity of the *secure by design* philosophy in such a degree that information system developers will be able to model, construct, test, deploy and maintain secure information systems through well defined and structured processes, which will motivate the consideration of security from the early stages of the development process, and with the aid of appropriate modelling languages. In such a vision, development is made even easier with the aid of computer-aided tools that enable to accurately track the security solution to the initial system requirements and therefore validate it against the security goals of the organisation where the system is deployed. Moreover, the research results are fed to a knowledge base that is used to educate users and professionals on all aspects of secure information systems.

7.3.1. *Vision, Principles and Practices*

7.3.1.1. *Vision*

Our vision is based on three main world view assumptions: (1) the development of secure information systems is a complex issue, which involves technical as well as social challenges; (2) processes, models, methodologies and automated tools can be employed to address the technical challenges and to assist in the development of secure software systems; (3) proper education of anyone involved in the development as well as in the usage of information systems is needed to support the outputs of research addressing the technical challenges and to contribute towards the achievement of the social challenges.

It is also important to define a set of principles and practices associated with a *secure by design* research area. Principles incorporate the world view and define the philosophical approach to solving problems. Practices are the methodologies, models, procedures and theories used to apply the discipline's knowledge base. Together, principles and practices form the foundation of the research area and promote further ordered study.

Therefore, the main objectives are the production of novel techniques, methods, processes and tools, which integrate security and information system development principles; the education of information system developers to use such techniques to analyse, design, implement, test and deploy secure information systems; and the education of information system users to ensure that the social implications of security are understood.

We argue that the *secure by design* philosophy is based on the following principles and practices:

7.3.1.2. *Principles*

- Security is a two-dimensional problem;
- Security considerations must be integrated in the development process;
- Information system users affect the social dimension of security and therefore they should be made aware of the implications and
- An adequate security-oriented education must be provided for information system users and professionals.

7.3.1.3. *Practices*

- Consider security from the early stages of the information system development;
- Define security requirements together with functional requirements;
- Maintain a separation of concerns between security and functional requirements;

- Ensure quality of security solution;
- Ensure security requirements satisfaction during the design stage and
- Develop curriculum based on the research results.

Although some of the above principles and practices are not novel, and they are based on the related areas of information systems and/or security engineering, the point is that current research and practice do not follow them.

7.4. Current Situation

7.4.1. *Independent Research Paths*

As we have discussed above, the consideration of security in the development of information systems raises a set of inter-twined issues in different research areas of computer science, such as the areas of security engineering and information systems engineering. However, these two research communities traditionally work independently. On one hand, information system development techniques and methodologies do not consider security as an important issue, although they have integrated concepts such as reliability and performance, and they usually fail to provide precise enough semantics to support the analysis and design of security requirements and properties [8, 26]. On the other hand, security engineering research has mainly produced formal and theoretical methods, which are difficult to understand by non-security experts and which, apart from security, they only consider limited aspects of the system.

7.4.2. *Lack of Appropriate Methodologies*

As indicated by current research [36, 14, 27, 29] information systems development methodologies do not create a security control environment early in the development process and modelling languages fail to include specialised handling of security requirements. However, there is a large number of work, which mostly has been developed over the last few years.

Initial work from the software engineering community produced a number of methods and processes for reasoning about non-functional requirements (NFR), including security. Chung [7] proposed the NFR framework to represent security requirements either as potentially conflicting or harmonious goals. From the security engineering community, Schneier [33] proposed attack trees as a useful way to identify and organise different attacks in an information system whereas Viega and McGraw [40] proposed 10 principles for building secure software. More recently, Anton *et al.* [2], proposed a set of general taxonomies for security and privacy, to be used as a general knowledge repository for a (security) goal refinement process. The pattern approach has been proposed by a number of researchers to assist security novices to act as security experts. Schumacher and Roedig [34] proposed a set of patterns, called security patterns, which contribute to the overall process of secure software engineering. Fernandez [12] specified security models as object-oriented patterns that can be used as guidelines for the development of secure information systems.

Although useful, these approaches lack the definition of a structured process for considering security. A well-defined and structured process is of paramount importance when considering security from the early stages of the development process.

On the other hand, a number of researchers model security by taking into account the behaviour of potential attackers. Van Lamsweerde and Letier [39] use the concept of security goals and anti-goals. Anti-goals represent malicious obstacles set up by attackers to threaten the security goals of a system. In addition, Van Lamsweerde [38] defines the notion of anti-models; models that capture attackers, their goals and capabilities. Similarly, Crook *et al.* [8] introduce the notion of anti-requirements to represent the requirements of malicious attackers. Anti-requirements are expressed in terms of the problem domain phenomena and are satisfied when the security threats imposed by the attacker are realised in any one instance of the problem. Lin *et al.* [21], incorporate anti-requirements into abuse frames. The purpose of abuse frames is to represent security threats and to facilitate the analysis of the conditions in the system in which a security violation occurs. An important limitation of all these approaches is that security is considered as a vague goal to be satisfied whereas a precise description and enumeration of specific security properties is still missing.

Differently, another "school of thinking" indicates the development of methods to analyse and reason about security based on the relationships between actors (such as users, stakeholders and attackers) and the system. Liu *et al.* [22] have presented work to identify security requirements, analysed as relationships amongst strategic actors, during the development of multi-agent systems. Giorgini *et al.* [14] have introduced an enhancement of Tropos [4] that is based on the clear separation of roles in a dependency relation between those offering a service (the merchant processing a credit card number), those requesting the service (the bank debiting the payment), and those owning the very same data (the cardholder). Moreover, Giorgini *et al.* [15] have proposed a PKI/trust management requirements' specification and analysis framework based on the clear separation of trust and delegation relationship. Although a relationship-based analysis is suitable for reasoning about security, an important limitation of these approaches is that they only guide the way security can be handled within a certain stage of the software development process.

Another direction of work is based on the extension of use cases and the Unified Modelling Language (UML). In particular, McDermott and Fox [25] adapt use cases to capture and analyse security requirements, and they call the adaptation an abuse case model. An abuse case is defined as a specification of a type of complete interaction between a system and one or more actors, where the results of the interaction are harmful to the system, one of the actors, or one of the stakeholders of the system. Similarly, Sindre and Opdahl [35] define the concept of misuse case, the inverse of use case, which describes a function that the system should not allow. They also define the concept of mis-actor as someone who intentionally or accidentally initiates a misuse case and whom the system should not support in doing so. Alexander [1] adds Threatens, Mitigates, Aggravates links to the use case diagram. Jurgens proposes UMLsec [18], an extension of the UML, to include the modelling of security-related features, such as confidentiality and access control. Lodderstedt *et al.* [23] also extend UML to model security. In their approach, security is considered by analysing security-related misuse cases.

An important limitation of all the use-case/UML-related approaches is that they do not support the modelling and analysis of security requirements at a social level but they treat security in system-oriented terms. In other words, they lack models that focus on high-level security requirements, meaning models that do not force the designer to immediately go down to security requirements.

On the other hand, a large amount of work has been devoted to security policies and the definition of security models. Various models[a] have been proposed based on mandatory access control (MAC), discretionary access control (DAC) and role base access control (RBCA). One of the first models was the Bell & Lapadula multi-level security model [3]. Another well-known model is the Chinese Wall model [5], according to which data is organised into three different levels.

The definition of security ontology is also an important area of research within the security engineering community. Initial efforts to define a widely accepted security ontology resulted in what is known as the Orange Book (US Department of Defense Standard DOD 5200.58-STD). However, work towards this standard started in the late 1960s and it concluded in the late 1970s. Therefore, important issues, raised by the introduction of the Internet and the usage of information systems to almost every aspect of our lives, are missing from the standard. More recently Kagal *et al.* [19] have developed an ontology expressed in DAML+OIL to represent security information, trust and policies in multi-agent systems, whereas Undercoffer and Pinkston [37] after analysing over 4,000 computer vulnerabilities and the corresponding attack strategies employed to exploit them have produced an ontology for specifying a model of computer attacks.

Although important and useful in many situations, the above work has a number of important limitations with respect to the integration to information system engineering practice. First of all, it mainly considers the later stages of the software development process. As it has been discussed in previous sections, it is important that security is considered from the early stages of the development process. Moreover, existing work is mainly focused on the technological aspects of security and it ignores, in general, the social dimension of security. It is important that security is considered within the social context and any social issues, such as trust and the involvement of humans, are taken into account.

[a] An extensive presentation and discussion of these models are out of the scope of this chapter and this book.

In previous work, the author has proposed a methodology, called Secure Tropos [26] that considers security from the early stages of the development process. Secure Tropos deals with the modelling and reasoning of security requirements and their transformation to a design that satisfies them.

The security-oriented process in Secure Tropos is one of identifying the security requirements of the information system, transform these requirements to a design that satisfies them, and validate the developed system with respect to security. In particular, the security-oriented process proposed by this research is mainly divided into five sub-activities: (1) identification of security requirements of an information system; (2) reasoning about different solutions; (3) development of a design that satisfies the security requirements of the system; (4) information system design validation and (5) attack testing of the information system under development.

The first step in the security-oriented process aims to identify the security requirements of the system. Security requirements are identified by employing various modelling activities that secure Tropos supports such as security reference diagram [26], security constraints [27, 29] and secure entities modelling [26]. In particular, possible security constraints that are imposed on the system and its stakeholders are identified and secure goals and tasks that guarantee the satisfaction of the identified security constraints are imposed on the actors of the system.

The second step in the process consists of identifying a design that satisfies the security requirements of the system, as well as its functional requirements. To achieve this, software agents are identified with the aid of the Tropos modelling techniques [27, 29] and secure capabilities that guarantee the satisfaction of the secure entities identified during the previous step are assigned to the agents. It is worth mentioning that at this stage, different architectural styles might be used to satisfy the functional requirements of the system. Therefore, an evaluation of how each of these architectural styles satisfies the security requirements of the system is needed. For this reason, an analysis technique [27, 29] is employed to enable developers to select among alternative architectural styles using as criteria the security requirements of the information system under development. The analysis involves the identification of specific security requirements and the evaluation of different architectural styles against these requirements. The evaluation results in contribution relationships from the different architectural styles to the probability of satisfying the security requirements of the system.

The development of a design that satisfies the security requirements of the system-to-be is a very challenging process. To assist developers, Secure Tropos includes a pattern language [30] consisting of security patterns based on agent orientation. The pattern language demonstrates the following important characteristics: (1) It contains security patterns that are based on agent-oriented concepts, such as intentionality, autonomy, sociality and identity and therefore it is suitable for modelling security issues of complex systems; (2) It explicitly defines each of the patterns of the language as well as it provides a precise definition of their relationships and (3) The structure of the patterns is described not only in terms of the collaborations and the message exchange between the components, but also in terms of the social dependencies and the intentional attributes, such as goals and tasks, of the agents involved in the pattern. The above characteristics allow for a complete understanding of the pattern's social and intentional dimensions, which are issues demonstrated by most of the current systems.

The fourth step of the process is the validation of the system's design. The Secure Tropos process allows for two types of validation: model and design. Model validation involves the validation of the developed models with the aid of a set of validation rules [26]. Validation rules are divided into two different categories: inter-model and outer model rules. Inner model rules assist the validation of each model individually, whereas outer model rules assist the validation of the consistency between the different developed models.

On the other hand, design validation aims to validate the developed solution against the security policy of the system. A key feature of Secure Tropos, which permits developers to perform design validation, is the fact that the same concepts used for analysis are also used for design and testing. As a result, the definition of these concepts enables developers to provide a direct map between the developed security solution, the security constraints and the security policy.

The fifth step of the process is the testing of the developed system at design time against possible attacks. A scenario-based security testing process, called Security Attack Testing (SAT) [31], is employed. Two sets of scenarios — dependency and security attack — are identified and constructed and security test cases are defined, from these scenarios, to test the developed design of the system against various attacks. In particular, the process aims to identify the goals and the intentions of possible attackers; identify through these a set of possible scenario attacks to the system; and apply these attacks to the system to see how it copes. By analysing the goals and the intentions of the attackers, the developer obtains valuable information that helps to understand not only how an attacker might attack the system, but also why an attacker wants to attack the system. This leads to a better understanding on how possible attacks can be prevented. In addition, the application of a set of attacks to the system contributes towards the identification of attacks that the system might not be able to cope and this leads to the re-definition of the agents of the system and the addition of new secure capabilities to the system to assist in the protection of those attacks. The SAT process consists of four main activities: (1) derive dependency scenarios where activity scenarios are identified that involve actors and resources of the system-to-be; (2) define security attack scenarios where security attack scenarios are constructed and validated; (3) define security test cases where test cases are identified and the security of the system against potential attacks is tested, using these test cases and (4) re-definition of system where extra secure capabilities are introduced to the system for each attack that it is identified that the system cannot prevent.

7.4.3. *Bad Practice*

In many cases, the inclusion of security on a system is driven by existing custom solutions (security mechanisms) rather than the system's real security requirements. Supporting the development of the security of the system on specific security mechanisms, as opposed to security requirements, prevents the consideration and choice of different and sometimes better solutions to satisfy the security requirements. For instance, imagine a system which requires identification and authentication. If the development of the system is based on some specific solutions to these requirements, such as username and password, then other solutions might be ignored, such as biometric identification and authentication, which in some cases could better fulfil the initial security requirements. Therefore, it is important that only the security requirements drive the development, as it happens with functional requirements, and not the well-known security solutions.

7.4.4. *Lack of Sharing Existing Knowledge*

As indicated above, information systems and security engineering communities mainly work separately. This separation results in restricted sharing of existing knowledge. For instance, different research events are organised by the two communities, different research publications and so on. Even widely used textbooks mostly concentrate in one part of the problem, either technical security issues or software engineering techniques, and they only contain, when they do, very limited information about the integration of the security and information systems engineering principles.

7.4.5. *Lack of Appropriate Education*

Professional training courses and university curriculum should help towards the solution of the above problem. However, they propagate it. Information systems and security engineering training as well as curriculum development in universities follows the separation of the two research areas. As a result, information systems engineering principles are taught separated from security engineering issues and vice versa. This means that information systems developers are not well educated regarding the security issues that might face during the development of information systems, and security engineers mostly are not familiar with current practices and issues surrounding information systems development.

7.4.6. *The Affect on Real-Life Scenarios*

All the above-mentioned problems affect the development of secure information systems as demonstrated by the following real-life scenarios:

- Requirements engineers do not usually receive appropriate training [13] in eliciting, analysing and specifying security requirements. As a result, they often confuse them with security mechanisms which are used to fulfil them. Therefore, they end up defining architectures and constraints rather than true security requirements [13].

- Information systems developers are faced with the development of secure information systems according to their security requirements. However, not all practitioners are security specialists neither they fully understand mathematical security models [25]. An information system developer without the appropriate security knowledge and without the support of practices that integrate security as part of the development process more likely will fail to develop the system according to its (security) requirements.
- Security engineers are often required to enhance the security of an existing system. However, current security models and methodologies used by security engineers neither fully analyse nor reason the implication that the addition of security components will have on the existing functionalities of the system. Without appropriate processes and methodologies to guide them, most likely they will fail.
- System developers are required to test, during design, whether the system under development satisfies its security requirements. However, the lack of appropriate languages and automated techniques makes such a task very difficult.

7.5. An Active Research Agenda

An active research agenda implies that hypotheses are being generated that address the fundamental question of the research area. The agenda should stand the test of time, with many researchers and practitioners in the discipline continually expanding the research that builds upon itself. The research agenda should be complex and substantial enough to be divided into sub-areas. Multiple sub-questions need to be formulated to guide the research necessary to contribute to the body of knowledge, which addresses the fundamental question asked by the discipline [20]. We believe that the following challenges are important for a research agenda for the *secure by design* philosophy.

7.5.1. *The Challenges*

Challenge 1: *Unify efforts to integrate security and information systems engineering.*

Although the need for such a unification has been recognised by various researchers, work on integrating security and information systems engineering is mainly carried out independently by members of each community. It is important to unify the efforts on the two fields. Only then we will be able to precisely identify the technical as well as the social issues that surround the development of secure information systems and produce solutions that truly work.

Challenge 2: *Consider the social dimension of security.*

Security is mainly considered as a technical issue by information system and security engineers alike. A mature solution that integrates security and information systems engineering should consider not only the technical dimension of security but also the social dimension. It is only when we consider both dimensions that we will be able to develop secure enough information systems.

Challenge 3: *Develop complete security ontology.*

The need for sound and complete security ontology is well recognised as an important issue for the development of widely accepted solutions on secure information systems engineering. Such an ontology will provide a firm and well-understood foundation to support the development of appropriate methods, processes and methodologies.

At present, work on defining such an ontology is carried out independently by the information systems engineering and security engineering research communities. This separation of work has resulted in an abstraction gap, which makes the integration and practical application of security issues in information systems engineering practices difficult. As an example, consider the term "security requirement". Although this term is fundamental; so far, it is used and interpreted differently by various researchers and practitioners [17].

Challenge 4: *Define a suitable exemplar.*

Typically, various information systems engineering approaches will be demonstrated using case studies, which are tailored to emphasise the key characteristics of the approach. However, such case studies often focus on specific problems. It is important, therefore, to define a suitable example problem (in software engineering community the term *exemplar* is widely used when referring to an example problem), which will emphasise the problems faced by the community and it will serve as a focal point for discussion and exchange of research ideas and results. In choosing such an *exemplar*, various criteria should be considered. For instance, the *exemplar* should be broad enough to cover all the possible issues, technical or social, that are associated with the development of secure information systems. Moreover, it should be generic enough as well as rich and complex enough to test the limits of any proposed approach.

Challenge 5: *Evaluate the different software engineering paradigms with respect to their appropriateness to integrate security*

Various software engineering paradigms exist such as model-driven, aspect-oriented and agent-oriented. All these treat software system development differently, using their own set of concepts and techniques. It is very important to identify the strengths and weaknesses of each of these paradigms when integrating security into the information systems development process.

Challenge 6: *Development of new techniques, methods, processes that consider security as part of the information systems development lifecycle.*

At present, most existing methodologies and models concentrate only on specific stages of the development process, such as security requirements engineering, or security design. It is vital, however, that security is considered throughout the development process and it is considered alongside the functional requirements and other non-functional requirements of the system-to-be. It is only then that we can consider security as part of the development process and not an isolated concept of the system. Therefore, it is important to develop new methods and techniques. These should support the formal (and simultaneous) modelling, reasoning and analysis of security and functional requirements, and the transformation of such (security and functional) requirements to a design that will satisfy them. Moreover, one of the main problems of considering security during the development stages of an information system is the lack of methods and techniques to trace the provided functionality to security requirements and also test the solution before the implementation of the system. Therefore, it is crucial to develop new methods and techniques to support traceability and validation of the proposed solution.

Challenge 7: *Tool support.*

Integrating security in the development process means adding extra activities in an already difficult task. Therefore, it is of paramount importance to produce tools to support the development process. A tool should not only support developers in modelling and reasoning about security (and functional requirements) during the analysis stage, but it should also help to transform the requirements to design, check the consistency of the proposed solution and also validate the security functionalities of the proposed solution against the security requirements of the system.

Challenge 8: *Transfer of security knowledge.*

Many information systems developers do not always have a strong background in computer security and lack expertise in secure information systems development. Nevertheless, in practice, they are asked to develop information systems that require security features. Development methodologies should consider that issue and provide methods and processes that allow even developers with minimum security expertise to analyse and design a system with security in mind. At present, security patterns seem to provide a right step into this direction. However, there is a need to enhance current pattern languages and provide a better integration with information systems engineering processes and methods.

Challenge 9: *Transit research results to mainstream system development.*

An important, long-term challenge is the successful transfer of research knowledge and best practice on developing secure information systems to industry. To achieve this, there is a need to make secure information systems engineering practice widely known (research and industry), standardise it and provide an agreed set of techniques, models and methodologies. This will ensure trust in the proposed methods and industrial confidence.

Challenge 10: *Provide suitable education.*

Education and professionalism are essential to the widespread recognition and deployment of any research area. The written record of knowledge and thought progression is valuable for future researchers and practitioners to reference when developing new theories and methodologies. Moreover, conferences and journals provide a forum for researchers and practitioners to exchange ideas, develop new knowledge and identify future lines of research. Separate curricula, professional societies and journals advance professionalism and are necessary for a separate discipline [24]. It is, therefore, essential to provide publications and events specifically focused on the secure information systems engineering to enable education and professionalism of researchers, practitioners and users of information systems.

7.6. Conclusions

This chapter argued that a *secure by design* philosophy should be employed in the development of information systems that require security. Such a philosophy is based on the principle that security is not only a technical problem and therefore a technical only solution will not result in the development of secure information systems. What is really needed is a mutual effort to consider security from the early stages of the development process to assist information system developers to fully understand the characteristics, principles and challenges that underlie the development of secure information systems.

This chapter discussed the need to develop a research area based on this philosophy and it outlined the aim and objectives of such a research area. Moreover, this chapter introduced its foundational principles and practices and it identified a number of limitations on current practice. This chapter is concluded by providing a list of challenges that are necessary for the advancement of the *secure by design* philosophy.

References

1. I. Alexander, Misuse cases: Use cases with hostile intent, *IEEE Software* **20** (2003) 58–66.
2. A. I. Anton and J. B. Earp, A requirements taxonomy for reducing web site privacy vulnerabilities, *Requirements Engineering* **9**(3) (2004) 169–185.
3. D. E. Bell and L. J. LaPadula, Secure computer systems: Mathematical foundations and model. The Mitre Corporation (1976).
4. P. Bresciani, P. Giorgini, F. Giunchiglia, J. Mylopoulos and A. Perin, TROPOS: An agent-oriented software development methodology, *Journal of Autonomous*

Agents and Multi-Agent Systems **8**(3) (2004) 203–236 (Kluwer Academic Publishers).

5. D. F. C. Brewer and M. J. Nash, The Chinese Wall security policy, *Proceedings of the IEEE Symposium on Research in Security and Privacy*, Oakland, California, 1–3 May 1989, pp. 206–214.
6. CERT Coordination Centre, Annual report (2006), available at www.cert.org.
7. L. Chung and B. Nixon, Dealing with non-functional requirements: Three experimental studies of a process-oriented approach, *Proceedings of the 17th International Conference on Software Engineering*, Seattle, USA, 1995.
8. R. Crook, D. Ince and B. Nuseibeh, Modelling access policies using roles in requirements engineering, *Information and Software Technology* **45**(14) (2003) 979–991 (Elsevier).
9. P. Devanbu and S. Stubblebine, Software engineering for security: A roadmap, *Proceedings of the 22nd International Conference on Software Engineering. Track on the Future of Software Engineering.* Limerick, Ireland, 2000.
10. DTI, Information Security Breaches Survey, Technical Report (2004), available at www.dti.gov.uk.
11. E. B. Fernandez, A methodology for secure software design, *Proceedings of the 2004 International Conference on Software Engineering Research and Practice (SERP'04)*, Las Vegas, NV, June 21–24, 2004.
12. E. B. Fernandez and R. Pan, A pattern language for security models, *Proceedings of the 8th Conference on Pattern Languages of Programs (PLoP 2001)*, Monticello, Illinois, USA, September, 2001.
13. D. G. Firesmith, Engineering security requirements, *Journal of Object Technology* **2**(1) (2003), ETH Swiss Federal Institute of Technology.
14. P. Giorgini, F. Massacci and J. Mylopoulos, Requirements engineering meets security: A case study on modelling secure electronic transactions by VISA and Mastercard, *Proceedings of the International Conference on Conceptual Modelling (ER)*, LNCS 2813, 2003, pp. 263–276 (Springer-Verlag).
15. P. Giorgini, F. Massacci, J. Mylopoulos and N. Zannone, Filling the gap between requirements engineering and public key/trust management infrastructures, In *Public Key Infrastructure*, (eds.) S. K. Katsikas, S. Gritzalis and J. Lopez, LNCS 3093 Springer, *Proceedings of the First European PKIWorkshop: Research and Applications*, EuroPKI 2004, Samos Island, Greece, June 25–26, 2004.
16. C. B. Haley, J. D. Moffett, R. Laney and B. Nuseibeh, Arguing security: Validating security requirements using structured argumentation, *Proceedings of the 3rd Symposium on Requirements Engineering for Information Security (SREIS'05) held in conjunction with the 13th International Requirements Engineering Conference (RE'05)*, Paris, France, 2005.
17. C. B. Haley, R. Laney, J. D. Moffett and B. Nuseibeh, Arguing satisfaction of security requirements, In *Integrating Security and Software Engineering: Advances and Future Vision*, (eds.) H. Mouratidis and P. Giorgini (Idea Group Publishing, 2006).
18. J. Jürjens, *Secure System Development with UML* (Springer-Verlag, 2004).
19. L. Kagal and T. Finin, Modeling conversation policies using permissions and obligations, In *Developments in Agent Communication*, (eds.) F. Dignum, R. van Eijk and M.-P. Huget *(Post-proceedings of the AAMAS Workshop on Agent Communication*, Springer-Verlag, LNCS), January, 2005.
20. D. H. Liles, M. E. Johnson, L. M. Meade and D. R. Underdown, Enterprise engineering: A discipline?, *Proceedings of the Society for Enterprise Engineering Conference*, June 1995.

21. L. C. Lin, B. Nuseibeh, D. Ince, M. Jackson and J. Moffett, Analysing security threats and vulnerabilities using abuse frames, *Technical Report* 2003/10, The Open University, 2003.
22. L. Liu, E. Yu and J. Mylopoulos, Security and privacy requirements analysis within a social setting, *Proceedings of the 11th International Requirements Engineering Conference* (IEEE Press, 2003), pp. 151–161,
23. T. Lodderstedt, D. Basin and J. Doser, Secure UML: A UML-based modelling language for model-driven security, *Proceedings of the UML'02*, LNCS 2460, Springer-Verlag, 2002, pp. 426–441,
24. H. B. Maynard, *Industrial Engineering Handbook*, 3rd Edition (McGraw Hill Book Co., New York, 1971).
25. J. McDermott and C. Fox, Using abuse care models for security requirements analysis, *Proceedings of the 15th Annual Computer Security Applications Conference*, 1999.
26. H. Mouratidis, A security-oriented approach in the development of multiagent systems: Applied to the management of the health and social care needs of older people in England, PhD thesis, University of Sheffield, 2004.
27. H. Mouratidis, P. Giorgini and G. Manson, When security meets software engineering: A case of modelling secure information systems, *Information Systems* **30**(8) (2005a) 609–629 (Elsevier).
28. H. Mouratidis, P. Giorgini and G. Manson, Modelling secure multiagent systems, *Proceedings of the Second International Joint Conference on Autonomous Agents and Multiagent Systems*, ACM, ISBN 1-58113-683-8, 2003, pp. 859–866.
29. H. Mouratidis, M. Kolp, S. Faulkner and P. Giorgini, A secure architectural description language for agent systems, *Proceedings of the 4th International Joint Conference on Autonomous Agents and Multiagent Systems*, Utrecht — The Netherlands, ACM, ISBN 1-59593-094-9, 2005b, pp. 578–585.
30. H. Mouratidis, M. Weiss and P. Giorgini, Modelling secure systems using an agent oriented approach and security patterns, *International Journal of Software Engineering and Knowledge Engineering (IJSEKE)* **16**(3) (2006) 471–479 (World Scientific).
31. H. Mouratidis and P. Giorgini, Security attack testing (SAT)-testing the security of information systems at design time, *Information Systems* **32**(8) (2007) 1166–1183 (Elsevier).
32. J. Saltzer and M. D. Schroeder, The protection of information in computer systems, *Proceedings of the IEEE* **63**(9) (1975) 1278–1308 (September 1975).
33. B. Schneier, *Secrets & Lies: Digital Security in a Networked World* (John Wiley & Sons, 2000).
34. M. Schumacher and U. Roedig, Security engineering with patterns, *Proceedings of the 8th Conference on Pattern Languages for Programs (PLoP)*, Illinois — USA, 2001.
35. G. Sindre and A. L. Opdahl, Eliciting security requirements with misuse cases, *Requirements Engineering* **10**(1) (2005) 34–44.
36. T. Tryfonas, E. Kiountouzis and A. Poulymenakou, Embedding security practices in contemporary information systems development approaches, *Information Management & Computer Security* **9**(4) (1997) 183–197.
37. J. Undercoffer and J. Pinkston, Modelling computer attacks: A target-centric ontology for intrusion-detection, *Proceedings of the CADIP Research Symposium*, 2002, available at: http://www.cs.umbc.edu/cadip/2002Symposium/
38. A. Van Lamsweerde, Elaborating security requirements by construction of intentional anti-models, *Proceedings of the 26th International Conference on Software Engineering*, Edinburgh, May, 2004, ACM-IEEE, pp. 148–157.

39. A. Van Lamsweerde and E. Letier, Handling obstacles in goal-oriented requirements engineering, *Transactions of Software Engineering* **26**(10) (2000) 978–1005.
40. J. Viega and G. McGraw, *Building Secure Software* (Addison Wesley, 2001).
41. E. Yu, L. Liu and J. Mylopoulos, A social ontology for integrating security and software engineering, In *Integrating Security and Software Engineering: Advances and Future Vision*, (eds.) H. Mouratidis and P. Giorgini (Idea Group Publishing, 2006).

Chapter 8

ONLINE TRANSACTIONS' SECURITY

JAMES KADIRIRE
Anglia Ruskin University

8.1. Introduction

There has been a sea change in the way that business is carried out in the digital age due to the proliferation of Internet technologies, which have made e-commerce more and more attractive. Merchants are able to reach out to customers all over the world where hitherto, it was only the preserve of multinationals. Consumers can shop online at their convenience, 24-hours a day, 7-days a week, order goods and have them delivered within a day or so, or even on the same day. The World Wide Web (Web), with its unprecedented exponential growth since the mid-1990s, is the main vehicle driving the popularity of e-commerce. A survey carried out by the Office for National Statistics [21] showed that people preferred to buy goods online for convenience and competitive pricing. The challenges in providing adequate security for online retailers are fairly steep ones. The threat to online shopping not only comes from securing the communications media like TCP/IP and the payment infrastructure, but also from internal fraud. Anyone can start up an Internet business and register as a merchant to trade online and according to Anderson [2], a survey carried out by the *Financial Times* in October 2000, found that 37% of dot-com executives have shady pasts, compared to 10% found in checks on the traditional companies. Online shopping is carried out via either notebooks/laptops, desktop computers or mobile devices like mobile phones and portable digital assistants (PDAs). The common denominator amongst all these devices is that they have access to the Web and the underpinning protocols like TCP/IP and HTTP, etc. This is usually facilitated via wireless local area networks (WLANs) using the IEEE 802.x protocol for wireless access, Global Systems for

Mobile Communications (GSM)/General Packet Radio Service (GPRS)/Enhanced Data GSM Environment (EDGE)/Universal Mobile Telecommunications System (UMTS)(3G) protocols for mobile devices and the Ethernet protocol, as well as WLANS, for desktop computers. The different protocols for accessing the Web have their own vulnerabilities and this chapter looks at some of the security issues that face online shoppers and the solutions proposed by researchers and being used by merchants. The introduction of the GPRS, EDGE and UMTS/3G networks, has made it possible for users to browse the Web and carry out online shopping. Pretty much, any retail company that is worth it is salt has some online presence. Mobile devices have truly become ubiquitous and pervasive, with over two billion mobile users worldwide [8]. This means that the potential for mobile devices being used for online shopping is also very high. However, according to Ahonen [1], the biggest hurdle that stops people from using their mobile devices online is people's concerns about security.

Weinraub [30], claims that for every 10 people who decide to make a purchase online, using either mobile devices, notebooks or desktop computers, only two actually go through with it. The main reason cited for this reluctance to purchase goods online was the fear of handing over credit card/e-cheque details over the Internet. That is a staggering 80% of potential sales which are going to be wasted because of security fears. When online shopping was introduced in the mid-1990s, there was a lot of anxiety about online transactions' security and this spurred the software vendors and banks to produce new protocols like the secure sockets layer (SSL) and secure electronic transactions (SET), which we will look at a bit later on. So, given all these anxieties about online transactions' security, why do we still bother about security? Well, predictions suggest that the market for internet shopping will continue to grow rapidly and according to the Office of Fair Trading [22], Verdict forecast retail sales in the United Kingdom, to nearly triple from £10.9 billion in 2006 to £29.1 billion by 2011. This means that online shopping would more than double its share of overall retail spending, from 4.0% to 9.3%. In addition to this, research from the Centre for Economics and Business Research predicts that online sales could comprise 40% of all retail sales by 2020, at a value of about £162 billion. This invariably means that security has to be at the heart of the enabling technologies. Despite all these fears about the lack of adequate security in online shopping, in comparison with normal credit card shopping and other forms of payment, online shopping is significantly more secure. Drimer *et al.* [11], distinguished researchers at the University of Cambridge demonstrated live on television how easy it is to breach the security of the much heralded chip and pin technology shown in Fig. 8.1, which is in widespread use today.

Figure 8.1 shows a typical chip and pin set-up and instead of signing a paper receipt to verify a card payment, the card owner enters a four-digit Personal Identification Number (PIN), just as in the case of using a cash point or automatic

Fig. 8.1. Chip and pin.

teller machine (ATM). Drimer *et al.* [11] demonstrated how the tamper proofing and the certification process of the pin entry devices (PEDs) is unsatisfactory. They implemented practical low-cost attacks on two of the widely deployed PEDs, the Ingenico i3300 and the Dione Xtreme, and successfully demonstrated, on the Watch Dog television programme, how by simply tapping inadequately protected smartcard communications, an attacker with basic technical skills can expose card details and PINs, leaving cardholders open to fraud. According to an article by Leyden [19], Lloyds TSB, one of the main high street banks in the United Kingdom admitted that flaws in the new Chip and PIN system recently introduced for debit cards in the United Kingdom open up the system to fraud. Conventional fraud may be down because of the system but crooks are still able to use cloned debit or credit cards in foreign ATMs or cash points. In spite of this, consumers feel more at ease using the Chip and Pin system than online shopping and even if they do not feel at ease with Chip and Pin, there is a large element of compulsion from retailers which leaves consumers with no choice but to use the Chip and Pin. When one looks at the actual statistics of online shopping, one sees that there is absolutely no evidence to support consumers' fears of buying things online. In a report by Weinraub [30], Gerry Sweeney, vice-president of Visa's online division states that:

> *"Of all credit card transactions, less than one-half of one percent are fraudulent. And of that miniscule number, less than one-tenth of one percent occur online. In other words, an online transaction is more than* 99.999% *secure."*

8.2. Security Threats/Attacks

Before we look at ways of securing online transactions, it is worth looking at some of the threats and attacks that online shoppers and indeed merchants are vulnerable to. A security threat has potential for security violations and a security attack compromises the information system security. According to Stallings [27], a security attack can be defined as any action that compromises the security of information owned by an organisation and is classified as either a passive or active attack. In a passive attack, the attacker only eavesdrops on the transmission channel(s) between the sender and the receiver with the sole purpose of gaining information about the transmission to be used later for some malicious purpose. An active attack involves active attempts on security leading to modification, redirection, blockage or destruction of data, devices or communication links. The most common active attacks are discussed as follows.

8.2.1. *Denial-of-Service Attacks*

These forms of attacks include flooding a Web server with very large number of transactions, which effectively render the Web server inoperable. This is effectively denying legitimate users e.g. potential customers, access to the merchant's Web server. Potential online shoppers using a WLAN/Wi-Fi at home can be prevented from using the Internet by a saboteur flooding an access point (AP) with network traffic. Users are unable to connect to the Internet because the AP is unable to cope with the volume of traffic, rather like receiving a busy signal when trying to use a phone line that is already in use.

8.2.2. *Replay Attacks*

These involve a hacker eavesdropping on a communications link and capturing some data that he or she can then re-use at a later date or time. A typical example of this attack could be a malicious user capturing some data, which may contain the legitimate user's password and then replays that message at a later time to gain unauthorised access to the system. If this happens at a merchant's database where the online shopper's confidential data is stored unencrypted, this could lead to the malicious user getting hold of shoppers' personal details like credit card numbers, home addresses, dates of birth, etc., which could be used for identity theft and all sorts of unlawful things.

8.2.3. *Masquerade Attacks*

Also known as "spoofing", this involves one computer on a network pretending to be another computer, usually with special access, in order to gain unauthorised access to some confidential data or restricted resource. One way of securing a Web server is to give access only to computers with a trusted IP address. Hackers, therefore, use a technique called IP spoofing to convince a computer on the other end of the connection that the computer they are using is assigned a trusted IP address. For home shoppers using a WLAN, a hacker can use a network sniffer to find out the service set identifier (SSID) of your access point and simply connect to it. If it is not properly secured, it will allocate him/her a valid IP address, which will allow him/her to masquerade as a legitimate user and get up to all sorts of mischief.

8.2.4. *Modification/Corruption of Data or Access Control Bits*

These attacks involve a hacker changing the exchanged data by insertion, deletion, re-ordering, delay or even changing the whole message.

8.3. Security Concerns and Requirements

The Internet is based on open network architecture, so information can be transferred freely and efficiently. This raises many security concerns for consumers wanting to shop online and some of the main concerns that plague consumers are:

(a) An online shopper transmits his/her confidential credit card details over the Internet. Can people other than the intended recipient e.g. the merchant or bank, read it?
(b) An online shopper agrees to pay £100.00, say, for his/her iPod. Will this payment information be captured and changed by some hacker on the Internet? and
(c) An online shopper is looking at a Website, which claims to be merchant X. Is it the real merchant X or a phishing attempt by some unscrupulous hacker masquerading as merchant X?

All these concerns are addressed by modern cryptographic techniques, which make e-commerce very secure. The security requirements used to address these security concerns are confidentiality, integrity and authentication.

8.3.1. *Confidentiality*

If ensures a message is kept confidential or secret so that only the intended recipient can read it. It is achieved via data encryption and is used in a security system to address security concern (a) discussed earlier. Confidentiality can be applied at two levels, i.e. the user data stored in a merchant's database and also while the data is in transit from the client to the server. The main attacks on data confidentiality are the passive attacks and an effective solution to combat this, is the use of cryptography so that even if a hacker gets hold of the data, it will be of no value to them as it will be encrypted.

8.3.2. *Integrity*

This is the process which ensures that if the contents of a message are altered in anyway while in transit between the sender and the receiver, it will be detected by the receiver and will therefore not be valid. Data integrity is typically achieved via the use of digital signatures and because digital signatures are so critical in online transactions security, we will give a brief explanation on how they work, shortly.

8.3.3. *Authentication*

This is the process of validating or verifying the credentials of an identifier (person) or entity presented to a system. Proper authentication relies on the identification and registration/credentialing processes which should not be compromised. It is essentially the associating of an identifier and credential bearer to that collection of information about someone or something in a particular context. When two parties are communicating over the network and one party wishes to establish the identity of the other, authentication must occur. This is a difficult process to accomplish securely, reliably and across a variety of systems and applications. The basic factors used to accomplish authentication are:

- Something a person has, e.g. a physical token like a key;
- Something a person knows e.g. a secret or a password;
- Something a person is e.g. physical characteristics particular to that person and
- Combinations of the above.

Authentication addresses security concern (c) discussed earlier and is typically achieved via digital certificates. In this chapter, we explore ways of how data integrity, confidentiality and authentication can be combined to provide security in online transactions.

8.4. Cryptography

This is a process that ensures that information is only read and used by its intended recipient using cryptosystems. There are two types of cryptosystems, namely, secret or symmetric key cryptosystems and public-key or asymmetric cryptography. Again, this chapter does not go into detail about cryptosystems, but just gives an overview.

8.4.1. *Symmetric Key Cryptography*

Symmetric key cryptography is governed by the encryption key, which in most cases, is just a binary number. It relies heavily on the secure distribution of the private key and cannot be easily deployed in a public network. Different encryption keys will produce different output messages, which are called the cipher text.

Figure 8.2 shows how secret-key cryptography works. Plain text and the symmetric cryptographic key are fed into the encryption algorithm. The output is the cipher text. When the recipient receives the cipher text, he/she inputs that into the decryption algorithm using the same symmetric cryptographic key and the output is the original plain text. Symmetric encryption follows two basic principles i.e. substitution and transposition. The Caesar Cipher is a good example of the substitution technique in which each letter of the alphabet is substituted with another letter of the alphabet nth places further down the alphabet. For example: "THIS IS A SECRET" (key $n = 3$) is substituted to

Fig. 8.2. Secret-key cryptography.

become "WKLV LV D VHFUHW". Transposition, which is a better technique than substitution re-arranges the positions of the letters as controlled by a key. For example, "THIS IS A SECRET" (key 4213) is transposed to IHSTSI S EAERTC. "THIS IS A SECRET" is divided into a block of four letters and the letters of each block are transposed according to a key "4213" such that the letters in the first, second, third and fourth positions are moved to the fourth, second, first and third positions, respectively. The reliability of an encryption algorithm is related to the key size or number of bits in the encryption key. Secret-key cryptography is very expensive in terms of its key distribution, so its use and uptake has been confined to governments and banks. For a sender and recipient to communicate securely using private-key encryption, they must agree upon a private key and keep it secret between themselves. If they are in different physical locations, they must trust a third party like a courier, the phone, e-mail, etc., to carry the secret key between the two parties and this is where the problem lies. Anyone who intercepts the secret key in transit can later read, modify and tamper with any of the information. Private-key cryptography is about 1,000 to 10,000 times faster than public-key cryptography, but because of the key management problem in symmetric cryptography, public-key cryptography is more widely used. In open systems, where the number of users is large, private-key cryptography does not scale well [25]. For example, if a cryptosystem has n users, each user must have $(n-1)$ keys, and the total number of keys is $\frac{n^2-n}{2}$ keys. For example, if $n = 1{,}000$, each user must have 999 keys, and the total number of keys in the cryptosystem is 499,500 [16]. Diffie and Hellman [10] came up with the idea of public-key cryptography to deal with the key management problem.

8.4.2. *Asymmetric Key Cryptography*

The primary benefit of public-key cryptography is that it allows people who have no pre-existing security arrangements to exchange messages securely. The need for the sender and receiver to share secret keys via some secure channel is eliminated and all the communications involve only public keys and no private keys are ever transmitted or shared. Each principal has a pair of keys called the private and public keys. The private key is kept secret or private and the public key is disclosed to the public, rather like someone's telephone number in the phone book. If someone wants to send you a message, they encrypt it using your public key. The message can only be decrypted with your private key. You can also use the private key to encrypt the data and only your public key can decrypt that data. A digital certificate functions much like a physical certificate and has information included with a person's public key that helps others verify that a public key is either genuine or valid. A digital certificate consists of three things i.e. a public key, certificate information i.e. identity information about the user, such as the user's name, user ID, etc. and one or more digital signatures. The purpose of the digital signature on a certificate is to state that the certificate information has been attested to by some other person or entity, like a certificate authority (CA). The digital signature does not attest the authenticity of the certificate as a whole, but it vouches only that the signed identity information is bound to the public key.

Public-key cryptography algorithms are built on a mathematical hard problem which makes deriving the private key from the public-key or the cipher text mathematically impracticable. The most popular public-key cryptosystems in use today are RSA, Diffie–Hellman and the Elliptic curve cryptography (ECC). The RSA cryptosystem [24], is built on the difficulty of factorising a huge integer prime number. Diffie and Hellman [10] is used for key exchange. It is built on the difficulty of computing discrete logarithms. The ECC is an emerging algorithm that implements public-key cryptography in new concepts that are built on the difficulty of discrete logarithm problems for elliptic curves [17]. The importance of the ECC is that it can provide the same level of security by using a shorter key and less calculation expense, compared to the other types of public-key cryptosystem [7].

8.5. Digital Signatures

A digital signature is an electronic means of ensuring data integrity. It can:

- Authenticate the identity of the sender of a message or signer of a document;
- Be used to ensure that the original content of the message is unchanged and
- Be automatically time-stamped.

One of the main problems with online shopping these days is that a person can order some goods on the Web from a merchant and if they are unhappy about any aspect of the transaction or the goods, they can always claim that they did not agree to that and ask for their money back either from the credit card company or the merchant. Digital signatures can be used for providing non-repudiation in online transactions i.e. they prevent the sender from claiming that he or she did not send the information. According to Menezes *et al.* [20], non-repudiation is a service which prevents an entity from denying previous commitments or actions. This ensures that the sender has an irrefutable proof that the receiver has received

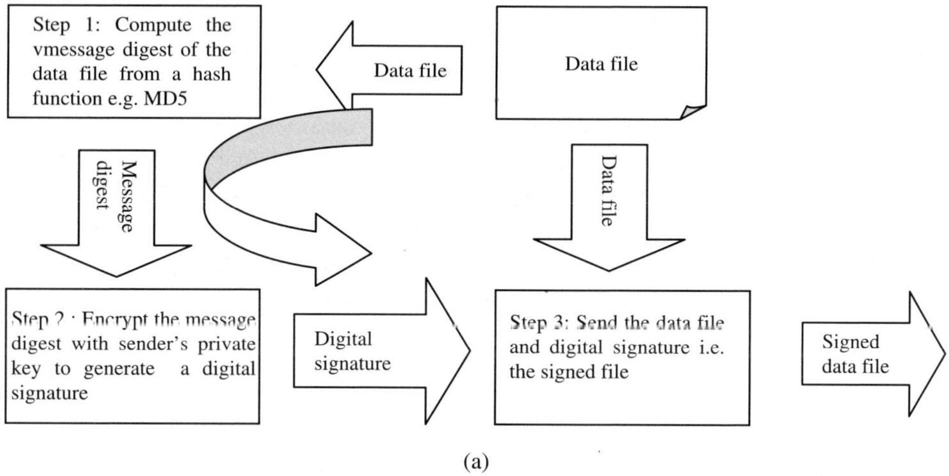

Fig. 8.3(a). Digital signature generation example.

the message and that the receiver has an irrefutable proof that the sender has sent the message.

Figure 8.3(a) shows how a data file can be signed, say by an online shopper i.e. the sender, using a digital signature and sent to the receiver i.e. the merchant. Here is how digital signatures work. The sender takes the file they want to send e.g. a data file with their personal details and computes a message digest (MD) from a hash function e.g. message digest algorithm 5 (MD5), or the secure hash algorithm (SHA-1), a widely used hash function with between 128-bit and 256-bit hash values. A MD is a fixed length string of digits or a single large number, usually between 128 and 256 bits in length, created from a text file using a one-way hash function [12]. The MD is then encrypted with the sender's private key to generate a digital signature. The sender then takes the original data file and the digital signature and these two are called the "signed data file", and sends the signed data file to the receiver e.g. the merchant.

Figure 8.3(b) represents the signed data file arriving at the receiver's or merchant's end of the transmission. When the signed data file arrives i.e. the data file and the digital signature, the same hash function used by the sender is used to compute the MD of the file received by the merchant. We will call this the expected MD for illustration's sake. The sender's public key is used by the merchant or receiver to decrypt the digital signature to get the actual MD of the data file. We will call this the actual computed MD for illustration's sake, as shown in Fig. 8.3(b). We then compare the actual computed MD with the expected MD to see if the file has been tampered within any way. If the two MDs are the same, the received file has been received with no loss to its data integrity and is therefore accepted as a valid transaction. If however, the two MDs are different, the data file's integrity has been breached and the transaction is rejected.

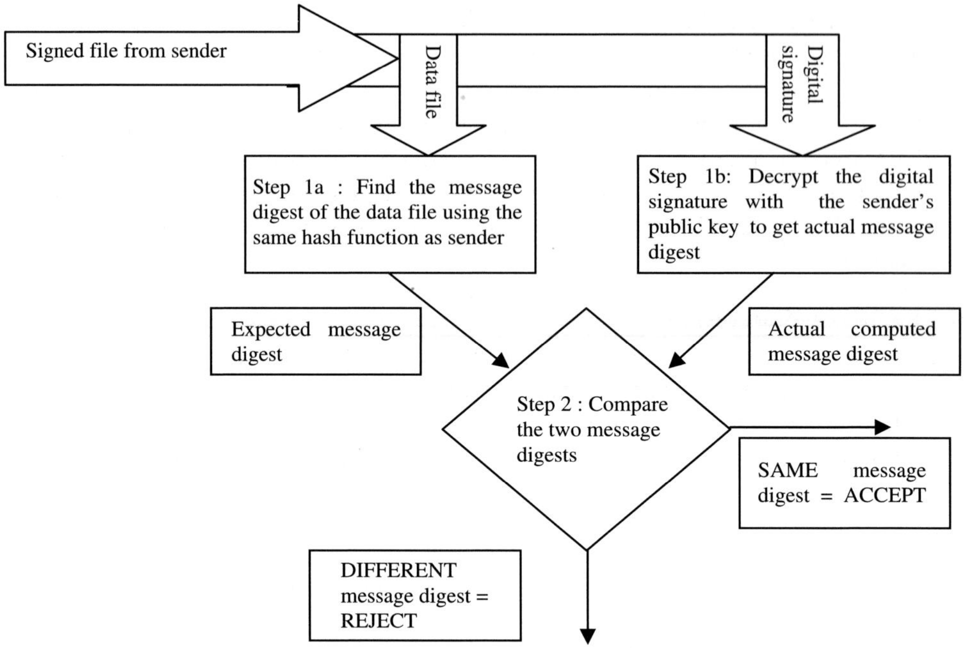

(b)

Fig. 8.3(b). Digital signature generation example.

8.6. Secure Sockets Layer (SSL)/Transport Layer Security (TLS)

The SSL [14] was designed by Netscape in 1994 and its main purpose is to encrypt the communication between a client and a server over the Internet that are both using Web browsers. A very similar protocol was subsequently developed by the Internet Engineering Task Force (IETF), called the TLS and usually SSL and TLS are used synonymously [9]. The current SSL version in use today is version 3.0. This chapter does not go into detail about the SSL/TLS protocol, but it just gives an overview. The SSL has two distinct entities, namely the client and the server. The client is the entity that initiates the transaction and the server is the entity that responds to the client and negotiates which cipher suites that are used for encryption. During an SSL session, the Web browser is the client and the Website server is the server. The SSL/TLS is made up of three protocols i.e. the Handshake Protocol, the Record Protocol and the Alert Protocol. The client authenticates the server during the Handshake Protocol. When the session is initiated and the handshake is complete, the data transfer is encrypted during the Record Protocol phase. If there are any alarms at any point during the session, then the alert is attached to the questionable packet and handled according to the Alert Protocol. The client always authenticates the server, and the server has the option of also authenticating the client. During the Handshake Protocol, the following important steps take place i.e. the session capabilities are negotiated. This means that the encryption (ciphers) algorithms are negotiated and the server is authenticated to the client. The SSL uses symmetric cryptography for the bulk data encryption during the transfer phase. However, asymmetric cryptography is used to negotiate the key used for that symmetric encryption. The client's Web browser always initiates a session by opening a connection to a server port and sending a request for a secure session to the server. This is the Client-Hello message which consists of the following parameters:

- The highest SSL version it supports;
- A nonce which consists of a 4-byte time stamp and a 28-byte random number. A nonce is a number or bit string used only once in a security session and in the context of SSL; it is used during the key exchange to prevent replay attacks;
- A session identifier (ID) which is used to update the connection parameters and
- A list of suggested cipher suites and compression methods that the client is able to support.

The server then responds with a Server-Hello message which consists of the following parameters:

- The server's public key;
- The session ID and
- The chosen version for ciphers and data compression methods i.e. the supported encryption methods.

The server may request a digital certificate from the client so that the client can also be authenticated, but this is optional and in practice it is rarely, if at all used. The server then sends a Server-Hello Done message, to signal that it is finished with the handshaking part of the protocol. The client then generates a secret session key, encrypts it and sends it with the server's public key. The session key then becomes the key used in the data transfer. The client then changes to the new Cipher-Spec and sends a Change-Cipher-Spec message to the server, which essentially tells the server that whatever the client sends from now on is encrypted. The client then sends a Finish message under the new algorithm and keys, to the server. The server changes to the new Cipher-Spec and sends a Change-Cipher-Spec message to the client. The server then sends a Finish message under the new algorithm and keys to the client. When this phase is complete, the Handshake Protocol is complete and the client and server can now exchange data in a secure transmission.

8.6.1. *Security Limitations of SSL/TLS*

8.6.1.1. *Client authentication*

Client authentication is optional in SSL/TLS, as it does not require the client to have a public-key certificate. Consequently, there is usually no client authentication.

The implications of this are that if a malicious user gets hold of someone else's card details, then he or she can impersonate the card holder and use the card holder's credit card to purchase goods online.

8.6.1.2. *Merchant authentication*

The client authenticates the merchant by his or her public-key certificate. However, SSL/TLS does not provide any means for verifying the merchant's legitimacy. Any malicious user with a Website and public-key certificate is able to establish an SSL/TLS connection and deceive the client. This opens up the client to the phishing scam which many unsuspecting Internet users have fallen victim to.

8.6.1.3. *Non-repudiation*

The SSL/TLS does not provide any means or proof of the transaction that has taken place between the merchant and the clients. That means the client can repudiate details of the transaction like prices, the goods' specification, etc. and in some cases either party can repudiate that the transaction has taken place at all.

8.6.1.4. *Data confidentiality*

The SSL/TLS makes available to the merchant, all the payment details of the clients to the merchant. That means if the merchant stores the clients' data in a local database, unencrypted, which some do or have done in the past, the client's personal data could be compromised if some malicious user breaches the database's security.

8.7. Secure Electronic Transaction (SET)

The SET is a protocol devised by a consortium of more than 10 big companies like Visa, MasterCard, Microsoft and VeriSign, etc., between 1996 and 1997, for securing credit card transactions over the Internet. Again, we only give a brief overview of the SET protocol here.

The SET participants are as follows:

Cardholder: The cardholder or the client, must have a public-key certificate from a certificate authority. The public key must be signed by the issuer bank, which guarantees the relation between the cardholder's key pair and the client's credit card.

Merchant: The merchant must have two public-key certificates for its use, i.e. one certificate for signing messages and the other for encryption. The merchant must also have the Payment Gateway certificate.

Payment Gateway: SET defines the Payment Gateway as a new participant in Web transactions. The main function of the Payment Gateway is to process

SET messages from the merchant and to forward SET messages to the acquirer bank.

Certificate authority (CA): The CA is not a direct participant in the transaction and can be described as a background participant. The main function of the CA is to provide public-key certificates for the cardholders, the merchants and the Payment Gateways.

Issuer: A financial institution that issues the credit card to the client. The issuer also works with the acquirer to transfer the transaction's funds from the client's account to the merchant's account.

Acquirer: A financial institution that acquires the transaction data from the merchant, checks this data with the issuer and works with the issuer to settle the transaction.

Here is a brief explanation of how SET works as described in the work of Anderson [2]. First, the customer sends the merchant server his/her certificate CC for his/her public key KC and a nonce, NC. The server replies with certificated public keys for the merchant (CS, KS) and its bank (CB, KB), plus a transaction sequence number S#. Then the customer sends a message containing an order, encrypted under the merchant's public key, and a payment instruction, encrypted under the bank's public key. Hashes of both of these are signed with the customer's private signing key. Next is an authorisation step, which can be performed online or deferred, as appropriate. The server sends the payment instruction to the acquiring bank, together with a summary of the order, which includes the amount payable, but not the exact description of the goods. The bank checks all this and refers to the card-issuing bank, if necessary. If everything is in order, then it sends the server an authorisation response similar to the traditional one (with an amount and an authorisation code), fortified with a signature.

C Æ S: C, NC, CC
S Æ C: S, S#, CS, CB
C Æ S: {Order}KC, {Payment}KB, sigKC {h(Order), h(Payment)}
S Æ B: {Summary}KB, {Payment}KB
B Æ S: sigKB {Auth_response}

8.7.1. *Problems with SET*

The SET protocol is a much more complicated security protocol than SSL/TLS and because the designers of this protocol were so concerned with providing security, they overlooked how easy it would be for the users to use it. According to Garfinkel and Spafford [15], this very complexity has been instrumental in the "failure" of SET as an online transactions' security protocol. Some of the reasons attributed to the apparent demise of SET are discussed below.

8.7.1.1. *Complexity*

The client must install special software i.e. client-side SET, have a pair of public/private keys and register with the Payment Gateway. This complexity restricts the client mobility.

8.7.1.2. *Public key infrastructure (PKI)*

The SET assumes the wide availability of PKI, which includes key generation, revocation, public-key certificates and private-key security, which is not the case.

8.7.1.3. *Expense*

The SET is too expensive for small payments.

8.7.1.4. *Client tracking*

The SET prevents the merchants from learning the clients' credit card numbers. Many big merchants use the credit card number as an index into a database for tracking their client's activities. This tracking is useful for marketing purposes. According to Paulson [22], because of this restriction by SET, the big merchants announced that they would not use SET unless they are allowed to use the clients' credit card numbers.

8.8. Disposable Credit Card Numbers (DCCNs)

Although SSL/TLS is the de facto protocol used in securing the communication between an online shopper and a merchant, it has its limitations. The SSL/TLS has many things going for it and it is implemented in pretty much all the major Web browsers on the Internet. However, as discussed in Section 8.6.1, SSL/TLS has its limitations and consumer concerns of online transactions' security still persist. In general, SSL/TLS does not authenticate the client and this leaves consumers' credit cards at risk of being used illegally by some malicious users. The credit card details are printed in clear text on the credit card and as a consequence, these details can be spotted by an attacker. Also, the credit card details are printed in clear text on the card holder's bank statements from the issuer and sometimes on the invoices from the merchants. If the client does not shred these statements and invoices carefully, an attacker can collect the credit card details from the client's bin. The most serious concern when using SSL/TLS is that, any merchant, who has a transaction with the client, has these details in clear text in his or her database. They may choose to encrypt the data after they have received it, but that is not part of the SSL/TLS protocol and is therefore a glaring security risk. One way of trying to deal with this serious security concern when using SSL/TLS in online transactions, is to add strong client authentication to the transactions. Therefore, the credit card details should not be enough to complete a transaction.

The problem of providing online transaction security is, by no means, trivial and the solution is not only in technological fixes. Researchers can come up with brilliant technologically sound security measures, but as SET has shown, air-tight security does not guarantee that the protocol will be adopted by online shoppers. Many banks are coming to the conclusions that the way forward lies with address verification rather than with cryptography [4]. Some radical thinking has to be adopted and the idea of disposable credit cards is one solution that is being investigated and indeed is in use by some institutions. Consumers have a real fear that their credit card information can be stolen when they are shopping on the Internet. American Express has been offering their customers single-use credit card numbers under the free service, called Private Payments [29]. A one-time credit card number, or DCCN is a technique whereby a temporary credit card number is sent to the merchant to complete a transaction. The DCCN is designed for single use only, after which the DCCN is invalid.

The DCCNs can either be generated online or offline. With online generation, there is still the issue of establishing a secure connection to the issuer during or just prior to the transaction. American Express's Private Payment and Shamir's SecureClick are examples of online generation schemes for DCCNs. With the SecureClick scheme [26], the client's Web browser contains a small plug-in, which is provided by the issuer during the registration. When the client decides to pay for goods or a service, he or she fills in a typical shopping form, but the credit card details stay empty. When the client submits the form for payment, the plug-in intercepts the form before it is sent to the merchant and initiates an SSL/TLS secure connection to the issuer. The client then authenticates himself or herself to the issuer by a pre-defined method from the issuer. This method can be a time-dependent password, a digital signature, a biometric device or any method, which can achieve strong authentication. After the authentication, the issuer generates a DCCN for the client. The client then fills in the DCCN in the appropriate fields on the shopping form. The plug-in forwards the shopping form to the merchant. Finally, the merchant processes the shopping form like any normal form. The generated DCCN has all the features of regular credit card details, such as a 16-digit length, the first digit being the card type and the last digit being the correct checksum. Furthermore, the merchant does not know that the credit card details are for a single-use card.

Assora *et al.* [3] have also proposed a new DCCN scheme based on off-line generation. The DCCN can be generated by devices such as smart cards, PDAs or mobile phones. This method has the added assurance; guarantees even, that the merchant is not able to generate a valid DCCN. The only person who is able to generate a valid DCCN is the person who has access to the client's secret. The scheme uses a pre-shared secret key between the issuer and the client to generate a hashed message authentication code (HMAC) for some selected data from the transaction. A HMAC is a message authentication code calculated using a specific algorithm involving a cryptographic hash function in combination with a secret key.

The scheme is secure, collision-free and adds minimum overhead on the issuer and client. Furthermore, the merchant does not have to change the actual system of performing the transactions because the generated DCCN has all the features of classical credit card details.

Any off-line generation requires the use of a cryptographic technique. The output of any cryptographic algorithm is mainly controlled by the encryption key and the algorithm. When using symmetric cryptography, the key must be at least 80-bits long and for asymmetric cryptography, the key length needs to be at least 160 bits, when using ECC [28]. In a typical transaction, when using a normal credit card, the client has to fill in a 16-digit credit card number, 4 digits for the start date, 4 digits for the expiry date and 3 digits for the Card Verification Checksum (CVC). According to the International for Standardization Organization [18], the length of the typical credit card number is 16 digits, although some credit card numbers can be of different lengths. There are 6 digits to identify the card type and the issuer and one digit for error detection. Consequently, there are only 9 remaining digits, which is approximately $\approx$ 30 bits; and this is not long enough to provide secure cryptography. Because a DCCN has to maintain the format of the actual credit card, the issuer does not have to decrypt the message from the client, but only needs to check the validity of the message. Therefore, the output of the algorithm, which generates the DCCN, is able to be truncated.

8.8.1. *The DCCN Based on a Hashed Message Authentication Code (HMAC)*

The main idea of the protocol is that the DCCN for a transaction is a hash for some selected data from the transaction, in addition to the client's secret key [3]. This hash is generated locally by the client during the transaction and sent to the issuer with the selected data. The issuer uses the client's secret key and the received selected data to check the validity of the hash and as a consequence, the validity of the DCCN. For optimum security, the HMAC, is used as a hash function to generate the DCCNs [5]. The HMAC function has the following properties:

- The output of the HMAC function is equal to the output of the approved hash function, which the HMAC is built on;
- The calculation expense is nearly the same as the approved hash function. Any device capable of calculating a hash function should be able to calculate the HMAC;
- The output of the HMAC can be truncated to fit the size of the normal credit card number and
- The HMAC is a standard for keyed-hash message authentication codes [13].

The protocol works as follows and is discussed in the following sections.

8.8.1.1. *Registration phase*

The client opens a bank account with the issuer and gets his/her credit card. In addition to the credit card, the client has a secret key associated with this credit card, which is used for any Web transaction with the credit card. If the client has more than one credit card, he/she must have a secret key for each credit card. The issuer must link every credit card with the corresponding key. The client must have an HMAC Creating Device (HCD) to generate the HMAC and to securely store the secret key.

8.8.1.2. *Transaction process*

When the client decides to pay for goods or a service from the merchant, he/she receives a payment form to fill in his/her credit card details. A DCCN is made up of the price, date, merchant ID and a nonce. However, each issuer is able to define their own parameters. The client enters the price, date, merchant identifier (mID) and a nonce on his/her HCD and calculates the HMAC for these values by using his/her secret key. The nonce can be one or two digits and is chosen randomly by the client. By using the nonce, the client is able to purchase goods with the same specification (price, date and mID) more than once and at the same time prevent the replay attack. The HMAC is generated as follows:

$$S = H(K \parallel m); \quad \text{where } m = (\text{price} \parallel \text{date} \parallel mID \parallel \text{nonce})$$

where H is the HMAC function, K is the client's secret key and S is the output of the HMAC hash function. The most widely used hash function today is the SHA-1 [13]. The output of SHA-1 is 160-bits long. Since, the output of the HMAC function is equal to the output of the approved hash function, if SHA-1 is used as the approved hash function, then S is 160-bits long.

S is the DCCN for the transaction. Comparing the size of S to the available digits in the typical credit card number (9 numeric digits), it can be observed that S is too long to be a typical credit card number. The output of the HMAC can be truncated from L bytes (the output of the hash function) to t bytes (the required length). The FIPS [13] imposes the following restrictions in order for a valid truncation to occur.

The application must make numerous trials impractical and t can be truncated to at least 4 bytes i.e. $4 \leq t \leq L$ (bytes) or in bits $32 \leq t \leq L$ (bits).

The first condition can be maintained because only the issuer, who knows the client's secret key, is able to check the validity of the DCCN. Normally, the issuer's data is well protected and thus the attacker cannot use this data for numerous trials. Online guessing attacks can be prevented by blocking the client's account after a few faulty attempts. The second condition is not attainable because the available space in the credit card number is only 30 bits.

Black and Rogaway [6] presented several methods to eliminate the output of block ciphers and at the same time keep the same level of security. The most suitable method for this DCCN generation is the Cycle-Walking Cipher.

$$S \text{ is 128-bits long (the output of the HMAC)}$$

t is the truncated HMAC, t must be at least 32-bits long and must be truncated from the most significant bit of S [13].

The problem is that the DCCN has only a room for 30-bits long. Therefore, the issuer must be able to check 32 bits of t while the client sends only 30 bits.

The solution is that if the issuer can calculate the value of these two bits from the received message from the client, then the client does not have to send them to the issuer. Here is a small algorithm to apply the Cycle-Walker Cipher to solve the problem.

$$S = H(K \parallel m);$$

While (true) begin

If $(S[L-31] = 0)$ and $(S[L-32] = 0)$ then break

Else $S = H(K \parallel S)$;

End {while}

Return S

where $S[L-31]$ and $S[L-32]$ are the bits number 31 and 32, respectively from S and the counting starts from the most significant bits. This algorithm finishes when the values of bits 31 and 32 of S are zero. Therefore, the client does have to send them to the issuer. If H is a good hash function, then the value S is a random value, therefore the probability of the bits 31 and 32 being zero together is 1/4. Thus, the algorithm above is repeated four times on average to get the desired result. Hence, it does not add too much calculation overhead on the client or the issuer. The remaining credit card fields i.e. the start date, expiry date, CVC and the most significant 6 digits of the credit card, must stay the same as the permanent credit card fields. The error detection digit, which is the last digit, should be calculated for the DCCN and added as a first digit to the DCCN. These fields should stay the same as the permanent credit card for many reasons. First of all, the merchant normally checks the validity of these fields such as the start and expiry dates and the error detection digit. Second, it is a small indication that the client knows the permanent credit card details. Third, the most significant 6 digits of the credit card are required to complete the clearance procedure.

8.8.1.3. *Transaction clearance*

When the merchant receives the payment information from the client, he/she forwards it to the issuer. The issuer can use the billing address, which is mandatory in the credit card transactions, to identify the client. The issuer uses the client's secret key (K) to check the validity of the DCCN. To validate the DCCN, the issuer follows the same procedure as the client to generate the DCCN and compares it with the received DCCN. If the DCCN is correct, the issuer authorises the transaction.

The billing address is used to identify the client. In the case of the client having more than one credit card, the issuer simply checks the DCCN with the entire client's secret keys. The correct secret key detects the intended credit card. The security of the system is established from the security of the HMAC. The HMAC is proven secure [5], even if the attacker knows any number of the generated DCCNs. The size of the key (K) is $\frac{L}{2} \leq K \leq L$ [13]; where L is the output of the hash function. If SHA-1 is used as a hash function, where $L = 160$ bits, K must be at least 80 bits. The recommended value for K is 128 bits. Another case in which the DCCN can be repeated is that of a replay attack, where the merchant simply submits the same DCCN and the same transaction details more than once. The issuer should maintain a small queue for every client, for example, the transactions for the last statement. This queue is useful for answering any query from the client. It is also useful to prevent any possibility of replay attacks from the merchant. Any received DCCN from a merchant should be rejected if it already exists in the queue from the same merchant. Any received DCCN from any merchant should be rejected if its date is older than the acceptable date i.e. the date of the last statement. The proposed protocol is able to achieve data integrity for the transaction. Any attempt from the merchant to change any information from the transaction details will be discovered promptly.

8.9. Summary

The rapid growth of the Internet with its enormous economic advantages manifested in e-commerce has revolutionised the way companies and consumers do business. A good percentage of traditional retail businesses, if not all, have a Web presence and online shopping is becoming increasingly common, but there are also security risks associated with that. Consumers are very reluctant to give out their credit card details online, as there is this perception that the risk of fraud or something going wrong is very high. It is true that there is some risk in online shopping, not least because some merchants do not encrypt their customer data once they have received it over an encrypted session using an SSL/TLS. Also, there are risks from eavesdroppers and malicious hackers when buying goods online. However, all the evidence has shown that there is more risk in using Chip and Pin and doing conventional shopping with one's credit card, than buying things online. Having said that, consumers buying things online should always be vigilant about who they are submitting their details to and it is always a good idea to check some of these companies to make sure they are genuine. Also, common sense dictates that one should not buy any merchandise as a result of unsolicited e-mails or spam, claiming to be legitimate merchants, as there is a lot of phishing going on. E-commerce is set to grow with the popularity of the Web and online security is always going to be an issue and researchers and banks are working together to devise increasingly better security protocols that are always easy for consumers to adopt. But for some people, the fear of online shopping is like the fear of flying and no amount of statistics proving how safe online transactions are, will ever convince them to shop online.

References

1. T. T. Ahonen, *m-Profits, Making Money from 3G Services* (John Wiley & Sons Ltd., 2002) ISBN 0-470-84775-1.
2. R. Anderson, *Security Engineering: A Guide to Building Dependable and Distributed Systems* (Wiley & Sons, 2001).
3. M. Assora, J. Kadirire and A. Shirvani, A web transaction security scheme based on disposable credit card numbers, *International Journal of Electronic Security and Digital Forensics (IJESDF)* **1**(2) (2007).
4. D. Austin, Flood warnings, In *Banking Technology* (July–August 1999), pp. 28–31.
5. M. Bellare, R. Canettiy and H. Krawczykz, Keying hash function for message authentication, *Advances in Cryptology–Crypto 1996, Lecture Notes in Computer Science* **1109** (1996) 1–15.
6. J. Black and P. Rogaway, Ciphers with arbitrary finite domains, *Cryptology ePrint Archive*, 2001, Report 2001/012.
7. S. Burnett and S. Paine, *RSA Security's Official Guide to Cryptography*, (Osborne/ McGraw-Hill, New York, USA, 2001).
8. Cellular Statistics, Latest mobile, GSM, global, handset, base station, & regional cellular statistics for April 2006, 2006. Retrieved March 26, 2008 from: http://www.cellular.co.za/stats/stats-main.htm.
9. T. Dierks and C. Allen, The TLS protocol, Ver 1.0, The Internet Engineering Task Force (IETF), 1999, RFC 2246.
10. W. Diffie and M. Hellman, New directions in cryptography, *IEEE Transactions on Information Theory* **IT-22**(6) (1976) 644–654.
11. S. Drimer, S. J. Murdoch and R. Anderson, Thinking inside the box: System-level failures of tamper proofng, *Technical Report Number* 711, University of Cambridge, Computer Laboratory, February 2008.
12. FIPS, Digital signature standard, *Federal Information Processing Standards*, FIPS 186-2, 2000.
13. FIPS, The keyed-hash authentication code (HMAC), *Federal Information Processing Standards*, FIPS 189, March 2002.
14. A. O. Freier, P. Karlton and P. C. Kocher, The SSL protocol, Version 3.0 Netscape Corp, 1996.
15. S. Garfinkel and G. Spafford, *Web Security, Privacy and Commerce*, Second Edition (O'Reilley, USA, 2002).
16. J. Graff, *Cryptography and e-Commerce* (Wiley & Sons, New York, USA, 2001).
17. IEEE, Standard specifications for public key cryptography, IEEE P1363 standard, 2000 Retrieved April 6, 2008 from: http://grouper.ieee.org/groups/1363/index.html.
18. ISO/IEC 7812-1, International Organization for Standardization (ISO) and International Electrotechnical Commission (IEC), Geneva. ISO/IEC 7812-1: Identification card — Identification of Issuers.

19. J. Leyden, Chip and PIN fraud hits Lloyds TSB, The Register 11th May 2006, 2006. Retrieved March 27, 2008 from: http://www.theregister.co.uk/2006/05/11/lloyds_tsb_chip_and_pin_fraud/
20. A. Menezes, P. V. Oorschot and S. Vanstone, *Handbook of Applied Cryptography* (CRC Press, 1996).
21. Office for National Statistics, Information and Communication Technology (ICT) (2006). Activity of UK businesses, 2004, amended 9th February 2006.
22. Office of Fair Trading, Internet shopping, 2007. "Retrieved March 26, 2008, from http://www.oft.gov.uk/shared_oft/reports/consumer_protection/oft921.pdf.
23. C. L. Paulson, Making sense of specifications: The formalization of SET (transcript of discussion), security protocols (2001). *Lecture Notes in Computer Science* **2133** (2001) 82–86.
24. R. L. Rivest, A. Shamir and L. Adleman, A method for obtaining digital signatures and public key cryptosystem, *Communications of the ACM* **21**(2) (1978) 120–126.
25. B. Schneier, *Applied Cryptography*, Second Edition (Wiley & Sons, USA, 1996).
26. A. Shamir, SecureClick: A web payment system with disposable credit card numbers, *Financial Cryptography (FC-02), Lecture Notes in Computer Science* **2339** (2002) 232–242.
27. W. Stallings, *Network Security Essentials Application and Standards*, Third Edition. (Prentice Hall, USA, 2007).
28. D. R. Stinson, *Cryptography: Theory and Practice* Second Edition (Chapman & Hall/CRC, USA, 2002).
29. M. Trombley, American Express offers disposable credit card numbers for online shopping, 2002, Computerworld Security, 7th September, 2000. Retrieved March 30, 2008, from: http://www.computerworld.com/securitytopics/security/story/0,10801,49788,00.html
30. J. Weinraub, Security Report, Retrieved March 24, 2008, from: http://www.educationjobs.com/security.htm.

Chapter 9

SECURITY AND E-ACCESSIBILITY OF ONLINE BANKING

HAMID JAHANKHANI, LIAQAT ALI and HOSSEIN JAHANKHANI
School of Computing, IT and Engineering
University of East London, UK

9.1. Introduction

The customers of today's global networked economy have higher demands and expectations from their financial institutions. Although online banking greatly increases the convenience of banking services, however, the issue of E-accessibility is becoming more stringent to avoid potential threats.

In the current world population, there are significant numbers that are visually impaired. Unfortunately, web developers and designers still think of web as purely a visual medium and are avoiding the fact that visually impaired people have the equal access rights. This chapter focuses on the E-accessibility level of online banking services and their security issues.

The World Wide Web Consortium (W3C) has provided the standard of the Web Content Accessibility Guidelines (WCAG 1.0), however, despite these standards, many websites still fall short of E-accessibility.

The security of online banking for visually impaired people is an important issue along with accessibility. Banks are using additional methods to achieve the extra layers of security but unfortunately not for everyone. In this chapter, a number of issues are highlighted according to the perceptions and experiences of the blind and partially blind people.

People with severe visual disabilities now have the opportunity to benefit from a wealth of information and services that was previously unavailable to them. With the help of synthesised speech and Braille display technology, even completely blind people can use the web. Blind people read the available information from the sense of their fingers called Braille.

Blind and visually impaired people can also read web pages by using software tools known as Screen Readers, for example JAWS and WindowsEyes. These Screen Readers are able to generate speech and/or refreshable Braille output for visually impaired people. However, the simplest web pages generally feature images and use tables to format their navigation menus and content, while many others use JavaScript, animations and other technologies hypothetically to make their navigation systems more user-friendly. This creates a number of problems for people with visual impairments, as they cannot see the images and their Screen Readers can have serious problems interpreting tables, animation and JavaScript. In many cases, this renders entire websites unusable for people with disabilities. Usability is paramount for the success of websites.

9.2. Web Accessibility

Web accessibility describes a person's ability to use a website over the Internet [4]. It is a phrase, which is customarily referring to the development and designing of websites that are accessible to all users regardless of any disability for physical limitations.

Tim Berners-Lee, W3C director and inventor of the WWW defines web accessibility as "*access by everyone regardless disability*".

A number of disabled people face a number of obstacles and problems when they are accessing websites such as poor contrast between text and background, fixed font size, missing and broken links, inadequate accessibility information, poor structural elements etc. Some of these obstacles and problems are relatively simple and easy to overcome but others require a little bit more consideration, more efforts and more thoughtfulness in the design area of web development.

Web accessibility is about adapting the design of the web content to the various disabilities. Therefore, we have to ensure that websites facilitate the use of assistive technologies.

The knowledge and understanding of E-accessibility is now rising at all levels of society. Traditionally, the term was allied only with the accessibility of buildings and in some cases with traditional media only, but, now is being extended to take account of online information and services. According to the Disability Rights Commission, "*One in seven people in the UK suffer some form of disability and many people experience some loss of sight and manual dexterity as they get older*" [9].

An accessibility standard helps web designers to identify and addresses these issues. The Federal Rehabilitations Act [6] and the WCAG 1.0 [10] are the two most important standards of the Web accessibility.

The WCAG 1.0 has three different priorities and 65 different checkpoints. Priorities are:

- ***Priority*** [1] ***or "A"*** *checkpoints are those that the developer of the web must satisfy to insure that the page itself is accessible.*

- ***Priority*** *[2]* ***or "AA"*** *checkpoints are those that the web developer should satisfy to ensure that certain groups will be able to access information on the web page and*
- ***Priority*** *[3]* ***or "AAA"*** *checkpoints are those the web developer may do to ensure that all content on the page is completely accessible.*

The US Rehabilitation Act, Section 508 contains 16 different standards.

9.3. E-Accessibility Barriers

The term "disability" is used very commonly and can direct several problems when accessing the Websites. People can have the combinations of different disabilities but all disabilities do not affect the way of E-accessibility. For example, a person using a wheelchair still can use a computer and access websites. Disabilities that really affect the E-accessibility level can be categorised as follows, [11]:

- **Visual** (Blindness, low vision and colour-blindness)
  The issue of visual impairment is one of the main barriers in the way of E-accessibility. There are several kinds of difficulties for people with visual disabilities such as multimedia, frames, forms, tables, navigation, complex notation, Java Scripts, image rendering and programming code.
- **Hearing Impairments** (Deafness).
- **Motor** (Incapability to use a computer mouse, slow response time, limited fine motor control).
- **Cognitive** (Learning disabilities, distractibility and the inability to remember or focus on large amounts of information).

It is difficult to say how many people in the world are actually disabled.

9.4. Access Technologies and Adaptive Computing

Most people with disabilities require assistive devices to access websites. Some of the alternative and adaptive strategies used by disabled people to improve the level of E-accessibility are as follows:

- Text-to-speech (synthetic voice)
- Digital audio
- Braille
- Scanning software
- Screen magnifiers
- Screen readers
- Speech recognition
- Text browsers
- Oversized trackball mouse

- Visual notification
- Tabbing through structural elements
- Sticky, bounce, toggle and mouse keys
- Voice browsers
- Eye tracking
- Mouth stick or a head wand
- Captioning
- Electronic pointing device and
- Alternative key boards and switches.

9.5. Accessibility Evolution Tools

The assessment and validation of websites is an art. Different tools are available for the assessment and validation of websites, which is itself a positive step towards E-accessibility. These accessibility tools play a critical role in ensuring the accessibility of the web and perform a static analysis of homepages or sites regarding their accessibility.

Testing and validation of a website is still very important. The two common types of HTML testers are validators and linters. The main differentiation among a validator and a linter is that a validator checks and validates a page against a published HTML specification for technical errors, while a linter checks a page for commonly and frequently made mistakes. It is often a good idea and technique to use both as they can sometimes find different types of problems [1].

A validator uses Document Type Definition (DTD), a language that describes the contents of Standard Generalised Markup Language (SGML) document, to check whether the page is syntactically correct or not. A validator is the only certain way of telling if your site is standards-compliant [13].

A key point to understand in regard of accessibility tools is that these tools can only partially check the accessibility of websites through automation and still require human judgement and checking or manual check of the website [3]. No automated accessibility evaluation tool can find all of your content's accessibility errors. Automated programs can only evaluate a few of the many possible accessibility issues that can arise in a particular website [12].

Till now, there are several different tools for testing, assessment and validation of websites, which are different from one another in several dimensions. Some of them do only testing while some other tools perform fixing of a page as well. They are different from each other in terms of effectiveness, cost and reliability. The important thing is to evaluate the quality of these tools. For a common web developer to develop and design a better and accessible website, the key role of these tools is very critical. By evaluation and comparing the accessibility tools, web developers and designers can act upon the appropriate selection and choice. This evaluation will also provide a competition between the tools manufacturers and will improve the tool's quality itself.

The automated tools identify different features of the websites that might cause a failure of the website in terms of its accessibility to disabled people. For example, if an image element in a website does not contain the Alt attribute, then the website will become an accessibility failure because the page cannot be accessed through the speaking browser. A complete list of these accessibility tools can be found at http://www.w3.org/WAI/ER/tools/complete.

9.6. E-Banking

The term "banking" can be applied to a large number of financial institutions who deals with financial data. Online banking is a system that allows bank customers to access their bank accounts and other financial information through a personal computer system via Internet.

Financial sectors including banks are the biggest users of information technology as compared to other industries. Online banking allows bank customers to communicate with the bank's internal network system and perform financial transactions through the bank's website. Simply defined, E-banking is "*an electronic channel used to provide retail and commercial banking products and services to customers*" [8]. It is one of the convenient ways of banking in the modern era. There are three types of internet banking: Informational, Communicative and Transactional.

Informational banking is the basic level of Internet banking with relatively low risk where bank only markets products and services. Using a stand-alone server, informational systems have no path between the server and the bank's internal network.

Communicative banking allows bank customers to interact with the bank's system in a limited form like sending mails and applications for loan. Compared to informational banking with communicative banking, the risk is a bit high because the server may have a path to the bank's internal networks. To monitor the network and to control the unauthorised access, appropriate actions must be there to keep the data secure from any outside viral attacks.

Transactional banking is the modern type of banking that allows bank customers to perform their financial transactions online. Transactional banking involves the high risk of security due to the connection between the server and bank's internal network. Customers can perform different types of operations such as checking their balance, paying utility bills, transferring money between accounts etc.

Internet banking is rapidly increasing and financial institutions are ever more alert to the challenges of developing an infrastructure to secure financial data. Security, authentication, trust and availability are, therefore, the major concerns in E-banking.

The issue of security in E-banking has different dimensions from logical to physical security. Some banks allow customers for direct dial-in-access to their system over a private network while others allow network access through the

Internet. Cyber laundering and phishing are the important issues to be considered. Counterfeiting of financial data such as credit card numbers, account usernames, password and social security numbers and hijacking banks' brand names are the major problems encountered in the E-banking industry.

E-banking breaches essentially fall into three categories [7]:

- Breaches with serious criminal intent e.g. fraud, theft of commercially sensitive or financial information:
- Breaches by casual hackers e.g. defacement of websites or denial of services causes websites to crash and
- Flaws with system design and set-up leading to security breaches e.g. general users being able to transact on other users' account.

All of these threats have potentially serious financial and legal implications.

It appears that the rate of security attacks is actually outpacing the growth of technology. Crackers and to some extent hackers represent a well-known threat in this respect and are responsible for a significant degree of disruption and damage to information systems. However, they are not the only criminal elements that have to be taken into consideration because evidence suggests that the technology is increasingly seen as a potential tool for terrorist organisations. The facts are that a class of elite intruders is now commonly able to attack sites, extract credit card account information and then cover their tracks by destroying digital evidence along their path. These intruders have become more brazen as they are more successful.

Hacking is defined as an electronic break-in into a company for the purpose of harvesting a wealth of information or may be with some non-malicious intention. Hackers are those people who do not have any bad intention but want to learn everything about a computer system, while crackers are the ones who are breaking into computer systems illegally. The early hackers had a love of technology and a compelling need to know how it all worked, and their goal was to push programs beyond what they were designed to do. The term "hacker" did not have the negative connotation as it has today.

Present-day hackers can be classified into different categories based on their knowledge and involvement in the field. There are some with in-born talent who really want to learn more of the computer systems and they also create software that helps to improve the overall technological infrastructure. Those who belong to this category with out any malicious intention are the highly respected computer pundits in the contemporary world. The second category hackers are not knowledgeable like the previous ones but they have experience with several operating systems and know how to exploit the vulnerabilities associated with them. Most of the security consultants fall into this category. Third and the last category, often called crackers are those with least knowledge and use the codes and methods developed by other people to do their cracking.

The attacks take place in several phases such as information gathering or reconnaissance, scanning and finally entering into the target system. Information gathering involves methods of obtaining information or to open security holes. It is just like the way in which the orthodox type of robbery is carried out. The robber will find out the whole information about the place that is to be robbed before making an attempt. Just like this, the computer attacker will try to find out information about the target. Social engineering is one such method used by an attacker to get information.

Various authors have provided definitions, such as:

> *"Social engineering can be regarded as 'people hacking', basically its hacker jargon for soliciting unwitting participation from a person inside a company rather than breaking into the system independently,* [1].
>
> *Social engineering is a hack that uses brains instead of computer brawn. Hackers call data centres and pretend to be customers who have lost their password or show up at a site and simply wait for someone to hold a door open for them. Other forms of social engineering are not so obvious. Hackers have been known to create phoney web sites, sweepstakes or questionnaires that ask users to enter a password,* [2].
>
> *Social engineering is an art of utilising human behaviour to breach security without the participant even realising that they have been manipulated,* [3]*."*

There are two main categories under which all social engineering attempts could be classified, computer- or technology-based deception and human-based deception. The technology-based approach is to deceive the user into believing that is interacting with the "real" computer system (such as pop-up window, informing the user that the computer application has had a problem) and get the user to provide confidential information. The human approach is done through deception, by taking advantage of the victim's ignorance, and the natural human inclination to be helpful and liked.

The scam, called "phishing" as a form of identity theft, is used to gain personal information for the purposes of identity theft, using fraudulent e-mail messages that appear to come from legitimate businesses. These authentic-looking messages are designed to fool recipients into divulging personal data such as account numbers and passwords, credit card numbers and social security numbers. Phishing is a two-time scam, first steals a company's identity and then use it to victimise consumers by stealing their credit identities. The term phishing (also called spoofing) comes from the fact that Internet scammers are using increasingly sophisticated lures as they "fish" for a user's financial information and password data.

According to Anti-Phishing Working Group (2008), the number of phishing cases reported in January 2008 was 29,284 and the number of unique phishing websites detected by the Anti-Phishing Working Group in January 2008 was 20,305.

Phishing becomes the most commonly used social engineering attack due to the fact that it is quite easy to be carried out; no direct communication between hacker and victim is required (i.e. a hacker does not need to phone their prey, pretending that they are technical support staff, etc.). Sending mass-mails to thousands of potential victims increases the chance of getting someone hooked. There are usually three separate steps for these attacks to work, which are setting up a mimic website, sending out a convincingly fake e-mail and finally luring the users to that mimic site.

The techniques that hackers use normally to obtain an unauthorised access are discussed in details in the following sections.

9.7. Trojan Attack

This can be accomplished by using a key logger program on a victim's computer systems. When users of computer system visit certain websites on their system via Internet and download different programs, key logger programs are also installed on their system without their knowledge. With installed key logger program, whenever a users visits to their bank's website, the information keyed in during that session will be captured and sent to the attacker. The Trojan is basically used here by the attacker as an agent. This agent can be used to steel information from the victim's computer system.

9.8. Use of Fake Websites

In this technique, an attacker tries to trap victim by disguising their identity to make it appear that the messages are coming from a reputed organisation and trusted sources. Once the fraudster gets success, the victim of data do not realise that they are accessing information from fraudster's website instead of the actual website. Information captured by attackers can be used later for further unauthorised transactions.

9.9. Password Compromise

Using password is an important element of online banking. Customers can access their banking information and can perform financial transactions via Internet by using an appropriate login ID and password to their accounts. Although it is an easier way of Internet banking for customers, it is important to understand that using of passwords does not offer ample protection against Internet fraud such as phishing. The major issue here is the compromising of passwords. The attacker can effortlessly seize the control of online transactions when a password is compromised. In such a scenario, passwords are not trusted mechanisms as an authentication token for online banking as no one can guess who is on the backhand of the system entering the same password.

Different security issues can be considered for a secure website over the Internet. However, the two major security issues are system security and information security. System security is about to make sure that no unauthorised person can change the content of the website while information security deals with the customer details, making sure that the sensitive information provided by customer like customer details, credit card numbers etc. are stored in a safe environment.

Encryption is an important element of the security in developing a secure website. Although it provides some of the difficulties for unauthorised people to intercept with the content and information of the website, but do not provide any mechanism to make sure that the information held on the server is secured.

9.10. Secure and Accessible Online Banking Challenges

People with visual disabilities have different perspectives towards the E-accessibility of online banking services. While, they are committed to the Internet services, however, there are a number of barriers for them to carry out secure online transactions. This is because, online banking industries are seeking to achieve the highest level of security by introducing new methodologies and security devices but this raises a number of issues of privacy, accessibility and security for visually impaired people.

The first and most important step for online banking is the login. The poor and inaccessible design of the home pages of these online banking websites makes navigation difficult for people with visual disabilities. Although in some cases it strengthens the security layer, it creates a number of complications in terms of accessibility when translated by the Screen Readers.

User names and passwords are the common methods for login. Security risks are increased when passwords are used carelessly, due to several reasons such as re-using a password which is common across different accounts. Thus, a password used for a low-security website can be easily compromised by an unauthorised person to gain access to a higher secured website. Usability is ever more important feature of security in general and password systems in particular. Different banks use different techniques for login, such as the following:

Method 1: The bank customer is required to enter their customer number which is basically the date of birth of the customer followed by another four-digit unique number identified by the bank. On successful completion of this stage, the customer can access the next page where the customer will be asked to provide their PIN and password information in a desired sequence.

It is likely that when blind people enter a desire character sequence for either date of birth, password or Personal Identification Number (PIN), they may make a mistake in that entry as the Screen Reader will translate those characters as they appear on the screen in the form of "$_*$" sign. Even, if, the Screen Readers translate the characters, then privacy and security issues are raised. In case of multiple wrong entries, the bank will hang the logging process.

Cognitive problems are also there and remembering a lengthy password is not an easy task.

Method 2: Very similar to Method 1, however, in the first step, bank customers are required to enter their surname and their 20-digit customer membership number provided by the bank. On successful completion of this step, the customers will be prompted to another step in which they are required to enter their five-digit pass code along with two characters from their memorable word in a desired sequence.

Method 3: This one-stage method is the easiest for people with visual disabilities as they do not need to translate the whole page using their Screen Readers in terms of login process. The bank only asks for the username and password to be entered by the customer. Usernames are basically account numbers.

Once successfully logged into systems, the next stage is the process of secure transactions. Unfortunately, many of the security and accessibility issues are there too. For example, blind and partially sighted people take longer than other people when they surf the Internet and carry out online banking; as a result, they are often logged out by the system because of time-out technique used in the online banking for security reasons.

Visually impaired people use different assistive technologies to overcome the barriers of accessibility. Some use Screen Readers and some use magnification software. Text-only Browser is also another assistive device used by visually impaired people but in case of banking websites it may not work due to their website security level of security, these devices will not be able to process JavaScript [8].

Online banking always demands customers to fill different forms if they want to apply for a particular facility such as loans, mortgages, applying credit cards etc. Sometimes, these forms require a very lengthy process to be completed even for sighted people. With some sites, filling in the application form is difficult because labels used in the forms are not logically associated with what they refer to and it is difficult to work out what is required in the specific field. For example, a wrong label may be read for a particular radio-button or checkbox used in the form. Sometimes, it is not even clear by just listening to Screen Readers that what is being asked for or what has to be done. Labelling is, therefore, very important in forms particularly for those users who are blind and use Screen Readers. Some Screen Reader like Jaws usually check the code that helps the system to find out what label is aligned with which edit field of the form. However, if the label and edit box are not concurrent in the code, then the Screen Reader will not correlate the label with the edit field and therefore the user will never know what the edit field is referring to [2].

Banks normally send PIN to visually impaired people in the normal format instead of Braille. Therefore, visually impaired people are dependent on their close relative/person to read out their PIN for them. Some banks are now sending these PINs in Braille format to their partially sighted and blind customers.

Bank cards including credit and debit cards are now widely used by many customers. Although the 16-digit number printed on these cards is in Braille format, but, the three digits access code of the card at the back is not in Braille format. This 3-digit access code plays an important role during online banking.

In the United Kingdom, almost every credit and debit card transactions are now required to input the 4-digit (PIN) when accessed and used. The ISO9564 explains the fundamental principles and techniques to provide the minimum security measures, which are required for the effective PIN management by the financial organisations. Banks and financial organisations are unfortunately failing to support blind and partially sighted customers over a choice of Chip and PIN cards [5].

Chip and PIN services involve different difficulties for blind and partially sighted people. Among them is the use of different varieties of the Chip and PIN machines used in the market. The availability of these different types of machines makes it difficult to cope with the transactions. Security itself is another issue of Chip and PIN devices. Blind and partially sighted people always feel insecure when entering a PIN in the machine. This involves both ATM and Chip and PIN machines. The big uncertainty is that someone can oversee their PIN. It is also possible that most of the visually impaired customers are timed out by the system when entering a PIN to a Chip and PIN machines; this is simply because they take longer to enter their PIN, [5].

Managing finances online using PIN and password is a safe and convenient way adopted by banks but banking industry is always looking to enhance security measures by introducing latest technologies, such as Card-Reader service, the PINs-entry device, codes generator, SafePass (that delivers a one-time 6-digit code as a text message to the customer's mobile) and many more, but, unfortunately, one way or the other these devices are inaccessible to visually impaired people.

Other accessibility issues like the use of Completely Automated Public Turing test to tell Computers, and Human Apart (CAPTCHA) images also create different concerns for visually impaired people (Fig. 9.1). The term CAPTCHA is now widely used for signing up to conduct secure online banking transactions.

If a website demands a visually impaired person to use a CAPTCHA technique or enter a number which was displayed in a graphic format for login, then this is

Fig. 9.1. CAPTCHA image example.

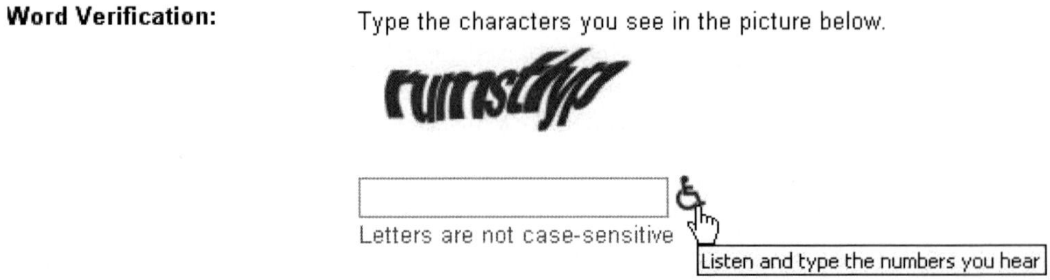

Fig. 9.2. Audio CAPTCHA image of Google search engine.

impossible without the help of a sighted person. This, of course, creates privacy and security issues.

Since 2006, Google is using the audio CAPTCHA for many of the Google services for blind and partially sighted people. In audio CAPTCHA, users simply have to click the link and type the number they hear (Fig. 9.2).

Banks all over the world should use the same technique to make sure that CAPTCHA is accessible to everyone. The Kiwi Bank of New Zealand is an example which provides this facility for their visually impaired customers.

9.11. Conclusions

People with severe visual disabilities now have the opportunity to benefit from a wealth of information and services that was previously unavailable to them. The awareness of E-accessibility is highly acknowledged in developed countries and now rising at all levels of the society. Standards have been established for the achievement of accessibility via technical specification and design interface by the W3C through a series of designed guidelines, WCAG 1.0, but, unfortunately web developers and designers ignore these standards and guidelines and still think of web as a purely visual medium and are avoiding the fact that visually impaired people have the equal access rights.

Online banking is more than an ideal environment for standard banking transactions because it provides customers an easy access to a wider range of services and allows them to deliver more timely and cost-effective services round the clock. People are motivated to the use a sophisticated online banking environment because of the expediency involved in online banking services. The popularity of online banking and web-based applications have grown tremendously but also presents a complex set of issues of security and accessibility for visually impaired people.

The security of online banking for visually impaired people is an important issue along with accessibility and goes hand-in-hand. Banks are using additional methods to achieve the extra layers of security but unfortunately not for everyone.

The challenge is to ensure that people with disabilities can take advantage of pervasive computing.

References

1. AnyBrowser, Accessible design tools, *Website Testing*, 2007, Website accessed on 27 October 2007, available at http://www.anybrowser.org/campaign/abtools.html.
2. BBA and RNIB, *Accessible E-Banking: Making Your Online Service Accessible to All*, (British Bankers Association (BBA) and Royal National Institute for Blinds (RNIB), October 2001), ISBN 1-874185-25-5.
3. C. Mancini, M. Zedda, A. Barbaro and M. G. Corsi, Users with different abilities and the Italian health websites accessibility: A survey, *Proceedings of the 9th EAHIL Conference Roma*, Italy, 2004.
4. H. Monaghan, Extending horizons for the whole community, Chichester District Council, Web Accessibility Policy, Information Technology, 2005, Electronic Government Department.
5. RNIB, Research: Statistics — Numbers of people with sight problems by age group in the UK, 2008, Website accessed on 26 February 2008, available at http://www.rnib.org.uk/xpedio/groups/public/documents/PublicWebsite/public_researchstats.hcsp.
6. Section 508, 508 Law, 1998, www.section508.gov, http://www.section508.gov/index.cfm?FuseAction=Content&ID=3.
7. C. Sergeant, E-banking: Risks and responses, 2000, Banks & Buildings Societies, Financial Services Authority.
8. F. Stakelbeck, China and E-Banking, Global Politician, 2008, available at http://www.globalpolitician.com/21348-china.
9. SystemConcepts, Usability-website accessibility, 2007, website accessed on 27 October 2007, Source: http://www.system-concepts.com/usability/accessibilitytesting/.
10. W3C, Checklist of checkpoints for web content accessibility guidelines 1.0, 1999, available at http://www.w3.org/TR/WAI-WEBCONTENT/full-checklist.html.
11. WebAim, Introduction to web accessibility, people with disabilities on the web, 2003, website accessed on 27 October 2007, http://www.webaim.org/intro/.
12. WebAim, The planning evaluation repair and maintenance process evaluating website accessibility, Web Accessibility in Mind, 2007, website accessed on 27 October 2007, available at http://www.webaim.org/articles/process/evaluate.php.
13. WebTips, Dan's web tips, validators: validators vs. linters: what's the difference?, 2006, website accessed on 27 October 2007, available at http://webtips.dan.info/validators.html.

Chapter 10

TOWARDS AN INTEGRATED NETWORK SECURITY FRAMEWORK USING THE Y-COMM ARCHITECTURE

GLENFORD MAPP*, JON CROWCROFT† and RAPHAEL PHAN‡

*Middlesex University

†University of Cambridge

‡Loughborough University

10.1. Introduction

Our current approach to network security has evolved in response to specific threats as they have been identified or in most cases as these threats have manifested themselves, often with painful consequences. We have been basically playing catch-up to the hackers as they find and exploit holes in our networking infrastructure or end-user systems. As a result, specific solutions have evolved such as the firewall which, in fact, can have detrimental effects on end-to-end performance. What is needed is an integrated approach to security as well as having security as a key part of any future communications systems. This chapter explores a new architecture for future telecommunications called Y-Comm in which security is integrated into the design.

10.2. The Future of Telecommunication Systems

Before we can propose a new framework for network security, we first need to look at a new architecture for telecommunications which will meet the needs and demands of its users. First, future systems must allow ubiquitous connectivity where users are always connected from anywhere and at any time. The need for continuous connection is being met by the development and deployment of a number of wireless technologies including 3G/HSPDA, WLAN with 802.11n being the latest network of this type that will be deployed, WiMax and satellite communications. There will also be new indoor technology such as Ultrawideband which can be used both as a location and communication technology.

However, the widespread deployment of wireless networks will have a significant impact on the evolution of the Internet. In the early Internet, end systems were primarily composed of Ethernet and Token Ring systems, which had a similar performance to the systems used in the core of the network. However, with the wide-scale deployment of wireless networks as end-systems, there will now be significant differences in network characteristics in terms of bandwidth, latency, packet loss and error characteristics. These developments mean that it will not be possible soon to think of the Internet as a single unified infrastructure. It would be better to view the Internet as comprising a fast core network with slower peripheral networks attached around the core. The core network will consist of a super-fast backbone using optical switches and fast access networks, which use ATM and MPLS. Most of these peripheral networks will make use of wireless technologies described above.

These developments also indicate that a new framework is required as they severely weaken the end-to-end arguments that were used when the Internet was designed [16]. The authors now believe that it is essential that a new architecture should also reflect this new reality. The authors, therefore, believe that it is necessary to think in terms of not one but **two** frameworks. The first framework deals with issues in peripheral networks and is called the **Peripheral Framework** while the second framework deals with issues in the core network and is called the **Core Framework**.

10.3. The Peripheral Framework

The Peripheral Framework is concerned with activities on mobile nodes and in the wireless networks to which they are connected. The peripheral network, therefore, deals with issues of connectivity — making sure that we are always connected to the Internet. One of the key features of this framework is the need to support vertical handover as the mobile moves around [12]. Vertical handover involves switching connectivity from one network to a completely different network. Vertical handover must be properly managed or else there will be an increased delay and packet loss experienced by connections associated with the mobile. Since different wireless networks have different characteristics in terms of bandwidth, latency, etc., vertical handover can therefore result in a significant change in the **quality of service** or QoS experienced by applications [8]. How these QoS changes are dealt with is also a key part of the Peripheral Framework.

Figure 10.1 shows the Peripheral Framework developed for heterogeneous networking. A more detailed explanation of the architecture is found in Ref. 11.

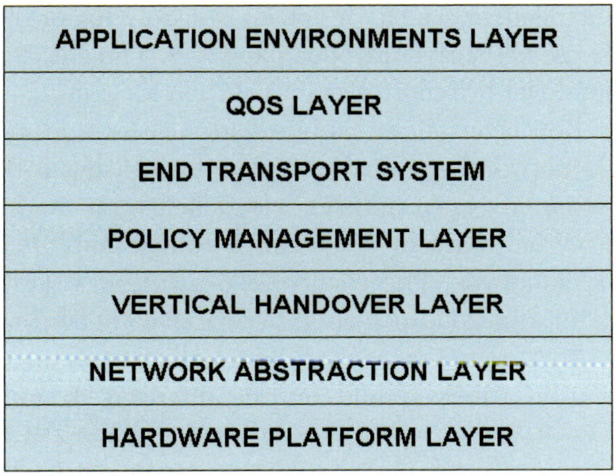

Fig. 10.1. The peripheral framework.

10.3.1. *Layer 1: The Hardware Platform Layer*

This layer is used to define the hardware components and technologies required to support a particular wireless network, including the electromagnetic spectrum, modulation techniques, Media Access Control (MAC) algorithms, etc. We can therefore represent each different networking technology as a vertical slice of the hardware platform layer. A key function of this layer is to determine which technologies are compatible and hence can be operated simultaneously.

10.3.2. *Layer 2: The Network Abstraction Layer*

This layer specifies a common networking interface which all networks employing this architecture must support. This interface is used to maintain and control the network on the mobile node. Different wireless device drivers must be written to map onto this layer. This layer therefore forms a bridge between the hardware interfaces of the wireless networks and the higher layers of the framework. Using this layer, the hardware generates events such as the availability of a given network, change of channel characteristics including signal strength as well as power being consumed by individual interfaces. This information is used by the upper layers to perform handover decisions, QoS algorithms as well as notify relevant applications. The higher layers also use this layer to control the individual network interfaces. Recent work done by the IEEE 802.21 Working Group [19] to develop a unified interface to control different wireless interfaces could be used as a starting point for the development of this layer.

10.3.3. *Layer 3: The Vertical Handover Layer*

This layer is concerned with the specification of mechanisms including state engines and triggers for vertical handover. There are two kinds of vertical handovers. The first is network-controlled and is managed and maintained in the core network. The second is client-controlled in which the client controls handover. Presently, most commercial systems use network-controlled handover. The authors, however, believe that client-based handover is a more elegant solution for a number of reasons. First, the network abstraction layer allows mobile nodes to have up-to-date information about their wireless networking interfaces, thus facilitating a better environment to support vertical handover. In addition, the client can take into consideration other factors such as the state of its network connections in its decision about when to implement handover. The client-based solution is also more scalable as it can easily access information on several networks compared to a network-based solution because network operators regard the topology and status of their networks as sensitive information which should not be divulged to competitors or third parties.

10.3.4. *Layer 4: The Policy Management Layer*

This layer is used to evaluate all the circumstances when handover should occur. The layer can be implemented by defining certain rules with regard to all the relevant parameters and their values which are evaluated with respect to handover. Policies can be essentially divided into two categories: reactive and pro-active policies. Reactive policies are triggered by changes in the condition of the networks to which the mobile node is connected. Such triggers are conveyed by the network abstraction layer. Pro-active policies attempt to know the condition of the various networks at a specific location before the mobile node reaches that location. Pro-active policies allow mobiles nodes to calculate the Time Before Vertical Handover (TBVH). This allows the mobile node to minimise the effects of vertical handover.

10.3.5. *Layer 5: The End Transport System*

This layer looks at moving data to and from the mobile node. Since most peripheral networks will be wireless, it is therefore important to ensure that network and transport systems operate efficiently so that applications running on the mobile node can receive a sustainable QoS. The TCP/IP, which is the Protocol suite used throughout the Internet, has been shown not to perform well in wireless networks [13, 20]. In addition, recent work has shown that TCP adapts very slowly to network conditions after vertical handover [5] without substantial help from the lower layers [18].

10.3.6. *Layer 6: The QoS Layer*

This layer helps to ensure that the QoS required by applications can be maintained as the QoS being offered by the networks is dynamically changing as the mobile node moves around. This framework defines two types of QoS support. The first is called **Downward QoS** which allows applications to specify the QoS they require and leaves the system to support such requirements over available network channels. The second type is called **Upward QoS** in which applications themselves attempt to adapt to the changes in network conditions. Current applications which cannot adapt to the changing conditions will employ Downward QoS while newer applications such as multimedia and networked games would need to use Upward QoS.

10.3.7. *Layer 7: The Application Environments Layer*

This layer specifies mechanisms and routines that allow applications to be built, which can use all the layers of the framework. An interesting approach to this problem is to adopt a toolkit approach which allows different types of application environments to be built. This approach is similar to that used in the deployment of X Window System [7]. Hence, various objects in the toolkit can be used to build a particular application environment. So, for example, using the toolkit, we could specify a context-aware, location-aware, application layer for high-mobile environments using Upward QoS mechanisms. Such an application environment will use the lower layers of the Peripheral Framework via the objects in the toolkit to build relevant applications.

10.4. The Core Framework

The Core Framework is shown in Fig. 10.2. A more detailed look at this framework is discussed in Ref. 6. It contains seven layers which are detailed below:

The first two levels of the Core Framework are similar in function to the first two layers of the Peripheral Framework. However, while the Peripheral Framework specifies software such as device drivers to support a given network on a mobile node, in the Core Framework, these layers represent the functionality that software modules need to support in the base-stations of a given technology. The network abstraction layer in the Core Framework allows the base-station and its resources to be managed from another location in the network.

10.4.1. *Layer 3: The (Re)configurable Network Layer*

This layer provides a control plane for (re)configuring networking resources in the core network. An efficient way of doing this is to virtualise hardware components to use a small number of virtual units such as switchlets or routelets. This layer would therefore make use of programmable and active networking techniques to control key infrastructural hardware, including routers and switches, base-stations, base-stations controllers, mobiles switching centres and GPRS/3G support nodes. Though there has been a lot of research done into virtualising routers and switches [17, 15], it is clear that more work needs to be done in virtualising mobile systems. In addition, in order to facilitate client-based handover, these interfaces must be made available to heterogeneous devices. The opening up of network infrastructure in this way raises security issues which must be addressed. For example, the possibility of mobile nodes demanding resources to which they are not entitled.

10.4.2. *Layer 4: The Network Management Layer*

This layer acts as a management plane that uses the programmable network layer to bring together hardware and software components to build enterprise class networks. Each network will have an operator that controls it. To do this, the layer must also provide Authentication, Access Control, Accounting and Charging (AAAC) systems [14]. It must also support the use of policy mechanisms that would allow operators to dictate which hardware components may be used on their networks. The policy management layer in the Peripheral Framework can interact with the network management layer in the Core Framework to help inform mobile nodes about network resources that can be used on specific networks. The policy management layer uses this information to tell the vertical handover layer on the mobile device about the network resources that can be used for a vertical handover.

10.4.3. *Layer 5: The Core Transport System*

This layer is about network addressing and transport mechanisms in the core network. Currently, TCP/IP is used in the core network and we are of the opinion that it should continue to be used, though a move to IPv6 is necessary to add enhanced network capabilities. Since the core network will be using switching technology, TCP/IP addressing is therefore being used to represent endpoints in the core network to which data is routed. A peripheral framework must, therefore, be connected to a core endpoint in order to use the core network. Core endpoints are therefore manifested as points in the access network to which several peripheral networks may be attached.

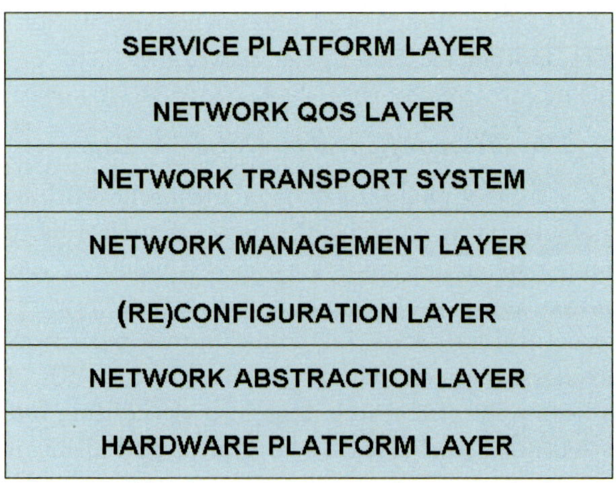

Fig. 10.2. The core framework.

10.4.4. *Layer 6: The Network QoS Layer*

This layer is responsible for QoS issues within the core network. It looks at how QoS is defined and the mechanisms used to establish and maintain QoS at different points in the system. The development of a very fast core and the adoption of wireless networks indicate that a lot of QoS issues have moved from the core network to peripheral wireless networks. We, therefore, believe that Internet QoS models such as IntServ [3] and DiffServ [9] need to be enhanced to deal with this new reality. A key concern is to prevent core endpoints (i.e. where the peripheral networks connect to the core network) from becoming overloaded. A suggested approach is to look at QoS not just in terms of individual flows or streaming classes but also on a network level. So, a peripheral network negotiates with the QoS manager in the Core Framework about the QoS being used by that network at the core endpoint to which it is attached.

10.4.5. *Layer 7: The Service Platform*

The service platform allows different agents to install and operate various services in a secure and controlled fashion. The service platform will provide the ability to install services in component form on several networks simultaneously, or on a single network. This will, therefore, allow the provision of both national and regional services to be easily constructed, e.g. traffic information in London would be a local service accessible to networks operating in London. When a service is installed, it must produce a Server Level Agreement (SLA) which specifies the minimum QoS that is needed to run the service. The SLA is given to the QoS manager in the core network, which checks to see if the core endpoint can sustain the required QoS.

10.5. The Y-Comm Framework

In the Y-Comm framework, the Peripheral Framework and the Core Framework are brought together to represent a future telecommunications environment which supports heterogeneous devices, disparate networking technologies, network operators and service providers. The two frameworks, shown in Fig. 10.3, share a common base sub-system consisting of the hardware platform and network abstraction layers. Both frameworks diverge in terms of functionality but the corresponding layers interact to provide support for heterogeneous environments. The shape of the structure can be represented by the letter Y; hence, the architecture is called Y-Comm.

10.6. The Y-Comm Security Model

It should be observed that none of the layers of the architecture described so far mentions security. This is because Y-Comm attempts to define a multi-layer

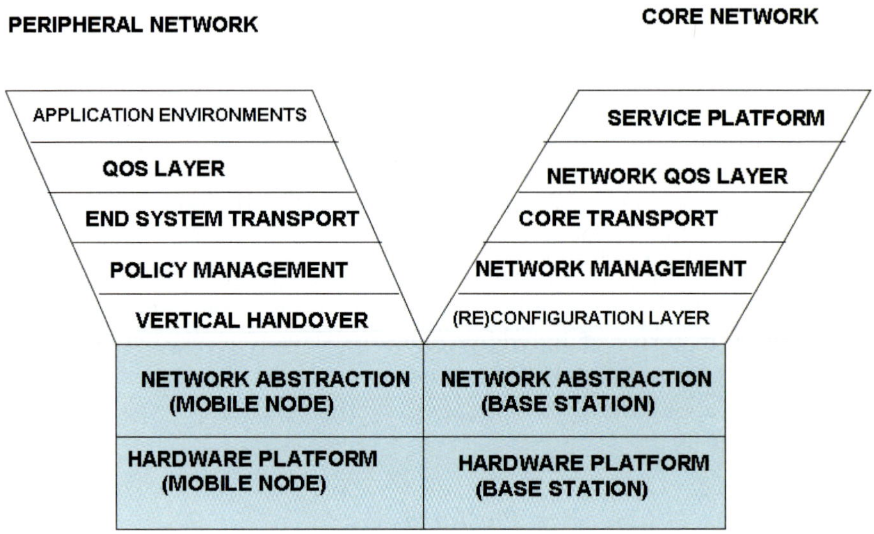

Fig. 10.3. The Y-comm framework.

security model which must be applied to both the Peripheral and Core Framework simultaneously to provide total security. The security layers must work together in as well as across both frameworks in order to be fully integrated with the new architecture.

The highest layer of security is at Layer 7 and is called **Service and Application Security** or SAS. In the Peripheral Framework, SAS defines the AAAC functions at the end-device and is used to authenticate users and applications. For example, when a user logs onto an end-device, it will be interacting with the SAS module which decides on the applications and programs the user is allowed to run on the mobile node. The SAS in the core network is used to monitor which services can be installed on a given network and which entities may use the service. Hence, SAS provides AAAC functions for services on the service platform in the core network.

The next security layer is at Layer 6 and is called **QoS-Based Security** or QBS.

Y-Comm also embraces the **Quality of Security Service** or QoSS approach [10] to security in which security requirements of applications are viewed as **part** of the overall QoS required by that application. Hence, dynamic changes in security are treated as changes in the QoS of an application. This level of security is therefore concerned with QoS issues and the changing QoS demands of the mobile environment. This may be due to several factors including the fact that mobile nodes are continuously changing their point-of-attachment as they move around. In addition, in order to meet their service-level agreements, servers may choose to replicate services closer to the current position of the mobile. So, it is necessary to ensure that core endpoints and peripheral networks are not overloaded or the security of these systems is not comprised; for example, if the security level of a proxy server is below the security requirements to connect to a secure network, then the QBS layer will not allow the service to be migrated to that network. In addition, the QBS layer also attempts to block QoS-related attacks, such as Denial-of-Service (DoS) attacks on networks and servers.

The next security layer is at Layer 5 and is called **Network Transport Security** or NTS. In the Peripheral network, NTS is concerned with access to and from end-devices and the visibility of these devices and services on the Internet. The NTS is important as it decides which end-user devices can connect and how that is done. Therefore, before any application or user can connect to any service or to any other end-user, NTS decides whether this connection should be established based on the various factors, including IP address, service name, the security profile of the user, etc. The NTS is therefore used to replace things like NAT in peripheral networks, etc. In the core network, NTS is used to set up secure connections through the core network. So, NTS in the Core Framework involves setting up secure tunnels between core endpoints using mechanisms such as IPSec to ensure that moving data across the core network is done in a fast and secure manner.

Finally, the fourth and last level of security is defined at Layer 4 but can also encompass Layer 3 and Layer 2. It is called **Network Architecture Security** or NAS. In the Peripheral Framework, this type of security attempts to address security issues involved in using particular networking technologies and the security threats that occur from using such a technology. A good example is the fact that wireless systems are inherently broadcast (over the air) so that security measures must be taken with respect to the transmission of packets compared to wired systems. The NAS is concentrated in Layer 4 as it is one of the key issues that must be taken into consideration when a vertical handover is activated. So, for example, if I want totally secure communication and I am doing a vertical handover between a LAN and a WLAN system, the system should ensure that link-level packets are also encrypted when a vertical handover is done from the LAN to the WLAN. In addition, because of the security threat posed by the wireless systems, NAS is used to define specific access-control policies with regard to devices and applications that may use any given network. So, when a mobile device wishes to use any given network, NAS is invoked to ensure that the user is authorised to do so. The NAS also ensures that the local LAN environment is as secure as it could be.

In the core network, NAS is used to secure access to the programmable infrastructure. The NAS in this context determines which switchlets, routelets or base-station resources may be used by the network management system. The NAS also prevents mobile nodes from acquiring resources reserved for other networks. The full Y-Comm architecture including its security layers is shown in Fig. 10.4.

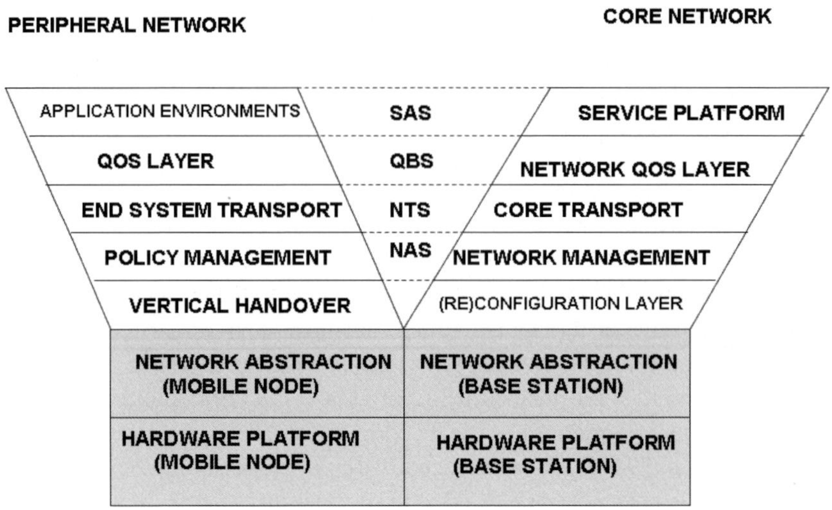

Fig. 10.4. The Y-comm framework with security layers.

10.7. Network Security Models in Y-Comm

Because these security functions are part of the communications architecture, it can be used to much greater effect than previous methods. This is because these modules will be started as part of the architecture and so will be invoked using normal communication routines. Y-Comm is therefore able to offer two distinct network security models. The first is called the **Connection Security** model in which security is applied when connections are being made. The second security model is called the **Ring-Based Security** model which is applied to restrict the accessibility of servers and services.

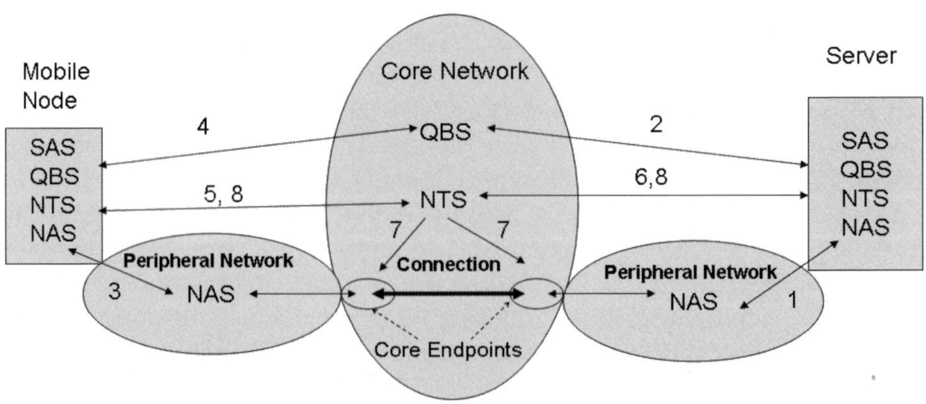

Fig. 10.5. Connecion_security_model.

10.8. The Connection Security Model in Y-Comm

In this model, the different security layers work together to establish a connection between a mobile node and a service being hosted at another site.

The Connection Security framework is shown in Fig. 10.5. We can show how the framework is used by looking at the interaction involved in setting up a connection. This is shown as a series of steps.

Step 1: The server is started. The NAS module in the server talks to the NAS module on the Local LAN to get access to its wireless infrastructure.

Step 2: The QBS security module on the server informs the QBS module in the core network about its Service Level Agreement (SLA) which contains the QoS associated with a connection to this service. When clients originally subscribe to the service (not shown Fig. 10.5), they are given the SLA associated with the service. Hence, they know the QoS requirements for connecting to the server.

Step 3: The mobile node is started. The NAS module in the mobile node contacts the NAS module in the peripheral networks to gain access to the wireless infrastructure.

Step 4: When the mobile node wants to use the service, the QBS module in the mobile node contacts the QBS module in the core network and asks for a connection with a given QoS to be made to the server. The QBS module returns two core endpoints which must be used to set up the connection.

Step 5: The NTS module on the mobile node contacts the NTS module in the core network and says that it would like a connection to the server, using the core endpoints, the QoS and security parameters.

Step 6: The NTS module in the core network contacts the NTS module on the server to signal an incoming call. At this point, the server can also check the security of the client as well as the security of the connection.

Step 7: If the server accepts the request, then the NTS module in the core network joins the two core endpoints.

Step 8: It then signals to both the client and server that a connection has been established.

10.9. The Ring-Based Security Model in Y-Comm

Ring-based security is an extension of "Off-by-Default", an idea introduced by Roscoe *et al.* [2]. The concept does not allow devices to be directly accessible over a WAN such as the Internet without initially interacting with the network infrastructure. A device, that wants to communicate with the outside world, must explicitly indicate to the networking infrastructure that it wants to be able to do so. When the system gives the server or user an IP address, this is registered in a dynamic DNS, making it possible to access the device or server over the Internet. The "Off-by-Default" concept is a new idea and merits consideration but it is impossible to implement using the present architecture. However, we believe that this concept can be significantly enhanced using Y-Comm framework because of its multi-layer security structure.

The Y-Comm structure uses four different modules to implement security: the SAS, QBS, NTS and NAS modules. The systems run in both the Core and Peripheral Frameworks. We would therefore like to use these systems to form rings of access, which guard access to certain services. The concept is similar to the rings of protection seen in the design of the Unix Operating System [1]. This concept was also supported on the 80286 and 80386 Intel microprocessors [4].

Access to machine specific servers and local services is guarded by the SAS module. So, servers that are local to the machine must register with the SAS module via the local loop interface when the server starts. The existence and state of the servers are not made known to any modules or applications outside the machine. So, the SAS servers ensure that only applications and users that are on the local machine are given access to the relevant service. Outside users cannot connect to these servers because these services are only known to the SAS module on that machine.

Servers that serve networks, such as DNS or DHCP, to provide LAN access must be visible at the network level. These servers must therefore register with both the SAS and NAS modules on the local machine. This NAS module will register the server with the NAS modules on the relevant base-station. This action will make these services accessible to other mobile devices on the local LAN. So, once a mobile device is cleared by its NAS module and registers with the NAS module at the base-station, it is informed about the location of all the relevant LAN servers that serve the local wireless networks.

Servers that wish to serve global clients must register with the SAS, QBS, NAS and NTS modules on the local machine. The NAS module will register the server with the NAS modules running on the relevant base-station. The QBS system on the local machine talks to the QoS manager in the core network. The core QoS manager needs to know the SLAs which the service needs and must ensure that the core endpoints by which the server will be accessed can deal with the QoS for that service. This information is stored in a Global DNS service which maps the service name to the core endpoints to which it is attached and the QoS required to maintain that service. So, when the QBM module is contacted about connecting to a given service, it checks the Global DNS to find out these parameters to work out which pair of endpoints should be assigned to the connection. If the service is not in the Global DNS database, then it will not be known to the QBS layer in the core and so no connection can be made. The arrangement is shown in Fig. 10.6.

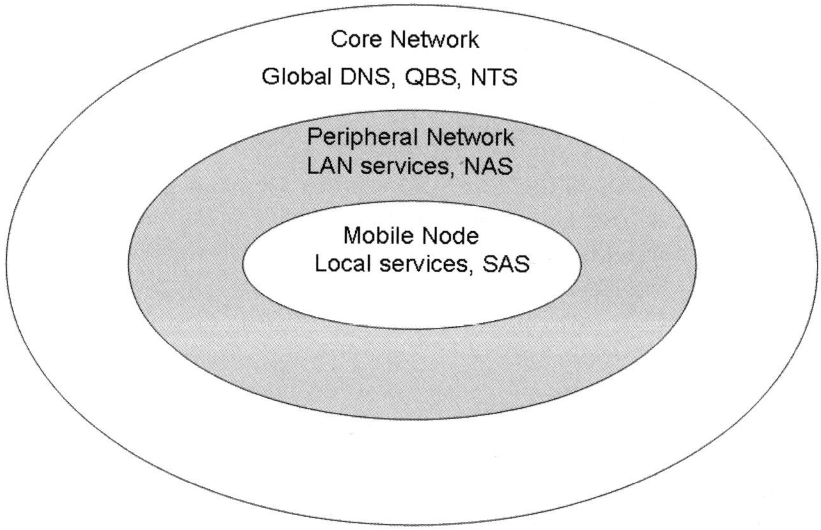

Fig. 10.6. Ring based network security.

10.10. Enhancing Communications Structures in Operating Systems

In order for ring-based security to work, we need to enhance some of the communication structures that are used in the OS. This can be done by using additional structures to the socket call framework which is currently used to manage Internet communications. First, the bind system call in the present socket library binds a socket to the IP address of a given interface or a set of interfaces. We believe that such a call should be augmented in the proposed architecture as the security system is in control of connectivity and we have just shown how to do security using these modules. What is therefore required for the system to work is that servers need to bind not only to an address but they also need to bind to a SLA which also includes its QoS and security requirements. This SLA will contain certain bits of information which can give us a clear indication of how these services are managed.

The most important of these is **scope**. A service can have one of three scopes: local, LAN or global. Local indicates that the services must run on individual machines and are under the control of the SAS module. Servers that have LAN scope are registered with the NAS module in the base-station. Again, these services are only available to other mobile devices on the same LANs. Finally, if the scope is global, then the service is registered with the Global DNS and is controlled by the QBS and NTS modules in the core network.

The other fields include the security measures which are also used to allow connections to be established with the server. These functions may include the use of access control lists, capabilities, or login/password checks. In addition, these functions are exported to the relevant security module which then applies these functions when there is an attempt to contact the server.

What is new here is that security checks are done even before a connection to the server is made. So for local services, the SAS module uses these security checks when clients attempt to connect to a local server. For LAN services, the security checks are executed by the NAS at the base-station when a request is made to connect to LAN services. Finally, for global servers, the security checks are executed by the QBS and the NTS systems in the core network. Hence, security is checked before a connection is completed but not after.

10.11. Conclusion

This chapter has looked at a new communications architecture for heterogeneous networking called Y-Comm which uses a multi-layer security model. The integration of the various layers in the security model as well as the fact that the model itself is closely integrated with the overall architecture make it possible to design new security solutions. We believe that the security models introduced using the Y-Comm architecture supersedes a lot of security techniques being used today, including firewalls, leading to a more secure but also a more efficient network infrastructure.

References

1. M. J. Bach, *The Design of the UNIX Operating System* (Prentice Hall Software Series, 1986).
2. H. Ballani, Y. Chawathe, S. Ratnasamy, T. Roscoe and S. Shenker, Off by default!, *Proceedings of the Fourth Workshop on Hot Topics in Networking (HotNets-II)*, College Park, MD, USA, November 2005.
3. R. Braden, D. Clark and S. Shenker, RFC 1633: Integrated services in the intranet architecture: An Overview, *Technical Report, IETF*, 1994.
4. B. B. Brey, *The Intel Microprocessors* (Prentice Hall, 2005).
5. D. Cottingham and P. Vidales, Is latency the real enemy in next generation networks? *Proceedings of First International Workshop on Convergence of Hetergeneous Wireless Networks*, July 2005.
6. J. Crowcroft, D. Cottingham, G. E Mapp and F. Shaikh, Y-Comm: A global architecture for heterogeneous networking, *(Invited Paper) Proceedings of the 3rd Annual International Wireless Internet Conference (WICON 2007)*, Austin, Texas, USA, October 22–24, 2007.
7. R. Scheiffler and J. Gettys, The X window system, *ACM Trans. Grap* **5**(2) (1986) 79–109.
8. A. Duda and C. Sreenan, Challenges for quality of service in next-generation mobile networks, *Proceedings of IT &T Annual Conference*, October 2003.
9. D. Grossman, RFC 3260: New terminology and classification for diffserv, *Technical Report, IETF*, 2002.
10. C. Irvine and T. Levin, Quality of security service, *Proceedings of the New Security Paradigms Workshop*, September 2000.

11. G. Mapp, D. Cottingham, F. Shaikh, P. Vidales, L. Patanapongpibul, J. Baliosian and J. Crowcroft, An architectural framework for heterogeneous networking, *Proceedings of the International Conference on Wireless Information Networks and Systems* (WINSYS 2006), August 2006, pp. 5–10.
12. J. McNair and F. Zhu, Vertical handoffs in fourth-generation multinetwork environments, *IEEE Wireless Communications* **11** (2004).
13. M. Meyer, TCP performance over GPRS, *IEEE WCNC*, 1999, pp. 1242–1252.
14. C. Rensing, H. Hasan, M. Karsten and B. Stiller, A survey of AAA mechanisms, protocols and architectures and a policy based approach beyond: Ax, *Report 111. Swiss Federal Institute of Technology*, 2001.
15. P. Rodriguez, R. Chakravorty, J. Chesterfield, I. Pratt and S. Banerjee, MAR: A commuter router infrastructure for the mobile internet, *Proceedings of the ACM Second*

Mobile Systems, Applications and Services Conferences (ACM Mobisys 2004), June 2004.
16. J. H. Saltzer, D. Reed and D. D. Clark, End-to-end arguments in system design, *ACM Transactions in Computing Systems* (1984) 277–288.
17. J. der Merwe and I. Leslie, Switchlets and dynamic virtual ATM Networks, *Proceedings of Integrated Network Management V*, 1997.
18. J. Scott and G. Mapp, Link layer-based TCP optimisation for disconnecting networks, *Computer Communications Review* **33**(5) (2003).
19. The IEEE 802.21, Working group website: http://www.ieee802.org/21/.
20. G. Xylomenes, G. Polyzos, P. Mahonen and M. Saaranen, TCP performance issues over wireless links, *IEEE Communications Magazine* **39**(4) (2001) 52–58.

Chapter 11

INTRODUCTION TO BEHAVIOURAL BIOMETRICS

KENNETH REVETT
University of Westminster, London, UK

11.1. Introduction

Biometrics is concerned with the scientific approach to user verification and/or identification. The focus will be on *behavioural* biometrics — based on the way they provide information to the authentication system. For instance, a person could be required to provide a signature, enunciate a particular phrase, or enter a secret code through an input device in order to provide evidence of their identity. Note that there is an implicit simplicity to behavioural biometrics — in that typically no special machinery/hardware is required for the authentication/identification process other than the computer (or ATM) device itself. In addition, the approaches prevalent in this domain are very familiar to us — practically everyone has provided a signature to verify their identity and we have one or more passwords for logging into computer systems. We are simply used to providing proof of identity in these fashions in certain circumstances. These two factors provide the foundation for the behavioral approach to biometrics. These modes of identification are substantially different from the other classes of biometrics: physiological and token-based biometrics. For instance, what is termed physiological (or biological) biometrics requires that we present some aspect of our physicality in order to be identified. Typical instances of physiological biometrics include iris scans, retina scans and fingerprints. Lastly, token-based biometric systems require the possession of some object such as a bank or identity card. Each class of biometrics is designed to provide an efficient and accurate method of verifying the identity (authentication) and/or the identification of an individual.

11.2. Types of Behavioural Biometrics

There are a variety of subdivisions/modalities within the behavioural biometrics domain. Each possesses unique characteristics in terms of ease of use, deployability, user acceptance and quality of the identification/verification task. In what follows, a summary of a variety of popular behavioural biometrics is presented, with an emphasis on the theoretical underpinning and practical implementation. The order is rather arbitrarily presented — and does not reflect any relative merits of the corresponding modality.

11.2.1. *Voice Recognition*

It is a venerable behavioural biometric in which users are requested to enunciate text as a means of identifying themselves. Voice can be employed for either speaker identification or speaker authentication. With respect to speaker identification, a person enunciates text and the speech patterns are analysed to determine the identity of the speaker. In the literature, this is referred to as speaker independent recognition (see [13]). This mode poses several interesting issues, such as what happens if the speaker is not contained within the database of speakers? As in all major forms of biometrics, any individual wishing to utilise the biometric device must at some stage introduce themselves to the system, typically in the form of an enrollment process. One of the principal tasks of the enrollment process is to register the person as a potential user of the biometric system (enrollment will be discussed further later in this chapter). In a speaker independent system, the user's voice pattern is analysed and compared to all other voice samples in the user database. The closest match to the particular voice data presented for identification becomes the presumed identity of the speaker. There are three possible outcomes: (i) the speaker is correctly identified, (ii) the speaker is incorrectly identified as another speaker, or (iii) the speaker is not identified as being a member of the system. Clearly, we would like to avoid the last two possibilities, which reflect the false acceptance rate (type II error) and the false rejection rate (type I error) as much as possible. When the speaker attempts an authentication task, the speaker has provided some evidence of their identity, and the purpose of the voice recognition process is to verify that this person has a legitimate claim to that identity. The result of this approach is a binary decision: either you are verified as the claimed identity or you are not.

The other major division within voice recognition biometrics is whether the enunciated text is fixed or free (Fig. 11.1)? That is, does the user enunciate a specific phrase (text-dependent), or are they allowed to enunciate any phrase (text-independent)? The speaker dependent version is easier from a matching perspective, in that the spoken text is directly matched to the information stored in the database. The text-independent approach allows the speaker to enunciate any speech they wish to. This approach requires a model of each

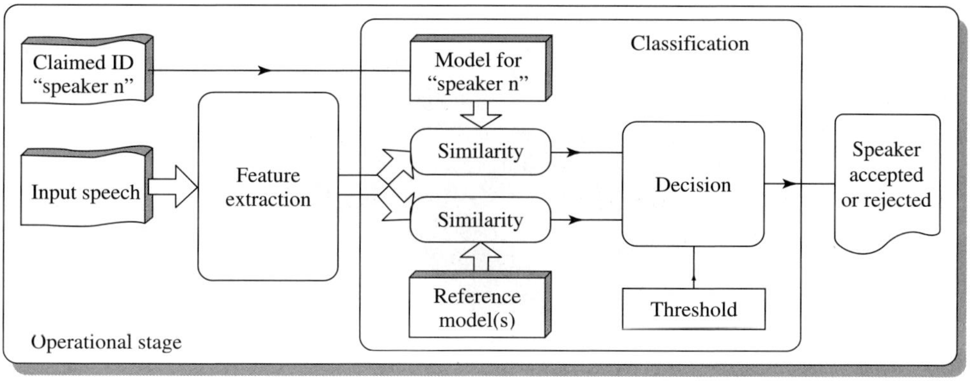

Fig. 11.1. An example of a voice recognition processing system. (Taken from Ref. 3.)

speaker, which is certainly more computationally expensive than the text-dependent approach.

11.2.2. *Signature Verification*

It is a behavioural biometric where users are required to present handwritten text for authentication. This is probably the most familiar of all biometrics — though currently not the most prevalent, due to the advent of computer-based passwords. There are two essentially distinct forms of signature based biometrics: on-line and off-line. With an on-line signature verification system, the signature characteristics are extracted as the user writes, and these features are used to immediately authenticate the user. Typically, specialized hardware is required, such as a pressure sensitive pen (a digital pen) and/or a special writing tablet. These hardware elements are designed to capture the dynamical aspects of writing, such the pen pressure, pen angle, and related information. In a remote access approach, where specialized hardware may not be feasible, the on-line approach is most suitable from a small portable device such as a PDA, where the stylus can be used for writing. The off-line approach utilises the static features of the signature, such as the length and height of the text, and certain specialized features such as loops (not unlike a fingerprint approach). Typically, the data are acquired through an image of the signature, which may be photocopied or scanned into a computer for subsequent analysis. As in all behavioural biometric approaches, a writing sample must be stored in the authentication database and the writing sample is compared with the appropriate reference sample before the acceptance/rejection decision to be made (see [2] for details). Again, there is the possibility of having text-dependent or text-independent signature verification. The same caveats that applied to voice also apply here — and really voice and signature are really very similar technologies — only the mode of communication has changed — which results in a different set of

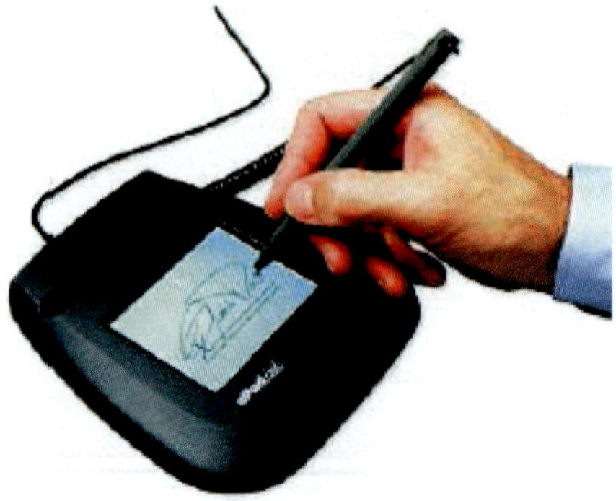

Fig. 11.2. An on-line signature verification system. (Interlink Electronics ePad.)

features that can be extracted. An example of an on-line signature setup is presented in Fig. 11.2.

11.2.3. *Keystroke Dynamics*

It is a behavioural biometric that relies on the *way we type* on a typical keyboard/keypad type device. As a person types, certain attributes are extracted and used to authenticate or identify the typist. Again, we have two principal options: text-dependent and text-independent versions. The most common form of text-dependent systems requires the user to enter their login ID and password (or commonly just their password). In the text-independent version, users are allowed to enter any text string they wish. In some implementations, a 3rd option is used — where a user is requested to enter a long text string — on the order of 500–1,500 characters. The user enrolls into the system by entering their text either multiple times if the short text-independent system (i.e. password) is employed, or typically once if the system employs a long text string. From this enrollment process, the user's typing style is acquired and stored for subsequent authentication purposes. This approach is well suited for remote access scenarios: no specialized hardware is required and users are used to providing their login credentials. Typical attributes that are extracted when a person types are the duration of a key press (dwell time), and the time between striking successive keys (di-graph if the time is recorded between successive keys). These features, along with several others are used to build a model of the way a person types [14, 15]. The security enhancement provided by this technology becomes evident if you leave your password written on a sticky notepad tucked inside your desk, which someone happens to find. Without this level of protection, possession of the password is required for a user to access your account. With the addition of a keystroke dynamics-based biometric, it is not sufficient that the password is acquired: the password has to be entered exactly (or at least within certain tolerance limits) the way the enrolled user entered it during enrollment. If

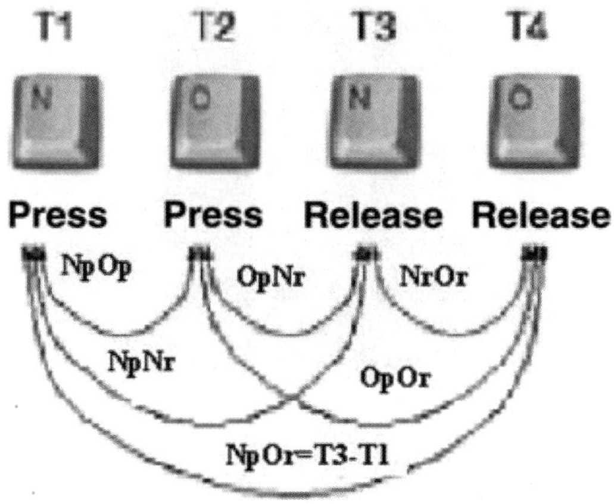

Fig. 11.3. The combinations of digraphs that can be generated from the character sequence "N" followed by "O". Note the subscript 'p' and 'r' indicate press and release, respectively.

not, the login attempt is rejected. An example of the notion of a digraph is depicted in Fig. 11.3.

11.2.4. *Graphical Authentication Systems*

They are employed as an alternative to textual-based password systems. There are issues with textual-based passwords regarding the strength, which refers to how easy it would be to guess someone's password, given free access to the computer system on which they are stored (an off-line attack). Studies have indicated that most people make their passwords easy to remember — such as their names, certain memorable places, etc. Generally speaking, the full password space is not utilized when people are allowed to select their own passwords. On a typical PC type keyboard, there are 95 printable characters, and for a password of eight characters, there are 95^8 (or 6×10^{15}) possible passwords that can be generated [19]. This is a relatively large search space to exhaustively explore, though not impossible in a realistic time frame with today's modern computing power (and the deployment of a grid of computers). But typically, most users explore a small fraction of this possible password space, and the possibility of a successful off-line attack is very real [17]. As indicated, the principle reason for the lack of a thorough exploration of password space is the issue of memorability. Here is where graphical passwords take over.

Graphical passwords are composed of a collection of images, each representing an element of the user's password. The images are presented to the user — who must select the password elements — possibly in a predefined order, but more often than not order is removed from the equation, depending on the implementation. A key difference between textual- and graphical-based passwords is that in the former, recall is required and in the later recognition is involved. The psychological literature has provided ample evidence that recognition is a much easier task than recall. In addition, it appears that we have an innate ability to remember pictures

Fig. 11.4. An example of the Passfaces™ graphical password authentication scheme. Note that on each page of faces, the user is required to select the correct face image — note that in this system there is an implied order to the selection process.

better than text. These two factors combined provide the rationale for the graphical password-based approach. An example of a classical approach, dubbed Passfaces™ is presented in Fig. 11.4. In this system, the user's password is a collection of faces (typically 4–6), that must be selected in order from a series of decoy face images.

11.2.5. *Mouse Dynamics*

It is a behavioural biometric approach designed to capture the static and dynamic aspects of using the mouse as a tool for interacting with a user interface, which contains the elements of their password, typically presented in a graphical fashion. The movement of the mouse over time and position is recorded, along with dynamical information such as the speed of the mouse movements is recorded, forming the features that are used to build a reference model of the user. Therefore, mouse dynamics is used in conjunction within a graphical password scenario, though the password may not consist of a collection of images to be identified. Instead, this approach is based human–computer interaction (HCI) features — how one interacts with an application is used to authenticate a user. Provided there is enough entropy in the game — enough possibilities for interacting with it, then one may be able to differentiate users based on this information. An example of a system developed by the Ahmed and colleagues [1] is presented in Fig. 11.5.

11.2.6. *Gait as a Biometric*

It is a behavioural biometric modality that relies on the walking pattern of a person. Even the great Shakespeare himself stated that "For that John Mortimer... in face, in gait in speech he doth resemble" (Henry IV/II). As Shakespeare himself intimated, there are subtle differences in the way a person ambulates. The results of a number of gait-based biometrics indicate that these differences are statistically significant — leading EER values on the order of 5% or less. There are two principal

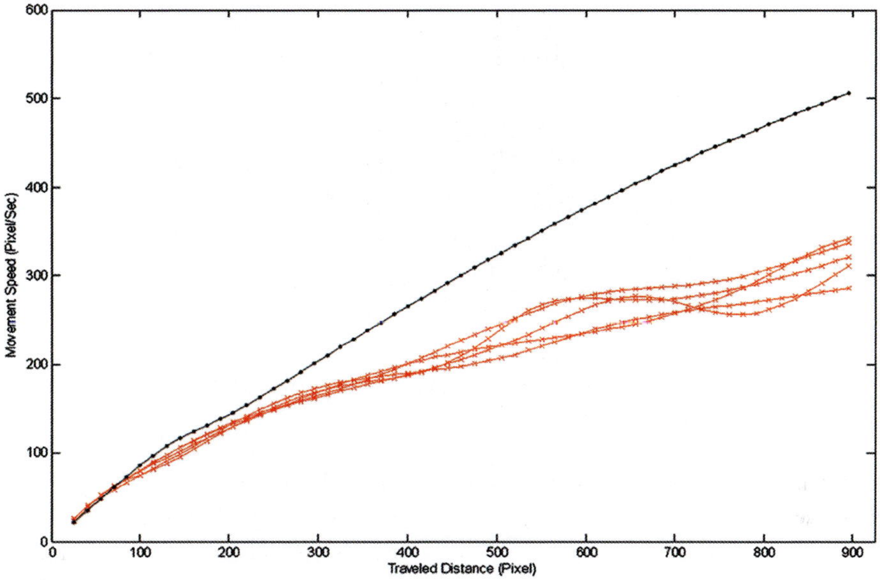

Fig. 11.5. A graph presenting the user profile solid top line versus a series of imposters based on average speed of mouse movements. (From Ref. 1.)

approaches to gait biometrics: machine-vision and sensor-based approaches. The former is the more traditional approach — and is suitable for scenarios where the authentication process must be mass-produced, such as at airports. In this scenario, a user can be scanned from a distance relative to the authentication point. Provided the data acquisition can occur quickly, this type of approach may be very attractive. The sensor-based approach (see Fig. 11.6 for an example of an accelerometer, the typical sensor used in gait analysis) acquires dynamic data, measuring the acceleration in three orthogonal directions. Sensor-based systems are quite suitable for user authentication — as it is obviously attached to the individual accessing the biometric device. Machine-vision-based approaches are more general and are typically employed for user identification.

The feature space of gait biometrics is not as rich as other technologies. This probably reflects the conditions under which the data are acquired — either a machine-vision approach with issues regarding lighting and other factors that typically degrade the performance of related biometrics such as face recognition. Even under the best of conditions (the gold-standard condition — see Appendix A for details), there are really only three degrees of freedom from which to draw features from. The current trend is to focus on dynamic aspects of walking — and the results tend to be somewhat better than static features when comparing EER values. When deployed in a multimodal approach, gait data, in conjunction with speech biometrics for instance, tend to produce very low EER values (see Appendix A for details). Research continues to find ways to enhance the feature

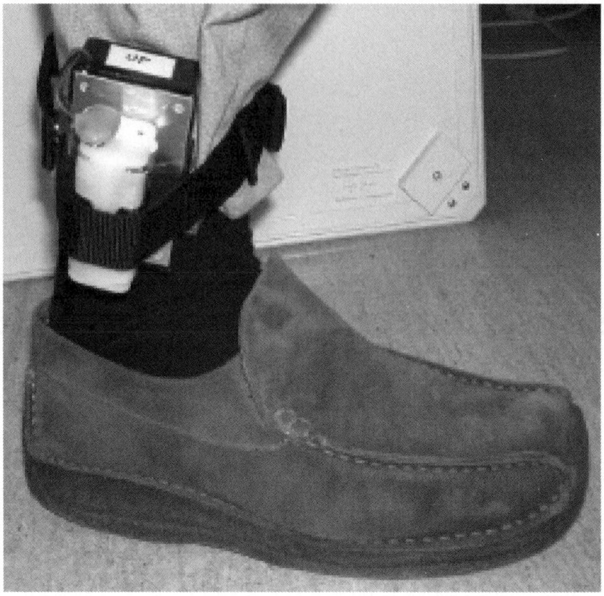

Fig. 11.6. A photograph of a subject wearing a sensor-based gait device termed an accelerometer. Note that it is capable of measuring acceleration in three different orthogonal directions. (See Ref. 5 for details.)

space of gait biometrics, but considering what is currently available, an EER of 3%–5% is quite respectable.

11.2.7. *Smile Recognition*

It is a behavioural biometric technique that uses high speed photography and a zoom lens generates a series of smile maps. These maps are composed of the underlying deformation of the relevant muscles and tiny wrinkles, which move in a characteristic fashion when a person smiles. In this approach, a collection of directional vectors is produced which form the contours describing the dynamical aspects of smiling. This approach requires further analysis to determine how effective it will be as a behavioural biometrics, as current results were produced from a small study cohort.

11.2.8. *Lip Movement Recognition*

It is a behavioural biometric based on lip movement recognition using shape similarity when vowels are uttered. In the method, we apply the mathematical morphology, in which three kinds of structuring elements such as square, vertical line, and horizontal line are used for deriving pattern spectrum. The shapeness vector is compared with the reference vector to recognise individual from lip shape (see [18]). According to experimental results with eight lips uttered five vowels, it is found that the method successfully recognises lips with 100% accuracy.

11.2.9. *Odor as a Biometric*

It is an often-overlooked class of behavioural biometrics, which is based on our sense of smell — olfaction-based biometrics. The human olfactory system is capable of detecting a wide range of odorants — using a relatively receptor system (see Freeman, 1991 for an excellent review). There are two principal processes involved in olfaction: stimulus reception and identification. There are questions regarding the specificity and sensitivity of the sense of smell. There are a number of professions that relay on a keen sense of smell — wine tasters, perfume experts and human body recovery are a few examples [16, 20]. It would therefore seem reasonable to assume that olfaction does have sufficient capacity to accurately identify a wide range of odors with high sensitivity. The question then shifts to whether or not humans exude sufficiently distinct odors such that we can be discriminated by them. Does the use of deodorant, colognes, and perfumes obfuscate our body odor beyond recognition? Lastly, how do we get a computer to perform olfaction?

The answer to the last question relies on the development of an artificial nose — the ENose [7, 20] — depicted in Fig. 11.7. It is composed of two modules: a sensor array and a pattern recognition system. The sensor array consists of a collection of sensors (typically 10–20) each designed to react with a particular odorant. The pattern recognition system is used to map the activation pattern of the sensor array to a particular odorant pattern. The sensor array can be designed from a variety of materials: conductor sensors:

- Made from metal oxide and polymers
- Piezoelectric sensors
- Metal-oxide-silicon field-effect-transistors
- Optical fiber sensors

each of these technologies can be deployed as the basis for the sensor aspect of an ENose system (for details, please consult [6, 8].

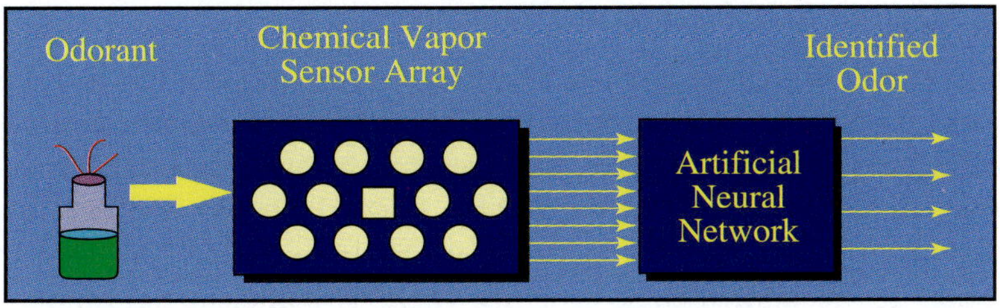

Fig. 11.7. The olfactory biometric scheme, highlighting the sensor array and pattern recognition components. (From Ref. 8.)

There are a number of pattern recognition systems that can be employed — cluster analysis, neural networks, and related classification algorithms can be employed with success. The current operation of the ENose system is essentially a 1:1 correspondence between sensor array number and odorants. Though the human olfactory system contains a great number of receptors (on the order of 1×10^6), they are used in a combinatorial fashion. That is, there is not a 1:1 correspondence between an odorant and the activation of a particular receptor. It is a distributed system — and ENose , if it is to succeed at all must adopt a similar approach. To date, there is not a clear direction in this area — it is really up to the neuroengineers to develop the required technology before it can be adapted to the biometrics domain. Though interesting, this approach will have to therefore wait for further parallel advancements in engineering before it can be considered a truly viable behavioural biometric — especially in a remote access context.

11.2.10. *Biological Signals as a Behavioural Biometric*

It is a novel approach that relies on the measurement of a variety of biological signals. These include the electrocardiogram (ECG), the electroencephalogram (EEG) and the electrooculogram (EOG) to name a few potential candidates. In the late 1970s, Forsen published a report that evaluated the largest collection of potential biometric technologies known at the time [4]. Included in this impressive list were the deployment of the ECG and EEG — quite prescient for 1977! The basic approach is to extract the signals from the user during the enrollment period, extract features and generate a classifier. When the user then attempts to log in, the particular class of signal is recorded, and a matching score is computed, which determines the decision outcome. This is really no different than any other behavioural biometric (and physiological for that matter) — the novelty here is the data that are acquired. In order to acquire biological signal data, specialized hardware is required. One of the tenets (or at least selling points) of behavioural biometrics is that no specialized hardware is required. It is anticipated that with the current rate of technological advancement, the amount of hardware required will be reduced to acceptable levels (see [12] for a nice discussion on this topic).

11.2.11. *ECG as a Behavioural Biometric*

The ECG is simply a recording of the electrical activity associated with the beating of the heart. A series of leads is positioned appropriately over the heart — which picks up the small electrical signals produced by various regions of the heart that generate electricity (i.e. the pace-maker or the sinoatrial node). The recording of the human heart-beat generates a characteristic profile (depicted in Chapter 8 of Fig. 4). The question to be addressed is whether there is sufficient variability between individuals such that this signal can form a reliable marker for any particular

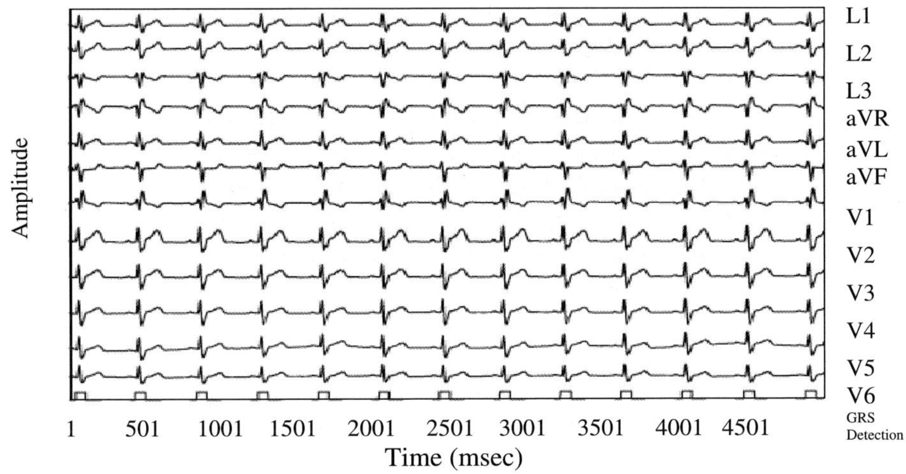

Fig. 11.8. A time series recording of ECG data and some pre-processing results. The x-axis is time and the y-axis represents the signals acquired from each of the 12-leads. The bottom row represents the SVM detection results. (See Ref. 10 for details.)

individual. The data presented in Chapter 8 of this volume indicates that there is plenty of evidence to suggest that *it is* sufficiently discriminating to produce a high degree of classification accuracy (near 100% in some studies). Figure 11.8 presents a typical authentication scheme employing ECG data (Taken from Ref. 10).

11.2.12. *EEG as a Behavioral Biometric*

The EEG is a recording from the scalp surface of the electrical activity of a collection of synchronously firing, parallel-oriented neurons. The EEG records the electrical activity of the brain — and as such is continuously active (even in patients in the locked-in-state condition, resulting from a stroke). Embedded within the on-going EEG activity are changes that occur in a correlated fashion with particular types of cognitive activities. The activities are typical cognitive functions such as thinking of a name, reading aloud, and listening to music. These signals can be isolated from the underlying background activity through a series of filtering and related techniques, which are discussed in some detail in Chapter 8 of this volume (see the references therein for more details). The goal in this approach is to associate particular electrical signatures that occur within the brain with particular cognitive tasks, such as entering a password to playing a video game.

The data obtained from EEG is sufficiently robust to generate a significant amount of inter-subject variability, and many studies have produced statistically significant classification results using "raw" EEG data (Fig. 11.9). In addition,

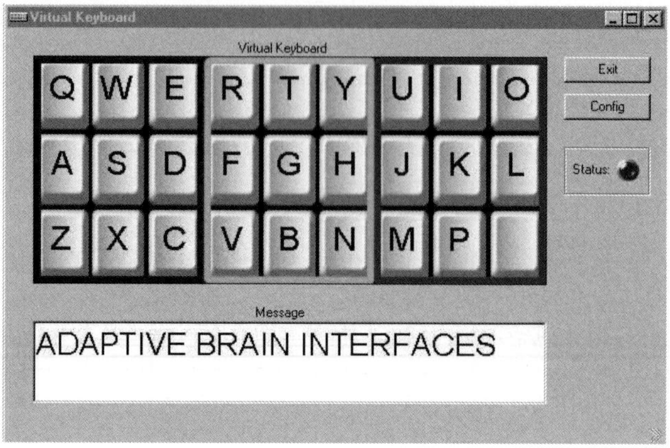

Fig. 11.9. An example of a virtual keyboard controllable through a variety of brain computer interface (BCI) systems.

through the process of biofeedback, a type of operant conditioning, people can control to some degree, the activity of the brain in response to particular tasks (Millner, 1969). This is the essence of the brain–computer interface (BCI), and forms the basis of an exciting area of research that is being applied to biometrics. For instance, users can control the movement of a cursor, type on a virtual keyboard and related activities.

That this technology can be used as an authentication system is receiving serious research efforts and the results appear to be quite promising, even at this early stage in the evolution of this technology. Again, there are the issues of the requisite hardware, which, as in the case for ECG technologies, can be expected to diminish with time. An example of a typical BCI protocol stack is presented in Fig. 11.10.

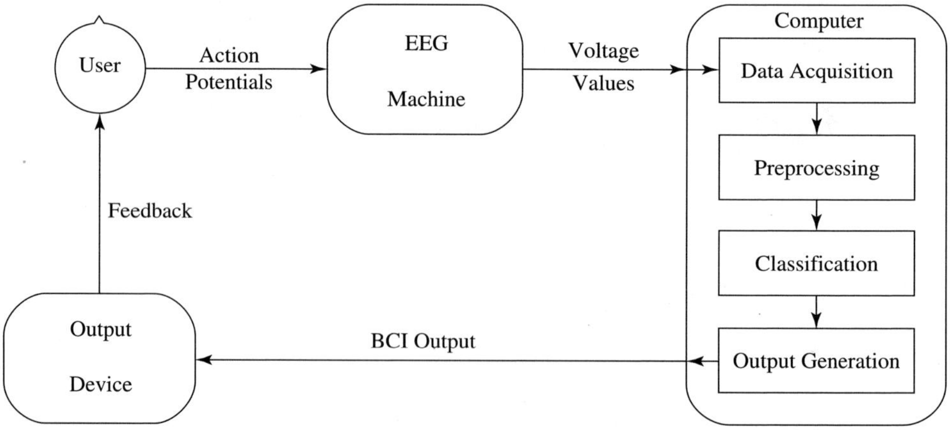

Fig. 11.10. An example of a typical BCI protocol stack, displaying the principal features and the feedback loop. (From Felzer, 2001.)

11.3. The Biometric Process

Virtually all biometric-based authentication systems operate in a standard triage fashion: enrollment, model building and decision logic. This set of processes is depicted in Fig. 11.11. The purpose of enrollment is to acquire data from which the other two modules can be generated. In addition, it serves to incorporate a user into the pool of valid users — which is essential if one wishes to authenticate at a later date. The enrollment process varies little across biometrics modalities with respect to the user's participation: provides samples of data. How much data are required depends on how the biometric operates. Typically, the inherent variability of a biometric modality will have a significant impact on the quality of the data obtained during enrollment. Issues of user acceptability — in terms of the effort to enroll — are a significant constraint and must be taken into account when developing the particular biometric. It is of no use if the system generates 100% classification accuracy if it is too invasive or labor intensive. This is an issue that distinguishes physiological from behavioural biometrics. Physiological biometrics is based on the notion of anatomical constancy, and individual variation. One would expect that in this situation, enrollment would be minimal. For instance, in a fingerprint-based system, once all fingers were recorded, there would be no need to repeat the process 10 times for instance. A fingerprint is a fingerprint? The same may not hold true for behavioural biometrics, where there is an inherent variability in the way the process is repeated. Our signatures are rarely identical — and the irony of it all is that the technology scrutinizes our behaviour at such a low level — that it is bound to find

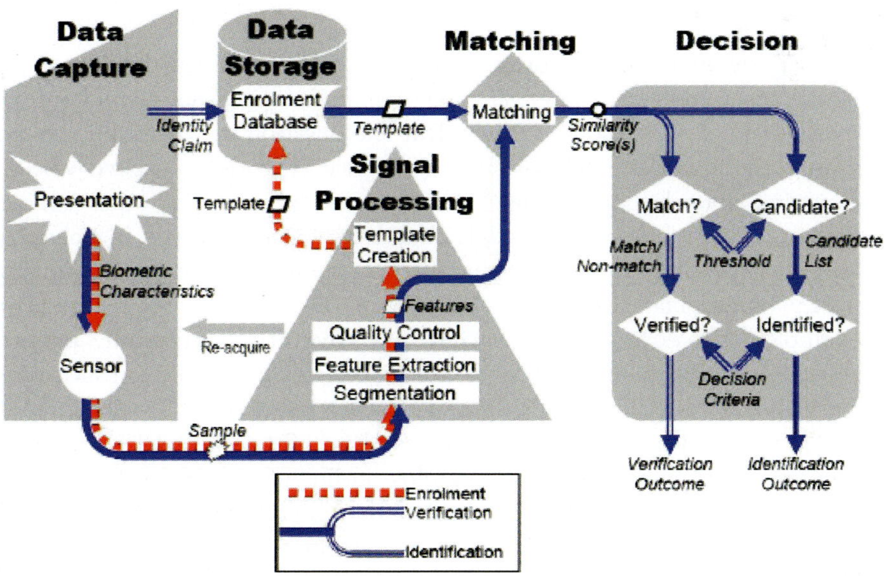

Fig. 11.11. The elements that comprise a complete biometric-based system, suitable for both verification (authentication) and identification.

some variation — even when we as humans, examining two versions of a signature produced by the same individual for example, find no clear differences.

There are two principal classes of features that can be acquired during enrollment, which can be categorized into static and dynamic features. Typically, static features capture the global aspects of the enrollment data. For instance, in signature verification, the static features capture the width/height ratios, and the overall time interval during which the signature was entered. Dynamic features include how the enrollment data change while they are being entered — such as the acceleration or the change in typing speed over time. One could envision that the static data are used as a gross approximation, with the dynamical features added in the event of a borderline decision. This presupposes that the static data are less informative than dynamical data. But at the same time, the issue of constancy might weigh static data more heavily than dynamic data, which tend to be more variable. Finding this balance is a difficult task — as it is not known in advance of the study. Typically, the results of the study are used to weigh the features — and different studies produce varying results — as the conditions are rarely identical between studies. There are also issues of data fusion — how does one incorporate a variety of features, which may operate on different time scales and differing magnitudes? These are important issues that will be discussed in Chapter 7, where multimodal biometrics is addressed.

Once these issues have been resolved, the ultimate result of the enrollment process is the generation of a biometric information record (BIR) for each user of the system. How do we transform the data that are collected during enrollment into a useful model? In part, this is a loaded question. On the one hand, one would assume that a model was available prior to collecting the data. But in reality, a lot of exploratory analysis is performed, where one collects all the data that appear possible to collect, and generate a collection of models, trying each to find out which provides the best classification accuracy. But the question is, where did the model come from in the first place? This is the way science progresses — so we proceed as normal barring any other indication.

There are a vast number of models that have been employed in behavioural biometrics. It is beyond the scope of this text to explore this area, as it would fill a number of volumes. The case studies that occupy the majority of this text provide some examples of a variety of approaches that have been successfully applied in this domain. Assuming that a BIR is created for each successfully enrolled person, a database is created with the BIR data. There are issues here as well. Should the data be encrypted to help reduce the success of an off-line attack? Generally, the answer is yes — and many systems do employ on-line encryption technology.

The decision logic is designed to provide an automated mechanism for deciding whether or not to accept or reject a user's attempt to authenticate. When a user makes a request to authenticate, their details are extracted and compared in some ways to the stored BIR. In order to decide whether to accept or reject the request, a decision process must be invoked in order to decide whether or not to accept the request. Typically, this entails comparing the features extracted from the authentication attempt with the stored BIR. There are a number of similarity metrics that have been employed in this domain. A factor that significantly impacts the matching/scoring process is whether or not the system utilizes a static or dynamic approach. For instance, in keystroke dynamics, one can employ a fixed text or a variable text approach to authentication. For a fixed text approach, a specific set of characters are typed — which can be directly compared with the BIR. This is a much easier decision to make than one based on a more dynamic approach, where the characters entered are contained within a much larger search space of possible characters. Of course, the ease with which the decision can be made is contingent upon the model building component — but none-the-less has a significant impact on the decision login. Given that a decision has been rendered regarding an authentication attempt, how do we categorize the accuracy of the system? What metrics are available to rate various decision models?

In part, this depends on the exact task at hand: is it a verification or identification task? Clearly an authentication task (also known as identification), the goal is to confirm the identity of the individual. This can simplify the match and scoring processes considerably — as it reduces the search task to a 1:1 mapping between the presumed identity and that stored in the database. The verification task is depicted in Fig. 11.12.

The task of identification is considerably more difficult than authentication in most cases. The entire database must be examined — as there is no information that could narrow down the search. As depicted in Fig. 11.13, the two process models are similar — barring the candidate list component, found only in the identification model. Another subtle distinction between these two approaches is depicted by the "adaptation" component present in the verification process model (Fig. 11.12). Adaptation of the BIR is a vitally important feature of a mature and viable biometrics. Take for instance a keystroke dynamics-based authentication system. After the user completes enrollment, and continues entering their password, the typing style might change slightly due to a practice effect or for other reasons.

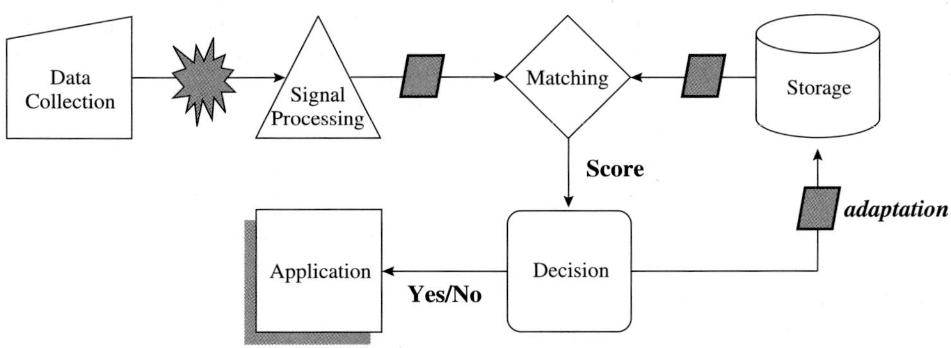

Fig. 11.12. A graphical depiction of the verification process model — indicating the principle elements and their potential interactions.

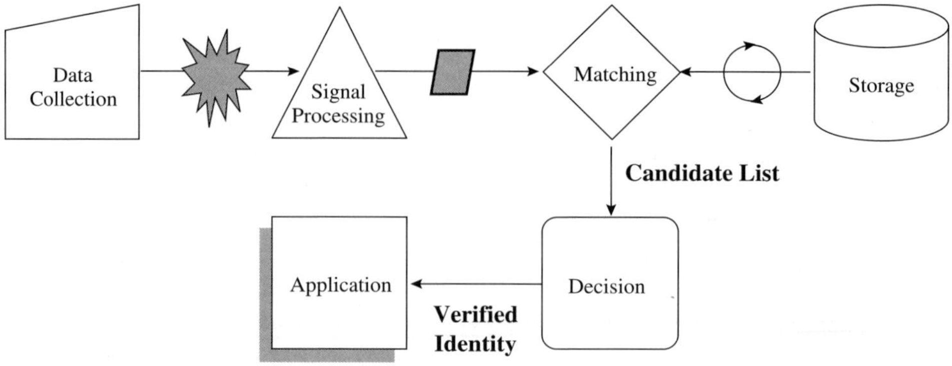

Fig. 11.13. The identification process model — depicting the principle difference between verification and identification — the candidate list element. (See the text for details.)

If the user is continuously matched against the enrollment data, the system may begin to falsely reject the genuine user. To prevent such an occurrence, the user's BIR must be updated. How the user's BIR evolves over time is an implementation issue. We tend to keep a rolling tally of the latest 10 successful login attempts, updating any statistical metrics everytime the user is successfully authenticated. This is possible in a verification task — or at least it is easier to implement. In an identification task — the issue is how does the system actually confirm that the identification process was successful? The system must only update the BIR once it has been successfully accessed — and this cannot be known without some ancillary mechanism in place to identify the user — sort of a catch-22 scenario. Therefore adaptation most easily fits into the authentication/verification scheme, as depicted in Fig. 11.13.

11.4. Validation Issues

In order to compare different implementations of any biometric, a measure of success and failure must be available in order to benchmark different implementations. Traditionally, within the biometrics literature, Type I (false rejection rate FAR) and type II (false acceptance rate FAR) errors are used as a measure of success. Figure 11.2 illustrates the relationship between FAR, FRR and the equal error rate (EER), which is the intersection of FAR and FRR when co-plotted. Note that some authors prefer to use the term crossover error rate (CER) as opposed to the EER — but they refer to identical concepts. When reading the literature, one will often find that instead of FAR/FRR, researchers report FAR (False Acceptance Rate) and IPR (the Imposter Pass Rate). The confusion is that this version of FAR is what most authors' term FRR, and the IPR is the common FRR. Another common metric prevalent in the physiological literature is the FMR (False Matching-Rate) and FNMR (False Non Matching-Rate). The FMR is used as an alternative to FAR (False Rejection Rate). Its use is intended to avoid confusion in applications that

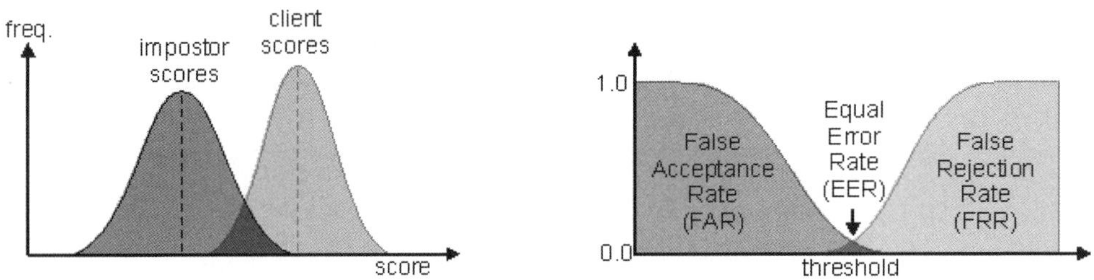

Fig. 11.14. When the FAR and FRR are plotted on the same graph, as a function of a classification parameter, the intersection of the two functions is termed the EER (equal error rate) or CER (cross over error rate).

reject the claimant (i.e. an imposter) if their biometric data matche that of an enrollee. The same caveat applies to FNMR as well.

A common result reported in the literature is the interdependence between the FAR and FRR (Fig. 11.14). Most studies report that one cannot manipulate one of the metrics without producing an inverse effect on the other. Some systems can produce a very low FAR — but this generally means that the system is extremely sensitive and the legitimate user will fail to authenticate (FRR) at an unacceptable level. From a user perspective, this is very undesirable — and from the corporate perspective, this can be quite expensive. If a user fails to authenticate, then their account is usually changed — and hence the user will have to re-enroll into the system. In addition, the help-desk support staff will be impacted negatively in proportion to the user support required to reset the users' account details. On the other hand, when the FRR is reduced to acceptable levels — then the FAR rises — which tends to increase the level of security breeches to unacceptable levels. Currently, there is no direct solution to this problem. One possible approach is to use a multi-modal biometric system, employing several technologies. This approach does not solve the FAR/FRR interdependency, but compensates for the effect by relaxing the stringency of each component biometric such that both FAR and FRR are reduced to acceptable levels without placing an undue burden on the user. The use of a multi-modal approach is a very active research area and will be discussed in some detail in Chapter 7.

In addition to FAR/FRR and their variants, it is surprising that the concepts of PPV and NPV, along with the concepts of sensitivity and specificity, often reported in the classification literature. PPV is the Positive Predictive Value and the NPV Negative Predictive Value of a classification result. The PPV provides the probability that a positive result is actually a true positive (that is a measure of correct classification). The PNV provides the probability that a negative result will reflect a true negative result. From a confusion matrix (sometimes referred to as a contingency matrix), one can calculate the PPV, NPV, sensitivity, specificity and classification accuracy in a straight-forward fashion, as displayed in Table 11.1.

Table 11.1. A sample confusion matrix for a two-class decision system. Note the TN, true negative, FP, false positive; FN, false negative; TP, true positive.

	Negative	Positive	
Negative	190 (TN)	10 (FP)	Specificity
Positive	10 (FN)	190 (TP)	Sensitivity
	NPV	PPV	Accuracy

The values for PPV, NPV, sensitivity, specificity, and overall accuracy can be calculated according to the following formulae (using the data in the confusion matrix):

$$\text{Sensitivity} = \text{TP}/(\text{FN} + \text{TP})$$

$$\text{Specificity} = \text{TN}/(\text{TN} + \text{FP})$$

$$\text{PPV} = \text{TP}/(\text{TP} + \text{FP})$$

$$\text{NPV} = \text{TN}/(\text{TN} + \text{FN})$$

$$\text{Accuracy} = (\text{TN} + \text{TP})/(\text{TN} + \text{FP} + \text{FN} + \text{TP}).$$

Furthermore, the use of specificity and sensitivity can be used to produce an ROC curve (see Fig. 11.15). The ROC curve displays the interplay between sensitivity and specificity — it quantifies the relationship between FAR/FRR, in

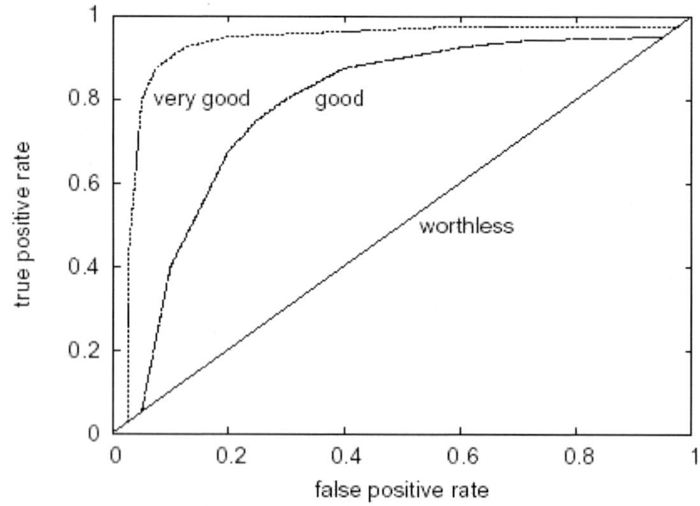

Fig. 11.15. An example of a Receiver Operator Characteristic curve, which displays the relationship between specificity and sensitivity (the x-axis is the false positive rate), and the y-axis is the true positive rate. The closer the curve approaches the y-axis, the better the result. Typically, one calculates the area under the curve to generate a scalar measure of the classification accuracy.

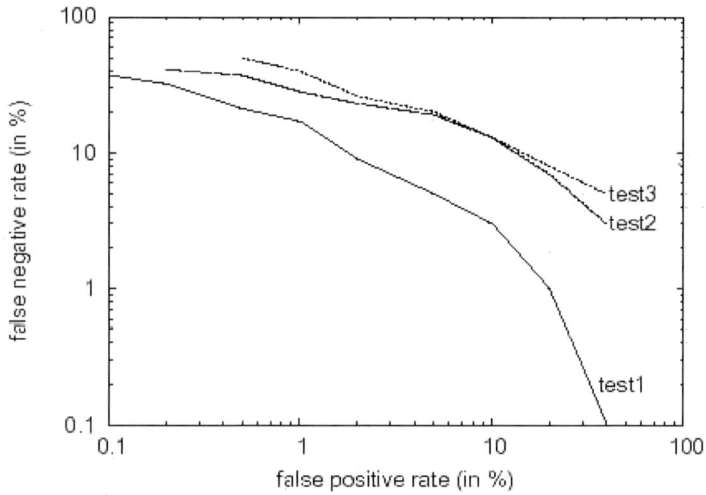

Fig. 11.16. An example of a DET curve (using the same data used to plot the ROC curve in Fig. 11.15).

a form that is more quantitative than a simple EER/CER plot. In addition, the likelihood ratio (LR) can be obtained simply by measuring the slope of the tangent line at some cutoff value. These measurements are very useful in assessing the quality of a classification result. They are used quite frequently in the datamining literature and related fields, but for some reason have not found a place in the biometrics literature.

In addition to the above metrics, the detection error tradeoff (DET) curve may be reported (see Fig. 11.16 for an example of a DET curve). To generate a DET curve, one plots the FAR (x-axis) against the FAR (y-axis). Typically, this plot yields a reasonably straight line, and provides uniform treatment to both classes of errors. In addition, by the selection of judicious scaling factors, one can examine the behaviour of the errors more closely than the ROC. Of course, the FAR/FRR values are obtained as a function of some threshold parameter. With a complete system in hand, we can now address the important issue of biometric databases — valuable sources of information that can be used to test our particular biometric implementation.

11.5. Conclusions

The chapter has highlighted some of the major issues involved in behavioural biometrics. A summary of the principal behavioural biometrics was presented (though the coverage was not exhaustive) — highlighting the principal techniques. The focus of this book is on a remote access approach to biometrics — and as such there is an implicit constraint that a minimal amount of hardware is required to deploy the system. One will note that in the list of behavioral biometrics, ECG,

EEG and gait were added. These approaches require some additional hardware over and above what is typically supplied with a standard PC. Their inclusion was to set the background for Chapter 8, which discusses the future of behavioural biometrics. Therefore, it should be noted that these technologies may not fall under our current working definition of a remote access approach — which can be defined as "a technique for authenticating an individual who is requesting authentication on a machine, which is distinct from the server which performs the authentication process". But if behavioural biometrics is to expand its horizons, we may have to consider other options from traditional ones such as voice, signature and keystroke interactions. Who knows what the future of technology will bring to us — which might make these possibilities and others a feasible option in the near future.

It is hoped that this text will highlight some of the advances of behavioural biometrics into the foreground — by highlighting some of the success stories (through case study analysis) that warrant a second look at this approach to biometrics. There are a variety of techniques that have been attempted — each very creative and imaginative, and based on solid computational approaches. In the final analysis, this author believes that behavioural biometrics — either alone or in conjunction with physiological biometrics — either is standard reality or in virtual reality — can provide the required security to enable users to feel confident that their space on a computer system is fully trustworthy.

References

1. A. A. E. Ahmed and I. Traore, A new biometrics technology based on mouse dynamics, Technical Report ECE-03-5, Department of Electrical and Computer Engineering, University of Victoria, P.O. Box 3055 STN CSC Victoria, B.C. V8W 3P6 Canada, 2003.
2. J. Coetzer, B. M. Herbst and J. A. du Preez, Offline signature verification using the discrete radon transform and a hidden Markov model, *EURASIP Journal on Applied Signal Processing* **4** (2004) 559–571.
3. T. D. Ganchev, Speaker recognition, PhD Thesis, University of Patras, Greece, 2005.
4. G. Forsen, M. Nelson and R. Staron, Personal attributes authentication techniques, ed. A. F. B. Griffin, *Rome Air Development Center Report RADC-TR-77-1033* (RADC, New York, 1977).
5. D. Gafurov, K. Helkala and T. Sondrol, Biometric gait authentication using accelerometer sensor, *Journal of Computers* **1**(7) (2006) 51–59 .
6. J. Gardner, Detection of vapors and odors from a multi-sensor array using pattern recognition, Part 1: Principle component and cluster analysis, *Sens. Actuator B* **4** (1991) 109–115.
7. P. Keller, *Overview of Electronic Nose Algorithms*, International Joint Conference of Neural Networks (IJCNN'99), Washington, USA, 1999.
8. Z. Korotkaya, Biometric person authentication: Odor, *Advanced Topics in Information Processing*, Lappeenranta University of Technology, Finland, 2003.
9. A. Martin, G. Doddington, T. Kamm, M. Ordowski and M. Przybocki, The DET curve in assessment of detection task performance. In Proceedings of EuroSpeech-97 (1997), pp. 1895–1899.

10. S. S. Mehta and N. S. Lingayat, Comparative study of QRS detection in single lead and 12-lead ECG based on entropy and combined entropy criteria using support vector machine, *Journal of Theoretical and Applied Information Technology* (2007) 8–18.
11. M. S. Nixon and J. N. Carter, On gait as a biometric: Progress and prospects, EUSIPCO, (2004), pp. 1401–1404,
12. R. Palaniappan, Multiple mental thought parametric classification: A new approach for individual identification, *International Journal of Signal Processing* **2**(1) (2005) 222–225.
13. L. R. Rabiner, A tutorial on hidden markov models and selected applications in speech recognition, *Proceeding of the IEEE* **77**(2) (1989) 257–286.
14. K. Revett, S. Magalhaes and H. Santos, Developing a keystroke dynamics based agent using rough sets, The 2005 IEEE/WIC/ACM International Joint Conference on Web Intelligence and Intelligent Agent Technology Compiegne, 2005, pp. 56–61.
15. K. Revett, I. Jokisz and H. Jahankhani, The effect of keyboard partitioning on keystroke dynamics based user authentication, ICGeS2007, Int'l Conference on Global e Security, UeL 18–20 April, 2007, pp. 292–297.
16. A. Teo, H. Garg and S. Puthusserypady, Detection of humans buried in rubble: An electronic nose to detect human body odor, Proceedings of the IEEE 2nd Joint EMBS-BMES Conference, Houston, TX, USA (2002), pp. 1811–1812.
17. J. Thorpe and P. C. Oorschot, Graphical dictionaries and the memorable space of graphical passwords, *Proceedings of the 13th USENIX Security Symposium*, San Deigo, USA: USENIX, 2004.
18. T. Wark, V. Sridharan, and V. Chandran, The use of temporal speech and lip information for mutli-modal speaker identification via multi-stream HMM, Proceedings of the International Conference on Acoustics, Speech, and Signal Processing, Istanbul (2000), pp. 2389–2392.
19. S. Wiedenbeck, J. Waters, J. C. Birget, A. Brodskiy and N. Memon, Authentication using graphical passwords: Basic results, *Human–ComputerInteraction International (HCII 2005)*, Las Vegas, NV, 2005.
20. A. Yamazaki, T. Ludermir and M. de Suonto, Classification vintages of wine by an artificial nose using time delay neural networks, *IEE Electronics Letters* **37**(24) (2001) 1466–1467.

Chapter 12

INFORMATION SECURITY MANAGEMENT AND STANDARDS OF BEST PRACTICE

THEO TRYFONAS
Safety Systems Research Centre, University of Bristol,
Bristol BS8 1TR, United Kingdom

Information (computer, intellectual property or on paper) is critical now more than ever within the business environment as businesses have expanded their services and have become more reliant on information technology (IT). As such, Information Security and its management is a critical element of the governance process within organisations. Information security has traditionally being viewed by management in terms of protecting information communication and technology systems and implementation of access controls. However due to a number of high profile organisational scandals such as the one of Enron and the legislation that was put in place to confront those (e.g., Sarbanes–Oxley Act in the United States), the Boards of Directors and Executive Management are compelled to examine their information security controls, ensuring that they are in place, adequate and operating effectively. Researches and professionals agree that information security is a multi-disciplinary, multi-departmental and multi-organisational issue which must get top managements priority. Information Security is no longer a dedicated function of the IT department but a business function of within the organisation which supports the viability of the organisation. In this respect, information security management is a critical component of organisational life. In this chapter we firstly review its core principles and subsequently explore some of the most influential international standards currently available to both business and technical managers as good practice for information security management implementation.

12.1. Principles of Information Security

Regardless of the sector in which they operate, many organisations are responsible for running increasingly diverse, complex and business critical information technology (IT) information systems and applications. They are also under increasing pressure to cut IT related costs, generate additional revenue, comply with appropriate legislative and regulatory requirements and deliver competitive advantage, while improving the speed and cost effectiveness of the service they provide to their clients. A single security breach, theft, unauthorised access or virus attacks can result in serious financial and reputational damages.

As organisations become more dependent on IT, there is an increased need for robust IT controls to ensure that the IT architecture is suitable for the business, the IT investment can be justified, the operational processes are optimal and that the IT strategy is clearly aligned with the business strategy. To provide this effective oversight it is essential that boards of directors and senior management understand their evolving roles in the governance of IT and apply the same level of management attention that they give to all major organisational assets. This can not be done on a piecemeal basis, rather it must be part of a formally prescribed IT governance framework appropriately supported and funded by management.

Business and IT governance, once considered separate, are now closely linked. Often IT was under the sole control of the IT Department and operated in isolation from other business functions with only occasional progress reports on specific issues. A consequence of this was often serious cost escalation, poor performance and dissatisfied users. However, various pieces of legislation, such as Sarbanes–Oxley, have meant that an effective IT governance structure is not just desirable; senior management have to sign off on assurances and they face financial and legal penalties if their organisation is not adequately controlled and secured. Information security is amongst others at the heart of IT governance and its management will contribute essentially into the successful steering and alignment of IT with business and organisational objectives.

The IBM Dictionary of Computing [39], states Information Security are "... concepts, techniques, technical measures and administrative measures used to protect information assets from deliberated or inadvertent authorised acquisition, damage, disclosure, manipulation, modification, loss or use". This definition speaks to IS as managerial function with an associated set of beliefs and practices that influences the status of security within an organisation.

Grobler [23] (p. 16) provides a broader definition of Information Security, "... the process of protecting information, systems and hardware, that use, store and transmit the information, from a wide range of threats in order to ensure business continuity, minimise business damage and maximise the return on investment by preserving confidentiality, integrity and availability of information and information assets". From Grobler's definition IS can be viewed from three different perspectives, these are (1) to protect the business from different threats in order to minimise the impact of threats, (2) to maximise return on investment and (3) to protect information assets.

Information Security is defined by Dhillon [15] as "the protection of information against unauthorised disclosure, transfer, modification, or destruction, whether accidental or intentional". BS7799/ISO17799 defines information security as "the preservation of confidentiality, integrity and availability of information". Raval and Fichadia [48] simply defined information security as "the protection of information assets". The information and the IT systems and networks that support it are important business assets. Their availability, integrity and confidentiality may be essential to maintain a competitive edge, profitability, cash-flow and respected organisation image. In general, security involves the protection against attack, failure or the occurrence of any other unwanted event. Security management has a set of defined objectives. These objectives form the basis for implementing security in any organisation. The objectives are discussed in the next section.

Security objectives define the end goals, or purpose of a security system [48]. There are five objectives of security management; these includes; Confidentiality, Integrity Availability, Authentication and Non-repudiation [15]. However, the widely agreed objectives in terms of definitions are; confidentiality, integrity and availability. These are sometimes called the "CIA" of information security [14].

12.1.1. *Information Security Objectives*

Confidentiality: This can be defined as the prevention of unauthorised disclosure of information [22]. Denning [14] stated that Confidentiality is the characteristic of information being disclosed or made available only to authorised entities at authorised times and in the approved manner. Confidential information means information that is "private or secret, carried out or revealed in the expectations that anything done or revealed will be kept private for a select group not available to the public" [6]. Confidentiality is sometimes called "secrecy" or "privacy" where privacy is the protection of personal data and secrecy is the protection of an organisations' data [45]. This ensures that computer-related assets are accessed only by authorised parties, which means that only those who should have access to something will actually get that access. Data may be encrypted to preserve confidentiality.

Integrity: Gollmann [22] defined Integrity as the prevention of unauthorised modification of information. This means that assets should be modified only by authorised parties or only in authorised ways [45]. Welke and Mayfield [49] recognise three particular aspects of integrity as, authorised actions, separation and protection of resources and error detection and correction. NCSC [42] clarifies by stating that integrity ensures that computerised data are the same as those in source documents that is; they have not been exposed to accidental or malicious alterations or destruction. Thus, there is some reasonable assurance that the information is accurate, can be relied upon to be factual and not modified or otherwise changed without going through a formal process to ensure integrity is maintained [6]. Krause and Tipton [33] are of the opinion that to produce information with high integrity, the entire system needs to function reliably. This means that inputs should be accurate and complete and only authorised transactions should be captured. Processing logic must be accurate and should process only in the manner specified and documented. Data must be stored in a secured manner so that no unauthorised changes are made and no loss of data occurs. Thus, information integrity is built into an information system through a collection of numerous measures working together [11].

Availability: This means that one is assured, with reasonable confidence and certainty, that the information and the information systems are always available when needed [6]. It can be said that a data item, service, or system is available if there is a timely response to requests, there is a fair allocation of resources, service or system can be used easily and in a way it was intended to be used and there is controlled concurrency, that is; support for simultaneous access, deadlock management and exclusive access as required [45]. Raval and Fichadia [48] interpreted systems availability as "the state of readiness of systems so that authorised users can access and use the system for their purposes and during expected times of operation". The very popular denial-of service attacks are to a large extent a consequence of this security requirement not adequately addressed. Information assets become unavailable due to intentional attacks or unintentional errors, therefore, it is necessary to monitor information assets to protect them from compromises on account of errors and attacks [8].

Authentication: This is the process of verifying the identity of a communicator [15]. The security requirement for authentication becomes important in the context of networked organisations. Authentication assures that the message is from a source it claims to be from. In today's net-centric world, a person can transact from virtually any place around the world. It is essential for businesses to verify if the person is who he claims to be. Raval and Fichadia [48] stated that proving the identity of a user is necessary for two reasons; firstly, to allow access to information assets to those authorised and secondly, to hold the person accountable for the act.

Non-repudiation: The importance of non repudiation as a security objective came about due to increased reliance on electronic communications and maintaining legality of certain types of electronic documents [8]. This led to the use of digital signatures, which allow a message to be authenticated for its content and origin [8]. Non-repudiation has been defined as a property achieved through cryptographic methods, which prevents an individual or entity from denying having performed a particular action related to data [45]. For instance, non repudiation in a digital signature scheme prevents person A from signing a message and sending it to person B, but later denies it was not him who signed it after all. Caelli *et al.* [8] stated that the core requirement for non-repudiation is that persons A and B have a prior agreement that B can rely on digitally signed messages by A, until A notifies B otherwise.

12.1.2. *Sources of Security Risks, Information Security Breaches and Associated Controls*

A threat is defined as a source factor that could result in a potential violation of security objectives, e.g., confidentiality of information. There are many types of electronic threats which an organisation faces daily which management need to be informed and aware of to implement appropriate security controls for organisational security. The types of threats to information security commonly reported include [18, 47]:

- Electronic theft, sabotage or wrongful disclosure of data or information
- Fraud
- Personal profit from computing resources
- Installation and use of unauthorised software, hardware and peripherals
- Illegal or illicit use of resources
- Abuse of computer access controls
- Physical theft, sabotage, or intentional destruction of computing equipment

It is important to note that no matter the source of the threat, it is the responsibility of those charged to direct the organisation to ensure that the threat is minimised vulnerabilities and implementing security controls. A control being defined as an action, device, procedure or technique that removes or reduces a vulnerability [45] (p. 7).

As can be identified from the threats, Information Security is only part of the bigger security picture of an organisation. Information security is a problem within the entire organisational structure. Hence it is no longer acceptable to view security as a technological solution, as threats may materialise from both external and internal sources. These threats may have a significant impact on the organisation [35]:

- Brand image and reputation
- Customer loyalty and retention
- Employee motivation and satisfaction
- Breach of legal or regulatory controls

Over 150 million United States customer data records have been compromised since January 2005 until July 2007, according to Privacy Rights Clearinghouse [46]. These security breaches have occurred over a cross-section of organisations from governmental agencies and universities to financial and healthcare institutions.

In the 2006 DTI *Information Security Breaches Survey* [18], which surveyed 1001 UK businesses, reported 62% had a security incident within the last year. The survey further states that the medium number of incidents suffered by respondent was roughly eight a year. Looking at the UK public sector specifically, the 2006 Society for IT Managers' (SOCITM) *IT Trends in Local Government* survey reported that only 4% of the responding local authorities reported identification of a breach of security [54].

The effect of incidents on the organisation can be devastating, beyond the effects of breaches to the customer privacy. The potential cost of security breach to an organisation can be direct or indirect. The 2006 DTI survey [18] reported that the average cost of a UK company's worst security incident of the year was roughly £ 12,000 (up from £ 10,000 two years ago). This is more when compared to the

2006 SOCITM survey reported average cost for local authorities was approximately £ 500. However the highest cost of dealing with a security incident was £ 25,000 for one local authority [54].

From these survey statistics, it can be seen that organisations are still failing to appreciate the impact of poor information security governance has on their internal and external customers. LogicaCMG [35] asserts, based on their own research, that "[organisations] underestimate the importance of information security governance and the impact of security breaches ...".

In the context of this discussion and after the definitions of threat and impact have been discussed we also need to formally define the term of "risk" that we have liberally used so far. Risk has been defined differently by different authors, though all such definitions refer to a key issue, that is, the presence of a threat to the system. Down *et al.* [17] (pp. 4, 5), for example, provides a general definition, referring to risk as "a set of circumstances that makes us nervous because we perceive the possibility of an undesirable outcome. When 'possibility' becomes 'probability', then 'a risk' becomes 'a high risk' and makes us feel very uncomfortable indeed". Carroll [10] defines risk as follows: "Risk is the probability that a threat agent (cause) will exploit a system vulnerability (weakness) and thereby create an effect detrimental to the system". Nosworthy [43] provide the following definition: "A risk represents the likelihood of a threat happening/causing a problem". Martin [38], on the other hand, refers to risk as follows: "The term risk is used to describe the possibility of a threat taking advantage of an asset's vulnerability". Pfleeger [45] defines a risk as "an unwanted event that has negative consequences". Walker [57] identifies four main general areas of risk: Strategic; Market, Credit; and Operational risk.

Dhillon [15] regards information system security at three levels of required control of technical, formal and informal nature respectively. Security policies strive to implement controls within these three levels. We will refer particularly to the role of security policies and their development within an organisational environment later in this chapter.

Technical Controls. At a technical level, the intent of security management is to secure the hardware, software and the data that resides in computer systems. Layton [34] pointed out the fact that an organisation needs to ensure that the hardware, software and data is not modified, destroyed, disclosed, intercepted, interrupted, or fabricated. Modification is said to occur when the data held in computer systems is accessed in an unauthorised manner and is changed without requisite permission [58]. Destruction is simply when the hardware, software, or data is destroyed due to malicious intent. Disclosure of data takes place when data is made available or access to software is made available without due consent of the individual responsible for the data or software. Unauthorised disclosure creates a serious impact on maintaining security and privacy of systems; however, this could be managed by instituting proper program and software controls. Interception is when an unauthorised person or software gains access to data or computer resources.

This could result to copying of programs, data, or other confidential information. Interception occurs when a computer system becomes unavailable for use. This may be a consequence of malicious damage of computing hardware, erasure of software, or malfunctioning of an operating system. This is sometimes referred to as denial of service. Finally, fabrication is said too occur when false communications are inserted into a network or records are added to an existing database.

Today's businesses are eager to grasp the idea of implementing complex technological controls to protect the information held in their computer systems. The focus of these controls is in the area of access control and authentication. However, before implementing technological controls, businesses should consider constituting a well-thought out baseline organisational controls like allocating responsibilities, awareness, etc.

Formal Controls. This form of control is a rule-based approach used to determine the consequences of misinterpretation of data and misapplication of rules in an organisation and also help in allocating specific responsibilities. In any organisation, development of the task force helps in carrying out security management and gives a strategic direction to various initiatives. An ongoing support is expected from the security managers in order to monitor these controls. Formal controls may address the hiring procedures during employment and the structures of responsibilities within organisational departments. A clearer understanding of the structures of responsibilities helps in attribution of blame, responsibility, accountability and authority.

Informal Controls. Increased awareness and an ongoing training programme is the most cost-effective control an organisation can consider. Training and awareness programs are extremely important in developing trusted employees. However, a focus on developing a security culture goes a long way in developing and sustaining a secure environment. The aim of the organisational subculture is to understand the intentions of management and to make members of the organisation committed to their activities. All this is possible by adopting good management practices. Forcht [19] is of the opinion that the first step in developing good management practices and reducing the risk of a security breach is by adopting some baseline standards. Later in this chapter we review two key approaches for the implementation of information security management, consolidated in the form of international standards.

12.2. Information Security Management and Its Components

Risk management, as defined by Blyth and Kovachic [6], "is the process of assessing risk, taking steps to reduce risk to an acceptable level and maintaining that level of risk" (p. 47). They also state that risks are analysed by managers for a number of aspects of their consideration, and that managers consider alternatives and implement plans to capitalise on returns on their investments. They add that a risk management process for information systems makes it possible for managers and their organisations to build an in-depth knowledge regarding their systems and how they are interrelated.

Blyth and Kovachic [6] (p. 47) identify five principles of risk management, including the following:

- Assess risk and determine needs.
- Establish a central management focus.
- Implement appropriate policies and related controls.
- Promote awareness.
- Monitor and evaluate policy and control effectiveness.

They also maintain that the successful organisation apply such principles through linking them into a cycle of activity which helps the organisation address risks in a continuous process. After senior executives assess the risks to their business operations, the organisation should:

- Establish policies and selected controls.
- Increase awareness of users to the policies and controls.
- Monitor the effectiveness of the policies and control.
- Use the results to determine if modifications of policies and controls are needed.

According to Barefoot and Maxwell [4], the primary purpose of a security manager is to lead the corporation in the protection of its assets by ensuring that all security functions are performed in a manner that is cost-effective, meets employee, corporate and customer's need and complies with the corporate policies, laws and regulations. Blount [5] stated that effective security management is based on a comprehensive and integrated strategy that includes three major components; Identity and Access Management (IAM), Security Information Management (SIM) and Integrated Threat Management (ITM). A complete security management solution must provide capabilities in each of these critical areas as stated by Raval and Fichadia [48]. These are discussed in details below.

Identity and access management (IAM): This area of control deals with creating and managing user identities, their accounts and access entitlements and enforcement of access policies across the environment. Today's businesses struggle with increasing numbers of applications and resources, coupled with the large number of users both within and outside the organisation. User's identities and their level of access to applications and data are at the core of most businesses and must be effectively managed. Identity and access management control helps to manage user identities and their access to critical IT resources. This provides an improved security for systems, applications, processes and data.

Security information management (SIM): One of the greatest challenges in an organisations security system is managing the flood of alerts generated by a growing number of multi vendor security devices and systems. Automation is required to isolate and prioritise the real security threats and the key to this automation is a security information management solution. A security information management system aggregates, filters and provides reports and analysis of all security-related events within an environment [56].

Integrated threat management (ITM): This approach allows for safe management and protection of electronic systems against a multitude of threats. These threats could be in the form of viruses, spyware, adware, hackers and spam publishers, etc. The ITM identifies and combats these electronic threats.

12.2.1. *Risk Measurement and Analysis*

Risk management comprises two stages, as illustrated in Fig. 12.1. Risk assessment involves identification and estimation of the risk (risk analysis) and comparing it against a given risk criteria to determine whether the risk is acceptable or not (risk evaluation). If the risk deemed acceptable was to be treated, there are measures to be implemented to modify the risk to an acceptable level (risk treatment).

The concept of risk assessment is a vital process for the development of appropriate defences [9], or the development of any product or application. Information risk assessment provides the means by which systems risks can be identified and assessed in order to select appropriate safeguards to implement [10]. In fact, the risk assessment process is a key component of security policy development [9]. Canavan [9] maintains that: "It is important to go through a risk assessment process to determined what you want to protect, why you want to protect it and from what you need to protect it". (p. 244). He also identifies six steps which are associated with risk assessment:

1. Identifying and prioritising assets
2. Identifying vulnerabilities
3. Identifying threats and their possibilities
4. Identifying countermeasures
5. Developing a cost-benefit analysis
6. Developing security policies

Pfleeger [45] maintains that security planning begins with risk analysis. He adds that risk analysis is a process to determine the exposures and their potential harm,

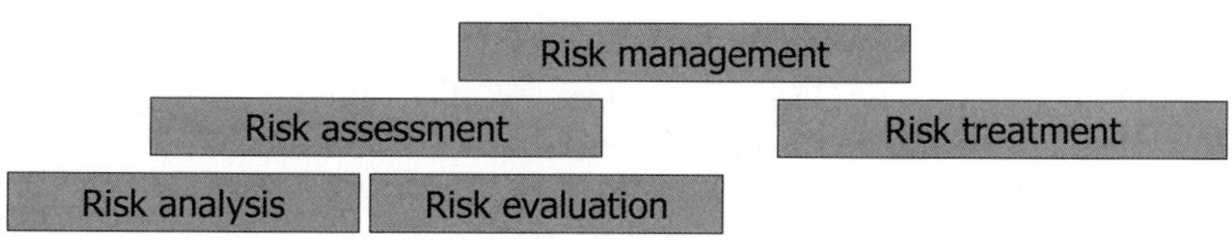

Fig. 12.1. Risk management task according to ISO 27001 [28].

maintaining the first of all, all exposures of a computing system are listed, followed by listing possible controls and their costs for each exposure. The last step of the analysis is a cost–benefit analysis; whether it is cheaper to implement a control or to accept the expected costs of the loss.

There are certain benefits of risk analysis. Some of these benefits are explained below [45]:

1. Improve awareness. Discussion of topics relating to security can boost the overall level of interest and concern among the staff.
2. Identify assets, vulnerabilities, and controls. Some firms are not aware of their computing assets and the vulnerabilities associated with those assets. Identifying assets and the determination of vulnerabilities, as will be explained later in this chapter, comprise the first two steps of risk analysis.
3. Improve basis for decisions. Controls may have a negative impact on productivity through increased overheads and inconvenience to users. Certain controls cannot be rationalised from the viewpoint of protection they provide, and some risks are so serious that they justify a continuing search for more effective controls. In both cases, the gravity of the risk affects the desirability of controls.
4. Justify expenditure for security. A number of security mechanisms are rather expensive without an obvious benefit, as it is difficult to determine the effectiveness of controls in the generic absence of incidents. In this respect a risk analysis can help to identify occasions which merit the cost of a major security mechanism, adding that it is frequently useful to identify the much larger risks from not spending for security.

Risk analysis, is a management process, adapted from similar practices in the filed of management. There are certain steps that need to be undertaken to analyse the security risks in a computing system, and different methods may refer to those in different terms, however in general there would be [45]:

1. Identification of assets.
2. Determination of vulnerabilities.
3. Threat consideration of estimation of likelihood of exploitation.
4. Computation of expected annual loss or other exposure indicator (level of risk).
5. Identify applicable controls and their costs.
6. Select appropriate controls for the environment under analysis.
7. Projecting annual savings of control.

In terms of identifying assets of a computing system, those can be grouped into categories, such as Hardware, Software, Data, People, Documentation, Supplies and logistics, etc. The first three of which are part of the computing system, whereas the remaining may not strictly speaking be a part of it, however they are important for its proper operation. This detailed list of assets forms a major step to start a risk analysis, and it is in fact an inventory of the system.

The next step of risk analysis is to determine the vulnerabilities of the assets mentioned above, which is a step that requires subject expertise, previous knowledge and imagination so as to predict what impacts or damage might occur to the assets and from what sources.

The third step is to determine how often each vulnerability could be exploited. There are different ways by which the likelihood of an event can be estimated, for example commonly may be used the following:

- Quantitatively, by referring for example to probability or possibility based on historical observation of incidents, either for a particular system if available, or for a more generic population.
- Qualitatively, such as for example in the Delphi approach, a technique in which several experts individually express their opinion and estimate the likelihood of an event.

Based on the previous information it may be subsequently possible to attempt to estimate the expected cost of each incident or a similar metric of *risk*. However, like the likelihood of occurrence, this value is also difficult to determine.

Thus those two steps largely differentiate the various approaches of risk analysis, as most would include an initial step of asset identification/inventorying and a subsequent step of identification of vulnerabilities and exposures of those assets. The adoption of a qualitative vs. a quantitative method to compute the level of risk results in the different approaches of risk assessment as discussed below.

Quantitative Approach. Quantitative risk analysis is a formal approach to understanding and analysing risks. It employs the probability of an event occurring and the likely loss should it occur [7]. It makes use of a single figure derived as a product of these elements, referred to as the Annual Loss Expectancy (ALE). Based on this it is theoretically possible to rank events in order of anticipated risk (i.e., the calculated ALE) and to make decisions based upon this.

Criticism of this approach is often associated with the potential unavailability, unreliability or inaccuracy of input data. Rarely can probability calculations be precise and, in some cases, a false sense of security could promote complacency. Furthermore, as events themselves are frequently interrelated the real level of risk could be different than the one estimated. In spite of the drawbacks however, a number of organisations have successfully adopted quantitative risk analysis [7].

Qualitative Approach. Qualitative approaches do not depend on the calculation of probabilities, as a subjective estimation of potential loss is derived instead. Parameter values are expressed using textual terms expressing intensity, for instances scales graded along a "high", "medium" and "low" spine. This approach takes into consideration significant subject knowledge and the judgment of the expert conducting the analysis.

The majority of qualitative risk analysis methodologies draw upon a number of interrelated elements: Threats, vulnerabilities, impacts and controls, as defined earlier in this chapter. A classic example of this approach is the CCTA Risk Analysis and Management Method, CRAMM [53].

Another possible view of risk analysis approaches is distinguishing them by whether they rely on previous knowledge for similar systems and cases, or each case of analysis is treated independently and with respect to its own particularities. In essence we can therefore distinguish between knowledge-based or model-driven approaches. In reality, many methods would use both models of the system under review and apply existing knowledge with regard to this type of system and its known and experienced security requirements.

Knowledge-Based Approach. This approach is based on reusing best practice and accumulated experience. It was particular popular in the past, where the number of assets and their vulnerabilities where limited, processing was usually geographically constrained and great emphasis was given to access control. A classic instrument of this approach is a security checklist, containing possibly typologies of assets, enumerating vulnerabilities and pinpointing to applicable security controls.

Model-Driven Approach. Not necessarily distinct from the previous two approaches, this approach employs systems modelling, usually object-oriented, to represent a system and analyse its relevant risks. Systems modelling can be also used in qualitative risk analysis (e.g., CRAMM), although the modelling follows usually a more traditional functional decomposition modelling approach. An approach which utilises object orientation is CORAS [16]. CORAS brings together aspects of various tools and methods and it provides detailed recommendations for the use of modelling techniques with the UML modelling language.

Whatever the approach used, systems modelling in risk analysis serves in general three key purposes:

- Describes the assessed system at an appropriate level of abstraction.
- Acts as a medium for communication, discourse and interaction between different groups of stakeholders involved in the risk assessment (experts, end-users, managers, customers, etc.).
- Provides concise documentation of the risk assessment results and the assumptions on which these results depend.

The fifth step is to identify applicable controls. Risk analysts usually adopt a worst-case scenario approach where they presume that no preexisting protection is available in the system and propose the appropriate controls in this respect. Afterwards a comparison with the existing controls can be made and the analyst could eventually conclude with recommendations for new controls or possible amendments of existing ones (e.g., reconfiguration of access controls).

An optional concluding step is to project potential savings. The cost of potential exposure as calculated previously minus the cost of new controls or reconfiguration of existing ones, gives an indication of the savings the information security programme has contributed towards. Such figures of course are largely subjective and this model is rather simplistic to be used as evidence of return of investment (ROI).

In spite of its widespread usage, there are arguments against the use of risk analysis, including the following, as summarised by Pfleeger [45]:

- The results represent subjective understanding of risk and are hence not precise.
- Use of risk methods may give a false sense of precision and subsequently immutability.
- Particularly the non-quantitative approached are perceived by many as of not having appropriate scientific foundations.

12.3. Challenges and Success Factors of Information Security Management

12.3.1. *Challenges of Security Management*

With the increasing awareness of security management, one would assume that organisations are mostly concerned with the events surrounding security issues. However, the mindset of business managers about investing in security is worrying. A research carried out by Accenture, conducted among senior IT professionals and board level executives shows that IT security investments are made with a view to implementing technology solutions to ensure compliance with external regulations [36]. Fifty-three percent of the survey respondents cited compliance as the biggest single driver of security investment. This suggests that information security managers are failing to convince the board of the business benefits of security investments. Warman [58] noted that the reasons for management of organisations to be complacent with security issues are firstly, the continuing perception that security is about blocking access and protecting assets rather than generating a return for the business. Another is that security is seen as primarily a technology issue, rather than an area where processes, people and other organisational factors are equally important and finally, security is widely perceived as a cost of doing business rather than a way of creating value.

Schneier [51] reported that such attitudes about security have caused many organisations to distance their security teams from other parts of the business. A research carried out by the U.S. Department of Homeland Security [11] reported that most executives view security as an operational issue and not a strategic issue. For example, when asked how well their company's security was aligned with business goals, 79% of high-ranking executives said the most effective alignment was in complying with government regulations and 74% said it was for protecting confidential information. Only 44% said security enhances the value of the brand,

and only 33% said it helps in managing the supply chain (ibid). The key problem is that most security managers do not know how to map their priorities to business objectives, and they are still failing in convincing the board of the business benefits of security investments. Again, most top managers do not understand how security fits into their business objectives. The IT security managers have a pivotal role in convincing the top level management of the rationale for investing in security.

The various challenges faced by security management in various organisations are based on the way security is viewed [55]. Warman [58] outlines some of these challenges as discussed below.

Management Complacency. Most business managers do not understand the ways in which IT security can influence their businesses, and part of the difficulty is that many managers, particularly at senior levels, are unable or unwilling to make use of computer systems themselves. This could be as a result of minimum training due to lack of time, or because of the uneasiness about technology. A now dated, still indicative, survey carried out by the business journal Management today and Microsoft (reported by Warman [58]) showed that 92% of managers were uncomfortable with computers, 58% have insufficient time to learn the systems and 31% had poor training or support. For this reasons, issues about IT security management are not considered as key issues in most organisations. Another key factor to this complacency is the fact that few IT security professionals have the communication skills and the business awareness required to give their directors a non-jargoned and concise insight on the state of information security in their organisation, and the potential benefits from investment in security. As stated earlier, CBR reported that most business managers invest in security only to maintain and achieve compliance [36]. The result of this is that funds available for security are shifted into compliance rather than enhancing security itself (ibid). Caelli *et al.* [8] agreed with the fact that very few managers would actually seek for insecurity in their businesses, but the problem of fitting the security paraphernalia into the operation of normal business often seems to be insurmountable.

Security as Overhead Cost. The perception that security is a cost of doing business has made security management to be viewed as an overhead cost. Expenditures receive much of the focus in organisations because they directly affect the organisation's bottom line. Protecting the financial condition and stability of an organisation is one of the most important issues in management. Security management is considered as an expense-driven activity that can directly affect an organisation's profitability and as such, it is considered as an expense on organisations balance sheets. The view of security as an overhead cost is an unfortunate outgrowth of the lack of inclusion of measurements and metrics as an essential element of security management. Most organisations fail to evaluate the losses that could result if appropriate security measures were not put in place.

Techno-Centric Approach to Security. In most organisations, security management is viewed as a technical issue rather than one that a business should embrace. The situation of the security professionals within the IT department makes it to be viewed by everyone as a technical issue. Krause and Tipton [33] pointed out that it is the business and not IT that owns the company data. IT only provides the technology and processes to implement the decisions of the data owners. Surely, most of the security controls are within IT because that is where most of the information is held, but the drivers are with the business. If the security of information is regarded as a business responsibility, and is traced back to business requirement, then funding becomes easier. Also, if business managers understand that IT does not own business information, then IT is not responsible for its protection.

Lack of Awareness. Although things are improving, there is still some serious lack of awareness about security issues in many organisations. In most organisations running a formal awareness program to promote information security and ensure that everyone knows their responsibility is still lacking. Due to this lack of awareness, people tend to take security unserious coupled with the view that security appears to be more of a hindrance than of practical business benefit. This at the long run has an adverse effect on a company's security system. Caelli *et al.* [8] stated that the basis for a successful security is people and many instances of data loss, corruption, or computer service unavailability can be avoided, if personnel are properly trained. When staff are aware of the threats to their organisation and the correct countermeasures that should be employed, only then is security effective. A comprehensive training and information security awareness programme is therefore a vital element for an effective security management.

12.3.2. *Requirements for Successful Security Management*

Management skills are fundamental to addressing the threats to computer and data security [58]. Not only have managers addressing functional and strategic issues to understand the need for security measures, but also recognise the value of active support and promotion of the measures [13]. For a security programme to become successful, some requirements are inevitably fundamental. These requirements are discussed below.

Management Commitment: Layton [34] stated that management commitment for information security is an element that must exist at the core of every information security program if it is going to be effective and successful in controlling the security risks. Management support goes all the way to the board of directors and the executive management. It is required that they take an active role in the information security program. Caelli *et al.* [8] pointed out that it is the responsibility of the information security officer or the highest-level position for information security within the organisation to develop and publish the information security policy documents, but it is the responsibility of the executive management team to ensure that the policies meet organisational, legal, contractual and regulatory requirements. Without the clear and active support of executive management, the information security program will not be as effective as it should and will likely fail at some point. Barefoot and Maxwell [4] consider management commitment as a key risk indicator to a successful security management because, without executive management support and commitment, the information security posture of the organisation would be at significant risk and likely to lead to devastating consequences for the organisation. According to Forcht [19], management must take initiative and view security issues as a part of an organisation's culture. Management should ensure that the employees understand the security plan. This could be achieved if management considers the following factors; leading by example, participating in the improvement process, knowing in details how to improve security and continual monitoring of the program.

In many organisations, senior managers may deliberately choose to ignore technical matters, or alternatively delegate the responsibility to more technologically competent junior staff. In doing so, they are losing control over their computer systems and their information resource [58]. Most junior employees cannot possibly have the wide scale appreciation of the necessary detail that constitutes a fully operational organisation. Numerous studies show that failure of security programs is attributed largely to management's lack of participation, misuse of improvement process, unwillingness to make a long-range commitments and failure to make the security program part of the business. Connolly [12] noted that responsible managers who show commitment to security programs by increasing their knowledge and extending this knowledge to their employees are fundamental to the implementation of a successful security program.

Security Policy: A primary step in securing an information system is by developing and implementing a security policy in the form of a dynamic document or set of documents [3]. The objective of a security policy is to provide management direction and support for information security [26]. The information security policy document provides a mechanism where the directors and senior management can lay down a clear statement of direction and 'rules' for the successful operation of the company. Gollmann [22] defines a policy as a set of rules, laws and practices that regulate how an organisation manages and protects resources to achieve its security policy objectives. Another view of policy is that it is a high-level management document that informs all users of the goals and constraints on using a system [45]. Goguen and Meseguer [21] argue that a security policy defines the security requirement for a system.

Implementing information security management involves developing policies and procedures that document the organisation's intentions to manage information with due care, throughout its lifecycle and keep it safe from unauthorised access and alterations. Gollmann [22] reiterated that policies provide a baseline for implementing security controls to lessen risk introduced by vulnerabilities.

Some researchers argue that it is theoretically impossible for an organisation to have a cohesive information security program that will appropriately protect the organisation's assets without having a written and approved information security policy [34]. Caelli *et al.* [8] stated that if the policy is created and implemented correctly, it will serve three main purposes within the company. Firstly; the policy will define the main security objectives which must be achieved and a security framework to meet business objectives. Next, the policy allocates responsibilities which include important definitions of ownership and custodian of data and systems. Without a clear definition of responsibilities, it is not possible to manage security and administration becomes unworkable. Finally, the policy statement can contribute directly to control. Examples of these controls includes personnel policies on recruitment, "safety first" policies and customer policies.

Security Awareness, Education, and Training: Security awareness, education and training are principles that must be implemented in every organisation. Forcht [19] stated that there is a clear difference between awareness, education and training. Awareness is typically directed at all users and tends to focus their attention on global security principles. Training, on the other hand, is much more in-depth and the message is directed at a specific group with an expected outcome. Education is another step beyond training where concepts and topics are covered in depth for the purpose of developing new skills and altering the outcome in some ways. These requirements are pre-requisites for a successful security management. Layton [34] argues that establishing a suitable information security policy and an effective information security awareness program will do more to protect an organisation than any firewall or piece of technology could offer.

Allocation of Information Security Responsibilities: Information security roles and responsibilities must be defined by management. It is impractical to assume that employees and users of the organisation's assets clearly understand their responsibilities for information security [8]. Lack of understanding constitutes a disaster to security management. According to Layton [34] information security responsibilities should be clearly defined and described within the information security policy document. The assumption is that if an asset has an assigned owner, the owner can be responsible for its protection and security. Barefoot and Maxwell [4] suggested that information security responsibility definition should start within the job description and then extended into each role as appropriate.

Reporting of security incidents: To achieve effective security management, there is a need to provide guidance on the actions that should be taken following any security incident, including procedures for reporting and responding to such incidents. All employers should be made aware of the procedures for reporting the different types of incidents that might have an impact on the security of organisational assets [12]. This should be included in the policy statement. Security violations may affect the entire business. Barefoot and Maxwell [4] suggested that management must be kept informed about these violations through progress and summary report.

12.4. Best Practice Advocated through Standards: ISO 27001 and COBIT

12.4.1. *General Principles of Good Practice*

The most important step in developing and deploying an IT security framework is to make an open and honest assessment of where the organisation currently is. Such an assessment should be conducted by reference to a broad cross-section of staff so that the overall assessment is not skewed by either the views of management or operational staff. Different stakeholders will have different views on what is important for the system and its security. End users for example will emphasise how important is to avoid system disruptions for shorter amounts of time, as it may prolong their working hours, whilst managers may be more relaxed on the same requirements.

A very useful method to perform this assessment is the use of Capability Maturity Models (CMM), which are widely used to measure overall levels of IT maturity within organisations. The CMM was originally introduced for software development within organisations and described specific steps and activities that would be required to move from one level to the next [25]. The idea was that it would help organisations improve the maturity of their software processes in terms of an evolutionary path. This path goes from ad-hoc, chaotic processes, towards mature and disciplined software processes. There were originally five levels in the CMM, as shown in Table 12.1, although many models now include a Level 0: Non Existent rating.

It can be seen from this table that there is a clear focus on continuous improvement and this is shown graphically in Fig. 12.2.

Regardless of the number of levels in a CMM model, perhaps the most important point is that the model chosen by an organisation best reflects their culture and processes and supports the principle of continuous improvement, and given the

Table 12.1. Typical stages in a capability maturity model.

Level	Description
1. Initial	Processes are ad-hoc, chaotic or poorly defined.
2. Repeatable	Basic processes are established and there is a level of discipline to stick to these processes.
3. Defined	All processes are defined, documented, standardised and integrated into each other.
4. Managed	Processes are measured by collecting detailed data on the processes and their quality.
5. Optimised	Continuous process improvement is adopted and in place by quantitative feedback and form piloting new ideas and technologies.

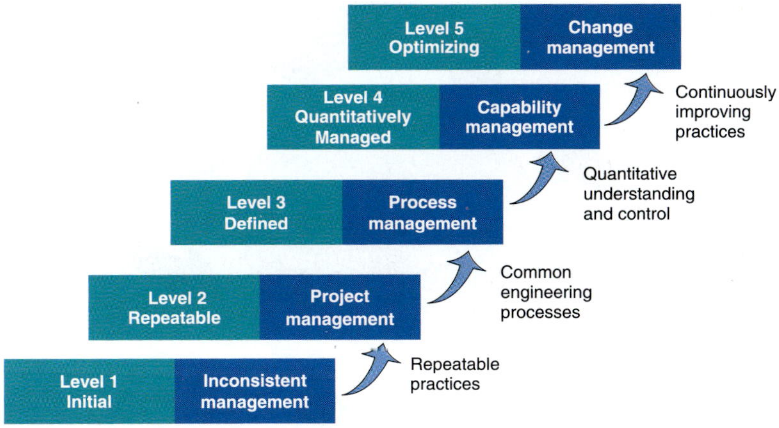

Fig. 12.2. Stages in the CMM model [20].

importance of senior management support detailed earlier in this chapter, it is a model that they are comfortable signing up to. Thus, a CMM can be adapted to provide a "route" to continuous improvement in other areas of IT delivery and IT governance. However, one of the issues with the CMM approach is that it focuses on "what" organisations should do in order to improve their processes and not "how" they should do it. In some cases there may be strict organisational requirements to follow and at other times it is left to individuals to determine what actions are required and how to execute these activities [32].

The Control Objectives for Information and related Technology (COBIT) have also produced a generic capability model that follows the "traditional" five level approach. This is a very interesting representation since it shows that different frameworks are designed for different purposes. It can be seen that COBIT is particularly useful for assessing high level, possibly business strategic issues, while others such as ITIL and ISO 17799 are designed to provide focussed advice in specific areas (service delivery an information security respectively). COBIT, ITIL and ISO 17799/27001 are the most quoted governance frameworks with indirect (COBIT, ITIL) or direct (ISO) relationship with information security. COBIT and ISO have been subsequently selected for in-depth analysis in turn.

A key feature of COBIT, ITIL and ISO 17799/27001, is continuous improvement and the concept of the *Plan, Do, Check, Act* model. The plan-do-check-act cycle as shown in Fig. 12.3 is a four-step model for carrying out change. Just as a circle has no end, the PDCA cycle should be repeated again and again for continuous improvement, hopefully achieving to progress the organisational area of concern (in this case information security) towards higher levels of maturity.

The model is particularly useful in the following circumstances:

1. As a model for continuous improvement
2. When starting a new improvement project

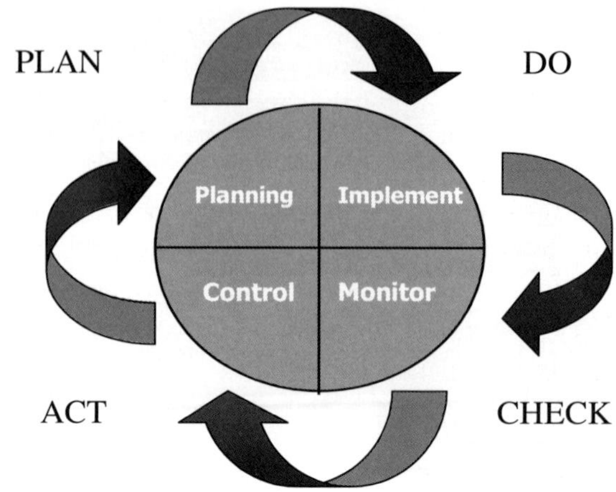

Fig. 12.3. Plan, Do, Check, Act Model of continuous improvement.

3. When developing a new or improved design of a process, product or service
4. When defining a repetitive work process
5. When planning data collection and analysis in order to verify and prioritise
6. Problems or root causes
7. When implementing any change

The procedure to implement this model should follow a set format:

1. **Plan.** Recognise an opportunity and plan a change.
2. **Do.** Test the change. Carry out a small-scale study.
3. **Check.** Review the test, analyse the results and identify what you've learned.
4. **Act.** Take action based on what you learned in the study step: If the change did not work, go through the cycle again with a different plan. If successful, incorporate what you learned from the test into wider changes. Plan new improvements, beginning the cycle again [2].

12.4.2. *ISO 27001*

This standard can trace it roots back to 1995 and the introduction of British Standard BS 7799. BS 7799 has undergone revision both in terms of structure and content since then. Part 1 became an international standard ISO 17799 in 2000, while Part 2 became an international standard, ISO 27001 [28], as part of the evolving 27000 series in 2005. ISO 27001 is seen by many as the de-facto international standard on establishing and improving an Information Security Management System (ISMS) for use in both the public and private sector.

BS ISO/IEC 17799:2005 ("ISO/IEC 17799"), the Code of Practice for Information Security Management [26] is a widely used reference and an exclusive model for information security. According to the DTI Information Security Breaches

Survey of 2006, 46% of the 1001 UK business participants had adopted the BS7799, of which ISO/IEC 17799 is the first part, at the time of the survey [18].

The purpose of the ISO/IEC 17799 as stated within the standard documentation is to give "recommendations for information security management for the use by those who are responsible for initiating, implementing or maintaining security in their organisation. It is intended to provide a common basis for developing organisational security standards and effective security management practice and provide confidence in inter-organisational dealings" (ISO/IEC 2005, p. 1). In other words, the ISO/IEC 17799 provides best practice security objectives which should be implemented and the associated controls (safeguards) that support organisational objectives [44, 52]. The ISO/IEC 17799 also provides guidelines for establishing an Information Security Management System within a selected unit or within the entire organisation. The ISO/IEC 17799 provides a subset of critical areas to be considered by those responsible for IS to facilitate their IS requirements (ISO/IEC 2005). As such the ISO/IEC 17799 provides a solid foundation for the development of information security practices and procedures within the organisation [26, 44, 50]. According to Myler and Broadbent the standard is good foundation for those who need to [41]:

- Create information security policies and procedures.
- Assign roles and responsibilities.
- Provide consistent asset management.
- Establish human and physical security mechanisms.
- Document communications and operational procedures.
- Determine access control and associated systems.
- Prepare for incident and business continuity management.
- Comply with legal requirements and audit controls.

The ISO/IEC 17799 contains 134 detailed information controls within eleven high-level security control objectives. The selection of security control objectives within each high level objective is determined by a combined approach considering control objectives' relevance and importance to the organisation's business processes and via risk analysis to identify areas of weaknesses (ISO/IEC 2005). Each control objective is divided into sub-objectives, which contains descriptions and explanations of activities to facilitate those charged with the selection of security controls to comprehensively consider various aspects of information security.

ISO 27001 is not that dissimilar to BS 7799, although new controls have been included to reflect technological and environmental developments. The standard consists of 11 sections that include 39 control objectives and 133 specific controls, as detailed in Table 12.2. The ISO/IEC 27001:2005 ("ISO/IEC 27001") is based on Part 2 of the BS7799, and is a management framework for the implementation of the selected controls of the ISO/IEC 17799. The ISO/IEC 27001 requires the organisation to establish a risk management framework to manage, review and

Table 12.2. The sections of ISO 27001.

Number (Standard Section Ref.)	Description
1 (A5)	Security policy
2 (A6)	Organising information security
3 (A7)	Asset management
4 (A8)	Human resources security
5 (A9)	Physical and environmental security
6 (A10)	Communication and operations management
7 (A11)	Access control
8 (A12)	Information systems acquisition, development and maintenance
9 (A13)	Information security incident management
10 (A14)	Business continuity management
11 (A15)	Compliance

improve information security and risk through the establishment an Information Security Management System (ISMS). The ISO/IEC 27001 mandates the use of ISO/IEC 17799 as a source of guidance for the selection and implementation of security controls. The ISO/IEC 27001 is therefore the governance tool which guides the deployment of selected security controls objectives based on the guidance of ISO/IEC 17799.

The ISO/IEC 27001 is constructed with a number of critical sections which specifies the conditions required to implement, maintain and improve an ISMS. In essence the ISO/IEC 27001 details information security concepts an organisation "shall do" while providing an auditing guide [37]. The standard emphasises continuous improvement, which forces management assigned responsibility for the ISMS to always think about security. To provide for continuous improvement the ISMS utilises the plan-do-check-act model, which in this case is implemented as

- Plan — Definition of the ISMS perimeter, formulation of the information security policy, risk assessment, preparation of the security action plan.
- Do — Implementation of the security action plan, improvement of information security awareness, and delivery of training to personnel.
- Check — Ensures the effectiveness of implemented security measures, the management of procedural control, the evaluation of data reliability and periodic auditing of the ISMS.
- Act — Implementation of the appropriate corrective, preventive measures and implementation of previously identified ISMS improvements.

A major requirement of the ISO/IEC 270001 to facilitate the implementation of the ISMS is the definition of the organisation's security perimeter. Security perimeter can only be defined if management known what information assets require protection and this decision is the responsibility of management. By defining the security perimeter, the implementers of the security controls, such as the IT

Department, are then able to create a roadmap to detailing information security strategies necessary [37]. At the end of the process, the implementation team is then able to produce a Statement of Applicability documenting what security measures will be implemented or not or where other controls address the identified risk.

12.4.3. *COBIT*

The Control Objectives for Information and related Technology (COBIT), were developed in 1996 by the Information Systems Audit and Control Association (ISACA), although they are now issued and maintained by the IT Governance Institute (ITGI). COBIT's mission, as stated on the ITGI website (www.itgi.org) is to "research, develop, publicise and promote an authoritative, up to date, international set of generally accepted information technology control objectives for day to day use by business managers and auditors". COBIT is an open standard and with the exception of the audit guidelines, are available free via the ITGI website. COBIT version 4 was issued in December 2005, with an update to version 4.1 in early 2007. The various iterations to COBIT have reflected its development from an audits tool in 1996 to much more of a strategic governance tool in 2005.

COBIT was designed by IT Governance Institute (ITGI) as a best practice to facilitate boards of directors, managers, users of IT services, IT auditors and executives to increase the value of IT and reduce related risks [29]. The COBIT framework examines IT controls within the business environment from a management perspective [31]. As such COBIT is considered by ITGI to be the tool for IT Governance, focusing on IT business processes through the use of IT resources in a controlled and measurable manner [29].

COBIT consists of four domains that contain the 34 control objectives and then 318 detailed control objectives. The four domains are

- **Plan and Organise (PO).** This domain covers strategy and tactics, and concerns the identification of the way IT can best contribute to the achievement of the business objectives. The realisation of the strategic vision needs to be planned, communicated and managed for different perspectives. A proper organisation as well as technological infrastructure should be put in place.
- **Acquire and Implement (AI).** To realise the IT strategy, IT solutions need to be identified, developed or acquired, as well as implemented and integrated into the business process. In addition, changes in and maintenance of existing systems are covered by this domain to make sure the solutions continue to meet business objectives.
- **Deliver and Support (DS).** This domain is concerned with the actual delivery of required services, which includes service delivery, management of security and continuity, service support for users and management of data and operational facilities.

- **Monitor and Evaluate (ME).** All IT processes need to be regularly assessed over time for their quality and compliance with control requirements. This domain addresses performance management, monitoring of internal control, regulatory compliance and governance.

Each CO is divided into a set of Detailed Control Objectives (DCOs), providing an outline and details on how CO must be managed and specific measurement requirements [31]. For example, within the CO "deliver and support" domain one of the DCO's is "ensure systems security". The CO is then mapped against the "Information Criteria" and "IT Resources" associated with the CO to indicate: the degree that control measures will satisfy different information criteria; and the degree to which control measures impact IT resources.

In addition the CO have associated management guidelines which detail the following [31]:

- Critical success factors (CSFs)
- Key goal indicators (KGIs)
- Key performance indicators (KPIs)
- RACI chart — Responsible, Accountable, Consulted and/or Informed

The COBIT framework can be seen graphically in Fig. 12.4. It shows how the four domains manage the IT resources to deliver information to the organisation according to business and governance requirements [27]. The 34 objectives each have a process overview that details the information that is obtained for each process, and this is shown in Fig. 12.5 below. One of the strengths of COBIT is that it is very audit orientated and provides very good checklists for the various aspects of IT within organisations [1].

Another strength of COBIT, especially in the most recent version 4, is the much greater emphasis on governance related issues, such as increased focus on IT regulatory and compliance requirements. This has been caused by events such as the enforcement of the Sarbanes–Oxley Act. There is also a greater emphasis on aligning the goals of the business to IT [40].

Indeed, version 4 of COBIT has quite a strong focus on governance and there is much correlation between the individual processes and the COBIT defined governance focus areas of

- Strategic alignment
- Value delivery
- Resource management
- Risk management
- Performance management

The COBIT framework, according to ITGI [30] ties business requirements for information and governance to the objectives of the IT Services function. ITGI

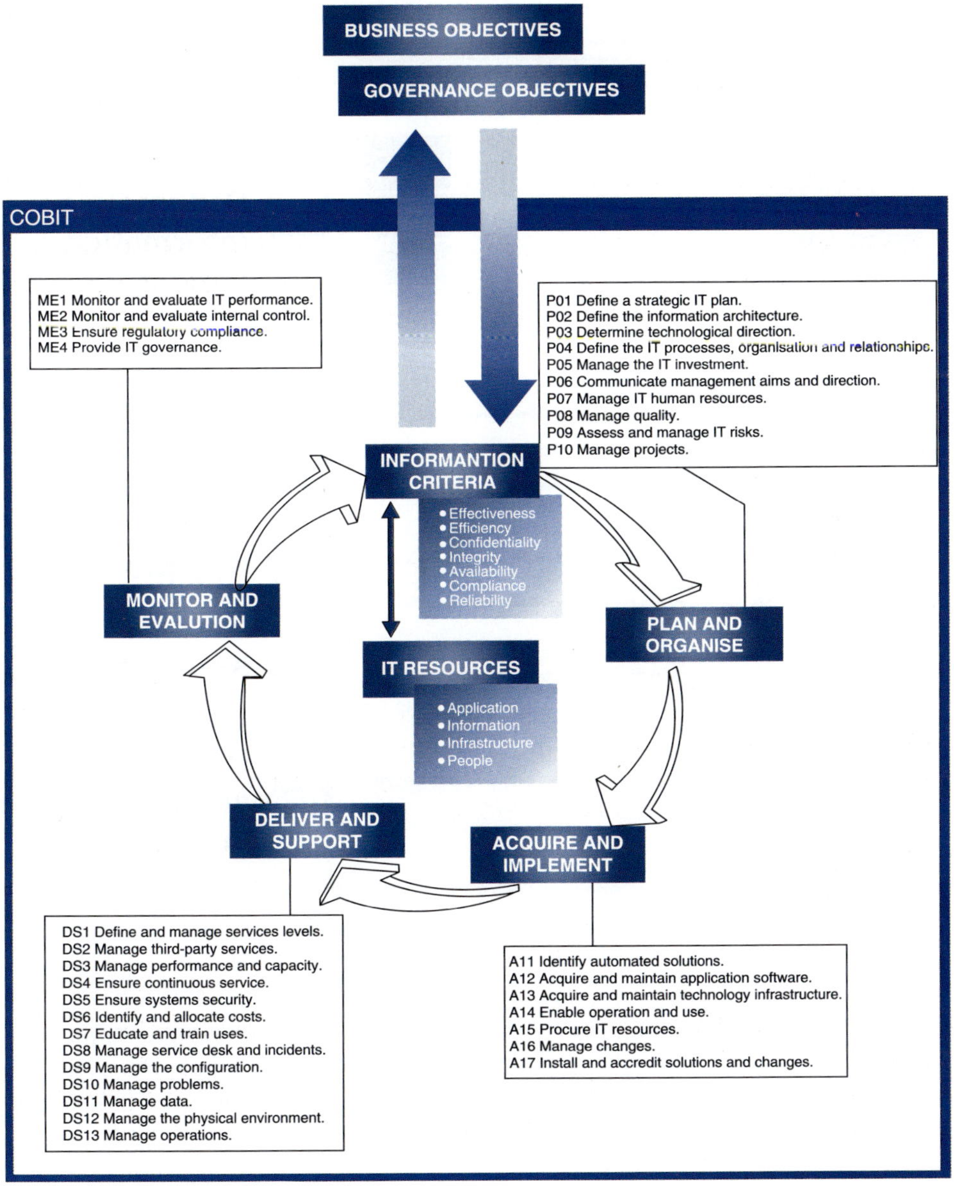

Fig. 12.4. The COBIT framework. (Reproduced from [27].)

further state that the COBIT process model enables IT activities and resources that support them to be managed and controlled based on COBIT control objectives and aligned and monitored using COBIT goals and metrics.

The COBIT principles can be visualised as having three dimensions: IT processes, IT resources and information criteria, as shown in Fig. 12.6 IT processes

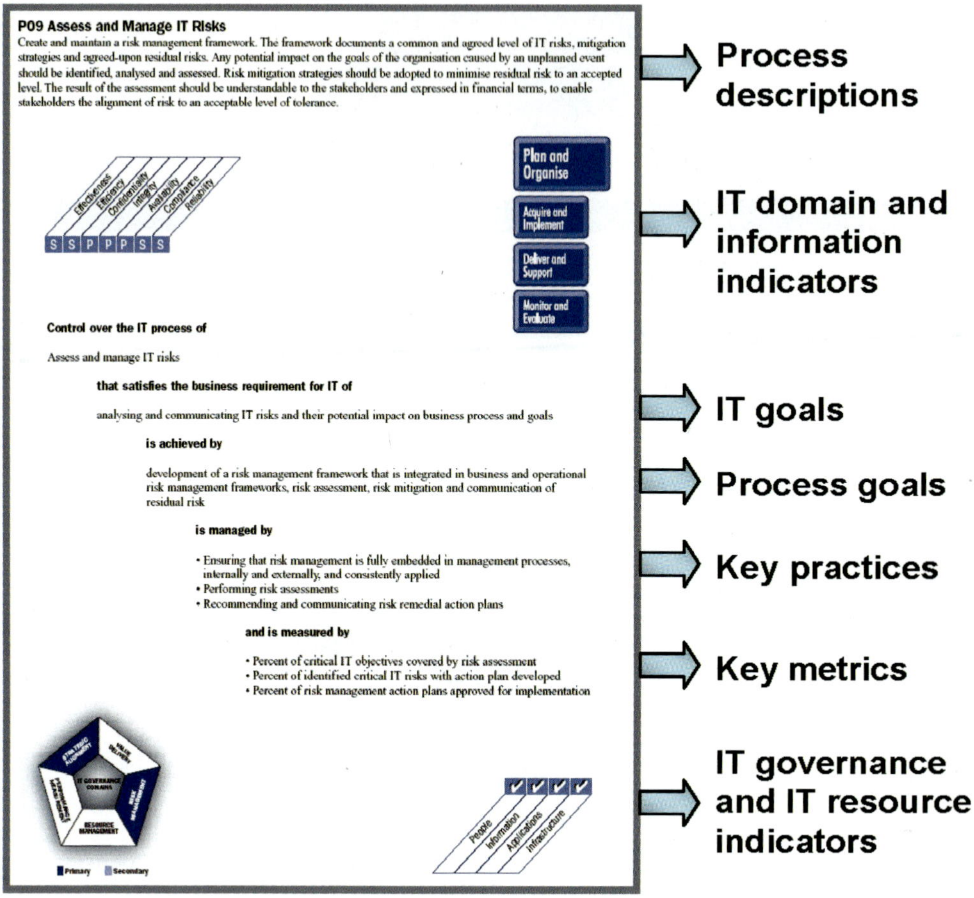

Fig. 12.5. The COBIT process with key components highlighted. (Reproduced from [27].)

include domains, processes and activities. IT resources include people, application systems, technology, facilities and data. Information criteria include the overriding concerns of quality, fiduciary, security, effectiveness, efficiency, confidentiality, integrity, availability, compliance and reliability.

An often quoted criticism of COBIT is that it sets out "what" should be done by providing a set of objectives and guiding principles, rather than "how" things could be done, and this had the potential to cause confusion when determining which framework to implement. However, this is being addressed and version 4 makes greater reference to the requirements of the other two main frameworks such as ITIL and ISO 17799/27001. Additionally, the COBIT mapping project is looking specifically at mapping COBIT with other frameworks such as the two mentioned above, but also frameworks such as Prince2 and PMBOK, that both relate to project management [24].

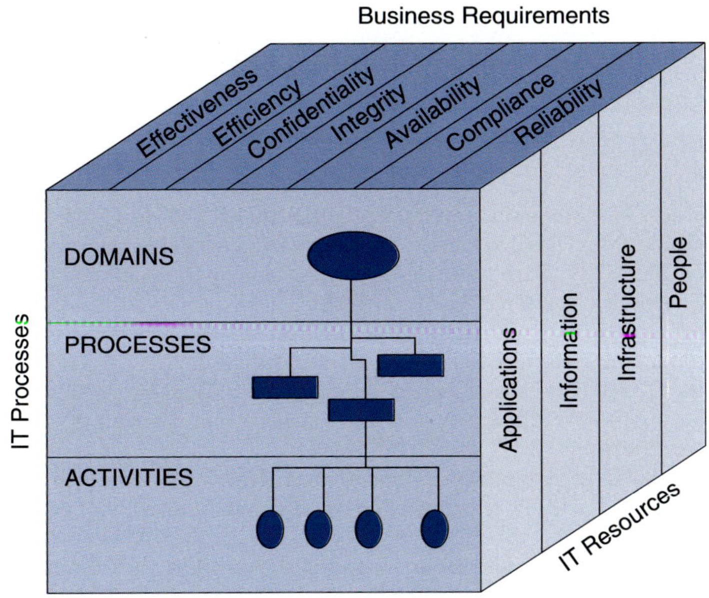

Fig. 12.6. COBIT Cube. (Reproduced from [27].)

12.5. Conclusion

Information Security has evolved from being a technical problem in the hands of IT personnel, to an organisational problem, as security issues emanate from both internal and external sources. Information is key asset for the success of many organisations and thus a governance structure is required to ensure that the correct executive management structures, policies and procedures are in place to direct the resources of the organisation to minimise the threats to information assets and maximise business benefit and added value.

Realisation of that fact will develop and mature over time, especially as legislation such as Sarbanes–Oxley, appears to be driving forward the "governance industry". Organisations may be best advised to examine all the various IT security and governance options that are available and develop their own approach that meets their needs. Significant input to this process may have the established and mature frameworks of information security and IT governance, particularly the two referred to in this chapter.

However, organisations should be aware that simply deploying existing frameworks will not 'solve' all their information security and IT governance requirements. Without, for example, the necessary senior management support and funding and staff awareness and training regime, so that all staff are aware of their responsibilities, frameworks are unlikely to deliver all of their intended benefits. Many implementations have failed because of a lack of a firm commitment to carry it through.

Acknowledgements

The author would like to thank Dawn Barrow, Mabel Ndigwe and Mike Parfitt for their input to this chapter.

References

1. G. H. Anthes, Model mania, *Computer World* **38**(10) (2004) 41–44.
2. ASQ, Adapted from Plan Do Check Act information available from www.asq.org, 2007.
3. A. I. Aton and J. B. Earp, *Strategies for Developing Policies and Requirements for Secure Electronic Commerce Systems* (2002).
4. J. K. Barefoot and D. A. Maxwell, *Corporate Security Administration and Management* (Butterworth-Heinemann, USA, 1987), 21–31, 109–114.
5. S. Blount, IT Security Management as a Business Enabler (2007). Available from http://ca.com/files/WhitePapers/sec_mgt_business_enabler_wp.pdf. [Last accessed 2nd August 2007].
6. A. Blyth and G. L. Kovacich, *Information Assurance: Security in the Information Environment*, 2nd edition (Springer-Verlag London Limited, London, 2006), 16–17, 96–97, 174–176.
7. C&A Security Risk Analysis Group, Introduction to Risk Assessment (2003) http://www.security-risk-analysis.com/introduction.htm. [Accessed 15th August, 2006].
8. W. Caelli, D. Longley and M. Shain, *Information Security Handbook* (Macmillan Press Ltd., USA, 1994), 1–9, 27–37, 75–85.
9. J. E. Canavan, *Fundamental of Network Security* (Artech House, Norwood, MA, 2001).
10. J. M. Carroll, *Computer Security*, Third Edition (Butterworth-Heinemann, Oxford, 1996).
11. T. E. Cavanagh, Navigating Risk — The Business Case for Security (2006). Report Number R-1395-06-RR. Available to purchase from http://www.conference-board.org/publications/describe.cfm?id=1231.
12. C. Connolly, Everybody's Business, *Security Management* **34**(1) (1990) 45–47.
13. D. Dalton, *The Art of Successful Security Management* (Butterworth-Heinemann, Worburn, 1998), 1–27.
14. D. E. Denning, *Information Warfare and Security* (Addison Wesley, 1999), ISBN 0-201-43303-6.
15. G. Dhillon, *Principles of Information Systems Security* (John Wiley & Sons, USA, Inc., 2007). 1–25, 100–108, 158–169, 221–229.
16. T. Dimitrakos, J. Bicarregui and K. Stølen, CORAS: A Framework for Risk Analysis of Security Critical Systems, ERCIM News 49 (2002). http://www.ercim.org/publication/Ercim_News/enw49/dimitrakos.html. [Accessed 7th August, 2006].
17. A. Down, M. Coleman and P. Absolon, *Risk Management for Software Projects* (McGraw-Hill Book Company, London, 1994).
18. DTI, Information Security Breaches Survey (2006) Department of Trade and Industry in association with PriceWaterhouseCoopers [online]. Available from: http://download.microsoft.com/documents/uk/security/downloads/DTI_2006_survey.pdf [Accessed 16th July 2007].
19. K. A. Forcht, *Computer Security Management* (International Thompson Publishing, USA, 1994), 370–450.
20. Gartner, MeasureIT. Special Edition 2001: Capability Maturity Model (2001). Available via www.gartner.com

21. J. A. Goguen and J. Meseguer, *Security Policies and Security Models, Lecture Notes* (1982). http://www.cs.purdue.edu/homes/ninghui/courses/Spring05/lectures/lecture08.pdf
22. D. Gollmann, *Computer Security* (Wiley & Sons, 1999).
23. C. Grobler, A Model to Assess the Information Security Status of an Organisation with Special Reference to the Policy Dimension, Mini-dissertation (2003) [online]. Available from: http://etd.rau.ac.za/theses/available/etd-08232004-100751/restricted/GroblerCP.pdf [Accessed 11th September 2007].
24. J. Heschl, An Introduction to COBIT 4.1 & Mapping COBIT to Other Frameworks and Standards. Presentation at ISACA e-symposium (2007) [online]. Available at http://www.isaca.e-symposium.com/archive300107.php (membership/login required).
25. W. Humphrey, Managing the Software Process. Software Engineering Institute. *The SEI Series in Software Engineering* (Addison-Wesley, 1989).
26. International Organisation for Standardisation/International Electrotechnical Commission, BS ISO/IEC 17799:2005 Information Technology-Security Techniques-Code of Practice for Information Management (British Standards Institute, London, 2005).
27. ISACA, COBIT 4.1 available as a free download from www.isaca.org (2007). Assessed May 2007.
28. ISO 27001, International Organisation for Standardisation/International Electrotechnical Commission, BS ISO/IEC 27001:2005 Information Technology-Security Techniques-Information Security Management Systems (British Standards Institute, London, 2005).
29. ITGI, Board Briefing on IT Governance, IT Governance Institute (2001). Accessed at www.ITgovernance.org.
30. ITGI, Information Risks — Whose Business are thy? IT Governance Institute (2005). Available via www.itgi.org/template.
31. ITGI, Information Security Governance — Top Actions for Security Managers (2007). Available via IT Governance Institute www.itgi.org
32. J. Jiang, G. Klein, H. Hwang, J. Huant and S. Hung, An exploration of the relationship between software development process maturity and project performance, *Information and Management* **41** (2004) 279–288.
33. M. Krause and H. F. Tipton, *Handbook of Information Security Management* (CRC Press LLC, USA, 1999), 353–465.
34. T. P. Layton, *Information Security* (Auerbach Publications, USA, 2007), 3–16, 55–69, 109, 126–130.
35. LogicaCMG, Information Security Governance: Board Briefing (2002) [online]. Available from: http://www.logicacmg.com/file/6265-581.3KB-LogicaCMG [Accessed 10th September 2007].
36. C. Louw, Boardroom Mindset Puts Security Second (2006). Available: http://cbr.co.za/article.aspx?pklArticleId=4198&pklCategoryId=378. [Last accessed 20th August 2007].
37. Lucent Technologies, Information Security Management Understanding ISO1779 (n.d.) [online]. Lucent Technologies. Available from: http://www.lucent.com/livelink/090094038006a83b_white_paper.pdf [Accessed 3rd November 2006].
38. N. Martin, *Simple Technique for Illustrating Risk* (SANS Institute, 2002).
39. G. McDaniel, *IBM Dictionary of Computing* (McGraw-Hill Inc., New York, 1994).
40. E. Mustafa and N. Robertson, Reframing the picture. Institute of Internal Auditors UK and Ireland Magazine (2006) 27–29.

41. E. Myler and G. Broadbent, ISO 17799: Standard for security, *Information Management Journal* **40**(6) (2006) 43–52.
42. J. D. Nosworthy, Implementing information security in the 21st century — Do you have the balancing actors? *Computers & Security* **19**(4) (2000) 337–347.
43. NCSC (National Computer Security Center), The Trusted Network Interpretation (1987) [online]. Available at http://www.fas.org/irp/nsa/rainbow/tg011.htm.
44. T. R. Peltier, Information Security Policies, Procedures, and Standards: Guidelines for Effective Information Security Management (CRC Press LCC, FL, 2002).
45. C. Pfleeger and S. Pfleeger, *Security in Computing*, 4th edition (Prentice Hall, United States of America, 2007).
46. Privacy Rights Clearinghouse, A Chronology of Data Breaches (2007) [online]. Privacy Rights Clearinghouse. Available from: http://www.privacyrights.org/ar/ChronDataBreaches.htm#Total [Accessed 21st September 2007].
47. T. Raschke, Leading by Example — Information Security in the Public Sector (2002) [online]. Nokia & IDC. Available at: http://www.nokia.com/NOKIA_COM_1/About_Nokia/Press/White_Papers/pdf_files/whitepaper_leadingbyexample_publicsector.pdf [Accessed 9th September 2007].
48. V. Raval and A. Fichadia, *Risks, Controls, and Security* (John Wiley & Sons, Inc., USA, 2007), 1–16, 50–55, 348–368.
49. J. E. Roskos, S. Welke, J. Boone and T. Mayfield, A taxonomy of integrity models, implementations and mechanisms, *Proceedings of the 13th National Computer Security Conference*, Washington, D.C., October 1990.
50. R. Saint-Germain, Information security management best practice based on ISO/IEC 17799. The Information Management Journal (July/August 2005), 60–66.
51. B. Schneier, *Why Management Doesn't Get IT Security* (2006) Available from http://www.schneier.com/blog/archives/2006/11/why_management.html. [Last accessed 21st August 2007].
52. K. Schwartz, ABC: An Introduction to IT Governance (2007) [online]. http://www.cio.com/article/111700/ABC_An_Introduction_to_IT_Governance [Last accessed 7 Feb 2008].
53. SIEMENS Insight Consulting, The Logic behind CRAMM's Assessment of Measures of Risk and Determination of Appropriate Countermeasures. Whitepaper (2005) [online]. Available at http://www.cramm.com/files/techpapers/CRAMM%20Countermeasure%20Determination%20and%20Calculation.pdf [Last accessed 7 Feb 2008].
54. Society for IT Managers, IT Trends in Local Government 2006/7 The Modernisation of Public Services, Executive Summary (2006) [online]. Available from: http://www.socitm.gov.uk/NR/rdonlyres/A9B3B46F-BFC2-41C9-9302-9047CA0C9AEF/0/IT Trends2006_ES.pdf. [Access on 22nd July 2007].
55. T. Tryfonas, On Security Metaphors and how they shape the emerging practice of secure information systems development, *Journal of Information System Security (JISSec)* **3**(3) (2007) 21–51.
56. B. Tsoumas and T. Tryfonas, From risk analysis to effective security management, *Information Management & Computer Security* **12**(1) (2004) 91–101.
57. S. M. Walker, *Operational Risk Management* (Connley Walker Pty Ltd., Victoria, Australia, 2001).
58. A. R. Warman, *Computer Security Within Organizations* (The Macmillan Press Ltd., London, 1993), 1–28, 71–96.

Chapter 13

SECURITY RISK MANAGEMENT STRATEGY

HAMID JAHANKHANI, MATHEWS Z. NKHOMA
and HARALAMBOS MOURATIDIS
University of East London, UK

The information and telecommunications services currently in use within local, national and global economies help facilitate the exchange of data at an unprecedented rate. The utility offered through information technologies has improved productivity and performance within for-profit organisations that have availed themselves of networked infrastructures that promote ease of access, remote access to data and enhanced speed and service. However, these assets also entail potentially damaging outcomes through threats to the network. Problems in network security have emerged from the different priorities that stakeholders have in respect to the data management infrastructure. These priorities are highly subjective and range from increased ease of use to increased limitations placed upon access and every point in between. These priorities are the result of two separate criteria that exist for users of a networked system: first, users want to enjoy efficient and pleasant experiences when using a system; second, users recognise specific risks and associate these with computer networks. Risk management is a significant component of network security, especially in respect to how the prerogatives of any network security system are defined.

13.1. Introduction

In recent years, information and telecommunications technology and services have expanded at an astonishing rate. The public and private sectors increasingly depend on information and telecommunications systems capabilities and services. In the face of rapid technological change, public and private organisations are undergoing significant changes in the way they conduct their business activities, including the use of wide area networking via public networks. These changes include mandates to reduce expenses, increase revenue and, at the same time, allow the participants to compete in the global marketplace.

Security management strategies have traditionally been framed as reactionary, where the outcome of the security management processes is defined according to goals set by the business hosting the security system. The decision to create or implement security within a specific company is done on the basis of security prevention of risk, but the risk is usually identified as the result of an existing flaw or a recognised problem. In this context, "risk-based decision-making and risk-based approaches in decision-making are terms frequently used to indicate that some systematic process that deals with the uncertainties is being used to formulate policy operations and assess their various distributional impacts and ramifications [16]. As a result, when networked security systems are put into place, these processes are designed specifically to counteract known risks or highly probable threats.

However, in the dialogue over successful network security policies, one perspective towards this outcome is that distinction must be made between imposing security policies that are based on logical assumption of risk and security policies that are based upon perceived risk. The study of how, why, and to what extent risk is embedded in network securities as a delimiting factor is therefore critical to determining whether a specific risk can be qualified and steps taken to prevent it, or whether the risk is non-qualitative but based upon emotional and contextual observations of other perceived threats.

In the study of network security, challenges have occurred in attempting to sell specific security strategies to management. Persons working in a management capacity not only have to evaluate and observe the impact that a security system will have in terms of economic costs, but also are keenly aware that problems associated with the adoption of new technology can harm appropriate operations among human users and in the attempts to integrate new technology into existing technology. Also, there is a knowledge gap between the skills that are traditionally held by managers and among those of network systems' specialists: while persons in management positions are neither ignorant nor uneducated concerning network security systems, this is typically not their area of expertise and persons in management may not accept the decisions made by security personnel. The converse is likewise true, where security specialists frequently do not understand the decision-making policies that management puts into place when approaching network security, as the security specialists are not trained in considering long-term logistical outcomes based upon implementation of a new or reformed security system.

In organisation's computer infrastructure, network security is implemented specifically to protect this infrastructure from known and unknown risks. Yet, no single component of an organisation is allowed to function without limitations, such as financial resources and the oversight of management and shareholders. The study of network security is therefore the study of risk assessment, where network security specialists identify specific risks, the potential impact of these risks, and whether the type and potential impact of a given risk qualifies it as a threat to the network.

Successful network security management is only achieved when the threat is avoided or, if the threat does come to pass, mitigated in terms of impact and outcome. However, risk assessment is a subjective process that is affected by multiple distinctive [32]. These variables include the priorities set by the network security supervisors, the capabilities of the security systems that have been implemented to target threats and the culture that arises within the organisation itself [16].

In order to implement an effective network security system, it is necessary to approach these distinctive elements and identify how sensible, appropriate risk assessment can be conducted. This indicates that all network security specialists must take into account the costs and the benefits of preventing one specific type of risk; the specialists must also identify the priorities within the organisation and provide risk management strategies to meet these. These requirements often cause the network security specialist to come into conflict with management when risk management strategies are determined that do not meet the expectations found within the organisational culture.

One venue in which this occurs is when changes need to be made to the network security system. Technology is finite; all technology tends to have limitations on how, why, and to what extent it can function. When a specific form of technology becomes outdated or obsolete, it needs to be replaced. Replacement of technology within security systems is an expensive and time-consuming process, and this is exacerbated by the need to implement widespread change when a single component needs to be changed; to clarify, when a server is outdated, the peripheral equipment attached to the server will also become outdated when the server itself is replaced. Also, older software and hardware are often not sufficient to manage new forms of risk within a networked system, suggesting that these may need to be replaced or updated even when they still function [22].

Analysis of risk is a new aspect of the balancing act that has thus far served as a normal state of operations for managers and network security specialists. Traditionally, "security management is a discipline in which making trade-offs is a continual activity: controls in the system versus controls in the environment, security control versus customer convenience and productivity, strong controls versus implementation and administrative costs and so on" [3]. In a risk assessment strategy, however, the network security specialist is faced with risks that are both recognised (e.g., hacking) and potential (e.g., using the company's logistical processes to transport illegal goods). This indicates that the network security specialist is faced with theoretical problems that could become risks, in addition to actual risks.

One concern that is increasingly mentioned by network systems specialists is that persons in management approach network security from a position of perceived risk as opposed to actual risk. When this occurs, the outcome is a separation between concrete risks that are seen as viable, legitimate problems that need to be resolved and theoretical problems that could potentially occur but have not yet manifested to any significant degree. Decisions to implement policies within networking are therefore grounded in questionable or dubious perceptions as opposed to specific themes or tendencies.

13.2. Focus on Priorities in Network Security

Risk management is the foremost concern of network security. Risks need to be identified, defined, described and strategies must be implemented to target these and resolve potential and actual problems. This indicates that when risk management is attempted, there is a subsequent dialogue that emerges from the processes of assessment. However, this dialogue is dominated by stakeholders within the security management process: stakeholders identify that specific outcomes have greater relevance to their own concerns and priorities, and subsequently seek to implement network security change that is in keeping with these concerns and priorities.

There are two specific negative outcomes that can result from the focus on priorities in respect to network security.

The first of these is conflict between stakeholders. General managers tend to be stakeholders in network security decision-making due to their control of financial resources and how to deliver news of change to the organisation and to those who are associated with the organisation (e.g., financial investors, etc.). General Managers also receive the feedback from outside sources and from persons and organisations, and integrate this feedback into the operational policies of the organisation. The general manager is more likely to be informed of perceived needs, especially how clients and customers need to be served by the organisation. Network security managers are stakeholders in that they determine how network security will meet potential threats [16]. The network security specialist is more likely to be informed of threats to the network: this awareness comes from information passed throughout the industry, as well as familiarity with the existing system and how threats to the system can manifest and potentially impact it [3]. Both sets of managers have the same goal when it comes to network security, which is to ensure that all potential risk is mitigated or avoided entirely. Yet, conflict occurs when both sets of managers place different — and often incompatible — priorities on how network security should define risk. The literature reports that general managers respond to external pressure from investors and from a desire to promote and preserve the brand name image of the institution. In contrast, network security managers are reported as responding to environmental concerns, especially in respect to threats that could occur within the organisation (e.g., internal service abuse by workers) or from external threats (e.g., hacks, worms, viruses, etc.). The degree of control that each set of managers has over network security differs according to the organisation and to the relationship formed between these two groups, but conflict does emerge when the interests of these two groups comes into conflict. The literature also suggests that these conflicting priorities can put the organisation at risk [16]. Mallery *et al.* find that responsiveness and adaptability are of paramount concern in implementing and promoting network security systems [23]. If time and energy is invested in conflict between stakeholders instead of successfully identifying and approaching shared solutions, the delay can place the organisation's network and digital information at risk for intrusion and exploitation.

The second negative outcome that can result from the focus on priorities in respect to network security is that of reactionary responses to perceived threat as opposed to actual threat. Risk analysis is derived from analytical management of data and the identification of how a specific risk can impact an organisation or a company. Scambrey, McClure, and Kurtz find that the degree of knowledge that needs to exist prior to involvement in appropriate decision-making processes for network security is not sufficiently met by those people who usually act as the end deciders in implementing networked security [30]. In a climate in which the presence and potential of risk and vulnerability are understood but the actual terms and conditions of the setting (e.g., the origins of the risk, the outcomes associated with a specific course of security) are vague or under-studied, this escalates the likelihood that risk management strategies will be implemented through reactionary processes instead of reasoned assessment of processes.

The literature on risk assessment and risk management has historically questioned the processes of how, why, and to what extent specific risks are framed as potential threats. Kemshall and Pritchard suggest that there is a necessary distinction made when a risk is classified in order to keep this risk as a manageable quantity: a risk is identified and quantified according to the degree of probability that it will occur within a specific context. In contrast, a threat is identified as a likely outcome based upon the environment in which assessment has occurred [19]. Risks and threats share similar properties, especially in respect to how these can change based upon the environmental factors demonstrated at the time. Despite these similarities, risks are not threats, and although threats can pose risks to the network or to the organisation, this is a different application of the term and should not be confused with risk in the defining process.

There is a distinction between risk analysis in network security and risk prevention in network security. The risk analysis phase is fundamental in identifying the types of risk that can potentially occur in a given system. When risk analysis occurs, the network administrator draws from outcomes that pose a risk to the system and classify these according to two traits: first, the network administrator has to identify the likelihood of a specific risk coming to pass; and second, the network administrator has to identify the degree of impact that a specific risk can have upon the network if a risk is manifested as a negative event. To clarify, a network administrator can draw up a table of outcomes based upon three specific known risks.

Risk management security programs also take into account normal, routine operational problems such as loss of access to the system or theft of computers. These risks tend to rank between *high* and *medium to high* in terms of probability of occurrence, and are far more likely to occur than security risks such as network intrusion. Yet even though these are more likely to occur and the damage can be significant, such as comes to pass when computers are stolen which contain the Social Security numbers of employees or clients, these are still mitigated in risk analysis and risk assessment. This should by no means imply that these risks are not considered at all, or that their outcomes go unrecognised by management and the information security staff alike. In recognising these risks, however, contingency plans are constructed in which solutions are integrated into the original network security management program. This suggests that the risks are recognised but are believed to be routine and can be prevented or, if these come to pass, actively reduced in terms of overall negative impact.

In studying how organisations manage potential risks to their networks, it is evident that some theoretical risks have a bare statistical chance of occurring but are treated as significant threats.

Preparing for a security threat that has an extremely low likelihood of probability is not done specifically to manage the threat itself. Schneier argues that the diversity of forms in which terrorist attacks can take place makes it statistically unlikely that any contingency plan will successfully identify and protect against damage to the computer systems [32]. Similarly, Simon Mitnick defines the "human element" as "the weakest link" in any security system, because the human component in any security system is likely to make decisions based upon the perceived effectiveness of the system as opposed to using his or her better judgment [25]. Thus, implementing expensive solutions for risks that have an extremely low probability of occurrence are done to provide a psychological "buffer" for stakeholders in the organisation [25], insurance companies are assured that the management is working to minimise risk; employees are assured that they will not suffer bodily harm; shareholders are assured that their investments are safe and so on.

Security analysts indicate that planning and implementation of security measures for risks that are not likely to occur is driven by an emotional investment in the minimisation of risk. Emotional issues and the exacerbation or distortion of security risks comprise a basic response to many threats. Security analysts identify this as an instinctual response to threat, wherein it is in an organism's best interests to respond to a threat as though the perceived risk were blown out of proportion to the perceived risk. When this occurs, the response is elevated and done in order to preserve the safety and integrity of the organism through providing a counter-threat that encompasses increased strength.

13.3. Business-Based Approaches to Network Security

The format and construction of the security system is a foremost consideration for businesses. "From a commercial standpoint, assurance of network security is a business enabler" [24]. As such, it is necessary to establish an effective security system and maintain it so that it continuously meets the needs of its owners, operators, and the customers of the business.

The foremost consideration when approaching any business-based security system is that both the business and the network security environments are constantly in a state of dynamic change. Historically, network security has been retroactive. Over time and with the gradual emphasis on interconnectivity that has been recognised within security systems, this was identified as a strategy doomed for failure because the existing designs often came into conflict with new or innovative design processes. When this occurred, conflict between systems or failure within one or more aspects of the systems occurred.

Recognition of failure and conflict has subsequently resulted in an emphasis on "ground-up" approaches to network security [16]. When a "ground-up" strategy is implemented, the business identifies security needs and puts strategies to target and resolve these into effect, literally building a new system from the ground up. There are liabilities within this strategy, especially in respect to the demands placed on a new network security system and the expense taken to implement and achieve desired outcomes.

Change is an important factor when identifying risk and planning specific security processes: network security is dependent on the criteria and components of the system involved, and these are in turn affected by the decisions made by users. As a result, all business-based-approaches to network security need to integrate these factors in order to effectively plan, program, implement, and maintain network security.

13.3.1. *Components of Network Security*

Security systems within any given organisation are defined and described by that organisation, which makes it difficult to formulate a single universal understanding of network security within a given context. With that said, all security systems have some manifestation of the following six components:

— "*Confidentiality*: The property of guaranteeing information is only accessible to authorised entities and inaccessible to others.
— "*Authentication*: The property of proving the identify of an entity.
— "*Integrity*: The property of assuring that the information remains unmodified from source entity to destination entity.
— "*Access Control*: The property of identifying the access rights an entity has over system resources.
— "*Non-repudiation*: The property of confirming the involvement of an entity in certain communications.
— "*Availability*: The property of guaranteeing the accessibility and usability of information and resources to authorised entities" [26].

Attacks against any one of these six components can be qualified as a form of risk, and the type of risk can then be identified and correlated based upon its source, its intent, and the component(s) that it targeted. Thus, while all forms of network security differ according to purpose and use, the system itself can be described using these components. Furthermore, any type of risk or threat can be contrasted against these systems: this shall be elaborated upon in the section entitled "decision-making and risk management."

Hardware and penetration testing are components of network security. These have traditionally played a more substantial role within network security due to the role played by these devices within risk assessment and respective targeting of same by those seeking to affect the security of the system.

Assessment of the effectiveness of a system tends to blend software, hardware, and other forms of security; if one component of the system fails, than the system as a whole is worthless. To ensure the appropriate functioning of the system, ongoing assessment and maintenance is required. McNab [24] found that the assessment of network security components tends to identify the effectiveness of components and also tests the limits on the system as a whole. These are accomplished through three testing channels that identify and target unique components of a specific networked system:

— "*Vulnerability scanning* uses automated systems (such as ISS Internet Scanner, QualysGuard, or eEye Retina) with minimal hands-on qualification and assessment of vulnerabilities. This is an inexpensive way to ensure that no obvious vulnerabilities exist, but it doesn't provide a clear strategy to improve security.
— "*Network security assessment* lies neatly between vulnerability assessment and full-blown penetration testing; it offers an effective blend of tools and hands-on vulnerability testing and qualification by trained analysts. The report is usually hand-written, giving professional advice that can improve a company's security.
— "Full-blown *penetration testing* [...] involves multiple attack vectors (e.g., telephone war dialing, social engineering, wireless testing, etc.) to compromise the target environment. Instead this book fully demonstrates and discusses the methodologies adopted by determined Internet-based attackers to compromise IP networks remotely, which in turn will allow you to improve IP network security" [24].

According to Kolodzinski [21], the primary goal then is to develop a scalable corporate security structure that is responsive to short- and long-term needs as well as shifts in technology. A basic tenet of such business driven computer network security planning is that senior management and its risk management function lead the charge on linking business strategies to computer network security and identifying where information is at risk. By knowing future needs, security planners can anticipate requirements for information protection with a view to making them able to expand or contract according to strategic actions that the company takes in pursuit of its targets and goals. Similarly, the planning process will be responsive to shifts in technology; needs must therefore be identified and systems put in place that allow for technology upgrades or add-ons.

There are also secondary goals after the initial infrastructure has been determined. While "it is not always possible to secure everything," it is also necessary to recognise that specific aspects of security can be "hardened" [1]. Here, *hardening* refers to the process of implementing and maintaining additional processes and protocols to reduce intrusion and negative effects. In their book, *Network Security Illustrated*, authors Albanese and Sonnenreich stress that "network hardening and network design are very closely intertwined processes" and that "network hardening compensates for practical network design compromises that networks need to make". Here, it is recognised that the initial design phase is necessary to effectively implement security protocols but there can be retrospective action taken by network security specialists to resolve problematic issues. This is beneficial because it ensures that corrections can be made to the system after it has been implemented: if change could not be made, then it would be necessary to continuously implement new security systems. With the ability to harden an existing system, changes can be made based upon recognised security flaws or new risks that emerge after the system has been implemented. Yet caution is also necessary, where "no amount of network hardening can compensate for poor network design" and network hardening technologies can do more harm than good if not properly utilised" [1].

13.3.2. *Decision-Making and Risk Assessment*

Boltz [5], suggests that risk assessment is necessary to critically implement risk management policies. Information security management protocols were established by the federal government in 1998 to ensure that government systems were effectively implemented to reduce risk and were continuously hardened over time to prevent emerging risks. Boltz [5], states that there is a "continuing cycle of activity" in respect to risk assessment, where the structure of risk assessment and risk management are dynamic; the data from one phase of the process is integrated back into another, and when system reform is made, this in turn generates new data sets that can then be bound over to another aspect of the risk assessment cycle. Author writes:

> Although all elements of the risk management cycle are important, risk assessments provide the foundation for other elements of the cycle. In particular, risk assessments provide a basis for establishing appropriate policies and selecting cost-effective techniques to implement these policies. Since risks and threats change over time, it is important that organisations periodically reassess risks and reconsider the appropriateness and effectiveness of the policies and controls they have selected [5].

Continuous assessment and analysis are therefore ongoing components of the risk management process. Inevitably, these processes will generate new information that demands responsiveness.

Other models of risk assessment and risk analysis integrate similar themes of knowledge generating informed outcomes, but these are not framed as cyclical processes. Instead, these are stage-centerd, where each phase of a given risk assessment and analysis process stems into the following stage. One component of how businesses approach decisions within network security is through risk management. "The risk management process minimises the impact of threats realised and provides a foundation for effective management decision-making" [10].

The use of risk management processes can integrate a modeling strategy in which specific known and unknown risks are categorised and qualified according to certain criteria. These are grouped within the Systems Development Life Cycle (CDLC) in which a five-stage process of risk management is attached to the five temporal phases of technology longevity within the workplace. These five stages are defined by Cole, Krutz, and Conley [10], as follows:

Stage One: Initiation — The need for the system and the purpose are documented. A sensitivity assessment is conducted as part of this phase. A sensitivity assessment evaluates the sensitivity of the IT system and the information to be processed.
Stage Two: Development/Acquisition — In this phase, which includes the development and acquisition activities, the system is designed, developed, programmed, and acquired. Security requirements are developed simultaneously with the definition of the system requirements. The information security requirements include such items as access controls and security awareness training.
Stage Three: Implementation — Implementation involves installation, testing, security testing, and accreditation. During installation, security features should be enabled and configured. Also, system testing should be performed to ensure that the components function as planned. System security accreditation is performed in this phase. Accreditation is the formal authorisation for system operation by the accrediting official and an explicit acceptance of risk.
Stage Four: Operation/Maintenance — The system performs its designed functions. This phase includes security operations, modification or addition of hardware or software, administration, operational assurance, monitoring and audits. These activities include performing backups, conducting training classes, managing cryptography keys, and updating security software.
Stage Five: Disposal — This last phase includes disposition of system components and products (such as hardware, software and information), disk sanitation, archiving files, and moving equipment. Information may be moved to another system, archived, discarded or destroyed. Keys for encrypted data should be stored in the event that the information is needed in the future. Data on magnetic media should be purged by overwriting, degaussing, or destruction [10].

While all five of these domains are important, the area of greatest specific interest here is in Stage Three, where it is recognised that implementation of the system functions as "an explicit acceptance of risk." The definition of risk is incorporated into the initial study of the network security system, as determined within Stages One and Two, and exists as described by Cole, Krutz, and Conley [10]. This suggests that risk management strategies make a distinction between "acceptable risk" and other types of risk. Gregg and Kim [15], suggest that risks are inherently quantifiable and that these risks can be further deconstructed and classified within the assessment process:

1. Acceptable risk — A term used to describe the minimum acceptable risk that an organisation is willing to take.
2. Countermeasure or safeguards — Controls, processes, procedures, or security systems that help to mitigate potential risk.
3. Exposure — When an asset is vulnerable to damage or losses from a threat.
4. Exposure factor — A value calculated by determining the percentage of loss to a specific asset because of a specific threat.
5. Residual risk — The risk that remains after security controls and security countermeasures have been implemented.
6. Risk management — The process of reducing risk to IT assets by identifying and eliminating threats through the deployment of security controls and security countermeasures.
7. Risk analysis — The process of identifying the severity of potential risks, identifying vulnerabilities, and assigning a priority to each. This may be done in preparation for the implementation of security countermeasures designed to mitigate high-priority risks [15].

The evaluation processes can further compartmentalise and define risk based upon these components and the relative impact that each individual aspect of risk can have upon specific procedural outcomes. A commonly-used term for this is assessment of risk through risk management planning; Cole, Krutz and Conley [10] refer to this as "defence-in-depth" planning. "Defence-in-depth is the practice of layering, like an onion, the defences and security countermeasures into zones, thus distributing the responsibility and accountability for information security over the seven areas of information security responsibility" [10]. Using the five stages of the SDLC cycle, it is possible to identify and predict specific policy based on the contextual setting constructed by the requirements of each individual stage.

13.3.3. *Outsourcing Security*

A final consideration in network security is that of outsourcing. Network security and network information are expensive processes to implement and maintain, and cost-conscious business organisation is likely to approach the expenditure in these areas as a necessary but expensive investment. Outsourcing is a strategy increasingly used by businesses to promote profitability through sending products or processes to other companies. These other companies are able to cut costs through producing these items or processes at reduced costs. It is now possible to outsource network security. However, as Axelrod [2] wrote in *Outsourcing Information Security,* moving information from in-house control to a third-party service provider carries with it numerous associated risks in addition to the potential cost benefits that could occur. Axelrod [2], wrote "outsourcing is as secure as you make it. There are multiple levels of security — both from a process perspective and a technology perspective — which companies can put in place to secure their business relationships, their data, and their intellectual property". He continues by noting that risks can be managed, where:

As companies allow business partners to access and process an increasing amount of proprietary data, applications, and intellectual capital, they are realising that not only must they get their business partners to commit to formalised security measures and policies, but companies must also take steps to protect themselves in the event that their business partners have a security breach [2].

This citation stresses a number of serious concerns for those companies who determine that outsourcing is an appropriate business strategy. First, the implication that agreement must occur for these partnerships to be successful exists. This agreement not only suggests compliance between platforms but also compliance with maintenance strategies and implementation of new technology as needed.

More importantly, however, this citation calls up the singular risk that occurs within outsourcing network security, where problems can befall the original company if the third-party service provider is breached. Companies that trade on client information, especially banks and medical organisations, therefore have the potential to exacerbate existing risk for their clients when they choose to outsource. It is therefore necessary to approach outsourcing from a cost-benefit ratio, as there is overlap between the outsourcing processes and the benefits obtained through security and security management.

13.4. Assessment and Analysis Models

Models of assessment and analysis are important to identify the perceptions of technology consumers and the decisions that consumers are likely to make in respect to technology. The Technology Acceptance Model, the Theory of Reasoned Action Model, and the Diffusion of Innovation Model help identify how technology is implemented within a specific user population and helps to clarify how these users will respond to its presence and effectiveness. To this end, it is necessary to approach these models as assessment tools that can clarify the roles of participants within the use of any technology-centerd system, including network security systems.

13.4.1. *Technology Acceptance Model*

The Technology Acceptance Model is used to predict the use of an information system, especially in respect to the ability to accept and diagnose "design problems before users have experience with a system" [14]. There are two specific criteria that affect the nature of TAM prediction, and these are *perceived usefulness* and *perceived ease of use*. The concept of perceived usefulness applies to the degree to which a potential user of technology approaches and surveys the status of a specific form of technology. The *perceived ease of use* refers to how easy the technology appears to be in terms of a user's overall familiarity with it or how quickly the user believes that he or she will adapt to its use.

TAM was first proposed by Davis *et al.* [12]. User acceptance of computer technology: a comparison of two theoretical models". This paper was a research experiment that modeled user behaviours of technology based upon the comparative ease of acquisition of information [12]. The researchers provided a sample population with a qualitative study and asked the respondents to clarify their personal perceptions of a specific form of technology. The researchers found that new users — those whom had never seen or experienced that specific form of technology before — were less likely to perceive it as useful and were less likely to perceive it as easy to learn than respondents who had prior experience with similar forms of technology. There appeared to be an increased willingness to interact with technology if the user recognised the potential utility or entertainment value imbedded within the device. Moreover, the attitude of the user appeared to predispose them to specific learning curves: users who perceived the technology as useless or non-important were less likely to adapt to its use at a rate comparable to that of a user who saw the technology as beneficial.

Davis *et al.*, [12] determined that these perceptions among users were fundamental concepts that the users brought with them when they sought to learn new technology. In this setting, TAM helped illustrate that a user who associated specific perceptions with a technology was more or less likely to use it, based upon the degree of resistance expressed in their self-reported perceptions. The researchers concluded that perception affects behaviour, which in turn affects the overall use of the technology.

Reliability and validity of TAM has been affirmed in multiple follow-up research studies. These studies help to affirm the feasibility and the overall usefulness of TAM. Most researchers concur that TAM is an appropriate strategy through which the perceived integration of technology into a specific organisation can be measured, but there are limits on TAM's overall usefulness. For example, Chau's [9], "An Empirical Assessment of a Modified Technology Acceptance Model" found that TAM was efficient, effective, and without peer as a modeling strategy, but that this was true only if certain criteria are met within the organisation and within the sample population [9]. This occurred because TAM measured perceived acceptance and did not measure actual acceptance: in short, TAM assesses the degree to which the sample population thinks they have adopted the use of a new technology and not the actual outcome of adoption.

13.4.2. *Theory of Reasoned Action Model*

The Technology Acceptance Model was developed as a component of assessment within psychological and motivational theory, specifically those ideas put forth through the Theory of Reasoned Action (TRA) was developed in 1967. This theory was used to assess human motivation based upon personal preferences, perception, and behaviour. The process of reasoned action in sociology is one in which the individual person or a close community of persons are assessed in terms of attitude and behaviours. If there is no barrier or limiting factor, the outcome is one in which the person will follow through on his or her subjective relationship between attitude and behaviour. To clarify, if a person prefers to wear the colour green, then it is highly likely that the person will choose to wear green clothing. Barriers that may prevent this are found in dissuading variables (e.g., criticism from a third party who asks why the person always wears green) or influencing variables (e.g., finding a great item of clothing that does not come in green but that the person still wants to wear).

13.4.3. *Diffusion of Innovation Model*

The diffusion of technology is a fundamental component of the assessment process. Diffusion of technology refers to the process through which technology is introduced to the population as a whole and is diffused therein; it is necessary to study diffusion as users adopt technology at different rates and demonstrate different perspectives towards technology and how it can be used. Diffusion is said to have occurred when specific individuals demonstrate functional awareness and abilities to use technology; these processes are then transmitted throughout other members of the population and "diffused" into their repertoire of use in respect to technology. When successful diffusion occurs, the population of users not only demonstrates basic knowledge of how the technology can be used but also displays a willingness to use it as directed, [28, 29]. To this end, diffusion of technology is a necessary topic of study in the exploration of network security and network protocols, as it helps to assess the degree to which security-specific technology is accepted and used by the population [33].

Rogers' [28], diffusion of Innovation model (DoI) is the standard model used to identify how new or alternative types of technology are diffused into the population. Rogers [28], defines new technology as an "idea, practice, or object that is perceived as new by an individual or other unit of adoption". The degree to which the technology is resisted or accepted by the population is characterised by diffusion into five specific population groups, each of which is distinguished specifically by how and to what extent the diffusion processes occur. The first group is comprised of *innovators*, who are the people who invent or install the technologies and are at the forefront of use: network security specialists fit into this population. The second group is comprised of the *early adaptors*, who identify that the new technology has desirable characteristics and makes changes in order to fit this technology into their lifestyles. Then the *early majority* approaches the technology and recognises its usefulness; these persons tend to receive psychological encouragement and validation for this decision after having watched the early adaptor group and their successes with the technology. The *late majority* receives the same encouragement from the early majority group and then adopts the technology as their own. Finally, the *laggards* acquire and use the technology after significant change has been made based upon the use of this technology; if given the alternative, most laggards would avoid adoption altogether [28].

The DoI is useful in studying network security as it helps identify how and to what extent the knowledge of usefulness moves through an organisation. For example, in a research study on the incorporation of information security tools to protect computer systems when the Internet was used, authors Neale, Murphy, and Scharl [27] studied how diffusion of these tools moved throughout a single organisation. It was noted that smaller companies in which the populations are small, centerd, and focused on network security are far more likely to adopt these types of network security than larger organisations with multiple participants and greater distance of communication between participants [27].

13.5. Models Used in Risk Assessment

There are multiple models that can be referred to as "risk assessment" strategies; some of these have already been identified as such within the context of this chapter. Eight common modeling approaches will be described to help clarify how risk assessment functions and the extent to which a specific approach is able to identify and resolve issues of risk.

13.5.1. *General Accounting Office Model*

Boltz [5], reports on the General Accounting Office (GAO) model. As was previously stated, this model of risk assessment approaches risk as a continuous, dynamic model in which continuous documentation of data is required. When documentation is achieved, this information can then be used to facilitate reforms in information security. Traits within the GAO model are achieved through framing risk in respect to "focal points", which are used to facilitate, coordinate, and execute "the business unit's risk assessment activities" [5].

13.5.2. *National Security Administration's INFOSEC Assessment Methodology*

The INFOSEC Assessment Methodology (IAM) is comprised of six core developmental and implementation phases. These include an analysis of risk within an organisation according to 18 baseline standards, an assessment plan in which risk is defined, an analysis of information and assessment, written reports on existing and proposed information, briefings with customers, and continuous organisational reviews.

13.5.3. *Shawn Butler's Security Attribute Evaluation Method*

Butler's [6], approach to security is done through the Security Attribution Evaluation Model (SAEM). The model functions on the premise that managers are "motivated to minimise security costs but maximise security benefits", a process that can have negative consequences even if high standards of care are maintained. "SAEM is a cost-benefit analysis process for analysing security design decisions that involves four steps: (1) a security technology benefit assessment, (2) an evaluation of the effect of security technologies in mitigating risks, (3) a coverage assessment and (4) a cost analysis" [6]. The third and fourth steps can be achieved simultaneously but must be separated for purposes of clarity. These processes are also dependent on assumptions of the general environment, such as that all "security products have been correctly installed, configured, and maintained" [6]. When these occur, the SAEM can provide a roadmap towards how certain risk criteria can impact an organisation and which aspects of risk analysis can be heightened.

13.5.4. *Carnegie Mellon's Vendor Risk Assessment and Threat Evaluation*

The V-Rate program by Carnegie Mellon functions on the assumption that vendors, such as those third-party software system designers, play a role within security. Often, threats emerge through compromising these third-party processes instead of through targeting the actual network infrastructure. Using the V-Rate program, organisations can reduce security risks and overall vulnerability through assessing the roles of the service providers and how the use of same can impact the organisation's internal security.

13.5.5. *Yacov Haime's Risk Filtering, Ranking, and Management Model*

The Risk Filtering, Ranking, and Management model proposed by Haimes [16] is an eight-phase program in which a program or a process can be critically examined in terms of risk. It is identified as a philosophical model instead of a mechanical model, as it does not promote structural reform but instead helps refocus on the goals and the setting in which goals are achieved.

— Phase I. In this, the scenario is identified and a model is used to represent success within a specific set of circumstances.
— Phase II. Risk is identified and similar models are conducted in which risk is manifest to varying degrees.
— Phase III. Bi-criteria filtering and ranking are used to identify the likelihood of specific risks and the outcome associated with these.
— Phase IV. Multicriteria evaluation occurs in order to identify and generate outcomes based upon the information contained in the previous phase.

— Phase V: Qualitative ranking is applied to classify which risks are the most likely to occur and identify under what conditions these are likely to occur.
— Phase VI: Risk management scenarios are generated to resolve these issues and classify specific directions if a scenario should happen to manifest.
— Phase VII: In this stage, mission critical items are framed within safeguards to ensure their preservation in spite of negative circumstances.
— Phase VIII: Finally, operational feedbacks are generated in this modeling scenario to help flesh out understanding of risk variables and suggest options for change or reform.

13.5.6. *Carnegie Mellon's Survivable Systems Analysis Method*

This model approaches risk as a criteria of survivability. The model stresses that risk needs to be framed as a threat the organization, especially in respect to how the organisation responds to specific influences. The model utilises three "key capabilities" within an organisation, which are "resistance, recognition and recovery":

Resistance is the capability of a system to repel attacks. Recognition is the capability to detect attacks as they occur and to evaluate the extent of damage and compromise. Recovery, a hallmark of survivability, is the capability to maintain essential services and assets during attack, limit the extent of damage, and restore full services following attack [8].

The criteria used to measure these three key capabilities are embodied in the definition of system capabilities, the definition of essential capabilities, the definition of capabilities in compromised scenarios, and an analysis model that uses these definitions to predict survivability within specific proposed and actual crises.

13.6. Organisations and the Need for Network Security

Computers have found their way into all areas of business, industry, education, and government. Increasingly far-reaching information networks linking computers and databases provide important benefits, including greater staff productivity and a sharper competitive edge [4]. The computer is the symbol of the modern, automated business. Its growing popularity plus the powerful business software, has resulted in an explosion of stand-alone data processing systems in many different departments of organisations throughout the world.

These computers are also increasingly distributed, wherein the processing power of a single machine is spread out through multiple machines within a network. Computer-based information systems within companies have played a significant role in improving communications and reducing processing time [18].

Similarly, information networks, which are similar to distributed computing in that the machines are linked, but differ because processing power and access to the computer are closely limited, are now solving many work and productivity problems. Networks promote information exchange by interconnecting distributed departmental computers and associated terminals, printers, and other devices with centralised computers so that all units function as part of a single, unified communications system. Within a single closed environment, such as is found in a company, the ideal result of these information networks and distributed processing systems is a single cohesive network in which authorised personnel can speedily and efficiently access computers and other system resources from any terminal or other device. Through networked computing, this can occur whether the devices are in the same room, building, city, or even located within the same country.

With this expansion of the importance and usefulness of information networks, important network security likewise becomes more important in turn. The Internet and the increased connectivity of computer systems operating on the Internet is the primary source of concern for security personnel: there are multiple other venues through which attacks can commence but the Internet is the easiest and therefore the most accessible for opportunistic exploits [17]. For those organisations that have a heavy investment in information, a well-planned network circulates this information efficiently and in a timely manner. Security within this type of system is critical to preserve and protect the integrity of data, but if inappropriately planned and executed these security systems can reduce network flexibility and still fail to prevent unauthorised access and information loss. Thus, the ability of a network to blend an advanced level of security with maximum operating flexibility must be considered carefully in any network plans. There are also other sources of significant financial loss, including computer viruses, stolen data, sabotage, network break-ins, network failure, software errors, and computer failures.

The cultivation of network security systems has been in effect as long as the use of computers themselves, but it appears as though the new format for network security systems is markedly different than that which governed prior systems, "security should be thought of as an art; it cannot be accomplished through the old 'tools and techies' model" [13].

Advanced network security has always managed to promote maximum network flexibility in addition to buffering and protecting against unauthorised computer access [4]. Yet network security has undergone multiple experiential phases in which certain aspects of security are isolated as targets for exploit — and are therefore more vulnerable — than others. "Computer security has focused on several issues over the years. In the initial stages, computer security focused largely on technical issues like encryption, access controls and intrusion detection systems exploits, yet as these issues were targeted for exploit and were in response targeted by new or improved security practices, other vulnerabilities were targeted in turn. Now, "economic, financial and risk management aspects of computer security have also become important concerns to today's organisations" [18]. However, these aspects of networked system are not primary goals of security: these are complements to, rather than substitutes for, the technical aspects of computer security.

Improving the focus on and the consistency of advanced security on multiple levels also makes it possible to include and impose an audit trail of network usage. This helps facilitate multiple outcomes within the security system, including assessment of resources and appropriation of new resources to the areas where they will be best received. Another benefit is that a user authorisation can be quickly and efficiently rescinded from the network. In general, this advanced security level can help reduce, if not eliminate, the need for costly additional security hardware such as data encryption devices.

Threats against corporate data continue to rise. More companies are storing increasing amounts of corporate data on information systems. Senior management expressed concern over the threat, but has done very little to counter the threat. Very few companies have established a dedicated information security staff. Most companies do not have a formal security policy. Many companies face problems in procuring skilled information security personnel. Senior managers fail to see information security as "value-added" contribution to "bottom line".

13.7. Perception-Driven Network Security

The association of perception with outcome within network security is, arguably, a concept that is fundamental to the basic study of network security. "It is well known that perfect security is very hard to achieve and usually the goal is to provide an acceptable security level, usually by trading security requirements with other functional and non-functional requirements of the system-to-be" [26]. Here, the emphasis on outcome and correlation between desired outcome and perceived assessment of risk is clear: certain standards that govern an acceptable level of risk are automatically accepted and integrated into expectations for security performance. The success or failure of a security system is therefore not dependent on whether it meets all risk and prevents all attacks, but rather whether it performs within the expectations of those responsible for determining its performance. Thus, all network security is inherently based within perception and its effectiveness evaluated based upon perceived outcomes.

While the need for network security continues to resolve itself within the networked culture, the literature on network security indicates problems in resolution between those aspects of security that require the greatest overall focus. One source comments that "Structure and dedicated resources breed confidence. And confidence, experts say, breeds better security. In a sea of data that fails to reveal relationships between security and best practices, the confidence factor is a welcome sight", and emphasises how the network security system of a company sets it apart as a recognisable entity among its less secure counterparts [18]. In this setting, it is clear that network security is not only identified as a needed component within a networked infrastructure but is actively used as an incentive to improve the image of the organisation.

"An organisation should not believe itself to be secure simply because it spends millions on security devices every year. The fact is that having an infinite budget and a large variety of security resources can often be more of a detriment than a benefit in many organisations. Organisations with vast resources at their command are very likely to try to solve security problems by implementing new security toys. [...] Security cannot be handled exclusively through expensive equipment, as many of us have been led to believe. Security is not a technology; it is a thought process and a methodology. Security within our technologies is nothing until security is in our minds" [13].

Williamson [34], suggests an approach to setting priorities for IT projects and some criteria for IT investment decisions that are potentially applicable to information security investment decisions. This approach consists developing a formal, quantitative way to assess the business value of proposed projects; engaging customers in a dialogue about the available resources and business needs throughout the year, not just at budget time; interviewing customers about their wants and needs; communicating frequently with customers about the Information Security (IS) department's achievements, current projects and short-term plans.

13.8. Risk Assessment and the Need for Network Security

The processes and the resources that are used to implement network security systems are fundamental concepts of the risk assessment process. As information and computer network security involves more than technology, companies are now spending more money and man-hours than necessary on cutting-edge technology. Inaccurate analysis of the company's needs can result in greater risk of information loss and higher frequency of security breaches. The literature demonstrates an almost universal accord concerning risk assessment and network security: it is recognised that risk assessment must be done in order to identify vulnerable targets and implement change based upon strategies that do achieve the greatest good.

According to Kolodzinski [20], analysing potential risk and the allocation of resources for computer network security and business continuity require strategic, long-term planning. Most companies tend to be reactive and respond with quick infrastructure solutions. A strategic approach to computer network security leads to a more efficient plan and a less expensive risk-management strategy. Aligning computer network security to corporate goals provides management with a framework for steering resources, whether it is toward infrastructure, improved controls, training, or insurance, based on a carefully thought-out process that analyses the level of risk the company is willing to absorb. Kolodzinski [20], states that this analysis leads to better computer network risk management. As a consequence, the company achieves higher levels of efficiency and cost-effectiveness essential to its profitable growth.

Yet while risk assessment must occur, the extensiveness and strategies used within the risk assessment processes are difficult to define and describe. Author Bruce Schneier uses the metaphor of a "fortress", where: "Internet security is usually described as a fortress, with the good guys inside the wall and the bad guys outside. Network owners buy products to shore up the barrier, on the logic that a stronger wall will give them better security. Flaws in the network are holes in the barricade, patches the mortar that closes them" [31].

The image of the fortress, Schneier continues, is inappropriate due to the motile composition of cyberspace. He finds that in the modern networked computer system, "there are too many of us, doing too many things, interacting in too many ways" to create a single barrier-type fortress [31].

There must be permissive ability to move between the various networked nodes, as well as throughout the Internet. Imposing single wall-like security features creates conditions where all participants are forced to define and redefine permissiveness on a continuing yet immediate basis. Schneier writes that "detection and response are far more effective — and cost-effective — than increased prevention", as the outcomes of mobility tend to be beneficial (e.g., increased website traffic) and it is only in rare conditions that the risk outweighs the benefits. In these instances, all security should be responsive. Responsive security differs from permissive security as it is identifies the processes that are ongoing and checks their validity; permissive security only allows access if the potential intruder meets predefined criteria for entry.

The image of the fortress, however, is more appealing for most network security specialists. Kolodzinski [21], reports that unprotected information and computer networks can seriously damage a business's future: the loss of classified or customer critical information, exposure of trade secrets, unacceptable business interruption, or lawsuits stemming from security breaches cannot only impact the processes as they occur but can also characterise the company as a security risk. A similar source reports that protecting company assets no longer takes the form of money or property management but is almost exclusively a digital venture [7]. Also, in *Security in the Boardroom*, it is frequently referenced that the network is only as secure as those who are responsible for it, and as upper management is asked to take control of network security is it increasingly framed through restrictions and exclusions on use [33]. *Security in the Boardroom: The Impact of Physical Network Security on Corporations and What Executives Need to Know and Do about it.* (New York: Outsource Channel Executives.) This latter source places a powerful emphasis on the use of security systems that are as controllable as possible; the image of an aggressively solid secure wall is dominant throughout the text.

Similarly, Gregg and Kim [15], frame the dialogue over network security as a program, a policy, and a desired outcome that must provide a barrier against intrusion. Their perspective is more flexible than that offered by Canavan [7], and Kolodzinski [21], where this barrier does not have to be a concrete but instead needs to be responsive and adaptive to transformed conditions. The authors clarify three core components of any risk analysis process:

— When a new program is developed, a risk analysis should be performed to establish the security state of the system. An analysis performed early on like this helps establish whether security problems exist. This is beneficial when new code or applications are developed for which problems can be found and fixed early on.
— An analysis of risk should be performed whenever changes are made to systems, processes, or programs. A risk analysis performed during this time is instrumental in uncovering vulnerabilities that occur as a possible side effect from the change.
— A vulnerability assessment should be performed periodically to examine the controls that have been implemented. It's also advisable anytime there has been a breach in security, an intrusion, or an attack. At this point, the assessment is critical because it can help uncover how the breach occurred and discover what problem in policy or system vulnerability allowed the event to occur [15].

The authors continue through stressing the relationship between information technology and network infrastructures. In their perspective, it is essential to focus specifically on the links between technology and security through assigning prerogatives to components and outcomes. These designate priories within the system, where:

After the IT systems, applications, and data assets are inventoried, the organisation must prioritise them based on importance to the organisation. This prioritisation is critical because many organisations do not have unlimited funds to implement proper security controls and security countermeasures to mitigate the identified risk from threats and vulnerabilities. This prioritisation is typically aligned to the organisation's business drivers, goals, and objectives. Then, assessing the risk of threats and vulnerabilities on an organisation's IT hardware, software, and assets can be done qualitatively, quantitatively, or via a hybrid approach [15].

With this in mind, Gregg and Kim [15], emphasise that the process of assessment and designation is a continuous, ongoing process that must form a barrier against intrusion but must also be flexible and responsive enough to function within a dynamic environment. This perspective is a combination between the ideas of Schneier [31], Canavan [7] and Kolodzinski [21]. In this respect, risk assessment processes are framed as the outcome of known and unknown factors, but the distinction between how these processes are managed characterises the status and outcome of risk assessment.

Risk assessment is also increasingly focused on the integration of the human element. Williamson [34], stresses that it is important to take egos and the need for validation into account when designing a networked security system. This includes multiple processes, such as working with committees structured to minimise the influence of any one individual or department; visiting with the business units;

communicate clearly how priorities are set so that people can anticipate project funding decisions; and developing a business case for every project, assessing its risks, its business value, and the cost of building or buying it. Additionally, Williamson [34], recommends that there needs to be a focus on interest in the constraints under which business customers operate, and staying on top of changes in the regulatory and competitive environment in which the business operates. Decision-makers within an organisation must be prepared to show how a proposed project fits with business goals: Williamson [34], does not clarify whether these persons are network security specialists or business managers, a point that would be valuable within the current literature analysis.

Most perceptions of risk assessment and the human element do not, however, concentrate on the potential value of the human element and instead pose the majority of human interactions as aggressive security threats [33]. The threat comes from the "inherent weaknesses" of those involved in the security system [33].

13.9. Management and the Perception of Network Security

Kolodzinski [21], presents potential grim scenarios for companies if they do not emphasise the importance of network security. An unprotected information and computer network means loss of data that are deemed crucial and confidential for the company's own development; loss of confidential third-party data; and business interruption or slowdowns that significantly impact the business as well as other parties. Kolodzinski [21] further stresses that any of these scenarios could result in loss of competitive advantage, lawsuit exposure, and unacceptable downtime (business interruption). Outcomes of this nature are contrary to successful organisational operations; the continued shift in power over the IT department by management indicates that these issues are recognised and that management has invested a personal stake in the processes governed by IT [33].

Because of the common technologies, inferences about IT are assumed to be applicable to information systems technology. Senior managers are becoming more and more aware of the need to address security and information technology investments within the context of the corporation's business goals. The literature on perception among management and network security indicates that there are conflicts between these two parties on the most effective ways to implement and control network security. It is without question that network security can and must occur, but there is a division in the literature at this point.

13.9.1. *Positive Perceptions of Management in Respect to Network Security*

In this domain, the leading argument is that organisations that are "extremely confident" in their network security profiles are often characterised as having greater overall security based upon the degree to which they invested in their security profiles. "That group tends to create far more structure around security within the organisation — in other words, making it a discipline and not something that happens as part of the IT group. They hire more security executives and give those executives more control over policy, spending and staff" [18]. When this occurs, the outcome is an increased focus on security as a component principle of the organisation itself; instead of being treated as a secondary or tertiary component of the company, network security is a fundamental component that helps promote the success of the organisation. In this setting, network security is no less important to success than are departments such as marketing or accounting.

Information systems (IS) executives are most concerned with ensuring that their technology goals are consistent with those of the overall business, believing that an effective organisation and usage of the company's data is a critical IS activity.

13.9.2. *Negative Perceptions of Management in Respect to Network Security*

According to the Computer Sciences Corporation [11], aligning IS and corporate goals is the primary challenge IS executives. In the survey, respondents focus on improving user productivity and collaboration through information networks. For example, the IT infrastructure improvement initiative currently under way in most respondents' companies is "enhancing or developing information networks", and the three most critical technologies being used in companies' infrastructure initiatives are wide area networks, local area networks and client/server. In addition, IS executives consider networks a key competitive tool. When asked to name the five most critical emerging technologies their companies will have to adopt by the year 2000 to remain competitive, respondents chose the Internet/World Wide Web, electronic commerce, groupware, broadband networks and network security tools [11].

13.10. Conclusions

Implementing effective network security requires planning and consideration of the risk variables involved. Using emotion-centerd decision-making, it is likely that risks that are less likely to emerge in a specific organisational setting will be given higher priority than risks that are more realistic or are more likely to manifest. However, emotion-centerd decision-making inspires system-wide change, and improves awareness of the need for planning, security analysis, and risk prevention.

There is a gap between management perceptions and security network specialist perceptions in respect to risk assessment and problems such as the need to placate investors and respond to socio-cultural events are also of interest and needs to be addressed.

Themes of conflict and problems in conflict resolution may make it difficult to prioritise the strategies that are necessary to effectively manage the network security systems. This is especially true in respect to threat identification and strategies that are intended to facilitate threat management. What is encouraging is that management and network security specialists share general — although not identical — conceptions of what constitutes a threat and why threats require intervention.

References

1. J. Albanese and W. Sonnenreich, *Network Security Illustrated* (McGraw Hill, 2004).
2. C. W. Axelrod, *Outsourcing Information Security* (Artech House, Boston, 2004).
3. D. Bailey, A philosophy of security management, *Information Security* (eds.) A. Abrahms *et al.* (2000) 98–111
4. J. M. Bailey, Effective information networks combine flexibility with security, *The CPA Journal* **59**(1) (1989).
5. J. L. Boltz, *Information Security Risk Assessment* (Government Accountability Office, Washington, DC, 1999).
6. S. Butler, Security attribute evaluation method: A cost-benefit approach, 232–240, *Proceedings of the 24th International Conference on Software Engineering,* Orlando, FL, May 19–25, 2002, ACM Press.
7. J. E. Canavan, *The Fundamentals of Network Security* (Artech House Publishers, 2001).
8. Carnegie Mellon University, Survivable systems analysis method, CERT, 2002. Retrieved 10 May 2007 from http://www.cert.org/archive/html/analysis-method.html
9. P. Y. K. Chau, An empirical assessment of a modified technology acceptance model, *Journal of Management Information Systems* **13**(2) (1996) 185–204.
10. E. Cole, R. Krutz and J. W. Conley, *Network Security Bible* (Wiley Publishing Inc, New York, 2005).
11. Computer Sciences Corporation, 1996. Retrieved 6 January 2007 from http://www.csc.com/
12. F. D. Davis, R. P. Bagozzi and P. R. Warshaw, User acceptance of computer technology: A comparison of two theoretical models, *Management Science* **35** (1989) 982–1003.
13. K. Day, *Inside the Security Mind: Making the Tough Decisions* (Prentice Hall, New York, 2003).
14. A. Dillon and M. G. Morris, User acceptance of new information technology: Theories and models, *Journal of the American Society for Information Science, New York* **93**(1) 31 (1996) 3–32. Retrieved 19 January 2007, from Joan, http://www.ischool.utexas.edu/~adillon/BookChapters/User%20acceptance.htm para
15. M. Gregg and D. Kim, *Inside Network Security Assessment: Guarding Your IT Infrastructure* (SAMS, New York, 2005) 47.
16. Y. Y. Haimes, *Risk Modeling, Assessment and Management.* 2nd edition (Wiley Blackwell, 2004).
17. H. Jahankhani, S. Fernando, M. Z. Nkhoma and H. Mouratidis, Information systems security: Cases of network administrator threats, *International Journal of Information Security and Privacy* **1**(3) (2007) 13–25, July–September 2007.
18. H. Jahankhani and M. Z. Nkhoma, Information security risk assessment, *International Conference on Information and Communication Technology in Management*, ICTM, 2005, 23–25 May 2005, Malaysia, pp. 426–437.
19. H. Kemshall and J. Pritchard, *Good Practice in Risk Assessment: Key Themes for Protection, Rights and Responsibilities*, Vol. 2 (Jessika Kingsley publishers, New York, 1998).

20. O. Kolodzinski, Cyber-insurance issues: Managing risk by tying network security to business goals, *The CPA Journal* **72**(11) (2002).
21. O. Kolodzinski, Aligning information security imperatives with business needs, *The CPA Journal* **72**(7) (2002) 20.
22. M. Malek, S. Ghosh and E. A. Stohr, *Guarding Your Business: A Management Approach to Security* (Springer, New York, 2004).
23. J. Mallery, J. Zann, P. Kelly, W. Noonan, E. S. Seagren, P. Love, M. O'Neill and R. McMullin, *Hardening Network Security* (McGraw-Hill Osborne Media, New York, 2005).
24. C. McNab, *Network Security Assessment* (O'Reilly, New York, 2004).
25. S. Mitnick, *The Art of Deception: Controlling the Human Element of Security* (Wiley, New York, 2002).
26. H. Mouratidis and P. Giorgini, *Integrating Security and Software Engineering: Advances and Future Vision* (Idea Group Publishing, 2007).
27. L. Neale, J. Murphy and A. Scharl, Comparing the diffusion of online service recovery in small and large organizations, *Journal of Marketing Communications*, **12**(2) (2006) 165–181.
28. E. M. Rogers, *Diffusion of Innovations.* 4th edition (Free Press, New York, 1995).
29. E. M. Rogers, *Diffusion of Innovations.* 5th edition (Free Press, New York, 2003).
30. J. Scambrey, S. McClure and G. Kurtz, *Hacking Exposed* (Osborne McGraw-Hill, New York, 2001).
31. B. Schneier, The enemy within: Walls don't work in cyberspace, *Wired* **11**(6) (2003).
32. B. Schneier, *Beyond Fear* (Wiley, New York, 2006).
33. M. S. Smith, *Security in the Boardroom: The Impact of Physical Network Security on Corporations and What Executives Need to Know and Do about it* (Outsource Channel Executives, New York, 2005).
34. M. Williamson, Weighing the NO's and CON's, *CIO*, April 15 (1997) 49.

Chapter 14

OPEN SOURCE INTELLIGENCE

DAVID LILBURN WATSON
Watson Business Solutions Ltd., UK

14.1. Definitions

Definitions of "open source intelligence" have varied over time and between organisations and nations. Most simply it is "unclassified information". Other definitions used are "information of potential intelligence value that is available to the general public".[a] The definitions of terms used in this paper are

Grey Literature — information, regardless of media, that cuts across scientific, political, socio-economic and military disciplines. This can be, but is not limited to, reports, of any type, studies, dissertations and theses, trade literature, market surveys, newsletters, etc.

Open Source — information that has been provided by any person or group without the expectation of privacy. That is the information, or the relationship between the author(s) and the information is not protected against public disclosure. Typically, this is information available for free or for which a fee is paid to access it.

Open Source Data (OSD) — the raw information from a primary source (Examples of sources are given below).

Open Source Information (OSIF) — data that can be deduced or edited from a number of sources that provide some filtering and validation and that information

[a]Dictionary of Military and Associated Terms, US Department of Defence, 2005.

is generic, produced in some digestible format and widely distributed. Examples of this include newspapers, books, general reports or briefings.

Open Source Intelligence (OSINT) — information that has been deliberately discovered, discriminated, distilled and disseminated to a select audience in order to address a specific question. This process applies the proven process of intelligence to information from a number of diverse sources and creates intelligence.

Publicly Available Information — information published or broadcast for general public consumption, lawfully seen or heard by a casual observer or made available to the general public at an open meeting. The information made available could be in any form or format. This includes, but is not limited to, data, photographs, maps, charts, reports, instructions, manuals, conversations, films, etc.

Validated OSINT (OSINT-V) — information to which a very high degree certainty can be attributed. It can be produced by an intelligence professional with access to classified material or assured sources for validation.

14.2. Some History

Governments, the military and organisations have been collecting, evaluating, processing and using intelligence from time immemorial. Each has done it in their own way, from government spies, through military Scouts, Exploring Officers or satellite imagery to competitive intelligence and "opposition shopping".

Some of these processes have used covert intelligence gathering processes, whilst activities like checking the prices of goods in neighbouring shops are overt.

From this point of view, OSINT has been used in one form or another for many years. In the commercial world, it is known as "competitor intelligence" and the Society for Competitive Intelligence Professionals (SCIP) has very strict rules about the legality and ethics of collection methods.[b]

OSINT has been practised by a number of different people and organisations, but really came to notice with Robert David Steele. Steele is a former US Marine Corps and Intelligence Officer for 20 years and has been called the "father of open source intelligence".[c]

In 1992, Steele wrote an article for the *Whole Earth Review* titled *E3i: Ethics, Ecology, Evolution, and Intelligence.* It presented an early view of his thinking on an alternative paradigm for national intelligence, one that stressed sharing and open sources instead of the traditional unilateral secrecy.

---

[b]www.scip.org.
[c]Microtimes — US Technology Magazine.

Between the 1993 and 2001 terrorist attacks in the United States, the Aspin-Brown Commission conducted a major inquiry into the state of the U.S. intelligence agencies and how well they were able to protect the United States against attacks and other perils in the United States and abroad. Steele testified at this commission about the tremendous value of OSINT.

The commission decided to put this theory to the test, and in March 1995 and set a contest between Steele and the U.S. secret intelligence community. The contest started on a Thursday at 17:00 and was to finish on the following Monday at 10:00. The target was to give a country analysis of Burundi.

On the Monday morning, Steele produced from

- Lexis-Nexis — the names of the top 10 journalists in the world reporting from Burundi.
- Institute of Scientific Information in Philadelphia — the names of the top 10 academics in the Burundi.
- Oxford Analytica — 22-page political–military summaries on Burundi in relation to the United States, to the UN, and to other regional issues.
- Jane's Information Group — one-page orders of battle for each tribe and one-paragraph summaries for each of the many news articles from the current Burundi crisis.
- East View Cartographic — Russian military 1:50 combat charts complete with contour lines and all cultural features.
- SPOT Image — confirmation that 100% of Burundi was immediately available in commercial imagery at the 1:50 level, cloud-free and less than 3 years old, all in the archives.

The CIA produced a regional — not country-specific — economic study and a PowerPoint chart of questionable value.

Even after this, the uptake of OSINT was minimal in the United States.

In the 2 years prior to 9/11, there were at least 15 books published by OSINT practitioner–authors calling for reform and focus on open sources. This, sadly, did not happen, but the United States started to pay attention to OSINT after that. In 2005, the "Commission on the Intelligence Capabilities of the United States Regarding Weapons of Mass Destruction" report stated that OSINT must be included in the all source intelligence process. There is now an open source centre in the United States.

Elsewhere in the world, there are other excellent OSINT cells. One of notes is in the Metropolitan Police (Specialist Operations Directorate — SO 11), where DC Steve Edwards has been awarded an MBE[d] for his work in this area. There are numerous others throughout the world, though availability of information on them in the United Kingdom is scarce.

[d]Member of the British Empire.

14.2.1. *The Great OSINT Debate*

Given the results of the Steele vs. "CIA and others" test, there is still much debate in professional intelligence circles as to the value of OSINT. In general, the intelligence professionals agree that open source data are useful and should be collected and analysed, just as data from classified sources are. The disagreement comes from the relative value of OSINT information compared to that of clandestinely obtained secret information. The result of this is that the amount of time and resources devoted to the collection and analysis of open source data is in dispute.

There are generally thought to be three differing views on OSINT and its value.

The first view holds that the policy makers actually prefer and believe that they gain more information and value from clandestinely collated intelligence than OSINT. They argue that the cost and process that is used to obtain this information means that it can be more trusted and that OSINT may be used to corroborate the findings. OSINT is not seen as being able to "get into the mind of" the target. Therefore, it is argued that the requirement is for "secrets" that can only be obtained by clandestine methods. It would be too cynical to state that the intelligence agencies around the world jealously guard their empires and budgets and like their *status quo*, without a "new kid on the block" showing the "professionals" how to do their job at less cost, less risk and faster.

The second view is that OSINT should be viewed as not only a supplement and complement to classified information, but also as a source of valuable information in its own right. This is the approach held by the supporters of Robert Steele.

The third approach is the middle ground approach. This view contends that whilst OSINT is important it may not provide all of the answers, but it will help to focus the requirements and provide insights into what needs to be found out clandestinely. This is using OSINT as the first filter.

Various estimates exist as to the public availability of information needed for the policy makers that is traditionally found by mainly clandestine sources. This varies from 80%[e] to more than 90%,[f] though as "Wild Bill" Donovan shrewdly pointed out, one needs to know where to go to find it.

14.3. Problems in Using OSINT

The professional intelligence analyst can face a number of problems in trying to effectively utilise open source data to produce OSINT, the main problem is lacking in the necessary skills to use it, being far more used to the traditional clandestine

[e] From Sherman Kent, often called the "father of intelligence analysis" in his days in the OSS and after whom "The Sherman Kent School for Intelligence Analysis" is named.

[f] Several people have suggested this including Lt. Gen Samuel V Wilson, former Director of the U.S. Defence Intelligence Agency (DIA) and Major General William Joseph "Wild Bill" Donovan, the "father" of the CIA.

intelligence processing. A close second to this can be the traditional intelligence analyst's (or their management's) automatic bias against the "newcomer" — the traditional "user resistance to change" situation. In addition to these two main problems, some others can include:

- Lack of training — to make effective use of available resources.
- Lack of access — to the Internet or Internet resources.
- Volume — the sheer volume of data that can be collected may cause problems, but on the other hand it can be used for validating what has already been collected and analysed.
- Tool — a lack of tools to effectively identify, collect analyse, visualise and present information.

14.4. Introduction to the OSINT Process

OSINT is distinct and separate from academic, journalistic or business research in that it represents the application of the proven process of national intelligence processing to a diverse number of global sources in order to produce tailored intelligence to the customer.

OSINT is the major new "weapon" in 21st century intelligence operations. It is not "new" as such, as organisations and nations have always understood the value of direct observation, reading of relevant material and legal purchasing of information services. What is new about OSINT is that there are three distinct areas that have become available. These are

- The use of the Internet as a tool for identifying, locating and sharing information.
- The explosion of published information that is available on such sources as the Internet as anyone, anywhere can publish information.
- The accessibility of many areas of intelligence source data that were formerly unavailable.

Critical to any OSINT strategy are four components:

- Sources
- Software
- Services
- Analysis

It is the analysis that is the key to successful integration of OSINT into a successful intelligence product and there are too few people with the right skills to perform this task.

Before the Internet, many people regarded the media as their prime open source. Some of these were private organisations, but a number were government sponsored such as the BBC[g] Monitoring Service and the US WNC.[h] The arrival of the Internet, as we know it, from 1992 changed forever and completely the way in which research can be carried out. It is estimated that by 2015 that half of the world will have access to the Internet.[i] The Internet is at the start of its life as a resource for intelligence professionals, it is only 15 years old and it will grow and develop over time.

14.4.1. *The Internet as an Intelligence Source*

The major problem with the Internet is that it has been oversold in the past, but too many sites are of dubious content or are biased. Anyone, anywhere, can publish any information that they want on the Internet. Sources of information are often out of date or rarely dated so that the researcher does not know if the information that they have found is current.

Another issue with Internet-sourced material is that is rarely consistently formatted, paginated, edited, filtered or verified. On account of this, if intelligence professionals do not show that they exploit the Internet, then "amateurs" will use it as a prime source, bypassing the professional intelligence analysts. They will then end up with intelligence that is from possibly dubious sources that has not been through the well-established intelligence processing cycle and probably end up with unverified intelligence of questionable accuracy and value.

That having been said, the Internet does open up multiple new sources of information and provides a means of rapid communication and information dissemination. Whilst rapid dissemination is possible, the sender has little or no control over the information once sent and the very act of sending it will leave a permanent "footprint" that can be recovered. See the section on "Anonymous Searching on the Internet", below.

Whilst these sources are opened up, it would be foolish to think that the whole of the Internet was available on the desktop after using a search engine. Often, it is necessary to search the "invisible web",[j] as there are pages on the Internet that are neither found nor indexed by the search engines. Most of this information is held in databases that can only be accessed by direct query with the results being displayed as Web pages. The size of the invisible or "deep web" is not known, but there are various estimates ranging from 50 times the size of the searchable Internet to several hundred times its size.[k] There are a number of specialised search engines that can investigate it.[l]

[g]British Broadcasting Service.

[h]World News Connection — formerly called the Foreign Broadcast Information Service (FBIS).

[i]From Dr. Vinton Cerf — acknowledged as one of the founders of the Internet.

[j]The expression was coined in 1994 by Dr. Jill Ellsworth to refer to information that was invisible to conventional search engines.

[k]Barker, Joe (Jan. 2004). Invisible Web: What it is, Why it exists, How to find it, and Its inherent ambiguity UC Berkeley — Teaching Library Internet Workshops.

[l]These vary in quality and several are currently unavailable, but an example is www.completeplanet.com.

One issue that is of serious concern to OSINT professionals is the volatility of the Internet sources. Internet sites come and go and information placed on the Internet is often out of date or never updated but still appears as a valid source for search engines.

14.4.2. *Anonymous Searching on the Internet*

As has been said above, anyone using the Internet will leave a footprint. Sometimes, this does not matter. Often with OSINT, the OSINT professional will not want anyone to know what they are doing and where they have been even if the information is all in the public domain.

The Internet is monitored by a variety of people and organisations and some of them may be linked to the requirement or deliverable. Alerting them to heightened interest may not be in the interest of the tasker.[m] There are a number of simple precautions that can be taken to make a smaller footprint, and these include:

- Use a firewall to protect the PC.
- Turn off cookies.
- Remove any Internet history stored.
- Delete cache files regularly.
- Consider storing all downloaded information on removable media.
- Consider using a dedicated Internet PC with a minimal-operating system on it.
- Consider using different PCs or Internet cafes or similar.
- Consider what details you are prepared to leave on a site where access is by registration.
- Consider logging on and off at regular intervals and using different ISPs.[n]
- Make your Internet connection as anonymous as possible.

These actions do not involve deception, but are prudent common sense.

It is worth noting that all Web sites have the ability to carry out traffic analysis to determine who has visited their site and what they have looked at. Whilst it is possible to obfuscate an identity, it is not possible to hide the fact that a Web site has been visited.

14.4.3. *Other Free Sources*

There are numerous sources of free data in addition to the Internet. Some of these are

- Newspapers
- Periodicals

[m] The person or organisation commissioning the research.
[n] There are a number of free ISPs in each country. If you use an ISP in another country, ensure that you set the clock on your PC to the time for the country so you do not stand out as a possible "outsider".

- Observation
- Radio
- Television
- Books
- Reports
- Journals
- Dissertation and theses
- Databases
- Libraries
- User groups, newsgroups and other special interest groups.
- E-mail lists (see Appendix).
- Academic or trade papers.
- Photographs and other image sources.

OSD is available from a number of different sources and is available in a number of different formats. Whilst many regard the Internet as the ultimate source for OSINT, it may be of surprise that the majority of the world's information is still in printed form. This is typically in libraries, universities or other similar repositories. The volume of information on the Internet is growing daily, but the volume of printed information is growing as well.

14.4.4. *A Comparison of Open Source versus Covert Source Intelligence*

OSINT and its counterpart, Covert Source Intelligence, can be compared at a high level as below:

Open Source Intelligence	Covert Source Intelligence
Open access	Closed access
Cheap	Expensive
Unknown reliability	Reliability known (?)
Independent	Dependent
Unstructured	Structured
Low risk	High risk
Least intrusive	Highly intrusive
Overt	Covert

As can be seen from the table above, there are distinct differences in the possible quality, accessibility and cost of obtaining raw intelligence between the two different sources.

14.4.5. *Commercially Available Sources*

There are numerous commercial sources (i.e., those that charge a fee for their services) available both on-line and off-line. They are typically formatted and structured and represent the edited and usually indexed offering of well-respected information providers. These can be used by almost anyone who wants to contract with the information service provider and there are numerous such service providers throughout the world. Some are generalist and others are highly specialised. Of the generalist service providers, probably the best known is Lexis-Nexis (used by Robert Steele and quoted above), but there are many others, some of the better ones are listed in the Appendix. If one wants to find a good service provider, the Association of Independent Information Professionals (AIIP)[o] is a good place to start. The charging structures for service providers varies between provider and this can make searching for information expensive unless the researcher is familiar with both the subject and the provider's service. There are advantages in using a specialised researcher if unfamiliar with the service provider's services.

In addition to on-line service providers, there is much available grey literature that is legally and ethically available through specialised channels or direct local access. Grey literature is usually understood as to be not available through commercial publishers and includes working papers, pre-prints, dissertations, theses, technical reports, specialised data sets and commercial imagery. There are many organisations and people producing grey literature and these range from governmental agencies, educational establishments, organisations creating documents for clients to a variety of formal and informal societies, clubs and associations.

Commercial imagery is now affordable, available and reachable to all in a way that it never was before with the availability of satellite imagery. The downside of this ease of availability is that the imagery is available to friend and foe alike — in fact usually anyone who has a credit card. In some cases, such as Google Earth, which is available even on the author's iPhone, it is free. The major downside with commercial imagery is the revisitation time allowing changes to take place between visits that are not always apparent. This limitation can be partially offset by the possible overlap of different satellites covering the same target at different visitation cycles and thus reducing the interval between successive images from a single satellite.

There are numerous sources of available information, they are of varied quality and all must go through the intelligence process to provide the final intelligence product. The ultimate intelligence expert available is the Eyeball, Human, Mk 1, where the user of the eyeball has direct experience in the subject matter that they are observing. The human is able to adjust quickly to their environment and search

[o]www.aiip.org.

out answers in a way that no other source or automated tool can. There are a number of methods of identifying the relevant expert for research, just as Robert Steele did (above). Some of these include:

- Social Scientific Citation Index (http://scientific.thomsonreuters.com/products/ssci/)
- Scientific Citation Index (http://scientific.thomsonreuters.com/products/sci/)
- Gales Ready Reference Shelf — which incorporates many databases of individual experts and information sources throughout the world (http://www.gale.cengage.com/servlet/ItemDetailServlet?region=9&imprint=000&titleCode=GAL9&type=4&id=111000)
- Dow Jones Factiva for searching the world's leading business news publications (http://www.factiva.com/)
- Specialised searches using search engines for the specific type of expert required
- Relevant professional associations from phone directories, local libraries or Chambers of Commerce

14.4.6. *Software*

If one has identified or collected raw data, it needs to be processed and analysed prior to final production. There are three main processes that need to be undertaken and these are

- Pre-processing of digital data, including conversion to digital media and language translation
- Collaborative work tools including referencing and indexing, visualisation and modelling
- Management and publishing of the intelligence product

There is some software that will perform some of the functionality defined above but no tools that can perform the whole process. The use of software will depend on the individual analyst's choice or that of his employers for each of the tasks that make up these three main processes.

14.4.7. *Services*

There are a number of service providers who specialise in OSINT services that are available to the intelligence researcher and analyst. These include:

- Collection Services that include, but are not limited to
    - Online collection
    - Surveys
    - Document collation
    - Reconnaissance
    - Surveillance

- Processing Services that include, but are not limited to
  - Data conversion
  - Hard copy to digital conversion
  - Language conversion
  - Database design and build
  - Modelling
  - Visualisation
- Analysis and Production Services that include, but are not limited to
  - Analysis
  - Interpretation
  - Report production

14.5. The OSINT Cycle

The OSINT life cycle is a defined process as shown in the diagram below and this is described below:

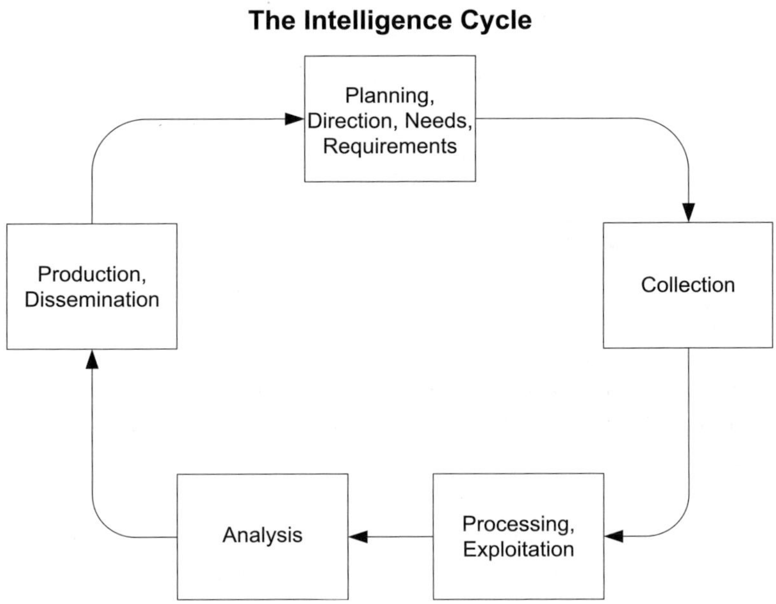

14.5.1. *Planning*

The most important stage in the OSINT process is the actual planning and requirements definition stage. If this is not correct and the requirements captured correctly, the results will almost certainly not meet the requirements of the tasker. The "tell me everything you know about..." approach is doomed to failure from the start. The tasker needs to be able to define in the most succinct way what they need to know and why. By understanding the needs and their context the OSINT professional can undertake their tasking in the most efficient manner. By getting this right and using feedback from intelligence products provided, the OSINT professional can maintain their focus and even sharpen it.

14.5.2. *Collection*

Collection of open source data is the process of turning the requirements definition from the tasker into an intelligence need. This requires the development of a collection strategy that will research available resources. The research that is carried out as part of the collection process matches the requirements definition to available intelligence and identifying new sources for them. The range and amount of open source data that is available can be enormous, so that precise planning and focussed collection are essential to prevent intelligence staff being overwhelmed with "information overload".

As much open source data can be of dubious value and may be out of date, one of the essential tasks in the collection process is the identification of subject matter experts who can validate sources.

As intelligence products have, necessarily, a short "time to market" additional time spent in collection and validation of open source data or sources will reduce the time available for analysis. Therefore, the OSINT professional should endeavour to not have to revalidate every piece of open source data or source each time it is collected but try to maintain a supply of known, trusted and validated sources and subject matter experts to validate any dubious ones.

Typically, source reliabilities will be assigned to sources, and these will range from "reliable" to "unreliable" and "unable to judge". In the same way as sources can be scored, the open source data itself can be judged. This is usually categorised between "reliable" though "unreliable" to "misinformation" and "deception" and of course "unable to judge".

It is also essential to ensure that copyright is acknowledged where appropriate as well as any relevant privacy legislation.

14.5.3. *Recording the Results*

Once the open source data have been collected, it is necessary to save that which satisfies the requirement. Saving the source data allows the OSINT professional to recover the information later as well as cite the sources. This may be done as a hard copy records but it is easier to store and process in electronic form. There are several methods used for this:

- Bookmarking favourites — using a browser's bookmarking or favourites facility.
- Saving content — either whole or part of a Web page — also to include as part of the record the URL and date downloaded.
- Downloading files — whatever sort of file it was, text, audio, video or other — also to include as part of the record the URL and date downloaded.

- Save a Web page — in whatever format required — also to include as part of the record the URL and date downloaded.
- Save the whole or part of a Web site — download part or all of a Web site — also to include as part of the record the URL and date downloaded.

14.5.4. *Processing*

Once the relevant source data has been collected, it needs an experienced OSINT professional to evaluate and sort the data. Much of the source data may be in non-digital form or in different languages and so paper evaluation and translation are essential in the processing part of the cycle.

Much of the source data that are available, unless it is from known and trusted sources, may well be subject to bias or fabrication by the author for their own purposes. Sorting fact from fiction, supposition or just fantasy is an essential task to be carried out as part of this process. The pedigree of the source is essential to understand and evaluate.

The criteria for evaluation of Internet open source data for reliability has been established and is outlined below.[p]

14.5.4.1. *Accuracy*

The following questions need to be answered:

- Is the information provided consistent with other sources?
- Can the information be compared to a validated source?

14.5.4.2. *Credibility*

The following questions need to be answered:

- Does the Web site clearly identify itself?
- Are there contact details on the Web site?[q]
- Do many other sites cite or link[r] to the source?
- Is the Web site hosted by a free or cheap service?
- Are there hit counters on the Web site?[s]

14.5.4.3. *Currency*

The following questions need to be answered:

- Does the material on the Web site appear up to date?
- Can material on the site be validated for currency?

[p]Based on "Web Site Evaluation Checklist", Joe Barker, University of Berkley, 2002 http://www.lib.berkeley.edu/TeachingLib/Guides/Internet/EvalForm.pdf.
[q]www.samspade.org is a favourite site for checking ownership of Web sites and many other related matters.
[r]It is possible to see who has linked to a Web site by using http://www.wholinks2me.com/
[s]Care should be exercised when using these counters as they may be manipulated.

14.5.4.4. *Objectivity*

The following questions need to be answered:

- Does the Web site have a bias?
- Who does the Web site represent?
- Does the Web site speak for a known organisation?
- Who else links to the site or does the site link to?

14.5.4.5. *Relevance*

The following questions need to be answered:

- Is the material relevant to the question to be answered?

14.5.5. *Production*

Typically, the deliverable of the OSINT process is the intelligence report. Reporting formats will vary between organisations, but there will be a degree of commonality for all reports. Each report should show the date of collection of source material, have a summary and use various sections to answer the requirements defined by the tasker.

Depending on the organisation, or the tasker's requirements, the report could be split by source type or have each of them mixed and cross referenced.

Typically, a list of sources is added to the report and gives the ranking of reliability of the source, a URL for the resource and a summary of the source.

14.6. Dissemination

The major difference between OSINT and other intelligence disciplines is that the latter are usually classified and the former can be shared with anyone. This also provides a feedback loop so that the intelligence product can be fine tuned if required.

14.7. Summary

Although many intelligence professionals have "turned their noses up" at OSINT, there has been a groundswell, albeit in pockets, to use OSINT as a recognised and valued intelligence source.

OSINT is not something everyone can do from day 1, but given appropriate training and tools it is possible to use OSINT both as a primary source and as a collaborator for other sources.

As has been seen from the examples above, there are people doing this today with great success, but they have all applied the intelligence cycle to the process to end up with focussed and targeted intelligence products.

14.8. Appendix

Some sources of online information:

Source Type or Function	Source Name	URL
Broadcast monitoring	BBC monitoring	http://www.monitor.bbc.co.uk/ and http://www.bbcmonitoringonline.com/mmu/
Broadcast monitoring	World news connection	http://wnc.fedworld.gov/ and http://wnc.dialog.com/
Commercial sources of aerial and satellite imagery	Virtual terrain project	http://www.vterrain.org/Imagery/commercial.html
Current awareness	Oxford analytics	http://www.oxan.com/
Defence monitoring	Jane's	http://www.janes.com/
Foreign affairs monitoring	Country watch	http://www.countrywatch.com/
Global risk monitoring	Political risk services	http://www.prsgroup.com/
Maps and charts	East view cartographic	http://www.cartographic.com/
Shipping information	Lloyd's register	http://www.lr.org/Industries/Marine/Services/Shipping+information/
Reverse phone look-ups (USA and Canada)	555-1212.com Inc	www.555-1212.com
Identity verification (UK)	GB Group plc.	www.gb.co.uk
Address verification	Experian	www.qas.com
Email, phone and address verification (USA)	Intellius	www.theultimates.com
Yahoo people finder (US)	Yahoo	http://people.yahoo.com/
Email lists		http://navigators.com/email_lists.html

Chapter 15

DIGITAL IDENTITY MANAGEMENT

ELIAS PIMENIDIS
School of Computing, IT and Engineering
University of East London, UK

Organisations are currently managing sensitive information of millions of individuals on an online basis. Increasingly people have little control over their own information and this is the real challenge facing organisations and governments in an era of exponential creation, networking and duplication of data, most of which is identifiable in nature.

Traditionally, identity management has been considered as the process of managing information for a person's interaction with an organization's information systems and assets.

Nowadays, there is a key fundamental business shift away from the traditional approach in which a user's identity was managed manually on a system-by-system basis.

Federated identity management, which supports multiple entities connected within a circle of trust, is one of the major initiatives growing out of Web services that will provide substantial benefits to corporations and consumers. Federated identity management (IDM) refers to the ability to establish trust relationships between various security domains to enable the passing of authentication, authorization and privacy assertions. This is a key aspect of identity management in the context of business-to-business integration, typically when Web services technology is used.

The European Union is actively pursuing an interoperable pan-European e-IDM since 2006. Under this initiative, EU governments have agreed to facilitate, as a matter of high priority, the establishment for mutual recognition of national electronic identities for public administration Web sites and services.

The key reason for promoting the interoperability of e-identification is the growth of demand for cross-border services, which needs to be met by equivalent cross-border framework of cooperation. Once this initiative reaches a good level of success and an acceptable maturity level, it could possibly prove the model for a worldwide interoperable e-identification system, increasing the efficiency of transactions, and significantly enhancing the effectiveness of border control systems.

15.1. An Overview of Identity Management

With the evolution of the Internet and its ever-increasing use by individuals organizations are managing sensitive information of millions of individuals on an online basis. Access to such sensitive details has to be provided and controlled with the outmost efficiency while protecting and promoting individual privacy. There is more personal information out there than ever before, and most of it is controlled by others. Increasingly people have little control over their own information. This is the real challenge facing organizations and governments in an era of exponential creation, networking and duplication of data, most of which is identifiable in nature.

Identification requirements are everywhere and increasing. Systems users have multiple identities which need to be managed. In the online digital environment, identification demands are becoming more frequent. Increasingly, more and more granular information is being collected about individuals by third parties, and these data are being used in novel ways, for novel purposes — not all of which benefit the individual.

There is a growing disjunction with the bricks-and-mortar world where individuals can often demonstrate our identity (or credentials) via an ID document offered for visual inspection. But in the faceless online world, identification credentials are often recorded in databases, compared or collated with other data, and stored indefinitely for further uses.

At the same time, the identity of other entities online is becoming harder to verify. Organizations often simply do not know who they are truly dealing with online, or how accountable they are with respect to the handling of personal information [1].

Identity management in the digital world does not have a clearly defined meaning, but technology-based identity management, in its broadest sense, refers to the administration and design of identity attributes, credentials, and privileges. Even the above attempt to shed more light into the context of the term does not resolve the issue. A single definition is not enough to provide full coverage of the issues and the topics under the overall umbrella of this significant topic. Below is a useful group of widely accepted explanations of terms that are closely related to digital identities and their management:

Identity management (IDM) — Systems and processes that manage and control who has access to resources, and what each user is entitled to do with those resources, in compliance with the organisations' policies.

Enterprise IDM — Enterprise identity management is understood as IDM that primarily serves the enterprise's needs. Control is exercised by the enterprise instead of by the individual (see in contrast user-centric IDM).

User-centric IDM — It is IDM that seeks to place administration and control of identity information directly into the hands of individuals.

Federated IDM — A relationship that allows the authentication of an entity verified by one identity authority to be recognised by other identity authorities in the federation [ISO/IEC 2005]. Federated IDM enables single sign-on.

Single sign-on — A form of software authentication that enables a user to authenticate once (identity is verified once) and be granted access to every application that the user has been authorised for [5].

Traditionally, identity management has been considered as the process of managing information for a person's interaction with an organization's information systems and assets.

Nowadays, there is a key fundamental business shift away from the traditional approach in which a user's identity was managed manually on a system-by-system basis.

In the past when a new employee was hired, accounts and permissions were set for the workstation, network, enterprise resource planning system and other corporate applications. Different administrators were responsible for configuring access to each of the different environments.

Today, there are a vast number of applications, platforms, services and systems that an employee may access. Users have a difficult time managing multiple usernames and passwords. As a result, security vulnerabilities occur when users select poor, easy-to-remember passwords or use the same password at a collection of independent sites.

The challenge is shifting in focus from supporting employees only, to providing access to external stakeholders such as customers, suppliers and partners. Organisations are looking at IDM to manage this complex process and to provide reliable, efficient and controlled access to resources. The goal is to provide the right people with the right access at the right time and prevent the possibility of identity fraud and theft. This involves establishing new processes and standards, a new level of relationship and trust, and new technologies—all of which cross organizational boundaries.

The basic components of IDM are

- Authentication — This is the process of verifying the identity of a person so access to protected resources can be properly granted or denied. Common approaches include passwords, digital certificates, biometrics, smart cards and smart tokens. These systems may be implemented as single sign-on systems.
- Authorization/access control — This is the process of ensuring that users are given access to applications or resources that they are entitled to review or use. Access control can be user-based, rule-based, role-based or a combination.
- Enterprise directory — This is a central data repository for holding and managing user identities and access privileges can also store rules and policies for the IDM architecture.
- User management — It includes a collection of systems that support the creation, maintenance, suspension, deletion and use of digital identities. It also features user self-service and the automation of the user management procedure [8].

15.1.1. *Contextual Identity*

The online world is a complex new environment that is constantly evolving offering users new possibilities, new ways of communicating, new ways of creating knowledge, sharing it and transacting with other individuals or organisations. Social structures online have to be established within a short time — unlike their real world counterparts. It is easy to consider that procedures based on personal contact or paper can be transformed into digital procedures for use online. But below the surface, more fundamental differences between the offline and the online world exist. Examples of these are the relative permanence of memories and the ease with which experiences can be shared between many actors across time and space barriers.

These differences are both qualitative (e.g., automated decision making) and quantitative (e.g., more data collected and stored for a longer period) in nature.

The speed of developments and potential irreversibility of their effects requires urgent attention on issues such as identity, trust, security and privacy.

The — sometimes conflicting — interests and issues that have to be reconciled are increasingly well understood. For example for such a conflict is an interest in identifying trading parties on one hand and providing anonymity on the other. The convenience of "portable" online identities is another example; users do not want to fill in similar forms for each service, yet there is the risk of disclosing more than is required. National security interests — sometimes positioned as overriding civil liberties in public debates — increases the need for proper data protection. And finally, while customer data are an important business asset, they can become a business liability in complying with data protection legislation [5].

In today's digitized world, personal information provided in different contexts varies. Identities may be used in or out of context. Identities used out of context generally do not bring desired results. For example, trying to use a coffee card to cross a border is clearly out of context. On the other hand, using a bank card at an ATM, a government-issued ID at a border, a coffee card at a coffee shop and an MS .Net Passport account at MSN Hotmail are all clearly in context.

In some cases, the distinction is less clear. An individual could use a government-issued ID at your ATM instead of a bank-issued card, but if this resulted in the government having knowledge of each financial transaction, many people would be uncomfortable. One could use a Social Insurance or Social Security Number as a student ID number, but that has significant privacy implications, such as facilitating identity theft. And you can use a .Net Passport account at some non-Microsoft sites, but few sites chose to enable this; even where it was enabled, few users did so because they felt that Microsoft's participation in these interactions was out of context.

Numerous digital identity systems have been introduced, each with its own strengths and weaknesses. But no one single system meets the needs of every digital identity scenario. Even if it were possible to create one system that did so, the reality is that many different identity systems are in use today, with still more being invented. As a result, the current state of digital identity on the Internet is an inconsistent patchwork of ad hoc solutions that burdens people with different user experiences at every Web site, renders the system as a whole fragile, and constrains the fuller realization of the promise of electronic services and digitized transactions [1].

15.1.2. *Internet's Problems or Identification Problems?*

The Internet, by design, lacks unified provisions for identifying who communicates with whom; it lacks a well-designed identity infrastructure. Instead, technology designers, enterprises, governments and individuals have over time developed a collage of isolated, incompatible, partial solutions to meet their needs in communications and transactions. The overall result of these unguided developments is that enterprises and governments cannot easily identify their communication partners at the individual level. Given the lack of a proper identity infrastructure, individuals often have to disclose more personal data than strictly required. In addition to name and address, contact details such as multiple phone numbers (home, work and mobile) and e-mail addresses are requested. The amount and nature of the data disclosed exceeds that usually required of real world transactions, which can often be conducted anonymously — in many cases the service could be provided without any personal data at all. Over the long run, the inadequacy of the identity infrastructure, that takes the above into account, affects individuals' privacy. The availability of abundant personal data to enterprises and governments has a profound impact on the individual's right to be let alone as well as on society at large.

Many of the problems facing the Internet today stem from the lack of a widely deployed, easily understood, secure identity solution.

A comparison between the bricks-and-mortar world and the online world is illustrative: In the bricks-and-mortar world, you can tell when you are at a branch of your bank. It would be very difficult to set up a fake bank branch and convince people to do transactions there. But in today's online world it is trivial to set up a fake banking site (or e-commerce site ...) and convince a significant portion of the population that it is the real thing. This is an enormous identity problem. Web sites currently do not have reliable ways of identifying themselves to people, thus enabling impostors to flourish. What is needed is reliable site-to-user authentication, which aims to make it as difficult to produce counterfeit services in the online world, as it is to produce them in the physical world.

Conversely, problems identifying users to sites also abound. Username/password authentication is the prevailing paradigm, but its weaknesses are all too evident on today's Internet. Password re-use, insecure passwords, and poor password management practices open a world of attacks, in and of themselves. Combine that with the password theft attacks enabled by counterfeit Web sites, and man-in-the-middle attacks, and today's Internet is an attacker's paradise.

The consequences of these problems are severe and growing. The number of "phishing" attacks and sites has skyrocketed. There are reports that online banking activity is declining due to account fraud and identity theft which are frequently the result of a single-factor, i.e., authentication exploitation.

15.2. The Strategic Need for Identity Management

The issue of authentication is a crucial one and both private organisations as well as government agencies have to develop their own strategy to ensure that authentication procedures for users accessing their systems and resources are fast, effective and efficient. To this effect organizations have to develop their own strategies for authentication and consequently identity management.

Banks are a natural partner in authentication strategies. Banks already provide authentication and guarantees to companies along the financial supply chain, and may be able (and can be urged by companies) to do the same along the physical supply chain. Around the world, banks must already authenticate identities, and verify fund flows, to meet various laws, such as those targeting money-laundering and terrorism. Banks can similarly endorse e-commerce and e-government activities, issuing digital certificates to validate identities and providing non-repudiation, or verifying the authenticity of the origin and delivery points in a transaction so that neither party can disavow their participation. In this way, banks assume the liability associated with e-commerce risk from participating companies, a process in keeping with their traditional role of assuming and managing risk. Furthermore, banks already collect most of the customer information they need to provide digital authentication, and unlike most non-bank authenticators, they are legally bound to collect, house and protect that data responsibly [2].

15.2.1. *Identity Authentication Strategy to Support Global Business Growth*

As the global network of suppliers, customers and payments becomes ever more complex, companies must be able to verify who they are dealing with — and be able to grant or deny them appropriate access to information and assets. It is therefore a strategic imperative for companies to design and execute a digital-authentication strategy. Those that discount authentication as an IT strategy, or wait for authentication standards and norms to reach best practice, will be at a competitive disadvantage.

Companies need identity authentication mechanisms that protect them properly against malicious use, not just monetary loss. Companies are rightly concerned about the gamut of risks associated with identity authentication, from phishing and pharming to monetary fraud. Identity authentication strategy must look beyond the potential monetary loss from payments transactions and make sure the company's reputation, data, and other intellectual property and assets are also protected.

Organisations must not underestimate the risks of improper authentication management having a major impact on their reputation. Hackers that manage to fraudulently portray themselves as a company can defame that company and steal usernames, passwords, credit card information and other personal information about its customers. While the intent of these thieves may actually be just to defraud the company's customers, irreparable harm will have been done to the reputation of the company itself, whether the hackers are successful in their attempts or not. Companies must therefore be proactive in protecting themselves against such breaches of corporate security [2].

The benefits of a successfully executed IDM strategy include:

- Minimised cost — An effective IDM solution can reduce the time users must wait to do their jobs by speeding the provision process for permissions and access rights.
- Better customer service — For example, the successful implementation of single sign-on to multiple applications can reduce the irritation of creating usernames and passwords for each application. IDM also addresses the issue of users selecting guessable usernames and passwords and repeating them at different sites.
- Improved security — When employees leave or change jobs, access rights need to be revised in a timely manner to make the organisation less vulnerable to risks.
- Reduced privacy risk — IDM solutions can reduce the risk of privacy breaches. Legislation requires companies to safeguard user privacy, guarantee the accuracy of corporate financial data and audit efforts to ensure compliance. Legislation includes the European Union Data Protection Directive (implemented through various laws and acts in the member states) and the U.S. Sarbanes–Oxley Act, Gramm–Leach–Bliley Act and Health Insurance Portability and Accountability Act (HIPAA).
- Personalization — IDM allows organizations to personalize content and delivery methods for the user and provide an improved self-service environment.
- Infrastructure — By providing an IDM architecture with reusable integration and security components, an organisation can reduce application development time and provide services more quickly.

15.2.2. *An Identity Metasystem*

A universal adoption of a single digital identity system or technology is unlikely to occur and a successful and widely deployed identity solution for the Internet requires a different approach — one with the capability to connect existing and future identity systems into an identity metasystem. A metasystem, or system of systems, would leverage the strengths of its constituent identity systems, provide interoperability between them and enable the creation of a consistent and straightforward user interface to all of them. The resulting improvements in cyberspace would benefit everyone, ultimately making the Internet a safer place with the potential to boost e-commerce, combat phishing, and solve other digital identity challenges.

An identity metasystem could make it easier for users to stay safe and in control when accessing resources on the Internet. It could allow users to select from among a portfolio of their digital identities and use them for Internet services of their choice, where they are accepted. A metasystem could enable identities provided by one identity system technology to be used within systems based on different technologies, provided that an intermediary exists that understands both technologies and is capable and trusted to do the needed translations.

It is important to note that the role of an identity metasystem is not to compete with or replace the identity systems that it connects. Rather, a metasystem should rely on the individual systems in play to do its work.

15.2.2.1. *Architecture of a proposed solution*

For a digital identity solution to be successful, it needs to be understood in all the contexts when an individual may wish to use it to identify himself. Identity systems are all about identifying oneself (and his/her properties) in environments that are not related to the individual. For this to be possible, all systems involved, i.e., those where the individual needs to digitally identify himself must be able to speak the same digital identity protocols, even if they are running different software on different platforms.

Such a solution, in the form of an identity metasystem, has already been proposed, and some implementations are well under way. The identity metasystem is based upon an underlying set of principles called the "Laws of Identity". The Laws are intended to codify a set of fundamental principles to which a universally adopted and sustainable identity architecture must conform.

By allowing different identity systems to work together in concert, with a single user experience, and a unified programming paradigm, the metasystem shields users and developers from concerns about the evolution and market dominance of specific underlying systems, thereby reducing everyone's risk and increasing the speed with which the technology can evolve.

The Laws of Identity seek to put users in control of their own identities, their personal information, and their online experiences. In the metasystem, users decide how much information they wish to disclose, to whom, and under what circumstances, thereby enabling them to better protect their privacy. Strong two-way authentication of identity providers and relying parties helps address phishing and other forms of fraud. Identities and accompanying personal information can be securely stored and managed in a variety of ways, including via the online identity provider service of the user's choice, or on the user's PC, or in other devices such as secure USB keychain storage devices, smartcards, PDAs, and mobile phones.

Further, the identity metasystem enables a predictable, uniform user experience across multiple identity systems. It extends to and integrates the human user, thereby helping to secure the machine–human channel.

Participants in the identity metasystem may include anyone or anything that uses, participates in, or relies upon identities in any way, including, but not limited to existing identity systems, corporate identities, government identities, Liberty federations, operating systems, mobile devices, online services and smartcards. Again, the possibilities are only limited by innovators' imaginations.

The metasystem would store no personal information, leaving it up to individual identity providers to decide how and where to store that information. The identity metasystem would not be an online identity provider for the Internet. It would provide a means for all identity providers to co-exist with and compete with one another — all having equal standing within the metasystem. And while Microsoft charged companies to use the original version of Passport, no-one will be charged to participate in the identity metasystem.

In fairness, the Passport system itself has evolved in response to these experiences. It no longer stores personal information other than username/password credentials. Passport is now an authentication system targeted at Microsoft sites and those of close partners — a role that is clearly in context, and one which users and partners are more comfortable. Passport and MSN plan to implement support for the identity metasystem as an online identity provider for MSN and its partners. Passport users will receive improved security and ease of use, and MSN Online partners would be able to interoperate with Passport through the identity metasystem [1].

15.3. Identity Management Tools, Techniques and Systems

Many companies use Internet-based digital certificates for transactions and communications, including government interactions. According to a survey by the EIU nearly one-half of all respondents already use Internet-based digital certificates for e-commerce transactions or communications. The certificates are also used for a variety of interactions with local or national government agencies. For example, 48% of companies use digital certificates in filing their corporate taxes electronically, and 44% use them to file sales taxes (e.g., value-added tax). Major reasons that non-users give for shunning the certificates include government restrictions on usage and the failure of certificates to interact with corporate systems [2].

Federation is the dominant trend in identity management. But many users still are not sure what federated ID management is, how it can benefit them or how they can implement it as part of the evolving new data centre architecture.

Essentially, federated ID management is a result of the modern world of distributed network services and refers to establishing trust relationships among decentralized security and policy domains. With a federated ID environment, a layer of abstraction is implemented over legacy identity and security domains. Using standardized methods, each domain can share its local identity and security information while retaining its own internal directory, metadirectory, account provisioning and public-key infrastructure services [4].

Federated identity management, which supports multiple entities connected within a circle of trust, is one of the major initiatives growing out of Web services that will provide substantial benefits to corporations and consumers.

This refers to the ability to establish trust relationships between various security domains to enable the passing of authentication, authorization and privacy assertions. This is a key aspect of identity management in the context of business-to-business integration, typically when Web services technology is used.

For example, if a company uses Web services to integrate its application with its suppliers, one approach will be to create additional user accounts for all eligible users in the various supplier companies. Managing the new accounts can create an administrative nightmare for the company. One alternative is to implement a delegated user system in which each supplier has one administrative account and is responsible for managing the accounts of its individual users. An issue with this approach is that delegated accounts become costlier as more suppliers are added. Also, the supplier has an additional burden of maintaining the accounts of its users who are accessing the system. Federated identity addresses this situation by enabling the suppliers to directly link the user information in each internal IDM system.

The primary features of a federated identity solution are

- A single sign-on (SSO) that operates across enterprises.
- The capability to link and unlink an account in one system to another. This identity mapping feature tells an application that Johnd is the same as Jday and that Jdoe is John Day and not Jane Day.
- A basis for trust between systems — having one company trusts the information and identity credentials it receives from another.
- A secure system for sharing and managing user authorization data between organisations.

Currently, SAML is in the lead and gaining momentum with a good adoption rate. SAML stands for Security Assertion Markup Language. SAML provides the basic definition and structure of security assertions, which are trusted statements used to communicate authentication and authorization information of a user to a remote service. These assertions contain information about end users, Web services or any other entity that can be assigned a digital identity. It is developed by the Organization for the Advancement of Structured Information Standards (OASIS).

SAML V2.0 uses XML Signature [XMLSig] to implement XML signing and encryption functionality for integrity, and source authentication. SAML V2.0 uses XML Encryption [XMLEnc] to implement confidentiality, including encrypted identifiers, encrypted assertions, and encrypted attributes [5, 7].

Federated Identity Management is mostly used for cross-domain single sign-on (SSO) to save user time and to eliminate costly password resets. With SSO, users

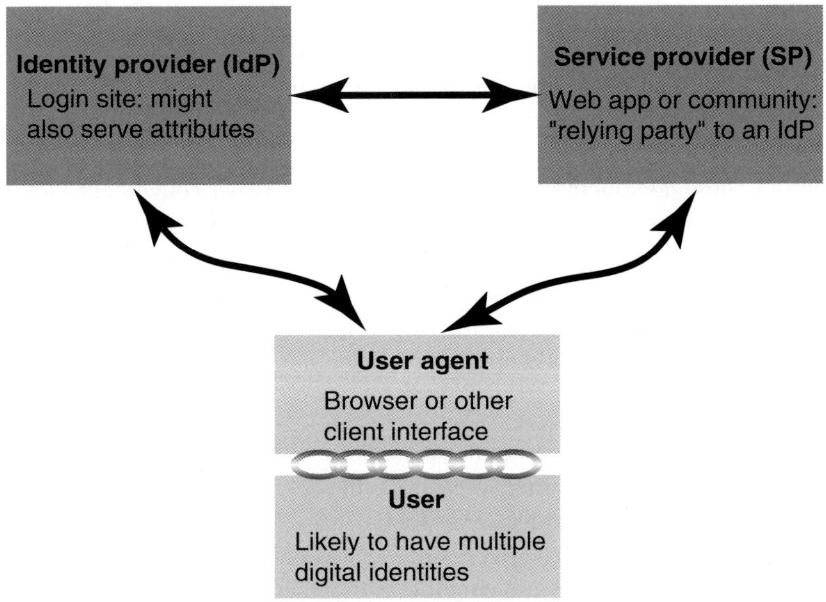

Fig. 15.1. SSO example.

can authenticate at one location and subsequently access protected resources at other locations. For example:

Enterprises that outsource functions, such as human resources and health insurance, turn to SSO (Fig. 15.1).

Many governments adopt SSO for citizen portals and cross-agency application access by civil servants.

Academic institutions often manage access to research databases at partner universities with SSO.

In all cases, the same players apply, namely:

Users and the *agents*, such as browsers, through which they communicate online.

Service providers (SPs) that offer Web applications and rely on an external source for crucial input into their authorization decisions.

Identity providers (IdPs) that authenticate users and manage attributes of common interest.

To protect the identity and application data being passed around, IdPs and SPs typically establish a *circle of trust* with technical and business agreements that define their respective responsibilities. Often, a circle of trust assumes a star-shaped topology with a single IdP and many SP spokes that trust the IdP but not necessarily each other. The IdP role is more complex because it must furnish the authentication infrastructure along with most of the security measures. SPs, looking to simplify by outsourcing to the IdP, expect to be offered toolkits that ease deployment.

Besides the commercial world and private organisations, governments are actively involved in digitising their services in a quest for improving citizens' transactions experience, efficiency of operations and effectiveness of their services.

A major function of government agencies is the issuing of licenses, permits, authorisations, registrations, certificates, and other official documents. Users who can print out these documents at home, work, or some other convenient location can avoid making time-consuming trips to government offices or waiting in frustratingly long queues. To prevent counterfeiting and fraud, however, there must be a way to print out original documents without changing their contents. The Korean Institute of Public Administration (KIPA) identified four key requirements for securely delivering more data and a wider range of common documents to users over the Internet:

- Most organisations preparing online services for citizens had existing information systems. New document-issuing systems should therefore be modularised for easy, cost-effective migration into legacy systems.
- Korean government agencies affix a unique seal on official printed documents to prevent their fraudulent use and copying. A comparable method of authenticating agencies that issue public documents online for printing on standard paper was needed.
- Hackers must be prevented from forging or altering the contents of public documents issued online.
- Various types of personal data pass between document-issuing organisations and users over the Internet, while information related to documents to be printed can be stored in computers. Illegal users must be prevented from accessing this data using functions such as printer spool control and access control to temporary files.

To address these challenges, KIPA has recommended that government agencies that currently issue or plan to issue documents online implement several technologies. For offline printing, these include

- High-density 2D bar codes that can store original documents and thereby preventing their forgery or alteration
- Digital signatures with public-key infrastructure (PKI) certification to authenticate organizations and documents
- Digital watermarking to protect the official seal of document-issuing organizations [3]

The above example is a case of a dual identity management challenge. The government agency responsible for issuing the document has to ensure that it manages the identity of the user (the citizen requesting the service) as well as managing and guaranteeing the authentication and authenticity of the printed document.

Identity management at e-government level is more complex than that of managing identities in the corporate world. One of the biggest challenges for e-government is the diverse number of agencies involved. Not only are citizens and local firms customers, but in many cases, the agencies can be seen as customers as well. But not all customers are equal with respect to level of information sharing; information sharing across networks between the police department and a city operated water department may be done at a different level than the police department and the public citizens of the area. One of the most successful implementations is that of e-government identification systems in Austria.

The Austrian citizen card is rather a concept that will show a variety of appearances. The health insurance card rolled-out to each Austrian citizen in 2005 is one of these appearances. The public identity card available as a smart-card in 2002 is another one. Further examples of the citizen cards are member cards of the Austrian computer society, or SSCDs shipped by CSPs that issue qualified certificates. Also bank cards for automated teller machines are following the citizen card concept since May 2005.

The citizen card provides electronic signature for authentication and electronic identity for identification and the principle can be applied to any smart card supported by a chip and PIN mechanism issued by any authorised organisation in Austria. The authorities use different personal identifiers derived from the source PIN of the natural person concerned and from the relevant procedural sector. This is an irreversible cryptographic derivation, which is used only once. This means that the source PIN cannot be retraced from the derived identifier. Similarly, it is impossible to derive a new identifier for another sector from an existing derivation.

Furthermore, Austria has gone a step further into integrating foreign e-IDs into its system by supplying a "Substitutional Source PIN" for foreign identity cards. This has so far been successfully implemented for the identity cards of Italy, Belgium and Finland. This is at the time of writing the most successful attempt to achieve interoperability of e-identification for government services across the European Union [9].

15.4. E-Identification Interoperability for e-Government Services

On the 25th of April 2006, the EU commission adopted the eGovernment action plan recognising the need for interoperable pan-European eIDM. Under this action plan EU governments have agreed to facilitate, as a matter of high priority, the establishment for mutual recognition of national electronic identities for public administration Web sites and services. The action plan foresees full implementation by 2010.

Electronic identities are fundamental for secure access to and convenient use of eGovernment services in Europe. Prior to the above date, a number of Member States have introduced electronic ID card (e-ID) schemes, whilst others are in various stages of implementation and planning. In researching the different approaches from governments and agencies across the continent and in particular the EU, it appears that there are a number of different technologies used to allow secure access to government services. In order to prevent these developments from creating new digital barriers across borders, a set of minimum requirements and common standards must be agreed to enable European e-ID solutions to interoperate.

One of the biggest challenges for e-government is the diverse number of agencies involved. Not only are citizens and local firms customers, but in many cases, the agencies can be seen as customers as well. But not all customers are equal with respect to level of information sharing; information sharing across networks between the police department and a city operated water department may be done at a different level than the police department and the public citizens of the area. It is important to look at each of these potential interactions as communication channels that need to be defined in terms of trust and content. These different entities each have a role in the community's response to a cyber-security event, and the exercise is structured to explore this aspect of emergency operations. Managing these multiple independent relationships is a challenge that grows exponentially with the number of channels. Thus the issue of creating secure and interoperable eIDM infrastructure becomes more complex.

e-ID schemes need to be able to authenticate users and to support a digital signature facility that can be consistent in an e-Transaction process. e-ID cards can incorporate advanced security features (such as biometric identifiers) for convenient proof of identity of a person. For governments, e-ID is therefore both a secure replacement for paper-based identity schemes and a reliable key to identify and authenticate users of e-enabled public services.

Similarly, as companies and organizations grow, they have multiple systems for managing and using digital identities. The complexities that result from having multiple identity systems generate high costs, management overheads and security vulnerabilities; e-ID can solve the identity and access management challenges, control environments and reduce complexity. However, the road to success is not an easy one.

The most common technology used in the applications reviewed is that of e-ID Cards. The basic concept of the e-ID card was introduced by TeleTrust in the TeleTrust Token 20 years ago. Since then advances have been made, but the basic concept remains: the e-ID card is designed to identify the card holder, to create a digital signature and to support routines for confidentiality. The card has two key pairs and certificates stored on the chip that forms the primary physical security feature of it. The card if of familiar appearance and usage to most residents of the Member States of the EU as its features and use are similar to those of a common credit or debit (charge) card that millions across the EU use in their everyday transactions with stores and services. However, recent research has demonstrated that identity cards had historically tended to be introduced by authoritarian regimes and were viewed as a repressive measure. As such they have been resisted by countries such as the United Kingdom and the United States (eema 2006). Such fears on the historical used of identity cards have been combined by the considerable cost of introducing, maintaining and operating such schemes (LSE 2005) that led the UK Government to recall its original plans and reconsider the whole scheme. This decision in relation to the e-ID scheme proposed for the United Kingdom is also compatible with the approach adopted by the United Kingdom towards eGovernment projects, where the aim is to develop cost efficient systems that would achieve value adding improvements to the services offered both to the government and the user.

However, nowadays the original purpose (authoritarian government use) has been lost and the identity card is becoming a useful tool. Since 9/11, resistance to the ID card in the United States had diminished and so has in the United Kingdom since July 2005 with the events of the London Underground bombings. The United States is moving ahead with an e-ID card scheme that would provide a single, secure credential that would be tamperproof and hard to forge, and could be used for physical and logical access. A recent example of the usefulness of such an ID is at the crisis of hurricane Katrina when it struck the city of New Orleans. At the time, 6,000 medical personnel wanted to go and help, but were unable to do so because no-one knew who they were. Thus the concept of a pan-European e-IDM card scheme might sound more plausible and more useful than ever before.

The key reason for promoting the interoperability of e-identification is the growth of demand for cross-border services which needs to be met by equivalent cross-border framework of cooperation. Much work is done at national level but is project-focussed and by definition time-limited, not allowing for the integration into the full European perspective. A further limitation at the time of writing is that no authority or governance mechanism at EU level is available for such an ambitious undertaking.

On a positive note though, substantial resources already available:

- Content, from complete eService applications through to data models and code list
- Containers, from databases, Web portals and CMS
- Registries, with agreed processes for managing components

The focus on their management is largely national and this limits the possibility of agreements on standards and protocols used and further attempts experiment with the integration of cross border services on interoperable authentication systems.

Once this initiative reaches a good level of success and an acceptable maturity level, it will form the basis for interoperability across services in other federal government systems (countries) and could possibly prove the model for a worldwide interoperable e-identification system, increasing the efficiency of transactions, and significantly enhancing the effectiveness of border control systems [9].

References

1. A. Cavoukian, Laws of Identity the Case for Privacy-Embedded Laws of Identity In the Digital Age, a White From the Information and Privacy Commissioner of Ontario, Canada. Available at http://www.ipc.on.ca/images/Resources/up-7laws_whitepaper.pdf
2. Economist Intelligent Unit, Digital identity authentication in e-commerce, The Economist, March 2007.
3. J.-W. Kim, K.-T. Kim and J.-U. Choi, Securing e-government services, *IEEE Computer* **39**(11) (2006) 111, 112.
4. J. Kobielus, Our Federated Future, Network World, 22 March 2004. Available at http://www.networkworld.com/supp/2004/ndc2/0322idmgt.html
5. R. Leenes, J. Schallaböck and M. Hansen, PRIME — Privacy and Identity Management for Europe, Project Whitepaper, 15 May 2008. Available at http://prime-project.eu/
6. E. Maler, Federated Identity through the Eyes of the Deployer, Sun Developers Network, 29 February 2008. Available at http://developers.sun.com/identity/reference/techart/deployment.html
7. OASIS Standard, Conformance Requirements for the OASIS Security Assertion Markup Language (SAML) V2.0, 15 March 2005. Available at http://docs.oasis-open.org/security/saml/v2.0/
8. L. Pang, A manager's guide to identity management and federated identity, Information Systems Control Journal **4** (2005).
9. E. Pimenidis and I. Savvas, E-identification technologies for e-government interoperability in the EU, International Journal of Electronic Security and Digital Forensics (IJESDF) **1**(2) (2007) 169–179.

Chapter 16

E-SECURITY AND CRITICAL NATIONAL INFRASTRUCTURES

HOWARD THOMPSON
NAS (Defense) Ltd., UK

16.1. Introduction

Since the end of the Cold War, the international landscape has been transformed. The ideological polarisation between two major power blocs, which determined world politics for over 50 years, has been replaced by a more complex and unpredictable set of relationships [17]. World-wide economic trends have broadened and global services are more evident than hitherto. In addition, advances in technology, particularly in communications, have done much to bolster the interrelationships between societies and individuals, and between businesses and economies. Geography and distance are growing less and less important as mankind enjoys the ability to travel farther, faster and cheaper than ever before. Moreover, ideas can be exchanged almost instantaneously; capital can be bartered, substituted, exchanged and traded around the world, and indeed beyond the world, at the push of a button in far away lands. And organisations are outsourcing processes and operations and locating them in central processing locations, often in countries other than their own.

These are promising and positive changes and are creating and strengthening new opportunities for governments, for businesses, for organisations and between nations.

But there are new challenges and obstacles which must be met. Protecting critical national assets and services in an increasingly complex and unpredictable interconnected world is becoming evermore difficult.

A nation's defence, public safety, the economy, and the quality of its national life have long depended on the efficient delivery of a number of essential services, among them telecommunications, energy, banking and finance, transportation, and vital human services such as the provision of food and water. These national essential services have, over time, become known as Critical National Infrastructure (CNI).

In recent times, there has been a growing recognition that the services that comprise a nation's CNI are increasingly dependent on an information environment, "that system of advanced computer systems, databases and telecommunications networks... that make electronic information widely and accessible... and which includes the Internet, the public switched telephone networks and cable, wireless and satellite communications systems [1]. In many counties, the loss of use of critical information systems would have a serious impact on the well-being of citizens and the proper functioning of government and industry. Many infrastructures need to be functioning at least at a minimum level for the public and private sectors to be able to survive: for example those concerned with the generation and distribution of electricity; the distribution of fuel; the supply of food and water; transport and communication systems; waste management; finance and insurance; and the information and telecommunication networks that link these together.

Although what we recognise as the traditional focus of national security has not disappeared, other considerations are now present. The threat of nuclear Armageddon has been largely replaced by other threats, mostly concerned with terrorism and similar malicious acts such as trans-national crime; but they are no less severe and have the potential to undermine wider national and international stability. The threats are aided by telecommunications and computer technologies only dreamed of even a short time ago.

The rapid growth and integration of a world-wide telecommunications infrastructure, based largely on the Internet, has brought critical infrastructures together in a manner which was hitherto unimaginable. Tracking dependencies has become more complicated and assessing criticalities more elusive. And to make matters worse, many now straddle both public and private sectors. In consequence this has created interdependence among them which it would now be hard to break. In the process of creating this interdependence, unprecedented threats have been created.

As information systems and their dependencies have become more and more valuable they have become more vulnerable to attack from the greedy, malicious or just plain curious.

The broad context of concerns has been outlined by the Organisation for Economic Cooperation and Development (OECD) It tells us that "globalisation, climate change, the transition to a modern technology intensive economy, demographic and social change, growing interdependencies... look set to increase the vulnerabilities of major systems during the 21st century. The provision of health services, transport, energy, food and water supplies, information and telecommunications, safety and security are all examples of vital systems which can be severely damaged by a single catastrophic event, a chain of events or the disastrous interaction of complex systems. There is growing concern that extensive disruption to, or collapse of, these systems could significantly impair future social development" [9].

Hardly a day goes by without we hear reports of hackers who have penetrated the security of computer systems and networks, caused vandalism to Web sites, disrupted command and control systems, stolen money from bank accounts, tampered with medical records or accessed sensitive data of one sort or another. The ever-present danger is that we could soon see rogue governments, terrorists and criminals making use of cyberspace to interfere with the essentials of national survival. The prevention of telephone communications, shutting down power generating stations, preventing the distribution of food and other essential daily needs, disrupting water supplies, hindering transportation, obstructing law and order enforcement, or undermining the many other essential services now taken for granted are some examples of what a nation might face [4].

In the United States, Europe and the UK, significant steps are being taken to safeguard information infrastructures and many initiatives are in evidence. However, much remains to be done if we are to fully achieve a safe environment for information age society.

In this chapter, we will consider the nature of a CNI and its importance to a nation's well-being and survival. A short consideration of the nature of information that underpins the operation of technologies, and its importance, is followed by a brief assessment of cyberspace, the threats and the potential for harm which lurks within. It is helpful to consider how those threats can be modelled and the impacts they may have; how the United Kingdom, the United States and the European Union (EU) approach this is outlined. Some thoughts on a strategic frame work are offered in conclusion.

16.2. The Value of Information in the CNI

It is a little more than five decades since the electronic computer became generally available to help organisations in their quest for more efficient control of processes. In that time we have seen a move from large mainframe systems needing a completely engineered environment and substantial human intervention, to the end-user revolution of the personal computer (PC) which fits neatly on the desk, interacting not only directly with the operator but also allowing almost instantaneous communication with the world at large through complex and continent-jumping communication networks.

Computer and communications technology now forms the basis of all CNI business decisions: its use, can improve dramatically the quality of those decisions; it abuse can disastrously undermine them. Information is the very lifeblood of a CNI.

Information is not, however, like other corporate assets which appear on the balance sheet and which can be assessed as such. It is intangible and its quality cannot be readily quantified. Moreover, it can be known to and controlled by two or more parties at the same time. It has a particular characteristic which makes it unique in this sense: it can be surreptitiously copied, transmitted and shared while remaining to all intents and purposes under the control of the owner. The very nature of electronic information means it can not readily be inspected; and it can be forged and manipulated in a manner which is not easily detected. The danger of disclosure, modification and denial are all constant bedfellows [18].

All CNI organisations rely heavily on information and communications technology for among other things, their administration, supply chain management order distribution or command and control of processes and machinery. If that information technology or the information it processes, stores or communicates is unavailable when it is needed, or if it is flawed or inaccurate, the consequences can be disastrous.

What has this meant for almost all nations whose very existence now depends on information-intensive computer and communications technology to manage and maintain the very structure of the nation?

The evolution of these technologies and the ability to influence remote infrastructure and machinery, often located far from the source, has had a profound effect on the ability of nations to control and maintain the availability of national services. In addition, almost all levels of an organisation's staff, and some outsiders, now have the ability to influence and affect the day to day operation of elements on a CNI by interacting directly with it through its information technology elements.

And not often considered in discussions about CNIs, is that such national infrastructure invariably refers to infrastructure that is critical to a nation, not necessarily infrastructure that resides within a nation. In today's world the CNI of any nation embraces complex interdependencies associated with a global reach; no nation is an island! There are many international issues to consider. Most governments are providing specialist advice through national security networks and warning systems and informing their citizens about the dangers; but the setting of standards for security and resilience in the CNI sectors is not yet international beyond cooperation between some nations.

Another factor that adds to the complexity of dependency is that many organisations and businesses that own parts of national CNI are now international. It is becoming increasingly less possible to confine national information to national boundaries without introducing levels of protectionism that would be considered impossible to implement. It may be that because of his, nations are unwittingly developing systemic vulnerabilities as they develop a new generation of networks, providing the malicious and the malcontent with the ability literally turn the lights out in a period of international crisis [3].

16.3. Cyber Attack — The Potential to Harm the National Interest

The term "national Information infrastructure" implies an analogy with the road or rail network; that it is a unitary, integrated, standardised infrastructure on which different vehicles (the services) are carried. This is a model which may apply in the future but does not, even today, apply universally. Rather, it is a plurality of networks and service providers such that any coherence in the infrastructure is derived from the different components working together rather than through a single unitary national network.

The problem is not made easier by three factors. Firstly, malicious acts of damage are not readily or easily predicable. Secondly, although the integration of infrastructure makes systems more robust by introducing redundancy, the inherent vulnerabilities in architectural elements provide a multitude of different means of attack avenues, physical, logical and human. And thirdly, the perceived need to increase productivity by restructuring, outsourcing and off-shoring generate additional pressures whereby "errors can occur, where responsibility is diffuse, where open access is mandated and hence where intruders can enter" [12]. In addition, each depends fully on the correct and continued use of information technology.

There are time relationships that affect the ability to resist and then limit damage will hinge on early identification of an attack and the implementation of defensive measures. A factor will be initial effectiveness of the attack, the technical capability of the attacker and the capacity to continue in the face of the adoption of countermeasures.

Nations face a range of covert and overt threats to their national security. While the many countries have faced a variety of threats in the past, a unique combination of factors, namely increasing global reach, capability, resilience, sophistication, ambition and lack of restraint on the part of attackers, place the current threat on a scale not previously encountered. This is the threat from cyberspace.

Cyberspace provides a new dimension in the ability of a government, organisation or individual to harm another country's national interest; and while information integration into large-scale infrastructures has many benefits, it has brought with it a darker side where the very attributes which make it so attractive, give grave cause for concern about the very functioning of the nation. A CNI is vulnerable to a range of constantly emerging forms of electronic exploitation. That exploitation may be on behalf of a government a terrorist organisation or a group of malcontent individuals. A threat can take the form of electronic attacks on vital information, communications, food and water supply and other critical systems. And electronic attacks can be carried out, directly or indirectly, by an "insider", or by someone with specialist knowledge or access. Threats or hoaxes intended to frighten and intimidate might not be uncommon.

The nature and magnitude of electronic attack against any CNI arises from a range of potential attackers, their motives and their capabilities. At one end of the scale are foreign states that have hostile intentions towards the nation it wishes to attack; at the other are what we call script kiddies, budding hackers who use publicly available tools to attack systems and cause damage and disruption. In the middle we have a mix of threat groups: terrorists, who may be seeking to add electronic attack to their existing capabilities; activists seeking publicity; criminals engaged in electronic theft; disgruntled employees, hackers, crackers and virus writers.

Foreign states are increasingly recognising the value of electronic attack. Any such attacks are likely to be carried out by specialists working for military or intelligence agencies actively seeking to acquire economic and financial intelligence, details of systems, including any vulnerability which, in time of crisis could be used to mount attacks designed to disrupt a nation's well-being.

Of note are that there are several terrorist groups who currently have the declared intention to damage national infrastructures. Although there is varied capability within terrorist groups, at the moment they seem to favour physical attacks rather than widespread and damaging electronic attacks. How long this will remain the case is a question on many people's lips.

Activist groups, also known as hacktivists, have been known to carry out denial of service attacks, Web site defacements and electronic sit-ins. There are many causes that excite a large number of disparate groups and some reasons for electronic attack have included such issues as global capitalism, the war in Iraq and pollution. Generally speaking, such protestors have not shown the intent or capability seriously to harm critical infrastructure, although the potential for this is ever present.

Hackers, Crackers and Virus Writers present a particular hazard. There is considerable disagreement about the meaning of the term hacker but in the main the term is used to describe a person or group of people motivated by a desire to analyse and explore other peoples systems. While this activity is likely to be illegal, this group is often separated from criminals because of the difference in motive. They are discrete from crackers who break into computer systems with malicious intent. Some hackers simply try to penetrate critical systems to gain prestige within the hacker community, or because it provided an interesting technical challenge. Others have an interest in deliberately damaging computer systems, for example because of a personal grudge or to protest against an organisation. There is a wide range of skill levels amongst the hacking community and at the upper end of the scale there is a high degree of capability. Virus Writers have different skills, not only creating the virus, but also in deploying new, creative and more effective means of damaging and destructive propagation.

There is a real threat, albeit difficult to quantify, from a disgruntled employee (or other contractor, visitor or consultant with privileged access) who, for whatever reason, could use individual expertise couple to a privileged level of access maliciously to damage systems or cause disruption. The type of attack and the potential to cause damage or disruption varies significantly from employee to employee. Hostile motivation could be for personal reasons or for an ideological aim.

Another common threat source is commonly described as the "script kiddie". These unskilled attackers scan the entire Internet in search of vulnerable victims, and then run hacking tools, which they have often downloaded from the Internet. Such individuals may succeed in penetrating office networks or in mounting limited denial of service attacks.

Threat groups have a range of electronic attack techniques or tools at their disposal. We can separate these attacks into categories: denial of service; network intrusion; viruses and worms; Trojan software; and malicious hardware. This is for ease of explanation — in reality the boundaries between the types of attack are often blurred. In the case of operational and process control systems (for example, SCADA), we can also add operator spoofing as an attack method.

Denial of Device (DOS) attacks are designed to render a system unusable, often without actually penetrating it. For example, attackers often cause denial of service by flooding target systems with unwanted data, effectively blocking legitimate use. In some cases an attacker will simultaneously flood a victim from multiple sources. This is called a distributed denial of service attack (DDoS). It is also possible to cause denial of service by exploiting well-known vulnerabilities in operating systems and applications.

Network intrusion is the term used to describe situations where an attacker penetrates a system remotely to cause a malfunction or damage, to scope the system and identify its vulnerabilities, or to remove data surreptitiously. Intrusion is commonly achieved by exploiting vulnerabilities in software on the target machine, but may also rely on techniques such as password guessing, wireless hacking or war dialling where the attacker dials a great many telephone numbers in an attempt to find a modem that will accept an inbound connection.

Viruses and worms cause damage and denial of service and are very common. The term virus is used to describe programs that infect individual files, while worms are stand-alone programs that copy themselves to victim machines. The most common virus works by attaching itself to an executable file and then infecting other executable files whenever the infected file is executed. The most common form of worm is a mass mailing worm. When executed by a victim, it emails itself to every other address in the victim's address book with a message to tempt these new victims to execute the worm and spread it further. Other kinds of worm do not rely on gullible users but instead automatically locate, exploit and copy themselves to vulnerable systems.

Trojan software, often hidden in e-mails or Web-pages for download, contains a hidden malicious payload. This payload can be executed by opening e-mails, attachments to e-mails, by being drawn into or automatically connected to infected Web sites. Often this payload opens a back door on the system to give the attacker access to the victim, but other payloads are possible. A Trojan could delete or change files or cause a serious malfunction to a system which might cause it to crash under certain pre-set conditions.

While viruses, worms and Trojans are often called malicious software, it is not often recognised that attacks can also be executed using malicious hardware. One example is a key logger. An attacker can covertly attach this small device to a target computer keyboard, where it will then record keystrokes. The attacker can then covertly remove the key logger and download the contents, revealing all of the passwords and usernames typed on the target machine.

Operator Spoofing is the term used to describe the operator being tricked into taking imprudent action based on spurious or false signals apparently from field devices. This attack is more complex as the attacker is required to access and modify the system to change the data points on the screen graphics in order to deceive operators and stimulate an event, e.g., an emergency shutdown. This requires understanding of the system and the process control (SCADA) software in use; the network address of the server(s), the ability to access and modify files on the server(s) including access through the company network (if necessary) plus necessary administrator privileges. Insider knowledge of the network would help significantly in this type of attack.

And there is an extra physical dimension to protecting assets held electronically. Often ignored is the fact that people are any infrastructure's most important asset; they are also its greatest danger. Cyberspace security is not a totally a technical problem; rather it is also people problem.

The technologies underpinning the use of the electronic tools used to control critical processes which are dependent on information are remarkably susceptible to a variety of events which can at best be troublesome and inconvenient, a worst catastrophic. If information is denied when it is needed, if it is inaccurate when presented or if its privacy has been compromised, any organisation may find it difficult to continue for long. What is certain is that people are becoming more computer literate and are able to involve themselves with the machines to a far greater extent than ever before. The knowledge and the opportunity to cause loss have increased dramatically.

The automation of the previously manual workplace, and the introduction of technology to every corner of the business environment, has caused far-reaching organisational changes. Traditional and long-accepted business roles and practices are changing as new, highly technical jobs are created in which skilled specialists deal mostly with machines. In addition to all of this, hardware and software may posses what can be termed artificial intelligence which can be turned against security protection designed into the system.

The nature of those able to attack systems has changed and the traditional list of attackers has been expanded to include computer professionals with the opportunity to commit attacks because of their close relationship with the infrastructure. Computer programmers, operators, tape librarians and engineers who function in the cyberspace environment are far more able to interact and interfere with CNI than their fellows of years gone by. Similarly, the methods and attack tool have also undergone a fundamental change. A new language has grown up to describe some of these: data diddling, Trojan Horses, logic bombs, salami techniques, super zapping, hacking, piggybacking, scavenging and data leakage are but some.

The timing of attack activity has changed in the computer environment and therefore in the CNI. In traditional crime scenarios, activity is measured in minutes, hours, days, weeks or longer. In the computer environment some attacks are perpetrated in mille-seconds. Moreover, geographically there are fewer constraints to the commission of attacks; it is now possible, through the medium of telecommunications, to be in one part of the world while at the same time directly causing the commission of a crime somewhere else in the world.

Equipment malfunction can be difficult to resolve and often prove a major problem. Many malfunctions are, however, brought about by attacks on other areas of the infrastructure, for example on environmental aspects such as air conditioning failure or power disruption. In addition, software is still an inexact science and it is not unknown for severe disruption to be caused by software malfunction; and human errors are frequent in computing where lack of thought, neglect, ignorance, ever-enthusiasm insufficient training and poor supervision are often responsible for system malfunction.

Even short strike action by employees is a threat to the continuance of operations, let alone major industrial action.

A major threat to any CNI from Cyberspace arises out of the vandal and hooligan element so prevalent in today's society. Even if such acts are not directed directly against the computer equipment, outside interference with communications and power supplies can cause serious disruption.

Any element of a CNI presents a hazard: even that which is not connected to the outside world is vulnerable to the predatory attentions of insiders whose intentions might be inimical to the functioning of the nation and its interests. This will be aggravated when interconnections are made, particularly when one infrastructure depends on the services provided by the other.

How can we prepare to detect, defeat and recover from cyberspace attacks?

16.4. Modelling Attacks on Information Infrastructures

To fight cyberspace attacks, it is necessary to assess the risk to the infrastructure from attacks from cyberspace, define consequence management and to determine how best to protect dispersed and distributed infrastructure elements within the CNI. In an information warfare scenario, it will be necessary to determine what counter offensive action against cyber attacks is possible or appropriate. To understand fully the nature of the threat and to be able to consider the response, there is a need to model the CNI, consider attacks (scenarios bearing in mind the interdependencies of individual elements of the overall infrastructure), consider how attacks can be recognised and defeated and last but by no means least plan for recovery and reconstruction if an attacks succeeds.

Foremost in CNI planning is fact that infrastructures have long attracted the attention of foreign and potentially hostile governments who may become involved with military action, of terrorists whose aims hostile to the CNI host nation, and criminals whose aim is to cause damage or to steal. When brought together, such infrastructures compound the business of both civil and military defence. Secondly, infrastructures are vulnerable to a wide range of vulnerabilities which are at best inconvenient and costly, and at worst catastrophic. Failures in interconnected structures can have severe knock-on effects elsewhere. Thirdly, experience has shown that cyber attacks, based on inherent weakness and vulnerability in software and other processes (such as poorly designed protocols), can very quickly lead to severe malfunctioning [13].

The potential attack groups who may threaten any CNI are often considered as belonging to one of four groups: state sponsored actors whose intention is to wage information warfare against a nation; insurgents whose aim is to wrest control of a country from a government; criminals whose goal is disrupt business or steal assets; and terrorists whose aim is to spread terror and dissent.

During the initial probing phase an attacker will undoubtedly do his utmost to remain undetected so as to prevent recognition of the existence of the probe, the identification of the source of the probe and the nature and potential impacts of the attack. As far as he is concerned, the success of the attack depends on preventing his early discovery and his attack plan. The attacker knows that this stage is very necessary if the attack is to be successful; intelligence collection and detailed attack scenario preparation will consume much of the attacker's time. At various times during this period the attacker may undertake live system testing, system discovery and attack practice. An attacker might use a wide variety of open-source intelligence to help him select CNI targets and build an attack posture to. Once the attacker has gained the information he needs, the attack can be swift and devastating and rapid denial of service can be achieved. It might be the aim of the attacker to ensure that the effect and consequences of an attack grows more slowly and is difficult to detect quickly and deal with, as it spreads across the infrastructures. Moreover, attacks need not necessarily be concentrated at one specific time: they may be dispersed and spread over time so that parts of the infrastructure may loose effectiveness individually, or en masse, causing significant disruption to national systems. If targets are chosen carefully, it could be possible to bring down the CNI with all that that entails. But how might an attacker proceed?

There are various methods an attacker might use from cyberspace to attack the target CNI and the systems that constitute the infrastructure services. Moreover, he will probably have in his possession a variety of widely and easily available tools to probe and then attack the infrastructure.

One armed with the information gleaned from his cyber intelligence gathering and depending on the nature of the target the attacker's objectives and the type of attack (i.e., denial of service), damage can occur rapidly. An attacker may use a wide variety of open source data and tools to select, and then attack, an infrastructure. The preparation for the attack may be opportunistic and short, or it may be prolonged and detailed. The actual attack may be extensive or over a very short (in relative terms) where the severity of the damage is heightened by the nature of the infrastructure attacked. The attacker will probe the infrastructure and then attack it, leaving the infrastructure owner to attempt to limit damage and then recover to a normal state of service provision. In addition, it is possible that an attacker may plan attacks such that a chain of attacks is maintained and perpetuated, particularly across differing time zones.

To cause the maximum effect attacks can be distributed over time if the attacker's plan is to reduce the effectiveness of the CNI by changing the specific targets over a period of days, such as an extended series of attacks on separate elements of the infrastructure. The operational use and effectiveness of the infrastructure may be reduced significantly in this scenario. Such attacks will exploit the most vulnerable areas of a distributed infrastructure; if attacks can be maintained over a time, and are resistant to measures to combat them, the time needed to reconstitute the services may be beyond the ability of the defender and consequently the infrastructure may be degraded beyond effective use. Moreover, if an infrastructure can be maintained in a semi or permanent state of damage, dependencies will be affected quickly and a critical will arise rapidly.

Rebuilding the infrastructures and reconstituting its vital services may prove very difficult depending on the nature of the attack and the prevailing national conditions. Reconstitution may be as simple as rebooting system elements or reverting to the status quo by the use of backup facilities. On the other hand, the complete shut-down and rebuilding of a complex system (such as an electricity supply or water supply) may be complicated by a many and diverse factors such as geographical location and physical distribution.

Similarly, reconstitution can occur at various rates. A central system can be rebuilt and services restored rapidly by rebooting it and reverting to its pre-attack state using backups, but this can be time consuming, labour intensive and costly. Moreover, some information assets may be lost. Other cases, such as the complete shutdown of an important central distribution system, can require a substantial amount of time for reconstitution. Thus, onset times, attack duration and recovery time depend to a great extent on the type of system attacked and on the ability of the attacker to maintain a continuous series of attacks, in effect to conduct a cyber campaign against a particular element of the infrastructure over time.

16.5. The UK Approach to Protecting Critical Infrastructure

For a number of years, the UK government has been striving to introduce a truly open data network (ODN), characterised as an "information marketplace" which enables "any company to provide any service to any customer and information services of all kinds, from suppliers of all kinds, to customers of all kinds, across network service providers of all kinds, in a seamless accessible fashion". Such an infrastructure is provided on a technology-dependent bearer service utilising high levels of application-level services such as e-mail, fax, remote login, database browsing, digital object storage and financial transaction services. The user of an ODN should be able to "access this capability as he/she moved from place to place; it should be scalable in the many dimensions of size, load, services, reach and utility;

should integrate a range of network technology and end-node devices; and should provide a framework for security." In short, the ODN should possess a number of characteristics. It should permit universal connectivity and be open to all users, not forcing users into closed groups or deny access to any sectors of society. In addition, it should be open to network providers, making it possible for any network provider to meet the necessary requirement to attach and become part of the aggregate of interconnected networks. Moreover, it should allow service providers an open and accessible environment for competing commercial or intellectual interests. And finally, it should be open to change by permitting the introduction of new applications and services. Of major importance is that it should not be limited to only one means of distribution but rather it should permits the introduction of emerging transmission, switching and control technologies as they mature and become readily available.

The UK government views the UK CNI as those assets, services and systems that support the economic, political and social life of the United Kingdom and whose importance is such that any entire or partial loss or compromise could: cause large scale loss of life; have a serious impact on the national economy; have other grave social consequences for the community, or any substantial part of the community; or be of immediate concern to the national government [7]. In the main, organisations that fall into the UK CNI definition are those concerned with the water, food distribution, finance, health, the emergency services, power distribution, communications, and transport. The impact of an attack on essential national resources and services could have serious national consequences and have "knock-on" effects and when the "golden dominoes", power, petrol and telecoms, are involved these escalate [14]. The services are based on capital-intensive facilities for which standby capacity is frequently difficult and expensive [13]. Most of these facilities rely on interconnection with computer and communications services, with operators in their turn reliant and dependent on software which is at best complex and not well-understood. If elements of the services malfunction, or are denied to the infrastructure, there is a strong possibility that of severe and possibly immediate impact on the well-being of the nation and its inhabitants. It is worth considering these in detail [6].

Communications. The UK's major telecommunications companies including BT, Cable & Wireless, NTL and THUS are all investing heavily in building "Next Generation Networks" and are collaborating on technical and commercial issues via the Ofcom-sponsored NGNuk.

The Emergency Services. All three major emergency services are now using a common "terrestrial trunked radio network" called Airwave from O2, to help them co-ordinate their activities in tactical situations. However, there are major concerns about the duration of its back-up power arrangements where a major power failure represents part of the emergency.

Energy Supply. In 1963, a power station in the United Kingdom became the first in the United Kingdom to manage all plant operations from a central control room. Since then, power stations have become increasingly dependent on IT and communications infrastructures to operate and deal with emergencies, often using Supervisory Control and Data Acquisition (SCADA) systems using radio links that monitor a plant and generate alarms if problems arise. These are open to disruption via jamming of the radio link, and have historically been poorly protected. SCADA systems also some times use the internet to communicate back to the command centre, opening a further range of vulnerabilities.

Finance Sector. Recent research undertaken for the Financial Services Authority (FSA), HM Treasury and the Bank of England established that the UK banking and financial sector in the City of London and beyond are now totally dependent on IT and telecoms systems for managing daily payment, clearing and settlement operations.

Food Distribution. According to a major UK Superstore, in its most recent operating and financial review, any significant failure in the IT processes of retail operations, such as barcode scanning or supply chain logistics, would impact its ability to trade. Furthermore, any major disruption to the transport infrastructure would cause significant damaging to the food distribution process, given the "just in time" nature of supply of food to most major modern supermarket chains from a small network of national supply hubs.

Government and Public Services. The majority of public services are now available online, and it is intended that the there should be a single central point of access to all online public services. Next generation public services are also being developed for technologies such as broadband internet, mobile phone and digital TV channels. The disruption of any of these communications channels could hamper access to critical public services.

Health Services. The UK Health Service spends many billions of pounds sterling on its IT infrastructures and has undergone the largest modernisation programme of its kind in the world. The infrastructure will provide, among other services, electronic care records, online appointment booking and electronic prescriptions, and the linking up around doctors and hospitals. Disruption here would have serious repercussions but short-term problems which could affect the health service would be those in other areas such as transport that might affect the ability of trained accident and emergency staff to move around.

Public Safety. In the event of an emergency, the public increasingly turns to technology for safety information. In the aftermath of the 7 July London bombings, there were 10 times as many calls as usual to the major mobile telephone networks. But many of these calls could not be delivered as the networks were designed for strictly limited resilience against overload. People also increasingly turn to the Web for information in a crisis, but again Web servers must be able to handle unusually large loads in the midst of any other problems that may be occurring.

Transport Infrastructure. The UK Highways Agency, responsible for England's motorways, has developed a single data network for the whole country. Traffic flow might be easily obstructed if electronic control and management systems were to become unavailable. This would have serious knock-on effects to let us say the distribution of food and fuel, which in turn start to affect other areas.

Water Distribution. As in the energy sector, SCADA systems using radio links are widely used in the water industry to manage water collection and distribution including the management and supervision of reservoirs and plants for monitoring and controlling water pressure, flows and reservoir levels. There is an obvious cyber security problem; these systems are open to technical attack which could quickly undermine there ability to function at all, let alone properly.

The idea of an infrastructure connotes the idea of a group of cohesive elements working within themselves, and together. On the face of it, the UK CNI seems to be a well-thought-out and cohesive infrastructure of organisations working towards the common aim of providing services which can be secured in the national sense and in keeping with the national security. But closer examination would suggest a different picture. John Ridley, a member of the UK Government Communications Headquarters (GCHQ) department the Communications Electronics Security Group (CESG), recently described the role of his department in supporting the Government's initiative to protect the UK CNI. He said, "Until recently, if you asked organisations' what they thought of the UK CNI, the chances are that you would be met by an embarrassed silence, not least because very little had been said or written publicly". The concept has been debated in the United States for some time; the United Kingdom is now "coming out". This is hardly surprising. Outside the deeply secretive government intelligence and security organisations, little seems so far to have been done to determine the nature and magnitude of the threat to the CNI. There is little evidence to show any major thrust to bring together the disparate elements that comprise the UK CNI, in an organised hierarchical manner. The concept of an "infrastructure" does not therefore appear to have been codified and treated as a whole; rather it is still a collection of separate elements. Neither has much work been published which shows how much work has been done to ascertain the nature and likely outcome of attacks against those organisations in their UK CNI role and whose inability to perform activities and provide functions may have an effect on the others.

Although the UK CNI, as a concept, shares a community of interest, it is nevertheless a collection of independent organisations whose business aims, cultures, products and dependencies are widely different. They are not always Government owned, and indeed in many cases constituent organisations may be in the hands of foreign owners. Their business aims and priorities are directed towards making profits for shareholders; generally they neglect intelligence and pay lip service to information security issues which could affect the national interest. In addition, they are not subject to a consistent or supervised level of information security in the management of assets for which they are custodians and which contribute to the UK CNI. There is, therefore, no directed implementation of information security and its management, notwithstanding the multitude of standards available. Moreover although the Centre for the Protection of National Infrastructure (CPNI) is responsible at Government level for advising on infrastructure security, there is no central sponsor who has executive authority to mandate security implementation levels.

Unlike disasters and other forms of terrorist or warfare attacks, where a great deal of planning has taken place to train for physical terrorist attacks and the consequent need to mange the scenario, this does not seem to be the case in terms of dealing with cyber attack on the UK CNI. This is notwithstanding that there having been severely damaging incidents which were real and very public!

In March 2004, a fire broke out in a British Telecom (BT) cable tunnel in Manchester and put 130,000 land lines out of action, affecting internet services and disrupting several parts of the emergency services communications network including Derbyshire and Cheshire police forces and the Greater Manchester ambulance service. Many bank cash machines in the area were closed since they make security checks over phone lines; local shops could not use credit and debit card machines for the same reason. At the same time, some organisations which had thought they had back-up communication routes in place should the BT services go down, found that these alternative routes used duct space in the same cable tunnel, and so were lost as well.

The December 2005 explosion at the Buncefield oil refinery in Hertfordshire, the largest peacetime explosion ever seen in Europe, offers another example of how damage to IT infrastructure can undermine services provided to the CNI. The offices of an IT services company adjacent to the oil depot were destroyed; the disruption of an automated admission and discharge system for a major hospital provided by the company was quickly affected. The company also ran the payroll system for a significant number employers, paying out billions of pounds each month. Significantly, however, as reported later by the company, good business continuity planning at the company ensured there was little disruption to these services [15].

There have been some information security surveys which suggest the concerns felt by organisations. The National High Tech Crime Unit[a] commissioned the National High-Tech Crime Survey (NTCS) in 2005. Of the businesses consulted during the survey, two-thirds opined that their greatest concern was that when assessing the impact of computer enabled-crime directed against them was the worry that their business functions would be preserved sufficiently for them to continue to operate in their markets!

[a]It became the Serious Organised Crime Agency 1n 2006.

The UK Office of National Statistics (ONS) 2007 report entitled "Focus on the Digital Age" [10] provides the most recent in depth study of the use of computer-based technologies — and by implication the UK CNI — in the UK. The report concludes that there has been an increase in electronic crime and that this has led higher investment in security in both the public and the private sector.

The United Kingdom has recognised that advances in digital technology and the consequent interdependence on computer-based communications have increased dramatically the ability of an attacker to undermine business security. The NHTCU, in 2004, was for some time at the vanguard of fighting computer-related crime.

The Department for Trade and Industry's (DTI) Information Security Breaches Survey (ISBS) (2006) concluded that more than one-half of businesses in the United Kingdom (52%) had experienced a malicious and pre-meditated e-security incident during 2005. The figure had, however fallen from a higher figure of 68% reported in 2003. According to the research undertaken by the DTI during the survey, slightly less than one-half of all businesses involved considered the worst impact to be significant disruption. The survey also suggested that large businesses (those with 250 or more employees) were particularly prone to becoming a victim of an attack with 84% experiencing a malicious and pre-meditated attack.

The ISBS identified, and then concentrated on, four types of malicious e-electronic security breach: virus infection including what it calls "disruptive software"; staff misuse of systems; unauthorised access by outsiders; and fraud aided by the use and misuse of systems. When compared with the previous year's survey results there were significant increases in attacks in 2003 over 2001 but in 2005 the figures fell or remained largely the same when compared with those of 2003.

The report suggested that the minimum cost of computer-crime experienced by UK companies with over 1,000 employees was in the region of £2.4 billion. For businesses with a staff of 100–1,000 employees, the figure was less at £177 million. Among the businesses with a staff figure greater than 1,000, more than a half of the cost of the incident(s) was as a result of dealing with virus infection, worms and Trojans (28%) which caused disruption and destruction.[b]

The ISBS survey suggested that, by the end of 2005, 40% of UK businesses had a formal e-business Security Policy. The 2005 figure was an increase over the 2003 survey findings which showed a little over 30% of respondents had developed a policy. Large businesses, unsurprisingly, were more likely to have implemented a policy.

The survey also suggested that only one person in ten in smaller organisations was aware of the contents and use of the British Standard 7799; this figure rose to one in four in the very large business organisations. In 2005, of those businesses that were aware of the standard, 67% claimed compliance, a rise of 8% on the previous year. Of the respondents to the survey in 2005, 97% believed that they had enjoyed benefit from implementation of the standard.

[b]Financial fraud was the next greatest cost (25%) but is not a feature of this enquiry.

In the report of the 2005 survey, it was suggested that 12% of the staff of UK businesses employed staff with specialist information security qualifications. The increase over the 2003 survey was only 1%. Large business concerns were shown to be more likely to employ qualified staff with 29%, representing an increase of 5% over the previous survey. The survey suggested that 6% of persons responsible for information security in business organisations were likely to have a formal qualification. However, most businesses — 88% — reported no having one with formal qualifications. Of the larger businesses reviewed during the survey, 98% reported hiring external security consultancy support to fill gaps in the need to protect assets.

The ISBS reported also that 40% of businesses had increased their spend on information security in 2005 with only 2% having reduced their spend. The average spend on information security was assessed as being in the region of 5% of the IT department's annual budget. Those businesses, whose senior management were more supportive of information security and considered it to be a high priority, were most likely to spend the greatest.

Business identified five areas of motivation for investment in information security: protection of customer information; maintaining data integrity; protecting business reputation; maintaining business continuity in a disaster situation; and compliance with legislation and regulations. The study suggested that less than 30% of businesses measured return on security investment during 2005, although this was balanced by a figure of 50% who had prepared a formal business case to quantify security benefits.

The United Kingdom to approach to defending the CNI has stressed that best way to combat many threats is to raise awareness of the dangers and how to avoid them. In 1999, the Home Office set up the National Infrastructure Security Co-ordination Centre[c] to identify threats, deliver alerts and briefings and promote good practice in electronic security. The National Infrastructure Security Coordination Centre (NICC) a government body set up to provide information security advice to CNI organisations drew on knowledge from across government, working in partnership with the owners of the systems that support critical services in both the public and private sectors. The NISCC has changed its name but little else has changed on the face of it. The Centre for the Protection of National Infrastructure (CPNI), its successor, provides information and advice but ensuring that appropriate safeguards are in place is still seen as the responsibility of individual organisations, and it is up to them to ensure the delay between identifying a threat and taking action is as short as possible.

Since 9/11, the UK view of its national security posture has broadened to include threats to individual citizens and to its way of life, as well as diverse but interconnected set of threats and risks, which affect the United Kingdom as an entity. The cyber threat has been recognised as significant and substantial. Public

[c]Now Centre for the Protection of the Critical National Infrastructure.

sector bodies are required to have a business continuity plan, co-ordinated and shaped through the Cabinet Office Civil Contingencies Secretariat. In the private sector, however, there are major gaps. The UK Department of Trade and Industry (DTI) produced useful guidance [20].

Unlike the United States which has deluged itself in Homeland Security Legislation, the United Kingdom has deemed it necessary to meet the threat through existing legislation supported by a small number of acts designed, in the main, to meet the terrorist threat in general. In 2001, the Terrorism Act 2000 replaced the Prevention of Terrorism Act 1974 and was followed quickly in the same year by the Anti-Terrorism, Crime and Security Act. In 2000, the Regulation of Investigatory Powers Act 2000 passed into law to be followed by two codes of practice in mid 2001. The Extradition Act 2003, (including the UK/US Extradition Treaty — strongly criticised by many leading UK commentators) became law. The Prevention of Terrorism Act 2005 passed into law in March of that year and was designed to meet a ruling by the Law Lords on a case brought under existing terrorist legislation. Other than the Computer Misuse Act 1990 and elements of the Data Protection Act 1998 and the Interception of Communications Acts 1985, no specific legislation directed solely at CNI protection is available (if the Official Secrets Acts are discounted of course).

But drafting adequate laws is only part of the picture: it is also necessary to have properly resourced agencies in place to enforce them. Previously, keeping track of cyber crime was the work of the National High Tech Crime Unit (NHTCU), but this role has since been taken on by the Serious Organised Crime Agency (SOCA). NHTCU was strong on forging international links but a lack of resources meant its regional coverage was not comprehensive across the UK, creating weaknesses in defences in some areas, particularly outside major population centres. It is too soon to gauge how effective SOCA will be. Partnerships between public and private sectors are essential for protecting the UK CNI. Most public sector bodies have implemented standby arrangements for re-routing communications if one part of the system goes down, and are also working together to buy spare capacity from alternative suppliers. What is key here however is for organisations to ensure that their standby plans remain solid as time goes on, that they are monitored and checked, and that organisations on whom any standby plans depend are maintaining full functionality of their systems following any possible maintenance or upgrade work.

16.6. The American Approach

In May 2003, Brainbench, the global leader in online skills assessment, and the Information Technology Association of America (ITAA), the leading association for the information technology industry, conducted a joint worldwide cyber security survey to explore the perception of cyber security preparedness [8]. The survey revealed major gaps in the preparedness of organisations to navigate safely through cyber space. Issues surrounding cyber security awareness and practices were found to be misunderstood and lacking in implementation. Almost 800 people worldwide were surveyed on a variety of topics considered to be vital to maintaining information security; more than half of the respondents were from the United States. The findings were considered to be compelling and represented a call-to-action if current and future cyber security threats were to be addressed successfully.

Specifically, the findings of the survey indicated significant cyber defence concerns caused by the human factor. Two-thirds of all respondents reported that they were aware of cyber security issues and were proactive in addressing them and were confident of their own abilities to deal with cyber security threats. However, these same people assessed their co-workers' and other companies' cyber security skills as low. At the same time, most individuals admitted that the information security knowledge had been acquired without any formal training.

The attitudes of non-United States and United States respondents were found to be closely aligned, which indicated that perceptions about cyber security cut across national and cultural divides and a central theme emerged from the study: that individuals believe that cyber security is someone else's problem, and certainly not their own. This attitude, cutting as it does across national and cultural boundaries, underlines the fact that the human factor in cyber security had been largely ignored. It indicated to US planners that that any plan of action to increase cyber security defences must assess the security threat posed by people lacking sufficient information security awareness as well as a consideration of technological issues.

In the United States, The National Strategy for the Physical Protection of Critical Infrastructures and Key Assets (NSPP) provides the major part of the overall homeland security strategy. Although the strategy provides a definition of "critical infrastructures" and "key assets". the meaning of "critical infrastructure" has been in a state of evolution for decades and is still being debated. Twenty years ago, "infrastructure" was defined primarily with respect to the adequacy of public works. By the mid-1990's, with the growth of international terrorism, American policy makers reconsidered the definition of "infrastructure" in the context of what became known as Homeland Security. In more recent times executive orders have refined and expanded the number of infrastructure sectors and the types of assets considered to be "critical" for purposes of homeland security.

The USA PATRIOT Act of 2001 provides the federal government's most recent definition of "critical infrastructure". The NSPP has updated the list of critical infrastructures and assets of national importance. The list is not considered to be static and may yet evolve as economic changes or geopolitical developments influence homeland security policy.

As the 1990s reached their mid point, the growing threat of international terrorism renewed federal government interest in infrastructure issues. The previous period had focused on infrastructure adequacy; in the mid '90s federal agencies grew increasingly concerned about infrastructure protection. The growing concern led to a renewed definition of "infrastructure" and presented it in a security context. In July 1996, President Clinton signed Executive Order 13010 establishing the President's Commission on Critical Infrastructure Protection (PCCIP). This Executive Order (EO) re-defined "infrastructure" as the "framework of interdependent networks and systems comprising identifiable industries, institutions (including people and procedures), and distribution capabilities that provide a reliable flow of products and services essential to the defense and economic security of the United States, the smooth functioning of government at all levels, and society as a whole". This had gone further than before by prioritising particular infrastructure sectors, and identifying specific assets within those sectors. It categorised "certain national infrastructures are so vital that their incapacity or destruction would have a debilitating impact on the defense or economic security of the United States."

In May 1998, President Clinton signed Presidential Decision Directive 63 (PDD-63) the goal of which was, within five years, to establish a national capability to protect "critical" infrastructure from intentional disruption. The "critical" infrastructures were "those physical and cyber-based systems essential to the minimum operations of the economy and government". This definition was noteworthy for its specific mention of "cyber" infrastructure. To help achieve its goal, PDD-63 directed certain federal agencies to lead the government's security efforts and identify private sector liaisons in specific critical infrastructure sectors. It also identified certain "special functions" related to critical infrastructure protection to be performed by federal agencies: national defence, foreign affairs, intelligence, law enforcement. The first version of a National Plan for Critical Infrastructure (also called for by PDD-63)13 defined "critical infrastructures" as "those systems and assets — both physical and cyber — so vital to the Nation that their incapacity or destruction would have a debilitating impact on national security, national economic security, and/or national public health and safety". 14 While the Plan concentrated on cyber-security of the federal government's critical infrastructure, the Plan refers to those infrastructures mentioned in the Directive.

But most everything changed on 11 September 2001. Following the terror attacks on the World Trade Centre and the Pentagon, there emerged new EO's relating to critical infrastructure protection. They were signed into law as Executive Order 13228,15 established the Office of Homeland Security and the Homeland Security Council. The list in E.O. 13228 is noteworthy for its specific inclusion of nuclear sites, special events, and agriculture, which were not among the sectors identified in PDD-63.

The President's Critical Infrastructure Protection Board was established in a separate Executive Order 13231. Its duties focused primarily on information infrastructure and made reference to the importance of information systems to other critical infrastructures such as "telecommunications, energy, financial services, manufacturing, water, transportation, health care, and emergency services".

Congress also passed the USA PATRIOT Act of 2001(P.L. 107-56). The PATRIOT Act was intended to "deter and punish terrorist acts in the United States and around the world, to enhance law enforcement investigatory tools, and for other purposes". While not being specific, the National Strategy does identify "cyber infrastructure" as being distinct from physical infrastructure and states that "DHS will place an especially high priority on protecting cyber infrastructure".

The Administration's National Strategy for the Physical Protection of Critical Infrastructures and Key Assets (NSPP) was released in February, 2003. The strategy defines three categories of what it considers to be "key assets". These range from the diverse array of national monuments, symbols, and icons that represent America's prominent historical attractions; centres of government and commerce; and facilities and structures that represent national economic power and technological.

In December 2003, The Homeland Security Presidential Directive 7 (HSPD-7) clarifying executive agency responsibilities for identifying, prioritising and protecting critical infrastructure were introduced. The Directive requires that DHS and other federal agencies collaborate with "appropriate private sector entities" in sharing information and protecting critical infrastructure.

16.7. A Pan-European Perspective

The pan-European dimension of critical infrastructure protection has slowly been emerging in recent years and threatens to add a further layer of convolution for policy-makers and practitioners. However, as usual, there seems not to be one answer or approach throughout the EU to the protection of CNI, not yet at least. The fact that critical infrastructures such as energy, telecommunications, transport and water in Europe are becoming increasingly interdependent creates more complexity and raises the risk of severe disruptions. The danger increases that a breakdown in one infrastructure in more than one country may cascade to other infrastructures, potentially at a European scale.

Currently, the understanding of the pan-European infrastructures with their broad range of geographic and sector-specific dependencies and interaction is still underdeveloped. Studying these complex infrastructure systems demands joint interdisciplinary efforts by researchers, industrial stakeholders and governmental organisations. The research dealing with complex infrastructure systems depends on the use of models and simulation environments as a tool because disruptions and mitigating measures, for obvious reasons, cannot be studied or tested in real world circumstances.

The initial call for member states to co-operate in infrastructure protection came in the aftermath of the Madrid bombings (March 2004), where deficiencies were seen in the sharing of intelligence on the threats to infrastructure [16]. There are proposals currently before the EU Commission member states whereby member states would be required to identify and designate all critical infrastructure components and undertake periodic security reviews. The results of those reviews would be coordinated by a central EU coordinating body which, in turn, would prescribe and monitor standards.

Attempting to standardise across the EU is fraught with difficulty. There are twenty-seven member states, each presumably with a particular definition of CNI, perceiving differing levels of risks and having different military, technical and political resources, to meet risks and to defend against them. It is probable that an EU-wide approach will need a degree of cooperation and information sharing much beyond what is currently acceptable to individual member states at the moment. For example, there are inevitable concerns about sharing sensitive national information.

The task and objective of the EU Framework 7 Project DIESIS (Design of an Interoperable European Federated Simulation Network for Critical Infrastructures) currently being launched is to develop a pan-European standardised modelling and simulation e-platforms for that are fundamental for the transnational exploration of safety aspects of the critical European and national infrastructures and the services they provide to European citizens and the European economy. Understanding the complex system of critical infrastructures, with all their dependencies and interdependencies is still immature. The study of these complex infrastructure systems demands joint interdisciplinary efforts of researchers, industrial stakeholders and governmental organisations to overcome all the difficulties involved. In order to address these challenges, DIESIS proposes to establish the basis for a European modelling and simulation e-Infrastructure based upon open standards to foster and support research on all aspects of critical infrastructures with a specific focus on their protection. This European e-Infrastructure will support full cooperation of the different partners in charge for studying (inter)dependencies of critical infrastructures, while preserving the confidentiality of the proprietary knowledge embedded into the different models and simulation packages. Indeed for a lot of economical, political, technological and social reasons, the European critical infrastructures, once isolated and autonomous, are becoming more and more coupled [19]. This project will design a modelling and simulation infrastructure for exploring the security of dependent critical infrastructures. While there are very good simulators for certain infrastructures available today, there are no suitable simulators capable of simulating the interaction of multiple dependent infrastructures. The platform will be provided within the scope of a European Infrastructures Simulation and Analysis Centre (EISAC) to be established later in the process. Potential users of EISAC comprise research groups in the relevant areas, public security offices, corporate research departments of operators of critical infrastructures and other industrial stakeholders, as well as European member governments [5].

16.8. Towards a Strategic Framework for Cyber Defence

Partnerships between public and private sectors are essential for protecting a national CNI. In a recent exercise designed to strengthen coordinated responses to what many perceive as a growing threat, business and government leaders from the United States, United Kingdom, Australia, New Zealand and Canada, as well as a number of other countries, simulated defending against a large-scale cyber attack. The exercises were designed to sharpen and assess participants' ability to respond to a multi-day, coordinated attack and better understand the "cascading effects" such attacks can have Participants of Cyber Storm II, which also included about 40 private-sector companies, enacted a scenario in which "persistent, fictitious adversaries" launched an extended attack using Web sites, email, phones, faxes and other communications systems. Cyber Storm II came two weeks after the Pentagon released an assessment of China's military might, warning the People's Liberation Army was intent on expanding its capabilities for cyber warfare. It also follows intelligence reports that utilities in several countries have sustained cyber attacks that caused power outages!

Traditionally, security has always been about protecting the physical assets of a business (and by extension a nation and its people) form harm. This has not changed in the information age; indeed the need to protect physical assets remains as strong as ever, particularly when it is realised that Cyberspace is totally reliant on the physical medium. But there is an additional pressing national need. Every facet of national business and service provision is now in some way connected with the CNI. The control, storage, movement, manufacture and distribution of assets crucial to the national interest, and much more besides, are represented electronically as information held as data. Security is important as never before.

Security is a cost to buy and manage. Insecurity is also a cost, either directly by actual loss of assets, or consequentially, for example the need to recover following an incident, investigation of security breaches, the effect of incidents on the morale of staff, the financial effect on the business or on public image of the organisation. In most organisations, it is the business risk that is paramount. It seems sensible that there must be common and consistent standards which help to balance cost-effective information security measures against the perceived risks. Senior management is ultimately responsible for security policy formulation and the implementation and enforcement of the rules within the organisation. It is agreed, however, that everyone at whatever lower level of the organisation also has a part to play in security. Management has a clear duty to ensure that information stored, processed and communicated by computers and communications bearers is secured at least to a level comparable with traditional security needs. There must be a soundly based system of controls which allow management to meet its obligations in an orderly and efficient manner thus ensuring adherence to policy and the safeguards it mandates. In summary, security management is fundamental to business success and by extension to the CNI.

But what about security management in this changed UK CNI-dependant world? Do security managers really understand the importance of what they do nationally rather than within a commercial or other business organisation? Along with the obvious advantages, it should come as no surprise that Cyberspace is used increasingly to support and promote illegal activities both at an individual level and state sponsored against national interests. But it seems that no two organisations agree entirely on what security management is or where the security manager lies within the organisation, let alone how this might be affected by membership of the CNI; some define the security manager by role, some by activity and some by professional placement. What is clear is that the nature of the security manager's role varies within the organisation served. More senior security managers will invariably be concerned with the strategy while those at a junior level will concern themselves with administering tasks of security implementation and control.

At national level there needs to be an analysis of the research work being done by researchers, industrial stakeholders, infrastructure operators, decision-makers and governmental organisations. This will allow the development of a common approach and terminology to aid definition of appropriate standards in critical infrastructure protection. In this way, it will be possible to develop a security strategy prototype of such an infrastructure as a proof of concept.

16.9. The Next Steps

We have outlined briefly why information is now almost universally considered an organisation's most critical asset, forming as it does the basis of all business decisions and processes, whether they are in the service sector or an industrial sector organisation. We have seen that the rapid spread of communications technology and its application in all business, commercial, industrial, social and political areas of daily life has inevitably added to the potential for committing criminal acts in which the computers is the central figure. For the most part, early criminal acts against the Information Super Highway, the precursor of the CNI, were directed towards fraud or theft of equipment for re-sale or damage to assets. Although computer equipment will always be attractive to thieves, the clamour for electronic commerce on the net has increased the potential for spectacular crime in other areas and for damaging attacks which could actually undermine a nation's ability to function. That threat has now grown to include the CNI of most, if not all, nations.

The CNI is seen as a major asset for any nation. Its damage deep from inside Cyberspace would be potentially serious. The information necessary for the operation of that infrastructure is therefore critical also, but is widely distributed, is held by a large number of organisations and continually changes. It also is a target for terrorists, criminals and hackers. The answers to the questions of who owns and protects the CNI are therefore vital to the normal functioning of the nation [12].

The protection of CNI is an international issue but the international legal framework to deal with the problem of trans-national attacks is less than well-developed. There are many advisory bodies but we need an international regulatory body and governments are providing specialist advice through networks of warning bodies. It is easy to argue that this should include mandating certain suppliers or nations whose products the national security authorities would endorse and implement in the CNI, while at the same time also prohibiting products from countries whose bona fides could not be guaranteed. There is no doubt that in developing new generation networks, nations may be developing systemic vulnerabilities by providing foreign powers with the ability to control its CNI in a period of international crisis.

A comprehensive international response to the challenge of cyber security requires action by all stakeholders in both public and private sectors. A broad policy for information assurance is necessary that builds on existing approaches and ensures that stakeholders are involved from the outset in initiatives so that policy is shaped properly and remains adaptive in the face of technological change. It is axiomatic that that the threats and vulnerabilities are as much global as national. Amongst the leaders of the Internet economy, nations are recognising that CNI protection can not be put solely in the hands of the private sector. Political leadership at national levels is necessary [11].

A new approach to civil defence is necessary, replacing old ideas with a civil protection network of CNI-enabled organisations, building on and strengthening local capacity to respond effectively to disruptions caused by e-security incidents. There should also be bi-lateral and multi-lateral cooperation on cyber attacks which focus on underlying drivers of insecurity. Most importantly, it should be recognised that governments can not meet all of the new challenges by themselves and that stronger partnerships with business and commerce are necessary.

Harnessing global technology brings massive opportunities but it also generates global threats. The interconnected nature of the challenges and their impacts demand attention.

References

1. Adapted from the definition in: US Senate Permanent Sub Committee on Investigations, Hearings on Security In Cyberspace, 5 June 1996.
2. BCS Thought Leadership Debate took place on 7 June 2006 http://www.bcs.org/upload/pdf/cnii.pdf accessed May 2008.
3. Critical National Information Infrastructure — Who owns it and how do we protect it? (http://www.bcs.org/upload/pdf/cnii.pdf accessed 29 April 08.
4. D.E. Denning, *Information Warfare and Security* (Addison Wesley, 1998).
5. http://idw-online.de/pages/de/news257725 accessed April 2008.
6. http://www.bcs.org/upload/pdf/cnii accessed April 2008.
7. http://www.cpni.gov.uk/ accessed may 2008.
8. http://www.itaa.org/infosec/docs/Brainbench.pdf.
9. http://www.oecd.org/department accessed May 2008.
10. http://www.statistics.gov.uk/focuson/default.asp#digitalage accessed April 2008.
11. Information Assurance Advisory Council — Protecting the Digital Society March 2002.
12. Lukasik, Goodman and Longhurst.
13. Lukasik, Seymour and Longhurst, *Protecting CNI Against Cyber Attack.* Oxford University Press Adelphi Paper, 359, 2003.
14. Parliamentary Information Technology Committee 2006 at www.pitcom.org.uk/modules.

15. PIT Comms: Briefings for Parliamentarians on the politics of information technology. November 2006: Critical National Infrastructure (http://www.pitcom.org.uk/briefings/PitComms1-CNI.doc accessed may 2008).
16. RUSI Monitor May 2008.
17. The National Security Strategy of the United Kingdom Security in an Interdependent World, March 2008.
18. H. Thompson, *The Manager's Guide to Computer Security* (Eurostudy Publishing, 1990).
19. www.diesis-project.eu/+DIESIS&hl=en&ct=clnk&cd=5&gl=uk accessed 6 May 08).
20. www.dti.gov.uk/files/file9952.pdf accessed April 2008.

Chapter 17

DIGITAL FORENSICS TECHNIQUES AND TOOLS

ROBERTO DI PIETRO and NINO VINCENZO VERDE
Università di Roma Tre – Dipartimento di Matematica
L.go S. Leonardo Murialdo n.1 – 00146, Roma – Italy

17.1. Introduction

Computer forensics is the science of digital crimes investigation. More likely than not, during investigative processes one has to collect digital evidences, which could be prevented from being altered. This is certainly true for computer related crimes, such as child pornography or electronic frauds. Further, the need to analyse a computer system may arise in non-electronic crimes investigation. As an example, one could control temporary Internet files of a user in a system to find out that words like "murder" and "accidental" have been searched through the Internet, proving the guilt of the suspected user in a murdering case and also proving the pre-meditation of the murder.

17.1.1. *The Need for Computer Forensics*

During the last two decades, the use of computer systems has become more and more widespread. As a consequence, today the world around us is full of computer systems and other electronic devices. People use to store sensible information in hard disks or other storage devices. Moreover, employers in organisations interact with computers every day.

For these reasons, a natural need in looking for digital evidences in computer systems arises; since sensible information about a user is probably stored in her computer system, there is a good chance to find evidence which correlates her with a crime by searching through files stored in hard disks.

Computer forensics may find a role even in organisations and companies; in fact, they normally state security policies which should not be violated by employers; if a user succeeds in bypassing the security mechanisms used to enforce that policy, expert personnel should exploit a way to prove that the user took actions to avoid security mechanisms.

However, computer forensics is not that simple. While searching for digital evidence, there are many obstacles an investigator has to avoid:

- Usually there is a large amount of files stored in computer system devices, and only few of them may constitute valid evidences. If the computer forensics specialist does not know where to find them, it could take too much time; to spend months to find an evidence is as bad as finding no evidence at all.
- It is not rare that information has been deleted from devices. If this is the case, searching inside files is useless, and other techniques should be developed.
- Computer systems, as well as files, are often protected by passwords. Investigators should find a way to read protected data in unauthorised ways. Since protecting data with password often involves cryptography, the more the cryptosystem used to encrypt files is strong, the more it will be difficult for a computer forensics specialist to read them; further, techniques used to break a code vary from cryptosystem to cryptosystem. Unfortunately, for some cryptographic protocol it has been conjectured it is computationally infeasible to break them; in presence of such encrypted files, other methods should be exploited to read them.
- It is a common mistake to make the assumption that information is stored in working devices. In some occasion there is the need of recovering data from broken devices.
- Each case is different from another. Techniques that should be used and actions that have to be taken, as well as what kind of digital evidence is needed, are mutable factors. Another issue is given by the computer environment, which includes the operating system, the type of storage devices used and the authentication methods. As an example, tools designed to retrieve deleted files from a FAT filesystem are not expected to work on an EXT3 one.
- Finally, when digital evidence is found, it should not be altered. If it cannot be proven that evidence has not been altered, it cannot be used as valid while persecuting a crime.

17.1.2. *Goals*

The main goal of a computer forensics specialist or team is to find digital evidences and to assure that they have not been altered. To achieve this goal, a computer forensics specialist must handle with the obstacles described earlier; this may suggest that a common methodology of investigation is required, as well as standard techniques for obtaining and preserving digital evidences should be developed. However, different cases lead to different investigation methods, and different methods require different technologies to be used.

Due to this consideration, there is a lack of standardisation in digital forensic methods of investigation. The steps to be taken in an investigative process should be described in an abstract way; description should not make assumptions about the system object of the analysis. Moreover, only steps that are common to each investigation processes should be described. Reith, *et al.* [19] proposed an abstract model to describe the main phases of a computer forensics investigation:

- **Identification** — First of all, one has to identify computer systems to be analysed. In incident response, this means recognising an incident and determining its type.
- **Preparation** — After computer systems have been identified, techniques to be used should be selected, and related tools should be prepared.
- **Approach Strategy** — The goal of this phase is to find a work strategy which is expected to maximise the amount of digital evidences found while analysing systems.
- **Preservation** — Digital and Physical evidences should be secured, preserved and isolated. The process of preserving digital evidences is referred to as the Chain of Custody.
- **Collection** — Physical scene of the crime could be related with data stored in the digital devices; physical evidences are also important while investigating on a computer crime. For example, finding the fingerprints of the suspected user on a computer keyboard proves that she used that computer. Physical scene has to be recorded, while digital evidences have to be duplicated.
- **Examination** — This is the phase in which digital evidences are to be discovered. It consists in locating and identifying potential evidences, providing a detailed documentation for next steps.
- **Analysis** — While Examination identifies potential evidence, the process of analysis determines significance of evidences, and draw conclusions from them. It is during examination and analysis that a crime theory should be elaborated. This is a quite difficult task, and often several iterations of Examination and analysis are required.
- **Presentation** — The elaborated crime theory should be presented and explained, providing the conclusions of the whole investigative process.
- **Returning Evidence** — Physical and digital evidences should return to the proper owner, eventually removing criminal evidences.

Using a model like the one described above, finding digital evidences becomes easier. However, this model does not say anything about what steps should be taken to carry out evidences from computer systems. Again, methodologies to be used vary from a case to another; there are, however, many common methods used to obtain digital evidences. They may consist in monitoring network connections and intercepting users' e-mail, in analysing data from physical devices and many others. Common methods to achieve the goal of finding digital evidences could be summarised in Data Locating, Data Seizure and Data Recovery.

17.1.2.1. *Data locating*

Physical devices are full of data, and searching the contents of the whole physical device could result in time waste. Luckily, depending on the system, there are common places where to search for sensible data. Data locating consists in making a selective research for sensible data in an electronic device. During this step one may consider the history of the system subject of analysis. If the system is connected over the Internet, looking for evidences in the browser history could be a good idea. Other techniques are based on the fact that sensible information, such as passwords and eventually everything was typed on the keyboard, were stored in the system's RAM. Since RAM is a volatile storage media, it is impossible to read data from the RAM. However, many operating systems often store data from RAM to mass memory devices, maybe storing users' passwords in a transparent way.

17.1.2.2. *Data seizure*

More likely than not, data should be seized from the suspected user. This includes passwords stealing, mail interception and network monitoring. Techniques to achieve this goal are very different, since data seizure depends from many factors: what kind of data is to be seized, where it has been stored (if it has been stored somewhere), if the computer forensics specialist has a direct access to the system. Moreover it is important to know when data can be seized. If a user log in everyday at 8:00 A.M., it is useless searching for the password in data recorded at 6:00 P.M. This leads to restrict the amount of data to be examined, and potentially eliminates time wasting. Generally, the more you know about the user, the better it is.

17.1.2.3. *Data recovery*

Sensible data could have been deleted from electronic devices; therefore, techniques for data recovery should be developed. There are many aspects to be considered in data recovering. In fact, physical condition of the device, its capacity and its nature are essential to data recovery. Recovering data from a broken device is quite different from recovering deleted data. Usually the work of repairing broken devices is made by specialists' teams, while deleted data can be easily recovered, if they have not been overwritten. If it is not the case, special instruments have been developed to retrieve overwritten data from physical devices.

Once data are recovered, they are usually stored in tape media. This work is often made by organisations specialised in data recovery. Usually organisations serve concurrently many clients, and use tape media to store recovered data. Since writing data to a tape media is a slow operation, and due to the fact that clients should be served concurrently, many obstacles arise during the process of data storing, and there is the need to define specialised architectures to handle the process which minimize waste of time.

17.1.3. *The Need for Tools and Techniques*

Working on a computer crime never is an easy job. The goal of the investigator is to find out every single digital evidences stored in devices, or at least an amount sufficient for building and supporting a crime theory. The more the suspected user is skilled, the more difficult it would be to accomplish to this task. Furthermore, it is possible for an investigator to make errors while performing specific actions, or omitting clues while looking for digital evidences.

Digital Forensics techniques have been developed to achieve the goals of Data Locating, Data Seizure and Data Recovery. It is to the computer forensics investigator to analyse a case and to select a sub-set of techniques to be used in the process of digital evidence discovery. Using known and tested techniques is more productive than taking improvised actions, which could only lead to waste time.

Nevertheless, performing actions required to use a technique manually could take a huge amount of time; manually looking at all the cluster in a hard disk could require months, or even years, of work. Moreover, particular tasks could not be performed without the help of specific software or dedicated instruments. A large variety of software and hardware have been developed to help computer forensics scientists in performing the process of examination of computer systems.

17.2. Basic Techniques

Computer forensics methodologies of work, as well as main goals in a computer system analysis, have been introduced. The next issue is to introduce techniques used to achieve these goals; since many of them require knowledge of hardware and operating system essentials, a little background will be provided when needed.

Techniques will be distinguished according to the area of interest among computer forensics goals introduced earlier. So there will be one category of techniques developed for data locating, one for data seizure and finally one for data recovery.

It is important to notice that an investigator could use more techniques, even belonging to different categories, to obtain single evidence. Suppose that sensible data were encrypted, so that they become unreadable by anyone but the owner, and then were hidden somewhere in a storage device. Techniques for data locating should reveal the data, but then it is necessary to decrypt them; in order to accomplish the last task, encryption key must be stolen, or data cryptanalysis, whenever is possible, could be made. Clearly the latter techniques belong to data seizure. Furthermore, it is obvious it is impossible hidden data if they have not been found. This example proves that the use of many techniques may be involved to obtain single evidence.

17.2.1. *Data Locating*

Data locating is the process of discovering sensible data stored in hard disks or other devices. While sensible information is often stored inside files, handled by the operating system of the computer system which the mass storage device is attached to, there are still many cases in which information is stored in the hard disk without the knowledge of the operating system. Since locating data among the files handled by the operating system could be a common and easy task, this would be not to look for data which are not; in fact these data have no structure at all, and they could seem like random sequences of bytes rather than sensible information that could be used as digital evidences.

It is preparatory to the use of specific techniques to know how operating systems use storage device for data management. Nevertheless, it should be understood that this topic is filesystem dependent, although basic principles are the same for almost every today's filesystems.

Storage devices are logically divided in one or more partitions, each one of which has at most one filesystem associated to it.

The mass memory of storage device is usually divided in blocks of equal size, which is determined when creating the filesystem: blocks are increasingly numbered, and the number associated with a block is known as the block address. Blocks are the basic logical unit handled by an operating system while handling storage devices; it follows that when an operating system tries to read (write) n bytes and the size of a block for the used filesystem is k bytes, then $\lceil n/k \rceil$ blocks will be read (written).

For being practical for users to manage data, the filesystem provides functionalities for data retrieving; files could fit in one or more blocks, not necessary with increasing addresses. The filesystem should maintain information about addresses of blocks used by each file, as well as the order in which these blocks are used by the file: information about file organisation is usually stored in the first blocks of a filesystem, and the way they are stored varies from filesystem to filesystem. For clarity, it will be assumed that the first blocks of a partition contain a file map table in which each entry corresponds to a file, including needed information associated to that file. Since the last part of a file rarely fit fully in a block, a special character is inserted at the end of a file to delimit its end.

17.2.1.1. *Cluster tips*

While blocks are logical entities which divide the space used by a partition, clusters are physical entities of a hard disk. Hard disks are usually divided in cylinders, and each cylinder is divided in separate clusters used to store information. It is common use to make the size of a block a multiple of the size of a cluster, so that in a logical block will fit an exact number of physical clusters. Many operating systems, like DOS and all Windows versions, share one rule, which could be summarised as: "one cluster, one file". That is, in each cluster will be inserted information belonging at most to a single file. As a consequence, when writing a file in a hard disk, some cluster would remain partially or fully unused. As the operating system can only write a full block, it follows that the unused space should be fit with some sequence of bytes; a common strategy is to select random data from the RAM, so that sensible information about applications' data stored in a volatile memory could flow in a mass memory device in a completely transparent to the user way. This obviously provokes a covert channel from the RAM to the storage device.

The cluster tips technique consists in reading clusters filled with data belonging to the RAM. Keeping in mind that these data are stored in a disk because of the operating system limit to write only a full block, they could be located by looking for an end-of-file character and then reading whatever follows. Even if defrag could limit the power of cluster tips technique, by minimising the number of unused clusters, it could not be completely avoided due to the "one cluster, one file" rule.

17.2.1.2. *Free space*

Since the filesystem should insert new files inside the storage device, it has to know if a particular block is free or not. Once again, the way in which this functionality is provided depends from file system to file system.

The operating system uses the free space only to store new data, since if there is no file associated to that block then there will be no information to handle. Data contained in free space will be transparent to almost all the applications. It is important to notice that operating systems provide functionality to read arbitrary blocks from the filesystem, whether they are marked as free space or not, so data stored in free space are not transparent to the operating system. A skilled user could hide sensible information inside blocks considered as free space by the filesystem; these data are not associated with any files, so they cannot be found by a file searching utility.

Another issue is how filesystem handle file deletion. For computational purposes, it is useless to overwrite stored data, as it is possible to mark the blocks previously associated with the deleted file as free space. Usually this operation consist in marking a specific bit associated with a file, while overwriting all the data contained in a file could result in a huge waste of time. Again, by searching through blocks marked as free space deleted file could be retrieved.

Although the free space technique allows reading deleted data, it is not a data recovery technique, since data were not really deleted from the storage device.

17.2.1.3. *Swap file*

The RAM is one of the main components of a computer system: running applications, keyboard inputs, data to be processed and processed data are stored inside the central memory. However, RAM size is really limited, compared to the capacity of storage device. Many applications need a huge amount of memory to work correctly. Further, if running applications saturate the available memory, the operating system cannot store any other application in the RAM until some space is made available; in practice, using too much memory often provokes the system to freeze.

A workaround used by recent operating systems is to swap data which are not believed to be used in a short time interval into a storage device. Windows operating system uses to store data from the RAM in a particular file, known as the swap file. It is neither not marked as free space, nor associated with a file in the file list; its only function is to provide space capability for the operating system to store data belonging to the RAM. In *nix operating systems family a different partition is used to store swapped data.

As for the cluster tips, the use of a swap file or partition creates a covert channel in which information flows from the central memory to a storage device. However, the swap file is considerably larger than a cluster, and there will be a better probability of obtaining digital evidence by looking inside it, rather than searching for unused space through clusters.

17.2.1.4. *Registry*

System and applications settings in windows operating systems family are usually stored in a database which takes the name of system registry. The system registry is physically stored inside the storage device, precisely it uses two files: `user.dat` and `system.dat`. The database which constitutes the system registry follows a hierarchic model: the main entities are keys, denoted by a string representing the name of the key. Each key can be linked to other keys (in this case they are called sub-keys) and with other entities called values. Values consist of three fields: a name representing the value, a type field which specifies what kind of data is represented in the value, and finally data associated with the value. This hierarchical model is similar to file organisation into directories; as for files, every key and every value have a unique path associated to them, which is the ordered list of keys to walk through for reaching the key or value, and which the last node of the list is the key or value itself.

Almost every operation in the windows operating system involves reading from and writing to the system registry. Searching for data through the registry is a good way to redevelop the history of actions taken by users of the computer system. Furthermore, the registry is a compacted and well structured database for which APIs are provided by the windows standard library; for this reason many applications prefer to store sensible data inside the register, rather than creating one or more file and defining specific formats for configuration files.

The physical organisation of the two files mentioned above is quite intricate, and it would be virtually impossible to read data without using specific applications; fortunately, Windows provide specific software to accomplish this task, which is called Regedit. Nevertheless reading from and writing to a file can be made only by sequential access, so it will be computationally inefficient to physically remove keys and values from the registry. If a command of value deletion is inserted, the system should move all the data following the value to be deleted a little earlier (exactly the number of bytes used by the deleted value); a similar process should be handled for updating. Moreover, the structure of the file should be preserved. Thus, it is more convenient to not physically remove deleted and updated keys or values; instead, they will not be displayed by the Regedit utility. Finally, the corruption of the registry files would result in a failure in booting the system. To assure the integrity of the register, the operating system preserves copies of it, so that a backup could take place when needed.

The registry technique consists in reading sensible data from the registry, even looking for deleted or still present data before they have been updated.

17.2.1.5. *History and temporary files*

Applications usually provide special functionalities to provide easy recovery of unsaved data after a boot crash. The way in which applications accomplish to this task is by saving temporary files periodically; if a system crash occurs, the most recent temporary file will contain the most part of lost data. Temporary files are also useful in Internet applications; users commonly visit some Web site more frequently than others. Storing information about these Web sites, such as images or hypertext documents, will speed up a browser, since data are present locally and should not be requested to the Web site for transfer. It is then obvious that temporary files may help in redeveloping the history of a computer systems: documents revisions, visited Web sites, and so on.

Another issue in looking for the visited Web sites comes from history files. These are files created by browsers and that contain information about visited Web sites, and the date when they have been visited. History files are used to help the user to find what he needs, e.g., making a list of the known Web sites matching a particular sequence of characters.

Although history and temporary files are useful for users, preventing the user to lose unsaved data and helping him to find what she needs, they represent an extremely dangerous threat which could be used by computer forensics specialist to obtain information about the actions taken by the user. To read temporary and history files could reveal a huge amount of information, and eventually relate them with a crime.

17.2.2. *Data Seizure*

Although data locating techniques allow an investigator to find sensible data, there are many situations in which the goal is not to find sensible information; rather it would be necessary to obtain information which is not stored in mass devices, such as passwords used to protect applications and data. Data locating techniques involved with the information flow from the RAM to the storage device, i.e., swap file and cluster tips, could accomplish this task, but the outcome is not sure; further using such techniques is possible whenever the investigator has physical access to the system, for example when the computer system is taken in custody.

Other techniques should be developed to capture and record user activities; such techniques fit in the Data Seizure techniques category; they involve the use of both specialised hardware and software to obtain information that is not usually stored inside computer systems.

17.2.2.1. *Keystroke loggers*

Users usually interact with the computer system by typing in inputs through the use of a keyboard. Such data input usually will not be stored in storage devices, unless there is a covert channel from the RAM to the storage device. For example, passwords used to authenticate users are never stored in the mass memory; rather the hash of the password is stored so that, even reading it using data locating techniques, it would be computationally infeasible to recover the original password, or another character sequence which authenticates the user.

However, in these case information flows from keyboard to central memory; keystroke loggers examine and record this flow of data and record it. They come both in hardware and software. While hardware devices are placed between the keyboard and the RAM (typically they look like keyboard cable extensions), software usually monitors keyboard drivers in Windows and keyboard device file in *nix; in both cases typed in inputs will be captured and recorded. Since hardware devices cannot interact with the system, they have a mass memory where to store recorded inputs.

Usually there are many premises before choosing to use a hardware or software keystroke logger. Software loggers may interact with the operating system, so could offer extra functionality which could not be performed by hardware devices, such as relating the keyboard input with the application the input was for, or sending recorded data through e-mail; however they are forever running processes, and their existence is known to the operating system; due to this reason, they could be easily detected and their effect could be avoided by killing the discovered process. Finally, the logging process has to startup before it begins recording keyboard inputs, and it could require administrator privileges to monitor keyboard activities.

Hardware devices solve the problems above: they simply need to be attached to the keyboard, and they will begin recording every input that flows from the keyboard to the RAM. Clearly these devices could not provide all the features of a software keystroke logger, and they have their own memory: compared to software logging utilities, they could store a considerably smaller amount of information. If the maximum capacity of recordable information is reached, keystroke loggers will begin overwriting less recent data. Even with these capabilities, hardware keystroke loggers are extremely useful; since they are completely transparent to the operating system, they will never be detected by protection applications.

17.2.2.2. *Wiretapping*

A requisite to use hardware or software keystroke loggers is the physical access to the targeted computer system; in many situations computer forensics specialists have to investigate on a system without being able to interact with it. A common technique used to seize data is to monitor network connections and traffic, also referred to as wiretapping.

Wiretapping comes in different forms, which differ from the network environment of the computer system which network traffic wants to be monitored; common features that determine how wiretapping should be performed are the kind of connection of the computer system (e.g., directly attached to the Internet through a phone connection, ore connected inside a Local Area Network), the medium used for data transmissions (coaxial cable, optic fibers or radio transmissions), local network topology and the nature of the connection to be monitored, which could be encrypted or not. In every case the goal is to read incoming and outcoming data, recording which are believed to be sensible.

The kind of connection determines where wiretapping should begin. If the computer system is part of a Local Area Network, then data will flow from the computer system to the LAN gateway before being forwarded to the Internet; in the case that the computer system is directly connected to the Internet, the data will first reach the Internet service provider of the active connection, and then they will be forwarded to the outside world. In both cases there is a checkpoint where data have to arrive before being forwarded, and usually there is a unique path from the computer system to this checkpoint.

If the computer lies inside a local network, the organisation of the LAN is also relevant; for example, if the computer is attached to an 802.3 Ethernet LAN, it is known that all the computer systems of the LAN are physically connected by a single wired cable. The access to one of this computer system will allow using a sniffer software utility to monitor all the traffic of the LAN. In the case of an unprotected 802.11 wireless network, data will be read by everyone who is listening on the frequency used by the access point to communicate with the computer systems; even if the wireless network is protected according to the WEP protocol, the network key can be easily obtained by using tools such as AirCrack and AirDump.

In every case the wiretapping requires a physical access to the connection media from the computer system to the checkpoint (Gateway or ISP systems), and hardware for data monitoring should be provided. Even if these two requisites are generally sufficient, many obstacles arise: data flowing through a network are often subjected to errors, in which case it could be requested to retransmit them. Moreover, it should be know how bits are encoded before flowing through the transmission media. Software utilities could be used to examine captured data, in order to redevelop the original data transmitted by the user.

Finally, another obstacle arises if the connections are encrypted. In this case a way to find the session key for the connection should be exploited, or cryptanalysis work should be performed on the encrypted recorded data; however, this is not usually an easy task, and for almost all the cryptosystem used today for securing Internet activities it is conjectured it is computationally infeasible to break them.

17.2.2.3. *Spyware*

Since wiretapping cannot handle encrypted connections, other ways should be used to track online activities of users. A useful technique is to use ad-hoc developed software such as spyware.

A spyware is a utility which, once installed on a computer system, monitors the online activities of the users of that computer, eventually seize data such as password or e-mails, and then sends the collected information to the investigator. Spyware are often used by companies and organisations, which track users' online activities and then sends them commercial spots based on the information recorded. Differently from wiretapping, spyware monitor online activities at an endpoint of the connection, so, in the case of an encrypted connection, outcoming data will be recorded before they will be encrypted, while incoming data will be recorded after they will be decrypted.

The main obstacle of spyware lies in the fact that they should be installed in the computer system before they can begin seizing sensible information. Typically spyware are hidden inside other applications, and the user will have no knowledge of the spyware; installing the main application will cause the hidden spyware to be installed. In other cases, the user is warned about the presence of the spyware inside the program. In this case the word spyware is no longer used; instead, the application will be said with ad-ware license. Finally, a tricky method to install spyware inside a computer system is to exploit browser vulnerabilities to build ad-hoc web pages which, if visited, will provoke the spyware to be installed.

17.2.2.4. *Electromagnetic radiations*

A last but useful idea in data seizure makes use of the Van Eck radiations. Monitors usually emit radiations which could be intercepted in a half mile range; in this case information flows from the monitor through a radio channel which could be revealed through ad-hoc hardware. Equipment for performing interceptions of the video display through the emanated radiations can be easily found in electronic stores; it consists of a broad band radio scanner, a good antenna and a TV set.

Using Van Eck radiations is a really powerful technique: although passwords are usually not displayed and cannot be used with this technique, many other sensible data can be seized; this technique allows capturing every activity performed by the user, and by the user point of view. However, there are many countermeasures to prevent this form of data seizure, based on the physical nature of the technique; for example, if the user is suspecting of being spied, it could decide to shield the room where the video display is, and so radiations will be blocked before they could leave the room. Of course shielding is quite impossible in the case of a laptop computer. Another way to prevent this technique to be successful is to make intercepting radiations useless, by creating interferences in the transmission, through the use of Electromagnetic Interference Shielding products.

17.2.3. *Data Recovery*

It is possible, for an investigator, to have to deal with a broken device; in this case the data could not be read by directly accessing them through a computer system, and other ways have to be found to successfully read data inside the storage device. The task of reading data from broken devices takes the name of data recovery [Vacca (2005)].

Data recovery is even used by companies and organisations; as they usually store a large amount of data among several storage devices, it is not so unexpected that a storage device will break in a defined time interval.

Techniques for data recovery usually depend on the nature of the storage device; usually data recovery is made by specialised organisations, which should provide a backup service to several clients concurrently. To achieve this goal networks are often used to store recovered data inside tape devices, which are accessed by one or more servers.

17.2.3.1. *Backup obstacles*

Although using networks allows handling several backup requests, there are many obstacles to deal with, which influence the performance of the backup operation:

- Backup window: this is the period of time when backup can be run. In companies, backup window is generally timed to occur during non-production periods, when CPU and network bandwidth are low.
- Network bandwidth: this is an important issue; if the network cannot handle the amount of data sent by users, the result would be a bottleneck in the network.
- System throughput: for a client system, it's a measure of the ability to push data into the Backup Server, while for the server it is a measure of the ability to accept data from multiple systems at the same time.

17.2.3.2. *Backup server architecture*

Several approaches for handling data recovery make use of a client server model; clients responsible for backup tasks forward recovery requests to a single server through a network; the server, which is referred to as the backup server, is responsible of managing and performing requests coming from the clients.

However, a drawback of this strategy is that if too many requests are concurrently forwarded to the backup server, it could be unable to successfully handle every single request; to overcome this problem, a slightly more complex design is used. To obtain a performance improvement, requests received from the backup server are scheduled and forwarded to many other servers, each one with a tape library attached to it: these servers, which take the name of tape servers, are responsible of storing data to their own tape library. In this way all requests coming from a client are received by a single backup server, thus obtaining a centralised design; furthermore, task requests are performed by several different tape servers, granting a performance improvement by distributing the load of work.

17.2.3.3. *Network backup designs* [9]

17.2.3.3.1. Direct-attached tape

Direct-attached tape is the simplest network design for data recovery. Basically, every server has a tape drive attached to him; this approach allows adding more capability to each server by simply attaching additional tape drivers to it. To automate this work, autoloaders are often used.

This strategy, however, has some serious drawbacks. As the number of server expands, the amount of time spent managing all of the tape backups grow proportionately; in this case a way to centralise the management of the backup is needed. Furthermore, every server should be provided with its own tape device and tape backup software; it follows that an auto loader is needed for each server. Since direct attached tape is the easiest strategy for data recovery, it could be used in small environments in which drawbacks do not represent a critical point.

17.2.3.3.2. LAN-based backup

Although using direct attached tape is usually effective in small environments, other strategies should be developed in order to grant an increase of efficiency in larger ones. In a LAN-based backup design a backup server, manages backup requests coming from the clients of the network; in this way, backup tasks are centralised on a single server, which schedules the requests and forward received requests to several tape servers, as established in Section 2.3.2.

Although in large environments LAN-based backup is more desirable than direct attached tape, backup requests, as well as all backup data, have to flow through a network in order to reach the tape servers; it is important to notice that the amount of data flowing through the network for backup operations is expected to be really huge: for this reasons, network bandwidth and backup window represent a crucial point in LAN-based backups.

17.2.3.3.3. SAN-based backup

Another way to provide backup services is to use servers and tape libraries to build a Storage Area Network (SAN); it follows that a tape library can be shared among several servers, and a server can be connected to different tape libraries. SAN-based backup takes advantage of the high performance of direct-attached tape and also the central management of LAN-based backup. Obviously, clients should be able to reach servers in the SAN, in order to forward their backup requests.

17.2.3.4. *Incremental backup*

Even if a suitable strategy is designed and implemented for backup tasks, in large companies, which should backup a large amount of information, obstacles presented above still represent a critical point; however, it is often the case for companies that backup operations are performed periodically, to assure data integrity; in this scenario it is possible to minimise the amount of data to be backed up, by storing in the tape libraries data that have been changed since the last backup operation was performed; this strategy is referred to as incremental backup. Two possible strategies are defined, depending on the data granularity level.

- **File level incremental backup:** New files, as well as files updated since the last backup operation was performed, are object of backup operations. If the size of files to be backed up is too large, file level incremental backup could not be an effective strategy.
- **Block level incremental backup:** If a file level granularity is not considered suitable for incremental backup, another strategy resides in checking only for the blocks which have changed from the last backup.

17.2.3.5. *Image backup* [11]

The goal, in an incremental backup strategy, is to backup only data that changed from the last backup operation. Another way to efficiently perform backup tasks is the one of Image backup; in this approach a copy or a snapshot of a filesystem is made at a particular point in time.

Image backups are faster than an incremental level backup, and provide the ability to easily perform a "bare bones" recovery of a server without loading the operating system; if an image backup strategy is used the backup software will operate at file system level, and eventually will update file system accounting data. By taking a snapshot of the filesystem it follows that even deleted data are captured, since metadata regarding to them are still present in the filesystem.

An important issue in imaging backup is how the backup is performed; an efficient approach is the Live Imaging, which consists in capturing a snapshot of a system while it is still functioning in a live environment. Live imaging is unavoidable in companies which provide digital services to the outside world by using systems subjected to backups: if the systems are shut down before a backup operation begins, then services provided by the system will be unavailable for at least all the backup window time. Live imaging is also necessary when judge instructs that evidence gathering must be conducted using the least intrusive methods available.

17.2.3.6. *Using MFMs* [7]

So far, only the way through which recovered data are stored in tape devices has been represented, while the question about how data are recovered from a particular device has not been answered yet. The way in which this process can be executed obviously depends from the nature of the storage device from which data want to be read; however, it is important to analyse at least a technique used to recover data from magnetic disks, since they are the most widely used devices. It is commonly quoted [4] that data inside magnetic disks can be recovered if they have been overwritten only once or twice, and that it actually takes up to ten overwrites to securely protect previous data. If a head positioning system is not exact enough, new data written to a drive may not be written back to the precise location of the original data. Due to this *track misalignment*, it is possible to identify traces of data from earlier magnetic patterns alongside the current track. As an example, when a bit is written in a location of the disk, the final result depends from the previous content of that location: If a 1 is going to overwrite a 0, the effect will be closer to obtaining a 0.95, while it will be 1.05 circa if the 1 is going to overwrite another 1. Normal disk circuitry is set up so that both these values are read as 1, but using specialised circuitry it is possible to work out what previous "layers" contained. It turns out that each track contains an image of everything ever written to it, but that the contribution from each "layer" gets progressively smaller the further back it was made. A more profound analysis of layers may be carried out with Magnetic Force Microscopes (MFMs) [15]. A MFM is an extension of atomic force microscopy (AFM) that images magnetisation patterns with sub-micron resolution. MFMs have recently become a very popular tool for characterising magnetic microstructure [13]. Cost, simplicity and the ability to look through non-magnetic overlayers with minimum sample preparation are all reasons why MFMs are routinely used by industry for magnetic materials characterisation. A major drawback of the MFM is that the rather complicated interaction between the magnetic tip and surface makes quantitative interpretation of the MFM images difficult.

A MFM images the spatial variation of magnetic forces on a sample surface. The system operates in *non-contact mode*, detecting changes in the resonant frequency of the cantilever induced by the magnetic field's dependence on tip-to-sample separation. MFMs can be used to image "naturally occurring" and "deliberately written" domain structures in magnetic materials.

17.3. Advanced Techniques

17.3.1. *Digital Watermarking*

While investigating on a system, a computer forensics team or specialist could find sensible data which have been altered by a user. In many cases, the altered data alone may not constitute legal evidence; therefore, a way to recover the original data must be used. Digital watermarking makes use of technologies to show if data are an edited copy of other data [Caloyannides (2004)]. If it is the case, watermarking technologies also determine if original data are protected by copyright, as well as if edited data are authorised or less. Furthermore, a way to rebuild the history of edited data is needed; this includes discovering the exact pathway from the owner to the unauthorised copy, as well as finding differences between original data and copied data. Digital watermarking techniques are widely used in detecting if a digital image has been altered or not.

Watermarking techniques should be robust, not "washable" with software of image elaboration as Photoshop, must resist to the compression of the image (in a format like JPEG for example) and imperceptible to our senses. There are two widespread techniques:

- Modifying part of file: for example changing the most significant bit of some pixel or the entire file by spreading the digital watermark over the entire file. A sophisticate technique is to put two digital watermarks in a file, one of which is simple to find, so that it could be believed that the watermark has been eliminated.
- Watermark Negation-schemes: an image can be divided in pieces not big enough to contain sufficient information about the existence of the watermark; these pieces are sent across Internet and then reconstructed at the destination. Other watermark-negation-schemes use tools which have the ability to remove digital signatures.

17.3.2. *Raid* [6]

RAID disks have become popular in recent years even in low end systems; concurrently to the widespread of RAID disks, forensics techniques have been developed to image data contained in them.

A simple and efficient way to accomplish this task is to read RAID disks data by using an operating system equipped with drivers which allow seeing the array of disks as a single logical disk. If this step is performed successfully, techniques for imaging a single disk can be used.

However, this technique makes assumption which could not be true in many cases; these are the use of a specific operating system with specific drivers equipped, and the possibility of accessing raid disks by that operating system. Moreover, this technique cannot be successfully used if RAID is done in software using a proprietary product, when headers on the disks are damaged, leading the array controller to refuse the use of the disks, or when is impossible to obtain the original controller and BIOS configuration.

In the last two cases, techniques to reassemble the array should be developed; below a method to reconstruct RAID 5 is used.

In order to reassemble the array, it is necessary to follow this rough order of tasks:

- Block size has to be determined: this is done by observing discontinuities in a disk: this is an easy task if text data are observed.
- The physical array period must be determined: this could be done by examining the parity blocks around the disks. It is important to notice that, when the disks

are full of random binary data, it is impossible to tell which disk carries the parity and which is used just to store binary data. From this point, it is necessary to tabulate where the parity is.

- Construct a Striping map by inspecting which physical blocks follow each other throughout the image. Reinforce this observation by looking at different physical blocks representing the same slot position.
- Reordering the disks will produce a much more obvious pattern in the striping map. This pattern will allow an investigator to guess the complete pattern without having to confirm each element in the map.
- Reconstruct the array into a single image, so that other forensic packages may be able to use it.

Since RAID 5 provides redundancy, if a disk is corrupted or even missing, it could be rebuilt on the fly by using the parity in the other disk; however, this process is manual and time consuming. Usually an analyst determines the map, and then re-assembles the array by copying the data into a single logical image; this process could take a very long time, particularly if the array is large. Once the image is reconstructed, if the so built map is correct the analyst will succeed in mounting the partition.

17.4. Tools

17.4.1. *Data Locating* (2.1)

17.4.1.1. *Swap file*

The Swap File is a Windows file that is used for virtual memory management. If a user is working on a file or document (even one that has been encrypted by a powerful engine), Windows can copy all or part of it in an open unencrypted form to the Swap file on a storage device. Encryption keys, passwords and other sensitive information can be "swapped" to the storage device too. Even if a user uses all the security features in the latest versions of Windows, simply investigating the Swap file in DOS mode with readily available tools may allow for significant data retrieval.

17.4.1.2. *EnCase* [23]

EnCase has the ability to acquire data in a forensically sound manner using software with an unparalleled record in courts worldwide. It can easily manage large volumes of computer evidence, (viewing all relevant files, including "deleted" files, file slack and unallocated space), as well as transfer evidence files directly to law enforcement or legal representatives as necessary. Review options allow non-investigators, such as attorneys, to review evidence with ease. Reporting options enable quick report preparation.

17.4.1.3. *Supershredder.exe* [33]

Supershredder is a windows tool which allows removing a single file from hard disks; it is important to remember that common file deletion only updates metadata information about the file in the filesystem, while contents of the file are not deleted. Supershredder provides a large variety of algorithms to successfully remove a single file in a storage device; the user may choose which strategy to use among the following:

- Single Pass: data area is overwritten once with "1" or "0".
- DoD 5520.22-M Standard: this method overwrites all addressable locations with a character, its complement, then another character and then verifies the result.
- Gutmann Method: the data area is overwritten 35 times. This method overwrites the drive taking into account the different encoding algorithms used by various hard drive manufacturers.

SuperShredder gives to users the ability to delete files using Gutmann and DoD [22].

17.4.1.4. *BcWipe:* [26]

BCWipe software provides the following main commands and options: delete and wipe a file or a folder, or a group of files and folders. Using this tool, users can completely remove all traces of previously deleted files; this tool also allows wiping unused portions of the Swap File. BCWipe shreds directory entries and MFT so that information cannot be recovered; it also removes lists of recently used files from the File Menus of specific programs, as well as users' Internet Cache, Cookies and History.

BCWipe allows user to encrypt the Swap file providing user with additional data security. It provides a hexadecimal File Viewer, so user can examine contents of files after wiping. This utility is useful for investigating the quality of the wiping process, for example when you use a custom wiping scheme. BCWipe v.3 allows user to wipe data using predefined DoD 5200.28-STD and Peter Guttmann wiping schemes; it also includes the Wiping Scheme Editor utility to view and edit the number of wiping passes as well as binary patterns used in every pass. Swap File Encrypting Utility added. BCWipe v.3 includes a "Task Manager" that allows the user to schedule some wiping tasks automatically. User can configure BCWipe Task Manager to remove all references to recently used files or only those in applications that he specifies. Users can configure BCWipe Task Manager to automatically wipe selected folders when they are not locked, i.e., when Windows starts up. To avoid a possible security leak, user can configure BCWipe to wipe the Hibernation File regularly. BCWipe v.3 recognises and shreds alternate data streams in files, created on NTFS disks in Windows NT/2000/XP operating systems and has the ability to Skip files and directories which do not require treatment during "Wipe file slacks" process.

17.4.1.5. *Eraser:* [24]

Eraser is an advanced security tool (for Windows), which allows users to completely remove sensitive data from a storage device by overwriting it several times with carefully selected patterns. It is free software and its source code is released under GNU General Public License. The patterns used for overwriting are based on [Gutmann (1996)], and they are selected to effectively remove magnetic remnants from the hard drive.

An important feature provided by Erase is that users can also define their own overwriting methods.

It is important to notice that Eraser cannot wipe the swap file, since it is opened at the startup with exclusive access, so that application cannot access it. Swap file could be cleaned only by accessing it from another operating system, or by disabling it from Windows system settings, so that the exclusive access to the swap file will be released; the last method obviously requires rebooting the system.

17.4.1.6. *Registry*

17.4.1.6.1. Regedit (Windows) [2]

Regedit is a Windows tool which allows reading the contents of the Windows registry; Due to the vast amount of information stored in Windows registry, the registry can be an excellent source for potential evidential data. For instance, windows registry contains information about users' accounts, typed URLs, as well as command history.

Windows 2000 and XP Registry Editor (regedit.exe or regedt32.exe) have an implementation flaw that allows hiding registry information, preventing users by viewing and editing them, regardless of their access privilege (Secunia, 2005). The flaw involves any registry value which name length is comprised between 256 and 259 characters. The overly long registry value (regardless of type) not only hides its own presence, but also subsequently created values (regardless of type) in the same key (Franchuk, 2005). This vulnerability allows malware to hide malicious code in .autorun entries such as HKLM\Software\Microsoft\Windows\CurrentVersion\Run. Any program or components specified in this key will be automatically run during system startup. However, Windows will execute these hidden entries successfully at startup (Wesemann, 2005). Some common malware scanners are not able to detect such maliciously crafted registry values (Gregg, 2005). Neverthless, Windows console registry tool (reg.exe) can display overly long registry values.

17.4.1.6.2. Kregedit (UNIX, KDE) [8]

Kregedit is a *nix utility developed by the KDE team, which allows to read Windows native registry files; however, only the NT registry format is supported.

17.4.2. *Data Seizure and Password Stealing*

17.4.2.1. *Keystroke loggers*

17.4.2.1.1. (Windows) KeyKey [34]

KeyKey is a tool which records recent keyboard activities in a computer system. A predefined output file is selected for the user convenience, although she can provide a different output file, called reporting file. KeyKey allows defining time intervals in which the logging functionality will be enabled; in addiction, a TimeStamp recording feature enables users to record the time of keyboard activity.

17.4.2.1.2. Spyware: (Windows) WinWhatWhere [16]

WinWhatWhere is a Windows tool that monitors and records user activities on a computer system; logged activities include e-mail traffic, visited Web sites, instant messaging communication, passwords, files and keystrokes. Logged information will periodically be sent through e-mail to a pre-defined e-mail address.

17.4.2.2. *Against adware*

17.4.2.2.1. Ad-aware SE enterprise console

The main functionality of Ad-Aware [27] is to prevent and detect malware attacks, removing malware applications when necessary.

Ad-Aware SE Enterprise Console is designed for corporate environments; this tool monitors network connection and allows a centralised management of the clients of the network through the use of a console. In order to accomplish this task, each client should have Ad-Aware installed. Furthermore, client updates may be centralised by the use of Windows Group Policies.

In order to detect malware applications, scan activities can be scheduled on clients through the use of Ad-Aware SE Enterprise Console.

17.4.2.3. *Ad-aware SE professional*

Ad-Aware SE Professional is another tool designed to prevent and detect malware attacks, and which provides the following functionalities:

- User-controlled spyware removal: Users are allowed to decide what if a malware has to be deleted.
- Custom scanning of RAM, registry, hard drives and external storage devices.
- The CSI (Code Sequence Identification) technology used by Ad-Aware SE Professional detects known variants of malware and then searches for similar codes to identify emerging, or unknown variants.
- An extensive Definition File library of identified and analysed spyware, which is continuously updated for new threats, is used.

- Ad-Aware SE scans and removes ADS (Alternate Data Streams), malware that secretly attaches itself to a program or file that a user downloaded.
- Browser hijackers are blocked from taking control of user's home page and re-routing Internet searches.
- Registry protection allows user to stop spyware/malware attempts to modify the start-up sections in the Windows registry.
- The Command line interface allows Ad-Aware to scan and remove spyware in the background, rather than showing the window on user's screen. Users can export or save scan reports in text or HTML format.

17.4.2.4. *Against spyware*

17.4.2.4.1. Spybot-search & Destroy [29]

Spybot-Search and Destroy detects and removes spyware. Data observed while scanning a computer system are divided in red entries, which represent spyware and similar threats, and green entries, which are usage tracks. Removing usage tracks is non-critical, and it depends on personal preferences. Red entries represent the real threats. Red entries are regarded as real threats and should be dealt with.

17.4.3. *Data Recovery*

17.4.3.1. *Incremental backup*

17.4.3.1.1. (UNIX) Dump [14]

Dump examines files on an ext2 filesystem and determines which files need to be backed up. These files are copied to a given storage device, and eventually manages remote backups. A dump that is larger than the output medium is broken into multiple volumes. On most media the size is determined by writing until an end-of-media indication is returned. On media that cannot reliably return an end-of-media indication (such as some cartridge tape drives), each volume is of a fixed size; the actual size is determined by specifying cartridge media, or via the tape size, density and/or block count options below. By default, the same output file name is used for each volume after prompting the operator to change media. A particular filesystem, as well as a list of files and directories, can be selected to be backed up as a sub-set of a filesystem. In the case of a mountpoint, either the path to a mounted filesystem or the device of an unmounted filesystem can be used. In the case of a filesystem, restrictions can be placed on the backup. Since making a dump involves a lot of time and effort for full dumps, **dump** checkpoints itself at the start of each tape volume. If writing that volume fails for some reason, **dump** will, with operator permission, restart itself from the checkpoint after the old tape has been rewound and removed, and a new tape has been mounted. **Dump** tells the operator what is going on at periodic intervals, including usually low estimates of the number of blocks to write, the number of tapes it will take, the time to completion and the time to the tape change. The output is verbose, so that others know that the terminal controlling

dump is busy, and will be for some time. In the event of a catastrophic disk event, the time required to restore all the necessary backup tapes or files to disk can be kept to a minimum by staggering the incremental dumps.

17.4.3.1.2. (Windows) backup premium [20]

Backup Premium is a windows utility for secure and reliable backup of user's valuable data. It works using the SFTP and FTP SSL protocols, so that the user can be confident in the security of data transferring process. This feature can be especially useful for business users who have very high requirements on the security of storing and transferring data. The program supports backup on fixed or network drives, removable media and FTP servers. Users can organise the automatic backup through the use of a built-in scheduler. Incremental approach to backup allows transferring only new and recently changed files, which decreases backup time and Internet traffic. It provides Blowfish encryption and the support of Windows authorisation system; moreover, it provides the functionality of compressing backed up data in a ZIP format. Finally, it provides functionality to backup registry information, user data from the Windows Profile and some widely used programs.

17.4.3.2. *Block level backup*

17.4.3.2.1. ByteBack (write block in forensics mode) [30]

This tool is considered the standard for forensics recoveries; starting from version 4, it provides support to UDMA, ATA & SATA devices up to two terabytes, as well as memory management, control of Partition and MBR manipulations. It implements the following techniques:

- Disk Cloning (mirroring)
- Forensic Mode (write block)
- Disk Compare (verification)
- Low Level Format (disk wipe)
- MBR, Partition and LDM backup
- Basic Partition Table Management (set active, partition hiding/unhiding).

17.4.3.2.2. Image backup

dd: [Rude (2000)] dd is thought for building an evidence file. It has special flags available to it that make it suitable for copying block-oriented devices, such as tapes. **dd** is capable of addressing these block devices sequentially. **dd** can be a powerful tool when acquiring and copying tapes for cases. dd gives a user the ability to make a complete physical backup of the hard disk; assuming that a user has an unknown tape to examine, if she is unsure of the block size used on the tape she could find the correct block size by using a ibs/iso flag; Determining the correct size trivially make the copying process faster. In this case a "count" flag is used so that only 1 block is read. By setting the input block size to 128 users can effectively find what the real block size. The output of the above command would most likely be an "error" message (which was user's intent) with the real block size revealed (say 1024, for example). **dd** can be used to chop up an image of a storage device in smaller pieces.

17.4.3.2.3. (Windows) Paraben's forensic replicator 4.0 [28]

Paraben's Forensic Replicator can acquire a wide range of electronic media from a floppy to a hard disk. Forensic Replicator's images can be compressed and segmented and easily read into the most popular forensic analysis programs. It constitutes a support for creating & viewing VHD (Virtual Hard Disk) and for viewing Linux EXT2 & EXT3 partitions. It Supports tableau write protection devices. Other features are NTFS image viewing, DoD standard wiping of media. It provides a Drive to Drive image option. **Forensic Replicator** allows compressing image files on the fly and encrypts data for secure storage of evidence-128 bit.

17.4.4. *Advanced Techniques [3]*

17.4.4.1. *Digital Watermarking* (3.1)*: IDMarc* [21]

Digimarc® IDMarc™ digital watermarking is a covert security feature for identity documents that enables trusted machine authentication of driver licenses and other IDs. Although it is imperceptible to the human eye, an IDMarc can be read by many commonly available document scanners equipped with special software. These scanners read and decode the watermark feature to detect counterfeits, photo swapping and data alteration. IDMarc is woven into the artwork of the secure ID so that it takes no additional real estate on the card, allowing implementation with both new and existing ID designs. IDMarc can carry unique data only for authentication purposes without using the cardholder's personal data. This protects citizen information and eliminates the risk of compromising cardholder privacy.

17.4.4.2. *Raid* (3,2)

17.4.4.2.1. BestCrypt v.8: [25] encrypting/accessing

BestCrypt creates and supports encrypted virtual disks, which are visible as regular disks with corresponding drive letters. This tool allows encrypting data with many encryption algorithms, each of which is implemented with the largest possible key size defined in the algorithm's specification. Starting from version 8, LRW Encryption Mode is implemented: this is designed for applications working on disk sector level and more secure than other popular modes used earlier, like Cipher Block Chaining (CBC) mode. Data stored on a BestCrypt disk are stored in a container file. Through the use of BestCrypt v.8 users can mount file-container as a sub-folder on NTFS disk. When the virtual disk is opened, users can read and write data as if it were a conventional removable disk. BestCrypt version 8 can be used for storing encrypted data in containers and accessing them through virtual drives; this is done by encrypting set of files into a single compressed and self-extracting archive.

17.4.4.3. *RAID recontructor* [31]

17.4.4.3.1. Raid data recovery
Runtime's RAID Reconstructor helps to recover data from broken

RAID Level 5 Array consisting of 3 to 14 drives.
RAID Level 0 Array (Striping) consisting of 2 to 14 drives.

Even if a user does not know the RAID parameters, such as drive order, block size and direction of rotation, RAID Reconstructor will analyse drives and determine the correct values. This tool implements the ability of creating a copy of the reconstructed RAID on a virtual image (.vim), an image file (.img) or a physical drive. It creates a copy of user's RAID at another location. It will collect sector by sector from each single drive involved and write these sectors in the correct order to the designated destination. This process is also called "de-striping". Because one drive is redundant in RAID 5, it is sufficient to have one less than the original number of drives (N) in the array. RAID Reconstructor can recalculate the original data from the N-1 drives. For a RAID-0 (striped) array user will need all drives. The RAID Reconstructor will recover both hardware and software RAIDs. It will recover from broken Windows Dynamic Disk sets.

17.5. Conclusions

In this article, some of the most representative technique and tools used in digital forensics investigations have been described; moreover, several tools have been presented for each technique: these tools are useful to investigator to produce and examine digital evidences. Digital investigation works on digital data both at physical and logical level and it is often supported by hi-tech instruments like MFM. Despite this, human intuition has a primary role.

17.6. Summary

The main goal of a computer forensics specialist or team is to find digital evidences and to assure that they have not been altered. In this chapter we have talked about the common methods to find these evidences: *Data Locating*, *Data Seizure* and *Data Recovery*. Data Locating is the process to search the sensible data of the whole physical device. May be that the device is full of data but, luckily, there are common places where to search; in data locating sections we have introduced the main techniques to find these data. In data seizure sections we have showed techniques like password stealing, mail interception and network monitoring, helpful to seize data from the suspected user and to find digital evidences. Moreover, sensible data could have been deleted from electronic devices; therefore, techniques for data recovery have been introduced in this chapter.

We have introduced first the concept and then the basics and the advanced techniques used today to achieve all this goals. Also, we have made a survey of tools used and useful to find sensible data, to capture and record user activities and to read data from broken devices. Following a standard techniques to find and preserve digital evidences may be helpful for computer forensics specialist. Certainly, working on a computer crime never is an easy job.

17.7. Glossary

Active Data: Active Data is information residing on the hard drives or optical drives of computer systems, that is readily visible to the operating system and/or application software with which it was created and is immediately accessible to users without deletion, modification or reconstruction.

Application: Software programs, such as word processors and spreadsheets that most users use to do work on a computer.

Archival Data: Archival Data is information that is not directly accessible to the user of a computer system but that the organization maintains for long-term storage and record keeping purposes. Archival data may be written to removable media such as a CD, magneto-optical media, tape or other electronic storage device, or may be maintained on system hard drives in compressed formats. This is very important in forensics since it can show recent changes that have occurred in the data on a hard drive.

ASCII (Acronym for American Standard Code for Information Interchange): ASCII text does not include special formatting features and therefore can be exchanged and read by most computer systems. Files that have a ".txt" extension are typical of ASCII files.

Backup: To create a copy of data as a precaution against the loss or damage of the original data. Most users backup some of their files, and many computer networks utilize automatic backup software to make regular copies of some or all of the data on the network. No one does it enough.

Backup Data: Backup Data is information that is not presently in use by an organization and is routinely stored separately on portable media, to free up space and permit data recovery in the event of a disaster. To see the backup data, you have to reload it onto a computer from whatever storage media it is on.

Bandwidth: The amount of data that can be transmitted in a fixed amount of time. For digital devices, the bandwidth is usually expressed in bits per second (bps) or bytes per second. For analog devices, the bandwidth is expressed in cycles per second, or Hertz (Hz).

Binary: Pertaining to a number system that has just two unique digits. For most purposes, we use the decimal number system, which has ten unique digits,

0 through 9. All other numbers are then formed by combining these ten digits. Computers are based on the binary numbering system, which consists of just two unique numbers, 0 and 1. All operations that are possible in the decimal system (addition, subtraction, multiplication, division) are equally possible in the binary system. We use the decimal system in everyday life because it seems more natural (we have ten fingers and ten toes). For the computer, the binary system is more natural because of its electrical nature (charged versus uncharged, or on versus off).

Bit: A measurement of data. It is the smallest unit of data. A bit is either the "1" or "0" component of the binary code. Eight bits are put together to form a byte.

Boot: (v.) To load the first piece of software that starts a computer. Because the operating system is essential for running all other programs, it is usually the first piece of software loaded during the boot process. Boot is short for bootstrap, which in olden days was a strap attached to the top of your boot that you could pull to help get your boot on. Hence, the expression "pull oneself up by the bootstraps". Similarly, bootstrap utilities help the computer get started. (n.) Short for bootstrap, the starting-up of a computer, which involves loading the operating system and other basic software. A cold boot is when you turn the computer on from an off position. A warm boot is when you reset a computer that is already on.

Burn: Slang for making (burning) a CD-ROM copy of data, whether it is music, software, or other data.

Byte: Eight bits. The byte is the basis for measurement of most computer data as multiples of the byte value. A "megabyte" is one million bytes or eight million bits. A "gigabyte" is one billion bytes or eight billion bits. A single character of ASCII code, such as a letter of the alphabet requires one byte of memory for a computer to use it.

Cache: A type a computer memory that temporarily stores frequently used information for quick access.

CD-ROM: (Pronounced see-dee-rom.) Short for Compact Disc-Read-Only Memory, a type of optical disk capable of storing large amounts of data — up to 1GB, although the most common size is 650MB (megabytes). A single CD-ROM has the storage capacity of 700 floppy disks, enough memory to store about 300,000 text pages.

Compression: A technology that reduces the size of a file. Compression programs are valuable to network users because they help save both time and bandwidth.

Computer Forensics: Computer Forensics is the use of specialized techniques for recovery, authentication, and analysis of electronic data when a case involves issues relating to reconstruction of computer usage, examination of residual data, authentication of data by technical analysis or explanation of technical features of data and computer usage. Computer Forensics requires specialized expertise that goes beyond normal data collection and preservation techniques available to end-users or system support personnel.

Cookie: Small data files written to a user's hard drive by a web server. These files contain specific information that identifies users (e.g., passwords and lists of pages visited). Cookies have gotten a lot of false bad press lately, and are not the all-present danger to security that some people believe.

DAT: Digital Audio Tape. Used as a storage medium in some backup systems. Kind of like the old 8-track tapes, but obviously a lot better. Data: Any Information stored on the computer system, used by applications to accomplish tasks, or available to users. Deleted Data: Deleted Data is data that, in the past, existed on the computer as live data and was been deleted by the computer system or by end-user activity. Deleted data remains on storage media in whole or in part until it is overwritten by ongoing usage or "wiped" with a software program specifically designed to remove deleted data. Even after the data itself has been wiped, directory entries, pointers, or other metadata relating to the deleted data may remain on the computer. Deleted data is where a lot of court cases based on Computer Forensics are won or lost.

Deleted file: A deleted file is a whole file, such as a Microsoft Word document, that has been deleted and the disk space it used to occupy has been designated by the computer as available for reuse. The deleted file remains intact until it has been overwritten with a new file.

Deletion: Deletion is the process whereby data is removed from active files and other data storage structures on computers and rendered inaccessible except using special data recovery tools designed to recover deleted data. Deletion occurs in several levels on modern computer systems: (a) File level deletion: Deletion on the file level renders the file inaccessible to the operating system and normal application programs and marks the space occupied by the file's directory entry and contents as free space, available to reuse for data storage. (b) Record level deletion: Deletion on the record level occurs when a data structure, like a database table, contains multiple records; deletion at this level renders the record inaccessible to the database management system (DBMS) and usually marks the space occupied by the record as available for reuse by the DBMS, although in some cases the space is never reused until the database is compacted. Record level deletion is also characteristic of many e-mail systems. (c) Byte level deletion: Deletion at the byte level occurs when text or other information is deleted from the file content (such as the deletion of text from a word processing file). Such deletion may render the deleted data inaccessible to the application intended to be used in processing the file, but may not actually remove the data from the file's content until a process such as compaction or rewriting of the file causes the deleted data to be overwritten.

Desktop: Usually refers to an individual PC, such as a user's desktop computer. It can also refer to the first screen presented after a Microsoft Windows Operating System has finished booting up.

Digital: Storing information as a string of digits — namely "1"s and "0"s.

Disaster Recovery Tape: Disaster Recovery Tapes are portable media used to store data that is not presently in use by an organization to free up space but still allow for disaster recovery. May also be called "Backup Tapes".

Disc (disk): It may be a floppy disk, or it may be a hard disk. Either way, it is a magnetic storage medium on which data is digitally stored. 'Disc' is often used for optical discs, while 'disk' generally refers to magnetic discs, but there is no real rule.

Disc mirroring: A method of protecting data from a catastrophic hard disk failure. As each file is stored on the hard disk, a "mirror" copy is made on a second hard disk or on a different part of the same disk. Also known as RAID 0.

Distributed Data: Distributed Data is that information belonging to an organization which resides on portable media and non-local devices such as home computers, laptop computers, floppy disks, CD-ROMs, personal digital assistants (PDA's), wireless communication devices (e.g., Blackberry), zip drives, Internet repositories such as e-mail hosted by Internet service providers or portals, web pages, and the like. Distributed data also includes data held by third parties such as application service providers and business partners.

Electronic Mail: Electronic Mail, commonly referred to as e-mail, is an electronic means for communicating information under specified conditions, generally in the form of text messages, through systems that will send, store, process, and receive information and in which messages are held in storage until the addressee accesses them.

Encryption: A procedure that renders the contents of a message or file unintelligible to anyone not authorized to read it.

Ethernet: Ethernet is a frame-based computer networking technology for local area networks (LANs). Ethernet is mostly standardized as IEEE's (see below) 802.3. It has become the most widespread LAN technology in use during the 1990s to the present, and has largely replaced all other LAN standards such as token ring, FDDI, and ARCNET.

File: A collection of data or information that has a name, called the filename. Almost all information stored in a computer must be in a file. There are many different types of files: data files, text files, program files, directory files, and so on. Different types of files store different types of information. For example, program files store programs, whereas text files store text.

File extension: A tag of three or four letters, preceded by a period, which identifies a data file's format or the application used to create the file. File extensions can streamline the process of locating data. For example, if one is looking for incriminating pictures stored on a computer, one might begin with the .gif and .jpg files.

File server: When two or more computers are networked together in a LAN situation, one computer may be utilized as a storage location for files for the group. File servers may be employed to store e-mail, financial data, word processing information or to back-up the network.

File sharing: One of the key benefits of a network is the ability to share files stored on the server among several users.

Firewall: A system designed to prevent unauthorized access to or from a private network. Firewalls can be implemented in both hardware and software, or a combination of both. Firewalls are frequently used to prevent unauthorized Internet users from accessing private networks connected to the Internet, especially intranets. All messages entering or leaving the intranet pass through the firewall, which examines each message and blocks those that do not meet the specified security criteria.

Floppy: A soft magnetic disk. It is called floppy because it flops if you wave it (at least, the 5-inch variety does). Unlike most hard disks, floppy disks (often called floppies or diskettes) are portable; because you can remove them from a disk drive. Disk drives for floppy disks are called floppy drives. Floppy disks are slower to access than hard disks and have less storage capacity, but they are much less expensive. And most importantly, they are portable.

Forensic Copy: A Forensic Copy is an exact bit-by-bit copy of the entire physical hard drive or floppy disk, including slack and unallocated space. Only forensic copy quality will hold up in court.

Fragmented Data: Fragmented data is live data that has been broken up and stored in various locations on a single hard drive or disk.

FTP: Short for File Transfer Protocol, the protocol for exchanging files over the Internet. FTP is most commonly used to download a file from a server using the Internet or to upload a file to a server (e.g., uploading a Web page file to a server).

GIF: Pronounced jiff or giff (hard g) stands for graphics interchange format, a bit-mapped graphics file format used by the World Wide Web. GIF supports color and various resolutions. It also includes data compression, but because it is limited to 256 colors, it is more effective for scanned images such as illustrations rather than color photos.

Gigabyte (GB): A gigabyte is a measure of computer data storage capacity and is a billion (1,000,000,000) bytes.

GUI: Graphical User Interface and pronounced GOO-ee. A program interface that takes advantage of the computer's graphics capabilities to make the program easier to use. Well-designed graphical user interfaces can free the user from learning complex command languages. The Windows desktop screen is a typical example of a GUI.

Hard disk: A peripheral data storage device that may be found inside a desktop or laptop as permanent storage solution. The hard disk may also be a transportable version and attached to a desktop or laptop.

HTML (Hypertext Markup Language): The tag-based ASCII language used to create pages on the web.

IEEE: The Institute of Electrical and Electronics Engineers or IEEE (pronounced as eye-triple-ee) is an international non-profit, professional organization incorporated in the State of New York, United States. It is the largest technical professional organization in the world (in number of members), with more than 360,000 members in 150 countries (as of 2004).

Image: In data recovery parlance, to image a hard drive is to make an identical copy of the hard drive, including empty sectors. (Akin to cloning the data.) Also known as creating a "mirror image" or "mirroring" the drive.

Instant Messaging ("IM"): Instant Messaging is a form of electronic communication that involves immediate correspondence between two or more users who are all online simultaneously. It is a conversation made up of typing rather than speaking words.

Internet: A global network connecting millions of computers. More than 100 countries are linked into exchanges of data, news and opinions. Unlike online services, which are centrally controlled, the Internet is decentralized by design. Each Internet computer, called a host, is independent. Its operators can choose which Internet services to use and which local services to make available to the global Internet community. Remarkably, this anarchy by design works exceedingly well. There are a variety of ways to access the Internet. Most online services, such as America Online, offer access to some Internet services. It is also possible to gain access through a commercial Internet Service Provider (ISP). The Internet is not synonymous with World Wide Web.

Intranet: A network based on TCP/IP protocols (an internet) belonging to an organization, usually a corporation, accessible only by the organization's members, employees, or others with authorization. An intranet's Web sites look and act just like any other Web sites, but the firewall surrounding an intranet fends off unauthorized access. Like the Internet itself, intranets are used to share information. Secure intranets are now the fastest-growing segment of the Internet because they are much less expensive to build and manage than private networks based on proprietary protocols.

IP address: An identifier for a computer or device on a TCP/IP network. Networks using the TCP/IP protocol route messages based on the IP address of the destination. The format of an IP address is a 32-bit numeric address written as four numbers separated by periods. Each number can be zero to 255. For example, 1.160.10.240 could be an IP address.

ISP: Short for Internet Service Provider, a company that provides access to the Internet. For a monthly fee, the service provider gives you a software package, username, password and access phone number. Equipped with a modem, you can then log on to the Internet and browse the World Wide Web and USENET, and send and receive e-mail. In addition to serving individuals, ISPs also serve large companies, providing a direct connection from the company's networks to the Internet.

JPEG (Joint Photographic Experts Group): An image compression standard for photographs.

Keyword search: A search for documents containing one or more words that are specified by a user.

Kilobyte (K): One thousand bytes of data is 1K of data.

LAN (Local Area Network): Usually refers to a network of computers in a single building or other small, discrete location.

Legacy Data: Legacy Data is information in the development of which an organization may have invested significant resources and which has retained its importance, but which has been created or stored by the use of software and/or hardware that has been rendered outmoded or obsolete.

Megabyte (Meg): A million bytes of data is a megabyte, the slang term is 'a meg.'

Metadata: Metadata is information about a particular data set that may describe, for example, how, when, and by whom it was received, created, accessed, and/or modified, and how it is formatted. Some metadata, such as file dates and sizes, can easily be seen by users. Other metadata can be hidden or embedded and therefore unavailable to computer users who are not technically adept. Metadata is generally not reproduced in full form when a document is printed. (Typically referred to by the less informative shorthand phrase "data about data", it describes the content, quality, condition, history, and other characteristics of the data.)

Migrated Data: Migrated Data is information that has been moved from one database or format to another, usually as a result of a change from one hardware or software technology to another.

Mirroring: The duplication of data for purposes of backup or to distribute network traffic among several computers with identical data.

MIS: Management Information Systems.

Modem: A piece of hardware that lets a computer talk to another computer over a phone line.

Network: A group of computers or devices that are connected together for the exchange of data and sharing of resources. A network can be as small as two computers, or as large as the public Internet.

Node: Any device connected to a network. PCs, servers, and printers can all be nodes on the network.

OCR: Optical Character Recognition is a technology that takes data from a paper document and turns it into editable text data. The document is first scanned, then the OCR software searches the document for letters, numbers, and other characters and attempts the conversion.

Offline: Term for a computer or node not connected to a network.

Online: Term for a computer or node that is connected to a network.

Operating System (OS): The software that the rest of the software on a computer depends on to make the computer functional. On most PC's, this is Microsoft Windows. Unix and Linux are other operating systems often found in scientific and technical environments.

PC: Personal computer.

PDA (Personal Digital Assistant): Handheld digital organizers. The most well known type of PDA is the "Palm" handheld computer.

PDF (Portable Document Format): A technology developed by the Adobe Corporation for formatting documents so that they can be viewed and printed exactly the same on any PC using the Adobe Acrobat reader.

Petabyte (PB): A petabyte is a measure of computer data storage capacity and is 2 to the 50th power (1,125,899,906,842,624) bytes.

Plaintext: The least formatted and therefore most portable form of text for computerized documents. ASCII files are often called plaintext files.

Pointer: A pointer is an index entry in the directory of a disk (or other storage medium) that identifies the space on the disk in which an electronic document or piece of electronic data resides, thereby preventing that space from being overwritten by other data. In most cases, when an electronic document is "deleted", the pointer is changed to a form that allows the document to be overwritten, but the document is not actually erased.

Private Network: A network that is isolated from the Internet. See Intranet.

Public Network: A network that is part of the public Internet.

RAM (Random Access Memory): The working memory of the computer into which application programs can be loaded and executed. The contents of RAM disappear(s) when the computer is switched off.

Residual Data: Residual Data (sometimes referred to as "Ambient Data") refers to data that is not active on a computer system. Residual data includes (1) data found on media free space; (2) data found in file slack space; and (3) data within files that has functionally been deleted in that it is not visible using the application with which the file was created, without use of undelete or special data recovery techniques.

Router: A piece of hardware that routes data from one local area network (LAN) to another, or from a LAN onto the Internet.

Sampling: Sampling usually (but not always) refers to the process of statistically testing a data set for the likelihood of relevant information. It can be a useful technique in addressing a number of issues relating to litigation, including decisions as to which repositories of data should be preserved and reviewed in a particular litigation, and determinations of the validity and effectiveness of searches or other data extraction procedures. Sampling can be useful in providing information to the court about the relative cost burden versus benefit of requiring a party to review certain electronic records.

Server: Any computer on a network that contains data or applications shared by users of the network on their client PCs.

Software: Coded instructions (programs) that make a computer do useful work.

Stand-alone computer: A personal computer that is not connected to any other computer or network, except possibly through a modem.

System Administrator: (sysadmin, sysop) The person in charge of keeping a network working.

TCP/IP: (pronounced as separate letters) Short for Transmission Control Protocol/Internet Protocol, the suite of communications protocols used to connect hosts on the Internet. TCP/IP uses several protocols, the two main ones being TCP and IP. TCP/IP is built into the UNIX operating system and is used by the Internet, making it the de facto standard for transmitting data over networks.

Terabyte (TB): A terabyte is a measure of computer data storage capacity and is one thousand billion (1,000,000,000,000) bytes.

TIFF (Tagged Image File Format): One of the most widely supported file formats for storing bit-mapped images. Files in TIFF format often end with a .tif extension.

VPN: (pronounced as separate letters) Short for virtual private network, a network that is constructed by using public wires to securely connect nodes. For example, there are a number of systems that enable you to create networks using the Internet as the medium for transporting data. These systems use encryption and other security mechanisms to ensure that only authorized users can access the network and that the data cannot be intercepted.

References

1. M. A. Caloyannides, *Privacy Protection and Computer Forensics, Second Edition* (Artech House, Inc., Norwood, MA, USA, 2004), ISBN 1580538304.
2. M. Cohen and D. C., Raid reassembly, A forensic Challenge (2005).
3. E. Eldridge, Choosing the right tape backup architecture, *Power Solutions* **2** (1999).
4. P. Gutmann, Secure deletion of data from magnetic and solid-state memory, in *SSYM'96: Proceedings of the 6th conference on USENIX Security Symposium, Focusing on Applications of Cryptography* (USENIX Association, Berkeley, CA, USA) (1996), pp. 8–8.
5. V. Ralevich, Persistence of memory how hard is it to erase data, Sheridan College (2006).
6. M. Reith, C. Carr and G. Gunsch, An examination of digital forensic models, citeseer. ist.psu.edu/reith02examination.html (2002).
7. T. Rude, Examples of using dd within unix to create physical backups, CISSP (2000).
8. C. H. Sobey, Recovery unrecoverable data, the need for drive-independent data recovery, *Data Recovery Labs, Inc.* (2004).
9. J. Tan, Forensic readiness, Cambridge, MA: @Stake (2001).
10. url01 (2008), http://math.nist.gov/ MDonahue/mfm.html.
11. url02 (2008), http://cnst.nist.gov/epg/Projects/MagNano/mfm_proj.html.
12. url04 (2008), http://www.guidancesoftware.com.
13. url05 (2008), www.ucl.ac.uk/cert/secure_disposal_guidelines.pdf.
14. url06 (2008), www.freewarereview.info/2005-11/supershredder_securely_erase_d.html.
15. url07 (2008), www.jetico.com/bcwipe3.htm.
16. url08 (2008), www.heidi.ie/eraser/.
17. url10 (2008), http://samba.org/ jelmer/kregedit/.
18. url11 (2008), www.valleysolutionsinc.com.
19. url12 (2008), http://www.symantec.com/security_response/writeup.jsp?docid= 2004-070118-4939-99.
20. url13 (2008), www.lavasoftusa.com/products/product-data-sheets/enterprise- df1.pdf.
21. url14 (2008), www.safer-networking.org/en/tutorial/index.html.
22. url15 (2008), http://linux.about.com/od/commands/l/blcmdl8_dump.htm.
23. url16 (2008), www.backup-premium.com/.
24. url17 (2008), www.softpedia.com/.
25. url18 (2008), www.paraben-forensics.com/catalog/product_info.php?cPath=25& products_id=374.
26. url19 (2008), www.digimarc.com.
27. url20 (2008), www.jetico.com.
28. url21 (2008), www.softpedia.com/get/System/Back-Up-and-Recovery/RAID-Reconstructor.shtml.
29. J. R. Vacca, *Computer Forensics: Computer Crime Scene Investigation* (*Networking Series*) (*Networking Series*) (Charles River Media, Inc., Rockland, MA, USA) (2005). ISBN 1584503890.
30. L. W. Wong, Forensic analysis of the windows registry, School of Computer and Information Science, Edith Cowan University (2006).

Chapter 18

IPOD, CELL AND SMART PHONE FORENSICS

M. MATTIUCCI, R. OLIVIERI, L. GIAMPIERI,
S. MONFREDA and G. FINIZIA
High Tech Crime Unit
Digital Forensics Department of Carabinieri
Scientific Investigation Division,
Military Police Force, Italy

The team of the High Tech Crime Unit of the Digital Forensics Department of the Carabinieri's Scientific Investigation Division, situated in Rome, Italy, has turned iPod, cell and smart phone forensics into one of its main activities. The investigative relevance of such evidences is immense when compared to the extraordinary large amount of this kind of devices spread around Italy and to the considerable presence of personal data, so often submitted to protection lack, stored in their memories. This work is meant to deal with some of the technical/scientific investigation hitches, on a tricky subject like this, together with several practical cases of PDA and cell phone analysis.

18.1. Cell Phones Towards Smart Phones

Smart and cell phone business community has lately recorded, both in Italy and in the rest of the world, a huge development that does not seem to reveal any stop signal. On the contrary, new digital services and features for the consumers as well as a broader and broader data transmission band, which involves wider data volumes and higher transmission speed to higher performance's advantage, are always pursued.

At investigating level, the influence of the digital communication and especially of the "mobile communication" is a crucial factor as to the various scenarios that could be sparked off, such as wire tapping, radio localisation and digital services' tracking and control (e.g., email, SMS, MMS, etc.).

Before continuing, the study field must be conveniently narrowed down as different investigation areas, bound to the mobile systems of communication, can be identified:

- *Wireless systems*: the digital communication systems typically identified as wireless systems use infrared or low-power radio waves to transmit data, often resorting to a point-to-point mode. Typical examples can be pointed out by the IrDA,[a] Wi-Fi[b] or Bluetooth[c] protocols, widely used both for digital devices' data exchange (e.g. cellular phone — cellular phone) and for proper local area networks implementations. There are many other examples such as: *cordless phones* which solely communicate with their PSTN connected "base station" and *private professional systems* whose peculiar tasks require privileged cellular networks (e.g. TETRA[d] for the police forces, hospitals, etc.).
- *Satellite phones*: they can connect directly to orbiting satellites that provide them with specific digital communication services. Different architectural features affect the satellite phones that are used as means of transmission on a large and very large scale.
- *Cellular phones*: usually referred as "mobile telephone". These devices connect themselves to a network made of cells placed on their operative areas which is obviously interconnected to the PSTN.

This article will be based on the electronic equipment and the services connected to the mobile telephony. In particular we will emphasise the new investigating possibilities regarding *smart phones*, simply defined in [2] as "*a PDA*[e] *that simultaneously provides the user with mobile telephony services, PIM*[f] *applications, schedule management, contacts, electronic documents, etc.*". In the forensic computing[g] the smart phone is then an *embedded system*[h] [5], *handheld*[i]

[a]IrDA: acronym of *Infrared Data Association*, protocol that defines the physical parameters for the short range digital data communication, by means of infrared light.
[b]WiFi: acronym of *Wireless Fidelity*, protocol licensed by "*Wi-Fi Alliance*" whose target is to describe the basic technology of the *wireless local area networks* (WLAN), based anyway on the IEEE802.11(g) family technical specifications.
[c]Bluetooth: this is an industrial standard for the accomplishment of *wireless personal area network* (PAN), standard IEEE 802.15.1. It allows to connect electronic agendas, cellular phones, PCs, printers, video cameras, videogames, etc. by means of a secure system and officially/legally acknowledged based on short waves.
[d]TETRA: acronym of *Terrestrial Enhanced Trunked Radio* this is a radiomobile transmission protocol used by police, hospitals, fire brigades and military corps.
[e]PDA: acronym of Personal Digital Assistant, often reported as electronic address book.
[f]PIM: acronym of Personal Information Management.
[g]Forensic Computing: "...the process of: identification, preservation, analysis and presentation of 'digital evidences' in a trial, by ensuring the admissibility" [9].
[h]Embedded System: special natured electronic system, word often used to distinguish an electronic device with specific tasks from a general purpose computer.
[i]Handheld device: such devices have been created for personal use with the fundamental peculiarity of the easy portability.

like, capable of managing: vocal and/or video calls, SMS,[j] MMS,[k] IM,[l] email, web surfing and/or WAP[m] and PIM, etc.

18.2. iPod and iPhone

An *iPod* is a special purpose digital device and, in particular, a *portable multimedia player*, that is a system capable of storing and let the user play multimedia files.

Designed and commercialised by Apple in 2001, different versions are available based on the use of flash memories or even hard disks for the aforementioned file's storage.

The iPod is an handheld device that can store any kind of files in its mass storage unit and there are more than 141 millions of such a units around the world nowadays (Wikipedia January 2008). Because of these issues the investigative relevance of this kind of equipment is huge.

Moreover, the iPhone device has integrated the functional requirements of the iPod to create a new type of very advanced smartphone. From this point of view, the *iPod forensics* is a target field for high tech crime investigation units that have to be studied in parallel to the smart phone forensics.

18.3. Mobile Forensics

Smart phone seizure on the crime scene, together with the subsequent forensic laboratory analysis, form a study section emerged a few years ago as an appendix of the forensic computing and now overbearingly in vogue as an independent sector due to the discrepancies with the mother subject originated by the following factors:

- *Portability of the devices object of analysis* which determines compact dimensions, the adoption of proper interfaces, batteries (usually more than one) and hardware alien to computers and servers.
- *The concept of the mass memory* as a temporary data storage, on optical and/or magnetic basis, that turns to be invalidated by the massive use of everlasting supplied volatile memory and/or flash memories.

---

[j]SMS: acronym of *Short Message Service* available on GSM (Global System for Mobile Communications).

[k]MMS: acronym of *Multimedia Messaging Service* comprised of multimedia items (pictures, audio, video, rich text) capable of expanding the SMS features. Such system has been standardized by 3GPP, 3GPP2 and Open Mobile Alliance (OMA) groups and is provided on GSM/GPRS (*General Packet Radio Service*) and CDMA (*Code Division Multiple Access*) like UMTS (*Universal Mobile Telecommunications System*).

[l]IM: acronym of *Instant Messaging*. This is a real-time textual communication, with/without multimedia items, between two or more users.

[m]WAP: acronym of *Wireless Application Protocol* is a standard for the Internet navigation by means of cellular phones and PDA (smart phone).

- The constant presence of *automatic "idle" states or hibernation*, against the waste of the howsoever limited energy supply of the batteries — such states determine even undesired changements of the data in the memories and sometimes they simulate stand-by states on the devices even though they are absolutely not (e.g. black screen, insensitivity to the keyboard, etc.).
- *The absence of a universal constructive standard for smart phones*, which unfortunately determines the existence of device families capable of providing approximately the same digital services, based on substantially different hardware, and the forensic analysis approach of which, then, must be different.
- *The continuous manufacturing of new smart phones*, in order to manage a full market into which creating new looks or particular services is the only way the companies have to survive.

This chapter is mainly oriented to the most popular platforms on which basis those systems are implemented, and so: Symbian, RIM,[n] Pocket PC (Windows Mobile), Palm OS and Apple iPhone. It can be said, with fine assurance, that those platforms include the majority of the smart phones actually in use, even though many cases of cell phones partially incorporate the typical PDA features of the above mentioned systems.

The principles of the forensic computing, that have to be followed in the seizure and the analysis processes of the smart phones, remain the following [1, 8]:

- "*No technical activity of the police forces must alter the data stored into the evidence*" unless obviously this does not result the only possible way to proceed and that so it legally turns into an unrepeatable examination.
- "*Only on exceptional circumstances the original data stored in the seized evidence can be directly accessed and in such case the operator must be in possession of proper skills in the sector in order to be always able to explain the reason and the implications of his actions*".
- "*The whole activity, both the seizing and the analysis must be subject to precise descriptive documentation that can eventually agree a third part examination of the processes and the tools as well as the re-obtainment of the same results*" described in the final report.

18.3.1. *Physical Seizure*

The initial hypothesis is that the smart phone must be exclusively seized on the crime scene without the need to immediately access it. The fundamental requirement, at this point, is *to prevent the device connection to the cellular network* as this would cause not completely predictable alterations of its data contents.

[n]RIM: acronym of Research in Motion Limited, Canadian company known for having produced the *BlackBerry* smart phone.

The most immediate solution could obviously be the device switching off but then several contraindications would come out:

- The system, once switched off, could then determine, in the phase of reactivation, the request of a PIN code for the SIM or USIM[o] or a security code for the PDA. This certainly will slow down the data mining as well as the analysis operations.
- The battery, albeit not used, will tend to slowly discharge until exhaustion and this, always in the analysis phase, corresponds to the need to proceed with the battery charger seizure (this fact is actually as trivial as complex at the same time given the great variety of existing smart phones).
- The batteries could be more than one and some of them may permanently power up "temporary" mass memories, thus it could be necessary to recharge the device, even several times, before and during the seizure and the analysis in order not to lose definitively the stored data.

The above mentioned suggests that the shutdown procedure, although in many cases used with excellent guarantees for everybody (it must be remembered that turn off and seal the device corresponds to prevent everyone from physically accessing and, eventually, making any damage), is absolutely not the best kind of approach and certainly not the one that provides the analysis results as quickly as possible. In the perspective to find an alternative solution the access to the operational smart phone by a specialised technician in the immediate would be optimal, provided that the action takes place in an *electromagnetic waves shielded room.* Unfortunately, such a screening system is usually difficult to find and requires at least the device to be carried in a laboratory, with the obvious issues concerning the transportation during which the smart phone is connected to the cellular network and rather it presumably connects to several cells with the consequent internal registry update.

Concerning the aforementioned issues several devices, very useful for the real-time analysis and the transport, have been lately implemented:

- *Jamming device*[p]: especially in "closed" crime scenes like rooms or premises, a jammer can determine the cell inaccessibility by the mobile phone and thereby obtain a kind of insulation of the evidence over cellular network. Unfortunately, the jammer's operational range is not simply definable (it is possible to easily interfere with other transmissions equipment in the area causing inefficiency and various problems) and its screening function is not 100% guaranteed, especially if the phone is a "smart" one and can work on different bands. To this it must be added that some countries require a legal authorisation notice for its use.

[o]SIM: acronym of *Subscriber Identity Module*, is a smart card (chip on plasticized support) which contains: the univocal provider access key and several user-related information (messages, contacts, etc.). A USIM is the equivalent of a SIM but working on UMTS protocol (*Universal Subscriber Identity Module*).

[p]A *jammer* is a device that emits a range of radio waves in order to prevent a mobile phone from connecting to the available cell.

- *Faraday's tent*: this is a field tent capable of screening the electromagnetic waves. This could be an ideal trick if it were not that is generally small, therefore it can accommodate one to two operators at most and a few equipment, and secondly the smart phone battery runs down faster, due to the continuous attempts to reconnect to the cellular network. An internal power supply could be very useful but the electric wires between the interior and the exterior of the tent are allowed within limits because they represent electromagnetic waves spread points.
- *Shielded cases*: these are shielded rigid suitcases, intended for the seizure, inside which different types of power plugs for smart phones are present. Sometimes equipped with their own power supply (long life) they can assure an optimal transfer mode of the seized evidence from the crime scene to the forensic laboratory where they are opened within the steady shielded room. In this way an excellent compromise solution between the data inalterability guarantee and analysis speed is achieved.

18.3.2. *Analysis and Differentiations*

The analysis methods of SIM-cards and smart phones are continuously updated and are implemented through both hardware and software. The direct interaction with the evidence in the shielded room is certainly a valid method from the point of view of the connection to the mobile network but offers few guarantees about the technician's activity. He indeed finds himself to work, in any case, on an alterable original evidence (event to avoid, when possible, in every computer or telematic investigation). It must then be pointed out that:

- Every smart phone has a proper interface the unfamiliarity with which could lead to analysis slowdowns.
- *Not all the data are made available to the user through the standard commercial interface of the device*, for instance the deleted items or the activity traces when detectable, these are crucial elements for the investigations.

For the purpose of the analysis, it becomes important to the analysis' goal to understand what kind of mobile phone we have to deal with and the consequent analysis methods and tools, fit for the purpose, to be used in order to quickly retrieve the most appropriate information. Concerning this the mobile devices [2] are fundamentally clustered into three categories:

- *Basic phone*: it implements voice calls and SMS services.
- *Advanced phone*: it implements basic phone services plus EMS,[q] a sort of SMS based chat session, the management of an email box through the connection to an email server and the WAP based navigation.

---

[q]EMS: acronym of *Extended Message Service*, SMS extension which allows to send messages containing simple images and longer texts.

- *High end phone*: it extends advanced phone services to the broader spectrum of the *full instant messaging* (IM) by supporting specific client applications, of the *multimedia messaging* with the MMS, of the electronic mail by supporting POP[r]/IMAP[s] and SMTP[t] protocols and also the chance to surf the Internet by means of the HTTP.[u]

Each of such phone categories require a different kind of analysis approach because of the very different detectable information, both evident and hidden. Smart phones are obviously included in the high end phone category and they require thus the most advanced level of analysis.

It must then be considered a further feature connected to the *SIM-card presence* into the smart phone, given that actually some CDMA[v] devices are natively comprised of the SIM card functionalities while others, always due to commercial issues, allow the insertion of different USIM-cards.

A SIM or USIM is a smart card equipped with a memory, a processor and a basic operating system, it is therefore useful to even consider it like an evidence split from the smart phone itself. A mechanical extraction of the card should then be planned (normally this is a 25 mm in length, 15 mm in width and 0.76 mm in height little board) and subsequently its insertion in a forensic SIM/USIM reader (hardware + software system) or the fixing of the little board on a common smart card compatible plastic support and then proceed with the reading of the data through a normal commercial smart card reader.

Finally, a correct smart phone analysis cannot be set aside from the removable media typology associated with it, i.e., non-volatile memory boards that, like the

---

[r]POP: acronym of *Post Office Protocol*, net protocol (application level) commonly used by email clients in order to download messages from remote servers through a TCP/IP connection (Transmission Control Protocol/Internet Protocol). The majority of Internet users, who use personal mailboxes on providers, usually access their mailbox by means of POP3 (POP version 3).

[s]IMAP: acronym of *Internet Message Access Protocol* (available versions IMAP or IMAP4, previously called *Internet Mail Access Protocol*), is a net protocol (application level) which, like POP, allows a local client to access the incoming email, stored on a remote server, through a TCP/IP connection. IMAP4 and POP3 are the most popular protocols for email retrieving on the Internet. Practically, every client and server should support them but the POP protocol is more diffused due to its major simplicity and less heaviness of execution.

[t]SMTP: acronym of *Simple Mail Transfer Protocol*, is the standard de facto for the email transmission from the client to the Internet. Nowadays the more performing and flexible ESMTP (*Enhanced o Extended SMTP*) is widely used.

[u]HTTP: acronym of *Hypertext Transfer Protocol*, is a net protocol (application level) which allows to transmit and receive information of the World Wide Web through HTML (*Hypertext Markup Language*) hypertext pages.

[v]CDMA: acronym of *Code Division Multiple Access*, multiplex technique (more digital signals run on the same transmission channel) which doesn't use successive times (as in the *Time Division Multiple Access* or several overlapping frequencies as in the *Frequency Division Multiple Access*) but an encoded algorithm which allows to obtain a digital signal that mathematically "assembles" more signals and from which is possible to re-extract the original signals by means of a related decoding algorithm. Widely used in different kind of modern digital cellular networks as, for instance, UMTS which use a particular variant known as W-CDMA.

SIM-cards, may be inserted or removed, in the specific case, in order to extend the device's storage capacity. The market offers a great variety of these media cards whose capacity range spreads from MB up to several GB. From this point of view they may contain an enormous amount of useful information for the investigation process and should be analysed with extreme care, taking into account the file system[w] they are based on. Some media card typologies are summarised as follows:

- *Multimedia Card (MMC)*: flash memory that is in fact a solid state hard disk (without any component in motion). The size of a postage stamp they can keep the information even if extracted from the device that hosts them and therefore subject to power supply shortage.
- *Secure Digital card (SD)*: similar to MMC it implements data protection features (e.g. accidental erasure prevention) and more performing transfer rates.[x]
- Memory Stick, TransFlash, etc.

Once the due clarifications on the technical features and everything that may be connected to smart phones are done we can move on pointing out the achievable analysis techniques.

18.3.2.1. *Removable media analysis*

It has been determined to start with media cards just because in most cases they support a FAT file system, typical of the earlier MS-DOS and Windows systems and therefore sufficiently archaic and notorious to be efficiently handled with the forensic computing consolidated products.

The availability of commercial products like EnCase, ILook, etc. allows to probe the contents of partitions, directories and files both noticeable and deleted. Refer to [8] for the technical/investigative standard procedure which, as a general rule, is not that far from the hard disks' one, if it were not that (e.g., non-standard media) the cards need for specific hardware readers to be used (compatible with the mechanics).

18.3.2.2. *SIM/USIM analysis*

Unlike media cards, SIM-cards are highly standardised units provided with uniform contents and a well known interface. For this reason some software tools, which perform by means of the card readers, have been created in order to copy the SIM-card low level data with the subsequent purpose of interpreting them and so providing the investigator with a useful information set.

[w]File system: file storage and management method on a mass memory. Classical examples are FAT (*File Allocation Table*) and NTFS (*New Technology File System*) typical of MS-Windows, HFS (*Hierarchical File System*) typical of Mac systems, EXT (*Extended File System*) typical of Linux systems, etc.
[x]Transfer rate: data transfer speed on digital bus.

Skipping in this issue the problems related to the circumvention of the PIN code, both for the topic width and for the usual use of the PUK supplied by the telephone provider as a key element for access to SIM-card, which data typologies the commercial forensic tools are able to provide is pointed out as follows:

- Basic data as
  - IMSI: *International Mobile Subscriber Identity*, unique number associated to the GSM or UMTS user. It is stored in the SIM/USIM and is sent to the cellular network by the mobile phone for identification purposes.
  - ICCID: *Integrated Circuit Card identification*, unique SIM/USIM identification number.
  - SPN: Service Provider Name.
  - LP: Language Preference.
  - PIM: in particular Abbreviated Dial Numbers (ADN), Fixed Dial Numbers (FDN), Last Numbers Dialed (LND), Voicemail Number, Administration Data, Service Dialing Number (SDN) and Capability Configuration Parameters with all the possible information they are concerned with.
  - SMS: text and correlated information, deleted entries and eventually retrievable texts.
- *Positional data* (if recoverable): which may provide more or less precise information about the SIM-card position when certain services are asked to the provider:
  - LOCI: *Location Information*, the code *Location Area Identifier* (LAI) of the phone's physical position (geographic area) last time it was linked to the cellular network.
  - PLMN e FPLMN: *Public Land Mobile Network* and *Forbidden PLMN*, network list the SIM-card is allowed/not allowed to automatically log on when searching for available networks (particularly important for the SIM-card that work abroad).
  - BCCH: *Broadcast Control Channel*, report channel that identifies the single connected cell.
  - CBMID: Cell Broadcast Message Identifier for Data Download.
- *EMS data if supported*: EMS longer than 160 characters recovery and reconstruction and eventually including simple multimedia data both noticeable and deleted.
- SMS and foreign language PIM data usually not managed according to the user's main language choice.

18.3.2.3. *PDA/cell-phone hardware analysis*

Mobile phone or smart phone hardware analysis, with the exclusion of SIM and media card, is the most complex activity but, at the same time, the one that provides on average a considerable amount of investigative results.

First of all the hardware/software forensic tool should connect to the device in order to proceed with *data dump*[y] and therefore interpret the image, thus obtained, to provide understandable information. As mentioned in the footnote (26) this is a real-time operation that makes it difficult to ensure the legal repeatability. Anyway, this currently represents the way that provides the best achievable guarantees.

By the way, the above mentioned data dump may not be necessarily complete as some data may not be accessible by this process or they may be even fundamental prior to its execution. Classic example is constituted by the system information (eventual passwords, memory availability, version, battery status, etc.) and/or the resident data management (ownerships, accesses, privileges, etc.). To solve this issue the only possibility is offered by the forensic tool that should provide a connection to the device in order to gain its remote control from a computer station, and this often requires the installation of a minimum server software in the device to the absolute detriment of the investigation repeatability.

A needful mention must be pointed out as for the smart phone's internal memory structure which can be in general distinguished in ROM[z] (often FLASH[aa] in order to make the device easily upgradeable) which contains software and data for the device's basic functioning and RAM[ab] for the software and user's data management. The previously mentioned data dump mainly concerns the RAM which is generally structured in such a way to distinguish a recursive and continuous access area, hence subject to uninterrupted updating, and a different area where data are accessed and modified more seldom [5]. The result is that the temporary data, including the deleted ones, tend to move in the heavily dynamic area and thus to disappear faster than they would into the equivalent mass memory of a PC. Ultimately, consistent erased data finding and rebuilding becomes complex and unlikely, even if not impossible.

The data types retrieved by this kind of analysis are varied and a brief mention is given as follows:

- PIN code or password recovery (cracking operation).
- Information spotting (noticeable and deleted) on PIM applications (like logs, calendars, contacts, meetings, etc.) which are obviously more complex than the ones detectable in the SIM-card.

[y]The *data dump* is the digital device's data download conceptually similar to the mass memory low level copy used for the hard disks. The substantial difference is that the dump of an active system like a smart phone is a "live" copy, that is a copy is created even if the resident data can change, without warnings, at any time during the copying.

[z]ROM: acronym of *Read Only Memory*, into which data permanently reside apart from electrical supply and they are unmodifiable.

[aa]FLASH: not volatile memory (similar to ROM from this point of view), whose contents can be electronically read and modified in blocks, from which a behavior apparently similar to the RAM (volatile memory) and the consequent slowness of access.

[ab]RAM: acronym of *Random Access Memory*, volatile memory (it only keeps the information until the power is supplied) which allows to run performing operations both on reading and on writing/deleting of data.

- Incoming, outgoing, not answered, deleted phone calls.
- SMS and MMS (evident and deleted), reminding that normally the smart phones maintain the majority of such communication system on their hardware and a minimum portion on the SIM-card.
- *Internet Messaging*: email and instant message sent, received and deleted.
- *Web application(s)*: Internet surfing traces (sites, filled in forms, sent information, web-mail) and other applications' use for communications on HTTP.
- Textual files, images, videos, sounds, compressed archives research (evident and deleted).
- *Misnamed file*: the research for files whose data collection is not coherent with their file extension and therefore the declared type (e.g., a ".bmp" saved with the file extension ".txt" in order not to show it is an image).
- *Hash*[ac] *of the device's memory* in order to compare it with the hash value of the image originated from the memory dump, in order to verify the adherence and for the integrity of the copy.

18.3.2.4. *Methodology of analysis*

From the legal point of view one is mainly interested in the modifications that the analysis process can perform on the device. In particular, in the ambit in study and connected to the interaction user-device, the following categories of analysis can be distinguished:

- *Invasive* analysis
- Semi-invasive analysis
- Non-invasive analysis

It is often necessary to switch from a non-invasive method to a partially or totally invasive one, due to the need to obtain particular information not easily extractable.

Non-invasive analysis occurs by acquiring the memory of the radiomobile terminal deprived of its SIM-card. This is accomplished by connecting the above mentioned terminal to a PC by means of the so-called "Flash" cable (serial + USB) and so the use of a specific programming software for the resident data reader unit. Ultimately, adopted hardware and software are in many cases similar to those used by the producers and by the specialised technical assistance laboratories for the programming/reprogramming of the radiomobile terminals. Such instruments can execute different tasks besides of the above mentioned data reading one: for instance the locking/unlocking of a mobile phone from the SIM lock or the operator lock, IMEI number modification, the operating system update, etc. It obviously deals with specific systems for device families that don't enjoy the interchangeability

[ac]Hash: a function or hash algorithm (e.g., MD5) is a mathematical method for creating a unique identification code of a string of bits. A good hash function is one way (one can't rebuild the string from the code) and creates very different identification codes even if the input strings differ from each other for a few bits.

even though on the market some tool-boxes, capable of connecting themselves and manage about 90% of the devices in study, are available.

It must be noticed that the radiomobile terminal, inasmuch embedded system endowed with a microprocessor, memory and operating system is subject to a startup phase very similar to the bootstrap of the personal computers from which, however, can differ as it can take place in two fundamentally different methods:

- Operating system and phone manager software startup following the pressure of the power-on key (usual user modality).
- "*Test Mode*" startup in which the terminal gets activated after being connected by means of a flash cable to a PC endowed with a software that defines a relationship of client-server type (PC – radiomobile terminal) towards the device.

The second modality allows the data seizure (copy) operations. The subsequent activity guarantees the perfect repeatability of the investigation and allows to obtain a great variety of information from the radiomobile terminal, as well as including those cancelled (not overwritten) or however not visible when the terminal is on (e.g., a few telephone address book entries, SMS, MMS, multimedia contents, etc.).

Unfortunately, the non-invasive analysis, commonly desired by all the investigators (police) and by the magistrates and judges cannot always be accomplished. As a matter of fact, many mobile phone models exist that are designed for the cable data connection. Sometimes it is difficult to locate the recovered information and understand its type, for example it may not be possible to establish if the retrieved information is visible when the terminal is on or if they belong to the cancelled ones. Finally, the requested period for the non-invasive analysis is very extended as the data copy, extracted from the terminal, is in binary format, thus basic and not elaborated, whereby is necessary a "reverse engineering" operation in order to highlight the contents to the operator. This last operation is usually not easy and sometimes not possible because of the lack of documentation about specific protocols, file systems, etc. One practical way is to proceed through the *individuation of the tags* which mark the beginning and the end of data blocks in such way to separate the various macro objects that compose the logic of the entire system, in particular:

- Assembly code that form the operating system
- The interpreted code that form the applications and the Java objects
- The address book
- The system files
- The multimedia files
- The SMS and MMS archives

The semi-invasive analysis is the easiest and most spread out and sometimes it can be confused with the only available cellular phone's forensic analysis. It is performed by activating the radiomobile terminal endowed with a SIM-card inside a proper shielded room in such way not to allow the connection to the cellular network.

Such an analysis can be performed when:

- One has the availability of a certified screened room
- The terminal is endowed with the SIM-card with which the last activation has occurred (particularly for the older models)
- The information of concern refers exclusively to those produced or received by the user bound to the SIM-card and that these have not been cancelled.

This because the analysis takes place through the pure vision of the contents via the terminal's display and/or across connections with data cable and forensic software (e.g. TULP2G, MobilEdit Forensic Edition, Mobile Toolkit, Oxygen) otherwise, sometimes, with software released by the manufacturer for the backup management, everything while the terminal is switched on.

The semi-invasiveness exists because the terminal switching on operative modality, albeit without the connection to the mobile network, involves some system data modifications, for instance last logon date and time, last access to the folders date and time in the case of external memories and more.

For investigation purposes such data may seem unnecessary or not interesting, however, from a formal point of view, their modification are a condition that implies the un-repeatability of the technical operation. Anyway it must be pointed out that *the activation of the terminal in a shielded environment does not determine the deletion of the data related to the user*, unless the activation takes place with a different SIM-card than the one used for the last switching on, situation in which significant data loss could evidently occur, such as the calls log and eventually the address book deletion, undoubtedly determining a unrepeatable investigation process.

The invasive analysis is the most long and complex to perform both for the technical activity difficulties that may emerge and for the particular equipment to be used. It consists of the direct reading of the data stored on the memory chips contained in the radiomobile device.

Five phases, in that sense, are identified:

- A first inspection to detect the device's memory chips *typology*
- A following phase concerning the *physical extraction of the chip* (unwelding or extraction by pins' micro-cutting)
- A delicate phase concerning the *interfacing* during which the memory chip is connected to a reading/programming hardware
- The *reading* of the information and its transfer on the PC
- The *reverse engineering* (interpretation) of the extracted data

It must be highlighted that the last two phases are undoubtedly the same of the non-invasive method, but what happens during the extraction and the interfacing is extremely critical and can very likely determine the investigation's unrepeatability (e.g. the radiomobile terminal's restoring to the original state is possible only in theory).

18.4. iPod Forensics

The iPod device implements substantially two operative modalities [6, 7]:

- *iTunes mode* (default): the software iTunes has to be installed on a computer that constitutes the main repository of the multimedia data. The iPod connects to iTunes through a USB hardware link and a software layer which places itself above the operating system, Windows or MacOS. This indicates that *the operating system does not directly interpret the iTunes protocol* but simply is a support to allow byte streams between the computer and the device. ITunes task is to convert the multimedia data in a format treatable by iPod in order to manage successfully the *synchronisation.*
- *Disk mode*: if iTunes is not taken into consideration it is anyway possible to directly access the iPod through the operating system, both Windows and MacOS. In such case *the iPod is recognised and, if necessary, mounted as a USB hard disk (or in general as a USB memory).* The file system that has been mounted and so that exists on the device, depends on the operating system of the computer onto which the iPod has been initialised (iTunes defrayable operation). *If it deals with a Windows system the implemented formatting will be a FAT32, on the contrary if the system is a MacOS X the pertinent formatting will be a HFS+.*

To be noticed that

- If the iPod is initialised on MacOS it can be also a *boot HD* for Apple machines.
- *When in disk mode the iPod can store every kind of files*, even multimedia ones but this does not mean that the internal player would be able to read or play them in general. Only the iTunes protocol allows to upload on the iPod multimedia files that can be readable and playable by the device.
- *FAT32 can be mounted by Mac OS with reading and writing permission while HFS+ cannot (in general) by Windows.* Such issue, together with the diffusion of Windows has made FAT32 file system one of the most used in the iPod environment.

18.4.1. *iPod Seizure*

Several different situations can be found on the crime scene:

- *Disconnected iPod not in play mode*: as a matter of principle the iPod can be managed as a portable hard disk when it comes to seizing, particularly if the player is inactive and nothing is connected to it. After all the data are stored on a semi-permanent memory which is not subject to alterations when the power supply is exhausted. Nonetheless the practical experience suggests to turn the iPod protection on by pressing the specific "hold" key and to connect the device to an external power supply in order to prevent the complete battery exhaustion. This because the totally exhausted battery can generate malfunctioning both of the battery itself (impossibility of recharging) and of the device on the reactivation process (access difficulty or partial data random loss).

An iPod can receive electrical supply directly from the data connection, therefore from USB or Firewire connection, but it is necessary to evaluate its power absorption. The most recent iPod models use, in many cases, USB interfaces, so it is important to have at one's disposal a 5 V–1 A power supply (sufficient for every existent iPod model) and, obviously, the iPod-USB cable. There's a full range of iPod's power adaptors that can support 100 V/240 V alternating current but what needed is a system self-supplied by a 10 A/h battery at least, as on the crime scene the public power grid cannot be taken into consideration and the transportation period can be long. The whole stuff (battery, connection and iPod) can be shut into a metallic case for transportation purposes.

- *Disconnected iPod in play mode*: in the event of a seizure, the images visualisation or the sound reproduction activities must be carefully shot through a video camera (acknowledging the process on the related minutes) as well as the stop action the operator must execute in the immediacy to return subsequently to the previous point. Multimedia files play does not technically invalidate the iPod contents but, on the other hand, affects the battery charge. It must be added that the evidence's direct access on the crime scene is always highly unadvised unless there is an immediate emergency (e.g., there's the certainty that some videos reproduce faces or important actions in order to hinder a criminal event); in such case it is important to carefully record every action, even by means of video shots, and the information acquisition minutes editing.
- *Recharging iPod not in play mode*: there are different chances to keep an iPod electrically supplied such as the use of specific power supplies with USB or Firewire ports, or docking stations that can allow further functionalities as, for instance, the acoustic broadcast, etc. In both the aforementioned situations it is necessary to furnish oneself with a universal power supply as the batteries charge may not be full. By always recording every action through a video camera the device must be unplugged from its power supply in order to connect it immediately to the one arranged for the seizure, and then closing the whole stuff into the above mentioned metallic box.
- *Recharging iPod in play mode*: the procedure is the same of the previous situation except that, obviously, the same video record becomes part of the seized data as it can collect sounds and images of investigating relevance, displayed, during the operations, on the device.

- *Computer connected iPod in iTunes mode*: when the iPod displays not to manually disconnect the device but to operate through iTunes it is important to accomplish that in order to prevent physical damages of the memory support. Once the iPod is regularly disconnected it is possible to operate as above reported. An objection could be that *if a synchronisation is in progress this will be accomplished before the iPod can be disconnected.* Such event is crucial from the forensic point of view if the synchronisation involves multimedia data removal. It is possible indeed to interact with the iPod in order to prevent the accomplishment of such deletion. The deleting process is anyway quite fast and thinking of efficiently interrupting it on the crime scene is not very likely. *In most cases it is suggested that the operator disconnects the iPod through the iTunes services in order to prevent considerable probable damages.*
- *Computer connected iPod in disk mode*: in order to prevent possible physical damage to the memory support it is suggested to disconnect the device by the unmount command the operating system provides (e.g. the recycle bin for Mac OS, the USB unmount for Windows, etc.). Once the iPod is disconnected it can be seized and analysed as above mentioned.
- *Standby mode*: the iPod, with different modalities depending on the model, turns from the common state of power saving to an extremely deep *stand-by state*, after a few hours it is not used (apart from its battery status). When the device is reactivated a sort of startup procedure takes place in which some system peculiar parameters must be set up. Though it is presumed that such parameters are solely stored in the system RAM it must not be excluded that some of them pertain to the disc data contents and therefore of the semi-permanent memory that is going to be analysed from a legal point of view. *It would be good then to avoid such stand-by mode and at least copy the data before this occurs.* However, there might be the chance that such state is the actual one at the moment of physical seizure. The condition to be taken into account are the same of the *disconnected iPod not in play mode* in which the stand-by mode has occurred. The "hold" key must then be activated and the device must be connected to a power supply. The result is that the iPod restarts prompting the user for the aforementioned parameters but the activated hold key does not allow to interact and so the procedure comes to a stop.

18.4.2. *iPod Data Cloning*

This paragraph deals with the physical copy of the data stored in the semi-permanent iPod memory through the interaction with a Windows XP SP2 operating system.

As the simple connection to the USB port can involve modifications of a device's data contents, in such an environment, one can operate in hardware or software mode in order to prevent such issues.

- *Write blocking software (only Windows XP SP2)*: apart from the obvious solution to adopt a write-blocking driver, it is possible, in such a simple and functional way, to modify the system *registry* through the use of the system application *regedit.exe*:
  - Locate the key *HKEY_LOCAL_MACHINE*.
  - Then the sub-key *SYSTEM*.
  - Then the sub-key *CurrentControlSet*.
  - Then the sub-key *Control*.
  - Verification of the presence of the key *StorageDevicePolicies*.
  - If non-existent it will be sufficient to create it.
  - In StorageDevicePolicies the DWORD value WriteProtect must be created.
  - A double click can set the value WriteProtect to 1.
  - Save the registry and the USB devices successfully mounted will operate in write-protect mode.
- *Write blocking hardware*: a write-blocking hardware is an electronic device capable of intercepting, announcing (eventually) and blocking the write commands that from the PC are sent to a peripheral device.

18.4.3. *Extension on the Concept of Write Blocking Related to iPod*

The iPod, as a substantially multifunctional device, allows to start the debate on the modality with which the write blockers operate and their limitations in principle.

The computer sends indistinctly to the write blocker the control signals Read and Write, and the blocker filters them by letting pass only the Read ones, providing, at the same time, an "illusory" feedback to the Write messages (otherwise continuous errors would happen). The data solely pass from the device to the blocker and then to the PC. Unluckily the matter is not that easy as the control signals are not pure Read and Write but must be interpreted depending on the communication protocol in use.

The write blocker has mostly the task to evaluate the device's typology and this happens when the device is connected to the blocker (e.g., USB).

The same device can then support several transmission protocols. When the blocker is connected, the computer, warned about the device's typology, will send some control signals corresponding to the service and the protocol that the active software intends to manage on the device. As a consequence the blocker will identify the protocol and the device and therefore apply the correlated interpretation rules of the control signals, already set up in its firmware, in order to prevent the peripheral's data contents modifications. The other possibility is that precise rules cannot be determined and so some blockers try to apply general protection rules while others simply communicate the impossibility of managing the device with precision (better method).

In the iPod case, for instance, *the blockers can quietly identify the activity in disk mode but they have troubles in understanding the iTunes protocol determining most of the times the denial to work in iTunes mode*, event moreover substantiated by the iTunes impossibility to run whatsoever synchronisation. It must be noticed that such write blocker behavior , though correct (iTunes cannot write on the device), is only the result of the identification of an unsuccessful possibility to operate.

18.5. Cell and Smart Phone Forensics Tools and Investigations

There are many tools useful to gain access, at a forensic level, to PDAs, smart phones, SIM and media cards but a fundamental issue is that none of them can ensure a comprehensive investigation, just because of the aforementioned variety of existent devices and of the combination of telematic services they can run.

In [2–4] one can notice that different tools helpful to the analysis of the hardware of the smart phone become useless when it comes to analyse the SIM-card and it can be noticed as well how specific tools for the SIM-cards have appeared on the market:

- *Infinity box* is one of the most recent solutions for programming and non-invasive analysis of Panasonic, Motorola, Bird (Fly), Siemens, NEC, Philips, Alcatel, BenQ, Nyndai, Newgen, VKMobile, Audiovox, Airness, KN-Mobile, Konka, DBTel, Emblaze, Vitel, Toshiba, Amstrad, etc. cellular phones. It runs on Windows operating systems (Win 98SE, ME, 2000, XP) and does not require any power supply as it takes advantage of the one provided by the USB port.
- *Redbox II+* is one of the most performing solution for programming and non-invasive analysis and it distinguishes itself both for the great variety of supported phones and for its performing speed. Alcatel, Panasonic, Vitel, LG3G, LG, Philips Sharp, BenQ, Vkmobile, Trium, NEC, Sony Ericsson Server Access & Nokia Server Access, Maxon PCB3, Toshiba, Samsung.
- *Alibaba Box* is an efficient solution for programming and non-invasive analysis, limited to about 200 phone models though (the most diffused on the Italian market).
- *PDA seizure, Pilot-link* and *BitPIM* support the copying, analysis and reporting of the device's hardware only.
- *SIMIS, ForensicSIM, Forensic Card Reader, SIMCon* support, on the other hand, the copying, analysis and reporting of the SIM-card only.
- *Cell/Device seizure, TULP 2G, GSM .XRY, MOBILEdit Forensic* support the whole aforementioned system functionalities, related to different levels and modalities of analysis.

It is necessary to reckon, at the same time, that the above mentioned tools are not perfectly interchangeable, even within the same point, as their sphere of activity on the various smart phone brands is not uniform.

As far as just pointed out it is evident that the technical survey on a smart phone is actually a technical activity not completely automatable. There are still many uncertainty factors and, on top of it, the investigations' repeatability is always on the edge. In any case, these investigations' results are always crucial and they will keep on being decisive due to handheld devices' spread and practicalness.

In the investigative standard procedure sphere the handheld devices' memory card analysis is always fructuous, for example with regard to pedopornographic digital stuff exchange and to criminally significant files occultation. The security a media card seems to be able to offer, due to the very small dimensions and the remarkable memory capacity, often leads the user to bring it along inside the pocket or the briefcase without even encrypt or protect its contents, occurrence that crucially speeds up the analysis procedures.

The large media card's storage capacity combined with the nearly usual presence of cameras (one or more than one) on the smart phones have been an exceptional coincidence from the legal point of view. Several investigations concern photos or videos shot by those devices that capture criminal events and in many cases they become crucial evidential elements. This occurs since the smart phone is always available as a video camera and does not give rise to most people.

Similarly, it must be brought to attention the investigative validity of the SIM-card data, especially related to LOCI and BCCH which more than once has allowed to verify the truthfulness of the suspect's statements as for instance "*...I was in this area, at this moment...*". Even if the positional information, retrievable from the SIM-card, are indeed not accurate as the ones of a GPS system, they can anyway be used whenever they point a location completely different from the one stated by the suspect.

Another large sector of technical investigation concerns SMS and MMS, actually incorporating a huge volume of interpersonal transmission on digital support, as they have combined the mobile phone practicalness to the reception delay and the unnecessity to immediately reply. SMS and MMS become true evidences of ideas, activities, individual preferences and as often happens they can confirm or exclude investigative hypothesis.

It must then not be forgotten that voice calls and SMSs can be used as trigger communications for mechanisms when the receiving cellular phone results properly modified. This may classically happen in the explosives' remote trigger mechanisms, investigative sector that has given a boost to research in the area of the cellular phone's damaged, both mechanically and thermally, digital circuits rebuilding.

As a matter of fact, however, the technical investigation of a smart phone's hardware is a crucial forensic area of study. The richness of Internet services and PIM available on the handheld give to this instrument a great appeal by those who have a remarkable necessity of communication and so: young/teenagers and managers of every kind. Meanwhile the smart phone has become an indirect instrument of possible control as always in close contact with the user (positional binding) who reposes extreme trust in it by having it always handy (low data and communication protection will).

The problem of *digital wire tapping* has then bossily come to the fore for the police forces. It can be just thought that if one wants to control the communications of a smart phone it would be necessary to put under control at least:

- Voice calls
- SMS
- MMS
- E-mail
- Instant messaging on TCP/IP
- Web browsing

A large flow of data, often correlated among them if not even hidden communication carrier (one can think of the odd VoIP on mobile). The classic wire tapping does not allow to manage all these correlated data and so lately new hardware/software systems, capable of seizing and, above all, integrating the digital wire tapping results, are under construction and testing. Very expensive and hardly manageable systems countered by a sly option achieved just by the extreme complexity of the smart phones. As they can run third party software, one can speculate the use of particular low-level programs (very similar to the Trojans[ad] that usually infect PCs), which by means of the same channels used by the device to access the Internet services could allow, in due time, a remote forensic operator to inspect the contents and the communication of the device at a proper distance.

This interception method, focused on the communication junction rather than on the transmission link, has already been tested in the PC ambit with remarkable success, especially with the presence of several protocols or cryptography. The fundamental issue that affects it concerns the installation of the control software which requires the device's availability for several minutes. It has been debated on the fact that such method is a mix between the forensic approach (after the event) and the pure investigative one (before the event) that should not be considered perfectly real-time like the wire tapping. Such issue unluckily subsists also with the vocal calls which, possessing a digital nature, are subject as well to delays typical of a digital network and so lose the features of the perfect real time. In any case the remote control of the smart phone is anyway an optimal way to intercept Instant Messaging, a tool which is pretty much used because of its simplicity, immediacy and power by a remarkable portion of youth in real and proper virtual communities. The remote control software in this case can replicate all the incoming and outgoing messages towards a forensic operator's client, who is free to save the communications (which are not necessarily point-to-point) and evaluate them in

[ad]Trojan horse: software which is illegally installed or imported together with legal software or data in order to gain the control of a target PC.

nearly real-time when that results necessary (it can be thought of the online lure cases).

18.6. Real Applications

Our high tech crime unit experimented a lot of tools and techniques on the field and the goal of this paragraph is to show some very interesting results that can be useful for the specialised technicians.

18.6.1. *A Real Case of a Smart Phone Non-Invasive Analysis*

We applied a non-invasive analysis method on a mobile phone *LG,model U8360*, that can legitimately be qualified as a smart phone, according to what we have seen.

The device has been initially deprived of its battery and USIM and then connected, by means of a Flash Cable for LG U8xxx, to a personal computer forensically set up. The software *VyGis v1.20* is installed on the PC. Once the connection both on the serial port and on the USB one is terminated, the battery of the radiomobile terminal can be replaced and the VyGis, properly addressed on the cellular typology, starts to run an activation session in which the device is managed as a slave and the PC as a master.

From this stage, though on the device's mini-screen nothing appears, it is possible to launch a dump of the whole memory EEPROM NAND of the hardware under examination, so obtaining a "raw" file which includes all the digital data blocks read from the device. Such bits array, apparently unstructured, must then be interpreted in order to highlight the data readable by the operator.

An interesting methodology consists in importing the file of interest in a forensic analyser like, for instance, *EnCase* and reacquire the copy obtaining de facto a format conversion from "raw" to "E0x" (owner format of EnCase). At this point one can proceed with the new file "*mounting*" as a virtual device and analyse it as if it was a mass memory in which to proceed to the research of deleted files or coherent pieces of them.

EnCase scripts or other specific software con be used for that purpose (e.g., *R-Studio,File scavenger,etc.*). The analysis can be obviously performed even with a hex-editor (e.g., *WinHex*) that require, as well known, great skill with the arrangement of the data used in the device.

Other tools like IDA-pro, a disassembler, can be useful in order to delimit the data areas from those containing executable code otherwise indistinguishable in the initial dump result. In the technical investigation one is naturally interested into the first areas as the code usually does not maintain alteration caused by the user action (unless to investigate on cracking actions inherent the device itself).

The hand made reverse engineering of the raw image of the smart phone memory is a fruitful methodology too. Even if it is very difficult to implement because of the need to visually understand tags and boundaries of the memory blocks from the hex dump, it is sometimes the only way to recover deleted data. We were able to identify and recover deleted SMS, JPG files, contacts, etc.

18.6.2. *iPod Seizure and Analysis: Practical Issues and Problems*

Regarding the data seizure or rather the iPod's semi-permanent memory dump have been taken into consideration several different generation devices with different storage capacity: a 20GB iPod classic, a 6GB iPod mini and a 80GB iPod classic, spanning from the third up to the fifth generation.

The highlighted attempts, conceived to obtain raw dumps (bitstream) have produced the same result as it has been possible to validate by the values of different calculated hash typologies that (exactly by typology) coincide. No particular problem has come out in a deterministic and repeatable manner even if an acquisition through software blocking (registry set up) under Windows, has attained an error that, by repeating the acquisition from zero, has not shown up again.

The various attempts under Helix v.1.9.07 with *dcfldd*, both in presence or absence of write blockers has always produces correct and coincidental results in absence of any technical problem.

In relation to the dump analysis by means of EnCase, FTK and SMART software, the situation to necessarily recover deleted files has been considered with different modalities:

- In *disk mode*, files of various typologies have been uploaded to then cancel them by acting through the file system and simply moving them to the recycle bin:
  - *Deletion by Windows on FAT32*: the files are marked as cancelled by the forensic software and recovered without any problem.
  - *deletion by Mac OS on FAT32*: the files are NOT marked as cancelled by the forensic software and their recovery is not necessary (they are simply moved into another folder).
  - *deletion by Msc OS on HFS+*: the files are not indicated neither as present nor as cancelled, file carving becomes necessary to an approximate recovery.
- In *disk mode*, files of various typologies have been uploaded to then cancel them by acting through the file system and bypassing the recycle bin (shift+canc) o by emptying it in Mac OS:
  - *Shift-canc deletion by Windows on FAT32*: the files are not indicated neither as present nor as cancelled, file carving becomes necessary to an approximate recovery.
  - *Deletion by Mac OS on FAT32 with recycle bin emptying*: the files are not indicated neither as present nor as cancelled, file carving becomes necessary to an approximate recovery.

— *Deletion by Mac OS on HFS+ with recycle bin emptying*: the files are not indicated neither as present nor as cancelled, file carving becomes necessary to an approximate recovery.

- In *iTunes mode*, multimedia files have been uploaded to subsequently cancel them always through the synchronisation:

  — It is possible in general to recover from a FAT32 formatted iPod even if the multimedia file name, once converted and registred in the iPod, is not anymore correspondent to the computer contents, basis of synchronisation, and this can render the researches more complex (the name is replaced by a code).
  — As to HFS+ formatted iPods the recovery can only be based on file carving.

- In disk mode *some iTunes libraries have been altered by appending foreign data.* The iPod keeps on visualising everything correctly, the iTunes does not realise of the modifications during the synchronisation and the forensic analysis does not show evident symptoms of alteration. Only the specific data carving can permit to identify the alterations, provided that the target of the research is at least partially known.

18.6.3. *The Forensic Approach to the iPhone*

We can really say that iPhone forensics could be a different field than smart phone forensics, this because the Apple iPhone device is something special. Its internal 32 bit microprocessor RISC Samsung S3C6400 533/667 MHz is able to manage several kind of digital memories and controllers, the operating system (Aloha OS X) is a small but complete Unix BSD implementation and customisation and everything is under control of the drivers and obviously the Aloha kernel.

There are no test-mode provided nor installed daemon for the memory dump and there's no way to directly access the mass storage unit (Flash or solid state disk) without iTunes. You cannot consider the iPhone as an USB removable hard drive such as for the iPod, the disk-mode is not allowed, so it is not possible to physically dump the mass storage unit.

The only way to access the iPhone memory is to install (under Aloha OS X) a server daemon for forensic purposes. To do it the first step is to install the standard installer.app layer via the safari browser and then the daemon together with a communication layer such as Open SSL. After that the daemon can be controlled by a remote session and/or you can directly browse in the file structure of the mass storage unit.

Based on the fact that we access by a remote session the operating system has to be active, so it is a kind of live analysis. Moreover our activity is obviously intrusive (it changes the stored data) so it is not exactly a sound forensic analysis.

18.6.4. *The Forensic Approach to the Blackberry and Windows Mobile 6 Phone*

Our practical studies about blackberry and Windows Mobile 6.0 mobile devices showed that the memory dump of the EEPROM NAND is possible but the situation becomes quite similar to the iPhone one. The reason is the need of an installed forensic daemon in the O.S. of the devices that can act as a low level cloner. The intrusiveness of the activity is clear even if the daemon is no more than 10 Kbytes and overwriting that area could destroy some concealed or deleted evidence. Anyway that compromise cannot be avoided at the moment.

18.7. Conclusions

The *digital forensics science* is growing and gaining tools, protocols and scientific publications day by day. We can see the sub-field of *computer forensics* as a consolidated area with its theories, methods and tools and the exploration of the world of stand alone systems' seizure and forensic analysis is about to be completed. On the other hand, the subfields of *ipod, cell and smart phone forensics* comes overbearingly in vogue as a research frontier for the technical investigation.

Stand alone systems' evolution towards a nearly ceaseless connection to other computers and, mainly, to the Internet has, at the beginning, defined indeed the *Network Forensics* sector, which academic and police organs have already been devoting themselves to for a few years, mostly for the Internet investigation. Afterwards, the unceasing miniaturisation of the devices, the commercial explosion of the mobile telephony services (which have been supporting for long time the Internet typical services in order to produce even new ones) combined with the availability of new functionalities such as the multimedia tools and players, has determined the more and more massive use of electronic handheld devices with a potential similar to the one of a PC, comprised of the UMTS telephony. This is the *smart phone's* advent, which promises to change our approach to interpersonal communication once again and therefore requires to revise the investigating techniques in the light of the new methods of digital interception and forensic analysis. The present article has been dealing with the theories and technical experimental problems in the *iPod, iPhone, Cell phone and PDA Forensics* field, with the idea to evaluate some of the new commercially available investigation tools. The need was to explore the new technical possibilities for the high tech crime investigation units about seizure, data cloning and analysis of mobile digital devices oriented to gain evidence for crime prevention and prosecution.

References

1. ACPO (Association of Chief Police Officers), *Good Practice Guide for Computer Based Electronic Evidence v3.0.*, U.K., 2003.
2. R. Ayers, W. Jansen, N. Cilleros and R. Daniellou, *Computer Security — Cell Phone Forensic Tools: An Overview and Analysis*, NISTIR 7250, Computer Security Division, IT Laboratory, NIST, Gaithersburg, MD 20988-8930, 2005.

3. R. Ayers and W. Jansen, *Guidelines on PDA Forensics* (NIST Special Publication 800–72, 2007).
4. R. Ayers and W. Jansen, *Guidelines on Cell Phone Forensics* (NIST Special Publication 800–101, 2007).
5. E. Casey, *Digital Evidence and Computer Crime*, Second edition (Elsevier Academic Press, 2004).
6. M. Kiley, T. Shinbara and M. Rogers, iPod forensics update, *International Journal of Digital Evidence* **6**(1) (2007), Purdue University Cyber Forensics Lab, Department of Computer Technology, Purdue University.
7. C. V. Marsico and M. K. Rogers, iPod forensics, *International Journal of Digital Evidence* **4**(2) (2005), Purdue University Cyber Forensics Lab, Department of Computer Technology, Purdue University.
8. M. Mattiucci and G. Delfinis, *Forensic Computing* (Rassegna Scientifica dell'Arma dei Carabinieri).
9. R. Mc Kemmish, *What is Forensic Computing*, Trends and Issues in Crime and Criminal Justice (118), Australian Institute of Criminology.
10. M. Strano, M. Mattiucci and R. Olivieri, *Manuale di investigazione criminale — Accertamenti tecnici su cellulari e smartphone* (Ed. Nuovo Studio Tecna, Rome, Italy, 2008).

Chapter 19

A METHODOLOGY FOR SMARTPHONES INTERNAL MEMORY ACQUISITION, DECODING AND ANALYSIS

ROSAMARIA BERTÈ, FABIO DELLUTRI, ANTONIO GRILLO,
ALESSANDRO LENTINI, GIANLUIGI ME and VITTORIO OTTAVIANI
Department of Computer Science, Systems and Production,
Faculty of Engineering, University of Rome "Tor Vergata", 00133 Rome, Italy

The role of mobile forensics in the crime investigation chain of next future is going to rise due to the market penetration of mobile devices and the enriched capabilities of smartphones. In fact, lots of data can be stored in the internal memory, providing a silent witness of private facts and crimes, not limited to digital crimes, but in the general scope of serious crime. Since the smartphone market is very huge and provides a great variety of manufacturers and models, there is a strong heterogeneity of the tools adopted to retrieve smartphone contents in a forensically sound way: in fact, in most cases, the mobile devices manufacturers implement their own (proprietary) protocols on the proprietary cable-jack and the proprietary OSs, causing the forensic operators to be overwhelmed by the one-on-one tools for every single mobile device. This paper proposes a new methodology and a tool, called MIAT, to acquire the data by using the removable memory cards (e.g., SD, mini-SD, MMC, etc.). After overviewing the current seizure methodology and its related problems when applied to the mobile device scenario, we will introduce an alternative methodology to seize, decode and analyze the data from internal memory, overcoming some problems of the traditional techniques.

Keywords: Mobile Forensics, Data Seizure, PDA, Pocket PC, Windows Mobile.

19.1. Introduction

The forensic process of mobile equipment follows the rules valid for general purpose hardware. As shown in (Peter Stephenson "Using a Formalized Approach to Digital Investigation") the general framework in Fig. 19.1, it can be assumed as valid to treat a mobile equipment. The only very big difference between mobile and general purpose forensics is related to the acquisition phase, since removing the internal memory of the mobile equipment is not a pluggable piece-of-hardware as for the general purpose hard disks. For this reason, the most delicate task in a mobile forensics investigation is to seize data from devices. In the case of a smartphone, it means to seize (with the content acquisition phase) data from the three non-volatile storage locations.

IDENTIFICATION	PRESERVATION	COLLECTION	EXAMINATION	ANALYSIS	PRESENTATION
Event/ Crime Detection	Case Management	Preservation	Preservation	Preservation	Documentation
Resolve/ Signature	Imaging Technologies	Approved Methods	Traceability	Traceability	Expert Testimony
Profile Detection	Chain of Custody	Approved Software	Validation Techniques	Statistical	Clarification
Anomalous Detection	Time Synch.	Approved Hardware	Filtering Techniques	Protocols	Mission Impact Statement
Complaints		Legal Authority	Pattern Matching	Data Mining	Recommended Countermeasure
System Monitoring		Lossless Compression	Hidden Data Discovery	Timeline	Statistical Interpretation
Audit Analysis		Sampling	Hidden Data Extraction	Link	
		Data Reduction		Spatial	
		Recovery Techniques			

Fig. 19.1. DFRWS digital investigation framework.

Data stored in SIM Card can be seized using a smart card reader and a forensics tool (e.g., TULP2G or SimBrush), using often a middleware (for example a PC/SC layer).

The data stored in Memory Card can be seized using an MMC or SD reader (USB or integrated) and a byte stream imaging tool (like DD): binary data are read from source, then stored as an image file, representing all the single bytes, including file system meta data. In this way, it is possible to analyze, e.g., the file allocation table to recover deleted data.

The major challenge is to seize data stored in the internal memory. Three memory types are usually present: ROM (read only) memory, storing OS boot image, RAM (volatile) memory, storing running processes data, Flash memory (non-volatile), storing user files and documents, logs, videos, sounds, etc.

The internal memory is usually a flash memory chip integrated into the mother board. Since this memory type can be erased and written a limited time number and safe transactions must be granted, a logging file system is adopted.

In order to forensically acquire the smartphone memory content, the tools recommended by the NIST standards adopt a remote-way procedure (connected-by-cable, or connected-by-wireless), which represents the major operative drawback. In a connected-by-cable scenario, the lack of an appropriate cable could inhibit the acquisition phase. On the other hand, in the connected-by-wireless approaches, the device needs to be correctly configured for data transmission, i.e., switch on Bluetooth or IRDA services.

Both methods may use a standard protocol like AT commands or OBEX to communicate with the device. Most of the protocol implementations contain proprietary variants introduced by manufactures (e.g., AT command variants exist for Siemens, Samsung, etc.). Obviously, these variants are specific for a single manufacturer and they are incompatible with other models, leading to a great unpredictability of the effectiveness of the tool in seizing operations in the crime scene.

Moreover, a protocol-based access could prevent the access to the whole filesystem. Finally, new releases of smartphone models imply constant updates in hardware/software interfaces. In short time (less than a year), a forensic seizure tool could become obsolete, if not upgraded frequently. This problem refers both to Symbian and Microsoft OS-equipped devices.

In this paper, we will present our alternative methodology, which is composed by following phases:

1. Acquisition phase, which drives the dump on an external memory card (par. 3); this phase is implemented in a tool available for both Symbian and Microsoft Windows Mobile environment. Details about those have been described in our previous work [2, 7].
2. Personal Data Decoding Phase, which interprets the seized data (such as the address book, call logs, SMS, MMS, TODO list, etc.) and convert them in a suitable format (par. 4).
3. Data Analysis Phase, which highlights the correlations among extracted information and helps the investigation in increasing its efficiency and effectiveness (par. 5).

19.2. State of the Art

Many standards for the forensic investigation are currently available for the general purpose computers ACPO (www.acpo.police.uk), IOCE (www.ioce.org) and IACIS (www.cops.org). Furthermore, recently the NIST released the guidelines on cell phone forensics (Jansen). The tools currently in use perform the acquisition of the mobile device internal memory in a remote way: a forensic tool is connected with the target device and, using the OS services, it extracts the data like SMS, MMS, TODO list, pictures, ring tones, etc. In order to achieve forensic seizure, closed and open source tools are available: the most important open source tool is TULP2G (tulp2g), which implements standard modem protocols (e.g., Hayes commands) and OBEX protocol to communicate with the device. Unfortunately, if a smartphone model does not implement these standards, the tool is unusable for the investigation. Among proprietary tools, the Paraben Device Seizure (paraben) is one of the most important; it implements specific, proprietary protocols (like D-Bus for Nokia smartphone) but, as for the TULP2G tool, unknown protocols make the acquisition impossible. Recently, a novel low-level approach for the forensic examination of flash memory can be found in JTAG as well as for the Paraben

Corporation, which released the "CSI stick" (csiStick) a portable data gathering and forensic tool, which allows to acquire data without using the forensic workstation. This solution, however, still relies on proprietary plugs (currently, Motorola and Samsung). The .XRY tool (xry) adopts a quite similar approach to Paraben, with remote acquisition via hardware-specific plugs. Furthermore, recently, Guidance Software added to Encase the mobile package, called Neutrino. For a more extensive review of tools available for Symbian, the reader could refer to Williamson.

(a)

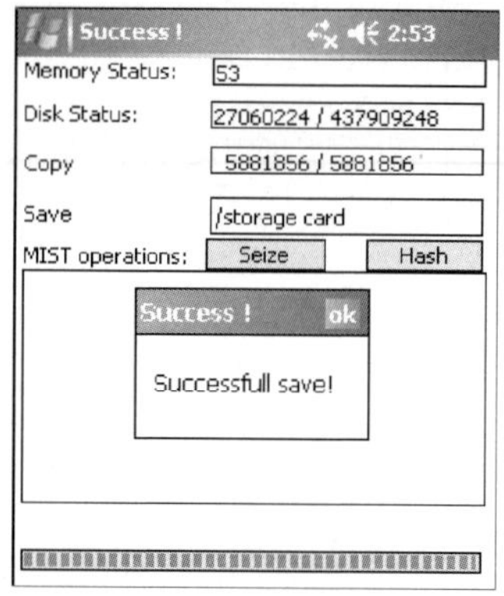

(b)

These figures show screenshots of several tools we developed. In (a) MIAT for Symbian. In (b) MIAT for Windows Mobile.

19.3. Acquisition Phase: MIAT

The Mobile Internal Acquisition Tool (MIAT) acquires data directly from the internal memory slot, spawning an acquisition application stored in the memory card (e.g., MMC or SD) held by the forensic operator. Even if the NIST guidelines say "to acquire data from a phone, a connection must be established to the device from the forensic workstation" [6], we believe that MIAT approach does not contrast those guidelines, but extends the forensic workstation concept to the removable memory where the MIAT executable resides.

The MIAT is an alternative way to seize the internal memory data relies on local execution of an application, which explores recursively the file system tree and copy each entry to a backup volume like an expansion memory card. During the acquisition process, files and directory are opened in read-only mode to preserve integrity: MIAT computes a digest in order to detect further corruptions. Firstly, in order to proceed with the acquisition, the device should be switched off or put in a Faraday Cage. Then, if the SIM and/or the memory card are inserted in the smartphone, we must remove them to collect them. We note that the SIM card is usually located under the battery, for this reason we must turn off the device. Once the SIM and the memory card have been acquired, we use the host memory card

Table 19.1. Files generated during the seizure process.

File	Contents
checksum.xml	File size, file tipology, file name, MD5 hash, seizure, duration, and creation, access and modification time
Info.xml	Information both about the device seized, like IMEI, device ID, platform type, model, manufacturer and about the seizure process, duration, battery consumption date of seizure
errors.xml	Information about errors that may happen during the process

(different from the original memory card found in the device, part of the seizure) for the internal memory seizure: the acquisition application on the memory card is split into two files, the executable and the installation file for Symbian OS (.SIS file).

A tool for seizing data is stored in a memory card and the acquisition is performed locally. In order to grant data integrity, MIAT performs the MD5 algorithm before and after copying each file and compiles a log file with all remarkable events (as shown in Table 19.1) in an XML format.

The main advantage of this approach relies on the use of the standard Symbian/ Microsoft APIs to access the file system (like Open, Read and Write): such system calls guarantee, e.g., Read, the integrity of the read cycles. If the data are seized on the removable storage support, we can access them with a common MMC o SD reader, and we are able to manipulate them with other tools described in paragraphs 4 and 5. The adoption of this methodology forces saving hardware tools like USB cables specific for each device or additional equipment like notebook PC to perform the acquisition; the forensic workstation is now the seized Mobile Equipment (ME) with a supplementary SD/MMC memory card with MIAT onboard.

In order to automatically identify the manufacturer and the model of the ME, we use the IMEI number (unique and available in every smartphone). In fact, the IMEI can be discovered by inspection of physical location under the battery: the crime scene operator sends a request (e.g., via his ME) containing the IMEI of the seized ME to a listening server and waits.

The server identifies the mobile phone model by checking the TAC number (first eight IMEI digits). Hence, the server sends the message back (e.g., via MMS), containing the ad-hoc compiled release of MIAT as attachment to the crime scene operator.

In Windows Mobile version, there is no need to perform this step, as there is only one version of MIAT for PPC and one for WM smartphones, both versions run for WM5 and WM6 devices.

The difference between MIAT and some other forensic tools (e.g., TULP2G) is that the former dialogues directly with the Operating System while the latter (e.g., Paraben device seizure) require an intermediary (located in the mobile phone) in charge of managing the messages sent by remote forensic tool to the mobile phone (as shown in Table 19.2).

Table 19.2. File last modification time changes.

File	Reboot	Acquisition	File	Reboot	Acquisition
100056c6.ini	B		101f6df0.ini	B	B
AlarmServer.ini	B	P	Applications.dat	B	M
backupdb.dat	B	P	btregistry.dat	B	
cbtopicsmsgs.dat	B		CntModel.ini	B	P
CommonData.D00	B	P	DRMHS.dat	B	
ECom.lang	B		HAL.DAT	B	
LocaleData.D05	B	P	nssvasdatabase.db	B	B
ScShortcutEngine.ini	B	P	smssmssegst.dat	B	
System.ini	B				

B, change happens for both tools.
P, change happens for PARABEN.
M, change happens for MIAT.

Since the intermediary code is generally closed, the second case makes impossible to verify how intermediary code is written.

19.3.1. *Symbian*

MIAT for Symbian, presented in Ref. 7, was developed to support and to test the methodology above described.

Symbian is an operating system derived from the Epoc operating system; Symbian OS supports a wide range of device categories with several user interfaces, including Nokia S60, UIQ and the NTT DoCoMo common software platform for 3G FOMATM handsets. The commonality of Symbian OS APIs enables development that targets all of these phone platforms and categories. In order to produce executable code which does not need of any other software layer (e.g., a JVM to interpret the bytecode), MIAT application was originally developed in C++, the native language of the Symbian OS. Note that since there are many versions of each combination OS/UI, there is a different SDK for each combination.

In order to build application for specific devices using the appropriate SDK of the correct version of the target phone.

The specific needed version of MIAT is recognised by the IMEI number as mentioned in the methodology above.

MIAT application uses OS APIs to explore and copy the entire internal memory file system to removable memory card; Fig. 19.2 shows how MIAT works, the scanning algorithm is iterative, starting at File System root and ending when all entries are seized.

Most relevant files are locked by system processes, while many files on the system are always opened and locked by system processes. For example, the file Contacts.cdb, which contains the database of contacts, is locked by PhoneBook, namely the address book process.

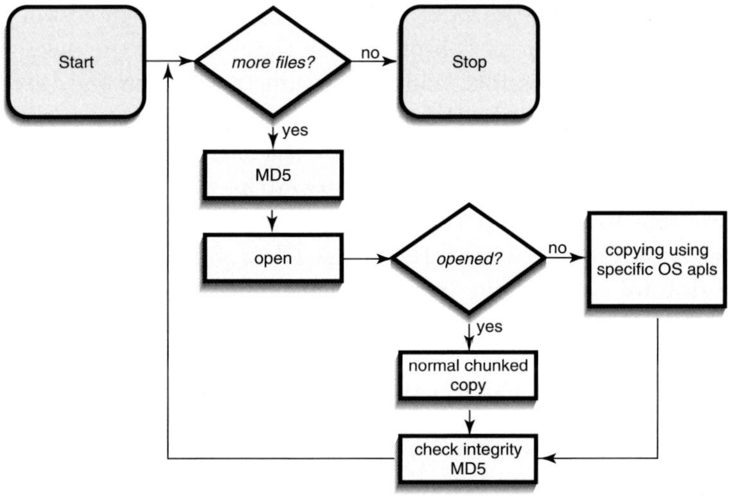

Fig. 19.2. How MIAT works.

In a recent work [3], MIAT adopted a further alternative way to get access to locked files. This way is accomplished by the Symbian RFs API method ReadFileSection that allows reading from a file without open it. By this method it is possible to seize the entire file system tree including files which have a persistent lock on; furthermore, this strategy preserves integrity because the access is established in read-only mode, guaranteed by the OS.

19.4. Personal Data Extraction

The main issue related to Personal Data Extraction phase is to overcome the heterogeneity of the available data; data set acquired from mobile devices manufactured by different vendors and also from different models of the same vendor are deeply different. The Personal Data Extraction (PDE) phase cover this issue interpreting the specific data formats and producing a set of XML files that enriche the input of the data analysis tool.

After a correct seizure with the MIAT extraction tool, we obtain a copy of the smartphone's filesystem on an SD memory, ready to be analyzed on a PC. A part of dumped relevant data (e.g., images, documents, etc.) are stored with a well-known format, like jpeg, doc, txt, etc., therefore they are accessible directly with standard tools like a file explorer. However, a large part of the mobile phone personal data (*SPPD*) are stored in some device databases, encoded by a proprietary file format and managed by specific OS functions. As second step of our methodology, in this paragraph we focus on retrieving such data and on converting them in a suitable format; at the same time, we will find a method to recover a part of deleted or invisible information as well.

The DBMS resources optimization strategy reduces the high-cost of DB's delete operations by flagging them as "obsolete": for these reasons the delete operations are scheduled as late as possible, and the circumstance when they are performed varies depending the kind of file. For instance, in the Symbian case, in *Contacts.cdb* the deleting operations are performed when the `Compress()` syscall is invoked. Operating system's tasks and third-party software as well can invoke this function, and they are able to know whether or not to perform compression by invoking `CompressRequired()` (see CContactDatabase Class, Symbian Developer Library). Let S the disk total space, F the free disk space, and W the amount of disk space wasted; the *Boolean* function returns true if:

$$(\mathrm{W} > 64\,\mathrm{K}) \vee (\mathrm{W} > 16\,\mathrm{K} \wedge \mathrm{W} > 1/2\,\mathrm{S}) \\ \vee (\mathrm{W} > 16\,\mathrm{K} \wedge \mathrm{F} < 1/20\,\mathrm{S}) \vee (\mathrm{W} > 16\,\mathrm{K} \wedge \mathrm{F} < 16\,\mathrm{K})$$

When the compression is completed, the contacts are rearranged, the space wasted by obsolete records is recovered and it is not possible to recover any deleted data. If the seizing operation occurs before the compression has been invoked, we will find a database file that will contain all data since last compression.

Personal information as address book, SMS, MMS, email, calendar, call and event logs are not accessible with standard tools. Even OpenSource and commercial software panorama do not offer any external tool able to parse such files and convert them in an open and suitable format, ready to be analysed. We used a Symbian case study to design a methodology for understanding the *SPPD* file formats, then we have developed a Java-based tool in order to parse these files and to convert them in an XML format. Before finding the best method, we looked for it by following two different approaches.

The first attempt was to import the seized files in a device OS emulator and to query DB data by native OS kernel DB functions, but we quickly realized that these functions are prevented to show obsolete-flagged records in query results.

As second way, for each relevant *SPPD* file, we tried how to understand its format through a low-level binary interpretation. So we developed a general-purpose methodology in order to help people in reverse engineer a binary file format. A comprehensive discussion about such methodology will be the subject of a future work. After applying the methodology on each *SPPD* file, we understood how to parse and how to convert them in an XML format. So, we implemented a parser for each file type and we tested them on a tested of 50 devices. Almost on all file types, we obtained a percentage of correct data extraction equals to 100%. As shown in Table 19.3, just in the case of *Contacts.cdb*, we reached 99.5% of average hit ratio. Now, we are working on such problem, reiterating over the methodology to find the cause. Moreover, as future work, we plan to apply such methodology on Windows Mobile device's filesystem.

Table 19.3. Symbian OS filesystem entries containing relevant SPPD. This table also shows the file recover hit ratio.

Entry	Description	Data recover (%)
Contacts.cdb	File containing phone's address book.	99.5
Logdbu.dat	File containing events log like: calls and video calls, received and sent sms/mms/mail trace, SIM changes, GPRS connections.	100
Calendar	File containing appointments, notes and anniversaries	100
Mail	Directory containing sent and received SMS, MMS, emails	100

19.5. Data Analysis

After the phase 2, all the smartphone's data are convenient for the Data Analysis phase.

The data analysis phase concerns with the process of looking and summarizing data with the intent to link useful information in order to support in developing investigative decisions. The data analysis tool was developed in order to capture the logical equivalent meta-information that data of a specific mobile device acquisition must contain (e.g., contacts information, calls details, ...). The common semantic expressed by the collected data helps to define the structure of the analysis tool.

The developed tool is composed by different views organized in a hierarchy; at the first level, we can find three different views:

The *Browser view* reorganizes collected data in a hierarchical way using a specific proximity measure. This view includes a standard and a file type view. The standard browser view exposes the seized files as they are on the target mobile device; in this case the adopted proximity measure is represented by the similarity to the original file system. The file type browser view clusters the collected files with respect to the content type (e.g., images, audio, etc.).

The *Logical Information view* allows to consult the collected data as they are recovered in the PDE phase; this view presents the information that are common to all the mobile devices, and linked with contacts, sms, events and calendar entries.

The *Analysis Engine View* allows to consult the collected data linking them in a specific way. This view is split into two different subview: *chrono-view* and *link view*. The former allows to consult a reconstruction of the mobile device event-history; entries in that history concerns with both human–device interaction (e.g., deleting a contact, saving an sms draft, ...) and device–device interaction (e.g., making a call, sending an sms, ...).

The latter allows to consult a reconstruction of the mobile device interactions. The seized device is represented as the central node of a star graph; the remaining nodes of the graph represent the mobile devices with which the seized device had at least one interaction of a specific type (e.g., send/receive a call, send/receive a sms, ...). The thickness of the edge linking each not-central node with the seized device node is determined as linear function of the number of different kinds of interactions, each one weighted appropriately. Figures 19.3 and 19.4 show

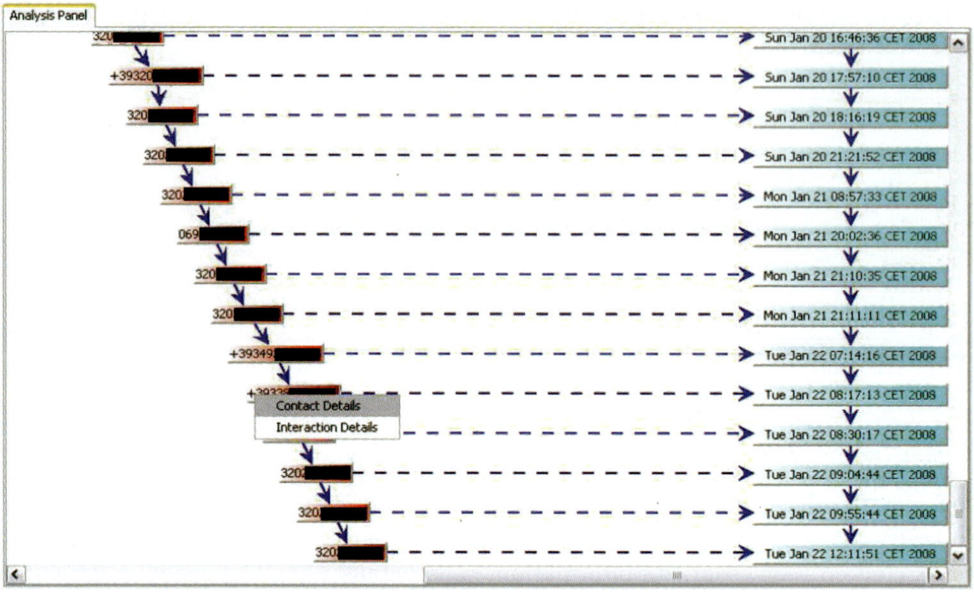

Fig. 19.3. Chrono-view.

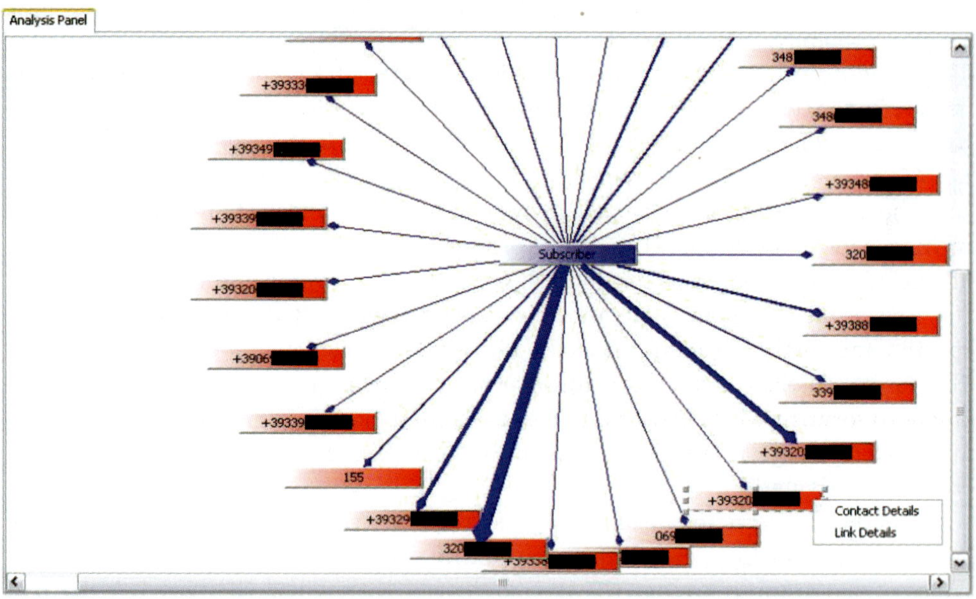

Fig. 19.4. Link view.

respectively the *chrono-view* and the *link view* GUIs of the implemented data analysis tool.

The realization of an integrated data analysis tool based on powerful automatic analysis and manipulation data techniques represents a core activity in the process of speeding up the investigative analysis. The need of automatic tools is strictly related to the size of the logic-dump obtained with MIAT that for very common devices (e.g., Nokia N Series) rises up to 32–64 MB (i.e., the size of the mobile device internal memory).

19.6. Conclusions

In this paper, we have presented an alternative methodology in order to seize the internal memory data of the Symbian smartphones. These devices have some peculiarities which heavily influence the investigation process. In fact they request a different approach for extracting evidence from their internal memory.

The tool MIAT, in both versions for Symbian and for Windows Mobile (not presented in this chapter), can lower the effort for Law Enforcement Agencies (LEAs), due to the non-technical skills requested for the forensic operators. This tool makes the acquisition phase independent from the plethora of one-to-one mobile phone cables, raising the parallelism for acquisition, and wiping out the cost-per-seizure acquisition (MIAT intent is to be released as open source software). For these reasons, it could automate and speed up the forensic process, especially when the amount of devices to analyze can overwhelm the high tech crime units.

Currently, the MIAT tool is experimented by an LEA, in order to verify its effectiveness over a large scale of Symbian models and to state the usability in the field of this software.

Current results show that MIAT for Symbian performs slower than Paraben in seizure times. Those are constrained by device type and filesystem density: as the mobile devices processor will speed-up, we believe that the performance between the MIAT and remote acquisition will be negligible.

Future developments of MIAT methodology will include (but are not limited to) the OS DRM mechanism, in order to maintain the highest level of device portability on Symbian phones and the development of a forensic analysis farm, where to forward all the seized images (from the crime scene) in order to provide to the forensic operator and investigator a real-time standard analysis of the mobile equipment.

Acknowledgments

We acknowledge Alessandro Di Stefano and Daniele Bocci for the contribution extracted by their degree thesis.

References

1. M. Breeuwsma, M. de Jongh, C. Klaver, R. van der Knijff and M. Roeloffs, Forensic data recovery from flash memory, *Small Scale Digital Device Forensics Journal* **1**(1) (2007) (http://www.ssddfj.org/papers/SSDDFJ_V1_1_Breeuwsma_et_al.pdf)
2. F. Dellutri, V. Ottaviani and G. Me, Forensic acquisition for windows mobile pocket PC, Proceedings of the Workshop on Security and High Performance Computing Systems, Part of the 2008 International Conference on High performance Computing & Simulation: HPCS (2008), pp. 200–205.
3. A. Distefano and G. Me, An overall assessment of mobile internal acquisition tool, Digit. Investig. (2008), doi: 10.1016/j.diin.2008.05.010.
4. ETSI and 3GPP, ETSI TS 123 003 V7.6.0 technical speci_cation (January 2008).
5. http://www.symbian.com/Developer/techlib/v70docs/SDLv7.0/doc_source/reference /cpp/ContactsModel/CContactDatabaseClass.html
6. W. Jansen and R. Ayers, Guidelines on cell phone forensics recommendations of the national institute of standards and technology. NIST (2007).
7. G. Me and M. Rossi, Internal forensic acquisition for mobile equipments', 4th International Workshop on Security in Systems and Networks (SSN), Proceedings of the International Parallel and Distributed Processing Symposium (IPDPS), 2008, IEEE Computer Society Press.
8. P. M. Mokhonoana and M. S. Olivier, Acquisition of a Symbian Smart phone's Content with an On-Phone Forensic Tool, Southern African Telecommunication Networks and Applications Conference 2007 (SATNAC 2007) Proceedings.
9. Paraben Corporation, Paraben device seizure. Paraben's Forensics Software www. paraben.com/, (2008).
10. Paraben Corporation. CSI stick, http://www.csistick.com/
11. Python for S60, Nokia Research Center, http://opensource.nokia.com/projects/ pythonfors60/
12. M. Santarini, NAND versus NOR, EDN, October 13, 2005 Symbian Developer Library 'Class CContactDatabase'.
13. P. Stephenson, Using a formalized approach to digital investigation.
14. TULP2G project website, http://tulp2g.sourceforge.net/
15. B. Williamson, P. Apledoorn, B. Cheam and M. McDonald, Forensic analysis of the contents of Nokia mobile phones, (2006).
16. XRY project website, MicroSystemation http://www.msab.com/en/, (2008).

Chapter 20

ANALYSIS OF E-MAIL HEADERS

ALESSANDRO OBINO and MASSIMO BERNASCHI
Institute for Applied Computing,
Italian National Research Council
Viale del Policlinico, 137 Rome, Italy

E-mail messages can be easily forged to show fake information about the sender or the time a message is created. However, a proper analysis of *headers* included in any e-mail message may provide valuable information. In this chapter, we describe e-mail *headers* organisation and present *MailMiner*, a tool we developed to ease the analysis by storing in a relational database the results provided by the tool itself.

20.1. Introduction

E-mail represents, along with Web browsing and *chat*, one of the most popular internetworking applications with billions of messages exchanged every day among friends, job colleagues and business partners. Despite common perception, the security of standard e-mail is very limited and information about the sender or the time sending that appears in an e-mail message can be easily forged. Hereafter, we recall the basics of e-mail protocols and format and show how the information found in message *headers* can be used to check its actual origin and the route followed to reach its destination.

20.2. E-mail Basics and Message Headers

The Simple Mail Transfer Protocol (SMTP) described in the RFC 821 and RFC 2821 [8] is a command-based text protocol (like HTTP albeit much older), used to exchange e-mail messages.

An example of SMTP-based communication (with command and response code in bold) is

RECEIVER **220** `receiver.yourisp.com SMTP Service at 23 Jun 2007 05:17:18 EDT`
SENDER **HELO** `sender.myisp.com`

```
RECEIVER   250 receiver.yourisp.com - Hello, sender.myisp.com
SENDER     MAIL From: <john@sender.myisp.com>
RECEIVER   250 MAIL accepted
SENDER     RCPT To: <paul@receiver.yourisp.com>
RECEIVER   250 Recipient accepted
SENDER     DATA
RECEIVER   354 Start mail input; end with.
SENDER     Date: Sat, 23 Jun 2007 13:26:31 EDT
SENDER     Subject: meeting
SENDER
SENDER     Let's get together Monday at 1pm.
SENDER     .
RECEIVER   250 OK
SENDER     QUIT
RECEIVER   221 RED.RUTGERS.EDU Service closing transmission channel
```

The RFC 822 and 2822 [9] define the syntax for text messages sent between computer users, within the framework of "electronic mail".

The process of sending an e-mail through the Internet involves, at least, four entities: mail client and mail server of both sender and receiver. Mail client, sometimes called Mail User Agent (MUA), is the software application used by the sender to compose the message; it is responsible of the generation of RFC822 compliant messages and of the SMTP communication with the sender's mail Transfer Agent (MTA) that is the server software application responsible of the delivery of the message. The receiver uses his own MUA to interact with the MTA of his organisation or Internet Service Provider in order to retrieve messages addressed to him. Protocols used in this final part of the communication can be either the Internet Message Access Protocol (IMAPv4 rev1 as specified in RFC 3501) or the Post Office Protocol (POPv3 as specified in RFC 1939) which are not relevant to the scope of the present work.

Besides the sender and receiver servers, most of the times an e-mail message sent through the Internet crosses other servers, called Mail Delivery Agents (MDA), in its path from source to destination.

Some of the most commonly used MUAs are Mozilla Thunderbird and Microsoft Outlook while examples of widespread MTAs are Sendmail, Microsoft Exchange and IBM Lotus Domino.

A message complaint with the RFC 822 is made of two sections separated by an empty line (defined as a sequence of Carriage Return and Line Feed, CR/LF). These sections are the header section and the body section (see Fig. 20.1).

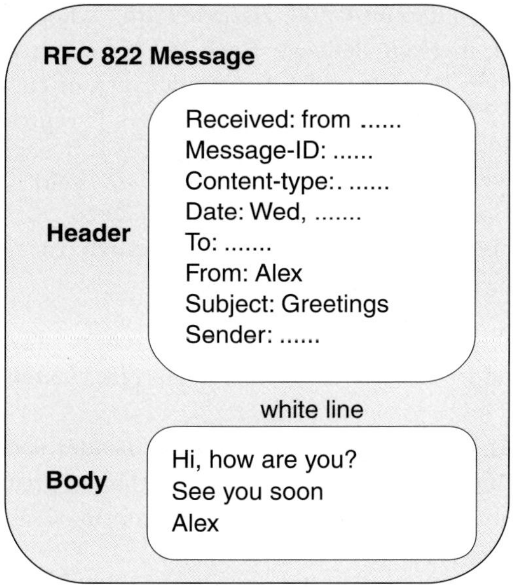

Fig. 20.1.

Most of e-mail users interact directly with the body section of a message whereas the header section is compiled by MUAs, using other information given by the user such as receiver address and message subject and MTAs. Each header field can be viewed as a single, logical line of ASCII characters, constituted by a field-name *(tag)* and a field-body separated by a semicolon followed by a blank space as shown in the example below.

```
Subject: this is the subject of an e-mail message
```

Tag is the literal identifier of the header. Headers may be classified in four main categories:

- Mandatory
- Optional
- Dynamic
- User defined

Mandatory headers are the minimum required (see Appendix 3 of RFC 822) to allow the delivery of a message. These are sender's e-mail address (header *From*), receiver's e-mail address (either *To* or *BCC*) and timestamp of sending (header *Date*).

Optional headers can be used to expand message distribution list (e.g. header *CC*) or provide additional information to receivers (e.g., header *Subject*).

Dynamic headers should be written by MTAs or MDAs to help reconstruction of message history. Within the dynamic headers, an important sub-class are the so-called *trace headers* (*Received* and *Return-Path)* whose values represent the routing information in message delivery. Each MTA dealing with a message writes a *Received* header while the last MTA before delivery often writes a *Return-path* header. An example of the structure of these headers is represented below:

```
Received: from serverX.domainX.com (XXX.YYY.ZZZ.WWW) by
          serverY.domainY.com (8.12.11.20060308/8.12.11) for
          receiver@receiverdomain.com with ESMTP id 17BGdbRM013174;
          Sat, 11 Aug 2007 18:39:38 +0200
```

The above example shows the *Received* header written *by* MTA (or MDA) server X receiving a message addressed to (*for*) *receiver@receiverdomain.com from* the MTA (or MDA) server Y.

Actually, although that format of the *Received* header is quite common and its syntax rules are specified in the RFC 822, the internal structure of its value can be customised in most mail server programs. Thus informations that header reports may vary.

As shown above the header usually contains the following elements:

- *From*: public name of sending MTA as presented in the **HELO** (or **EHLO**) command of the SMTP and IP address as determined by the TCP/IP connection
- *By*: public name and software version of receiving MTA (responsible of content write)
- *With*: transmission protocol used
- *Id*: spool identifier of the temporary file used by the receiving MTA to store the message
- *For*: e-mail address of the final recipient
- *Timestamp* of reception

In most messages there is a *stack* of *Received* headers organised as follows:

```
Received: from server_{n-1}.domain_{n-1}.com (x_{n-1}.y_{n-1}.w_{n-1}.z_{n-1}) by
        server.rcvrdomain.com for receiver@rcvrdomain.com with ESMTP
        id 8FDSK1F3013174; Sat, 11 Aug 2007 18:46:12 +0200
...
Received: from server_2.domain_2.com (x_2.y_2.w_2.z_2) by server_3.domain_3.com
        for receiver@rcvrdomain.com with ESMTP id 35HKTFXY546455
        Sat, 11 Aug 2007 18:39:25 +0200
Received: from server_1.domain_1.com (x_1.y_1.w_1.z_1) by server_2.domain_2.com
        for receiver@rcvrdomain.com with ESMTP id 17BGdbRM013174;
        Sat, 11 Aug 2007 18:36:38 +0200
Received: from client.domain_1.com (x_0.y_0.w_0.z_0) by server_1.domain_1.com
        for receiver@rcvrdomain.com with ESMTP id 75FBdxNQ786590;
        Sat, 11 Aug 2007 18:35:23 +0200
```

where $x_i.y_i.w_i.z_i$ is the IP address of the i-th mail server.

Note that the first header in the list is written by the nth MTA on the path from source to destination (i.e. the receiver's MTA) while the last one is written by the first MTA receiving the message (i.e. the sender's MTA) from the client (i.e. the sender's MUA). The RFC 2821 explicitly states that (i) an Internet mail program must not change the value of a *Received* header that was previously added to the message header; (ii) SMTP servers must prepend *Received* headers to messages; (iii) they must not change the order of existing lines or insert *Received* headers in any other location. Thus, in a regular message, the *by* server of the $(i-1)$th *Received* header should be equal to the *from* server of the ith header. The so-called *anonymous remailers* intentionally violate those rules to make difficult the tracing of the route followed by a message. The internal consistency of each *Received* header is also important to determine whether a message is genuine or not. Most Internet mail programs carry out a reverse DNS lookup on the IP address by which a message is received and show in the *Received* header the name corresponding to the IP address if it is different from the name used in the **HELO** (or **EHLO**) command. For instance:

```
Received: from mailserverX.domainX.com (realhostX.domainX.com
        [XXX.YYY.ZZZ.WWW])
```

In this example a server presents itself as `mailserverX.domainX.com` but the receiver finds that the name corresponding to the server IP address is `realhostX.domainX.com`. That does not necessarily mean that there is something wrong. Many hosts acting as mail servers have one or more *canonical* names (basically an alias of the official name registered in their **A** record of the DNS). If the mail program uses a canonical name, then the reverse lookup of the IP address will show a different name (corresponding to the official name registered in the **A** record). An inspection of the DNS records of the server can easily point out such situations.

Attention is also required looking at the timestamp of the *Received* headers. Apparently it looks reasonable to expect that the timestamp of $(i-1)$th header always precedes the timestamp of the ith header. Actually, small deviations can be due to a misalignment of the systems' clocks and should not be considered an indication of header forgery. A more reliable test can be done using the spool ID also reported in the *Received* header. If it is possible to access the log files of the server that apparently wrote the header, then the spool ID can be used to search for a matching entry in the mail program log file. An example of entry written by Sendmail follows:

```
Mar 23 09:02:57 mailhost sm-mta[28266]: m2N82n6I028266:
from=<abc@myisp.com>, size=618, class=0, nrcpts=1,
msgid=<01c88cdd$b442d900$af1440d9@myisp.com>, proto=ESMTP,
daemon=MTA, relay=[XXX.YYY.ZZZ.WWW]
```

where `m2N82n6I028266` is the spool ID.

If an entry with the spool ID reported in the *Received* header is not present or if the information it reports (e.g. the timestamp) are not consistent with those reported in the *Received* header, then it is likely that the header has been forged.

The spool ID is a piece of information completely local to the server which manages temporarily a message during its trip from the sender to the receiver and should not be confused with the Message-ID that is the value associated with the *Message-ID* header. The *Message-ID* header is added to each message by the first mail server that handles it and there should be only one *Message-ID* header in any message. An example of *Message-ID* header is

```
Message-ID: <3A2EDAFA.F4735272@myisp.com>
```

The value of the header contains a single unique message identifier. The uniqueness of the message identifier is guaranteed by the host that generates it. The RFC 2822 suggests to put the domain name (or a domain literal IP address) of the host on which the message identifier was created on the right hand side of the "@", and put a combination of the current absolute date and time along with some other currently unique (perhaps sequential) identifier available on the system (e.g. a process id number) on the left hand side. The angle bracket characters are not part of the Message-id; the Message-id is the sequence of characters between the two angle bracket characters.

Each mail server program uses its own rules to build the left hand side of the Message-id. For instance, Sendmail uses, since version 8.12, the following format for the Message-id: `<$t.$i@$j>` (where $\$t$ is the time, $\$i$ is the spool identifier of the message and $\$j$ is the domain name). The time is codified as follows: `YMDhmsNNPPPPP` where: `Y` is the year since 1900, mod 60; `M` is the month (January = 0, ... , November = A, December = B); `D` is the day (1–31); `h` is the hour; `m` are the minutes; `s` are the seconds; `NN` is the number of the *envelope;* `PPPPP` is the process ID (at least five characters). The characters used for coding the date are [0–9][A–Z][a–z]. According to these rules, `jBFE6F53022319` corresponds to `2005 December 15, 14:06:15` since

```
Y:j ⇒ (2005 -1900) mod 60 = 45 ⇒ j; M: B ⇒ December; D: F ⇒ 15;
H:E ⇒ 14; m: 6 ⇒ 06; s: F ⇒ 15;
NN: 53; PPPPP ⇒ 022319.
```

User defined headers represent an extension of the protocol. Each entity involved in the delivery of e-mail messages has the chance to define additional headers whose tag usually starts with the sequence $X-$. Those headers are widely used by MUAs, MTAs and other services, such as antivirus and antispam systems, to trace the result of different manipulations applied to the message. Few user defined headers, originally introduced by a single vendor, became standard de-facto (e.g. the *X-Mailer* header written by the sender MUA to identify type, version and operating system of the mail client).

20.3. Analysis of E-Mail Headers

The basic factor of insecurity in e-mail messages is that the SMTP does not require the verification of sender's identity (loss of authentication). Although a number of extensions of the SMTP protocol allow to negotiate security levels, these are not commonly used and actually they are mostly ignored by mail server administrators.

The forensic analysis of an e-mail message aims at discovering the history of a message including the time of sending and the identity of all involved entities (MUAs, MTAa, MDAs). This task is made difficult by the loss of authentication in SMTP and by the potentially malicious behaviour of one or more MTAs in the path from source to destination. *Header forgery* is one of such malicious activities. It can consist of both

- Writing of fake headers
- Manipulation (and also removal) of headers written by other entities

with the objective of hiding the real origin of the message or preventing the tracing of the route followed by the message to reach its destination. The first objective can be achieved by removal of the sender's identity (sender looks anonymous) or by manipulation of the sender's identity in a way that message seems to be sent by someone other than the sender (a real mail user or not). Note that an anonymous e-mail is legitimate in principle as it happens for the ordinary mail (obviously except for contents) but it represents an explicit violation of protocol specifications. The so-called Anonymous Remailers are MTAs which *anonymise* the sender by using header forgery, along with other techniques.

The substitution of sender's identity with another existent identity is also known as *e-mail spoofing*.

Self-sending messages are a special case of e-mail spoofing where the sender's e-mail address appears as the spoofed e-mail address of the receiver that actually receives the message in Blind Carbon Copy. Header forgery is widely used by spammers (the sending of spam messages is considered illegal in many nations) and by cyber criminals making phishing or social engineering. They use e-mail addresses of well-known organisations (e.g. banks or government institutions) to get trust from potential victims and grab them sensible information such as credit card numbers on bank account credentials. E-mail spoofing is often realised through malicious Open Mail Relays, Open Proxies or poorly configured MTAs, MDAs and HTTP Proxies which allow to send or forward messages by users that are not part of their respective domains.

A classic example of forgery (available from http://www.rahul.net/falk/mailtrack.html) follows:

```
From webpromo@denmark.it.earthlink.net Tue Jul 8 13:05:02 1997
Return-Path:
From: webpromo@denmark.it.earthlink.net
Received: from denmark.it.earthlink.net (denmark-c.it.earthlink.net
```

```
[204.119.177.22]) by best.com (SMI-8.6/mail.byaddr) with ESMTP
idNAA21506 for ;
Tue, 8 Jul 1997 13:05:16 -0700
Received: from mail.earthlink.net (1Cust98.Max16.Detroit.MI.MS.UU.NET
[153.34.218.226]) by denmark.it.earthlink.net (8.8.5/8.8.5) with SMTP
id NAA12436; Tue, 8 Jul 1997 13:00:46 -0700 (PDT)
Received: from adultpromo@earthlink.net by adultpromo@earthlink.net
(8.8.5/8.6.5) with SMTP id GAA05239 for ; Tue, 08 Jul 1997 15:48:51 -
0600 (EST)
To: adultpromo@earthlink.net
Message-ID: 199702170025.GAA08056@no-where.net
Date: Tue, 08 Jul 97 15:48:51 EST
Subject: Hot News !
Reply-To: adultpromo@earthlink.net
X-PMFLAGS: 12345678 9
X-UIDL: 1234567890x00xyz1x128xyz426x9x9x
Comments: Authenticated sender is
Content-Length: 672
X-Lines: 26
Status: RO
```

The To: line is a forgery; the actual recipients list was hidden, probably with a blind carbon-copy (*Bcc:* header). The Message-ID: line is an obvious fake. The second Received: line shows this inconsistency:

```
from mail.earthlink.net(1Cust98.Max16.Detroit.MI.MS.UU.NET
[153.34.218.226])
```

That is, the machine that delivered the mail to `denmark.it.earthlink.net` identified itself as `mail.earthlink.net` (as argument of the **HELO** or **EHLO** command) but is was actually named `1Cust98.Max16.Detroit.MI.MS.UU.NET`. The third `Received:` line is completely bogus. If the mail came from a dial-in customer at Uunet, there would not be any more `Received:` lines. If the mail was being relayed from Uunet, this `Received:` line would indicate Uunet, not Earthlink. Further, this `Received:` line contains e-mail addresses, not machine names. So, this e-mail was forged to make it look like it came from Earthlink but was actually injected from Uunet. Whether this was by an Earthlink customer or some other Uunet customer is impossible to tell without further information available, in this case, only to Earthlink.

Another example (from the same site) is the following:

```
Received: from cola.bekkoame.or.jp (cola.bekkoame.or.jp
          [202.231.192.40]) by srv.net (8.8.5/8.8.5) with ESMTP id
          BAA00705 for ; Wed, 30 Jul 1997 01:15:27 -0600 (MDT)
From: beautifulgirls585@aol.com
Received: from cola.bekkoame.or.jp (ip21.san-luis-obispo.ca.pub-
ip.psi.net [38.12.123.21]) by cola.bekkoame.or.jp (8.8.5+2.7W/3.5W)
```

```
with SMTP id OAA11439; Wed, 30 Jul 1997 14:35:50 +0900 (JST)
Received: from mailhost.aol.com(alt1.aol.com(244.218.07.32)) by
aol.com (8.8.5/8.6.5) with SMTP id GAA00075 for <"">; Tue, 29 Jul
1997 22:19:42 -0600 (EST)
Date: Tue, 29 Jul 97 22:19:42 EST
Subject: You can have what you want...
Message-ID: <574857638458.HWF39862@aol.com>
Reply-To: beautifulgirls585@aol.com
```

Here, the second `Received:` line indicates that `cola.bekkoame.or.jp` received the mail from a machine which identified itself as `cola.bekkoame.or.jp`, but was actually `ip21.san-luis-obispo.ca.pub-ip.psi.net`. This message was probably forged from a Psi.net dial-in account. Moreover, the IP address mentioned in the third Received: line cannot be matched via *whois* or *traceroute*. It certainly does not match AOL, indicating that the line is bogus.

As last example, we show a comparison of the headers of a forged and a regular e-mail (source http://www.emailaddressmanager.com/tips/headers.html).

SPAM HEADER

```
Return-Path: <ydcddlhanqz@yahoo.com>
Received: from mail.fx.ro (mail4.fx.ro [193.231.208.4])
        by fx.ro (8.12.7/8.12.7) with ESMTP id
i2OAVxGs024789; Wed, 24 Mar 2004 12:31:59 +0200
(EET)
Received: from mailv.fx.ro (localhost.localdomain
[127.0.0.1])
        by mail.fx.ro (8.12.11/8.12.3) with ESMTP id
i2OAVxaA004610;
        Wed, 24 Mar 2004 12:31:59 +0200
Received: (from root@localhost)
        by mailv.fx.ro (8.12.11/8.12.3/Submit) id
i2OAVxh1004609;
        Wed, 24 Mar 2004 12:31:59 +0200
Received: from 206.85.220.156 by 217.225.143.240;
Message-ID:
<VHUCXEYVIXPEUNUKOJEW@hotmail.com>
From: "Julianne Lloyd" <ydcddlhanqz@yahoo.com>
Reply-To: "Julianne Lloyd" <ydcddlhanqz@yahoo.com>
To: boby_con@fx.ro
Cc: bodistvan@fx.ro, bogdan.micu@fx.ro, bogdan@fx.ro,
bogdans@fx.ro
Subject: Get viagra over night - no prescription needed
Date: Wed, 24 Mar 2004 08:31:16 -0200
X-Mailer: AOL 9.0 for Windows US sub 740
MIME-Version: 1.0
Content-Type: multipart/alternative;
        boundary="--05917340466547820851"
X-Priority: 3
X-MSMail-Priority: Normal
X-IP: 162.238.92.104
X-RAV-Bulk: RAV AntiVirus classifies this e-mail as spam
(accuracy medium)
X-RAV-Signature:
250F0FB03547C3C93609D82815AB3746
X-RAVMilter-Version: 8.4.3(snapshot 20030212) (mail)
X-UIDL: l+/"!l-H"![JK!!^\J"!
Status: RO
```

REGULAR EMAIL

```
Return-Path: <bogdan@fx.ro>
Received: from srv01.advenzia.com (root@localhost)
        by emailaddressmanager.com (8.11.6/8.11.6) with
ESMTP id i2OApwQ14083
        for <support@emailaddressmanager.com>
X-ClientAddr: 193.231.208.29
Received: from corporate.fx.ro (corporate.fx.ro
[193.231.208.29])
        by srv01.advenzia.com (8.11.6/8.11.6) with
ESMTP id i2OApvs14078
        for <support@emailaddressmanager.com>; Wed,
24 Mar 2004 10:51:57 GMT
Received: from mail.fx.ro (mail3.fx.ro [193.231.208.3])
        by corporate.fx.ro (8.12.11/8.12.7) with ESMTP id
i2OAtxBr025924
        for <support@emailaddressmanager.com>; Wed,
24 Mar 2004 12:55:59 +0200
Received: from localhost.localdomain (corporate2.fx.ro
[193.231.208.28])
        by mail.fx.ro (8.12.11/8.12.3) with ESMTP id
i2OAtoQe006624
        for <support@emailaddressmanager.com>; Wed,
24 Mar 2004 12:55:50 +0200
Date: Wed, 24 Mar 2004 12:55:50 +0200
Message-Id: <200403241055.i2OAtoQe006624@mail.fx.ro>
Content-Disposition: inline
Content-Transfer-Encoding: binary
MIME-Version: 1.0
To: support@emailaddressmanager.com
Subject: How to read email headers
From: bogdan@fx.ro
Reply-To: bogdan@fx.ro
Content-Type: text/plain; charset=us-ascii
X-Originating-Ip: [80.97.5.101]
X-Mailer: FX Webmail webmail.fx.ro
X-RAVMilter-Version: 8.4.3(snapshot 20030212) (mail)
Status:
```

Fig. 20.2.

Many tools are available in Internet, both free and licensed, to carry out an analysis of the header section of e-mail messages aimed at pointing out possible manipulations of the message. Some of the most interesting features available in these tools are:

- Hop analysis: analysis of single trace-header values and cross-analysis of the trace-header sequence to identify possible inconsistencies
- Port scanning: verify the real existence of mail servers on single hops and possibly search for other services available (ports in *listen* state) on the sender's MTA
- Open Relay check: verify that a message did not cross mail servers known as open relay or open proxy trough queries on Web databases
- Tracing of the route to the sender's MTA
- Verification of sender's identity
- Abuse reporting (to the sender's ISP or government institution)
- Information gathering: search for sender's e-mail address through Web search engines, newsgroups and forums, also by looking for posts made by the sender

We carried out an evaluation, whose results are summarised in Table 20.1, of some of the most common tools available on Internet. In the rest of this section we briefly present the tools we analysed.

E-MailTrackerPro is a commercial program available in standard and advanced edition. The standard edition can be used to trace received e-mails allowing to find out the geographic location of sender and its Internet Service Provider. This can be useful in spam or *phishing* e-mails that frequently contain malicious viruses, threats and scams that can result in the loss of data and confidential information that can be used for identity theft. The tool allows checking if the e-mail has been misdirected (altered to make it look as if it came from somewhere else) or if it is legitimate or even if it has originated from a high risk part of the world such as China or Russia. Advanced edition includes an online mail checker that offers the preview of all e-mails on the server before delivery. This stops high risk and suspicious e-mails from reaching and infecting the host. The tool also includes an e-mail client that supports white/black lists on sender and recipient mail address, sender's IP address, sender's country, message size and subject.

Table 20.1. Main features of some, widely used, mail messages analysers.

Tool	Check of origin	MUA	Hop analysis	Port scan	Abuse report	Info gathering
E-mail Tracker Pro [4]	No	Yes	1-by-1	Yes	Yes	No
Net.daemon [6]	Yes	No	No	Yes	No	No
Spamx [11]	No	Yes	No	No	Yes	No
Adcomplain [1]	No	No	All	No	Yes	No
WebTracer [13]	Yes	No	1-by-1	Yes	No	Google & Newsgroups
Mail Tracker [5]	No	No	All	No	No	No
Spam Identifier [10]	No	No	Sender only	No	No	No

The e-mail address trace feature helps identifying the original route to the sender's network. The tool creates an HTML report organised in four main sections (general information, route map and hop analysis, sender's network/domain owner details, listening HTTP, HTTPS, FTP and SMTP services on sender's MTA).

The e-mail message trace feature requires copying the e-mail headers in a text area. User gets warned that *spammers* and *phishers* use to modify or add fake information. To prevent this kind of behaviour and find out the real sender's identity the tool performs some — not well-specified — checks. E-mailTracker also allow sending an automatic report to the sender's MTA abuse reporting e-mail address.

Net.Demon is a complete suite of Internet utilities integrated into a single user-friendly interface. Instead of the tabbed dialog box used by most Internet toolkits, Net.Demon uses the Windows MDI (multiple document interface). This means multiple tools can be run within the same session. It also provides a built-in text editor. Main features of the tool follow:

- Finger: get information about an user from a system
- Whois: find information about a server in the Internic database
- Traceroute: trace the route an IP packet takes to reach a remote host
- Ping: ping a host to see if it's alive
- IP Resolve: perform forward and reverse lookup on IP addresses
- Address Scan: scan a range of IP addresses
- Identd Server: provide IDENT services for other systems
- URL Reader: read the raw HTML data for a Web page
- Time Sync: query time from a remote server, or set your clock to it
- Verify E-mail: contact a mail server to see if an e-mail address is valid
- System Info: displays local Winsock info as well as IP address and host name
- Protocols: list protocols known by local Winsock
- Port services: list port services known by local Winsock
- Netstat: get the network status from a remote server
- Sysstat: get the system status from a remote server
- Port watcher: detect attempted connections to local host
- Keepalive: generate Internet traffic to keep ISP from dumping
- Terminal: connect to any TCP port
- Port Scan: scan all ports at a given IP address

Actually, the tool development seems to be abandoned since last version available is quite outdated (April 2001). Attempts to test the features of our interest ended in a failed — error 500 — HTTP query to Spamhouse Block List, probably due to URL redirection.

Spamx is a Java tool available for (virtually) any operating systems. The tool is not much more than a mail client including white/black list rules configuration, a spam checker, an automatic abuse reporting tool and a DNS lookup tool.

We were interested only in its spam checker functionalities to understand what kind of tests the tool performed. We tested the 3.0.2 version and found it to be quite unreliable even for messages that were quite clearly forged (spam messages).

Adcomplain is an interesting Perl script for reporting inappropriate commercial e-mail and Usenet postings, as well as chain letters and "make money fast" messages. It can automatically analyses an e-mail message, compose an abuse report and send the report to the offender's Internet Service Provider. The report is displayed for approval prior to mailing. Adcomplain can be invoked from the command line or automatically from many news and mail readers. Even if the last version is not quite up-to-date (1999), the tool provides a valid header analysis. Some examples of the results we obtained analysing a spam message follows:

- "Received:" header has suspicious text: "may be forged"
- "Received:" header's HELO name significantly differs from *peer* name
- Site in "Message-Id:" does not match "Received:" origin

This kind of details can be included in an automatic spam report to U.S. Federal Trade Commission.

Webtracer is presented as a professional software suite to perform Internet related forensic research. Its purpose is to retrieve as much information on an Internet resource as available, using a wide range of existing protocols and tests. The tool consists of five modules:

- Deep Internet resource analysis
- Bulk analysis
- E-mail header analysis
- Log file analysis
- Web traps

Webtracer uses publicly available Internet protocols and databases to discover relationships between resources. Resources include Internet domain names, Web site addresses (URL's), host names (server names), e-mail addresses and IP addresses. The relationships are searched recursively in a search tree. Each time a resource is clicked in the tree, related resources will be searched using the available protocols and databases. The goal is to come across a resource that leads to the required information (such as the identity of the sender's of an e-mail).

Mail Tracker is available online at http://www.theinquirer.net/email_tracker.htm and should allow to track down e-mail originators; it also tells how to complain about received spam. Here are some interesting results provided by forensic tests performed on trace headers by this tool:

- Data elements are missing from the *from* section
- The reverse lookup name does not match the *from* name
- The *by* name does match the *from* reverse name in the header above
- The elapsed time between headers is more 5 min

Table 20.2. Information about licensing and supported platforms of the analysed tools.

Tool	Licensing	Platform
E-mail Tracker Pro	Licensed per installed copy (15 days free demo)	Microsoft Windows
Net.daemon	Shareware	Microsoft Windows
Spamx	Freeware	Platform Independent
Adcomplain	Open source	Linux, Windows, Os/2
WebTracer	On line version: registered users only	-
	Client version: licensed per installed copy (15 days free demo)	.NET framework
Mail Tracker	On line/free	N/A
Spam Identifier	On line/free	N/A

Spam Identifier is another online tool (www.spamid.net) that traces the sender of an e-mail message and helps in preparing an abuse report (but it does not send it automatically) (Table 20.2).

The use of such tools is a good starting point in a forensic analysis but there are some recommendations to follow.

- Never trust the results of a single tool since they often provide very different answers; moreover some features are offered only by a sub-set of tools.
- Tools are often oriented to spam detection rather than evaluating the reliability of the message through the check of its headers: header forgery could be there even if the message body seems to be perfectly legitimate.
- Always check what kind of analysis the tool carries out on trace headers: if the analysis is made *per single hop*, it is important to verify which hop is analysed and analyse every hop anyway, even if it means to do that manually.
- Never trust automatic abuse reporting as it risks to transform, in turn, a victim in a spammer.
- Make appropriate use of port scanning features.
- Always use the results of past analysis: this helps identifying possible common elements among different cases under study.

As a matter of fact, none of the tools we analysed is fully compliant with all the above recommendations and in particular with the last one since no one provides a repository to store useful information found in the analysis phase. To address this issue, we developed a new tool named *MailMiner*.

20.4. MailMiner

MailMiner is written using the Perl language and makes uses of already existing packages (e.g. Mail::Header, Mail::MboxParser, etc.) available from the CPAN [2] to parse all e-mail messages found in a mailbox that uses the *mbox* format. Messages in an *mbox* mailbox are concatenated and stored as plain text in a single file. The beginning of each message is indicated by a line whose first five characters consist of "From" followed by a space. A blank line is appended to the end of each message. For messages stored in Outlook mailboxes (files with *.pst* extension) it is possible to resort to tools like *readpst* [7] to convert them to the *mbox* format before processing them.

MailMiner allows performing Data Mining on mail messages by

- Archiving useful information extracted from the header section of e-mail messages
- Building a series of indexes of non-compliance with the RFCs

Data extracted from the headers are stored by using S*qlite* [12] which implements a self-contained, server-less, zero-configuration, transactional SQL database engine.

The archiving repository is organised according to an Entity-Relation model built on entities *message* and *header* by using the following basic rules:

- Every message is made of a set of headers and a body.
- Every header has a type and a category.
- Trace headers are structured.

Figure 20.3 shows the organisation of the data repository. *no_compliance_i* is a Boolean value; some possible indexes of no-compliance are header forgery, trace forgery, invalid sender's e-mail address, e-mail spoofing, open relay and so on.

MailMiner populates the database according to a simple algorithm that we describe in pseudo-code:

```
For each header h1,...,hm
    Let vj be the value of hj
    Perform all non-compliance checks
```

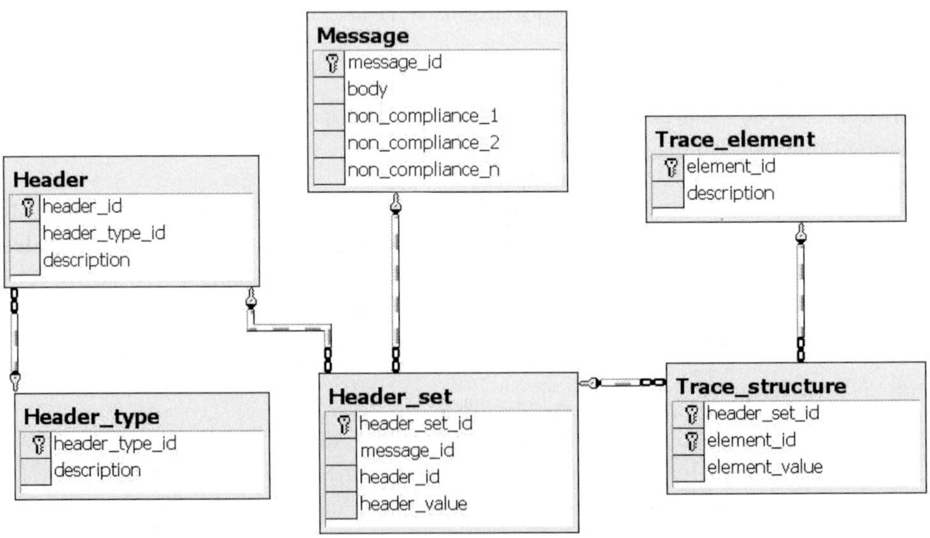

Fig. 20.3.

```
If hj is a trace header
   Archive the values of vj substructure in table
          Trace_Structure
Else
   If hj is not in table Header
       Add a new record to table Header for hj
       Archive vj in Header_set
End
```

Among the compliance tests there are all those made possible by queries to the DNS (for which we use the Net::DNS::Resolver package available from the CPAN).

As an example of the usage of *MailMiner*, we report some results of a test carried out on about 10,000 messages. The number of different headers found was 136 with the distribution reported in Fig. 20.4.

Such a high percentage of User-defined headers is symptomatic of the activity performed on a message by the different entities involved on its delivery. Figure 20.5 reports a distribution of User defined headers found during our analysis, showing that most of them are the result of antivirus and antispam filters; some others are introduced by MUAs for message marking (read, replied, forwarded, etc.).

About 10% of the analysed messages presented a wrong sender's e-mail address (that is an address that could not be used by a genuine MUA). Moreover there were, on average, 4 *Received* headers per message. This means that direct delivery from the sender MTA to the receiver MTA is not very common with the consequence that a message can be manipulated in several places along its route to destination.

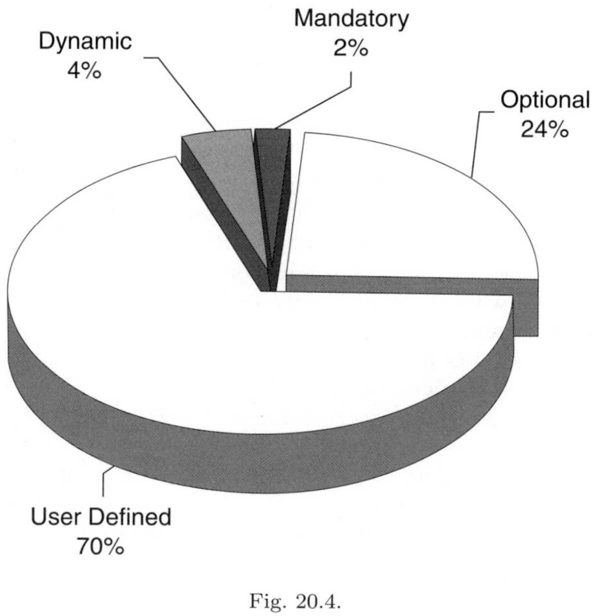

Fig. 20.4.

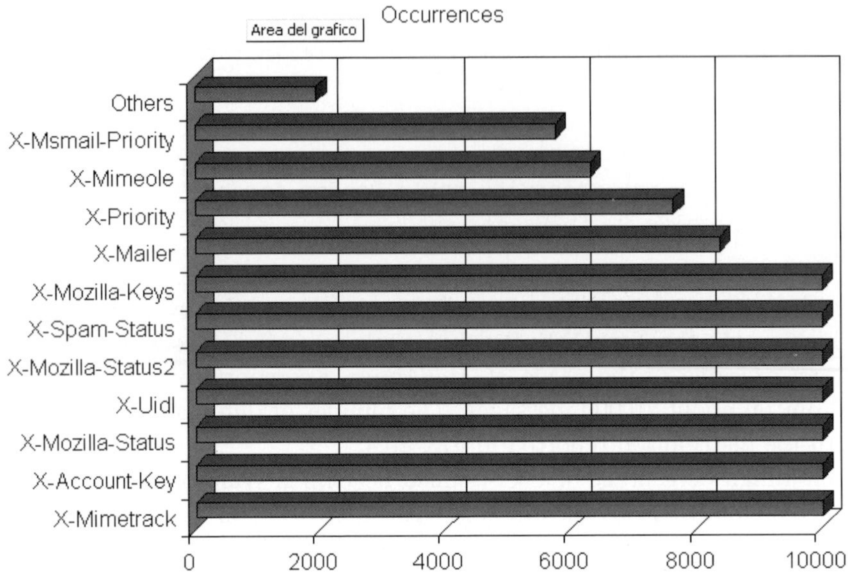

Fig. 20.5.

However, the most interesting result is that only for a very limited number of messages (~1%) all the information reported in the trace-headers passed the consistency tests so that the origin and the route followed by the message could be determined in a fully reliable way.

For the future we expect to develop a version of *MailMiner* that could work as plug-in of mail client/server programs so that the database is updated in real-time during message delivery.

MailMiner is available on request by sending an e-mail to *mailminer@iac.cnr.it.*

References

1. AdComplain: http://www.rdrop.com/users/billmc
2. Comprehensive Perl Archive Network: http://www.cpan.org/
3. David Wood: *Programming Internet Email*, O'Reilly, 1999
4. Email Tracker Pro: http://www.emailtrackerpro.com/
5. Mail Tracker: http://www.theinquirer.net/email_tracker.htm
6. Net.Demon: http://www.netdemon.net/
7. Readpst: http://alioth.debian.org/projects/libpst/
8. RFC 2821: http://www.ietf.org/rfc/rfc2821.txt
9. RFC 2822: http://www.ietf.org/rfc/rfc2822.txt
10. Spam Identifier: http://www.spamid.net
11. Spamx: http://www.spamx.com
12. Sqlite: http://www.sqlite.org/
13. WebTracer: http://www.forensictracer.com/analysis.php?action=mailheader

Chapter 21

DIGITAL EVIDENCE MANIPULATION USING ANTI-FORENSIC TOOLS AND TECHNIQUES

HAMID JAHANKHANI and ELIDON BEQIRI
School of Computing, IT and Engineering
University of East London, UK

Criminals are exploiting now digital communications to commit a wide range of crimes such as identity theft, online piracy, financial fraud, terrorism and pornography distribution. Computer forensics is the discipline that deals with the acquisition, investigation, preservation and presentation of digital evidence in the court of law. Whereas anti-forensics is the terminology used to describe malicious activities deployed to delete, alter or hide digital evidence with the main objective of manipulating, destroying and preventing the creation of evidence. The aim of this chapter is to present three of the current anti-forensic approaches and focusing only on memory-based anti-forensics known as anti-forensic live CDs. Memory-based bootable live CDs are specially built Linux operating systems that boot directly from the CD drive into the Random Access Memory (RAM) area. These packages do not load into the hard drive, do not change files or alter other variables in the target system unless specified by the user. Live CD's are used mainly for penetration testing and other security-related tasks and they include a variety of software packages used for anti-forensic purposes.

21.1. Introduction

The advent of Information Technology (IT) and personal computers has transformed significantly our way of living. Most of our day-to-day activities rely heavily upon the use of electronic devices and digital communications. More people are relying on these technologies to learn, work or entertain. In 2003, USA Census Bureau estimated that 62% of the households had access to a personal computer while 55% had access to the Internet [5]. Without doubt, digital communications can be considered as one of the greatest inventions of the last century because of its impact and benefits on the society.

On the other hand, digital communications have provided new opportunities for criminals and shaped the ways they commit crime [23]. Criminals are exploiting now digital communications to commit a wide range of crimes such as identity theft, online piracy, financial fraud, terrorism and pornography distribution. Furthermore, the incidences of some of types of crimes increased significantly with the introduction of digital communications and personal computers. For example, Internet communications have escalated the problem of child pornography by increasing the amount of material available, the efficiency of its distribution and the ease of its accessibility [29].

According to Schneier, a well-known security expert, electronic crime is flourishing because of three main reasons: a) automation, b) action at distance and c) technique propagation [21].

(a) Automation: Software packages are used to perform repetitive tasks and cross-reference more and more data.
(b) Action at distance: We live in a global digital communication era. Criminals perform electronic crimes in distance and with a high rate of anonymity and
(c) Technique Propagation: Successful electronic-crime techniques and malicious software is propagated easily through the Internet.

Law enforcement agencies have started dealing with crimes involving electronic devices and communications since the 1970s when these technologies were introduced. These were coined as electronic crimes since electronic devices and digital communications were used to commit them; while electronic evidence was defined as information or data of investigative value that is stored or transmitted by electronic devices [1].

The most important input of a computer forensic investigation is the digital evidence. Digital evidence can be envisaged as the counterpart of fingerprints or DNA in the digital world. Therefore, criminals will attempt to cover the traces of their malicious work by using amongst all anti-forensic methods to manipulate and tamper the evidence or interfere directly with the process [11].

Anti-forensics is the terminology used to define the activities of hackers or other cyber criminals aiming to undermine or mislead a computer forensic investigation. There are no well-established definitions regarding this discipline since it is quite new and it is yet to be explored [12]. Peron and Legary define it as "*...four categories of evidence destruction, evidence source elimination, evidence hiding and evidence counterfeiting...*", [11], while, Grugq [20], defines anti-forensics as "*the attempt to limit the quantity and quality of forensic evidence*".

Anti-forensics is not yet established as a discipline; however, the variety of related techniques and available tools online is an indicator that this might be the case in the near future. Furthermore, anti-forensic tools and techniques are developed not merely by hackers but also by well-known security experts who are fascinated by the subject. Grugq seems to be one of the most dedicated anti-forensic researchers so far. With more than five years of anti-forensic studies, he ended up losing his job after publishing "*Art of Defiling: Anti-Forensics*" [20].

The Metasploit Anti-Forensic project by Liu is part of the Metasploit project which targets audiences interested in penetration testing. Liu's presentation titled "*Bleeding-Edge Anti-Forensics*" which was co-presented with Brown for an Infosec World Conference was the most descriptive work of what he did so far about anti-forensics.

There are number of techniques that are used to apply anti-forensics. These techniques are not necessarily designed with anti-forensics dimension in mind. For instance, folder shielders have been designed in order to primarily provide a level of security and privacy, but they can be used as an anti-forensic tool since they can hide data. The others are as follows:

- **Digital media wiping:** A proper wiping of the media that contains the digital evidence, will simply disappear the evidence.
- **Steganography:** Someone can use Steganography to hide a file inside another and make the investigator unable to take advantage of the evidence.
- **Privacy wipers:** These are tools aim to delete any privacy traces from operating systems (OSs), applications or both. If properly used, the investigator might find no evidence at all inside the digital media.
- **Rootkits:** Rootkits can subvert the OS kernel and even react to forensic acquisition processes by hijacking the way the OS uses areas like process management or memory management to extract the evidence.
- **S.M.A.R.T anti-forensics:** This kind of technology can be used by an attacker to suspect if a hard drive has been taken out for a forensic duplication process.
- **Homographic attacks:** Such an attack can mislead an investigator since some letters that look similar to the human eye can be replaced with others in such a way to make a malicious file look legitimate.
- **File signature modification attacks:** Someone can purposefully change the file signature of a file to make it look something else.
- **Encryption:** This can be used almost in every anti-forensic stage in order to obscure and make unreadable and unusable the evidence.
- **Metadata anti-forensics:** Information about data (metadata) can be altered in order to hide user actions.
- **Slack space anti-forensics:** Someone can hide malicious software in areas that OS might not use, like slack space, because they might be considered as reserved or empty.
- **Secure digest functions (MD4, MD5, etc.) collision generation:** Someone can alter a file and then use anti-forensic software to make this file having the same MD4 or MD5 value like before the alteration, thus bypassing a forensic integrity check.
- **Digital memory anti-forensics:** There are programs that are able to hide processes or other evidence from memory.
- **Misleading evidence:** Someone can leave evidence in such a way to mislead the forensic investigation.

- **Packers/Binders:** Someone can use such a program in order to transform a file by changing its structure, thus it can bypass security mechanisms that searches for malicious behaviour patterns inside files.
- **Forensic tools vulnerabilities/exploits:** There are already implementations available to show that some of the computer current forensic tools can be either bypassed or exploited.
- **Resource waste:** To purposefully leave traces in a big network in order to make the forensic investigator waste valuable resources and time.
- **Forensic detection:** Someone can install a mechanism to be triggered after any computer forensic-related presence.
- **Anonymous actions:** It includes every action that can be done by a fake or an unknown identity. The result from the investigator is to fail to trace back the malicious activities.
- **Anti-forensics in flushable devices:** Someone can take advantage of devices that can be flashed (like PCI cards or BIOS) and install malicious code inside them, thus they can remain unnoticed.

21.2. Memory-based Bootable Environments

Memory-based anti-forensics techniques rely heavily on the use of open-source security live CDs. There is a plethora of "live CDs" — commercial and freeware — available in the markets that are tailored to meet particular user's needs such as : data recovery (SystemRescue CD, ERD Commander), security (BackTrack, NST), PC benchmarking (StressLinux, Ultimate CD), gaming (LLGP, Freeduc-games) or even alternatives for fully functional OSs (Knoppix, Kanotix). Frozen Tech (www.frozentech.com) provides an almost complete list of available open source live CDs that are freely distributed online [4].

A live CD is nothing more than a compact disc, DVD or USB drive, which contain an OS image file and a boot loader program, used to start or boot a computer system. An image file is a single compressed file that contains the entire OS programs and files. Bootable CDs, also known as LiveDistros, are mostly available freely open-source license agreement. According to this agreement, "anyone can modify and redistribute the original OS without asking for permission of retribution from the author" [18].

The concept behind using removable media for storing OSs is not new. In the early introduction of personal computers, OSs (such as MS-DOS) were loaded into the memory from removable media (usually floppy discs). With the advent of mainframes (considered as the first generation of computer systems), the instructions to hardware components were given by punched cards, which although did not constitute an OS in per se did introduce the concept of OS. It is worth mentioning though that punched cards were not effective as live CDs since extensive processes required hundreds of them [9].

Mainframes were not the only computer systems that used removable media for storing instruction programs or OSs. For example, discless computer systems do not have OSs installed; instead, they load from a copy of the OS located in a network server. Either OSs such as MINIX are distributed mainly in removable media (CD, floppy etc.) because of its extremely small size; MINIX kernel counts only 4,000 program lines whereas other OSs rely on millions of lines of code [15].

Although live CDs are the preferred tools of trade in conducting memory-based anti-forensics, most of them were designed for security testing purposes. Good collections of security testing tools are distributed with these portable media and are usually used by computer security professionals to troubleshoot their computer systems and networks. Unfortunately, even malicious users are taking advantage of these specially built packages to perform illegal activities, amongst all anti-forensics.

Live CDs manage to recognise and work with a variety of hardware components thanks to a device manager named "udev" [19], which is the device manger for most of UNIX/LINUX-based systems. Having a device manager that interacts with most of hardware devices promotes the inter-operability and portability of live distros.

Network services are fully accessible using a livedistro. Connections to Internet or local computer systems are easily implemented permitting the user to perform most of the tasks available in OSs that run from the hard disc.

Memo-based anti-forensics techniques are difficult to beat for a variety of reasons. The most important advantages of using these techniques are:

(1) Lack of digital evidence;
(2) Compatibility, flexibility and portability;
(3) Anonymity;
(4) Availability of tools and
(5) Freely distributed.

SecurityDistro.com, a site dedicated to memory-based security tools, lists over 40 memory-based packages [22]. Most of the packages offer similar tools and interfaces. Among these packages are: Backtrack, Anonym.OS, Helix, Penguin Sleuth and Auditor collection, which have a wide range of security tools that might be used to deploy anti-forensic activities.

Backtrack live CD currently provides an excellent sophisticated collection of security software that can be used to perform anti-forensics. Amongst all, Backtrack offers the user the opportunity to use a well-established security-focused framework named "metasploit" [14]. Metasploit framework is a collection of security tools used to test the security side of computer systems, penetration testing and exploitation. In addition, metasploit contains a special module called anti-forensics, which is a collection of anti-forensic tools (example: timestomp, sam juicer, slacker, transmogrify etc.) that can be loaded and used directly from the live CD aiming at manipulating digital evidence in a local or remote target system.

According to Liu, a well-known anti-forensic researcher, these tools are designed to tamper with or break well-recognised industry tools such as Encase, NTFS and PGP desktop, with the final objective of manipulating the digital evidence and compromising the investigation findings [13].

Once Backtrack is downloaded, the Linux-based OSs can be burned in a bootable CD and be ready for use. Instructions on how to accomplish this task are available on various sites online. By using a memory-based live CD such as Backtrack, a malicious user can easily manipulate these digital evidence attributes: file extensions and signatures, timestamps and hiding data in slack space. Windows Security Accounts Manager (SAM) hashes can be copied and dumped in RAM without leaving a single trace in the window's accounting logs, therefore, manipulating the forensic investigation process.

21.2.1. *Modifying File Extensions and Signatures*

Computer system files are identified by two attributes: file extensions and file signatures. "A filename extension is a suffix to the name of a computer file applied to show its format" while a file signature (known also as the magic number) is a set of characters stored at the beginning of the file. For each file format, there is a unique file signature, for example, executable files in Windows are identified by file signatures starting with the letters MZ etc. [13]. Therefore, to hide a file in a computer system suffices to change its extension and add the letters MZ at the beginning of that file. By using this technique, files containing pornographic material, for example, can be masqueraded as system files and go undetected by computer forensic tools.

Memory-based anti-forensic tools such as Backtrack can be used effectively to manipulate file extensions and signatures. Metasploit anti-forensics package includes a tool named "Transmogrify" that allows a user to masquerade malicious files by altering file signatures or extensions. Transmogrify is able to alter file extensions and headers without being detected by forensic tools like Encase [14]. Computer forensic tools usually compare file extensions to their related headers (magic numbers) to determine if files were altered or tampered with. Each file extension (such as:exe, bitmap, doc etc.) has got a unique magic number (hexadecimal), which is stored at the very beginning of each file. If a forensic tool finds a mistmatch between the file extension and the magic number in the header of the file, a red flag is raised and the file is marked as tampered. Transmogrify manages to alter the file extension and the magic number so that a perfect match is created. The file altered will not be detected by forensic software since both the extension and the header looks legitimate. Both Encase and Forensic Toolkit (well-established computer forensic software packages) fail to detect this type of anti-forensic actvity performed by using Transmogrify.

21.2.2. *Modifying Timestamps*

From a computer forensic investigation point of view, file timestamps are very important because they provide the necessary evidence to prove if certain anti-forensic activities occurred at a certain moment in time or whether a user was logged in a computer system. For this reason, malicious users might attempt to modify timestamps in order to eliminate compromising evidence. Timestamp is the data appended to a file that shows when a file is created, accessed, modified or entry modified. These file attributes are also known as MACE (Modified–Accessed–Created–Entry Modified) attributes. Anti-forensic tools attempt to modify these data parameters in order to mislead computer forensic investigators.

Backtrack again provides the perfect tool to modify timestamps. The tool is called "Timestomp" and is included in the metasploit framework. Timestomp is a program developed by metasploit project, which gives the user the opportunity to modify all New File Technology System (NTFS) timestamp parameters [14]. The NTFS is the proprietary file system of modern Windows OSs including NT, 2000, 2003, XP and Vista [16]. Windows MACE attributes can be really important from a computer forensic point of view, since they can be crucial in determining who accessed or created certain files, their access times etc. With Timestomp, all four file attributes (MACE) can be altered permanently; forensic tools (Encase, FTK etc.) will consider these values as legitimate timestamps.

Timestomp uses only these Windows system calls: NtQueryFile () and NtSetInformationFile (); the Setfiletime () call is not used to modify timestamps making difficult to detect the alteration. Setfiletime () call is a well-documented Windows API call used to modify file timestamps easily detectable by forensic tools. However, by using the unpublished Windows system calls (NtSetInformationFile () and Setfiletime ()), all four MACE attributes can be tampered with safely [27]. Timestomp not only alters timestamps but also can set these values to zero confusing forensic Windows explorer or forensic tools like Encase.

Timestomp can be used also as a stand-alone program to modify timestamps; however, its potential is fully explored when used within the metasploit framework. Meterpreter is a program (started from Backtrack's metasploit 3 framework) that permits a user to connect remotely to a target computer without directly accessing the hard disc or leaving traces in the registry. This is achieved through an advanced connection technique called direct memory injection [13].

From within this module, Timestomp can be executed to modify file timestamps. Since all the operations are conducted in temporary memory (RAM), no digital evidence is left in the systems to indicate traces of anti-forensic activity. Certainly, it will be almost impossible for a computer forensic investigator to notice timestamp modification since its parameters will look legitimate.

Timestomp provides specifically an option (switch-b) tailored to confuse Encase. The MACE values are modified in such a way that Encase will only be able to display blank values or consider them legitimate. Timestomp MACE modification will make all file timestamps useless and compromise computer forensic cases since the digital evidence will be considered unreliable.

21.2.3. *Hiding Data in Slack Space*

Slack space is the preferred hard disc area used by malicious user for storing illegal software, documents or pictures because files stored in it are not seen or accessed by Windows explorer; "data is hidden in unallocated or unreachable locations that are ignored by the current generation of forensic tools" [6]. The user is completely unaware of the existence of such files.

A variety of malicious programs might be used to hide data in the slack space; however, "Slacker" is one the most proficient tools used to perform such activities. Slacker, which is named after the slack space, is developed by the metasploit team and is released as a module within the anti-forensic package [14]. Slacker uses a sophisticated technique to hide programs, files or any other type of data in the slack space. It takes the data, fragments it into thousands of pieces and then distributes it across the slack space in the hard disc. This program mainly stores the data in stable file such as system files (Windows/system32 files), which are not examined by computer forensic tools. Slacker's main features include file splitting and slack space hiding; these features make slacker very hard to trace.

If a computer forensic tool is used to analyse the data in the slack space, no evidence will be discovered since individual fragments of data will not help to construct the true nature of the hidden file; for the forensic tool data is so diffuse that it looks like random noise [3]. Only slacker can recompose the fragmented pieces of data to create the original file. Slacker have proven to be successful also against PGP desktop, a security tool that includes some tools claming to wipe out completely the slack space. Metasploit researcher Liu has proven that data written in the slack space with slacker cannot be wiped out even when PGP desktop tool is used [14].

21.2.4. *Dumping SAM Hashes without Leaving Traces*

Backtrack can be used by malicious users to steal SAM hashes from local or remote target systems. Windows OSs store password information locally in a system file called SAM; otherwise known as SAM file.

This file is very important from the security point of view since it contains all system-user passwords in an encrypted format. Encryption of user passwords is performed by Windows using a proprietary encryption utility called system key which uses "strong encryption techniques to secure account password information that is stored in the SAM file". In a computer system running Windows OS, the system key utility (program) is located at this logical address: C:/Windows/system32/config/system. The system key program also contains the key used to encrypt the passwords stored in SAM.

Usually, access to SAM is restricted since it is a system file. Even if the user manages to copy SAM in a portable media device, it will be difficult to unmask the hidden passwords since a key stored in the system utility is needed to decrypt SAM. The key must be extracted first from the system key utility. However, with Backtrack, a user can extract quite easily user passwords hashes stored in SAM, without leaving digital evidence since all the operations are performed in RAM. On the other hand if a user attempts to access or copy SAM without Backtrack from within, windows OS, data will be added to the system log file.

Backtrack is also used to recover SAM password hashes remotely without leaving digital evidence in the target system. An exploit named "lsass_ms041_011" is used to connect remotely to vulnerable computer systems. Once a connection is established remotely, "meterpreter" is used to fully explore the target. Thereafter, another special metasploit anti-forensic tool named "Sam juicer" can be used to copy password hashed from remote SAM. Sam juicer performs the task without accessing SAM file, the registry or writing any files in the remote computer system hard disc since it uses direct memory injection to perform the task [13].

21.3. Discussion and Evaluation of the Memory-based Anti-forensic Tools and Techniques

Memory-based anti-forensic tools and techniques interfere substantially with the investigation process by altering or hiding digital evidence. Because memory-based anti-forensics techniques are0000 deployed directly in temporary memory (RAM), the defence strategies must be focused at this memory area. Slack space must be scrutinised and analysed as well. These defensive strategies must be implemented to mitigate to a certain extent memory-based anti-forensic activities:

a) Improving signature analysis: Memory-based anti-forensic tools manage to modify file extensions and signatures. In the meantime, some of the most-used computer forensic tools (Encase, Forensic Tool Kit (FTK) etc.) fail to detect such changes. Encase checks only the first two characters of a file signature, which can be easily modified (e.g. MZ for executable files). Therefore, automated tools will not be able to detect file signature modifications achieved by using memory-based anti-forensic tools. In this case, manual investigation should be conducted provided that the investigator identifies the suspicious files. Forensic tools need to be redesigned to tackle file signature modification. A good way forward would be the redesign of the searching process so that files are checked from top to bottom for patterns of data. This method might produce good result since particular patterns of data might be associated to certain files; if these patterns are not present, then further investigation can be conducted.

b) Capturing data in RAM: Memory-based live CDs load, operate and store data in memory, unless specified otherwise by the malicious user. For this reason, the main area to look for digital evidence is the memory of the local or remote computer system used for anti-forensic activities. Running processes, ports, uploaded or downloaded files might indicate the occurrence of anti-forensic activities.

Memory-based anti-forensics relies mainly on volatile memory, while traditional anti-forensics is deployed in the secondary memory storage area (hard disc). For this reason, memory-based live CD activities are hard to detect since volatile memory is unstable and easy to erase. Computer forensic investigators might be able to collect digital evidence only if the perpetrator's computer device is seized and is not shut down; however, if the user has removed the live CD and turned off the system, the evidence is lost permanently since memory is volatile. Forensic tools such as Memparser [26], or Windows Memory Forensic Toolkit [28], can be used to collect valuable data in memory provided that the system is still running.

c) Improving timestamp analysis: Windows systems record data about individual files in a file named Master File Table (MFT). In the MFT table, NTFS stores Standard Information Attributes (SIA) and File Name (FN) attributes for each file. The SIA records standards timestamp information (MACE attributes) while FN attributes stores information regarding the name of the file. Interestingly enough, FN attribute values are updated only when the file is created or moved while SIA attributes can be changed even later on. Currently, forensic tools like Encase check file timestamps only through the SIA values on MFT. The FN attributes are not checked or compared against SIA values in order to certify its authenticity. For example, if FN attribute "created" is smaller than SIA attribute "created", the file timestamp is tampered with. Forensic tools must compare both SIA and FN attributes to mitigate this anti-forensic technique.

d) Statistical analysis of slack space: Memory-based anti-forensic tools like slacker use intelligent space selection techniques to distribute file fragments. It will be difficult to detect this activity since the new fragmented files are stored amongst older data that was residing previously in the slack space. Statistical analysis of the slack space probably is the best counter technique to be used against slack space hiding. If fragmented files are identified probably, the slack space is used to store data probably using slacker or similar software packages.

e) Improving computer forensic tools: Some memory-based anti-forensic tools (e.g. Timestomp, Transmogrify) tackle computer forensic packages. Timestomp provides a specific option to trick Encase; by using switch-b timestamps are set to blank. Encase and FTK, two prestigious computer forensic tools do not recognise Timestomp changes [13]. This is a clear indication that these tools must be improved (or re-written) in order to properly detect timestamp alterations, particularly because modified timestamps can compromise the success of a computer forensic case in the court of law. On the other hand, Encase and FTK do not detect file extension or signature modification achieved with Transmogrify; even in this case, these tools must be improved in order to detect traces of anti-forensic activity. Computer forensic tool designers should examine carefully how memory-based live CDs interact with the system in order to improve their future released tools; the idea behind this is to use the same anti-forensic tools to defeat them.

21.4. Acting Anonymously

From a forensic scope, anonymity can be considered as a major anti-forensic approach. Below are some of the tools that are used and they are discussed in detail.

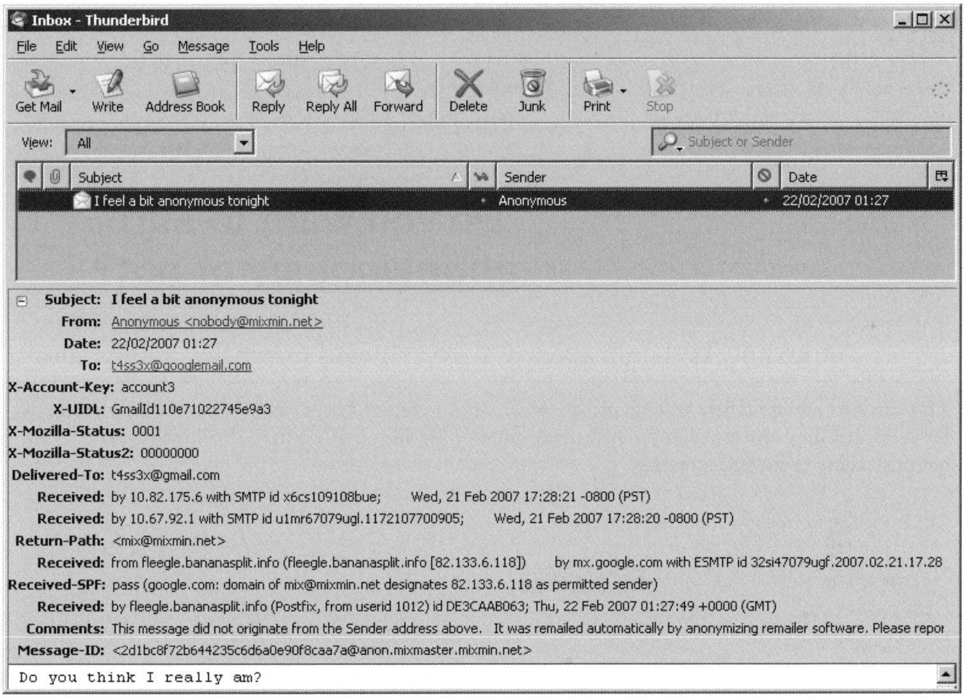

Fig. 21.1. Anonymous mail details.

21.4.1. *Anonymous Mail Accounts*

These are accounts that are created using services available on the Internet that facilitate anonymous mailing. This will make the process of e-mail tracking more difficult as the mail headers are altered and no IP address details will be available [10].

Figure 21.1 shows that the sender is an anonymous authority — in this case "bananasplit.info". There is also an e-mail address (abuse@bananasplit.info not visible in the picture) where more information about the sender's IP could be requested by the forensics team.

21.4.2. *Anonymous Proxies*

Nowadays, there are plenty of anonymous proxies on the Internet with a significant number of them being free [7, 2]. Although these proxies promise Internet anonymity, they do not always talk about the level of anonymity service they provide.

Below is the result of a test on a high anonymity service in order to find out the amount of information on the user's identity. The anonymity has been checked against a website (anonymitytest.com) that shows the IP address of the visiting address along with a service ("whois") that aims to trace back the IP.

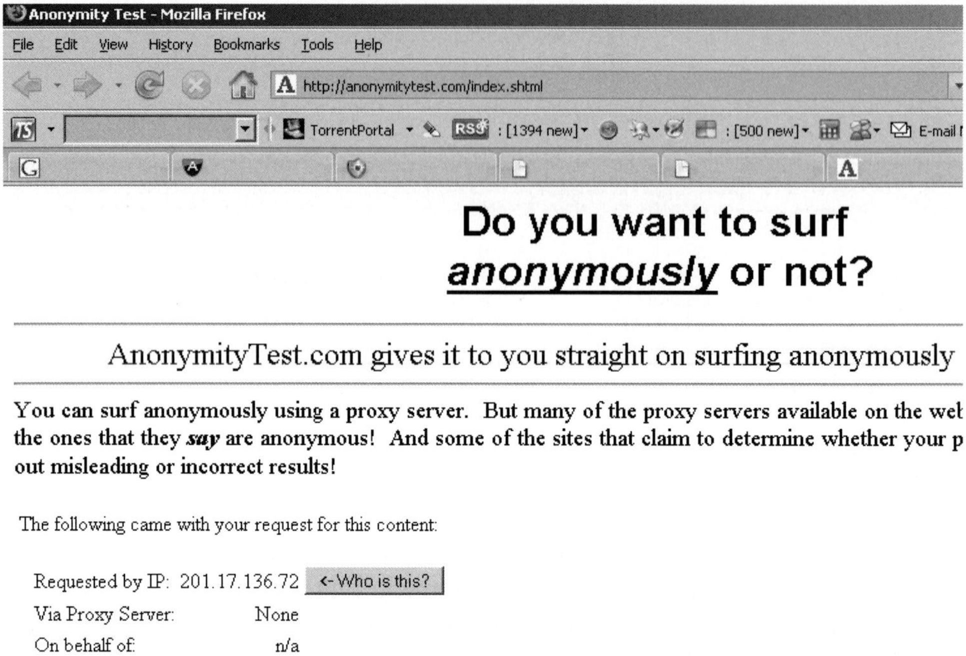

Fig. 21.2. Anonymity service.

It is important to note that in special cases even a high anonymity server can reveal all the information regarding its users. All someone has to do is to monitor and analyse the traffic patterns coming to and from that proxy [8].

Figure 21.2 shows that high anonymity proxy puts its own IP number to the visited web page in order to keep the client anonymous.

An attempt to trace back the address will come up with the details of the proxy server and not the user's (Fig. 21.3).

In order to get more information about the real visitor's identity, the anonymous proxy provider has to be contacted. Here are the problems someone might face:

(1) There are cross-boarder legal issues. In this case, the domain ends with "br" which means that the proxy owner is located somewhere in Brazil and
(2) The anonymous proxy provider — no matter the geographical location — might claim that all the logs are deleted and nothing is saved regarding their clients (anonymous visitors). In that case, only a government regulation which enforces IP logging would provide a connection to the client's IPs.

It is important to note that someone can be a part of an anonymous network — like Tor — in order to achieve anonymity. In this case, it is not even feasible for the governments to totally follow an IP or a packet, since the information is going

```
nic-hdl-br:   DSS30
person:       Diego Santos Soares
e-mail:       virtua@virtua.com.br
created:      20000424
changed:      20070128

nic-hdl-br:   RII19
person:       Ricardo Ide
e-mail:       ricardoide@globocabo.com.br
created:      20011010
changed:      20011010

remarks:      Security issues should also be addressed to
remarks:      cert@cert.br, http://www.cert.br/
remarks:      Mail abuse issues should also be addressed to
remarks:      mail-abuse@cert.br

% whois.registro.br accepts only direct match queries.
% Types of queries are: domains (.BR), BR POCs, CIDR blocks,
% IP and AS numbers.
```

Fig. 21.3. Proxy server details.

through a variety of interconnected nodes with some encrypted links through several countries. This is a more efficient way to keep the users anonymous.

The only way for someone to monitor an amount of Tor's traffic is to set up a fake Tor server and monitor the traffic of some other servers as well. In a crackdown of a recent crime investigation in Germany, police seized 10 Tor servers for suspicion of a child-porn investigation [17].

21.5. Wireless Anti-forensics Methods

How about if someone launches an attack using multiple access points from the roof top of a high building in the middle of a crowded city with the help of a strong directional antenna?

Siles [24, 25], in his excellent two-part article "Wireless Forensics: Tapping the Air" unveils some "*de facto*" and some new wireless anti-forensics methods. Some of the major approaches are [24, 25];

- The use of illegal channels, like channel 14 in the United States and Europe.
- The use of strong layer-2 encryption.
- The modification of the 802.11 specification (Raw Covert, MadWifi patches) and
- Wireless MAC spoofing.

While in theory, the forensics investigator should monitor every single packet of every channel available around the suspect, in reality the post-incident response could end up quite dramatically. This could be due to: ignorance regarding the channels and access points used, legal barriers between the access point and the forensics acquisition, non-cooperative ISPs etc. The forensic process should be enhanced with security mechanisms which would upgrade the post-incident reaction to real time. The real-time acquisition tools should have capabilities of capturing activity of all the wireless point within a respectable distance.

21.6. Conclusion

Anti-forensics is a reality that comes with every serious crime and involves tactics for "safe hacking" and keeps the crime sophistication in a high level. Computer forensic investigators along with the forensic software developers should start paying more attention to anti-forensics tools and approaches.

If we consider the computer forensics as the actions of collection, preservation, identification and presentation of evidence, anti-forensics can affect the first three stages. Because these stages can be characterised as "finish to start" between them from a project management point of view, the failure of one of them could end up as a failure of the lot. Thus, there is a high impact of anti-forensics to the forensics investigations.

Officially, there is no such thing as anti-forensic investigations because the anti-forensic countermeasures are still part of the investigator's skills.

References

1. J. Ashroft, Electronic crime scene investigation: a guide for first responders, 2001, Available at: http://www.iwar.org.uk/ecoespionage/resources/cybercrime/ecrime-scene-investigation.pdf [accessed 22 June 2007].
2. Anonymous INET, Fast proxy server list, 2007, http://www.anonymousinet.com/, cited on 28 February 2007.
3. S. Berinato, The rise of anti forensics, 2007, Available at: http://www.whitehatsec.com/home/resources/trade/07tradenews/062607CSO.html [accessed 16 June 2007].
4. N. Brand, Frozen tech, The LiveCD List, 2006, Available at: http://www.livecdlist.com/ [accessed 11 August 2007].
5. Census Bureau, Computer use in 2003, 2003, http://www.census.gov/population/pop-profile/dynamic/Computers.pdf [accessed 21 June 2007].
6. Forensicwiki.org, Anti-forensic techniques, 2007, Available at: http://www.forensicswiki.org/wiki/Anti-forensic_techniques [accessed 5 August 2007].
7. Free Proxy, www.FreeProxy.ru, Free PROXY servers, 2007, http://www.freeproxy.ru/en/free_proxy/, cited on 16 February 2007.
8. S. Gibson, Gibson research corporation, Security now — transcript of Episode 70 — Achieving Internet Anonymity, December 14, 2006, http://www.grc.com/sn/SN-070.pdf, cited on 3 February 2007.
9. L. Gichioco, Computer technology: From punch cards to clustered supercomputers, 2004, Available at: http://tle.geoscienceworld.org [accessed 28 July 2007].
10. T. C. Greene, The register, Net anonymity service backdoored, August 21, 2003, http://www.theregister.co.uk/2003/08/21/net_anonymity_service_backdoored/.
11. R. Harris, Arriving at an anti-forensics consensus: Examining how to define and control the anti-forensics problem, 2006, Available at: http://www.dfrws.org/2006/proceedings/6-Harris.pdf [accessed 12 April 2007].
12. H. Jahankhani, B. Anastasios and K. Revett, ECIWS, digital anti forensics: tools and approaches, 2007, Available at: http://academic-conferences.org/pdfs/eciw07-booklet.pdf [accessed 21 June 2007].

13. V. Liu and P. Stach, Defeating forensic analysis, CEIC 2006 — Technical Lecture1, 2006, Available at: http://stachliu.com/files/CEIC2006-Defeating_Forensic_Analysis.pdf [accessed 18 August 2007].
14. Metasploit.com, 2007, Available at: http://www.metasploit.com [accessed 12 June- 29 July 2007].
15. Minix, 2007, Available at: http://www.minix3.org/ [accessed 4 June 2007].
16. NTFS, 2007, Available at: http://www.ntfs.com [accessed 15 August 2007].
17. J. Oates, The register, German police seize TOR servers, September 11, 2006, http://www.theregister.co.uk/2006/09/11/anon_servers_seized/, cited on 27 January 2007.
18. Opensource.org, 2007, Available at: http://www.opensource.org/ [accessed 21 July 2007].
19. Qlogic.com, Persistent naming using udev in Linux environment, 2007, Available at: http://www.qlogic.com/documents/datasheets/knowledge_data/whitepapers/SN0130979-00.pdf [accessed 3 August 2007].
20. Ruxcon, The art of defiling, 2004, Grugq, www.ruxcon.org.au/files/2004/13-grugq.ppt, cited on 16 February 2007.
21. B. Schneier, Secrets and lies, in *Digital Security in a Networked World* (John Wiley and Sons Inc., USA, 2000).
22. Securitydistro.com, Security distros, 2007, Available at: http://www.securitydistro.com/index.php?option=com_weblinks&catid=11&Itemid=4 [accessed 15 August 2007].
23. D. Shinder, Scene of the cybercrime, in *Computer Forensics Handbook* (Syngress Publishing, USA, 2002).
24. R. Siles, Security focus, Sebek 3: Tracking the attackers, part one, January 16, 2006, http://www.securityfocus.com/infocus/1855, cited on 12 February 2007.
25. R. Siles, Security focus, Wireless forensics: Tapping the air — Part Two, January 08, 2007, http://www.securityfocus.com/infocus/1885/2, Cited on 16 February 2007.
26. SourceForge, Memparser, 2007, Available at: http://sourceforge.net/projects/memparser [accessed 11 August 2007].
27. E. Van Buskirk and V. Liu, Digital evidence: challenging the presumption of reliability, 2006, Available at: http://risk-averse.com/index_files/JDFP.pdf [accessed 11 August 2007].
28. WFTK (Windows Memory Forensic Toolkit), Digital investigations, 2007, Available at: http://forensic.seccure.net/ [accessed 12 August 2007].
29. R. Wortley and S. Smallbone, Child pornography on the internet, 2004, Available at: http://www.cops.usdoj.gov/mime/open.pdf? Item=1729 [accessed 21 June 2007].

Chapter 22

HIDDEN DATA AND STEGANOGRAPHY

DAVID LILBURN WATSON
Watson Business Solutions Ltd., UK

Research into data hiding has grown dramatically. A large number of techniques have been developed to hide data in recent years for a variety of reasons ranging for secret communications to digital watermarking to discourage copyright theft. Data hiding is not new, and this chapter looks at the history of data hiding from earliest times, through codes and ciphers into the physical hiding of electronic data and the use of modern computing techniques to assist in data hiding.

22.1. A Potted History of Data Hiding

One of the first recorded instances of hiding data was when Histiaeus did not like living in Susa, and made plans to restore his power in Miletus by instigating a revolt in Ionia.

In 499 BC, he shaved the head of his most trusted slave, tattooed a message on his head and then waited for his hair to grow back. The slave was then sent to Aristagoras, who was instructed to shave the slave's head again and read the message, which told him to revolt against the Persians. Aristagoras, who was disliked by his own subjects after an expedition to Naxos ended in failure, followed Histiaeus' command, and with help from the Athenians and Eretrians, attacked and burned Sardis. According to Herodotus, this was the entire cause of the revolt, although this is very unlikely [2].

At around the same time, Herodotus reports how Demeratus needed to advise Sparta that Xerxes was about to invade Greece. So, he sent a message on the table underneath the wax of the wax tablet so that the wax tablet appeared blank [2].

A large number of techniques for early data hiding were invented or reported by Aineias the Tactician [9], including letters hidden in messengers' soles or women's earrings, text written on wood tablets and then whitewashed and notes carried by pigeons. He also proposed hiding data by changing the heights of letter strokes or by making very small holes above or below letters in the text.

The early Chinese used to write messages on thin strips of silk and encase the silk in balls of wax, called "*la wan*" and then hide the balls of wax in, or on the person of the messenger [4].

In the Yuan Dynasty, when the Chinese were ruled by the Mongolian Empire, they wanted to overthrow the Mongolians and return to self-rule. The rebel leaders decided to use the upcoming Moon Festival for the uprising. At the Moon festival, it was traditional to eat Moon cakes and the rebels baked their attack plans in the Moon cakes. The cakes and the plans, were freely distributed to the rebels who successfully overthrew the Mongolians. Moon cakes are still eaten in celebration of this event [4].

The ancient Romans used to write between the lines using invisible inks based on readily available substances such as fruit juices, urine and milk. When heated, invisible inks would darken and become legible. Ovid suggests using milk to write invisible [6].

An early researcher into steganography and cryptography was Johannes Trithemius (1462–1516), who was a German Monk. His "Steganographia" described systems of magic and prophecy but also contained a complex system of cryptography. It contained details of hiding messages in seemingly innocuous text, for example, every other letter in every other word:

Padiel aporsy mesarpon omeuas peludyn malpreaxo

which reveals "prymus apex" [4]

Another was his invention of the "Ave Maria" cipher. The book contained a number of tables, each with a number of words, one per letter. To code a message, the message letters are replaced by the corresponding words. If the tables are used in order, one letter per table, then the coded message appears to be an innocent prayer.

In 1550, Girolamo Cardano (1501–1576), proposed a simple grid for hiding messages in a seemingly innocuous letter. A piece of paper has a number of holes cut in it and when it is placed over the seemingly innocuous letter, the real message appears in the holes in the piece of paper. This method is still used today for some multiple choice exam-making processes when they have to be marked manually.

Francis Bacon (1561–1626) has long been rumoured to have be the author of a number of Shakespeare's plays. Proponents of this idea point to the fact that there are a number of hidden texts — steganographies — hidden in the text. These contain the name "Bacon" [5].

Steganographia was only published posthumously in private circulation as Trithemius feared the reaction of the authorities. It was finally published in 1606.

Giovanni Porta described how to conceal a message inside a hard boiled egg by making an ink from a pint of vinegar and an ounce of alum. The writer would write a message on the shell, which being porous, would allow the ink to permeate the shell and leave the message on the hardened egg white. The message could only be read when the egg shell was cracked.

Porta was also famous for categorising cryptography into three types:

(1) Transposition;
(2) Substitution by symbol and
(3) Substitution by letter [7].

In 1641, John Wilkins (later, the Bishop of Chester), produced a treatise [10] on cryptography that was used in the English Civil War for hiding messages. He proposed a number of schemes ranging from coding messages in music and knots to invisible ink and described the basis of cryptanalysis by letter frequencies.

The earliest actual book on steganography [8] was written by Gaspari Schott in 1665. The book was published in four volumes and was mainly based on the works of Trithemius.

Auguste Kerchoff, in 1883, produced "Cryptographie Militaire". This evaluated previous cryptographically systems and provided a solution to the use of cryptography when using the telegraph. He proposed a number of rules for this and also made the observation that it was the cryptanalyst not the cryptographer who could determine the strength of a cryptographic key.

The 20th century saw the greatest growth in steganography. Lord Baden Powell used to map the Boer gun positions but to ensure that if he was caught that the Boers would not know what he had been doing, he built the maps into drawings of butterflies.

In World War 1, the Germans used a variation of the Cardano Grille for sending messages.

During World War 2, null ciphers were used to hide secret messages. The null cipher appeared to be an innocuous message, so would not arouse suspicion and would those not be intercepted. This is a message reportedly sent by a German spy [3].

Apparently neutral's protest is thoroughly discounted and ignored.

Isman hard hit. Blockade issue affects pretext for embargo on by-products,

ejecting suets and vegetable oils.

Taking the second letter in each word to decode the message reveals:

Pershing sails from NY June 1

When photography was sufficiently developed, it was possible to microfilm messages. The Germans used microdots in World War 2 (and World War 1).

Captured prisoners put on television from the Vietnam were reputed to have used blinking to send messages and the US POWs in the Hanoi Hilton were reputed to use a variation on the Cardano Grill to tap messages to each other.

With the advent of computing, it was possible to store and hide data in ways never thought of before the computing age.

This chapter discusses some methods of data hiding.

22.2. Physical Hiding of Computer Media

In many cases, the physical computer media has been hidden from view or disguised in some way or another. Some typical examples of this type of data hiding include:

(1) An USB thumb drive which contained a complete forgery kit including masters for passwords, driving licenses and other identity documents. USB thumb drives now contain up to 32 Gb of data. Other similar devices available are wristwatches with USB storage in the strap or wrist cameras that can take up to 100 pictures (Some examples of "small media" are given in Fig. 22.1).
(2) Spare disc drives inside the computer being seized that are not connected at the time of seizure. These can often contain the "goodies" and are connected only when there is no one else around so that the user can access the contents in private and
(3) Hard discs still in their wrappers stored on the shelf that on close inspection have had their anti-static bags slit open, but the discs appear to be empty (i.e. formatted).

22.2.1. *Other Simple Methods of Hiding Data*

There are many other simple methods of hiding data; these are usually only limited by the imagination of the person trying to hide the data. Two common methods include:

- Hiding data in the middle of a long file and
- Deleting a document and then undeleting it (using either DOS or Windows)

In the examples below, Windows has been used, many of the techniques can be adapted for other operating systems (OSs).

22.3. Changing the File Name

File naming is a process where the file has a unique name that the user remembers and associates with the file contents. The file may be put in a directory or folder which may be similarly named.

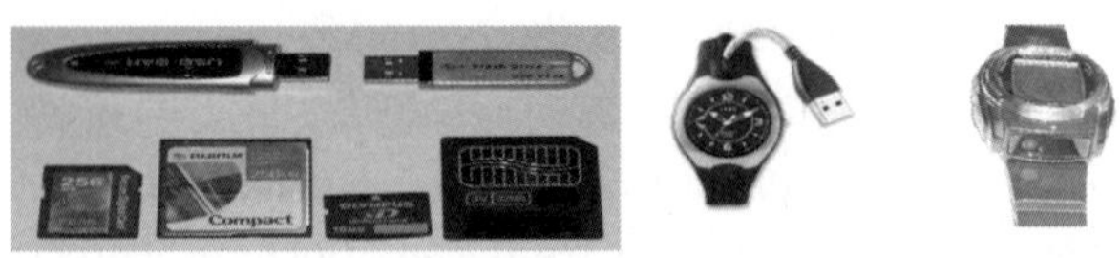

Examples of thumb drives and small memory cards | Wristwatch with USB storage | Wristwatch with inbuilt camera

Fig. 22.1. Examples of small media.

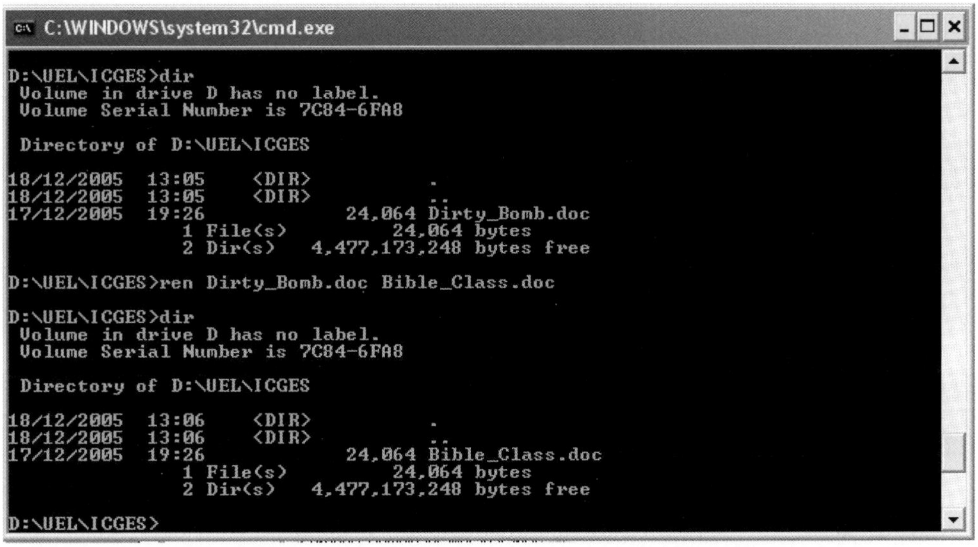

Fig. 22.2. Changing the file name (an example).

So, for example, if the file name was "Dirty_bomb.doc", this may be a clue as to the contents, but if this were changed to "Bible_Class.doc", the meaning would be much less obvious. This is shown in Fig. 22.2, where a listing of the directory shows the file "Dirty_Bomb.doc" as being 24,064 bytes long. This is renamed using the "ren" command to "Bible_Class.doc" being the same length and showing the same date and time. It is, in fact, the same file with just its name changed.

Text searches would identify the file if such terms as "bomb" or "dirty bomb" were used but there is no easy way to determine if the file name has been changed.

22.4. File Attributes

It is possible to "hide" a file so that its presence is "hidden" or its presence is suppressed from listing or identification. Each file has a number of file attributes that can be assigned to a file as shown in Fig. 22.3.

If the hidden attribute is applied, then the file will be suppressed from directory listings as shown in Fig. 22.4. This shows a directory listing showing the presence of the file "Dirty_Bomb.doc". Using the "Attrib" command, it is possible to see that the file has the "A" attribute (the "Archive file" attribute). By applying the hidden attribute to the file (attrib +H Dirty_Bomb.doc) and running a directory listing shows that the file is no longer listed. If the "attrib" command is run again, it shows that the file now has the "Archive" file and "Hidden" attributes.

It is possible to actually view the file if the "dir" command is used with the "attribute" switch set to display hidden files, as shown in Fig. 22.5.

Using the Tools | Folder Options | View option in Windows Explorer will allow the hidden files to be viewed.

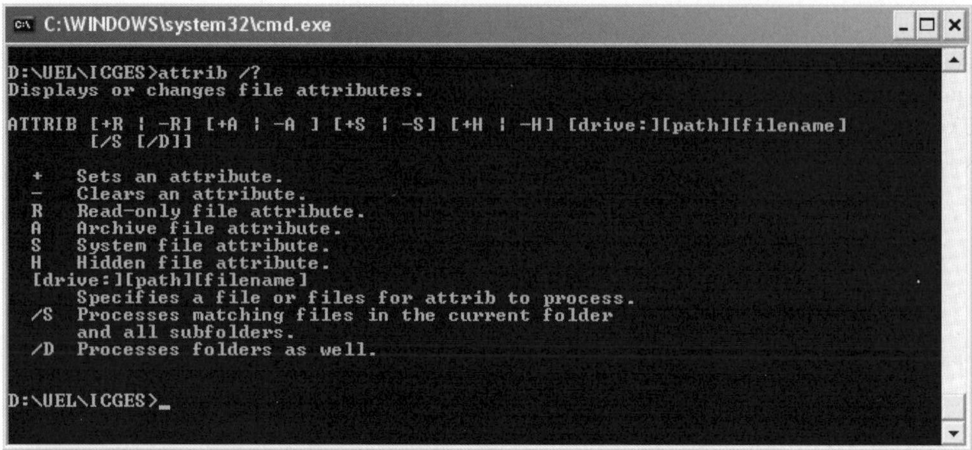

Fig. 22.3. File attributes.

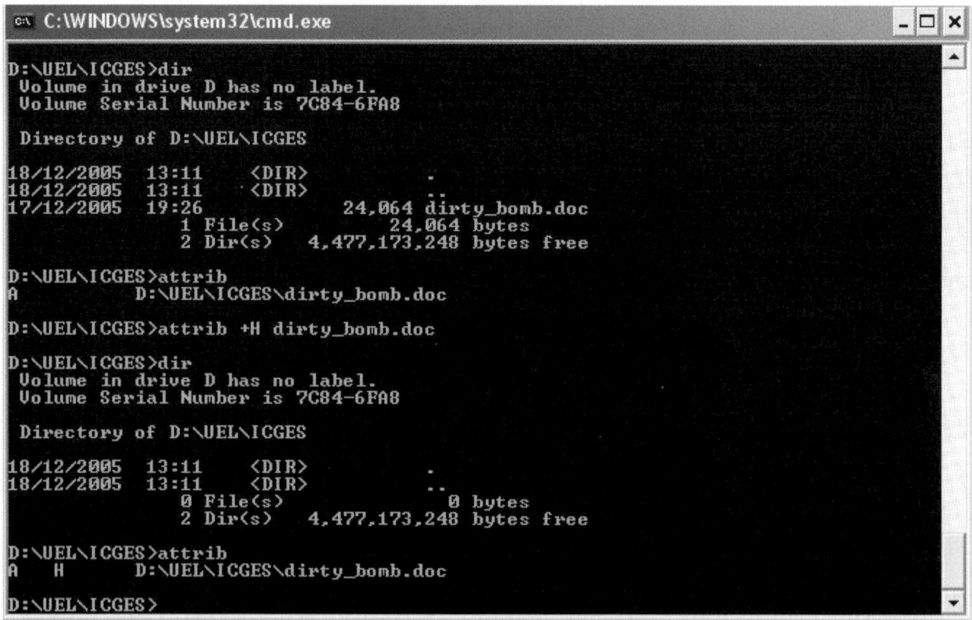

Fig. 22.4. Hidden attribute to a file.

22.5. Changing the File Extension

Changing the file extension is similar to changing the file name, except the three-letter extension (or file suffix) is changed.

Whilst there are naming conventions for system files, many user-created files, on the other hand, can have any three-character alphanumeric extension and be

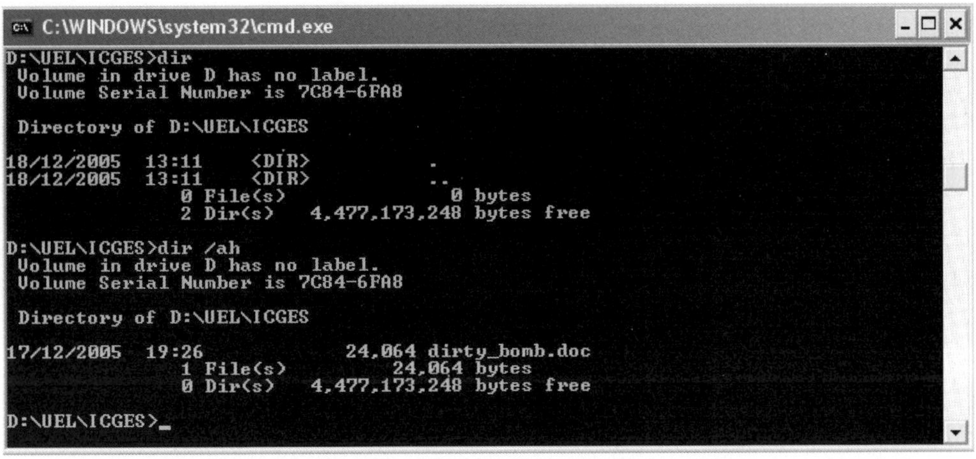

Fig. 22.5. Display of hidden files.

considered a legal filename. Therefore, a user can create any file extension on any file in an attempt to disguise it.

So, for example, if a ".doc" file was to be hidden from a search of all document files, it could be changed to being a ".abc" file and would therefore not appear in a "*.doc" file search.

The same process as shown in Fig. 22.2 would be used, but only changing the file extension and not the file name.

Text searches for the contents of the file would recover it as would a full-file listing — assuming you knew the name of the file being searched for.

Remember, just because a file is not reported in the search does not mean that it is not present — merely the search did not report its presence.

It is also possible to change the file extension association so that a different file to the one expected could open the file or that it just may not be opened at all as shown in Fig. 22.6.

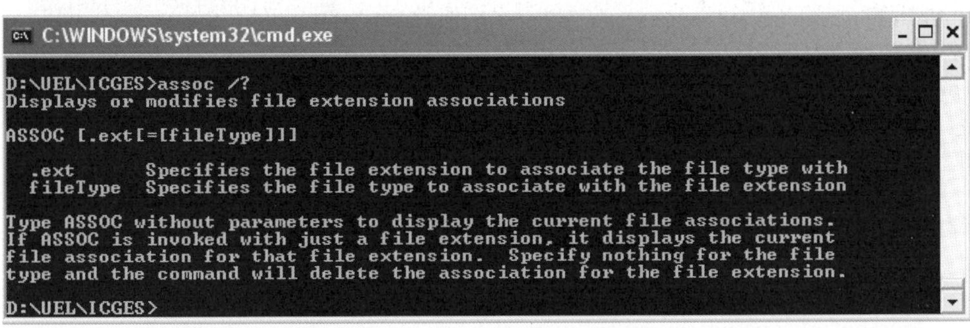

Fig. 22.6. File extension association.

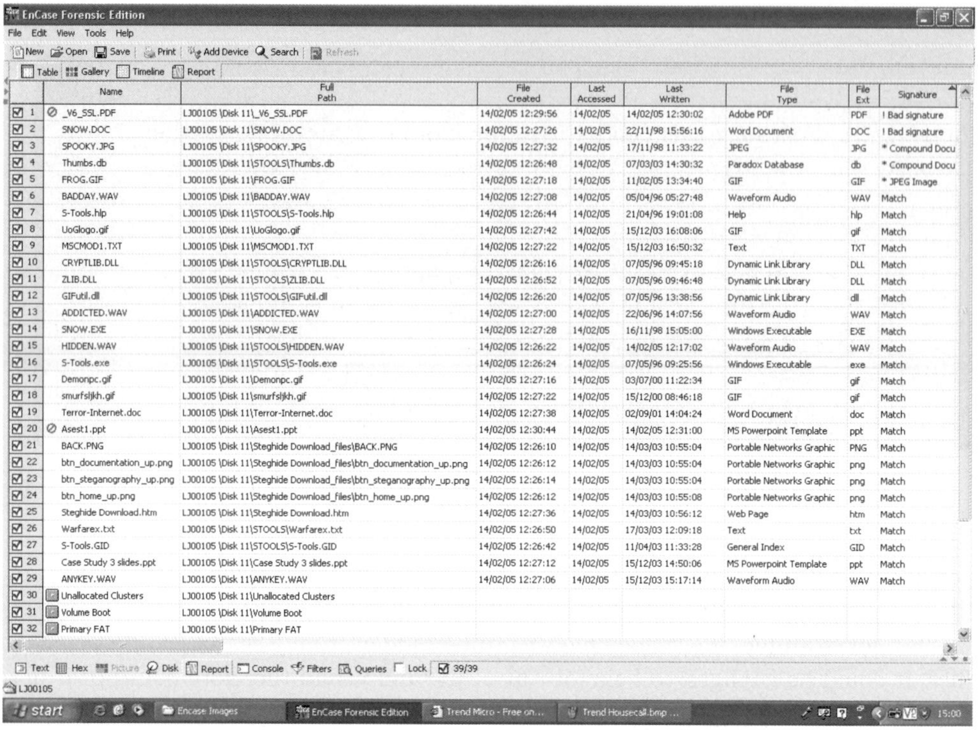

Fig. 22.7. Encase: an example.

Changing the file extension is much easier to detect as each file has an identifier at the start of the file that allows the program to determine whether it can open it or not. Checking this information against the file extension is called "signature analysis".

A number of forensic tools can identify files that have been modified in this manner. An example of this is Encase, as shown in Fig. 22.7.

As it can be seen, the first two files show that the file signature has an entry "! Bad Signature" indicating that the file contents do not match the file extension held in the Encase File signature Table for the given extension.

The third and fourth images are shown as "*Compound Document" indicating that the file has been renamed.

The fifth one shows a "*JPEG Image" indicating that the ".GIF" file is in fact a ".JPEG" file.

The remainder of the files show "Match" indicating the header matches the file extension.

Other specialised tools can produce similar results in their own way. Finding examples like this indicate that the file may deserve some attention.

It is not always necessary to know the list file signatures (or "magic numbers" as they are sometimes referred to), as they are usually present in forensic tools.

22.6. File Splitting

Large or even small files can be split into a number of component parts. In the early days of PC computing, this was typically used to split large files to fit them onto floppy discs, but today this can be used for more nefarious purposes.

If the file extension is changed and if the file name is carefully chosen, then it should reveal nothing about the contents of the file. Viewing the component parts of the file will often reveal nothing of note.

This sort of approach is detected by determining the file names of the split file and their types. Typically, such a web site as www.filext.com is used to determine what the extension means. Another clue is in the programs that are located on the system by taking a listing of the programs currently installed. Another clue is the series of filenames in sequence as shown in Fig. 22.8.

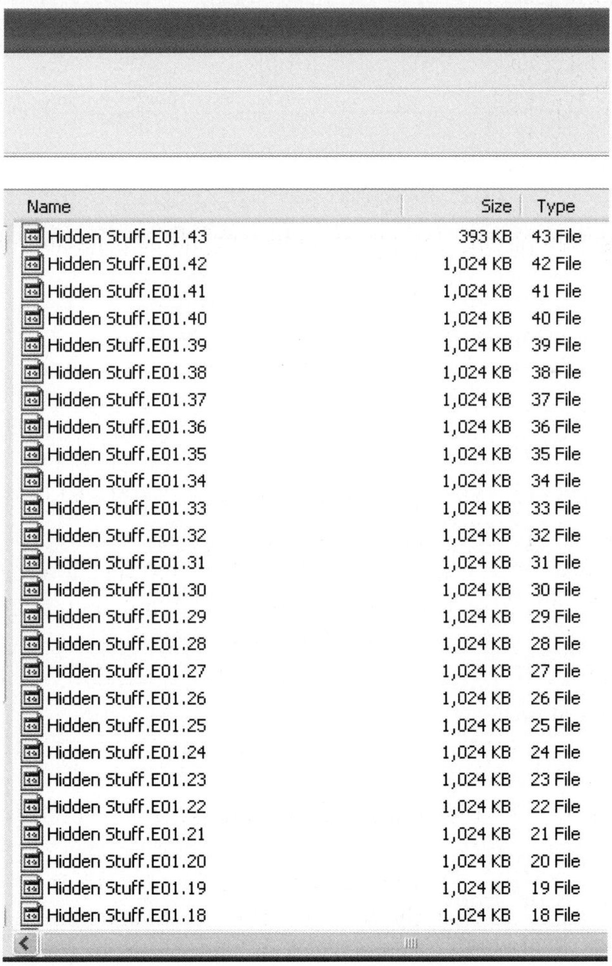

Fig. 22.8. Series of filenames in sequence.

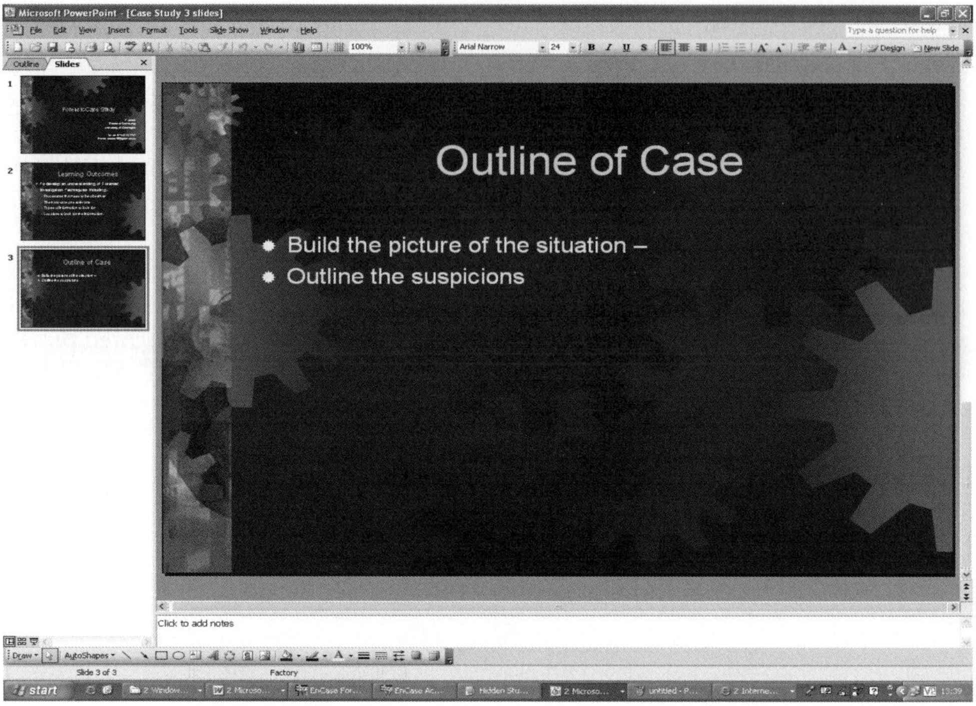

Fig. 22.9. Normal slide.

22.6.1. *Using the Same Colour Text as Background*

It is possible and relatively simple to use the same colour text as the background effectively hiding the data. An example is shown below, the "normal slide" being shown in Fig. 22.9 and the "hidden" message being shown in Fig. 22.10.

Again, to detect this type of data hiding, it is necessary to have some idea that there is actually hidden text present.

In this case, the tool "MetaData Assistant" has been used to scan the file for any metadata and the report that it produces is below:

<snip>

Comments:

no comments

Font matching background:

Blocks of font matching background: 1

Slide: Slide 3 Location: Text-Box 4 Text-box rectangle

Small font (size 3 or smaller):

No text or shapes with small font

<snip>

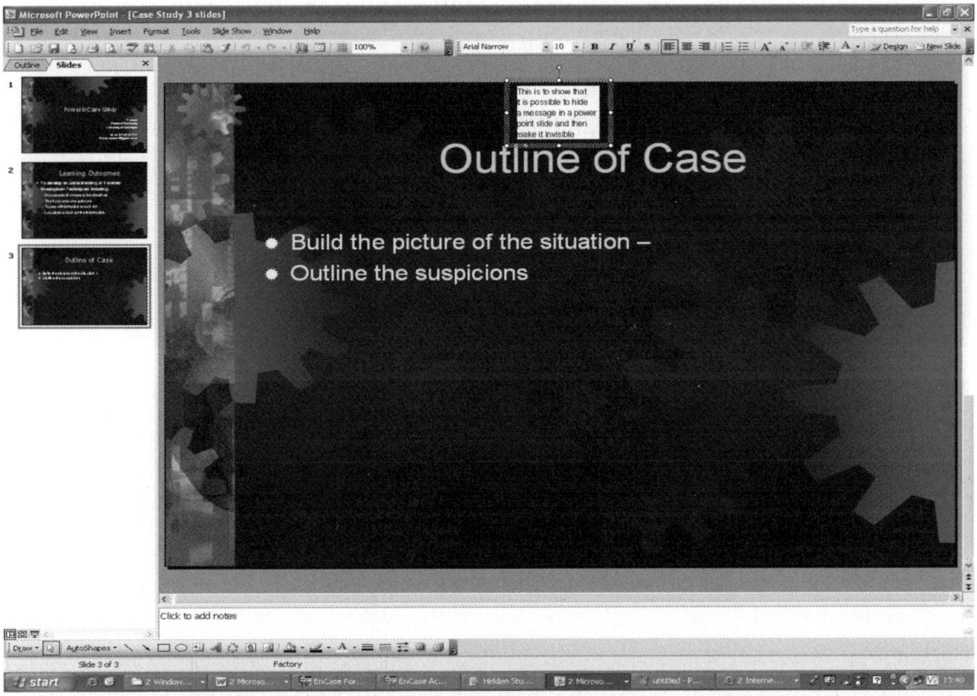

Fig. 22.10. Hidden message.

Note: Figures 22.9 and 22.10 reproduced with thanks to Drs Andrew Blyth and Iain Sutherland (University of Glamorgan).

22.7. Alternate File Streams in NTFS

The NTFS, the file system used by Windows NT, Windows 2000 and Windows XP has a feature that is not well documented and is unknown to many developers and most users.

This feature — Alternate Data Streams (ADS) — allows data to be stored in hidden files that are linked to a normal visible file. Streams are not limited in size and there can be more than one stream linked to a normal file.

This can allow an unscrupulous person to store any data that they want in the ADS or to execute programs in it.

The easiest way of demonstrating this is as shown in Figs. 22.11–22.15 using standard Windows and DOS features.

Figure 22.11 shows where I have copied calc.exe into a directory on its own. It shows a size of 112-Kb and a date modified of 29/08/2002.

In Fig. 22.12, the ADS is appended to "calc.exe" using "notepad.exe" as shown below.

Examination of "notepad.exe" shows it is 68-Kb as shown in Fig. 22.13.

Looking again at "calc.exe", we see that the size has not changed but the date modified has, as shown in Fig. 22.14.

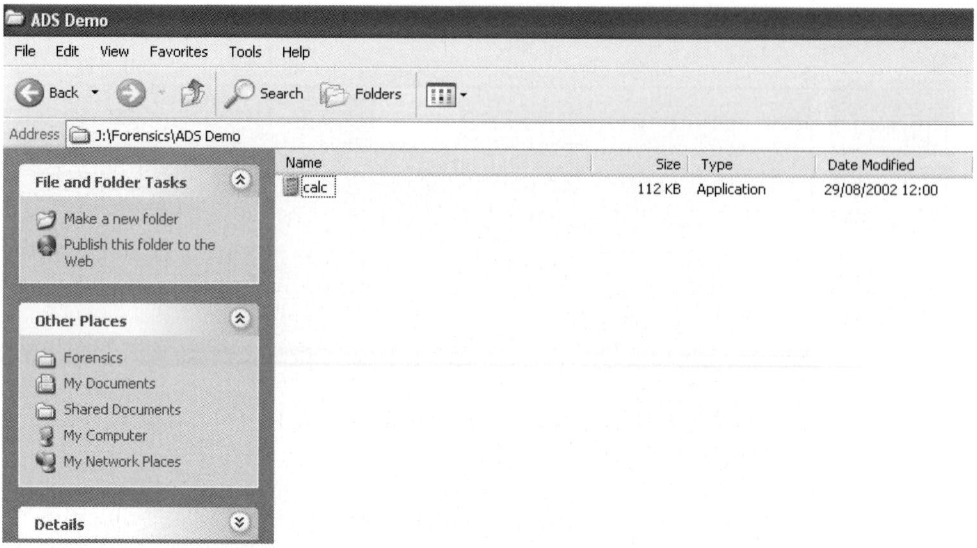

Fig. 22.11. ADS demo of forensics.

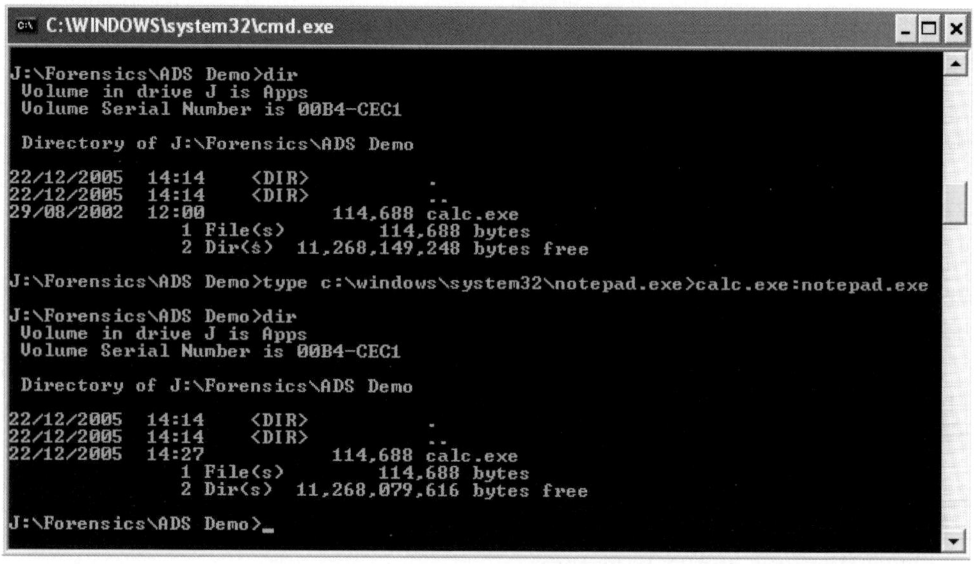

Fig. 22.12. ADS appended to calc.exe.

It is now possible to execute the new ADS notepad.exe using the standard command start as shown in Fig. 22.15.

This starts up “notepad.exe” as can be shown in the result in Fig. 22.16(a).

However, if the Windows Task Manager is examined, it shows that “calc.exe: notepad.exe” is running, as shown in Fig. 22.16(b).

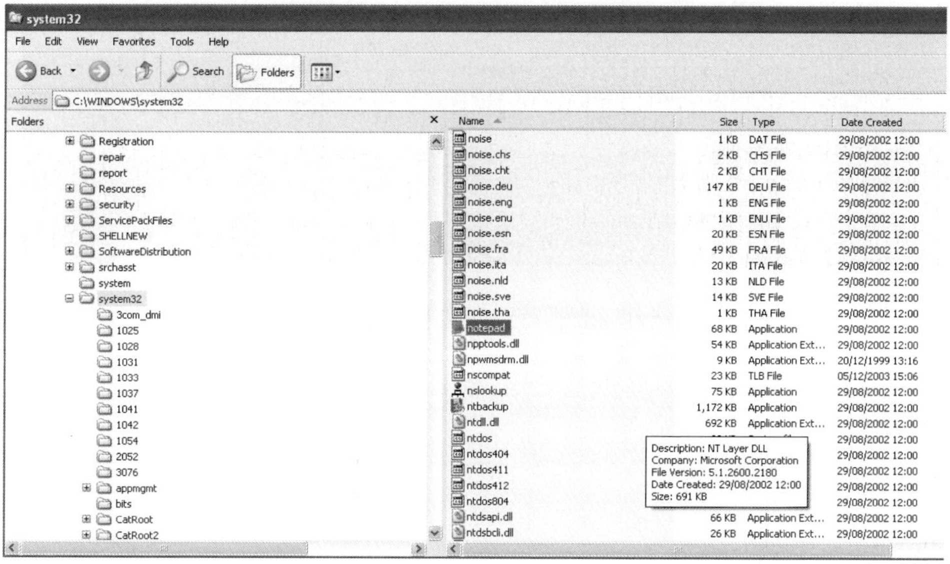

Fig. 22.13. Examination of notepad.exe.

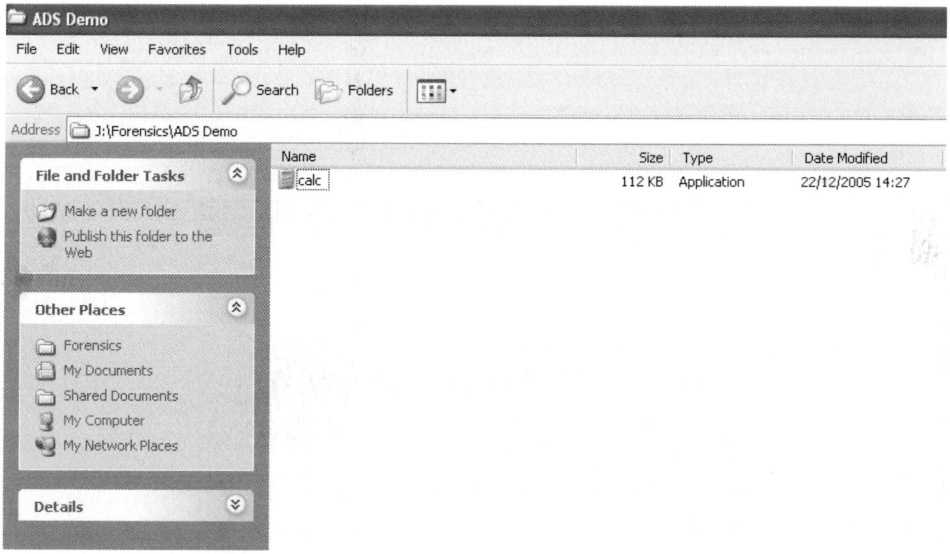

Fig. 22.14. Modification of date.

In earlier versions of Windows, the ADS was not shown, just the "calc.exe", with XP, the ADS is shown as well.

There are a number of forensic tools that can detect alternate data streams in use, the most well-known being LADS.exe. Figure 22.17 shows LADS running on the directory where the ADS was created.

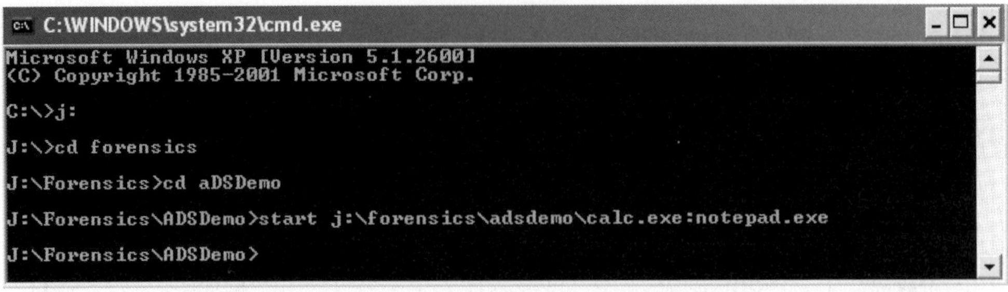

Fig. 22.15. Execution of new ADS notepad.exe.

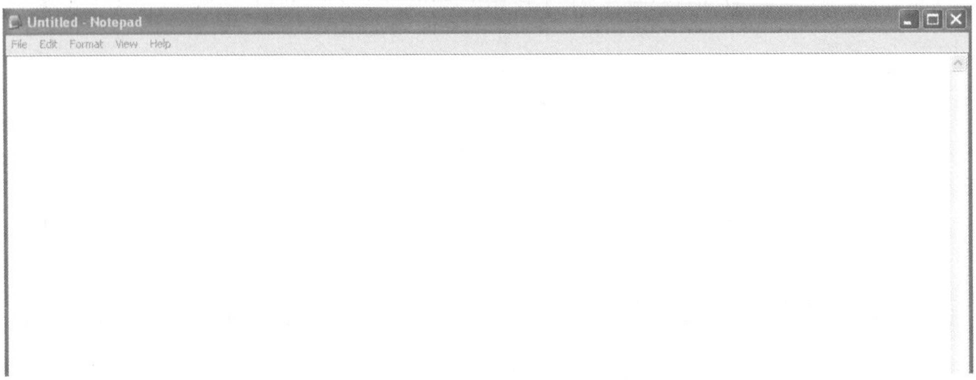

Fig. 22.16. (a) Result of notepad.exe.

Fig. 22.16. (b) Examination of windows task manager.

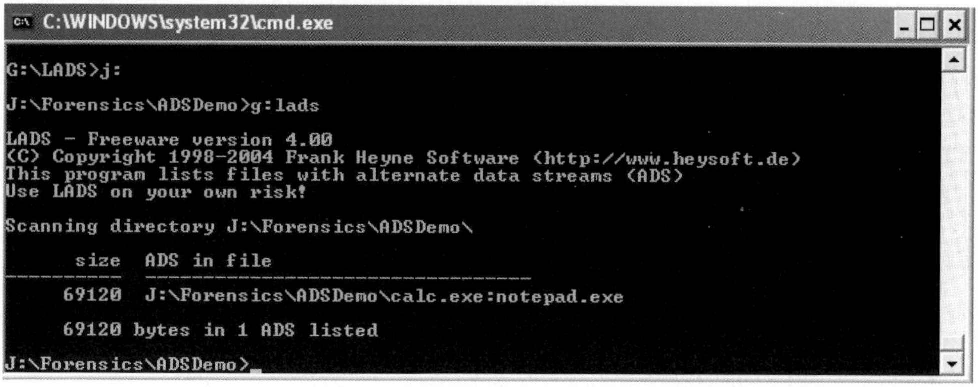

Fig. 22.17. LADS running on the directory.

22.8. Password-Protecting Files

One of the simplest ways of protecting a file from casual examination is the application of a password to the file. Using Forensic Toolkit (FTK) from Access Data, it is easy to detect the encrypted files in a case as shown below in Fig. 22.18.

However, FTK comes with serious password cracking tools so that the recovery of such passwords (even using ALT key sequences) is a relatively trivial matter.

There are numerous password-cracking tools available.

22.9. Cryptography

Cryptography is derived from two Greek words — "krypto" — hidden and "graphie" — to write, so literally it is hidden writing.

Data that can be read and understood by all is called "plaintext" or sometimes "cleartext". The method of disguising plaintext to hide its contents is called "encryption". The result of applying encryption to plaintext is called "ciphertext". The process of returning ciphertext to its original plaintext is called "decryption". Whilst cryptography is the science of securing data so that only the intended recipient can read it, "cryptanalysis" is the science of analysing and breaking the ciphertext to recover the plaintext.

Cryptography is the solution used when there is a requirement to transmit a message to someone and there is a possible adversary that wants to intercept that message and read it. It is the science of mathematics used to encrypt and decrypt data.

It has been thought that cryptography in some form or other has been around since writing was discovered.

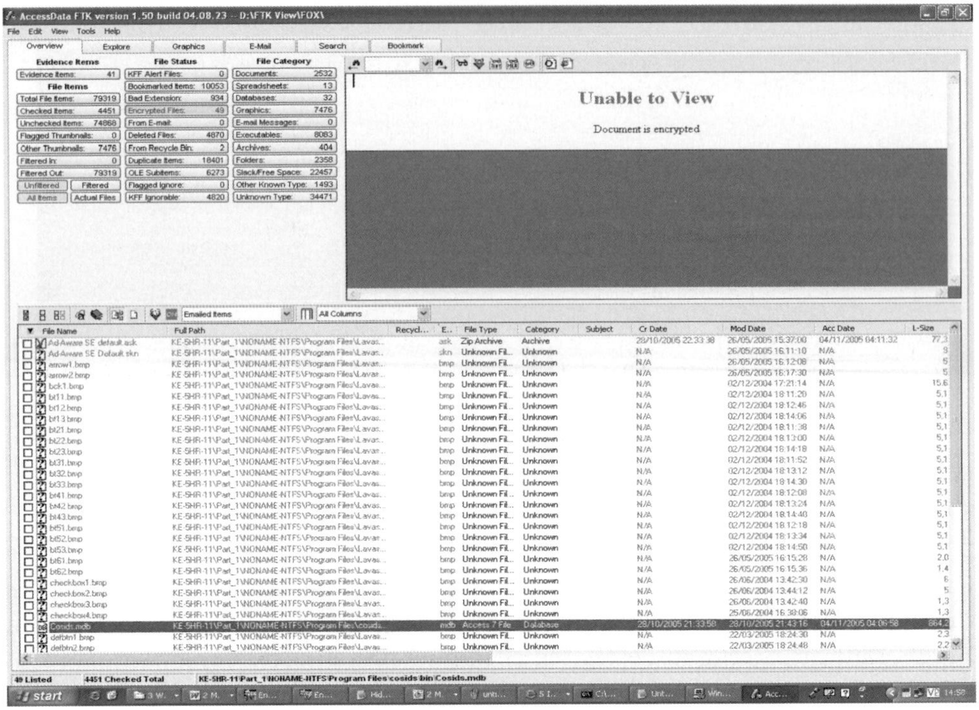

Fig. 22.18. Detection of encrypted files.

The most often cited first use of cryptography was the "Caesar Code" named after Julius Caesar. When he wanted to send messages to his generals, he sued messengers, but he did not trust them, so he made sure that if the message was intercepted, they could not be read by encoding them unless one knew the secret of the code.

To do this, he replaced every "a" in the text by the letter "d", "b" by "e" etc.

If the person trying to read the message did not know the "shift by 3" rule, they could not make sense of the message.

In today's world, there are four main requisites for sending secure communications over insecure channels such as the Internet:

- **Authentication** — the process of proving one's identity;
- **Confidentiality** — ensuring that no one apart from the intended recipient can read the message;
- **Integrity** — assuring the recipient that the message received has not been altered during transmission and
- **Non-repudiation** — proving that the sender really did send the message.

Note: This is a high-level overview of cryptography and its application; many other books deal with the subject in detail.

22.9.1. *What Cryptography Can Do*

Cryptography is not the universal panacea to all communication problems. Typically, cryptography can provide:

- Confidentiality for transmitted or stored messages;
- Prove the integrity of a message (i.e. it has not changed in transit) and
- Authenticate the sender (and so provide non-repudiation).

22.9.2. *What Cryptography Cannot Do*

Cryptography can only protect data when it is encrypted and remains encrypted. Before it is encrypted and after it is decrypted, it is not protected by encryption. Whilst it may seem obvious, it is worth stating that encryption can only protect data, it cannot protect physical assets apart from data. Encryption can not also protect against processes designed to subvert the encryption process that do not actually decipher the encryption process. Such processes could include keyboard loggers, filmed actions and theft of decrypted ciphertext or unencrypted plaintext.

22.9.3. *The Caesar Code Revisited*

The Caesar Code is known as a single transformation. It takes the original plaintext and makes a "single transformation" of it to the ciphertext. Typically, a page has the alphabet in the first column on a page and in the second column is the transformed alphabet. If the details of the transformation (i.e. the "shift by three") are stolen, or even copied, then the code is compromised.

Now, if a book has many pages in it and each page has a different transformation on it and all the pages are numbered, then all one has to disclose is the page number of the transformation used to encrypt the message. In this way, the page number becomes the "key". This is called "multiple transformations".

In this case, even if the notebook were disclosed, the actual key (the page number used to encrypt the plaintext) would not be known and a brute-force attack of approximately half the pages would be needed to decrypt the ciphertext.

The single transformation is simple to decrypt for an attacker. What we may think is a strong cipher may not actually be one.

If we take the example of multiple transformations and assume that there are 256 pages in it, then there are exactly 256 keys available (the number of pages). In binary terms, this is represented by 100,000,000 or 2^8 or an 8-bit "key space".

Over time, a random sample of keys would take about 128 different keys before hitting the right decryption key.

If the notebook had 65,536 pages, instead of 256, then this gives a 16-bit key space. This means that the number of attempts needed to find, on average, the right key to decrypt the ciphertext is 32,767 — giving a stronger cipher than the 8-bit one.

This is the same idea as used for modern ciphers.

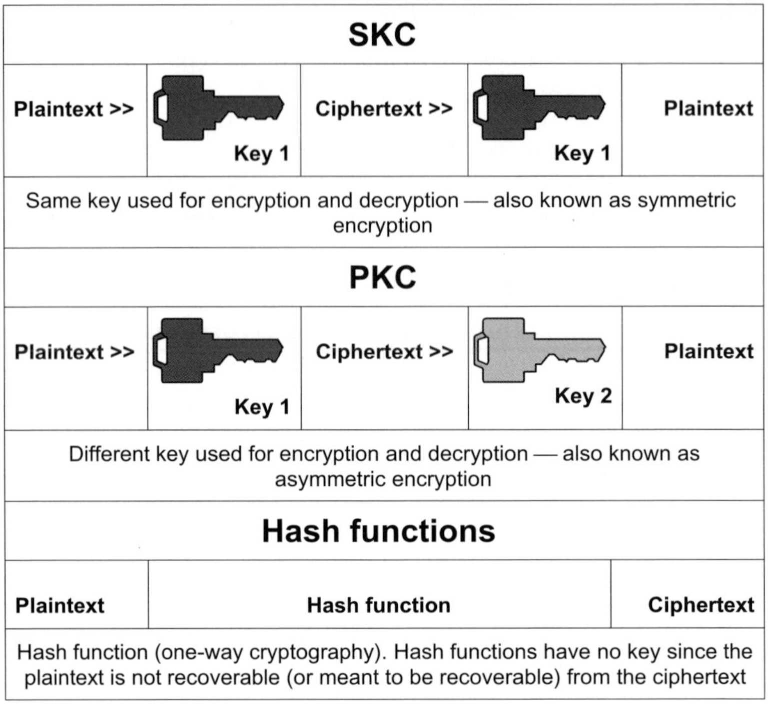

Fig. 22.19. Different types of cryptographic algorithms.

A 56-bit key space has about 7×10^{16} different keys, which sounds a strong cipher, but such a cipher can be very quickly broken with specialised hardware. Whilst 128-bit keys were once thought of as the ideal, now either 256- or 512-bit keys are recommended if you want to keep your secrets safe.

22.9.4. *Types of Cryptographic Algorithms*

Three different types of cryptographic algorithms are discussed (Fig. 22.19). These are:

- Secret Key Cryptography (SKC) — using a single key for both encryption and decryption;
- Public Key Cryptography (PKC) — using one key for encryption and another for decryption and
- Hash functions — using a mathematical transformation to irreversibly "encrypt" information.

22.9.4.1. *SKC*

With SKC, a single key is used for both encryption and decryption. As shown above, the sender uses a key (K1) to encrypt the plaintext and sends the ciphertext to the receiver. The receiver applies the same key (K1) to decrypt the message and recover the plaintext. Because a single key is used for both functions, SKC is also called symmetric encryption.

In this form of cryptography, the key must be known to both the sender and the receiver and that is the secret. The greatest problem with this form of cryptography is the distribution of the key.

The SKC schemes are generally categorised as being either stream ciphers or block ciphers. Stream ciphers operate on a single bit at a time and implement some form of feedback mechanism so that the key is constantly changing. A block encrypts one block of data at a time using the same key on each block.

22.9.4.2. *PKC*

The PKC is stated to have been first described publicly by Stanford University Professor Martin Hellman and graduate student Whitfield Diffie in 1976. Their paper described a two-key crypto system in which two parties could engage in a secure communication over a non-secure communications channel without having to share a secret key. There is evidence though, that it was invented in the United Kingdom in 1970 but kept as a military secret [1].

The PKC employs two keys that are mathematically related although knowledge of one key does not allow someone to easily determine the other key. One key is used to encrypt the plaintext and the other key is used to decrypt the ciphertext. The important point here is that it does not matter which key is applied first, but that both keys are required for the process to work. Because a pair of keys is required, this approach is also called asymmetric cryptography.

In PKC, one of the keys is designated as the public key and may be advertised as widely as the owner wants. The other key is designated as the private key and is never revealed to another party. It is straight forward to send messages under this scheme. Suppose Alice[a] wants to send Bob a message. Alice encrypts the message using Bob's public key; Bob decrypts the ciphertext using his private key. This method could also be used to prove who sent a message; Alice, for example, could encrypt some plaintext with her private key; when Bob decrypts using Alice's public key, he knows that Alice sent the message and Alice cannot deny having sent the message (non-repudiation).

22.9.4.3. *Hash functions*

Hash functions, also called message digests and one-way encryption, are algorithms that use no key. Instead, a fixed-length hash value is computed based upon the plaintext that makes it impossible for either the contents or length of the plaintext to be recovered. Hash algorithms are typically used to provide a digital fingerprint of a file's contents used to ensure that the file has not been altered by an intruder or virus.

[a]Traditionally, Alice and Bob are the names used for the sender and recipient.

It is often thought that there cannot be two files with the same hash value. This is not correct, but the difficulty is in finding two files with the same hash value. It is very difficult to create a file with the same hash value as another file, which is why hash vales are used in information security and digital forensics.

Hash libraries are sets of hash values corresponding to known files. A hash library of known good files, for example, might be a set of files known to be a part of an OS, while a hash library of known bad files might be of a set of known child pornographic images used in digital forensics.

22.9.5. *Differences in Encryption Algorithm Types*

There are three main types of encryption algorithms as they each perform different functions.

The SKC is ideally suited to encrypting messages. The sender can generate a key for each message to encrypt the message. The recipient needs the same key to decrypt the message.

The PKC can be used to encrypt messages but it can also be used for non-repudiation. If the recipient can obtain the key for the message sent (the session key) encrypted with the sender's private key, then only this sender could have sent the message.

Hash functions are used for ensuring data integrity because any change made to the contents of a message will result in the recipient calculating a different hash value than the one placed in the transmission by the sender. Since it is highly unlikely that two different messages will yield the same hash value, data integrity is ensured to a high degree of confidence.

22.10. Steganography

Steganography is also derived from two Greek words — "steganos" — covered and "graphie" — to write, so literally it is covered writing.

With cryptography, the data is present but hidden. It is possible to detect encrypted data quite easily, but decrypting it is the problem. Steganography is art-and-science of having a covert communication channel whose existence is unknown as the plaintext is hidden in an unremarkable cover media that does not arouse suspicion or show that a hidden message is present.

The examples given earlier of Histiaeus and Demeratus are classical cases of steganography. Today, the aim of steganography is still the same as it was in those days, which is keeping the presence of the message undetected. They are detectable but are not as easy to detect as an encrypted file in a folder.

The simplest method of hiding information within a file is to replace all of the least significant bits (LSB) within the file. This change can barely be seen by the human eye, if one knew to look for it. This method does not work when an audio file is used for the cover for the hidden message as changes to the LSB adds "noise" to the output that can be detected in the quiet periods of sound playback.

Steganalysis, the art-and-science of detecting steganography, can easily detect this.

The most common method of hiding a message is to hide the message in an image. There are many ways of doing this from the LSB replacement to the complex patchwork algorithm. This randomly selects pairs of pixels of a given image. The brighter of the pair is made brighter and the darker is made darker. This change is so subtle that it is undetectable to the human eye, even at high levels of magnification. The contrast change between the two changed pixels forms part of the bit pattern for the hidden message. To ensure that there is little chance of detection, there is a limit on the number of such changes that can be made. The number of changes will depend on the actual source image. The first program to do this was JSteg, a freeware program developed by Derek Upham. This did not encrypt the hidden message, just hide it. A later version of the, JStegShell, offered the possibility of encryption. There are a number of freeware and commercial tools available for hiding messages within images.

An example of steganography using Andy Brown's S-Tools is shown below in Fig. 22.20, demonstrating how easy it is to hide data in an image.

S-Tools is started and the recipient or host to hold the hidden image is dragged into the S-Tools program area. S-Tools only works with gif or bmp files. Transparentliaglogo.gif has been dragged across (LIAG are the Land Information Assurance Group — the Army's own Cyber Warriors).

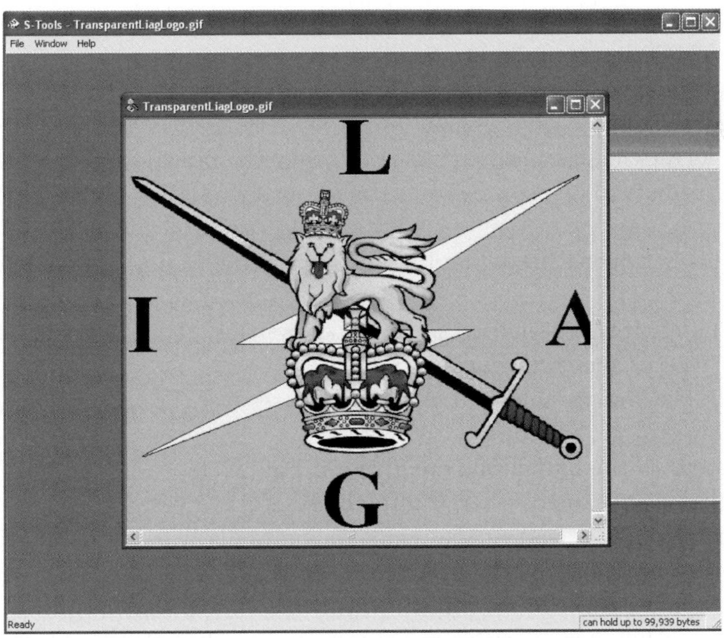

Fig. 22.20. Steganography using Brown's S-tools.

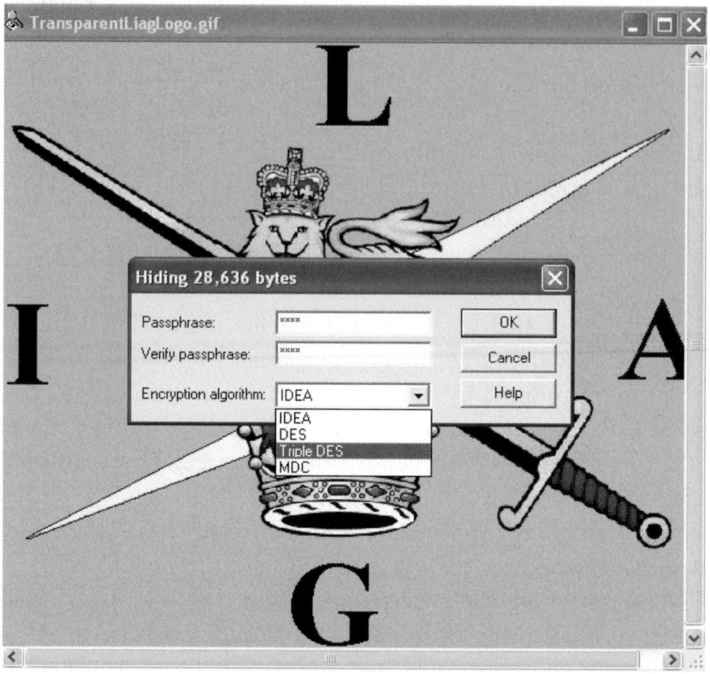

Fig. 22.21. Image of UK.gif.

This shows that in the bottom right-hand side of the screen that the image can hold up to 99,939 bytes of data.

A second image, UK.gif, was dragged across to the S-Tools active window as shown in Fig. 22.21.

This asks for the passphrase (twice) to insert the image to be hidden and what crypto algorithm is to be used. S-Tools has options for IDEA, DES, Triple DES and MDC as can be seen.

Once this has been selected and input, the picture-hiding options are chosen as shown in Fig. 22.22.

Once the choices are made, the image shows that there is hidden data in the image displayed as shown in Fig. 22.23.

By right clicking the image, it is possible to recover the hidden image as shown in Fig. 22.24.

The details of the hidden file are given in Fig. 22.25.

There has been much research into the digital watermarking use of steganography. This mars either an image, video or audio file with a digital identity. The digital identity is unique to the owner of the file and so can be used to copyright the file, image, video or an audio file.

There are a number of service providers offering digital watermarking services for copyright owners.

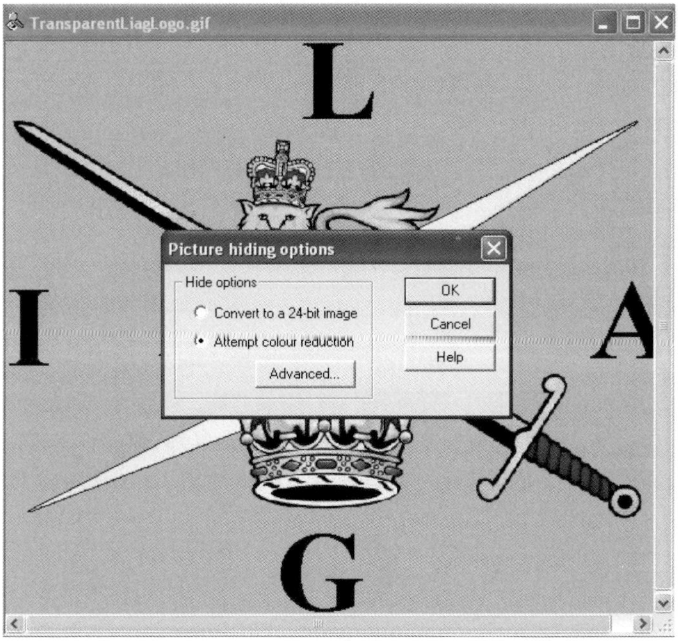

Fig. 22.22. Picture-hiding options.

Fig. 22.23. Hidden data image.

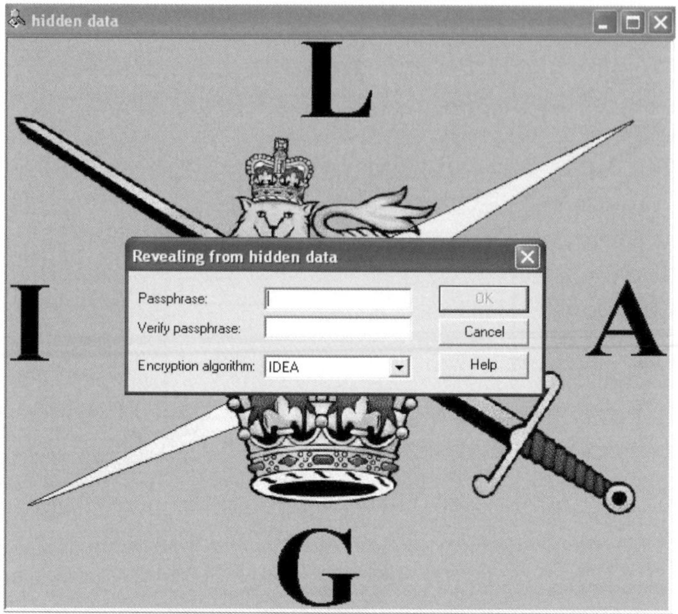

Fig. 22.24. Recovery of the hidden image.

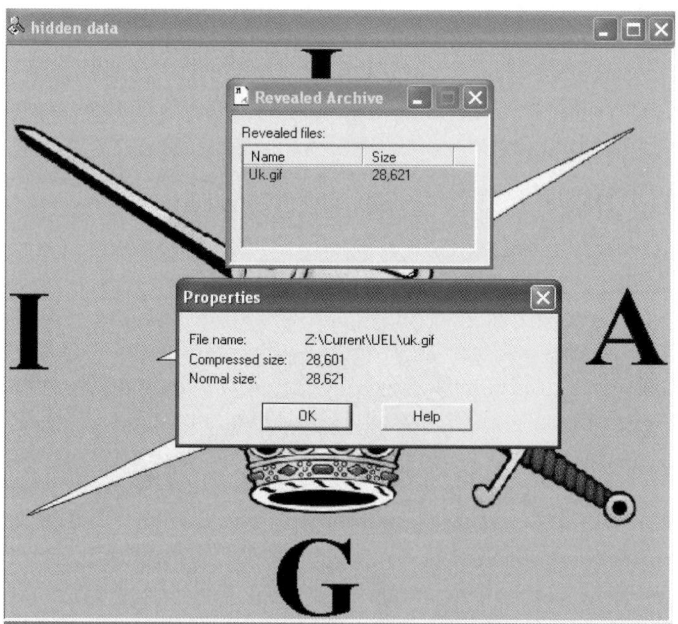

Fig. 22.25. Details of the hidden file.

For those who want to hide a message in an audio file, one of the most commonly used ones is MP3Stego. This has a command line interface allowing the user to encode and decode a file.

If one wants to place hidden.txt into the file innocentmusic.wav, and use the pass phrase "secret", one would enter the following command.

encode –E hidden.txt –P secret innocentmusic.wav innocentmusic.mp3

This compresses the innocentmusic.wav and hidden.txt into innocentmusic.mp3 using the pass phrase "secret".

To decode this:

decode –X –P secret innocentmusic.mp3

This uncompresses innocentmusic.mp3 into innocentmusic.mp3.pcm and saves the hidden text as innocentmusic.mp3.txt.

It is also possible to hide data in the slack space in a cluster on a disc drive. Slack space is where the actual data being stored requires less storage than the cluster size, an entire cluster is reserved for the file. The unused part of the cluster is called the slack space.

It is also possible to hide data in transmission protocols such as TCP and IP where data can be hidden inside certain header fields.

Hiding data in text files is a more challenging option because there is less redundant data to replace with a hidden message. Text-based steganography is very fragile in that the hidden message can be altered by anyone who actually alters the text in the original text in which the hidden text is to be inserted.

There are a number of methods used to hide text in text-based steganography; the most common are:

- Open space methods;
- Syntactic methods and
- Semantic methods.

Open space methods can use extra spaces in the text after a period mark, at the end of a line or by the use of right justification. For example, a single space can signify a "0" whilst a double space can signify a "1". It is also a possible change of the vertical height of certain letters to indicate "1"s or "0"s.

Syntactic methods use punctuation and the structure of the text to hide data without actually changing the meaning of the message.

Semantic methods assign two synonyms different values (1 and 0).

In the example below, a program called "snow" is used. It makes use of open space methods using the end of line spaces.

To demonstrate this, a text file "infile.txt" was created containing the text "innocent message" in it as shown in Fig. 22.26.

The command shown in Fig. 22.27 will conceal the message "I am Lying" in the file infile.txt, which will be encrypted and compressed with the password "test".

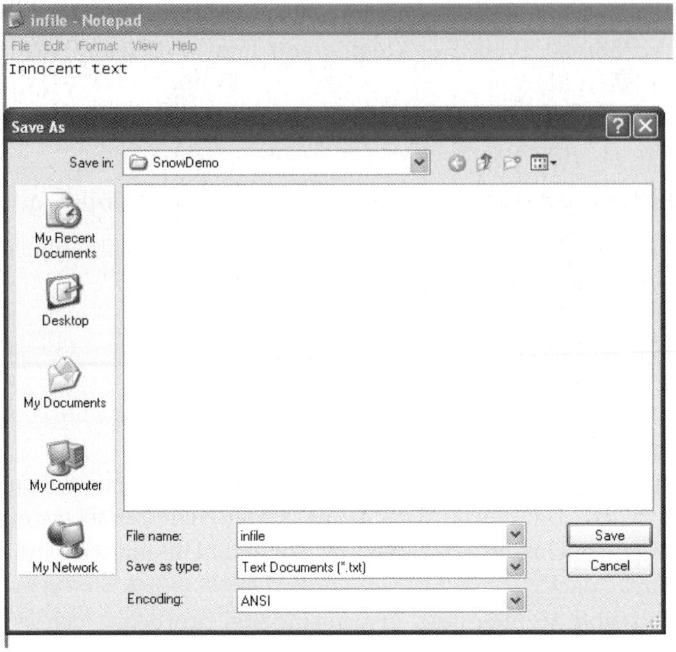

Fig. 22.26. Creation of a text.file "infile.txt".

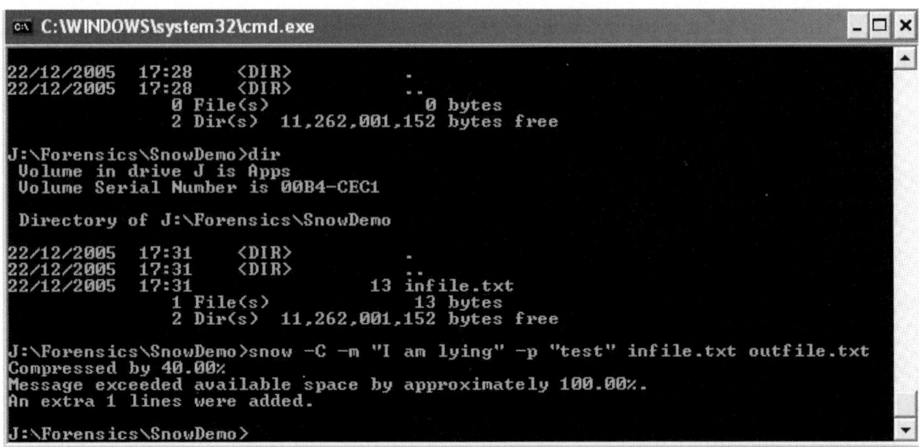

Fig. 22.27. Command.

To extract the message, the process in Fig. 22.28 is performed.

22.11. Summary

The need for hiding data has been around for many more years and will be for many more years to come. The methods used have become more sophisticated but so have the methods for its detection.

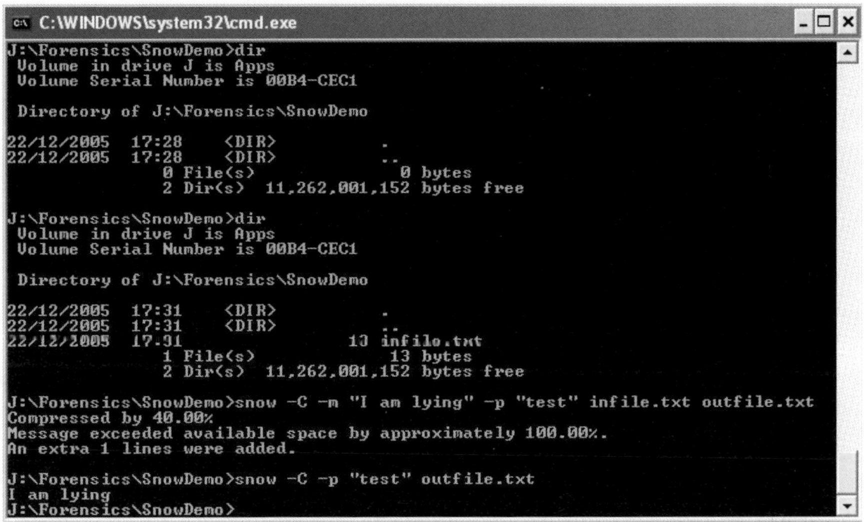

Fig. 22.28. Extraction of the message.

With the onset of computing, the scope for data hiding has grown immensely.

References

1. J. H. Ellis, The possibility of secure non-secret digital encryption, CESG Report, 1970.
2. Herodotus, *Histories of Herodotus* (Penguin Classics, 2003).
3. N. F. Johnson and S. Jajodia, Exploring steganography, *Seeing the Unseen, Computing* **31**(2) (1998).
4. G. Kipper, *Investigators Guide to Steganography* (Auerbach, 2003).
5. P. Leary, http://home.att.net\~tleary (is the most recent).
6. Ovid, *Art of Love* (Modern Library Classics, 2002).
7. G. Porta, *De Furtivis Literarum Notis*, 1563.
8. G. Schott, *Steganogaphica*, 1665.
9. A. Tacticus, *How to Survive Under Siege — Aineias the Tactician* (G. Duckworth and Co, 2002).
10. J. Wilkins, *The Secret and Swift Messenger*, 1641.

Chapter 23

CYBERSPACE AND CYBERCRIME

VAGELIS PAPAKONSTANTINOU
Attorney At Law, PK Partner, Athens

23.1. Cyberspace and Cybercrime

Ever since it escaped the pages of science-fiction books and entered the real world, cyberspace developed a strained relationship with security. This should have caused no surprise to those close to the new medium. First, a level of security compromise is to be expected in any field of human activity that is entirely new — things have to settle down for the regulator to take positive and long-lasting measures. Second, cyberspace per se was coined at the time as the last border of freedom, a space where humans may do as they please — this inevitably involves some tolerance when it comes to security. Whichever the reasons may be, the fact remains that cyberspace did and still does present a number of security issues that, nevertheless, are increasingly considered unacceptable in contemporary e-commerce-profiting societies.

Cyberspace, at least in a form relevant to the purposes of this analysis, is a relatively recent addition to our everyday lives. Widespread public use of the new medium does not date much before the 1990s. Of course, cyberspace was a term well-known and a space well-visited before that time, but its use was restricted to certain academic or other *avant-garde* circles much too scarcely populated to matter. In order for security to become a consideration, widespread use is necessary — and, such widespread use only happened more or less during the 1990s.

However, the lack of any substantial history does not mean that the new medium did not have enough time to complete its development cycles. On the contrary, cycles that would have taken years to complete were expedited at the astonishing speed that later came to characterise the new era (that is, the Information Society). Cyberspace was conceived and originally implemented as a borderless new space, transcending physical borders and formal legal rules, that would constitute the ultimate frontier for human freedom (and, discussions today on whether we should tax the Internet prove that this line of thinking still survives, *albeit* misguided). This afforded a certain level of freedom to its users at first. As more and more people were entering cyberspace, it was established that not all of them adhered to the same benevolent principles. In addition, cyberspace suddenly became commercially meaningful. Transcending its own original borders, that destined it to become a communications tool among individuals (preferably academics), cyberspace discovered e-commerce. From that point on, developments are known to everybody; within only a handful of years a previously unknown (if not, unheard of) tool was turned into an inseparable, self-evident part of doing business. Today, the commercial value of cyberspace simply cannot be measured. And, where there is money, crime inevitably follows.

Cybercrime thus emerged in cyberspace. Cybercrime comes in two forms: first, crime committed in cyberspace that was previously unknown to humanity (including regulators). Second, crime committed with the assistance of the Internet. Each case presents a number of particularities.

The self-evident cybercrime category refers to crime that was previously unknown, but only takes place in the Internet. All of the e-commerce-related crimes, as known today, may serve as examples: P2P (copyrighted) unauthorised file exchanges, libellous blogs and blogging and other Internet Service Provider-related crimes, e-commerce identity theft or fraud, all constitute new crimes that would have been impossible without the Internet. Some of these cases will be elaborated in this or in the following chapters in detail, a remark, however, of general application refers to their evolving character. Because such cybercrimes are connected with human creativity, as expressed mostly in ways of doing business on-line, they cannot be anticipated, but are only dealt with in retrospect. Web 2.0, that will be analysed later, constitutes an excellent example; once users had the opportunity of creating content and transmitting it to the millions, a new series of security risks emerged. The same is true with regard to social networks operating online. The law has no way of knowing from which direction the next flow of rights' infringements will come. In all these cases, it has to act in retrospect, sometimes only temporarily addressing the "loophole", until the conditions are mature enough for formal regulation. However, in all these cases, as is true with all new fields of human activity, an increased security risk is to be expected.

The second category of cybercrime is probably more widespread, and refers to new ways of committing crime, using the Internet. Exactly as the Internet has enabled new ways of doing business, it has also enriched the potential opening to criminals. In this case, older and well-known (and, dealt-with) crimes project themselves into cyberspace: pornography becomes on-line pornography, gambling becomes on-line gambling, credit-card fraud becomes on-line credit-card fraud. In all these cases, the law does have an answer; all said crimes (and any crime-making use of the Internet) are well-regulated in the real world. Difficulties arise, first, when the on-line circumstances gravely affect the conditions that the law required for a crime to be committed until its on-line version emerged, and, second, due to the inherent international nature of cyberspace. An example of the first refers to on-line gambling: in countries of state gambling monopoly, is it a crime for their citizens to bet in on-line betting Internet sites whose servers are located abroad? The second source of difficulties is quite obvious: crimes committed through the Internet are hard to trace and fight in a world based on state-sovereignty and state-restricted crime prevention mechanisms. The Internet known no physical borders and this conflicts with any criminal law, as known so far.

This chapter will therefore attempt to throw some light into the relationship between cyberspace and cybercrime. In order to do this, first, a brief security-focused analysis of the Internet, and in particular its Web 2.0 form, will be attempted, before approaching specific cybercrime-prone on-line activities, in order to demonstrate how assessments of the first part of this analysis apply in practice (other activities may be found in the following chapters as well).

Two definitional clarifications first: For the purposes of this analysis, cyberspace and the Internet (the world wide web (WWW)) shall be hereinafter used as synonyms. Regardless of the (perhaps deplorable) arbitrary character of this decision, the fact remains that, being firmly in a Web 2.0 environment and perhaps planning for the immediate future, definitional clarity, as opposed to general public use, is of secondary importance. The second clarification pertains to the legal background: that will be unavoidably European- (and indeed, EU) centred. When possible, regulations and case law from other jurisdictions (mostly American) shall be provided, but the legal basis upon which the following analysis shall build is that provided by the European Commission.

23.1.1. *The Internet as a Living Space*

In order to assess the security implications of the Internet today, be they "cybercrimes" or not, we first need to examine what it encompasses, at least from a user perspective. Cyberspace, or the Internet, is an environment in constant development and human behaviour relating to it, that gives birth to such security issues, inevitably follows. Both the main characteristics and contemporary trends of the on-line world shape the "cybercriminology" background.

As regards the Internet's main characteristics, these relate to such common and well-known issues such as its "distributed" or its "borderless" nature. On their analysis, there is no need to over-expand. By now, it is common knowledge that the Internet is an uncentralised, borderless (virtual) space. Although some form of governance is sometimes necessary (for instance, ICANN), there is no such thing as a central authority (state or other) that monitors and regulates its use. This lack of centralised control creates a number of security issues: rules are applied selectively (if at all), individuals find no (easy way for) redress, organisations are constantly set up and dismantled in a virtual world. State regulations, as already noted, are generally useless over the Internet. Unless Internet organisations refer to well-established real world (even listed in stock exchanges) enterprises, there is really no way of imposing penalties or regulating effectively Internet companies (see, for instance, the struggle of the music industry, even after Grokster, to being down Internet piracy organisations by blocking their IPs). The same is valid for individuals as well — once they decide to set their blog on an off-shore server, there is very little the criminal courts of their country may do against their potentially unlawful postings [8].

Both the "distributed" and the "borderless" nature of the Internet are issues known for quite some time. Nevertheless, their full potential has probably not been properly assessed yet, if placed under the "ubiquitous computing" or "intelligent environments" light: "*on the road to the realisation of the Ambient Intelligence (AmI) vision, physical space becomes augmented with computation, communication and digital content, thus transcending the limits of direct human perception. An Intelligent Environment consists of a set of technologies, infrastructures, applications and services operating seamlessly across physical environments (e.g. neighbourhood, home, car), thus spanning all the different spheres of everyday life. Their inhabitants, humans and agents, will carry out tasks, most of which will be very similar to those that we do today, only their activities will be very different. The introduction of ICT and its applications in order to support these activities (and improve the efficiency of tasks) will change many of their parameters and properties, especially those related to space and time*" [3]. In other words, the term "ubiquitous computing" describes living conditions whereby individuals will never really exit a computing environment — information about them will flow around them through invisible computing mechanisms (for instance, RFID), in order to facilitate mundane tasks. Obviously, such information shall not stay isolated; information must flow, and the only accommodating network is the Internet. Broadband connections and Wi-Fi technologies are already in commonplace. Eventually, once all this information is placed on-line, the security and regulatory issues shall be formidable; if today contemporary legal schemes find it hard to regulate cyberspace, although only a handful of human activities are projected on-line, we can imagine what the situation will be like once individuals project their digital personal on-line, into such "intelligent environments".

On the other hand, the new medium needs public trust in order to develop. E-commerce, by today, is a substantial part of the world economy; some of the most expensive companies around the world are but Internet search engines (Google, Yahoo, Baidu); others are on-line service or goods providers (Amazon). These evaluations evidently need public trust in order to be supported. However, cyberspace always aroused public suspicion, both by its ardent users and by those who are not at ease with new technologies. Again here Internet's "distributed" and "borderless" nature is to blame. Experienced users who are aware of the difficulties of controlling the new medium need increased security in order to entrust their money in it. On the other hand, people with partial exposure to new technologies find it hard to comprehend a medium not connected to the physical world; this, together with public (justified or not) fears about on-line crime has called for a possibly secure and well-regulated environment.

E-commerce has thus imposed in practice the need for increased security standards in cyberspace. Equally, e-commerce has probably led to "cybercriminality", at least in its most serious forms, it could be maintained therefore that the same phenomenon gave birth to the problem and makes its solution imperative. At any event, the quest for "cybersecurity" shall be an endless one, given the basic characteristics of the Internet. This, coupled with the full-force-speeding-ahead trend towards ubiquitous computing (slowly making itself felt through Web 2.0 applications that shall be immediately discussed) only increases the level of efforts required, sometimes making obvious the shortcomings of contemporary legal schemes when put to the test.

23.1.2. *Web 2.0: Some Security Considerations*

Web 2.0 is the talk of the day (notoriously, Time's Man of the Year for 2007 was You — as seen through a mirror ingeniously published on its cover — due to the power Web 2.0 has awarded individual users). The term is used to describe what constitutes a, perhaps if seen from a distance self-evident, development of the new medium; in its original form, the Internet was used for one-way communication — each website communicated information to the public who, for the most cases, could only but read them passively. Web 2.0 has taken the next step, making the Internet a two-way communication tool; now users replied back, or even transmitted information themselves. The Internet's role changed (or was enhanced) from informative to communicative.

Web 2.0 is evidently based on wide public participation. User-generated content, blogs and blogging, social networks are all based on an as increased basis of Internet users as possible. Efforts have been undertaken to take the Internet out of the computer context and into TV sets, making it thus even more accessible to the last ones who refuse to acquaint themselves to it. The word, and the world's agenda, is "penetration": the level of countries or societies in more indexes than purely technology-connected ones is by now estimated according to its Internet connections.

The fact, however, remains that the Internet and cyberspace per se remained unchanged: the emergence of Web 2.0 applications and mentality did not necessarily mean that regulatory issues were resolved. Difficulties connected to their "borderless" and "distributed" nature continue to plague public trust, at a time when it is most needed.

In this context, Web 2.0 did nothing to appease public concerns (because after all it was not within its scope to do so). Quite on the contrary, its identifying characteristics (public participation, user-generated content and social networks), its market importance (Web 2.0 — in effect only — Internet sites are being sold and bought for billions of dollars) as well as its business models (that keep pressing at the border of, the said, on-line projection of the human persona) have only exacerbated the "cybercriminality"-related issues.

As regards the identifying characteristics of Web 2.0 applications, as experienced after all by each one of us who has basic exposure to the Internet today, public participation is perhaps its most dominant trend. Practically, all Web 2.0 applications (as embodied into Internet sites) ask from users to participate into something (update their profile, upload their content, tell others what they are up to right now) in ever-increasing numbers — indeed, the more the merrier. Users necessarily interact among themselves. Not only do they "do something" but their actions interact with those of other users (for instance, while making friends in social networking sites, rating videos etc.). And, unavoidably this worldwide, wide interaction is not worry-free. Exactly because it relates to millions of humans, each with their own disposition, temperament and even agenda, conflicts are bound to exist while interacting; this can be very much true both in the real world (when real-life individual rights are infringed) or in cyber-world (when a new flow of "cybercriminality" that only takes place in cyberspace and does no harm to physical objects or persons is emerging).

User-generated content is a more than obvious source of conflicts. As it is by now known to everybody, there is a great commercial value in facilitating exchanges of users-created content (mostly pictures and videos). Such facilitators (in the form of Internet sites such as YouTube) are being traded for billions of dollars. Evidently, when millions of individual users create billions of separate pieces of content (videos, music, pictures etc.), conflicts are bound to appear. These conflicts may either concern interpersonal relationships (users infringing each other's rights) or mass, "institutional" infringements (see, for instance, the attack on the fundamentals of Intellectual Property (IP) Law launched by P2P networks or videos uploaded in YouTube).

Social networks constitute a more subtle source of security concerns. Social networks are intended to acquaint among themselves as many of their users as possible. A lot of effort, in the form of complicated algorithms, marketing and social science has been devoted in making this successfully; indeed, the most valuable networks are those that have created the most links among their millions of users. The standard way of accomplishing this is to become as intimate as possible: each user creates the so-called personal profile, in which his or her preferences, thoughts and realities are laid down with as much detail as possible (in order to attract as many compatible friends as possible). Security problems are plain for everyone to see, and can range to anything from sexual harassment of minors to fraud or even crime collaboration (suicidal tendencies included). From a security perspective, each Internet personal profile (and many users have more than one) is a source of risk, being effectively the equivalent of an individual being exposed to social, real-life interaction. Never mind that all this takes place over the Internet: crimes, either Internet-assisted or Internet-enabled, usually tend to take a very real-life format at the end.

Risk sources shall probably not cease to emerge or even diminish in the near future, mostly due to Web 2.0 business models. In order to create a successful Web 2.0 application, that shall hopefully sell for billions, entrepreneurs need to assimilate or create as much as possible human-interaction situations. In other words, questions need to become as intimate as possible: in the, not so far, past it has been "tell us a bit about yourself", at the time these lines were written it is "tell us what exactly you are doing now". Virtual worlds also need to be as real-life as possible in order to become convincing (and, thus, attractive): a multitude of virtual worlds are made available to users, be they equipped with, virtual, swords and weapons or be they reality-like recreations (virtual money and contracting included). All these situations create security concerns that there is no practical way for regulators to address effectively. Because Web 2.0 is found at the avant guard of human (commercial) creativity, a level of security compromise is to be expected. On the other hand, because Web 2.0 is addressed to millions of users, an otherwise expected security glitch could affect individuals at an unprecedented scale. The main risk posed by Web 2.0 is that it has managed to bridge the unthinkable: living at unchartered waters in millions.

23.2. Certain (Contemporary) E-Commerce Security Highlights

Because e-commerce, be it in its Web 2.0 or in its more traditional format, is intricately connected with human ingenuity, a definitive analysis of its legal aspects is impossible. Apart from certain self-evident aspects (for instance, on-line contracting, that shall be elaborated in one of the following chapters), all other of its instances unavoidably have to be examined on a per-case basis. This is not only due to their unexpected form, reflecting some of the most ingenious human creations, but also due to their ever-changing content. E-commerce applications change along with their users at an unprecedented speed: a traditional book-selling website may record its users' preferences in order to enhance its book suggestions (adding thus privacy legislation to its list of relevant fields of law), protect its sales processes through patents and imposing them against competitors if necessary (adding also IP Law to the picture), and even try to artificially fragment cyberspace by creating country-specific shops and sell its wares only to residents of the same country (completing the mix of laws with some unfair competition or even EU, if in Europe, Law). E-commerce is a dynamic part of the market that shows no signs of settling down; as long as it re-invents itself every second or third year, adequate and comprehensive regulation of its many aspects is plainly impossible.

Security concerns and "cybercriminality" unavoidably follow this scene of continued developments. No one can regulate effectively risks whose full extent has not unfolded (or will never unfold because their sources will have been replaced within a couple of years since they first appeared). Legislation may come in very broad terms (as are, for instance, the fundamental data protection principles). On the other hand, the need for public trust is as pressing as ever; the more Internet businesses trade in billions of real-life dollars, the more the public needs to trust and use as much as possible the Internet. The two trends are obviously conflicting: e-commerce develops and keeps creating new sources of risk, while millions of individuals have to use them in order to keep the world economy going. Security balances and checkpoints, industry self-regulation and watchful regulators are all necessary, but there is only so much they can do by definition.

It is under this light that the following analysis should be read. What is effectively attempted is to address, from a security perspective, certain e-commerce-related issues in their contemporary form. Risks, "crimes" and their regulatory responses for each one of them tend to change constantly; while the analysis shall focus on certain basic e-commerce aspects that are thought to be as fixed as possible in the on-line context, readers should be aware that on-line notions, issues and solutions tend to outrun traditional, off-line publishing.

23.2.1. *Cyber-Enterprising*

Cyber-enterprising lies at the heart of the "cybercriminality" issue. As already said, e-commerce is inseparably connected to the most innovative and creative ways of doing business. Practically, millions of people around the world are thinking up of new ways to make money out of cyberspace; once they have identified an opening they storm in, in order to capitalise on their findings as quickly as possible before the next on-line trend makes their own obsolete. E-commerce, particularly Web 2.0 business models only have a life span of a few years. Within this time, their owners either make it big (whereby a major sale is in order) or they quit for the next wave. Even the same applications have to change constantly in order to keep relevant: on-line social networks a couple of years ago afforded different functionalities to their users as compared to today (and only the future knows how they develop in their effort to create real-life income).

Another point to be taken into consideration (that was too analysed above, under Section 23.1.2) is mass participation. By now virtual enterprising is not addressed to a handful of people in a few technologically advanced societies; rather than that it is addressed to the whole wide world. The only measurement of success today for any e-commerce application (essentially, website) is the number of individual visits ("hits") to its webpages — mass participation is thus pursued at any cost. This only exacerbates the security problem. Infringements to individuals' rights now come in waves and indeed may originate from anywhere in the world.

The above two factors were indeed analysed above (under Section 23.1.2). What could perhaps constitute a useful perspective while analysing virtual enterprising refers to highlighting the, typical by now, "cyber-enterprising legal process". This process has become time after time and Internet "phenomenon" after Internet "phenomenon" typical when it comes to e-commerce. It certainly builds upon the above two factors (need to innovate and mass participation) and it also takes into consideration certain financial factors as well: the short life-span of most e-commerce applications and the need to sell. In this context, the typical "cyber-enterprising legal process" is comprised of three stages: first comes a certain disregard for contemporary laws, second an exacerbation of the problem while the (successful) Internet application explodes through mass participation, and, finally, an arbitrary *ex post* solution not always in the best interest of either themselves or individuals.

The disregard for contemporary laws is inherent to virtual enterprising, at least during its conception stage. Innovators usually do not bother to ask their lawyers, and, even if they do, they tend to ignore their opinions. Indeed, there is no other way to explain P2P networks or users' video exchanging sites like YouTube or even the original iTunes deal offered to users. There is no way that any competent attorney would have counselled the first P2P network facilitator that basing its marketing strategy on affording users to exchange copyrighted material in millions would constitute a legitimate enterprise. There is no way that the owner of YouTube was not aware ever since its launching date that users, when creating their videos, invariably step into well-established and protected IP rights and his website made profit out of this. And, there is no way that no one told Apple that binding users through its iTunes to its iPod would not ultimately stand a chance (well, in Europe at least). And, nowadays, it is highly improbable that no one is advising on-line social networks on the privacy implications of certain policies they implement. Nevertheless, all of these projects got at the time the green light to be implemented at a mass scale. It could be because innovators feel that they need to risk in order to reap profit. Or because they feel that cyberspace affords them different rules than traditional real-life distribution channels. Or because they simply feel that contemporary laws need to change. Whichever the case may be the fact remains that practically all ground-breaking on-line projects present serious legal issues, at least when examined under the law then in effect: it seems that after all cyber-enterprising includes a certain level of cybercriminality by definition.

Successful on-line projects evidently exacerbate the problem. A lot of e-commerce applications do not meet public acceptance and eventually die out — their legal shortcomings never thus come to affect us. Those of them, however, who do appeal to the public, increase the problem into unexpected dimensions. Once P2P networks became successful, millions of users were logged in at any time exchanging millions of songs. When YouTube was sold, it came packed with millions of videos all including some form of IP infringement (be it in background music or using extracts from copyrights videos). When iTunes had to withdraw, its lock on iPod millions of users had already paid it under the previous terms and conditions (indeed, Apple is into a First World War barracks-type pitch fight to protect other equally obvious shortcomings, such as its country-specific sale of content through its local "stores", clearly infringing EU law). At any event, during this second stage of the "cyber-enterprising legal process", the problem is blown up but not resolved or even acknowledged. Millions of users see their rights infringed, various watchdogs complain, regulators start thinking that something should be perhaps done, but on-line enterprises seem to just wait for the problem to simply go away.

What on-line enterprises and entrepreneurs patiently wait is for the third stage of the "cyber-enterprising legal process" to take place, that is, for the final, big settlement. Once it is firmly established that mass infringement of rights does take place and that something must be done about it the same enterprises that caused the problem are ready to discuss. However, by now, they are big enough to negotiate favourable terms. Regulators generally show understanding while imposing fines for past sins to major players (and taxpayers) in their economies (with the exception perhaps of Microsoft). A settlement is thus reached that may include payment of some amount but is rather addressed to the future, adjusting the situation to legal requirements (P2P networks had to shut down but P2P television of telephony thrives; YouTube had to settle through payment of an arbitrary amount to content providers; Apple has to change its Sale Terms and Conditions from time to time). This settlement not always serves the best interests of the public, or even of the business itself, but is seen as a remedy than a solution.

The above typical "cyber-enterprising legal process" is necessarily cyber-criminality-prone. The disregard for legal requirements in its first stage means that the possibility of crimes being committed through the new applications is assessed and, ultimately, accepted. If crimes or security loopholes do make themselves evident during the second phase they are neglected, with the hope of acquiring in the meantime a base from which to favourably negotiate. The final settlement is the, winning, exit for the original perpetrators, leaving society to face with the problem. Although it could be supported that these stages are met in other dynamic fields as well (for instance, finance), they tend to constitute the rule when it comes to cyber-enterprising.

23.2.2. *Blogs, Blogging and Cyber-Opinioning*

The issues relating to blogs and blogging are well known by now: blogging has become such a popular trend that very few of us do not own or do not have sometimes owned or even regularly contributed to a blog. Using the Internet as a two-way communication tool has not been a recent idea (certainly not a Web 2.0 contribution), but its widespread, almost unanimous use has only been a recent addition. Before the time of blogs on-line *fora* or bulleting boards served the purposes of user interaction. These options were available since the early days of the Internet; what is new, is the unprecedented scale of today's opinion expressing over the Internet.

Mass participation means mass influence as well. By now millions of people visit and learn or entertain themselves from blogs. Blogs have thus developed towards two, interesting from a security point of view, directions: First, mostly news-related, blogs have become small news agencies of their own, employing several people and creating substantial income. Second, blogs are the preferred way to bring to the public unpopular or shocking news or even to organise acts of opposition. Each of these categories presents different crime-related issues.

The first category was perhaps a foreseeable development. The most successful blogs, that indeed started out by individuals, had to develop in order to survive. Particularly, those blogs that offered information on niche fields (see, the discussion on the long tail of the Internet) had to keep gaining in depth in order to keep users connected. They thus developed into small news agencies, employing several people or expanding overseas. Nevertheless, news blogs never lost their character, meaning that they never intended to be possibly impartial news agencies, but rather ways to express conceptions and ideas of their, individual, creators. These same creators also participated in the market or field they covered through their blog. When money also came into the picture, conflicts became inevitable. Blogs and bloggers may infringe rights of third parties mentioned in their blogs (indeed, several blogs have as central purpose to identify "bad" participants in the market they cover); they also may misguide public opinion (and perhaps, shares' value) to their own benefit. Bloggers are also frequently operating through nicknames, and are hard to find (and sue, if applicable). Given the "borderless" nature of the Internet, users are rather advised to exercise caution, than to file later for damages. Regardless of the latter recommendation, however, the fact remains that the development of blogs as known today constitutes a continued source of risk both for users and unsuspecting third parties who may find themselves mentioned in them.

Blogs are also the preferred way of self-organisation when it comes to acts of public opposition. This may be in the form of publishing shocking news (for instance, photos), or organising events or posting breaking news from sites of upheaval or repression. All of the above have been used in more than one instances until today all around the world. Bloggers are sometimes identified and prosecuted; most of their actions, however, are successful, at least from a raising public awareness perspective. The immediate nature and the, at first, anonymity that the new medium affords have made it indispensable to similar causes. Nevertheless the security risks potential is obvious for everyone to see — whether we should live with it or take positive measures to abolish it is a totally different discussion.

Cyber-opinioning came at the time the Internet was invented and has accompanied it ever since. It has developed, taking advantage of enhancements afforded by new technologies, but it has remained in essence the same, affording the option to individuals to express themselves and interact, sometimes anonymously. Misunderstandings, and "crimes" therefrom stem from a misguided perception by authors that expressing themselves over the Internet differs from expressing themselves in the streets or in the traditional press, as well as, by a reader's award to their readings of more value than they are really worth. Although this situation has often led to serious difficulties, abolishing or even policing it more effectively hardly appears the best (if at all possible) way forward.

23.2.3. *Framing and Deep-Linking (and Associated Practices, Including GoogleNews)*

Practices such as "framing" or "deep-linking" may be categorised, according to the distinction above (under Section 23.1), as infringements of rights that take place in cyberspace and were previously unknown both to humanity and regulators. They both relate to e-commerce and in particular to the so-called cyber-enterprising. In order to properly explain their function, certain clarifications need to be made with regard to the commercial use of the Internet. Because since the first days of the Internet until today no effective business model has been devised to make users pay money (and e-commerce companies incur income) from service provided on-line (indeed, the trend of "free" has gained exponentially in strength [14]) advertisers' money is the obvious alternative. In fact, today the biggest companies over the Internet (and some of the biggest, as least according to their Stock Exchange evaluation, in the world) are solely based on advertising income. However, in order for advertisers to spend their money on Internet sites, they need proof that their clients' webpages are indeed visited by individual users. Individual visits per webpages (or, "hits") have become thus the Holy Grail of the Internet, at least from its business perspective, these days. In fact, "hits" (a term that shall be used here invariably, regardless whether it refers to individual visits or repeated downloads of the same page by one user or in any of its other, technical, distinctions) are the standard measurement of a websites' success. It is according to its number of daily, weekly or monthly hits that its owners ask for advertising spending, mostly in the form of banners affixed onto one or more of the same website's pages. It is according to their number of hits that bloggers count their readership, product- or service-selling websites their potential clientele, search engines their use (and penetration) to the public.

It becomes therefore evident that whoever wants to make money (or even a difference) out of the Internet needs to generate as much as possible traffic to his website, in order to then ask for adequate advertisement spending. The more the hits, the merrier.

In this continued struggle for hit-dominance, it would have been quite extraordinary if a number of deviant practices did not arise, aimed at directing hits towards webpages that do not deserve it. Under this category fall the various offsprings of cyber-squatting (most common today in its form of typo-squatting), that shall however not be analysed in this chapter because of the relatively stable environment accomplished by ICANN to-date, as well as, such tricks as "framing" or "deep-linking".

"Framing" broadly refers to the practice whereby specific webpages are "boxed" or "incorporated" into the websites of third parties other than their owners'. For instance, a newspaper article is included as a whole into a blogger's website; or, real-estate classified ads of a certain website are copied-pasted into another Internet site even under a different or even better presentation method; or, government information webpages are "boxed" into a search engine's results to a query relevant to their content. Evidently, around such "boxes", the advertisements and the Internet site of the perpetrator appear to Internet users.

The obvious result of "framing" is the loss of hits for the original owner of the webpage and the, unworthy, increase of hits for the "framer" 's website. Evidently, in the above example, the blogger will have diverted to his/her Internet site people who want to read that same article but who would have otherwise visited the newspaper's Internet site. The site that copied the classified ads will have increased its number of hits based however on content provided by the original, even cruder, Internet site. And, the search engine will have found a new way for generating hits, diverting them from the official government site from where the relevant webpages were taken. In all these cases, the conflict of interests and rights is plain for everyone to see: the "framing" site increases its income with content that does not really belong to it, while the content-owner loses Internet traffic (in equal or other numbers).

On the other hand, "deep-linking" refers to the practice whereby that internet traffic for the "victim" website is diverted to its internal webpages, where it does not matter that much. In e-commerce websites, mostly their homepage has come to matter; it is this page that attracts the most hits, and it is the hits of this page that receive attention; for instance, a newspaper will count hits on its first page and not necessarily on each webpage containing a single article. By deep-linking to its internal webpages, even without copying-pasting them and framing them into another website, still damage is caused to their owner, because users avoid the homepage (and thus their hit is missed) and visit directly the webpage that interests them.

Websites that engage into "framing" or "deep-linking" customarily claim that they do offer an added-value service to the public, organising information better. People have too little time, and by sorting out huge Internet sites (such as those of newspapers) and guiding them to the exact webpage that interest users is a worthy cause. Additionally, deep-linking at least does not cause much harm, because after all it is normal, and lawful, "linking" (the equivalent of referencing to the academic world), only to the exact webpage and not the homepage (again, the equivalent in the real world is referencing to a page rather than the cover page).

Regardless of the merits of such reasoning, the fact is that by now both "framing" and "deep-linking" have been found unlawful around the world. In more than one jurisdiction, it has been established by courts that these practices constitute unlawful infringements of the IP rights of the original webpages' rightsholders. The legal grounds may vary, ranging from traditional IP (copyright) law to (EU-specific) database protection legislation (the *sui generis* database right). There is thus not much meaning in continuing the above discussion.

What does however merit discussion is the contemporary forms (or not) of similar practices, the most well-known of which today is GoogleNews (http://news.google.com). As by now everybody knows, Google has established a service free to its users, whereby more than 4,000 news websites are scanned daily, and the news appearing thereon are indexed with a 3-line summary to Google's website. Users may visit Google's site, read the news and the summaries and click on the article, if they wish to, in which event they are guided to the actual webpage (not the homepage). Additionally, users may store keywords in their personal profile, and Google shall alert them whenever a relevant article appears in one of its indexed sources (again guiding them to the actual webpage). From this point of view, the practice is nothing more than a typical deep-linking example, with an information aggregator oranising information and guiding traffic first on his/her own Internet site and then to the internal webpages of its sources. Newspapers that participate in the GoogleNews "programme" evidently lose on hits from their webpages. Nevertheless, such is the power of Google today (or the amount of hits created to its featured articles anyway) that from all around the world, including some of the biggest and best news agencies, only the press from Belgium objected (and, evidently, succeeded). The rest have not reacted; the situation is obviously found at its second stage of the "cyber-enterprising legal process" described above, under Section 23.2.1 — the outcome shall probably depend on the popularity of this new service.

23.2.4. *On-Line Auctioning*

Although auctioning is by no means an activity previously unknown to humanity, the Internet has taken it at a whole different level. What was restricted in the past to expensive assets (real estate or machinery) or art and was fragmented and difficult to participate, which is still the case in the real-world, has nowadays become available to the millions who are now bidding for anything from the most mundane and trivial to the most exclusive and expensive. On-line facilitators have afforded Internet users this possibility. Internet e-commerce sites are offering to sellers the opportunity to upload their goods and the respective asking prices and to bidders the opportunity to participate in a simple but secure on-line auction. Money again is made through advertisement. Facilitators normally do not accept any responsibility for the professionalism of their users — most of the times they are not even aware what is being auctioned through their Internet sites.

It is exactly these unique characteristics of on-line auctioning that have caused various cyber-criminality issues. Sales by individuals to individuals around the world have frequently led to fraud. Unmonitored auctioning has led to crimes being committed whenever yet another vulgar auction (for instance, human-body parts, nazi memorabilia) takes place unnoticed (or, noticed too late).

And, whenever there is profit to be made, various schemes are put to work (for instance, automated-bidding web-bots) that thrive at the borders of either the law or contract. It is these three sources of risk (seller-bidder relationship, subject-matter auctioned and system operation) that taunt contemporary on-line auctioning providers.

Seller-bidder relationships are bound to include some fraud when the numbers rise to the millions; fraud is, expectedly, the greatest security concern when it comes to on-line auctioning. It may come in many forms: sellers may dispatch to successful bidders goods that are far from what was promised on-line. Credit card details that are used for payment by bidders may be put to other, unauthorised, uses as well. The anonymity frequently afforded to both sellers and bidders by on-line facilitators only aggravates the problem. The international element, when combined with the insubstantial value of most transactions, means that any attempt to legal recourse makes no sense, at least from a financial perspective. What we are effectively left with is with millions of individuals being helplessly engaged into international transactions whose proper execution they are in no condition to secure.

This fundamental problem of lack of trust has been addressed in the best way possible by on-line auctions facilitators. Websites making their money out of the number of auctions that take place through them (and the hits and advertising income realised therefrom) are absolutely interested in providing their users with a possibly secure, worry-free environment. Various systems have been put to this end: on one side of the auctioning system secure electronic-payment systems (such as PayPal) are offered to users in order to avoid credit card fraud. At the other, more difficult to regulate, side, scoring systems have been devised in order to generate public trust. According to these systems, sellers are graded each time they successfully complete a sale through an on-line auction. Users are to trust those sellers with the highest scoring. Public trust is thus gained upon a trial-based, past-experience system that is said to resemble as much as possible as the real world, whereby the most experienced and well-known merchants make the most sales.

The subject-matter auctioned in on-line auctioning systems is the second source of risk. With millions of auctions taking place simultaneously some unlawfulness is unavoidable. This may be in the, relatively harmless, form of auctioning prohibited goods (for instance, medicine) to the vulgar (and perhaps hard to believe) recorded cases of human-body parts or nazi memorabilia auctioning. Website facilitators defend themselves on the basis of the fundamental e-commerce legal principle, that providers are not held liable for their users' actions unless they known them (or are notified accordingly). Whenever therefore a prohibited auction takes place, the facilitator shall invariably claim no knowledge of it; if he/she is notified in time, he/she must disrupt it. This legal treatment, unavoidable and ultimately fair as it may be, may save on-line auctioning entrepreneurs from their responsibility for what is actually being auctioned on their webpages but does, however, little good to the credibility of the system altogether.

Finally, the system itself may be bent: auctioning facilitators are, as seen, equipped with secure electronic-payment systems and scoring systems and the law on their actual liability and are hoping for the best, but security concerns are far from resolved. Whenever there is money to be made, various schemes are put to work. Scoring systems may be cheated; in fact, sellers knowing that unless they are able to show some history and positive scoring they will not make any sales will most probably make their first sales to themselves, thus adding up to their profile. Accordingly, any elaborate bidding system can be cheated; because users are expected to bid until the last minute but technical (on-line) restrictions do apply when submitting, web-bots and other applications have been developed that will perform this task for them according to their instructions (for instance, "make my best offer one minute before the auction's ending time). In fact what on-line auctioning is sometimes is a fight between web-bots for last-minute bid submission. Naturally, this "bends" or even blatantly breaks the terms of service of the Internet site hosting the auction, but it is hard to prove (and the site's owner will ultimately prefer not to disturb his/her clients).

On-line auctioning systems are ultimately connected to the issue of trust over the Internet. Cyber-enterprising, in this case, simply refers to facilitating simple transactions between individuals around the world. Some cyber-criminality is bound to emerge somewhere in cyberspace and, because cyber-entrepreneurs are not (could not, as well) be held liable for it, it ultimately is up to individuals to protect themselves. However, public trust needs to be vested in the new medium in order for it to develop; whether by scoring systems or secure payment systems or any other idea that may come up in the future on-line auctioning systems' continued well-being actually depends on it.

23.2.5. *On-Line Gambling*

On-line gambling belongs to the type of cyber-criminality that, as discussed above under Section 23.1, is committed with the assistance of the Internet. In fact, the Internet is ideal for on-line gambling. Its "borderless" and "distributed" character means that gambling sites may be set up anywhere in the world, where they are lawfully allowed to operate, and still make sales to individuals in countries that may have a legislative gambling prohibition. On the other hand, individual users find that they no longer need to travel to casinos or have access to betting shops (or be adults, for the same purposes) in order to gamble; with the assistance of a multitude of gambling Internet sites, they can do so as the comfort of their living rooms. From this point of view, it is a win–win relationship.

Those who do lose out of this are legislative prohibitions and society ethics. A state ban on gambling may easily be circumvented; servers (and their sponsoring companies) may be setup off-shore, individuals who gamble on them are hard to trace. The situation is even worse for those countries that have a state monopoly (such as the case in may EU Member States): in this case, the Internet has enabled gambling outside state-sponsored channels and governments find it increasingly difficult to explain why these sites should be banned, at least based on reasonable argumentation (if state-sponsored gambling is allowed after all, accepting thus gambling risks to individuals, why is it wrong for private parties to engage in a profitable activity?). Society ethics also suffer: those societies that have chosen not to afford the option of gambling to their members, out of fear for what gambling can do to them, find it impossible to apply their decision to those of their members with Internet access.

Naturally, legislative and other solutions do exist: on-line gambling facilitators may be held liable for breaking the laws of one country although their servers are off-shore and their sites are addressed to the whole wide world, when, according to what has become a basic Internet principle, out of the content of their webpages it can easily be concluded that all or part of them is addressed to residents of that particular country. This, for instance, can be established through flag-enabled special webpages for these users, the languages an Internet site is offered into, its terms of service and other similar case-specific circumstances. At any event, the fact remains that infringement by a website of the laws of a particular country may be established in courts regardless whether it has a domain in .com or country-specific; acquiring thus court protection is possible after all — applying it is a totally different issue.

When application of the law or a court decision is the issue at hand, more radical measures are needed. Because gambling sites will normally operate behind off-shore companies, located in jurisdictions, that is, that are broadly negative to the option of losing valuable tax income, usually international judicial co-operation agreements for the enforcement of court decisions are of no use. This is why countries, such as the United States, have recently launched a two-front, practical thought-of attack on on-line gambling. First, they attacked their payment systems; then, they attacked the big players, those of on-line gambling sites that were audacious enough to become big enough to be listed in Stock Exchanges and behave like multinational enterprises.

The first attack was easy enough: all on-line gambling is obviously based on credit card payments. And, credit card payments are easy to trace and are also carried out by well-established companies that want no trouble with the law. What America, therefore, did what to make all on-line gambling-related credit card payments illegal. Credit card companies had to comply — on-line gambling sites were suddenly deprived of their source of income (at least from Americans).

The second offensive included the arrest of certain high-profile CEOs or (on-line gambling) company owners on American soil, demonstrating thus that no one is safe, no matter how big his/her company is (regardless whether it has been until recently lawfully operating in America too, and whether it continues to do so in others, equally sophisticated, jurisdictions across the Atlantic).

Reactions in Europe towards on-line gambling greatly vary. Many Member-States sponsor a state monopoly on all gambling activities, a practice that may conflict with more than one field of basic EU Law. Those Member-States profiting from a very strong gambling industry that has fully taken advantage of cyberspace (for instance, the United Kingdom) are obviously positive and would like the markets of other Member-States, profiting from a state monopoly, to open. The latter have reacted in various ways in order to protect their internal markets: sometimes, in the cases for instance of Greece and Italy, it has gone as far as prohibiting or attacking Internet cafes (being popular points of access to individuals). A number of court decisions both at Member-State level and even of the European Court of Justice have thrown some light into the matter. However, until the time this analysis was written, no final decision had been reached for opening up the markets; in principle, however, it is evident that the existence of a state monopoly per se deprives governments from any argumentation on whether gambling (in its on-line or off-line format) is good for their citizens — discussions have shifted by now on how to preserve tax income and Stock Exchange (government-owned) corporations' evaluations.

At any event, the fact remains that countries are broadly at war with on-line gambling, sometimes winning and sometimes losing battles. Regardless whether opposed to the idea itself, as is the case in America, or fighting to preserve profitable state monopolies, as is the case in the EU, the issue of on-line gambling raises eyebrows among legislators and security officials in both sides of the Atlantic. Users, on the other hand, are not as negative: on-line gambling sales are peaking. Users also enjoy an unprecedented freedom to gamble (never before was it so easy to any European to gamble on American sports) and to compete — once poker went on-line its international champions came in the unlikely forms of 19-year-olds who, instead of having to spend their whole life in bars, spent a lot of time on the Internet playing with other players around the world and improving their technique. Choice is ultimately good for them.

The internet has therefore enabled a question of ethics to be repeated: is gambling to be allowed or not? Societies that thought they had already answered that had better think again — the Internet has changed all their data. If the society's answer to the above question was an unequivocal "yes", then its member may only rejoice at the limitless options that are opened to them by the Internet. If the answer to the same question has been a (hypocrite's) "no, as a general rule, but yes when made by the government", then the Internet has emptied this society of its arguments: no more is gambling unsafe, connected to crime in shabby-looking places — the on-line Internet sites are a polished and secure way of doing business. This society had therefore better think again whether tax income has not blurred its judgement criteria. Ultimately, it is only those societies that prohibit gambling altogether that face the greatest threat by the on-line environment; because controlling it in cyberspace is more or less impossible that they might need to re-evaluate their original decision. The Internet has in fact been a catalyst in the case of on-line gambling, depriving hypocrites of their arguments and confronting general bans against an individual choice.

23.2.6. *On-Line Advertisement*

On-line advertisement has of course nothing to do with crime. Quite the contrary: it is in effect because of on-line advertisement that the internet developed in the way we know it. As already explained (under Sections 23.2.1 and 23.2.3), practically all e-commerce models have failed to identify until today any serious source of income other than advertising revenue. Based on the number of individual visits per webpage, advertisers are willing to pay its owner in order to place advertisements (in the form of banners) on them. Search engines (most notably, today Google, but others too) have devised similar methods to commercialise on users' searches: each time a term is searched, users are served apart from the results with, sponsored, links to goods or services that might be of interest to them. Software industries are trying to make models of on-line advertisement as effective as possible. As long as advertisement income continues to flow into the Internet, its existence (and even its development into Web 2.0 or Web 3.0) is secure.

On-line advertisement, in itself, has therefore nothing to do with cyber-criminality. Those few legal issues that were posed by it (for instance, potential infringements by AdWords to trademarks) have been resolved by now. The reason why it shall be briefly elaborated here pertains to the technical means it uses or the business decisions it makes from time to time in order to increase its effectiveness, and the effect the latter have on the law.

Banners constitute intrusive technology. As all of us know by now they tend to pop-up anywhere on webpages, sometimes obstructing our view or even refusing to go away until they have displayed their message in full. What is perhaps unclear to most of us is that banners are elaborate computer programs. Apart from their impressive graphics, they have code behind them that carries out a series of functions, from displaying properly depending on our system configuration to transmitting information back to their owners. Banners are seldom one-way communication tools that, once incorporated onto webpages, their job is to simply stay there and make themselves visible. Under normal circumstances, they are expected to keep communicating with their owners long after they have been attached onto webpages. The type of communication and the depth of information transmitted is the object of some, security and legal, controversy.

To begin with, banners and other on-line advertising methods are invariably security loopholes. Because of their need to communicate information back to their owners, they need an open communication channel from each computer they play on. An open communication channel is, understandably, very bad for security network. This is the reason why most contemporary Internet browsers come equipped with security settings that by default prohibit pop-ups from showing; users may disable them at their peril. On the other hand, on-line advertising in fields such as porn may have even worst implications for those unlucky users who visit relevant Internet sites: apart from all the above, it may come bundled with malicious software (for instance, dialers) that will attempt to defraud unsuspecting visitors and non-experts of their money.

Evidently, any malicious actions by banner-covered software onto individual computers are crimes committed by their owners (or even the site owners, if they were aware of it). Most countries have by now implemented computer crime legislation that normally covers any unauthorised actions taking place on computers. The problem in this case is not that much identifying the crime but rather applying the sentence; the "distributed" nature of the Internet means that perpetrators may reside off-shore, while users will normally prefer to format their hard drive than take the expenses and time to pursue them.

Even if users are not bothered by two-way communication happening on their computers, in return of watching these fancy videos, they would probably be interested to find out that on-line advertisement schemes in their more aggressive forms systematically make profiles on them. This is a business choice by on-line advertisers: in order to improve the effectiveness of their practices, they try to profile users. In fact, the Internet advertises exactly on this added-value functionality, as opposed, for instance, to television. While on television, broadcasting of commercials is random and advertisers have no way of knowing if they were broadcasted to interested parties or if individuals really saw them, the Internet is more accommodating. Mechanisms do exist to ensure that ads showing on users' computers are relevant to their taste and preferences; the same mechanisms let advertisers know whether users have viewed their ads (by clicking on them) or not. All these are accomplished by more or less making profiles on users. User preferences, IPs, sites visited etc. are all meticulously recorded, by way of installing relevant software onto a computer (through, for instance, banners or cookies), and are then transmitted back for further processing. Once profiles have been prepared, users are only shown those banners that are thought to be more relevant to them. Although this practically breaches every known data protection notion and principle, at least in the EU, perpetrators are hard to track because they tend to operate from friendlier jurisdictions. Here again, the costs required to do this are prohibiting.

On-line advertising is thus interesting from a security point of view when it comes to its business models and the decisions behind them. Because of the struggle for (on-line) survival, aggressive decisions are frequently taken that disregard both the law and fundamental security requests by users. The situation is aggravated by contemporary e-commerce trends. Once it has been established that advertisement income is the only sustainable solution for Internet services that users want to be given out to them for free [14], the only way forward is for on-line implementations to increase in number and improve in quality. This will unavoidably lead to increased attacks on individual privacy, and may even create some additional loopholes in computer security. Although each one of us should be grateful to on-line advertising and the money it has spent in order to make the Internet what it is today, we should all at the same time be watchful when it asks for more privacy and security compromises in order for it to thrive.

References

1. L. Edwards, *The New Legal Framework for E-Commerce in Europe* (Hart Publishing, 2005).
2. T. Hingston and S. Adam, Click-through banner advertising: A technical review, available at http://ausweb.scu.edu.au/aw2k/papers/hingston/paper.html.
3. http://conferences.theiet.org/ie06/index.htm.
4. I. Lloyd, *Information Technology Law*, 4th Edition (Lexis Nexis UK, October 2004).
5. M. I. Melnik and J. Alm, Reputation, information signals, and willingness to pay for heterogeneous goods in online auctions, February 2003, Available at SSRN: http://ssrn.com/abstract=452820.
6. OUT-LAW News, Google proposes 'crumbled cookies' in privacy pledge, 1 October 2007, http://www.out-law.com/page-8511.
7. C. Reed and J. Angel, *Computer Law*, 5th Edition (Oxford University Press, 2003).
8. *The Economist*, Leaks and lawsuits, 6 March 2008.
9. *The New York Times*, Italian case may open up crossborder gambling, 07 March 2007.
10. The Register, Google news faces 1m fine in Brussels, 18 September 2006.
11. The Register, Google loses Belgian news appeal, 22 September 2006.
12. The Register, US restricts online gambling, 12 July 2006.
13. P. Todd, *E-Commerce Law* (Routledge Cavendish, 2005).
14. WIRED, Free, 16 March 2008.
15. WIRED, Online poker: Is it legal?, 5 August 2006, http://www.wired.com/techbiz/media/news/2006/08/71547.
16. WIRED, Refusing to fold, Online poker players bet on prohibition repeal, 21 May 2007, http://www.wired.com/politics/law/news/2007/05/gambling_laws.

Chapter 24

INTELLECTUAL PROPERTY RIGHTS: THE SECURITY PERSPECTIVE

VAGELIS PAPAKONSTANTINOU
Attorney At Law, PK Partner, Athens

24.1. Introduction

Although Information and Communication Technology (ICT) practically left no field of law unaffected, very few cases have been recorded where they actually threatened the very fundamentals upon which such a particular field was constructed; Intellectual Property (IP) Law is however one of those fields that holds such an enviable distinction — once ICT fully released its potential, no principle of theoretical construction was left unharmed.

This chapter attempts to highlight only some of the numerous ways in which ICT affected, and continues to affect, IP Law: the emphasis this time is on cybercrime and security issues. In this context, a brief elaboration upon the reasons of change shall be undertaken before embarking upon the presentation of selected fields, where the most noteworthy issues have been raised (content, software, databases). The case of peer-to-peer (P2P) networks shall attract particular attention, not so much on account of their legal treatment, an issue more or less resolved by now, but because new technologies (for instance, IPTV or even VOIP) have brought them onto the surface in totally new contexts. Finally, the EU Database Right shall be briefly presented, again from a cybercrime and security enhancement point of view.

In short, the relationship between ICT and IP rights is, to say the least, challenging. First, there are several occasions for conflict: holders of traditional IP rights (namely, the content industry[a]) have emerged as self-appointed keepers of the, traditional, IP scheme and are thus opposed to any ICT that challenges traditional IP norms (as controlled by them): for instance, P2P networks are to be judged illegal, unless they are equipped with Digital Rights Management (DRM) systems. Then, there are several occasions for profit while exploiting previously unchartered areas: ICT opens new sources of income for the Content Industry and others. New e-commerce methods add up to existing distribution channels for the benefit of those players in the market who can identify them; however, such "unchartered areas" usually come with a price on security. Finally, ICT constitutes in itself a difficulty with regard to its own legal treatment: software is still after several decades in the middle between copyrights and patents, while in some (European) jurisdictions databases profit from then own customised legal protection. In this contradictory and evolving environment, stakeholders change sides frequently and try to bend the rules to their own favour.

For the purposes of this analysis, however, IP Law should be perceived as inseparably connected by now with ICT and e-commerce models. This is an inherently risk-taking process — and risk gives birth to security issues. These issues are invariably technology-related and may be tackled with either in traditional legalistic ways or through technology itself. For those cases that have enough history of implementation to form "cybercrimes", the answer lies in formal legal methodologies. On the other hand, for those technologies and individuals that are involved with state-of-the-art developments, an increased security risk is to be expected, to which traditional legal action may not (or even, could not) always be of much use.

A single clarification to be made in advance relates to the, essentially, EU focus of the analysis that follows. Intellectual Property Law does fall within the First Pillar and is indeed more or less harmonised among all EU Member States. In the same context, the EU approach may not always be compatible with third-country approaches (for instance, in the case of copyright). Despite the substantial unifying efforts of international organisations (mostly, the WIPO) not every country uses the same rules when it comes to the protection of IP. The analysis that follows shall be primarily EU-focused, paying particular attention to EU regulation and case law (and legislative particularities as is the Database Right itself) while approaching the security and cybercrime issues brought by ICT on IP Law.

[a]The term "content industry" shall denote in this chapter "an umbrella term that encompasses companies owning and providing mass media, and media metadata. This can include music and movies, text publications of any kind, ownership of standards, geographic data and metadata about all and any of the above" (see Wikipedia at http://en.wikipedia.org/wiki/Content_industry).

24.2. ICT and Intellectual Property (IP) Rights

The legal scheme for the protection of IP is one of the few whose very premises have been challenged by the emergence of ICT. Because, as it will be immediately demonstrated, ICT changed the physical mechanisms used by legal theory ("works" and distribution channels), the protection of works of the intellect will never be the same. The attack on IP by ICT has been two-fold: not only have the exploitation methods of traditional assets ("works") been challenged, but also newcomers in the field (most notably, software but also databases, multimedia etc.) have demonstrated the limits of the IP rights system.

The legal response so far to these challenges has taken shape in a struggle, first, as regards already known "works" (music, text, image) to maintain well-known notions and mechanisms, by "cybercriminalising" new methods afforded by technology, and, second, as regards newcomers such as software or databases to avoid creating new fields of law but rather to incorporate, admittedly with little success so far, them into already existing categories of "works" in traditional IP law sense.

24.2.1. *The Basics of Change*

The conflict between the legal system for the protection of IP and ICT became unavoidable once the latter effectively altered the nature of the former's protected subject-matter (by digitising information) and reversed century-long practices (by adding telecommunications networks to traditional distribution channels).

24.2.1.1. *The digitisation of information*

The digitisation of information achieved by the, then, emerging Information Technology (IT) signaled the first difficulties for the copyright scheme [16]. Until that time, the copyright system for protecting IP had worked relatively successfully for over two centuries. It was first developed in the United Kingdom back in 1710; only at that time did mankind realise that works of the intellect could be of an economic value, based on their "use" by others, and therefore constituted "property" of their author (or right-holder). In this sense, the system that was then developed, and is still largely in use today, focused upon protection of the "work" of the intellect against unauthorised reproductions (copyright being essentially the right to copy). The author of a protected work under this legal scheme was entitled to compensation for each and every use (reproduction, copying) of his/her work by others.

The digitisation of information challenged the practical (not theoretical) parts of this scheme, by altering the nature of its subject-matter. Until then "works" came out essentially in the form of texts or music or drawings (or even movies). In these forms, however, reproductions (copies) of any "work" were relatively easy to control (and thus, ask for compensation): books had to be printed and sold on bookshelves, music had to be copied into vinyl and sold on record stores. All these actions of reproduction involved cost (and thus could not be undertaken by anyone), and were controllable because of the relatively restricted distribution channels (shops) and the fragmented market (international commerce meant totally different things at the time). The digitisation of information managed the first blow to this scheme: once texts and music and pictures became digital, anyone could reproduce them at minimum cost. Evidently, the 17th century scheme, whereby any act of copying would confer money to the author of the work, automatically became obsolete: copying became so vast that the content industry could no longer control it as effectively as it did in the past. Even new (and technically challenging) "works" (for instance, movies) did not escape this global trend; once it was established that there was a market in copying and storing them in users' computers, it was only a matter of time before the appropriate (copying) technologies were unleashed to the public.

24.2.1.2. *Distribution over networks*

The emergence of networks, and in particular the Internet, managed the second, and crucial, blow to the legal scheme for the protection of IP, essentially by increasing exponentially the number of distribution channels. Until the interconnection of (home) computers was made possible, the digitisation of information alone, regardless whether annoying in itself for the Content Industry, remained inevitably "computer-isolated": any user could store tons of protected material in his/her computer, but use essentially was confined to his/her computer alone. Because networks did not exist (at least outside the academic or employment environment), any exchange of protected works with other users had to be performed physically, by means of copying onto a disc and carrying the disc to another computer in person. Consequently, even at that time the Content Industry was not particularly discomforted: although its property was digitised and copied massively, user-isolation meant that purchases of originals were not substantially affected.

Once user networks and, ultimately, the Internet emerged this was no longer the case: connected users were suddenly able to exchange "files" (incorporating unauthorised copies of protected material) without moving from their homes, at a single press of a button and at a marginal cost. Traditional distribution channels (i.e. shops) were shattered. No longer was it necessary at least for some users to purchase the original in order to digitise the work in it — the, vast, Internet community made sure that once a single user in the whole wide world purchased the original and digitised it everybody could then have it for free through a simple download [16].

To make things worse, e-commerce systems emerged that enthusiastically facilitated the exchange of files among users, namely peer-to-peer (P2P) networks; the analysis of the demise of the first attempts and their subsequent forms shall follow (see under Section 24.3). At any event, the Content Industry now had to face new e-commerce systems that demonstrated new ways of exploitation of its works.

The traditional IP system, based upon control of the "uses" of a work through control of its distribution channels, quickly became obsolete.

Nevertheless, at that point an interesting approach was adopted from a law-making point of view: rather than devising new theoretical and practical schemes to adopt to the new, irreversible, conditions, law-makers (supported by the Content Industry) stuck to the traditional IP scheme: in effect, control over distribution channels was attempted to be regained through Digital Rights Management (DRM) security-enabling technologies and the "cybercriminalisation" of new distribution channels such as P2P networks. At the same time, new ICT by-products (for instance, software or databases) were struggled into existing legal schemes (through an appropriate patching-up whenever needed) rather than introducing new legal tools: the limitations of this approach are making themselves obvious practically on a daily basis.

24.2.2. *The Case of "Content" (or "Works")*

Evidently, the changes presented above affected the traditional assets ("works") of the IP scheme, in other words, content (texts, music, audio, images etc.). New-type IP assets, such as software (see under Section 24.2.3) or databases (see under Section 24.4), presented other challenges to the traditional system for the protection of IP.

It is hard to overestimate the importance of content in contemporary markets. Content is an extremely valuable asset in itself (just imagine how much a pop hit or a bestseller is worth) and the reason of existence of all networks (the Internet included). Once we have found ways to communicate, we evidently need information to exchange: digitised content is practically the blood that runs in the telecommunications veins. Without, it, the Internet and much of contemporary television, to name only a few, would be void and irrelevant.

From this point of view, the conflict between ICT and traditional IP forces, as represented by the Content Industry, was inevitable. IT enabled the digitisation of content and telecommunications made the distribution of such digitised content easy to anyone. Unauthorised reproductions (copies) and distribution of protected material were bound to become extremely popular among Internet users (as it will be seen under Section 24.3, peer-to-peer networks). The Content Industry, whose income was severely affected by these developments, launched a counterattack to the technologies that threatened its interests (P2P Networks) and introduced technologies that would warrant application of the law (DRM systems).

Evidently, the field of law under discussion when it comes to the legal treatment of content is copyright (or, for some Continental European countries IP Law, as opposed to Industrial Property Law, essentially patents). It was therefore basic copyright legislation that ICT threatened, and it was copyright legislation that the Content Industry wishes to secure through its own ICT implementations (DRM systems).

When the first ICT technologies emerged and were put to public use the copyright system's reason of continued existence was loudly questioned by several proponents of "information is free". Under the new circumstances of an interconnected world, evidently a 17th century system whereby each copy of a work brought income to its author was no longer relevant [13]. Evidently, at the other end stood the Content Industry, a multi-billion infrastructure that came to secure a legal system absolutely necessary for its own survival. To keep the analysis short, it seems that the traditional IP system will not be abandoned, or changed, after all: the very same ICT technology that threatened its existence shall be put to its rescue.[b] While the Content Industry shall continue to sell its products and charge per "use" assisted by DRM systems (see under Section 24.2.2.1), the once proponents of "free information" are allowed to implement their ideals through the use of "open" licenses (see under Section 24.2.2.2) — the latter, however, being yet untested do present a series of security concerns.

24.2.2.1. *DRM systems as IP rights enforcement supplements*

Digital Rights Management (DRM) systems is the copyright response to challenges that are from time to time presented to it by IT. In short, DRM systems purport to ensure that reproductions of a "work" by a user conform to the respective lawfully acquired (and paid for) license. The equivalent in other industries would be for instance, a speedometer in cars pre-programmed not to go over the national speeding limit at any circumstances. The DRM systems ensure that users enjoy the copies of works they purchased to the exact extent they purchased them, by pro-actively prohibiting any circumvention of these rules. In practice, this is achieved through a two-step process: first, each individual piece of content is "tagged" with appropriate additional bits of information invisible to the user. Second, a system is implemented in the copying process that controls the reproductions of such, tagged, content (either by "locking" the copying machines or by reducing the quality of the content once it is copied in an unauthorised way). The DRM systems ought to be considered outside of the on-line environment: their predecessors first made their appearance in VCRs during the 80s and are still present in the off-line world, for instance, having divided the world into "zones" for DVD-related purposes.

The DRM systems are, therefore, the Content Industry's IT solution to the problems caused to it by that same IT.

Intellectual Property Law embraced DRM technologies. Ever since the digitisation of information made copying easy, even before networks aggravated the situation, appropriate amendments were made to the law in order to secure the Content Industry's interests. However, this is done in an indirect way. As far as IP Law is concerned, only "*technological measures that restrict acts unauthorised*

[b]In one yet confirmation of Lessig's "code is law" (in Code v2, available at http://codev2.cc/).

by authors", the so-called Technical Protection Measures (TPMs), are explicitly acknowledged and protected. In this context, according to the WIPO Copyright Treaty,[c] "*contracting parties shall provide adequate legal protection and effective legal remedies against the circumvention of effective technological measures that are used by authors in connection with the exercise of their rights under this Treaty or the Berne Convention and that restrict acts, in respect of their works, which are not authorised by the authors concerned or permitted by law*".[d] The TPMs are thus passively protected in the wording above, by means of an explicit recognition of their existence in the Treaty. This is, nevertheless, not the only layer of protection afforded to TPMs: in addition to their passive protection, they are also actively protected against (in the case of e-commerce, at least) "hackers": "*contracting parties shall provide adequate and effective legal remedies against any person knowingly performing any of the following acts knowing [...] that it will induce, enable, facilitate or conceal an infringement of any right covered by this Treaty or the Berne Convention: (i) to remove or alter any electronic rights management information without authority; (ii) to distribute, import for distribution, broadcast or communicate to the public, without authority, works or copies of works knowing that electronic rights management information has been removed or altered without authority*". The TPMs (regardless whether in the e-commerce context or other) are thus protected two-fold: first, their existence is explicitly acknowledged in the text of law; second, anybody who tempers with them or anybody who passes along content whose TPM have been tempered with, shall be persecuted.

The obvious question now relates to the relationship between TPM and DRM. Despite the fact that certain views have highlighted their differences (in most of cases with an ultimate aim of justifying attempted circumventions of DRM systems[e]), judging even from their wording their actual relationship becomes clear: TPMs are the technical measures upon which DRM is based. In other words, TPM corresponds to the first of the two-step process described above: the "tagging" of individual works with additional bits of information in view of their later use in DRM systems. And, by the same token, if the act of attaching TPM onto content is recognised and protected by law, most certainly the introduction of DRM rules for the use of such (TPM-enriched) content is also, indirectly, equally recognised and protected. The use of DRM systems, regardless whether off- or on-line, is therefore, technically, lawful under IP Law.

[c] Its provisions, with regard to DRM at least, have been implemented in the United States through the Digital Millennium Copyright Act (DMCA), and in Europe through Directive 2001/29/EC of the European Parliament and of the Council of 22 May 2001 on the harmonisation of certain aspects of copyright and related rights in the information society, commonly known as the EU Copyright Directive (EUCD) or the Information Society Directive (Infosoc).

[d] Art. 11 WCT.

[e] See, for instance, the, largely legalistic, argumentation whether TPM that can be circumvented can be "effective" or not, in order to infer whether they are protected by the WCT (see, however, Art. 6 par. 3 of the EU Copyright Directive).

This is more or less the situation by now both in Europe and in the United States. In Europe, the Copyright Directive[f] devotes a whole Chapter (Chapter III) to the "*Protection of Technological Information and Rights-Management Information*". In this context, the Directive's Articles 6 and 7 repeat, in effect, the WIPO Treaty's provisions seen above. In the United States, Section 103 of the Digital Millennium Copyright Act (the so-called "*anti-circumvention provisions*") effectively implemented the same WIPO Treaty provisions. It is therefore safe to say that by now the lawfulness of DRM systems according to IP Law and as far as its scope is concerned should be taken for granted, both in the off-line and in the online environment.

The above by no means evidence the lawfulness of contemporary or future DRM implementations. The DRM systems constitute systems that essentially reflect the business practices of their owners. The IP Law is only allowed to go as far as to validate the lawfulness of the existence of DRM technologies per se. The law cannot say whether a particular, essentially e-commerce model, implemented by a corporation is lawful, in other words, whether the rules for the use of content implemented by DRM technologies conform to IP Law provisions. This is something that should be assessed each time on an ad hoc basis, while taking into consideration not only IP Law but other fields of law that might be potentially relevant (for instance, Data Protection Law, Competition Law etc.).[g]

24.2.2.2. *"Open source" initiatives: the case of the creative commons license*

At the other end of strict DRM rules and heavily regulated content provision stands the, relatively recent, "open source" movement. Proponents of the notion of "freedom" inherent in the new medium (the Internet, but all ICT as well) have undertaken formal initiatives to assist their ideas once it became evident that the copyright scheme is here to stay, despite the assault unleashed upon it. Naturally, the most popular initiative in this context relates to the open source software, that will be discussed later in this chapter (see under Section 24.2.3.2); here, its equivalent in relation to content shall be briefly elaborated.

As far as content is concerned, the path to its free delivery to users is paved these days mostly by the Creative Commons initiative. A number of parallels may be drawn between it and the software open source movement: in the case of software

[f] Directive 2001/29/EC (the EU Copyright Directive).

[g] See, for instance, the numerous legal adventures of the, dominant in the market for the time being, Apple's iTunes. Having started out with a business model of "closed" DRM technologies that only allowed reproduction on iPods and prices that vary in different countries (not allowing inhabitants of one country purchases from its on-line store in another), after successful attacks by European organisations they have, at the time these lines are written, changed into a more open DRM system, allowing reproduction on other mp3 players and even selling (for a higher price) non-DRM content, while their pricing differentiations at least within the EU shall most probably need to be abandoned soon, after scrutiny by the competent EU authorities has begun.

the General Public License (GPL, currently in its version 3) is released by the Free Software Foundation for use by all software developers who wish their programs to be made available free (or, at least, outside a number of copyright restrictions) to the public. In the case of content, a set of (altogether six) Creative Commons Licenses are released by the Creative Commons Organisation for use by all content rightholders who wish that it is made available outside a number of copyright restrictions (depending on the type of license) to the public.

Seen under this light, the Creative Commons Licenses came to fill a gap in the (on-line) market. New artists or even established ones that do not care much about the Content Industry are given the chance to make their works available to the public under fewer copyright restrictions, exactly in the same way as software developers have been able to do for years. Nevertheless, the Creative Commons system suffers from an inherent difficulty that may ultimately threaten its existence altogether: content is far more complex than software. Content may include anything from text to images or music; it can be a single work but the norm is that it will be a composite work incorporating more than one work in it (and respective sets of rights). For instance, a picture of friends uploaded on a social networking website includes the rights of the photographer but also the rights of the persons appearing in it. Accordingly, a typical video uploaded on-line shall include images as well as music, all bearing different sets of right. In all these cases, the rightholder who makes available a work under the Creative Commons License must warrant to all those who are prompted to use it that all rights in the work are secured — a far from straightforward task, given the complexity of a typical work.[h]

At any event, the Creative Commons organisation is set, in its own words, to define "*the spectrum of possibilities between full copyright — all rights reserved — and the public domain — no rights reserved. Our licenses help you keep your copyright while inviting certain uses of your work — a "some rights reserved" copyright*" [5]. In this context, it has released a set of six different licenses to be used by authors depending on the uses they would like their works to be put to; technical guidelines are intended to assist them while incorporating these licenses (naturally, on an "as is" basis) onto their works. The first Creative Commons license is named "Attribution" and only retains the author's moral rights upon his work, while users are free to do whatever they want with it. The second Creative Commons license is named "Attribution Share Alike" and, on top of "Attribution", asks that all new creations are licensed under identical terms (in effect this type of license is paralleled to the GPL software license). The third Creative Commons license is

[h]This is after all one of the first conflicts caused by a work used under a Creative Commons License: a photographer made a picture of a girl available under a CC license and the photograph was used in an advertising campaign by a big company, only to find out that the girl's parents objected to their daughter's picture being used in this way. The case was eventually settled, but is typical of difficulties inherent in content uses that may come in the future (see http://lessig.org/blog/2007/09/on_the_texas_suit_against_virg.html).

named "Attribution No Derivatives" and, evidently, allows only for the distribution of a work on an "as is" basis. The fourth Creative Commons license is named "Attribution Non-commercial" and, equally evidently, prohibits any commercial use of a work. The fifth Creative Commons license is named "Attribution Non-commercial Share Alike" and it combines the obligation to distribute only for non-commercial purposes with the obligation to distribute under exactly the same terms (in effect abolishing any commercial uses even in licensed derivative works, an option available under the "Attribution non-commercial" License). Finally, the sixth Creative Commons license is named "Attribution Non-commercial No Derivatives" and is expressly the most strict of them all, effectively allowing mere re-distribution of a work (serving thus mostly advertising purposes) [6].

Unlike the Free Software Foundation, Creative Commons has developed a strong international orientation, with "porting" processes for its licenses (practically, however, only one of them) developing in parallel in several countries of the world. Nevertheless, given the complexity of content per se and, perhaps, the complexity of the 6-license system (that a non-legal person shall probably struggle to come in terms with), it remains to be seen whether they shall eventually become as successful and established as their GPL equivalent.

24.2.3. *The Case of Software*

Software per se is probably the most important addition to the copyright scheme during the last century (and, allegedly, the most controversial one too). Notwithstanding whether a "work" or "product", software emerged during the second half of the 20th century and came to constitute an invaluable asset within a very short period of time. Its legal treatment became thus imperative: two were the obvious options, one referring to its inclusion within already existing legal schemes and the second pertaining to creating a new legal tool. Without over-expanding on the subject, the first option was finally chosen: software was to be protected as "literary work" within the context of copyright; the limitations of this approach are making themselves obvious every day [19].

Software has always been an uncomfortable guest in the copyright legal system. First of all, rather than fitting-in, a series of specialised adjustments had to be made in order to accommodate some of its "functional" characteristics [19]. Then, the level of protection afforded by the copyright scheme to software was always considered inadequate by the software industry: in fact, only the interface (the "expression") of a computer program may be copyrighted — the source code is largely not protected and the algorithms underneath are most certainly left out of the copyright scope. These limitations, coupled with its market value, have turned the major players in the software industry towards patents, an approach that created even more problems (as will be shown under Section 24.2.3.1).

On top of said difficulties when it comes to protecting software under the copyright regime comes the multitude of its forms. Software has since its first appearance exited the computer environment, to become present all around us. Today, apart from the inside of computers, software may be found in mobile phones, DVDs, cars, homes etc. The notion of "ubiquitous computing", when computing shall happen constantly around us without us taking notice any more, is steadily leaving research environments to enter mainstream vocabulary. And, evidently the building component of such ubiquitous computing will be software.

In this context, it appears had to perceive software still as "*literary work*", according to its legislative treatment under the copyright scheme. Rather than a "*work of the human intellect*" software has become an everyday tool divided in components and sold as a whole or as subsets performing a single function, sold off-the-shelf or bespoke, incorporated in machines or intended to be traditionally "run" on computers, residing locally or in the cyberspace. Accordingly, its functions practically encompass by now the whole of human activities, including work, home and entertainment environments. The common factor of all the above is, evidently, more a "product" than a "work", more an industry than art.

Within this environment the security connection is inevitable. Even if one is not ready to acknowledge that "code is law" [12], the more software is used in domains of human life the more security add-ons shall be enabled in it in order to make this, socially, possible. E-commerce systems, for instance, would have been void of any (consumer) interest if appropriate security mechanisms for tracing fraud and executing a lawful transaction were not in place. Such systems evidently include mechanisms both for prevention of mishappenings and for detection of the culprit if, despite all measures to the contrary, a breach is identified. Adequate security systems, that will not only be robust but also look that way to the public, are of central importance to any software implementation regardless whether used for profit (e-commerce systems, web banking systems etc.) or other purposes (privacy protection, state security etc.). Public trust in new technology implementations is only gained through adequate security measures installed therein.

Apart from add-ons on software applications intended to increase or create security and public trust, one must not forget that software per se is an asset of great value that is, unfortunately, easily copied. Security measures are thus attached in it too in order to discourage this possibility (in the same way that DRM systems operate for content). Again such mechanisms should aim both at prevention and at detection against what essentially constitutes "*software piracy*".[i] Software piracy, despite appearances, has not always been at the top of the industry's agenda,[j] because, especially back in the 80s, distributing inadequately protected software was a shortcut for securing a wide customer basis. This largely explains the mediocre level of security with which the software industry even today protects

[i]For a definition of "software piracy", see http://www.bsa.org.

[j]The Business Software Alliance organization ("*the foremost organization dedicated to promoting a safe and legal digital world*") was only established in 1988 (see http://en.wikipedia.org/wiki/Business_Software_Alliance).

its products from unauthorised copying (if, for instance, compared with aggressive DRM technologies used by the Content Industry). Security concerns thus, when it comes to software per se, are mostly addressed by legal means, when a breach of the respective End-User License Agreement (EULA) is indeed identified and the offender is called to indemnify the infringed party.

Finally, the making of unauthorised copies is not the only way software may be involved with crime. Software is frequently used as a tool to commit crime, in or out of the computer environment. Two distinctions may be drawn in this case: crimes that are committed with the assistance of software developed especially for this purpose, or crimes that are committed on software. The first category will generally refer to any technologies designed to, for instance, steal credit card numbers, make unwanted calls, open holes on security systems etc. The second category involves unauthorised access to computers and tampering with their software (for instance, computer hacking) regardless whether for profit or not. Here again, security measures are expected to be aimed both at preventing such actions and, if committed, in assisting detection of the culprits.

It, therefore, becomes clear that software is called to assume a multitude of roles. First, it has to adequately protect itself and its economic value. Second, it has to adopt in various social and technical environments, securing public trust through adequate prevention and detection mechanisms. Third, software often becomes the crime itself, when either put to this cause or "suffering" from the crime. The complexity of this situation is only marginally assisted by software's current legal treatment under the copyright scheme, as a "literary work", a legal solution nevertheless that seems to have missed the need for change.

24.2.3.1. *Patents*

Patents, when related to copyright, stand at the opposite side of the IP spectrum. In principle, the two systems were never meant to intersect (and, indeed, in Continental Europe, at least copyright falls within IP Law and patents fall within Industrial Property Law, two entirely different fields): copyright was the devised mechanism to award creativeness when it comes to "works of the intellect" (books, pictures, music etc.), whereas patents were themselves invented in order to protect machines indented for industrial use. Software blurred the distinction, because, despite of the fact that it is placed under the copyright scheme as "literary work", it presents elements that not only constitute a "work of the intellect" but are also used as "tools" in everyday life. Additionally, copyright notoriously protects very little out of a typical computer program (in effect, only its interface), whereas patents, if awarded, go as far as protecting the algorithm (evidently, for that particular use). It is in view of the above that, first, a number of software or software-like inventions have been awarded patents despite the rules to the contrary, and, second, many regulatory initiatives have been undertaken towards patentability of software, that have been, nevertheless, all unlucky so far.

From this point of view, any security issues concerning software per se when it comes to the dispute between copyright and patents will inevitably follow the above, or any future, developments. As long as software continues to be protected by copyright, as is currently the situation, great use for digital (computer) forensics tools shall lie with regard to copyright infringements: because copyright does not protect the algorithm and only a substantial portion of the source code, minor modifications among competitors that essentially launch the same product are to be expected. All these cases present an evident security interest while, for instance, evidenced in court. The same applies to patent-protected software: any unauthorised use will have to be supported by sufficient digital (computer) forensics-provided data. On the other hand, the software industry shall continue its efforts to equip their software tools (for the purposes of this analysis, DRM technologies etc.) with patents in order to better protect them against competition [15].

24.2.3.2. *Open source software*

The notion of open source software, being by now in its version 3 of its most popular EULA (the GPL) need not be presented here; here it is enough to be noted that open source software is the response (of, mostly, the developers' community) to proprietary software — rather than locking the code of a computer program, a typical open source alternative provides access to its source code and the freedom, under certain conditions, to edit it. Open source software affects the electronic security field in more than one way: first, it is a potential field of digital forensics implementations in itself, while, for instance, establishing adherence to the EULA (GPL) terms and conditions. Second, open source tools have been made available for the digital forensics field, whose effectiveness, however, needs to be evidenced. Finally, the notion of the open- or open-source domain and advanced search options have already posed previously unknown security/intelligence issues.

As far as open source software per se is concerned, the freedom-to-use principle has already given birth to a number of disputes, particularly with regard to adherence to the inevitable respective conditions. The fact that a computer program is made available on an open source basis does not necessarily mean that conditions for its use do not apply as well. For instance, the GPL license asks that any software using software provided under its terms and conditions is, in turn, made available to the public under the same, GPL, license (Section 2). This principle, the so-called inheritance principle, constitutes one of the most widely disputed issues relating to the open source scene. While the software market and the software industry challenge its validity, because of the obvious restraints it creates for their business development, the open source proponents (mostly, the Free Software Foundation) are ardent supporters of its fair character and justified use in practice. It is in this context that software made available to the public under any version of the GPL license carries this burden. Nevertheless, once in a while proprietary software emerges that allegedly uses GPL-licensed software without observing the inheritance principle (that would, in effect, make it as a whole or part of it open source as well). It is in these cases, where the actual use and the extent of incorporation, if any, of GPL-licensed software in proprietary software is to be established, that digital (computer) forensics tools inevitably hold a central role.

The digital forensics field is in turn affected by open source tools. Open source tools are claimed to have a legal benefit over closed source tools because they have a documented procedure and allow the investigator to verify that a tool does what it claims [8]. In these cases, however, the effectiveness of such tools, when used in certain conditions, needs to be validated [3].

Finally, the notion of the "open source" domain, where information is posted on-line in a simple way for everyone to use, when combined with powerful search tools that are also freely distributed over the Internet create a powerful outcome with serious security implications from an intelligence gathering and processing point of view [18]. Again, in this case, procedural and substantial measures need to be undertaken in order to monitor the flow of information from security-sensitive fields in the free, on-line environment.

24.3. P2P Networks and Their Effect on the Law

As already seen, the distribution of content over networks gravely affected, if not threatened, the premises of IP Law: once digitised content was exchanged by users over networks (the Internet), the traditional scheme for the protection of IP and the Content Industry itself were never going to be the same. The story of the first P2P networks, and their ultimate judicial demise, is by now known to everybody, therefore here it shall be briefly elaborated: P2P networks emerged once the Internet found its way into the homes of users and constituted a, then novel, e-commerce method: the network facilitator only made its infrastructure available to individual users who exchanged (mostly copyright protected) content and profit was made through advertisement. This e-commerce business method in effect created a mass copyright infringement, whereby millions of users exchanged millions of, then, songs around the world. The battle between the Content Industry (as represented by its music branch) and new e-commerce players (P2P networks) was fierce and lasted a decade. Although a number of cases were initiated by music labels against P2P networks, the one that finally did make it to the US Supreme Court was the one by Metro-Goldwyn-Mayer against Grokster. The question at hand was whether a P2P networks facilitator could be held indirectly ("secondary") liable for the uses of its software by its users (that is, for the unauthorised exchange of copyrighted material among its users). At first[k] the P2P operators seemed like they could get away with it: courts were confused with the, then relatively recent VCR cases, where VCR manufacturer Sony was not held responsible for copying of TV shows

[k]Court of Appeals (Ninth Circuit) 380 F 3d 1154.

performed by users using its sets,[l] and drew the analogy between this case and P2P networks facilitators. Nevertheless, the US Supreme Court held otherwise[m]: based on quantitative and qualitative criteria (for instance, 90% of content stored on P2P networks is copyrighted material, 100 million users exchanged more than 1 billion files on a monthly basis, plus the fact that P2P operators actually advertised this aspect of their systems) it decided, in short, that P2P networks operators are ultimately liable for the actions of their users, and thus made the continuation of their operation in their then form no more viable.

The, American, verdict on Grokster unavoidably affected the way all countries around the world viewed P2P networks, "cybercriminalising" in effect all their known at the time forms. By the time this analysis was written, P2P facilitators, at least in their previous form, were more or less abolished from the Internet. In the meantime, legal implementations, for instance Apple's iTunes, dominated the Internet, providing users with the opportunity to download and use content in a legitimate way.

P2P technology, however, was not abolished as well. Quite on the contrary, it constituted and continues to constitute a widespread technology used for a multitude of purposes. The only part of it that was judged unlawful in the Grokster case was the encouragement of users to trade unlawfully copied content; if, however, ways were devised that such downloads be made legal, then the technology would be perfectly lawful. Additionally, other fields, such as Internet telephony (for instance, Skype) may use the same technology at its basis.

Nevertheless, the very nature of P2P networks makes them a security nightmare. Their "shared" character, whereby each user uses them in any way he/she pleases and the facilitator has practically no way of centrally controlling or even monitoring such uses, inevitably leads to serious security implications. P2P networks, under their contemporary legal treatment and using contemporary legal schemes, are bound to present a series of security threats, at least when seen from, for instance, the Content Industry (when it comes to music or video sharing) or the state (when it comes to Internet telephony) perspective.

As far as their file exchange function is concerned, P2P networks may (or may not) have struck the correct balance between user exchange of content and copyright adherence through the implementation of DRM (e-commerce) systems. As already seen, these systems constitute the technical, but not necessarily legal or even business, reply to widespread Internet distribution of content through P2P networks. As soon as it is established that such exchanges are lawful, through the exchange of DRM-protected material, P2P networks may continue to thrive based on their technological advantages. It is in this context that, for instance, WebTV initiatives such as Joost have been released: P2P technologies equipped with DRM security for content aim at the Internet public [7].

[l] *Sony Corp. of America v. Universal City Studios Inc.* 464 US 417 (1984, BETAMAX case).
[m] *Metro-Goldwin-Mayer Studios Inc. et al. v. Grokster Ltd., et al.*, 27 June 2005.

The limitations, however, of using DRM technologies over P2P networks are already making themselves felt. DRM technologies are thought to be too much restrictive in the Internet context,[n] and the Content Industry has been accused of not understanding the new medium. Once the Grokster case has established that P2P facilitators could no longer operate in the way they used to, the Internet community pressed for continued use of P2P networks in a lawful way. The Content Industry (music, but ever increasingly video as well) is pressed to devise ways of distributing content making use of P2P networks or other on-line channels in an efficient and lawful way for users; its existence itself is said to depend on its ability to adopt into the new reality. For the time being, such voices have not been heard, and DRM-equipped P2P (or other) networks appear to be the only contemporary lawful solution for the on-line provision of content; the situation may, however, change drastically in the future.

As regards other implementations of P2P technologies, Internet telephony (VOIP) has attracted particular attention. P2P technology is the basis upon which such popular solutions such as Skype are built.[o] However, regardless whether over the Internet or not, telephony remains a telecommunications service from a state/public security perspective. In Europe, various regulatory initiatives have been adopted for the processing of telecommunications data, the most well-known of which is probably by now the Data Retention Directive.[p] Such a Directive asks that, particularly with regard to Internet telephony, a series of personal data (user ID, telephone number, IP address etc.) be kept by the service provider for periods from six months up to two years.[q] However, here again the inherent "shared" nature of P2P networks, upon which VOIP is based, present a series of security issues: in effect, governments find it increasingly hard, if not impossible, to monitor Internet phone calls, even if they have the lawful right (and have followed the lawful procedure) to do so [4].

It therefore becomes evident that P2P technology is found at a constant conflict with the law. Its "shared" nature presents inherent security difficulties that are not dealt with efficiently under contemporary legal schemes. Under the regulatory framework in effect, user sharing of content over P2P networks shall continue to be illegal, while VOIP shall continue to grant criminals with intolerable potential as to their secure communication. "Cybercriminalising" appears thus to be the legal

[n] Indeed, for instance, Apple's iTunes are already making DRM-free content available through their service for an increased fee.

[o] See http://www.skype.com/help/guides/voip and Salman A. Baset and Henning G. Schulzrinne, An Analysis of the Skype Peer-to-Peer Internet Telephony Protocol, available at http://www.eecs.harvard.edu/~mema/courses/cs264/papers/skype-infocom2006.pdf.

[p] Directive 2006/24/EC of the European Parliament and of the Council of 15 March 2006 on the retention of data generated or processed in connection with the provision of publicly available electronic communications services or of public communications networks and amending Directive 2002/58/EC, OJ L 105/54, 13.04.2006 (available under http://europa.eu.int/eur-lex/lex/LexUriServ/site/en/oj/2006/l_105/l_10520060413en00540063.pdf).

[q] Art. 5 and 6 of the Data Retention Directive.

response to P2P networks so far. On the other hand, contemporary legal schemes that were devised at times when networks did not exist or were at least state-restricted and controllable (such as, for instance, the standard telephony networks) may be by now obsolete, crushed under wide Internet connections available to everyone for a marginal cost and individuals' inherent need to communicate globally. It remains thus to be seen whether in the future P2P networks shall be subdued to contemporary (IP and other) legal restrictions, or whether the need for a novel regulatory approach shall be identified and adopted, in order to better accommodate a newcomer that, for all its judicial troubles, has proven extremely resilient and adopting and is evidently here to stay.

24.4. The EU Database Right

The *sui generis* Database Right is an EU law exclusivity. Back in 1996, when it was established that the European database industry was lagging behind its American and Japanese competitors, it was felt that an adequate regulatory boost would be given by introducing a new, special (*sui generis*) right particular to the needs and particularities of databases[r]: the outcome was the Database Directive,[s] implemented by now into the national law of all EU Member States. The Database Directive essentially establishes that databases are indeed protected by copyright, under IP Law; those of them, however, that do not qualify for such a protection may profit from the new, *sui generis*, Database Right. And, quite a few of them may not profit from copyright protection: because copyright is awarded to "works of the intellect" that must present some originality and uniqueness (even at such low levels as computer programs are allowed to), most databases, whose added value lies most on the breadth of their contents and not their creation will generally fail the test. In addition, even those that do pass the test may still find it hard to abide by copyright rules, whereby fees are due for each complete reproduction of a work rather than access to part of it. At any event, the Database Directive became European law at a time, in the late 90s, when the online and, broadly speaking computer, environment[t] made a new market, that of databases, possible.

Given that the Database Directive was expressly not to be confined within the computer-automated environment,[u] the breadth of its scope was, originally at least, astonishing. Indeed, under the Database Directive such collections as dictionaries, single webpages cataloguing information or even web links, exhibition catalogues, the newspaper classifieds, or even the public tenders published in the financial press may be considered as a protected material. The only substantial limitation to widespread application of the Directive came later, through case law of the

[r]See Recitals 11 and 12 of the Database Directive.
[s]Directive 96/9/EC of the European Parliament and of the Council of 11 March 1996 on the legal protection of databases, OJ L 077, 27/03/1996.
[t]See, for instance, Recitals 9, 10 or 13 of the Database Directive.
[u]See Recital 14 of the Database Directive.

European Court of Justice (ECJ). In a well-known case,[v] the ECJ stressed upon the importance of the distinction between "creation" of a database and "obtaining, verification and presentation" of its contents. In effect, in order for a database to qualify for the *sui generis* database right, its owner must have spent substantial resources while either obtaining or verifying or presenting its contents — merely spending money for its creation will not suffice. This ruling left out of protection such cases as phone books or, as was the disputed subject-matter, betting listings.

Regardless of the above restriction in its scope, the fact remains that the Database Directive constitutes a powerful tool in the hands of owners of compilations, both in the online and the off-line environment. Particularly over the Internet, where substantial work may have been put to compiling webpages and publishing freely (with an aim to make profit through other means), Internet site owners may find themselves at a disadvantage, being squeezed between strict copyright requirements and broad unfair competition provisions. Practices, for instance, such as framing or deep-linking may effectively be dealt with under the database, *sui generis* right, regulatory framework (see Chapter 35, ...). On the other hand, although beneficial to the arsenal of lawyers, it remains to be seen whether the database right shall indeed at some point in the future serve its proper purpose, that is to boost the European database market,[w] or whether it shall remain a legalistic alternative to a series of more or less already known and accounted for e-commerce practices.

References

1. D. Bainbridge, *Intellectual Property*, 6th Edition (Longman, 2006).
2. L. Bently and B. Sherman, *Intellectual Property Law*, 2nd Edition (Blackstone Press, 2004).
3. B. Carrier, Open source digital forensics tools: the legal argument, available at http://www.digital-evidence.org/papers/opensrc_legal.pdf.
4. A. Gavras, Security weakness of VoIP, 2007, available at http://www.eurescom.de/message/messageMar2007/Security_weakness.of-VoIP.asp.
5. http://creativecommons.org/about/.
6. http://creativecommons.org/about/license/.
7. http://www.joost.com/forums/f/p2p-technology/.
8. http://www.opensourceforensics.org/.
9. P. Leith, *Software and Patents in Europe* (Cambridge University Press, 2007).
10. L. Lessig, *Free Culture: The Nature and Future of Creativity* (Penguin Books, 2005).
11. L. Lessig, *Code: Version 2.0* (Basic Books, 2006).
12. L. Lessig, Code is law, Code v2, available at http://codev2.cc/.
13. L. Lessig, Free culture, available at http://www.free-culture.cc/.
14. I. Lloyd, *Information Technology Law*, 4th Edition (LexisNexis UK, 2004).

[v]European Court Justice, *British Horseracing Board v William Hill*, C 203/02, [2005].
[w]See DG Internal Market and Services Working Paper, First evaluation of Directive 96/9/EC on the legal protection of databases, 12 December 2005, available at http://ec.europa.eu/internal_market/copyright/docs/databases/evaluation_report_en.pdf.

15. Microsoft patents digital audio DRM watermark, 2007, PC Advisor, 12 September 2007, available at http://www.pcadvisor.co.uk/news/index.cfm?newsid=10696.
16. P. Petrick, Why DRM should be cause for concern: an economic and legal analysis of the effect of digital technology on the music industry (November 2004), Berkman Center for Internet & Society at Harvard Law School Research Publication No. 2004-09. Available at SSRN: http://ssrn.com/abstract=618065.
17. C. Reed and J. Angel, *Computer Law*, 5th Edition (Oxford University Press, 2003).
18. D. L. Watson, Stealing corporate secrets using open source intelligence (the practitioner's view), *Int. J. Electronic Security and Digital Forensics*, **1**(1) (2007) 71.
19. R. Widdison, Software patents pending?, *The Journal of Information, Law and Technology (JILT)* 2000 (3). <http://www2.warwick.ac.uk/fac/soc/law/elj/jilt/2000_3/widdison/>.

Chapter 25

LEGAL ISSUES ON THE PROTECTION OF PERSONAL DATA ON THE INTERNET

EVI CHATZILIASSI
Data Protection Authority, Greece

25.1. Introduction

In the early years of the use of Internet, it was fashionable to claim that a space of absolute freedom and beyond the reach of traditional laws was developed, where everybody could access information and exchange views with no limitation. It was soon discovered that due to its distributed nature (as a network of computers communicating with each other on the basis of the Transport Control Protocol/Internet Protocol) and consequently due to the unlimited transborder flow of information, images, video and sounds, the use of Internet threatened some of the foundations of our democratic society, such as the right to privacy.

Privacy is acknowledged in Europe as a fundamental right and freedom and it is viewed as an essential provision for a dignified existence.[a] As data protection constitutes a part, the most visible and tradable part of privacy, Article 8 of the European Charter of Human Rights recognises that everybody has the right to protection against the uncontrolled collection, processing and distribution of their personal data.[b] Every Internet user reveals, either intentionally or unintentionally, a series of personal data in each Internet session and therefore the handling of the data collected raises many different data protection issues. Personal data are collected and processed not only by natural or legal persons that publish these data on the Internet, for example, on a website, in a discussion forum or in an online journal, but also by legal persons (companies), which provide to individuals and companies

[a] Article 8 § 1 of the European Convention for Human Rights and Article 7 of the European Charter of Fundamental Rights.

[b] For the differences between "privacy" and "data protection", see Nicola Lugaresi, Principles and Regulations about on line Privacy: "Implementation Divide" and Misunderstandings in the European Union,available at http://ssrn.com/abstract=333440.

access to the Internet (Internet Access Providers — IAPs) or/and other services on the web, such as web hosting (Internet Service Providers — ISPs). This chapter aims at offering an approach of on-line data protection in Europe.

25.2. Legal Framework

25.2.1. *Directive 95/46/EC and Directive 2002/58/EC*

On October 24, 1995, the European Union (EU) adopted Directive 95/46/EC on the protection of individuals with regard to the processing of personal data and on the free movement of such data (hereinafter referred to and as "data protection Directive)[c] and the member states were required to enact its provisions until October 24, 1998. Although the purpose of the data protection Directive was to keep a balance between the fundamental right to privacy and the free flow of information in the EU, in order to avoid restrictions for privacy reasons, the aforementioned Directive provides strengthened remedies to data subjects for breaches of its requirements. Compared to national legislation that existed until the data protection Directive's adoption, the Directive extends controls over the transfer of personal data to third countries and grants data subjects (natural persons with regard to whom personal data are collected and processed) three substantial rights: the right of access, the right for correction and deletion and the right to object on legitimate grounds to further processing of their personal data. In 1997, the EU adopted Directive 97/66/EC concerning the protection of personal data and the protection of privacy in the telecommunication sector,[d] which was repealed and replaced by Directive 2002/58/EC concerning the processing of personal data and the protection of privacy in the electronic communication sector (hereinafter referred to and as "ePrivacy Directive"),[e] in order to provide an equal level of protection for the users of publicly available electronic communications, regardless of the technology used.[f] The member states had to bring into force the provisions necessary to comply with this Directive until 31.10.2003.

It is essential to note that both Directives apply in principle to personal data processed on the Internet. Directive 2002/58/EC particularises and complements the general data protection Directive in the electronic communication sector by establishing specific legal obligations and technical provisions. According to Article 3 § 1, this specific Directive applies to the processing of personal data in connection with the provision of publicly available electronic communications services in public communication networks in the community. As Internet services are electronic communication services, the Internet forms part of the public

[c]Official Journal L 281 of 23.11.1995, pp. 31–50, Article 29 of the Directive 95/46/EU established a Working Party, which is an independent EU Advisory Party on Data Protection and Privacy and consists of representatives of all data protection authorities in the EU.
[d]Official Journal L 24 of 30.1.1998, pp. 1–8.
[e]Official Journal L 201 of 31.7.2002, pp. 37–47.
[f]See Recital 4 of Directive 2002/58/EC.

electronic communications sector. On the other hand, the general data protection Directive applies to any processing of personal data falling in its scope, irrespective of the technical means used and therefore it applies to all matters that are not specifically covered by the ePrivacy Directive, such as the obligations of the controllers, the rights of the data subjects and the processing of personal data in non-publicly available electronic communications services.

The most significant provision of the ePrivacy Directive is that of Article 5, which provides that Member States must ensure the confidentiality of communications (by means of a public communications network and publicly available electronic communications services) and of the related traffic data. The Member States must, *inter alia*, prohibit the storage of that data by persons other than users without the consent of the users concerned. The only exceptions foreseen relate to persons lawfully authorised in accordance to Article 15 of the directive and to the technical storage necessary for the conveyance of a communication.

The fact that the data protection Directive applies to personal data processed on the Internet was recognised by the European Court of Justice (hereinafter referred to and as "ECJ") in the Linqvist case.[g] Mrs Lindqvist, after following a data processing course, set up a webpage containing details about members of a Parish Church, including information about a member, who had injured her foot. Lindqvist did not obtain the consent of the individuals before posting the information on the website and also failed to inform the national (Swedish) Data Inspection Board about the publication of sensitive data regarding the health of the members of the Parish church on the website. The ECJ held that the act of referring, on an Internet page, to various individuals and identifying them by name or by other means, falls in the scope of the data protection Directive, as it constitutes the processing of personal data wholly or partly by automatic means within the meaning of Article 3 § 1 of the above-mentioned Directive.[h]

25.2.2. *Exemptions of the Applicability of Directive 95/46/EC*

There are two types of data processing in which the data protection Directive does not apply. According to Article 3 of the Directive, its provisions do not apply to the processing of personal data in the course of an activity, which falls outside the scope of Community law, and in any case to processing operations concerning public security, defense, state security and the activities of the state in areas of criminal law. The same exemption is foreseen in Article 1 § 3 of the ePrivacy Directive.

The second exemption refers to the processing of personal data exclusively for personal or household purposes. The objective of this provision is to prevent everyday activities of natural persons from falling within the scope of this Directive and subsequently being subject (the individuals) to the obligations foreseen for the controllers. The growth of social networking websites and on-line personal journals,

[g]Case C-101/01, Bodil Lindqvist, [2004], ECR 2003, pp. I–12971.
[h]See Recital 27 of the ECJ ruling in Case C — 101/01, as above.

which has as an outcome that individuals assume a central role in the collection, processing and distribution of personal data, raises new issues relating to the extent to which individuals may be able to benefit from this exemption. In its decision in the Lindqvist case, the ECJ held that this "exemption must, therefore, be interpreted as relating only to activities which are carried out in the course of private or family life of individuals, which is clearly not the case with the processing of personal data consisting of publication on the Internet so that those data are made accessible to an indefinite number of people".[i] The ECJ by making a distinction between private and public access on the Internet, adopted a narrow approach to the interpretation of Article 3 § 2 of the data protection Directive as applied to the Internet. It, consequently, placed an onus on individuals to limit access to their webpages to a defined group of people through appropriate technical measures (such as applying an obligatory password and blocking the pages from search engines) in order to be able to benefit from the exemption of processing data for personal or household purposes.[j,k]

Data processing on the Internet exclusively for journalistic, artistic or literary purposes could be regarded as a third exemption, since Article 9 of the data protection Directive requires that member states should foresee exemptions for the processing of personal data exclusively for the above-mentioned purposes, in order to keep a balance between these two fundamental human rights, the right to the protection of personal data on one hand and the right of freedom of expression on the other.

25.2.3. *International Application of EU Data Protection Law*

The main issue raised is whether the provisions of the data protection Directive could apply to the processing, and especially the collection, of personal data by websites which are based outside the EU. Article 4 § 1c of the Directive, regarding the applicable national law, states that: "Each Member State shall apply the national provisions it adopts pursuant to this Directive to the processing of personal data where... c) the controller is not established on Community territory and, for the purposes of processing personal data makes use of equipment, automated or

---

[i]See Recital 47 of the ECJ ruling in Case C — 101/01, as above.

[j]For the issues raised by the ECJ's interpretation of Article 3 § 2 of the data protection Directive with regard to the social and technical innovations encountered in the Intenet, see R. Wong and J. Savirimuthu, All or nothing: this is the question?: The application of Art. 3 (2) Data Protection Directive 95/46/EC to the Internet, available at SSRN, http://ssrn.com/abstract=1003025.

[k]The Dutch Data Protection Authority, following the ruling of the ECJ in the Lindqvist case, considers that the the data protection act does not apply to the maintenance of a weblog or a website, where access is limited to a restricted circle of family members and friends, through appropriate technical measures (such as applying an obligatory password and blocking the pages from search engines). See Dutch DPA, Publication of Personal Data on the Internet, December 2007, p. 13, available at http://www.dutchdpa.nl/documenten/en_pb_2007_privacy_legislation_internet.shtml?refer=true.

otherwise, situated on the territory of the said Member State unless such equipment is used only for purposes of transit through the territory of the Community".

The objective of this provision is to ensure that individuals are not deprived of the protection to which they are entitled to as regards processing of personal data that is taking place within their country only because the controller is not established in Community territory.[l] Due to the fact that there is a great chance that controllers could locate their establishment outside the EU, in order to bypass the application of the EU data protection law, Article 29 Working Party adopts a wide interpretation of the terms "makes use of equipment, automated or otherwise, situated on the territory of the said Member State". According to its opinion, the existence of two elements is essential for the application of the national law of a Member State which enacts the provisions of the data protection Directive: (a) that the controller undertakes some kind of activity and (b) that the controller intends to process personal data. However, the power of disposal of the controller should not be confused with the property or ownership of the equipment. Therefore, the Article 29 Working Party considers that the use of cookies (text files) by a controller, established outside of the Community territory, which are placed on the hard disc of the Internet user's computer (who is established inside the Community territory irrespective of its nationality), while a copy is kept by the website or a third party, would lead to the application of Article 4 § 1c of the data protection Directive and of the national data protection law concerned. Use of JavaScripts or banners by websites or advertising companies, would also have, as a result, the application of EU data protection law, in the case that the technical means are used for the collection and processing of personal data.[m]

25.3. Data Protection Directive's Basic Definitions

25.3.1. *Personal Data*

The definition of personal data is provided in Article 2 sub-section (a) of the data protection Directive and reads as follows: "Any information relating to an identified or identifiable natural person; an identifiable person is one who can be identified, directly or indirectly, in particular by reference to an identification number or to one or more factors specific to his physical, physiological, mental, economic, cultural or social identity".

From the above-mentioned definition, it becomes clear that even IP addresses, meaning the Internet addresses used by computers to communicate their identity on the Internet and attributed to the Internet users computers by ISPs, can be

[l]See Recital 20 of Directive 95/46/EU.

[m]See Article 29 Working Party, Working document on determining the international application of EU data protection law to personal data processing on the Internet by non-EU based websites, 30.5.2002, available at http://ec.europa.eu/justice_home/fsj/privacy/workinggroup/wpdocs/2008_en.htm, G. J. H. Smith, *Internet Law and Regulation*, 3rd Edition (London, 2002), p. 372, 7-014.

considered as personal data for the following reason. The ISPs are able without disproportionate effort to trace the IP address back to the natural person. Not only the static IP addresses (which are attributed to individuals with permanent Internet connection by ADSL or via video cable) are considered as personal data but also the dynamic ones (which are attributed to individuals using a modem or ISDN for the duration of each connection), since the ISPs systematically keep records of the date, time, duration and dynamic IP address given to the Internet user.[n] In order to decide whether an item of data could indirectly lead to the identification of a natural person and therefore be considered as personal data, it is important to examine whether the controller or a third person can, by using reasonable means, identify the said person.[o]

25.3.2. *Sensitive Data*

The data protection Directive makes a distinction between "normal data" and "data of special categories". According to Article 8 of the aforementioned Directive, the member states should prohibit the processing of personal data relating to a person's racial or ethnic origin, political persuasions, religion or philosophical beliefs, membership of a trade union, health and sexual orientation, unless the data subject has given his/her express consent or unless another legitimate ground for the processing (from those foreseen in Paragraph 2 of Article 8) exists.

Processing of sensitive data is subject to a more stringent regime as it results to a deeper invasion in an individual's private life. The term "explicit" consent refers to the quality that the consent should have in order to legitimise the procession. The consent must be freely given, specific, informed and not implied. Therefore, the express consent cannot be replaced by providing the opportunity to stop the processing at a later point (opt-out system). The second legitimate ground that could be used in order to legitimise the publication of sensitive data on the Internet, is the one foreseen in Article 8 § 2e, which refers to sensitive data that have been consciously published by the data subject himself.

25.4. Basic Principles of Data Processing

25.4.1. *Fair and Lawful Processing*

The first basic principle that should be respected by controllers is that personal data should be fairly and lawfully processed (Article 6 § 1a of Directive 95/46/EC). The term "process" as defined in Article 2 subsection b of the data protection Directive includes obtaining. In order to obtain personal data lawfully from the data subject,

---

[n]See Dutch DPA, as above, p. 10, Article 29 Working Party, Privacy on the Internet — An integrated EU Approach to On-line Data Protection, November 2000, available at http://ec.europa.eu/justice_home/fsj/privacy/workinggroup/wpdocs/2000_en.htm, Controversial, G. J. H. Smith, *Internet Law and Regulation*, 3rd Edition (London, 2002), p. 375, 7-023.
[o]See Recital 26 of Directive 95/46/EC.

the latter should be informed about: (a) the identity of the controller and, where applicable, the representative appointed, (b) the purpose(s) of the processing, (c) the recipients or categories of recipients of the data, (d) the obligatory or optional nature of the information to be provided and (e) the existence of and the conditions for exercising the rights to access, to rectify and delete the data concerned (Article 10 of the data protection Directive).

The main issue raised by the application of this principle to the processing of personal data on the Internet, is that during an Internet connection different kinds of processing operations are taking place, which are invisible to the data subject. Cookies which are placed on an Internet user's hard disc and contain information about pages viewed, advertisements clicked, user identification number or automatic hyperlinks to third parties are typical examples of such an invisible processing. The Article 29 Working Party expressed its great concern about the risks for privacy inherent to the processing of personal data, in the case that data subjects are completely unaware of the processing that takes place and declared that even the "invisible" and automatic processing is subject to the same terms, conditions and guarantees as any other processing of personal data.[p] At a later point, the Working Party recommended that the data subjects should, also, be informed, prior to any collection of personal data, about the existence of automatic data collection procedures. In particular, the data subject should know the purpose of these procedures, their period of validity, whether their acceptance is necessary to visit the site and, finally, the option to object to their use and the consequences of deactivating such procedures.[q]

25.4.2. *Finality Principle (Specified and Legitimate Purposes)*

According to Article 6 § 1 subsection b of the data protection Directive, personal data should be collected for clearly defined, explicitly specified and legitimate purposes and their further processing may not be incompatible with these purposes.

In the case that personal data were collected for a specific purpose, their further publication on the Internet should be consistent with the initial purpose in order to be legitimate. To judge the compatibility of the publication with the original purpose of collection, the controller should take into account, *inter alia*, the connection between the initial and the new purpose, the nature of the data in question and the consequences of this processing for the data subject. Due to the open nature of Internet, where information is accessible by a large and unknown audience, and to the fact that there are no means of controlling the purpose for

[p]See Article 29 Working Party, Recommendation 1/1999 on Invisible and Automatic Processing of Personal Data on the Internet Performed by Software and Hardware, 23.2.1999, available at http://ec.europa.eu/justice_home/fsj/privacy/workinggroup/wpdocs/1999_en.htm.

[q]See Article 29 Working Party, Recommendation 2/2001 certain minimum requirements for collecting personal data on line in the EU, 17.5.2001, available at http://ec.europa.eu/justice_home/fsj/privacy/workinggroup/wpdocs/2001_en.htm.

which data may be further processed by third parties, is highly unlike that such a publication is legitimate. Even the re-use of personal data already published on the Internet does not legitimise the specific processing. The controller is obliged to have an independent legitimate ground for the publication and the new purpose should be compatible with the one of the initial publication.[r]

25.4.3. *Adequacy, Relevance, Necessity — Accuracy*

The third basic principle for legitimate processing of personal data is that data should be adequate, relevant and necessary (not excessive) for the accomplishment of the purposes stated (Article 6 § 1 subsection c of the data protection Directive). From the above-mentioned provision, it is made clear that the purpose of processing is not permitted to be vague or liberal because in that case it is impossible to check whether the data collected and processed are actually necessary for the accomplishment of that purpose.

In accordance to subsection d of the aforementioned article, data being processed should also be accurate and up-to-date. The controllers are obliged to take all appropriate measures in order to ensure that data are consistent with the truth and that inaccurate or incomplete data are rectified as soon as possible.

25.4.4. *Limited Retention*

The last basic principle of processing personal data (Article 6 subsection e) is that the retention period of data should be limited to the absolute necessary for the purposes for which data were collected or for which they are further processed. Data retention is an increasingly contentious issue as it is highly relevant to the data retention practices of ISPs and others involved in the transmission of data through the Internet. According to Article 6 of the ePrivacy Directive, providers are obliged to erase traffic data (data processed for the purpose of the conveyance of a communication on an electronic communications network or for the billing thereof) relating to subscribers and users or to make such data anonymous when they are no longer needed for the purpose of the transmission of a communication or for billing purposes.

The principle of restricting the data retention period evidently conflicts with the desire of law enforcement authorities for longer periods of data retention by private sector entities. This is the reason that led in several member states to the adoption of national legislations (according to Article 15 § 1 of the ePrivacy Directive, setting out the conditions for restricting the obligations foreseen in Article 6), which provide for the retention of data by service providers for the prevention, detection and prosecution of criminal offenses. The legal and technical differences between national provisions led to the adoption of Directive 2006/24/EU

[r]See Dutch DPA, as above, endnote 10, pp. 19–20, G. J. H. Smith, as above, endnote 10, p. 381, 7-040.

on the retention of data generated or processed in connection with the provision of publicly available electronic communications services or of public communications networks (hereinafter referred to and as "data retention Directive"),[s] which aimed to bypass the obstacles created to the internal market for electronic communications, since service providers were faced with different requirements regarding the types of traffic and location data to be retained and the conditions and periods of retention.[t] Given the importance of traffic and location data for the investigation, detection and prosecution of criminal offenses, the data retention Directive foresees that different categories of data (Article 5) should be retained for a period that varies from six months to two years from the date of communication (Article 6) and that competent national authorities have access to the data retained. However, the national law determining the conditions under which the competent authorities gain access to the retained data, should be in accordance with Article 8 of the European Charter of Human Rights and with the principles of necessity and proportionality deriving from this article.

25.5. Grounds for Legitimate Processing of Personal Data

Controllers that wish to lawfully process personal data on the Internet must require the consent of the data subject concerned unless another legitimate ground of those listed in Article 7 of the data protection Directive exists.

25.5.1. *Consent*

Consent should be the legitimate ground for the greater part of data processing on the Internet. Consent is the specific (in the case of sensitive data "explicit"), freely given and informed indication of his wishes by which the data subject signifies his agreement to personal data relating to him being processed. From the above-mentioned definition derives that the principle "silence lends consent" does not apply and, therefore, the controller is obliged to seek for the data subject's consent for every specific type of data processing.

The data subject is entitled to withdraw his consent at any time, action that renders the processing unlawful, unless the controller can justify the data processing by means of another legitimate ground. In order to avoid unlawful publications on the Internet, as far as these publications are based on consent, controllers should take technical measures that allow for the active deletion of personal data in case that a data subject withdraws his consent.[u]

The data protection Directive does not entail any specific provisions about the processing of personal data of minors meaning young people under the age

[s] Official Journal L 105 of 13.4.2006, pp. 54–63.
[t] See Recitals 4–6 of Directive 2006/24/EU.
[u] See Dutch DPA, as above, endnote 10, pp. 21–22.

of 18.[v] Young people, though, tend to reveal detailed information about themselves and their friends on the Internet, for example, in social network environments. The controllers, therefore, are obliged to obtain the consent of the young person's parent or legal representative before any publication on the Internet and to inform the data subjects of their rights and obligations in a way that is clear and understandable to the target group.

25.5.2. *Other Legitimate Grounds — Legitimate Interest*

As it is mentioned previously in this chapter, besides consent the data protection Directive comprises five other legitimate grounds for processing personal data (Article 7). Each of these legitimate grounds pre-supposes that the processing is necessary for: the performance of a contract to which the data subject is party, the compliance with a legal obligation to which the controller is subject, the protection of the vital interests of the data subject, the performance of a task carried out in the public interest or in the exercise of an official authority and finally for the pursue of the legitimate interests of the controller or of a third party to whom the data are disclosed except where such interests are overridden by the interests for fundamental rights and freedoms of the data subject which require protection.

It is essential to note that there is a clear distinction between the collection and processing of personal data, which is necessary for the compliance of the controller with a legal obligation or for the performance of a task carried out in the public interest and the publication of these data on the Internet. The existence of an obligation to collect and process personal data for the aforementioned reasons does not mean that their publication on the Internet is lawful. The publication itself, as a form of processing, should be necessary for the compliance of the controller with a legal obligation or for the performance of a task carried out in the public interest in order to legitimise the publication of personal data on the Internet.[w]

An interesting issue that relates to the fifth legitimate ground of processing personal data, is that this particular ground is usually mentioned by search engine providers in order to legitimise the processing. According to Article 7 subsection f of the data protection Directive, the processing of personal data is lawful, if it is necessary for the purposes of the legitimate interests pursued by the controller or by the third party or parties to whom the data are disclosed, except where such interests are overridden by the interests for fundamental rights and freedoms of the data subject which require protection.

Several search engine providers claim that the processing of Internet user's personal data is indispensable in order to improve their services, to optimise their

[v] The Article 29 Working Party has recently (on 18 February 2008) adopted the Working Document 1/2008 on the protection of children's personal data, available at http://ec.europa.eu/justice_home/fsj/privacy/workinggroup/wpdocs/2008_en.htm.
[w] See Dutch DPA, as above, endnote 10, pp. 22–23.

system security, to prevent and detect fraud, to provide personalised advertising and, finally, to comply with legal requests. The Article 29 Working Party, recognising that search engines have become a part of the Internet user's daily life, had recently issued an opinion on data protection related to search engines.[x] Throughout this opinion, the Working Party is trying to strike a balance between the legitimate business needs of the search engine providers on one hand and the protection of the personal data of Internet users on the other.

Search engine providers claim that the storage of the content of user queries in their server logs is absolutely necessary for the improvement of their services, which is the outcome of the analysis of these data (for example, the analysis of the kind of queries that people make, the way in which they choose to refine those queries and the search results that they choose to follow). The Article 29 Working Party, however, considers that search queries do not need to be attributable to identified individuals in order to contribute to the improvement of the search engine providers services, and, therefore, the processing of personal data cannot be considered legitimate. As search engine providers have a legitimate interest in maintaining the security of their systems and in preventing or detecting fraud, they have adequate grounds for processing personal data for the aforementioned purposes. This processing, however, should be subject to strict purpose limitation, in order to be legitimate, meaning that data stored for security purposes may not be used for another purpose such as the improvement of their services.

The claim of the search engine providers that the storage of the content of user queries is indispensable in order to provide to Internet users personalised advertising, cannot legitimise the collection and processing of personal data. The wish of search engine providers to increase their revenues by providing personalised advertising, could only be considered legitimate if it takes place on the ground of the consent of the data subject and not on the ground of the pursue of a legitimate interest of the data controller (in this case of the search engine provider). Finally, their obligation to comply with legal orders, resulting in the disclosure of personal data, cannot justify the processing of such data solely for this purpose, as the legal obligation to comply with a court order should not be considered as an obligation or justification for collecting and processing personal data.

25.6. Data Subjects Rights

Processing of incorrect or incomplete personal data may have severe consequences for data subjects, the natural persons with regard to whom personal data are being processed. The data subject is even more affected if the processing of his data relates to the use of the Internet or to the publication of his data on the Web, as Internet

[x]See Article 29 Working Party, Opinion on data protection issues related to search engines, 4.4.2008, available at http://ec.europa.eu/justice_home/fsj/privacy/workinggroup/wpdocs/2008_en.htm.

is an open environment, which encourages browsing. The data protection Directive has, therefore, granted data subjects three substantial rights: the right of access, the right for correction and deletion and the right to object on legitimate grounds to further processing of their personal data.

25.6.1. *Right of Access*

According to Article 12 subsection a of the data protection Directive, the controllers should, without constraint and without excessive delay or expense, inform the data subject as to whether or not data relating to him are being processed as to the purpose(s) of the processing, the categories of data concerned, and the recipients or categories of recipients to whom the data are disclosed. The controllers, also, have the obligation to communicate to data subject in an intelligible form the data undergoing processing and any available information as to their source. The right of access could be considered as the core of data subject's rights since it is indispensable in order to submit a specific request for correction/deletion or a specific objection for the further processing of personal data.

Because of the fact, that data processed in publications on the Internet are usually publicly accessible and free of charge, the data subject has direct access to his personal data and does not need to submit a formal request to the controller in order to gain access to these data. The right of access, therefore, is of particular importance regarding publications to which access is restricted.[y]

25.6.2. *Right for Correction and Deletion*

Pursuant to Article 12 subsection b of the above-mentioned Directive, data subject has the right to ask the controller to rectify, erase or block the data, the processing of which does not comply with the provisions of the data protection Directive, in particular because of the incomplete or inaccurate nature of these data. The majority of national laws which enact the provisions of this Directive, foresee that the refusal of a controller to satisfy the request of a data subject for correction or deletion of his personal data should be reasoned.

The legitimate ground on which the processing of personal data, with the form of publishing data on the Internet, is founded is of great importance for handling this kind of requests. As we have previously mentioned in this chapter, data subjects can always withdraw their consent. Therefore, in the case that the publication is founded on the consent of the data subject, the controller should always comply with a request for deletion and should take this possibility into account even when designing the technical system. If the processing of personal data is founded on another legitimate ground of those referred in Article 7 of the data protection Directive, the data subject may request the deletion or correction of his data in the

[y] See Dutch DPA, as above, endnote 10, p. 39.

event that data are incorrect, incomplete, irrelevant or excessive for the purpose for which they are processed. In the case that the request is justified, the controller is obliged to comply.

25.6.3. *Right to Object*

In addition to the right of access, the right for correction and deletion, the data protection Directive grants data subjects the right to object on legitimate grounds to the processing of data relating to them. According to Article 15 of the aforementioned Directive, this right applies in the case that the processing is justified by a legitimate ground of those stated in Article 7. The right to object, therefore, applies to processing that is lawful but by virtue of the data subject's personal circumstances may be unlawful in relation to the specific data subject. If the objection is justified, the data controller may no longer continue the processing of the relevant data.

25.7. Transfer of Data to Countries Outside of the EU

The transfer of personal data to countries outside of the EU is, in principal, prohibited. It is considered legitimate only when the third country or the recipient, established in a third country, complies with regulations that offer an adequate with the European level of protection. The European Commission, following the procedure provided in Article 31§ 2 of the data protection Directive, decides whether a country meets that level. Examples of countries that offer an adequate level of data protection are Switzerland, Canada and Argentina. Specific agreements have, also, been made with the United States in order to legitimise the transfer of personal data to companies that apply the Safe Harbour regulations.[z]

Transfer of personal data to third countries should not be confused with the accessibility of Internet publications to an indefinite number of people all over the world. That was pointed out in the decision of the ECJ at the aforementioned Lindqvist case. The court stated that there is no transfer of data to third countries, in the meaning of Article 25 of the data protection Directive, where an individual loads personal data onto an Internet page, thereby making those data accessible to anyone who connects to the Internet, including people in a third country. The previously mentioned provisions should only apply when the intention of the controller to export personal data to third countries is explicit.[aa] Following the judgement of the ECJ, many member states of the EU adopted the same interpretation of the definition "transfer to a third country".[ab]

[z] The countries that offer an adequate level of data protection are available at http://ec.europa.eu/justice_home/fsj/privacy/thirdcountries/index_en.htm.

[aa] See Recital 56–71 of the ECJ ruling in Case C — 101/01, as above.

[ab] See Dutch DPA, as above, p. 49, CNIL (French DPA), Deliberation no 2005-276 of 17 November 2005, available at www.cnil.fr.

25.8. Access to Traffic Data by Third Parties (Directive 2002/58/EC)

A very interesting issue that is related to the interpretation of the ePrivacy Directive and is confronted by courts in almost every member state[ac] is the following: could third parties have access to traffic data retained by Internet Service Providers (ISPs) in order to ensure their effective protection of copyright in the context of civil proceedings? The ECJ had the chance to address this issue when a Spanish court made a reference for a preliminary ruling, asking whether the non-profitmaking organisation, Productores de Música de España (Promusicae), acting on behalf of its members who are holders of intellectual property rights, had the right to access the personal data relating to the use of the Internet by means of connections provided by Telefónica de España SAU ('Telefónica') in order to effectively protect the intellectual property rights of its members.[ad]

The ECJ in its judgement tried to reconcile two different fundamental rights, namely the right to respect for private life on one hand and the rights to protection of property and to an effective remedy on the other.[ae] The ECJ, interpreting, in principal, Article 5 (confidentiality of communications and of the related traffic data) and 15 (exceptions to confidentiality that could be adopted by member states) of the ePrivacy Directive in connection to Directives 2000/31, 2001/29 and 2004/48, concluded that member states are not required to adopt an obligation to communicate personal data to third parties in order to ensure effective protection of copyright in the context of civil proceedings.[af] The authorities and courts of the member states should, though, interpret their national law in a manner that is not in conflict with any of those fundamental rights (the right to respect for private life and the rights to protection of property and to an effective remedy) or with other general principles of Community law, such as the principle of proportionality.

25.9. Conclusion

Although Internet is considered nowadays as a very important element of our day-to-day life, it undoubtedly raises many different and complicated issues with regard to the processing of personal data. The right to protection against the uncontrolled

[ac] See, for example, Chiara Garofoli, Cri 6/2007, pp. 182–185, who presents the recent jurisprudence of the District Court of Rome, which, following the intervention of the Italian DPA, concluded that the Internet users' interest in protecting their privacy and keeping their personal data confidential is not outweighed by the interest of copyright holders in pursuing unlawful acts, such as making copyright-protected material available on the web.

[ad] Case C-275/06, Productores de Música de España (Promusicae) v Telefónica de España SAU [2008], available at http://curia.europa.eu.

[ae] The fundamental right to property, which includes intellectual property rights such as copyright, and the fundamental right to effective judicial protection constitute general principles of Community law, see Recital 62 of the ECJ ruling in case C-275/06 and further jurisprudence, as above.

[af] See Recital 70 of the ECJ ruling in case C-275/06, as above.

collection, processing and distribution of personal data is confronted with other fundamental rights such as the right of freedom of expression, the right to protection of property and the right to an effective remedy. The EU has adopted two Directives that aim at keeping a balance between the right to privacy on one hand and the free flow of information in the EU on the other (Directives 95/46/EC and 2002/58/EC). Both Directives apply to personal data processed on the Internet and they foresee basic principles that should be respected by the controllers, and specific legitimate grounds for the processing of data. Data subject's consent should be considered as the main ground for lawful processing of personal data on the Internet due to its distributed nature. The data protection Directive, also, grants the data subjects three substantial rights: the right of access, the right for correction and deletion and the right to object on legitimate grounds to further processing of their personal data.

The jurisprudence of the ECJ on matters of on-line data protection is trying to reconcile the exercise of this right with other fundamental rights and recommends that the courts and the authorities of the member states should interpret their national data protection law in a manner that is not in conflict with any other fundamental right or with any general principle of Community law such as the principle of proportionality. Bearing this in mind, the national authorities should address new issues of data protection on the Internet that will definitely rise due to the rapid progress of technology, by trying to maintain a balance between the confronting rights.

References

1. Article 29 Working Party, Opinion on data protection issues related to search engines, 4 April 2008, available at http://ec.europa.eu/justice_home/fsj/privacy/workinggroup/wpdocs/2008_en.htm.
2. Article 29 Working Party, Privacy on the Internet — an integrated EU approach to on-line data protection, November 2000, available at http://ec.europa.eu/justice_home/fsj/privacy/workinggroup/wpdocs/2000_en.htm.
3. Article 29 Working Party, Recommendation 1/1999 on invisible and automatic processing of personal data on the Internet performed by software and hardware, 23 February 1999, available at http://ec.europa.eu/justice_home/fsj/privacy/workinggroup/wpdocs/1999_en.htm.
4. Article 29 Working Party, Recommendation 2/2001 certain minimum requirements for collecting personal data on line in the European Union, 17 May 2001, available at http://ec.europa.eu/justice_home/fsj/privacy/workinggroup/wpdocs/2001_en.htm.
5. Article 29 Working Party, Working document on determining the international application of EU data protection law to personal data processing on the Internet by non-EU based websites, 30 May 2002, available at http://ec.europa.eu/justice_home/fsj/privacy/workinggroup/wpdocs/2008_en.htm.
6. Article 29 Working Party, Working document 1/2008 on the protection of children's personal data, 18 February 2008, available at http://ec.europa.eu/justice_home/fsj/privacy/workinggroup/wpdocs/2008_en.htm.
7. CNIL, Deliberation no 2005-276 of 17 November 2005, available at www.cnil.fr.

8. Dutch DPA, Publication of personal data on the Internet, December 2007, available at http://www.dutchdpa.nl/documenten/en_pb_2007_privacy_legislation_internet.shtml?refer=true.
9. ENISA (European Network and Information Security Agency), Security issues and recommendations for on line social networks, October 2007, available at http://www.enisa.europa.eu.
10. C. Garofoli, Cri 6/2007, pp. 182–185.
11. IWGDPT (International Working Group on Data Protection in Telecommunications), Report and guidance on privacy in social network services, "Rome Memorandum", available at http://www.berlin-privacy-group.org.
12. N. Lugaresi, Principles and regulations about on line privacy: "Implementation Divide" and Misunderstandings in the European Union, available at http://ssrn.com/abstract=333440.
13. G. J. H. Smith, *Internet Law and Regulation*, 3rd Edition (London, 2002), 372, 7–014.
14. R. Wong and J. Savirimuthu, All or nothing: This is the question?: The application of Art. 3(2) Data Protection Directive 95/46/EC to the Internet, available at SSRN, http://ssrn.com/abstract=1003025.

Chapter 26

DOWNLOADING MUSIC: LEGAL FACTS

HAMID JAHANKHANI and CHRIS O. FOLORUNSO
School of Computing, IT and Engineering
University of East London, UK

This chapter looks into available data regarding downloading of free music on the Internet and looks at views on this problem from the government and recording industry, as well as legal cases that have sprung up around this issue, focusing on principles of copyright law. The topic presents both sides of the issue and pays attention to how the issue of downloading music for free over the Internet has affected both internal and external business environments in terms of the ability to adapt to change.

In very recent developments [8], Virgin Media, one of UK's top Internet Service Providers sent out about 800 letters to customers warning them that they should not be downloading illegal music files via file-sharing sites and threatened its customers with disconnection if they persist. This move is guaranteed to highlight the issue in homes especially the ones where parents are oblivious to their children's actions online.

26.1. Introduction

Google your favourite artistes e.g. "Beyonce + mp3 + download" and you are presented with thousands of different websites offering free downloads to Beyoncé Knowles' copyrighted music files for free.

Why should Beyonce and other artists continue to write and make new songs and put them to Compact Discs (CD) or sell on sites like iTunes if they are not going to be paid for their efforts? Certainly, no one would like to go to work if at the end of the month they do not end up with being paid for their sweat.

Market analysts say "Digital downloading will continue to soar in popularity, with spending expected to hit £ 600million by 2012" [28].

Downloading music illegally instead of paying for the CD is starting to have reaching effects on our society [1, 24]. This will, in turn, discourage artists from producing new material and it will teach future generations that stealing is okay as long as one does not get caught.

The words in the street within teenagers about downloading music online according to a report [22], are:

"Pretty much everyone that I know downloads music on their computer",

> "It's stealing from the musicians",
>
> "For me, it feels like you're not stealing when you're just downloading something off the computer",
>
> "I mean, downloading a song or two of a particular artist is not going to hurt them that much",

According to a report by Jupiter Research, [21], "The digital youth of today are being brought up on a near limitless diet of free and disposable music from file-sharing networks.

But who is to blame? Levine [18], gave a perfect breakdown of why the blame should be for the record companies for looking like "conglomerate ogres" in the book "*the art of downloading music.*" Levine [18] writes:

> "People who file-share think they are ripping off the record company, not the artist, and they don't care about stealing from a big conglomerate, which is not the right attitude, but the record companies brought that on themselves to a certain extent".
>
> "They may be earning the lion's share of the royalties, but when you steal a track, the poor artist's smaller percentage gets lost as well" [18, p. 31].

Downloading illegally is stealing and stealing is wrong. Shergill-Connolly [23] thinks that lawsuits might be enough to bring parents attention to their kids as most parents do not even know that their kids can download music from the Internet, let alone the fact that their kids are downloading illegally so maybe parents out to share in the blame and hopefully if they are aware of this, they would be able to teach or explain to their kids and bring in to sync the consequences of stealing, which they would have been speaking to them about by now (hopefully) and also the legal consequences of piracy.

Unauthorised music downloading has hit record levels in the United Kingdom and the main reasons, according to Entertainment Media Research, is that people have lost the fear of being prosecuted. This, in turn, is continuing to have a massive impact on CD sales [1].

Before the Internet and the broadband revolution, to take ownership of music to play whenever and however you like as a consumer was basically limited to two options:

(1) Recording directly from the radio station which involves buying a blank cassette and waiting for hours for that famous track to be played. Of course, you had to be content with the radio jingles and DJ voice-overs and
(2) Buying the LP (7/12-inch record) or CD directly from record shops which always involves taking a trip to and from the high street.

These days, with technological advances — both in hardware and software, copying music has become much easier and accessible. All you need is an adequate computer with a reasonably recent operating system (OS), a good broadband connection coupled with up-to-date antivirus and software to connect you to the millions of music files out there on the Internet, this can be the authorised e.g. ITunes or one of the many unauthorised ones like Limewire or Kazaa.

The sales volume of the music industry has decreased significantly in the last few years. According to Smith [24], "CD music sales are down 20% from the same week, a year ago (March 2006)..." and that "the seven year decline in CD sales doesn't look to be turning around anytime soon".

One of the main reasons, according to the report, is the massive illegal sharing and downloading of music on peer-to-peer (P2P) networks. These networks enable consumers to share, distribute and copy digital content, such as music, very quickly, easily and freely.

The copyright industries, including the music industry's response was to introduce Digital Rights Management (DRM) which can be simply defined as the use of "access control technologies used by publishers and other copyright holders to limit the usage of digital media or devices" [19].

The DRM is now being used by most of the commercial online stores, such as Apple's ITunes which has dominated the market. ITunes sold more than 1 billion of such downloads in the first few years of operation, and currently selling over 5 million songs a day with, around 58 songs every second [4], shows that consumers are still willing to pay for legal downloads despite the fact that copies are freely and easily available.

The use of DRM has been controversial. Advocates argue that it is necessary for copyright holders to prevent unauthorised duplication of their work to ensure continued revenue streams whilst opponents argue that the use of DRM is a big mistake or that DRM violates local laws and so calling for a move to DRM-free music.

Without a doubt, one of the most widely discussed topic these days in the digital technology world is the move to DRM-free music. Apple's Jobs [15], talked previously on his "Thoughts on Music" about the disadvantages of DRM, stating "he would surely switch to DRM-free music in a heartbeat, if only the music labels would let him". Two months later, EMI Music, one of the big-record label, Jobs was referring to responded to his speech by launching DRM-free superior quality downloads across its entire digital repertoire and made Apple the first online music store to receive the new premium downloads [13]. It is only a matter of time before other labels follow if we judge by the result of a recent survey, which found that almost two-thirds of music industry executives thought that removing digital locks from downloadable music would make more people buy the tracks [16].

All these seem to prove that DRM might soon be a thing of the past unless something better is introduced to replace this unworkable solution as echoed by Bill Gates [6], who said "DRM is not where it should be. In the end of the day, incentive systems (for artists) make a difference. But we don't have the right thing here in terms of simplicity or interoperability".

But there is one question which remains unanswered in this topic of DRM i.e. what are the implications of a DRM-free (music with no digital restrictions) society? Who wins and who loses? Will it be Apple Inc., the average consumer or the record companies?

These issues have not been discussed very broadly. Most papers dealing with digital content and piracy have focused on abolishing the technology rather than suggestions on what could be done or what other idea could be exercised to combat the rise in unauthorised music downloads, [5, 11, 14], the most notable of these fights against DRM in the United Kingdom was the petition signed by 1,400 people and sent to No. 10 Downing Street [12], which was later rejected with the response "copyright owners should be able to continue to protect their content in that way".

Those focusing on the need for DRM look for ways to convince the public that they are on their side and against it [15].

A recent poll on Pocket-lint [20], reported that 62% of people are prepared to pay extra for music that is DRM-free. That surely must be good news to Apple if we recall Jobs' announcement that Apple's music download store will be offering DRM-free music.

Thompson [26], does not seem to believe this; he says "If Apple switched off FairPlay, then they would probably sell a lot more songs, on which they make very little money, and a lot fewer iPods, on which they make a lot".

Apple's downloaded files come with restrictions on their use, enforced by FairPlay, Apple's version of DRM. Of course, Apple's format, AAC, combined with FairPlay-encoded files is not compatible with any music devices other than Apple's media players. Therefore, dropping FairPlay could mean no more interoperability problems with players from other manufacturers, although Jobs did say in his recent manifesto [15], that he had no problems with this idea, the "Lack of interoperability" meaning that iTunes purchased music will no longer be exclusive to the iPod, giving consumers the freedom to use ANY device they wish to play their media.

In reply to speculations and statements like this, Jobs replied his critics at the EMI presentation announcing its DRM-free music rollout, Dalrymple [10] stating;

> "Some doubted Apple's sincerity when we proposed this solution to the interoperability problem this year saying that, as the number one digital music store and the number one maker of digital music players, we [Apple] had too much to lose by breaking the proprietary bonds between the iTunes music store and the iPod music players."
>
> "Hopefully, by our actions here today and over the coming months they will conclude that we are continuing to do exactly what has earned us these number one positions — doing the right thing for the customer, and the right thing for the customer going forward is to tear down the walls that preclude interoperability by going DRM-free."

Apple's iTunes is now faced with some competition with Amazon.com joining the online music sales business with unrestricted DRM-free downloads. Amazon has a massive 2 million songs by over 180,000 bands and artists from big music labels like EMI and Universal [2].

A DRM-free system could also have some effects on download and player technologies, therefore manufacturers will be looking to add AAC support to their MP3 players to capitalise upon Apple's popularity work and to make transition easy for existing iPod users. This will expand the potential market for AAC files to almost all portable players from iPods and Microsoft' Zune, because Zune "does support unprotected AAC files much in the same way that iPod supports unprotected WMA (Microsoft proprietary) music files" [17]. With that change, a lot more music services will consider using the AAC format either instead of or in addition to MP3.

For consumers, the big advantage of DRM-free music is that it can be played on any device they own, and there will be no arbitrary limit on the number of different devices that can be used. But there could be implications when purchasing DRM-free files.

Before going DRM-free, downloaded files could only be played on computers and MP3 players authorised to play them, but that is all changed now. It would be a bad idea for consumers to share any media purchased from the iTunes Store or Amazon on a P2P network simply because its source would be easily traceable. There have been talks about personal data being embedded within files, this could the purchaser's name, Apple ID if from ITunes or e-mail address, if this turns out to be completely true, then users will need to be really careful with what they do with the files.

26.2. Downloading Music, the Act

There have been a lot of news stories and publicity around music downloads and there is hardly a day we do not hear about the hefty penalties, lawsuits, warnings about losses and inventions designed to halt this so-called free gifts.

The British record industry began a campaign back in 2004 warning those swapping copyright-protected works that they are breaking UK copyright law and risk being penalised (BPI) but there are still some grey areas about when it is illegal to download music online.

Swapping files using file-sharing software over P2P networks are illegal in most countries and purchasing music from online stores like HMV, ITunes and OD2 is perfectly legal.

26.2.1. *Music Downloads a Brief History*

Wikipedia defines downloading as "the transferring of a music file from an Internet-facing computer or website to a user's local computer".

The most popular of downloads is music. This term covers both the legal and illegal downloads of copywritten material with or without permission or payment if required.

The first phase of this terminology came into existence as a problem within Copyright Law, where music was being downloaded from the Internet without the owner's permission. This is generally blamed on the introduction of technologies such as Peer to Peer (P2P) file sharing, which would be later popularised by Shawn Fanning who in 1999 launched Napster. But [18], one must also put some of the blame on the move away from slow dial-up Internet connections to high-speed broadband service and the proliferation of recordable CD equipment, which is now within the financial reach of the average consumer. The combination of these technologies has been convenient and innovative fun for everyone and a growing disaster for the music industry. As evidenced by recent reports about the "high levels of unauthorised downloads [1].

Legal music downloads typically involves the purchase of a song or an album available for downloading on the Internet. This is the second phase of music downloads, the online music store, whereby songs can be downloaded at a price; these include iTunes, URGE, Napster and MSN's Music store. Some even allow free downloads to its users by selling advertising spaces on their websites to pay from the music downloads; the latest example of these is SpiralFrog. SpiralFrog is a Web-based, ad-supported music experience, combining music discovery with the free acquisition of audio and music video files, licensed from major and independent record labels and publishers.

With this new phase, we are now seeing a change in the way music is sold and charts rated in different countries. In the United Kingdom for instance, there is now an additional way of measuring how the sales of singles or albums are going in the constant weekly race for the top spot; the UK Number 1 spot. History was made when "Crazy" by Gnarls Barkley reached the top spot in the UK Singles Chart based on download sales alone back in May 2006. Because until that month and according to the BBC "download sales could only count towards a chart position if the song could also be bought in shops, and under new rules, downloads can be counted as long as physical copies go on sale the following week" [7].

Downloading music, first became popular with file-sharing technologies such as P2P networks, with people breaking copyright laws by not paying for any of it. The Recording Industry Association of America (RIAA) claimed that this practice was damaging the music industry, and a series of law suits led to many of these networks being closed down. The most popular case of these was that of Napster which has now been re-branded and turned into a legal, pay-per-song music-download site.

26.2.1.1. *Napster*

Napster was a file-sharing service that paved the way for P2P file-sharing programs such as Kazaa, Limewire, iMesh, Morpheus and BearShare, which are now used for downloading music, pictures and other files. The popularity and repercussions of the first Napster have made it a legendary icon in the computer and entertainment fields and when discussing the topic of Music Downloads.

Napster was founded in 1999 as a P2P music file-sharing service by 18-year-old Shawn Fanning with the help of his uncle and friends. Napster did not store the music itself in its servers, but provided an index to files in other people's computers. More than 60 million users took advantage of the service, and it quickly became one of the most controversial ventures on the Web, because much of the music being shared was copyrighted material.

The music industry sued the company, claiming losses of millions in royalties. Napster lost the case in 2000 and was about to be shut down, except for a last-minute stay from the Circuit Court of Appeals. Because Napster could not reach an agreement with major record companies, it filed for Chapter 11 in 2002 [3]. Chapter 11 is a chapter of the US Bankruptcy Code, which permits reorganisation under the bankruptcy laws of the United States.

Napster was a different way of distributing MP3 files. Instead of storing the songs on a central computer, the songs live on users' machines. When you want to download a song using Napster, you are downloading it from another person's machine, and that person could be your next-door neighbour or someone halfway around the world.

Therefore, in summary, to use the Old Napster you needed:

- A copy of the Napster utility installed on your computer;
- A directory on your computer that has been shared so that remote users can access it and
- Some type of Internet connection.

The provider of the song needed:

- A copy of the Napster utility installed on his/her computer;
- A directory on his/her computer that has been shared so that someone else could access it;
- Some type of Internet connection that was "on" and
- A copy of the song in the designated, shared directory.

At its peak, Napster was perhaps the most popular website ever created, it went from 0 to 60 million visitors per month in less than a year [9].

Napster made P2P sharing popular and hence created a market; the computer program that made this possible was refined and rewritten by others, which led to more companies offering similar illegal services sharing music collections with other computers all over the world.

26.3. P2P Networking

The most important of them all is the P2P networking. The term P2P refers to "peer-to-peer" networking.

Webopedia defines P2P as "A type of network in which each workstation has equivalent capabilities and responsibilities. This differs from client/server architectures, in which some computers are dedicated to serving the others".

Traditionally P2P means, when, computers are in a P2P network and typically situated physically near to each other and run similar networking protocols and software. The modern view of P2P computing, P2P networks stretch across the entire Internet, not just a home local area network (LAN).

A P2P network implements search and data transfer protocols above the Internet Protocol (IP). To access a P2P network, users simply download and install a suitable P2P client application and there are lots of P2P networks and P2P software applications in existence.

26.3.1. *Vulnerabilities of P2P File-Sharing Networks*

P2P file-sharing systems have become the single most popular class of Internet applications of the new decade, but, we need to be aware of its vulnerabilities.

The P2P applications such as BearShare, Warez, Morpheus, BitTorrent, iMesh and KaZaA, make it easy for users to exchange files with each other over the Internet. But while these programs are a good way of sharing information, they are not entirely harmless and can cause problems for your personal system.

- *Copyright Issues*: A file-sharing application makes it easy to share music and other files. However, unless the user has the explicit permission of the copyright owner to possess or distribute the material, the user may be in violation of copyright laws.
- *Network Capacity*: Most P2P applications will usually be configured so that other users can access your hard drive and share your files all of the time. This constant file transfer can degrade your computer's performance. Therefore it is best to disable file-sharing access on one's computer to maximise performance and also to prevent access which leads to a breach in privacy.
- *Privacy*: If you are running a file-sharing application, make sure that you know which files and data the program can access and open up to others. You may be inadvertently sharing personal information, such as e-mail messages and credit card information with others.
- *Security*: The security of a computer running file-sharing applications can be at risk to worms and viruses. If these, any of these malicious codes infects your computer, apart from destroying all data stored on your computer, it can also spread to millions of computers on the Internet.

26.4. Legal Facts

The question is, downloading from the Internet, is it legal? For that many people though, it is as simple as going online with one of many P2P file-sharing programs or visiting one of the hundreds of virus-infected websites such as www.simplemp3s.net, selecting the tracks, downloading and burning it to a CD-ROM or straight to the MP3 player. But what are not so simple about downloading music are the copyright protection laws that is broken everyday by downloading some music tracks off the Internet. To make things even more difficult and confusing, these laws vary from country to country and the fact that music can be lawfully downloaded.

The fact is, if you use file-sharing services such as Kazaa or Limewire to download commercial tracks, and perhaps even share what music you already have, you are more than likely breaking copyright law. The British Phonographic Industry (BPI), now successfully bringing and winning lawsuits against file-sharers in the United Kingdom.

According to BPI, "Illegal peer-to-peer file-sharing has already had an enormous effect on British music sales; with an estimated £ 1.1bn in revenue lost in the last three years as a direct result. Its members cannot hope to continue investing in new music if people do not pay for it." (BPI Article)

"Ignorance is not a defence," [7] said Judge Justice Lawrence Collins. At the hearing, two men were found liable for unlawful file-sharing in Britain back in 2006 and ordered to pay thousands of pounds in damages.

One of them was ordered to make an immediate payment of £ 5,000, with total costs estimated at £ 13,500 and final damages were expected to take the bill even higher. The other was ordered to make an immediate payment of £ 1,500, pending final determination of costs and damages and there have been more cases since then.

According to the BPI, in October 2004, 23 settlements arose out of the 26 cases.

26.4.1. *Copyright Law*

Laws regarding the sharing and downloading of music on the Internet vary from country to country.

For example, Canadians have the ability to download media copyrighted files legally; this started back in December 2003, when the Canadian Copyright Board ruled that downloading files of a P2P network was legal, but uploading those files was not. This has made Canada the country with the greatest number of file sharers per capita in the world (OECD, 2005). However, this is all about to change with changes to Canada's copyright laws.

The Italian parliament in 2004 voted in favour of imposing jail sentences of up to three years on anyone caught uploading or downloading unauthorised copyright material to and from the Net [25].

The Copyright, Designs and Patents Act 1988, is the current UK copyright law. It gives the creators of literary, dramatic, musical and artistic works the right to control the ways in which their material may be used.

The rights cover: Broadcast and public performance, copying, adapting, issuing, renting and lending copies to the public. — UK Copyright Service.

Therefore, in the United Kingdom, a person must have permission to make a copyrighted work (such as a sound recording) available for download on the Internet and doing so i.e. uploading without permission of the copyright owner will be against the law.

In the United States, the law is much stricter and also deems copying of copyrighted music as illegal. The US Code protects copyright owners from the unauthorised reproduction, adaptation or distribution of sound recordings, as well as certain digital performances to the public. In more general terms, it is considered legal for you to purchase a music CD and record (rip) it to MP3 files for your own use. But uploading these files via P2P networks would constitute a breach of the law.

In the European Union (EU), the 2001 EU Copyright directive, which implemented the 1996 WIPO treaty ("World Intellectual Property Organization Copyright Treaty"), prohibits P2P, claiming that it is a violation of the directive [27]. However, it is worth noting that not all European member states have implemented the directive in national legislation.

26.5. Conclusions

Unauthorised music downloading has hit record levels in the United Kingdom and the main reasons, according to Entertainment Media Research, is that people have lost the fear of being prosecuted. This have in turn is continuing to have a massive impact on CD sales. The British record industry began a campaign back in 2004 warning those swapping copyright-protected works that they are breaking UK copyright law and BPI but there are still some grey areas about when it is illegal to download music online.

In this chapter, the problem of unauthorised and illegal music downloads from the internet and looking into what the music industry was/is doing to combat the problem was presented.

By taking views from a users' perspective, who has the option of either buying or copying digital music; the effects and impact of new and emerging technologies are having on consumer behaviours and the responses from the legal system, law enforcement and the music industries' and its underlying penalties, as well as rights and usage restrictions enforced through DRM in their attempt to reduce and hopefully put an end to this act.

With the record labels now gradually rolling out DRM-free music downloads, will help drive illegal downloader's to buy their music the right way, this plus the fact that there have been a lot of publicity around the actions of trade groups that represents the recording industries both in the United Kingdom (BPI) and the United States (RIAA) showcasing the successful lawsuits which they have brought against illegal downloaders, consumers who download music from legal file-sharing websites will start to outnumber those using illegal services.

References

1. N. Anderson, Unauthorized music downloading hits record levels in UK, July 2007. [Online] Available: http://arstechnica.com/news.ars/post/20070731-unauthorized-music-downloading-hits-record-levels-in-uk.html (accessed 28/10/07).

2. Amazon, Amazon.com launches public beta of Amazon MP3, September 2007. [Online] Available: http://phx.corporate-ir.net/phoenix.zhtml?c=176060&p=irol-newsArticle&ID=1055053 (accessed 28/10/07).
3. Answers.com — [Online], 2007. Available: http://www.answers.com/ (accessed 28/10/07).
4. Apple, 100 Million iPods Sold CUPERTINO April 9, 2007. [Online] Available: www.apple.com/pr/library/2007/04/09ipod.html (accessed 28/10/07).
5. Apple Classic, 'Hold everything' Apple Overview, August 2007. [Online] Available: www.apple.com/ipodclassic/ (accessed 28/10/07).
6. BBC, BBC News gates: digital locks too complex, December 2006. [Online] Available: http://news.bbc.co.uk/1/hi/technology/6182657.stm (accessed 28/10/07).
7. BBC, BBC News 'Court rules against song-swappers', January 2006. [Online] Available: http://news.bbc.co.uk/1/hi/entertainment/4653662.stm (accessed 28/ 10/07).
8. BBC, Warning letters to 'file-sharers', July 2008. [Online] Available: http://news.bbc.co.uk/1/hi/technology/7486743.stm (accessed 29/07/08).
9. M. Brian, How Gnutella. [Online] Available: Works, http://computer.howstuffworks.com/file-sharing.htm (accessed 28/10/07).
10. J. Dalrymple, Apple, EMI offer higher-quality DRM free downloads, April 2007. [Online] Available: http://playlistmag.com/news/2007/04/02/drmfree/index.php (accessed 28/10/07).
11. C. Doctorow, Greetings fellow pirates! Arrrrr!, June 2004. [Online] Available: www.kokogiak.com/thatboxinthecorner/drmcutup.asp (accessed 28/10/07).
12. 10 Downing Street, E-Petitions, Ban the use of digital rights management, February 2007. [Online] Available: www.number-10.gov.uk/output/Page11020.asp (accessed 28/10/07).
13. EMI, EMI music launches DRM-free, April 2007. [Online] Available: www.emigroup.com/Press/2007/press18.htm (accessed 28/10/07).
14. S. Granneman, The big DRM mistake, March 2006. [Online] Available: www.theregister.co.uk/2006/03/03/the_drm_mistake/ (accessed 28/10/07).
15. S. Jobs, Thoughts on music, February 2007. [Online] Available: www.apple.com/de/hotnews/thoughtsonmusic/ (accessed 28/10/07).
16. Jupiter Research, Music execs criticise DRM systems, February 2007. [Online] Available: http://news.bbc.co.uk/2/hi/technology/6362069.stm (accessed 28/ 10/07).
17. J. Layton, How stuff works 'How does Zune compare to iPod?, 2007. [Online] Available: http://electronics.howstuffworks.com/zune-ipod.htm (accessed 28/10/07).
18. S. Levine, *Art of Downloading Music* (London, 2004, Sanctuary).
19. Microsoft, Windows media DRM partners, 2007. [Online] Available: www.microsoft.com/windows/windowsmedia/forpros/drm/9series/providers.aspx (last accessed 25/ 10/2007).
20. S. Miles, Poll finds music lovers will pay extra for DRM free music, April 2007. [Online] Available: www.pocket-lint.co.uk/news/news.phtml/7298/8322/users-happy-pay-extra-drm.phtml (accessed 28/10/07).
21. M. Mulligan, Free music still beats legal in Europe, November 2005. [Online] Available: www.mp3.com/news/stories/2520.html (accessed 28/10/07).
22. R. Seith, Downloading copyrighted music, March 2003. [Online] Available: www.connectwithkids.com/tipsheet/2003/115_mar12/music.html (accessed 28/10/07).
23. S. Shergill-Connolly, Explain the legal consequences of pirating music, March 2003. [Online] Available: www.connectwithkids.com/tipsheet/2003/115_mar12/music.html (accessed 28/10/07).

24. E. Smith, Sales of music, long in decline, plunge sharply, March 2007. [Online] Available: http://online.wsj.com/public/article/SB117444575607043728-oEugjUqEt-To1hWJawejgR3LjRAw_20080320.html?mod=rss_free (accessed 28/10/07).
25. The Register, Italy approves 'jail for P2P users' law, May 2004. Retrieved on 15/10/2007. [Online] Available: www.theregister.co.uk/2004/05/20/italy_p2p_law/ (accessed 28/10/07).
26. B. Thompson, Why I don't believe Steve Jobs, February 2007. [Online] Available: http://news.bbc.co.uk/1/hi/technology/6353889.stm (accessed 28/10/07).
27. UER-Lex, Directive 2001/29/... harmonisation of certain aspects of copyright and related rights in the information society, June 2001. [Online] Available: www.ivir.nl/legislation/eu/copyright-directive.doc (accessed 28/10/07).
28. Verdict Research, Digital downloads rise, October 2007. [Online] Available: http://news.yahoo.com/s/nm/20071012/wr_nm/britain_downloads_dc (accessed 28/10/07).

Chapter 27

THE USE OF ELECTRONIC SURVEILLANCE IN CONDUCTING CRIMINAL INVESTIGATIONS ON THE INTERNET

MURDOCH WATNEY

Professor and Head of the Department of Criminal Law and Procedure
University of Johannesburg, South Africa

Cybercrime undoubtedly threatens the global growth and future of the Internet. Governments cannot ignore the abuse of the Internet and must address cybercrime that includes terrorism.

Many governments have elected to utilise electronic surveillance as an investigatory method in addressing the prevention, detection, investigation and prosecution of crime in an electronic medium such as the Internet.

Although the use of surveillance may not be a new investigatory method, surveillance on the Internet differs from surveillance conducted in the physical world. It is for this reason that the use of electronic surveillance has evoked a lot of debate. In this chapter, the author attempts to address the justifiability of the use of electronic surveillance in criminal investigations on the Internet and to counteract some of the criticism leveled against electronic surveillance without negating the risks inherent to electronic surveillance.

The emphasis in this chapter will, from a global perspective, primarily be on the legal regulation of the use of Internet surveillance as an investigatory method bearing in mind that surveillance is made possible by means of technology.

Surveillance technology is ever-evolving and allows increasingly for non-obvious but invasive technological access to information of Internet users. Criminal and intelligence investigations in most countries face the same dilemma: legal regulation of surveillance technology is required to allow the pursuit of justice whilst maintaining a human rights culture and to prevent governments becoming police states with all Internet users seen as guilty until proven innocent. It is a dilemma that requires all stake-holders to carefully monitor the surveillance methods employed in criminal and intelligence investigations conducted on the Internet.

27.1. Introduction

Crime is as old as humankind. However, as the society evolves, the types of crime and its methodology change. Criminal investigation has to keep abreast with these changes to ensure the apprehension of perpetrators and institution of criminal prosecutions.

Law and more specifically those laws governing the criminal justice system traditionally developed in a physical world. Prior to the development and implementation of computing technology and the Internet, many technological developments and changes brought with it advantages as well as disadvantages such as providing new tools for the commission of crime. For example, criminals could use the telephone to plan their activities and the record player and radio increased the incidence of copyright infringement [6]. In most instances, the "traditional" laws governing the criminal justice system were flexible enough to accommodate the investigation of crime resulting from these "new" communication technologies.

The biggest challenge to the "traditional" laws governing the criminal justice system resulted from the decision of the United States of America (US) in 1992 to commercialise the Internet.

The history of the origin of the Internet is well known. The Internet originates from the early 1960s in the United States as a result of a project referred to as ARPANET. This project aimed to ensure a nation-wide computer network that would continue to function even if a large portion of it were destroyed by a nuclear attack [8].

The commercialisation of the Internet had many far-reaching consequences, some not anticipated by the United States and other Internet-connected countries. Most of these consequences impact on and challenge the "traditional" laws governing criminal investigations and intelligence gathering. Intelligence gathering, in this context, refers to the gathering of evidence in a criminal investigation.

The following question comes to mind: why and how do governments provide for criminal investigations by means of electronic surveillance on the Internet? Interrelated to this question, the following issues are addressed:

(a) The differences between criminal investigations in a physical and an electronic medium such as the Internet;
(b) The effect of globalisation on the Internet and crime, including terrorism;
(c) A brief summary of the reasons why governments apply electronic surveillance to the Internet in gathering information with specific reference to the legal regulation of the Internet in addressing crime and now also terrorism;
(d) A discussion of what is understood with "surveillance" and how surveillance is used in the collection of information as part of conducting criminal investigations on the Internet;
(e) Using the only treaty on cybercrime, the Council of Europe Convention on Cybercrime as a benchmark to evaluate international Internet surveillance laws;
(f) A brief overview of the electronic surveillance laws of the United States and EU;
(g) The difference between search and seizure and Internet surveillance and
(h) Whether the use of electronic surveillance of information on the Internet as an investigatory method is justifiable. Many criminal investigations in the physical world employ surveillance as an investigatory method. Although the use of surveillance may not be a new investigatory method, surveillance on the Internet

differs from surveillance conducted in the physical world. It is for this reason that the use of electronic surveillance has provoked a lot of debate.

This chapter focuses on the use of electronic surveillance as an investigating method in conducting criminal investigations and intelligence gathering on the Internet. The aim is not to discuss the collection, analysis and presentation of electronic information as evidence in a court of law, although the information should be collected in such a manner that it would be admissible in a court of law.

The discussion proceeds from a global perspective with reference to various legal aspects pertaining to electronic surveillance and specifically the Internet. Although surveillance is made possible by means of computing technology, the focus will not be on the technological aspects of electronic surveillance, but on the legal impact of information and communication technology on criminal investigations and intelligence gathering and the legal regulation (governance) of Internet surveillance.

27.2. Differences Between Criminal Investigations in a Physical and an Electronic Medium

27.2.1. *Introduction*

The laws governing the criminal justice system developed in a physical world and the question arises whether these "traditional" laws can accommodate the electronic medium or whether the "traditional laws" should be adapted or whether new laws should be implemented? A meaningful answer can only be provided once criminal investigations in a physical medium have been compared with that of an electronic medium.

27.2.2. *Criminal Investigations in a Physical Medium*

The laws governing criminal investigations and investigative methodology were developed for a physical medium and may be characterised as follows:

(a) The object of the crime is mostly tangible in nature;
(b) The main perpetrator is physically present during the commission of the crime;
(c) Crime is predominantly investigated within the borders of a country and jurisdiction as well as the choice of law is seldom issues of dispute. Countries mostly have territorial jurisdiction regarding the investigation of a crime within a physical medium;
(d) Law enforcement agencies conduct criminal investigations and ensure the enforcement of law;
(e) The traditional procedural approach to criminal investigation, including the investigatory methods, is predominantly reactive: once a crime has been committed and brought to the attention of the law enforcement agency, an investigation commences and
(f) The criminal and procedural laws are aimed at the detection and investigation of a crime resulting in a prosecution.

27.2.3. *Criminal Investigations on the Internet, an Electronic Medium*

The commercialisation of the Internet and specifically the introduction of the World Wide Web (WWW) in 1995, resulted in the rapid integration of the Internet into the global society. The Internet presented several challenges to the traditional criminal justice system, such as:

(i) The use of computer technology introduced a new medium, namely an electronic medium that now co-exists with the physical medium. The Internet expanded the electronic medium to include a global borderless 24-hour, 7-days-a-week communication and information system.

(ii) The Internet caused conduct prohibited in the physical world, such as child pornography, fraud ("identity theft") and theft, to move to an electronic environment. Many of these traditional crimes can now be committed faster and in some instances, with more serious consequences, within an electronic medium. The Internet also introduced new methods of criminal conduct, the so-called "Internet" crimes such as denial of service (DoS) and hacking. Initially, some countries did not specifically criminalise this conduct but it could also not be accommodated under the definitions of the traditional crimes (see Section 27.3 hereafter). The non-regulation of the so-called Internet crimes resulted in legal uncertainty.

(iii) Cyber crimes are committed without the physical presence of the perpetrator at the time and place of commission of the crime. Where more than one perpetrator is involved, it is possible for the perpetrators to communicate online without meeting face-to-face. The crime can also be committed against more than one victim. One-on-one victimisation is not typical of cybercrime, as cybercrime can be automated, unlike real-world crime. With automation, perpetrators can commit thousands of crimes quickly and with little effort and one-to-many victimisation could be seen as the default assumption of cybercrime [13].

(iv) The Internet introduced the information age, which evolves around the generation, exchange, receipt and storage of intangible information. A crime is committed in respect of information. The investigation of cybercrime involves the gathering of information. Information is converted into evidence when it becomes admissible as evidence in a court of law (see Sections 27.3.1 and 27.5.1). Electronic evidence can be defined as the electronically stored information that can be used as an evidence in a criminal case. When an investigator is collecting information from computing environments, the ultimate aim is to obtain evidence that is admissible as evidence in a court of law, and to preserve its evidential weight optimally [13]. In the discussion of electronic surveillance, information and electronic evidence will be used interchangeably.

(v) The nature of the Internet is one of many factors that contributed to globalisation (see Section 27.3.2). For purposes of this discussion, it is important to define the meaning of the term "globalisation". Globalisation can be defined as "(t)he free flow of technology, persons, data, images, pests, information, waste ideas and, now terrorist networks that both constitute and characterise globalisation are very hard to slow down or to stop. The world of the Internet, with its built-in capacity to seek ways around obstacles and to continue working even when some nodes are taken out, typifies these global flows" [12].

It encompasses a world in which things are increasingly done at a distance and is characterised by the quick and easy, and in many instances, cheap transfer of technology, data, information and now terrorist networks [12]. The terrorist attack on the World Trade Centre and Pentagon in the United States on 11 September 2001 (referred to as 9/11) was facilitated by globalisation [11, 12] (see Section 27.3.1). After 9/11, it became clear that the attacks were prepared in part in Western Europe, particularly in Germany, and communicated by means of the Internet [5].

(vi) Although the Internet introduced a 24-hour, 7-days a week, faceless, borderless and global information and communication medium that facilitates the commission of crime any time from anywhere in the world, the activities take place somewhere and virtual spaces are downloaded and accessed in particular places in the physical world [11]. It is therefore clear that although an Internet user has borderless access to the Internet, the crime originates from a place in the real world and the result thereof is experienced in a place in the real world. It is also possible to link the crime to a person in the real world. Investigators can establish the identity of the perpetrator by means of investigating the information (data) available on the Internet. If a crime originates from outside the borders of a country, for example, money laundering or a paedophile syndicate the investigators have to rely on international assistance and co-operation in gathering the information and sharing the information.

From these challenges, the following problems regarding the investigation of cybercrime are identified:

(a) Some countries have inadequate criminal and procedural laws to detect, prevent, investigate and prosecute cybercrime. Even if a country regulates conduct on the Internet within its territory, crimes may be committed from outside the country's borders and originate from a country that does not regulate Internet conduct or enforce such a regulation. It is therefore important to have an international treaty on cybercrime, which regulates the collection of evidence in respect of criminal investigations. If Internet-connected countries ascribe to such an international treaty, one will have harmonised criminal and procedural laws in respect of criminal investigations.

(b) Crime investigators must act quickly as the electronic trail may otherwise go cold. Inadequate foreign cooperation and assistance in investigating international crimes and sharing of evidence result in investigators not being able to locate the criminal [13]. An international treaty can go a long way in assisting foreign co-operation and assistance.

(c) Although the emphasis in this chapter is on the collection of information by means of surveillance and not on the admissibility and reliability of the evidence, it is useful to note that the criminal procedural laws of most countries recognise, either explicitly or implicitly, that electronic data can qualify as an evidence in criminal proceedings. However, electronic evidence is different from paper-based documentary evidence in a number of ways. Computer evidence is more fragile than paper documentation. Like any other evidence, electronic evidence must be authentic, accurate, complete, convincing and admissible. The criminal procedural laws and/or laws of evidence of Internet-connected countries must address the admissibility and reliability of such an evidence [13].

(d) The traditional procedural approach to law enforcement has been re-active. This approach is not successful in the detection and prevention of crime. As terrorism is an ongoing threat to some countries, it is important that information is gathered to detect and prevent such attacks (see Section 27.4).

(e) A law enforcement agency cannot effectively investigate a crime on the Internet without the direct or indirect assistance of a third party, namely the Internet Service Provider (ISP). The evolution of Internet legal regulation in addressing crime (see Section 27.4) highlights the changed role of the ISP as a mere conduit of information.

(f) Compliance with legislation within an electronic medium is not easy to enforce. For example, in most countries, the distribution of, access to and possession of Internet child pornography is prohibited but who ensures compliance? The question arises whether the ISP should carry this obligation.

(g) Criminal investigation within an electronic medium such as the Internet may be hampered by the use of technology, such as peer-to-peer file-sharing, steganography (information hiding) and anonymous remailer that make it possible to "hide" crimes.

(h) Information and communication technology is ever evolving. Investigators must not only keep up with the technological changes, but also have the necessary technical ability and skills to investigate cybercrime and gather information. It is also important that the law governing criminal investigations is flexible enough to accommodate technological development.

27.3. The Internet and Globalisation of Terrorism and Crime

27.3.1. *Defining Cybercrime, Cyber-terrorism and Information Warfare*

Universal definitions for cybercrime, cyber-terrorism and information warfare do not exist [18]. Some definitions may be too wide and others too narrow. However, for the purpose of this discussion, it is relevant to conceptualise the different terms.

Cybercrime is defined as any unlawful conduct involving a computer or computer system irrespective of whether it is the object of the crime or instrumental to the commission of the crime or incidental to the commission of the crime [18]. Information warfare and cyber-terrorism resort under the concept, cybercrime.

The term "cyber-terrorism" was coined in 1996 by combining the terms "cyberspace" and "terrorism." The term became widely accepted after being embraced by the US Armed Forces [9]. Much of the interest in cyber-terrorism and information warfare originates from the events of 9/11 and the subsequent discovery of information and communication technology tools used to plan and coordinate those attacks. After the events of 9/11, reports surfaced that Al Qaeda had been transmitting hidden data over the Internet. Data hiding refers to the act of taking a piece of information and hiding it within another piece of data. Hidden maps of terrorist targets and instructions were posted in sport chat rooms and pornographic sites that could be accessed by anyone with an Internet connection [18].

Information warfare (IW) can be defined as the actions taken to infiltrate, corrupt, disrupt or destroy the information systems of an adversary, or to defend one's own information systems from such attacks. The IW by itself could be used as a means of mass disruption rather than mass destruction. However, if a nation is going to be physically attacked, the attackers may well use any means at their disposal, including IW [9, 18].

Cyber-terrorism can be defined as the unlawful attack or threat of attack on computers, networks and the information stored therein for the purpose of intimidation or coercion of a government or its people for the furtherance of a political or social goal [9, 18]. It appears that cyber-terrorism is not a serious threat at present, as a terrorist can achieve more terror and fear by means of a physical attack. A cyber-terrorism attack may be combined with a physical terrorist attack. The biggest threat lies with the use of the Internet for terrorist-related activities, such as secure communications between terrorists utilised, for example, for the exchange of ideas, information and plans related to the planning and targeting of terrorist attacks, canvassing of financial support, spreading of propaganda as well as recruitment. Although a cyber-terrorism attack might not be an immediate threat at this stage, there are some who are of the opinion that cyber-terrorism will increase in frequency and that it will be a continuous threat during the 21st century [9, 18].

Closely linked to cyber-terrorism and information warfare is other cyber crimes aimed at achieving specific objectives similar to that of cyber-terrorism and IW. Some of these cybercrimes may not be destructive but might be disruptive and result in economic damage and loss. These cybercrimes may include website defacement (also referred to as web graffiti), the launch of malicious codes (malicious software or code refers to viruses, worms and Trojan horses, logic bombs and other uninvited software designed to disrupt a computer's operations or destroy files) and DoS attacks. Other cyber crimes such as money-laundering and organised crime may also be used to finance terrorism.

27.3.2. *The Effect of Globalisation on Crime and Terrorism*

The US Justice Oliver Wendell Holmes remarked that "(w)e must study the history in order to understand the path of the law" [17]. If history provides us with the background to legal developments, it is worthwhile to compare the Gunpowder Plot of 1605 with the 9/11 US terrorist attacks. The Gunpowder Plot was not the first example of terrorism, but for the purpose of this discussion, a correlation can be drawn between the Gunpowder Plot and 9/11.

Approximately 400 years ago, an event that would become known as the so-called Gunpowder Plot occurred in England [10]. On 5 November 1605, Guy Fawkes and his co-conspirators planned to blow up the houses of parliament at Westminster. The purpose of this terrorist act was to overthrow the government and to kill King James 1 and as many members of parliament as possible. Fawkes and his colleagues hoped to restore the power of the Roman Catholic Church in Britain. Their plot, when discovered, led to state trials in the court of Star Chamber, multiple executions and renewed legal disadvantages for Catholics in Britain.

Although this happened over 400 years ago, it interestingly enough has parallels with the events of our time. Who can forget the terrorist attack of 11 September 2001?

> "The images of the 11 September terrorist attacks on the institutions that symbolised American economic, military and political power have been forever seared into the collective memory of humankind. History will sombrely mark the day two passenger planes, hijacked by Islamic terrorists, flew into the twin towers of New York City World Trade Center, which collapsed hours later with a thunderous roar. On the same day a third passenger plane, also hijacked by Islamic terrorists, attacked the Pentagon and a fourth airplane, intended for an attack on the White House or the Capitol, crashed en route to Washington, DC" [5].

What do the Gunpowder Plot and 9/11 have in common besides being examples of terrorism? [10]. The following are common denominators: a plot against the state, devotees of a religious minority believing that God was on their side, the planned use of weapons of mass destruction to secure their objectives, civil fear and outrage and calls for the use of extreme retaliatory measures including atypical and unfair trials and extreme punishments. Most importantly, irrespective of whether it was the detection of the Gunpowder Plot or the investigation of the 9/11 terrorist attacks in 2001, information is crucial in the prevention, detection, investigation and prosecution of a crime such as terrorism.

However, much has changed over 400 years. One of the biggest changes of the 20th century is the coming into existence of globalisation. The term "globalisation" only gained widespread use in the 1990s [12]. Former US secretary of state, Colin Powell [12] correctly stated in 2001 that "terrorism is the dark side of globalisation".

Although the Internet brought about many advantages, many disadvantages were to follow as well. From the perspective of cyber criminals such as terrorists and those involved in for example organised crime, the Internet has the following advantages: It allows for reduced transmission time in communications, enabling members of an organisation to coordinate tasks internationally. Information and communication technology also reduce the cost of communication. In the past, terrorists had to centralise their major activities to reduce detection, often associated by direct travel and telephone communication. This made them more vulnerable to discovery and their operations could effectively be wiped out with just one major raid. However, the Internet now enables terrorist organisations to disperse throughout the world resulting in the decentralisation of their operations and consequently easier safeguarding.

> "At the start of the new millennium, contemporary transnational terrorism takes advantages of globalisation, trade, liberalisation, and exploding new technologies to perpetrate diverse crimes and to move money, goods, services and people instantaneously for purposes of perpetrating violence for political ends. Terrorist groups live and operate in a borderless world" [23].

The 9/11 US terrorist attacks stunned the world, especially since nobody thought that an attack of such magnitude could be carried out using aeroplanes as weapons of major destruction. The attacks were followed by a perplexing question: was the information pointing to the possibility of such attacks available prior to the attacks and if affirmative, why had the information not been analysed to prevent the tragedy? This question has resulted in various global responses and we can now distinguish the position prior to 9/11 and the position after 9/11, as new laws and technology aimed at the prevention, detection, investigation and prosecution of terrorism followed the attack.

Although 9/11 was a watershed occurrence that set in motion various forms of governmental response, the response has not been limited to terrorism but also encompass other forms of cybercrime. Whereas terrorism and crime were traditionally confined to the boundaries of a specific country, the Internet enables borderless commission of both cyber-terrorism and cybercrime.

It is important to evaluate the responses of governments in combating, detecting and investigating cyber-terrorism and other cybercrime within the ambit of a human rights culture. Guaranteed human rights *inter alia* include rights such as the right to fair treatment before the law, the right to privacy, freedom of expression and the freedom to hold religious beliefs. From an international perspective, the United Nations Declaration of Human Rights of 1948 upholds various rights such as the right to liberty, life and security of person and most western countries recognise various human rights. Although a government must protect its citizens against crime and terrorism, it must also uphold the protection of human rights applicable to all citizens, including a suspect, irrespective of the crime committed.

> "No one doubts that new dangers are presented to contemporary society by fanatics of all religions, by suicide bombers; and by the access of the combatants to new and powerful weapons of death and destruction which endanger individuals and frighten peaceful civilian populations. Against such dangers, democratic societies are certainly entitled to protect themselves. But they must do so in accordance with the norms of constitutionalism" [10].

Governments are increasingly confronted by problems such as cyber-terrorism and cybercrime that cannot be solved by individual nation-states alone. Although individual governments have jealously guarded their sovereignty, solutions to these new borderless problems necessitate new forms of cooperation on a global level and the creation of new global institutions aimed at the investigation of cybercrime.

27.4. Motivation for the Use of Electronic Surveillance on the Internet

27.4.1. *Introduction*

The evolution of Internet legal regulation illustrates the motivation and reasons for the use of electronic surveillance as an investigatory method in obtaining information available on the Internet [22; see Section 27.5.1].

However, are the evolution of laws that regulate terrorism and crime on the Internet known and understood?

> "... This is the Law. How could there be a mistake in that?'
> 'I don't know this Law', said K.
> 'All the worse for you', replied the warder" [7].

The evolution of Internet legal regulation in addressing crime makes for interesting and in some instances controversial reading. It illustrates how governments grapple with finding solutions in addressing crime committed in an electronic medium such as the Internet but at the same time try to uphold a balance between security and protection of human rights.

This evolution may be divided into three phases, namely self-regulation, conduct regulation and the extension of conduct regulation to include control of information on the Internet by means of electronic surveillance.

27.4.2. *First Phase: No Governmental Regulation of the Internet*

When the United States commercialised the Internet in the early 1990s, it left the regulation of the Internet to the Internet community (consisting of the inventors and users of the Internet). Initially, the role of the law was perceived as irrelevant in respect of the Internet. It was felt that as the Internet was created by technology, it should therefore be regulated by technology.

What many may not realise is that the first phase was fraught with tension between some of the inventors of the Internet and users on one hand and the US government on the other hand in respect of Internet control, initially not in addressing cybercrime, but regarding issues such as the root authority. As the Internet became more commercial in the 1990s, the US government, who had funded the development of the Internet and had acted as passive and absentee custodian of the Internet from the 1970s through much of the 1990s, realised the relevance and importance of governmental control of the Internet [6].

Goldsmith and Wu [6] discuss the early years of self-governance and the making way for governmental control. In their discussion, the ongoing tension between users and intensified governmental control, is illustrated.

As Internet usage increased, the exploitation of the Internet by means of online crimes increased. The Internet was never created with security as its main objective, but had been designed to be open with distributed control and mutual trust among the users. The wake-up call for legal regulation came with the release of the "I love you" virus in 2000 [8]. Although the Federal Bureau of Investigation (FBI) and the Central Intelligence Agency (CIA) traced the origin of the "I love you" virus to the Philippines within 24 hours after its release, the conduct was not a crime in the Philippines at that stage. The perpetrator could therefore not be prosecuted or extradited to the United States to stand trial [4]. This illustrates the effect of globalisation on crime, as crime is increasingly committed outside an affected country's borders. Furthermore, even if an affected country has legislation in place that criminalises such a conduct, the legislation is meaningless if the perpetrator resides in a country where such a conduct is not also criminalised. Cybercrime can therefore only be effectively addressed if the various Internet-connected countries have harmonised legislation and provide assistance in the investigation of the crime and sharing of information.

Self-governance in addressing crime on the Internet proved unsuccessful. In 1988, the first worm, the Morris worm (named after its creator) was released [17]. It is interesting to note that this worm was released prior to the commercialisation of the Internet when the Internet was used by a small trusted group of people, mostly from a military and educational environment. The release of the Morris worm by a member of this trusted group was therefore met with not only dismay, but also with a growing realisation that self-regulation is ineffective to enforce compliance of ethical conduct on the Internet. Countries such as South Africa did not initially regulate conduct on the Internet but argued that the "traditional" criminal and procedural laws were flexible enough to accommodate criminal conduct on the Internet. The non-regulation of conduct during the period 1993 (commercialisation of the Internet in South Africa) to August 2002 resulted in legal uncertainty since the "Internet crimes" could not be accommodated by the South African traditional criminal law and could not be investigated as the conduct was not prohibited as a crime.

Governments realised that they needed a solution in addressing crime on the Internet, or differently put, in securing the Internet. Western governments also acknowledged that there should be an international treaty, which outlines guidelines in addressing cybercrime aimed at establishing harmonised legislation in the various Internet-connected countries. This would assist internationally with combating and investigation of cybercrime.

In 2001, the Council of Europe Convention on Cybercrime introduced a treaty on cybercrime, which was signed by all the Council of Europe member countries and four non-European member countries namely Japan, Canada, the United States and South Africa [21; see Section 27.6]. This is currently the only international treaty on cybercrime.

27.4.3. *Second Phase: Legal Regulation of Conduct*

Legal regulation of conduct on the Internet brought about the second phase in this evolutionary process. This solution was only partially successful. The application of the traditional law enforcement methods, tools and approach to crime within an electronic medium hinders the effectiveness of conduct regulation. Traditionally, a reactive approach was applicable to criminal investigations. This approach fails to address the deterrence, detection and investigation of crime on the Internet and/or prosecution of the perpetrator.

A crime such as "identity theft" illustrates the shortcomings of a reactive approach. By the time the commission of the crime is detected, it is difficult to establish the identity of the perpetrator, since the information (evidence) identifying the perpetrator are, in many instances, not available anymore. Information is vital in crime prevention and detection.

Traditionally, the police is responsible to investigate a crime as well as ensuring compliance with and enforcement of laws. The Internet, however, brings about challenges that necessitate the involvement of third parties, such as the ISP. For example, a government may place a statutory obligation on ISPs to prevent access to child pornography. If the ISP does not have such a statutory obligation, law enforcement becomes very difficult.

Furthermore, countries make laws applicable within their borders but this does not address the challenges of criminal investigations outside the territory of a country. International cooperation and assistance as well as the harmonisation of laws are the most effective methods of successfully addressing the challenges of cross-border cybercrime. Across border, investigations rely on the availability of and access to information.

Due to the problems experienced with the traditional law enforcement methods, tools and approach, governments were already prior to 9/11, looking for a more effective solution to address criminal investigations on the Internet. Effective investigation does not only rely on the prohibition of conduct as a crime but also on the collection of information on the Internet, because without collection of an evidence, a criminal prosecution cannot be conducted successfully.

27.4.4. *Third Phase: The Extension of Conduct Regulation to Include Laws Aimed at State Control (or Governance) of Information Available, Accessed and Distributed on the Internet, or Differently Put, the Availability of and Access to Different Types of Information on the Internet*

The event of 9/11 served as a catalyst to move from conduct regulation to the third phase of Internet legal regulation, namely extending the laws regulating conduct to include laws aimed at state control (governance) of information available, accessed and distributed on the Internet, or differently put, the availability of different types of information and access to different types of information on the Internet.

The third phase, which is now in the early stages of development and refining, is characterised by governments facing the dilemma of deciding which form such an Internet state control should take.

The following forms of governmental control of information on the Internet exist:

- No access to the Internet as practiced by countries such as Cuba; or
- Electronic surveillance: most western countries such as the EU member countries, the United States and other countries such as South Africa apply surveillance technology or are in the process of enacting legislation regulating the use of surveillance technology (bearing in mind that different surveillance methods may be used) or
- Censorship: practiced by countries such as China, Saudi Arabia and Singapore. These countries have adopted technological apparatus to monitor Internet messages, censor websites and prosecute those who speak out against government policies as undermining "state security" [3].

Although this chapter focuses on electronic surveillance, it is important to draw a distinction between state surveillance and censorship. Censorship is an ultra form of governmental control and includes surveillance of the Internet user as well as information available, distributed and accessed on the Internet such as monitoring the content of websites. Contrary to state surveillance, censorship affects the free flow of all information and access to all information globally. The purpose for extensive censorship is in some instances wider than addressing crime and may be politically or ideologically motivated [6]. Censorship in respect of specific information may be, in some instances, justifiable, for example a governmental imposition of a statutory obligation on the ISP to prevent the distribution of and access to child pornography differs [22].

27.5. Understanding the Use of Surveillance as an Investigatory Method on the Internet

Before referring to the surveillance laws of other countries that govern criminal investigations on the Internet (see Section 27.7), it is important to understand what is meant by the term "surveillance" and whether there exists any international treaty in respect of Internet surveillance that can serve as a benchmark (see Section 27.6).

"Surveillance" means in its broadest "to watch over". Information gathering by means of electronic surveillance can be conducted by means of the use of non-communication devices such as biometrics, RFID and video cameras. In respect of communication devices such as the cell phone and the Internet, the information has to be gathered on the communication medium.

"Surveillance" of the Internet is an umbrella term that refers to the collection of different types of information on the Internet by means of surveillance methods (or procedures) (Table 27.1). The aim of electronic surveillance conducted on the Internet is the gathering (collection) of electronic information (evidence) to investigate a serious crime (which includes terrorism and terrorism-related activities). Since Internet surveillance is invasive, it is limited to serious crimes and terrorism.

When investigating a crime on the Internet, investigators seek the collection of the following types of information:

Investigators gather content or traffic data by employing different surveillance methods or procedures, namely interception, monitoring and data retention or data preservation (Table 27.2).

Table 27.1. Type of information.

Type of information	Possible definition
1. Content data	Content information is the equivalent of a letter inside an envelope [2]. Content data is not defined in the cybercrime convention but it is understood as the meaning or purport of the communication or the message or information being conveyed by the communication. The collection of content data is also perceived as more invasive as collecting traffic data.
2. Traffic data	Traffic data is the addressing and routing information, equivalent to what one would learn from reading the outside of a sealed mail envelope without being allowed to open it, whereas content information is the equivalent of the letter inside the envelope [2]. Traffic data is information that is automatically generated when a criminal uses the Internet and can be useful to those investigating crime as it is similar to the physical DNA or fingerprints that are left at a physical crime scene. Although there does not exist a uniform definition for traffic data, most of the definitions have similarities, such as traffic data refers to data indicating the origin, destination, duration, termination, duration and size of the communication (Goemans and Dumortier, 2003) or that traffic data refers to the records kept by the ISPS when a user engages in online activity (Edwards and Howells, 2003). The Convention on Cybercrime defines "traffic data" as any computer data relating to a communication by means of a computer system, generated by a computer system that formed a part in the chain of communication, indicating the communication's origin, destination, route, time, date, size, duration or type of underlying service.

Table 27.2. Surveillance methods.

Surveillance method	Possible definition of such methods
1. Monitoring	Monitoring is the listening to or recording of the content data at the time of the communication.
2. Interception	Interception is applicable to the collection of content data of communications in the course of its transmission. It is the acquisition of the communications by someone other than the sender or the receiver.
3. Data retention	Traffic data retention is the blanket retention of traffic data (not content) for a certain period of time of all users irrespective of whether the user is suspected of committing a crime (including terrorism). Data retention is the process of storing data which means to keep data that is currently being generated (real-time) in one's possession in the future [13].
or	or
4. Data preservation	Data preservation ("quick freeze") is the preservation of specific traffic data of an identifiable Internet user for a specific criminal investigation for a limited period of time. It refers to data that already exists in a stored form and that must be protected from anything that would cause its current quality or the condition to change or deteriorate. It requires that data must be kept safe from modification, deterioration or deletion [13].

The surveillance method used will depend on the type of information required and the format of the information at the time of gathering. The information (content and traffic data) on the Internet may be collected either by the ISP (indirect surveillance) or by the investigating agency (direct surveillance). Interception and monitoring is applicable to obtaining content information that is fluid and in movement at the time of gathering. The content is gathered during the transmission of the communication. Interception relies on a suspicion in advance of a criminal act. The collection and "storage" of traffic data is aimed at traffic data that is static, recorded and stored at the time of the gathering. Traffic data may be stored either by means of traffic data retention or traffic data preservation. In many countries, data retention or data preservation is a new legal procedure or power that never existed in the physical medium. The ISPs in some countries may have an obligation to retain the traffic data of all users for a specified time, the so-called blanket traffic data retention. Other countries may provide only for data preservation where the ISP is ordered to preserve ("freeze") the traffic data of a specified user in respect of a specific criminal investigation.

The format of electronic information on the Internet is not always clear. For example, an unopened e-mail waiting in the mailbox of an ISP until the addressee downloads it to her computer, may be considered either as electronic information in transit or in storage. If the unopened e-mail is considered as electronic information in transit, interception will be applicable whereas if it is considered as information in storage, search and seizure will be applicable (see Section 27.8).

Surveillance methods must be distinguished from other information-gathering methods. Surveillance methods are procedures used to ensure the access to and availability of the information on the Internet (see Section 27.8).

27.6. The Council of Europe Convention on Cybercrime

The Convention on Cybercrime (Cybercrime Convention) is a multi-lateral instrument aimed specifically at addressing crimes committed in an electronic medium (computing environment) such as the Internet. As of date, 43 countries have signed the Convention and 21 states of the Council of Europe have ratified the convention.

The aim of the Cybercrime Convention is to combat cybercrime by requiring signatory countries to establish certain substantive offenses and adopt domestic procedural laws to investigate cybercrime; it furthermore addresses criminal and procedural law on an international level to ensure the harmonisation of laws governing the criminal justice systems and to provide international cooperation and assistance in the collection of electronic evidence pertaining to specific criminal investigations.

Articles 14–21 contained in Section 2 of the Cybercrime Convention, provides for the following information-gathering methods:

(i) The expeditious preservation of stored computer data at the request of the nation where the crime caused damage (Article 16). The specified traffic data may be preserved up to a maximum of 90 days to allow the competent authorities to seek its disclosure;
(ii) The expeditious preservation and partial disclosure of specified traffic data (Article 17);
(iii) Disclosure of specific stored computer data or subscriber data by means of a production order (Article 18). Subscriber data is information about the use of the service and the user of that service;
(iv) Search and seizure of a computer system or part of it to obtain stored computer data (content and traffic data) (Article 19);
(v) Real-time collection or recording of traffic data in respect of specified communications (Article 20) and
(vi) Interception of content data of specified communications in respect of serious offences (Article 21).

It is important to note that the Cybercrime Convention provides for preserved traffic data and not for retained traffic data (Articles 16, 17 and 20). This may be explained within the context of which the Cybercrime Convention was drafted. The negotiation and drafting began in 1997 and the treaty was completed and opened for signature about two months after the 9/11 US terrorist attacks. It was not drafted against the background of terrorism but with the focus on specific cybercrime investigations.

The preservation of computer data is applicable to data that already exists in a stored form to be protected from anything that would cause its current quality or condition to change or deteriorate. The wording "to order or similarly obtain" in article 16(1) is intended to allow the use of other legal methods of achieving preservation, including production order and search and seizure warrants that simultaneously facilitate the disclosure of the data to law enforcement agencies [13].

The production order in terms of Article 18 is only applicable to stored, existing data and not to traffic data or content that has not yet come into existence. Article 18 does not impose an obligation on ISPs to keep records of their subscribers or to ensure the correctness thereof. "Subscriber information" is defined in the Cybercrime Convention as any information, contained in the form of computer data or any other form, that is held by a service provider, relating to the subscribers of its services, other than traffic or content data by which can be established: (a) the type of the communication service used, the technical provisions taken thereto and the period of service; (b) the subscriber's identity, postal or geographic address, telephone and other access number, billing and payment information, available on the basis of the service agreement or arrangement.

The Cybercrime Convention refers to the collection of traffic data as "real-time collection of traffic data" and to the collection of content data as "real-time interception". The purpose of this distinction is to assist countries that distinguish between real-time interception of content data and the real-time collection of traffic data. The Cybercrime Convention also uses the term "collect or record" (Articles 20 and 21), since some states do not differentiate between the collection of traffic data and the interception of content data.

The Council of Europe acceded that the meaning of "interception" is problematic. The Council of Europe proposed that the procedure of interception can be distinguished from the procedure of search by looking at the state or format of the information, namely whether the information is in transit or inert [4]. Interception would be applicable to data moving between computers or storage files whereas the search procedure would be applicable to static information stored in one machine or one file store [4].

Although the cybercrime convention constitutes the current internationally agreed upon benchmark, it does not mean that the Cybercrime Convention is above criticism or flawless [13]. The Cybercrime Convention has been criticised as being extremely law enforcement-orientated without providing sufficient human rights protection and limitation of governmental use of powers. The Convention on Cybercrime only provides for international cooperation in prosecuting cybercrime but makes no provision in securing networks [1].

Despite this criticism, the Cybercrime Convention fulfills an important aim, namely to unite countries in fighting cybercrime, which is important when crime on the Internet is investigated.

27.7. Brief Overview of Internet Surveillance Laws

27.7.1. *Introduction*

Although surveillance as an investigatory method existed prior to 9/11, 9/11 had a global impact and resulted in the extension and intensification of electronic surveillance and the implementation of surveillance laws by granting law enforcement and intelligence agencies with more surveillance powers.

When evaluating the surveillance law of a country, the following aspects should be taken into consideration:

(a) The only international treaty on cybercrime, the Cybercrime Convention, serves as a yardstick to establish which countries were signatories to the Convention or have acceded to the Convention, whether the signatory country has ratified it and the extent to which a signatory country comply with the Cybercrime Convention.

The surveillance laws of those countries that were not signatories or did not accede to the Cybercrime Convention, may be compared to the Cybercrime Convention to see how their Internet surveillance laws differ from the Cybercrime Convention;

(b) Although the Internet was not designed as a single entity with a single authority that governs the legal development and use of the Internet, dominant "powers" have emerged in respect of the Internet legal regulation, such as the United States, EU and China [6]. It is important to take note of the surveillance laws of the United States and the EU; and the extent of the influence of the dominant powers on the surveillance laws of other countries and

(c) Countries can also learn from each other by taking note of the deficiencies in the surveillance laws of a country and try to prevent following suit, for example, some shortcomings have been identified in the surveillance laws of the United States (see Section 27.7.2.1).

27.7.2. *Foreign Internet Surveillance Laws*

27.7.2.1. *US surveillance laws*

After 9/11, the United States was the first country to implement comprehensive electronic surveillance legislation in the form of the Uniting and Strengthening of America by Providing Appropriate Tools Required to Intercept and Obstructing Terrorism Act of 2001 (hereafter referred to as the US Patriot Act). This legislation provides the US government with additional tools to help track, prevent and combat terrorism. It is also applicable to serious crimes such as organised crime. In September 2006, the United States ratified the Cybercrime Convention, which came in force in January 2007.

The US Patriot Act was signed into law on 26 October 2001. It was passed overwhelmingly with a vote of 357-6 in the House of Representatives and 98-1 in the Senate. It is a substantial piece of legislation with many components. It consists of 342 pages and amends over 15 different statutes. Relevant for purposes of Internet surveillance, is Title II of the US Patriot Act that establishes the operational guidelines by which information can be gathered. Title II covers all aspects of the surveillance of suspected terrorists, those suspected of engaging in computer fraud or abuse, and agents of a foreign power who are engaged in clandestine activities. In particular, the title allows government agencies to gather "foreign intelligence information" from both US and non-US citizens.

When evaluating the US Patriot Act, it must be done within context, namely the loss of approximately 3,000 lives as a result of the 9/11 terrorist attacks. Due to the particular circumstances at the time of the terrorist attacks, the US Patriot Act was passed soon after 9/11. The US Patriot Act was enacted to protect the people of the United States against further terrorist attacks.

Prior to 9/11, surveillance laws existed in the United States. The effect of the US Patriot Act was to expand the authority of the US law enforcement agencies for the stated purpose of fighting terrorist acts in the United States and abroad. This expanded legal authority is also used to detect and prosecute other alleged potential crimes (Pikowsky, 1994). The US Patriot Act makes numerous amendments to pre-existing statutes governing covert electronic surveillance pursuant to the investigation of domestic crimes as well as investigations concerning foreign intelligence and terrorism.

The US Patriot Act provides for the collection of content and traffic data by means of the following surveillance methods: interception, monitoring and data preservation. It also provides for indirect surveillance (where the ISP gathers the information) or direct surveillance (where the law enforcement or intelligence agency gathers the information on the Internet themselves).

Some of the provisions of the US Patriot Act have been criticised and the following shortcomings have been identified, namely weak judicial oversight, lowering of requirements in applying information and intelligence-gathering methods and weak privacy protection. The shortcomings may result in the possible abuse of government power.

When discussing the US electronic surveillance laws, it is important to note that the pre-existing 9/11 laws were drafted at a time of fixed location circuit-based switching systems. These laws were primarily aimed at the telephone that uses a circuit-switched network based on transmitting voice. One cannot overlook the different technical architecture of the Internet and the telephone regarding the amendments made by the US Patriot Act to the existing surveillance laws [2]. Some of the amendments update the laws to reflect that we live in a digital age and this should be welcomed from an investigatory perspective. For example, a warrant that is applicable in all jurisdictions in the United States is welcomed, since it gives effect to the borderless nature of the Internet.

However, in some instances, it can be dangerous to merely apply laws that were primarily designed for the telephony architecture to the Internet, as it may have adverse consequences such as privacy violations. This is illustrated in respect of pen registers and trace-and-trap devices. By allowing the use of packet sniffers on the Internet as analogue to the pen register and trace-and-trap devices used for the telephone, one finds that the privacy rights of Internet users may be violated due to the differences between the Internet and telephony architecture [2]. Using packet sniffers as an analogue to pen registers, may reveal content information and not only traffic data as is the case with a telephone network [2]. On a telephone network, transactional and content information are separated by time and space, whereas traffic data and content data co-exist in time and space on the Internet due to the Internet architecture. Every communication sent over the Internet is broken down into smaller components called "packets". The packets are sent separately across the Internet and re-assembled at their destination. Each packet has two components: a header and payload. The header contains the information necessary for the network to deliver the packet to its destination for re-assembling the message (this information can be called traffic data), whereas the payload section of the packet contains the actual communication (this information can be referred to as the content). Unlike the telephone network, the Internet data does not travel along a single path. This means that traffic data and content co-exist in time and space within the packet on a packet-switched network when collecting traffic data. The latter explains why the use of packet sniffers may result in serious privacy implications.

Since the 9/11 US terrorist attacks, terrorism has been identified as a serious and ongoing threat against the United States with the consequence that the United States is at present in the process of increasing surveillance powers by means of legislation.

27.7.2.2. *European Union*

On 14 December 2005, the European Parliament approved the Traffic Data Retention Directive. The Traffic Data Retention Directive 2006/24/EC is applicable to traffic and location data that are generated or processed on European level in the course of the supply by the providers of publicly available electronic communication services or of a public communication network. It is not applicable to search engines as they are providers of information services and the keeping of the search terms would amount to the retention of content data. It provides for indirect surveillance by the ISP that provides a communication service.

Article 1(1) states that the purpose of the Directive is to ensure that the data are available for the detection, investigation and prosecution of serious crimes. Article 1(1) determines that each member country may itself define serious crime in their national law. Article 1(2) furthermore determines that the Directive shall apply only to traffic and location data on both legal entities and natural persons and to the related data necessary to identify the subscriber or registered user. It does not apply to the content of electronic communication including information consulted using an electronic communications network. Article 5 outlines the categories of data to be retained as follows: (a) data necessary to trace and identify the source of a communication; (b) data necessary to identify the destination of a communication; (c) data necessary to identify the data, time and duration of a communication; (d) data necessary to identify the type of communication and in respect of Internet e-mail, the Internet service used and (e) data necessary to identify users' communication equipment or what purports to be their equipment and in respect of Internet access, Internet e-mail and Internet telephony, the calling telephone number for dial-up access, the digital subscriber line or other endpoint of the originator of the communication. Article 6 provides for the compulsory traffic data retention by the ISP of all users for a period not less than 6 months but not exceeding 2 years from the date of the communication (indirect surveillance). EU member countries had until 15 September 2007 to implement national legislation providing for compulsory traffic data retention subject to the condition that countries could postpone compliance with the Directive until 15 March 2009, which many EU member countries had done. Each country will determine the crimes to which data retention will be applicable.

By 15 September 2010, the European Commission (EC) will evaluate the success of traffic data retention in the prevention, detection, investigation and prosecution of crimes involving the Internet. Although traffic data retention originates from the security part of the European government, the data retention law is counterbalanced by the EU data-protection regime.

Most of the member states of the EU signed the Cybercrime Convention. However, the EU deviated from data preservation as provided in the Cybercrime Convention. The main reason would be the continuous threat of terrorism, with terrorism being described as the threat of the 21st century.

27.7.2.3. *Affect of the United States and EU surveillance methods on other countries*

It has been said that where the United States goes, others will follow [12]. It is therefore relevant to examine closely what is happening in the United States and to establish if the lead of the United States is followed and if affirmative, to what extent and degree it is followed elsewhere.

Today, most western countries provide for surveillance. The question is which surveillance methods should be used and which other information-gathering methods should be applicable. The impact of the EU's decision to impose a compulsory traffic data retention obligation on ISPs will most probably have a rippling effect with many non-European countries following suit (see Section 27.9.5).

Non-European countries such as South Africa need EU member countries and the United States as online trading partners and therefore developments in the EU and the United States are closely monitored.

It is essential that although there does not exist a global surveillance law, countries should harmonise their surveillance laws to provide for cooperation and assistance in respect of investigations (see Section 27.9.2).

27.8. Difference Between Search and Seizure and Internet Surveillance Methods

Internet surveillance, search and seizure and production orders may be employed as investigatory methods in gathering information in criminal investigations. Search and seizure, however, is not an Internet surveillance method. An Internet surveillance method is the method used to ensure access to and availability of different types of Internet information. The gathering of information by means of interception, monitoring, data retention or data preservation are generally secretive and the physical presence of the law enforcement agency is not necessary. The surveillance can be carried out by the ISP (indirect surveillance) or by the investigating agency (direct surveillance).

Search and seizure is only applicable to traffic and content information that is static, recorded and stored, in other words to information that is already available. It is normally gathered in the presence of the law enforcement agency and is not secretive. The gathering of the data takes place at a single moment in time, in other words, the period of the search is in respect of data that exists at that time and is not ongoing as is the case with traffic data retention [13].

27.9. Some Considerations Regarding Surveillance as an Investigatory Method on the Internet

27.9.1. The question is how can cybercrime be prevented, detected, investigated and a suspect be prosecuted?
Conduct regulation does not assist in the gathering of information on the Internet. To be able to investigate crime and prevent the Internet from becoming a "lawless frontier", the investigator needs to gather information. Information gathering on the Internet can only be done by means of surveillance. After 9/11, most countries apply surveillance to the Internet;

27.9.2. Investigating a crime needs a harmonised global approach to assist in cross-border investigations. Although the global Internet community does not speak with one voice in respect of information gathering, there should be some consensus amongst the Internet community on the surveillance methods used to gather information;

27.9.3. Surveillance is made possible by means of technology that is increasingly becoming more refined and sophisticated. Surveillance may be non-obvious, but it is intrusive. Surveillance technology must be regulated, otherwise it can be abused. It is important that surveillance laws provide for judicial oversight and compliance with pre-requisites.
Former US President, James Madison remarked:

> "If men were angels, no government would be necessary. If angels were to govern men, neither external nor internal controls on government would be necessary. In framing a government which is to be administered by men over men, the great difficulty lies in this: You must first enable the Government to control the governed; and in the next place, oblige it to control itself."

27.9.4. Surveillance is not only used for investigating a crime, but it may be seen as a security measure in protecting users' information against crime;

27.9.5. EU countries are in the process of implementing traffic data retention legislation. Other non-EU member countries will most probably follow the EU's approach and implement blanket traffic data retention. In the United States, there is growing interest and pressure to apply traffic data retention to investigate, for example child pornography.

Data retention has evoked a lot of discussion and the following aspects may be considered:

(a) Traffic data retention results in many debatable questions. For example, why is blanket retention of traffic data and not data preservation needed for law enforcement or security intelligence purposes? Traffic data retention may violate the legal presumption that one is presumed innocent until proven guilty, the user's privacy right and to a lesser extent, freedom of expression. It is argued in favour of blanket traffic data retention that it may provide the only clue to the identity of the perpetrator. Every piece of traffic data may be described as a piece of jigsaw puzzle and the more data there are to cross-check, the harder will it be to put the law enforcement agencies on the wrong track. Although all forms of traffic data may be altered or masked by sophisticated criminals, investigators may be able to establish the information. Also frequently asked: why are the traffic data of all Internet users stored for a certain period of time? The answer is that the crime may not be detected at the time of the commission or someone who had not been identified as a suspect may be identified as a suspect at a later stage. The retention period of the traffic data is questioned. The EU has proposed a maximum period of 2 years but the EU countries may extend the retention period. Some criticism has been leveled at the retention period arguing that the traffic data may not be relevant after 2 years, since the Internet constantly changes with some websites and IP addresses becoming obsolete after a period of 2 years;

(b) The ISPs carry not only the responsibility of traffic data retention but also the responsibility of securing the collected traffic data against unauthorised access to the stored data;

(c) Blanket traffic data retention of all users is very intrusive and infringes the right to privacy. One has to balance the security benefits derived from the collection of traffic data, such as the prevention, detection and investigation of a crime and prosecution of the suspect with the privacy violation. It is not only surveillance that infringes the individual's right to privacy, but the crime also infringes the privacy right of the user. Content data has been perceived as more privacy infringing than traffic data but it can be argued that traffic data may reveal more of an Internet user's habits than content data and may therefore be more invasive;

(d) Countries that apply traffic data retention, should counterbalance it with data protection laws. It is, for example, not easy to distinguish between traffic data and content data on the Internet. In practice, content and traffic data are often generated simultaneously resulting not only in revealing data, which is necessary for the conveyance of an electronic communication (namely traffic data) but which also shows elements of the content indicating the interests or habits of the user;

(e) Governmental abuse of powers may occur in countries that do not ascribe to a data protection regime. The EU has strict data protection laws in the format of the EU Directives 95/46/EC and 2002/58/EC and

(f) Data retention laws are pro-active policing. This means that it aims at addressing the security risk or the crime commission risk before it is actually committed. This is commendable as policing and investigation is normally re-active. The question remains whether intelligence agencies and law enforcement agencies will be able to identify a threat before its actual commission, especially since such a huge amount of traffic data will be stored and time is normally of essence when investigating, for example, terrorism.

27.9.6.1. Investigations on the Internet rely on the assistance of the ISP either by means of direct or indirect surveillance. The threat of crime and now terrorism has changed the role of the ISP from mere conduit of information to the gathering of information by means of surveillance. It may be argued that it is not reasonable to impose such a burden on the ISP;

27.9.6.2. Criminal investigation must be conducted in such a manner that it complies with procedural safeguards;

27.9.6.3. George Orwell in his book, 1984, feared the coming into existence of the so-called "big brother" surveillance state with the state watching over the personal lives of all people [12]. The implementation of surveillance laws regulating the use of surveillance technology has contributed to the existence of a surveillance state. There is a fine line between a surveillance state and a police state and to prevent a country from slipping into a police state, the powers of investigatory and intelligence agencies must be subjected to judicial scrutiny;

27.9.6.4. When gathering information in the interest of criminal investigations or national security, one must be sensitive to the political, sociological, economical and ethical implications of surveillance and

27.9.6.5. The final word on surveillance laws has not yet been spoken. Surveillance laws may still be challenged in national constitutional courts and regarding EU member states, it may also be challenged in the European Court of Human Rights.

27.10. Conclusion

> "A complete free society — if it is to survive — requires citizens who exercise self-restraint and who are willing to accept the consequences of failures of that self-restraint. At some threshold of failures, however, citizens demand of their government protection from each other. At some point, such protection curtails the freedom of citizens and the citizens find themselves in a police state. Thus, the pendulum swings between anarchy and totalitarianism, between unbridled freedom and censorship, between anonymity (i.e. no accountability) and Big Brother (i.e. no privacy). To achieve the balance of costs and benefits, we must first understand the problems we hope to solve" [16].

The problem, governments wish to address is the challenges inherent in the prevention, detection, investigation and prosecution of crime, terrorism and IW to ensure that the benefits of the Internet are maximised.

Goldsmith and Wu made the following contentious statement: "The greatest danger for the future of the Internet come not when governments overreact, but when they don't react at all". All will undoubtedly agree that action on governmental level is required to ensure the growth and prosperity of the Internet but how far can government action pursue justice before it amounts to an abuse of powers?

Many governments of Internet-connected countries have elected surveillance of information on the Internet as a solution to address the problem of cybercrime and terrorism. Thus, the pendulum has swung from no Internet regulation to conduct regulation and extending conduct regulation to include control of information by means of surveillance. The question is whether the benefits of surveillance are in balance with the prejudice suffered, e.g. privacy infringement and possible abuse of powers. At present, most governments are of the opinion that the use of surveillance as an investigatory method is justified.

References

1. D. Alexander *et al.* (eds.), Immunizing the internet, *Harvard Law Review* **119** (2006) 2445–2463.
2. R. Berkowitz, Packet sniffers and privacy: why the no-suspicion-required standard in the USA patriot act is unconstitutional, *Computer Law Review and Technology Journal* (2002) 2–8.
3. K. Bowrey, *Law and Internet Culture* (Cambridge University Press, USA, 2005), pp. 8–9, 194–197.
4. I. Carr, Anonymity, the internet and criminal law issues, *Digital Anonymity and the Law*, C. Nicoll, J. E. J. Prins and M. J. M. van Dellen (eds.) (T M C Asser Press, The Hague, 2003), pp. 161–188.
5. C. Fijnaut, J. Wouters and F. Naert (eds.), *Legal Instruments in the Fight Against International Terrorism: A Transatlantic Dialogue* (Martinus Nijhoff Publishers, Leiden, 2004), pp. 1, 5.
6. J. Goldsmith and T. Wu *Who Controls the Internet* (Oxford University Press, USA, 2006), pp. 73, 81, 84, 103, 145, 166–167.
7. C. Gringas, *The Laws of the Internet* (Cromwell Press Limited, UK, 2003), p. 1.
8. J. S. Hiller and R. Cohen, *Internet Law and Policy* (Pearson Education, Inc., USA, 2002), pp. 75–76, 95, 98–100, 169–171.
9. L. Janczewski and A. Colarik, *Managerial Guide for Handling Cyber-terrorism and Information Warfare* (Idea Group Publishing, USA, 2005), pp. 10, 43, 222–225.
10. M. Kirby, Symposium: the judiciary in a constitutional democracy, *South African Journal of Human Rights* (2006) 21, 43, 45.
11. R. Larry, *Globalization and Everyday Life* (Routledge, UK, 2007), pp. 7, 110.
12. D. Lyon, *Surveillance After September 11* (Polity Press, Cambridge, UK, 2003), pp. 13–15, 29, 89, 109–112.
13. A. Nieman, *Search and Seizure, Production and Preservation of Electronic Evidence*, Unpublished LLD Manuscript (University of North-West (Potchefstroom Campus), South Africa, 2006), pp. 3, 36–41, 49, 51–53, 96, 99–140.
14. R. A. Pikowsky, An overview of the law of electronic surveillance post September 11, 2001, *Law Library Journal* **94**(4) (2002) 601–616.
15. J. Podesta, USA Patriot act — human rights magazine, 2002, Available at: https://www.abanet.org/irr/hr/winter02/odesta.html.
16. R. S. Poore, Computer forensics and privacy: at what price do we police the Internet, in *The Privacy Papers Managing Technology, Consumer, Employee, and Legislative Actions*, R. Herold (ed.) (CRC Press LLC, USA, 2002), pp. 33–34.
17. M. Rustad and C. Daftary, *E-Business Legal Handbook* (Aspen Publishers, Inc., USA, 2002), pp. 5–6.
18. R. W. Taylor, T. J. Caeti, T. J. Loper, E. J. Fritsch and J. Liederbach, *Digital Crime and Digital Terrorism* (Pearson Education, Inc., USA, 2006), pp. 9–15, 27, 43, 378–379.
19. D. Van der Merwe, Computer crime, *Journal of Contemporary Roman Dutch Law* (2003) 33.

20. D. Van der Merwe, Information Technology crime — a new paradigm is needed' 2007, *Journal of Contemporary Roman Dutch Law* (2007) 311.
21. M. M. Watney, State surveillance of the internet: human rights infringement or e-security mechanism, *International Journal of Electronic Security and Digital Forensics* **1**(1) (2007a) 42–54.
22. M. M. Watney, The evolution of internet legal regulation in addressing crime and terrorism, *Proceedings of the Conference on Digital Forensics, Security, and Law* (Arlington, Washington DC, 2007b), pp. 19–29.
23. Zagaris, International judicial cooperation and counterterrorism, in (2004) *Legal Instruments in the Fight against International Terrorism: A Transatlantic Dialogue*, C. Fijnaut, J. Wouters and F. Naert (eds.) (The Martinus Nijhoff Publishers, Leiden, 2004), pp. 39–50, 94.

Chapter 28

THE LEGAL CONFLICT BETWEEN SECURITY AND PRIVACY IN ADDRESSING TERRORISM AND OTHER CRIME ON THE INTERNET

MURDOCH WATNEY
*Department of Criminal Law and Procedure,
University of Johannesburg, South Africa*

Internet security aimed at addressing terrorism and other crime is not only a technological issue as it invariably impacts on the legal system and the rights it contain such as the right to privacy. Many information security professionals focus on security and pay little or no regard to the privacy rights of the Internet user whereas the opposite can be said of privacy activists. Internationally countries face the challenge of applying new approaches to the Internet in the prevention, detection and investigation of terrorism and other crime and prosecution of the perpetrator. Securing the Internet against terrorism and other crime result in a conflict between security, a technical issue and privacy, a legal aspect. A perplexing question is whether an Internet user can expect online privacy or whether globalisation and the approaches in combating crime and especially terrorism, have not resulted in an online environment that is incompatible with privacy rights.

The impact of Internet security on privacy is not automatically justifiable on the basis that Internet security aims to protect against criminal actions. The method used in securing the Internet and the impact thereof on Internet privacy must be subjected to scrutiny by all stakeholders to ensure compliance with legal principles to prevent abuse of governmental powers. This chapter investigates to which extent Internet privacy rights may justifiably be restricted in the interest of law enforcement and national security.

28.1. Introduction

The Internet was introduced enthusiastically to various countries by their governments and without paying much regard to the effect the Internet could have on the legal system.

The Internet confronts all countries on a global level with the same problems. International cooperation is required to dismantle terrorist networks, fight the growth of child pornography Web sites, combat cybercrime such as organised crime and protect their cultural industries against piracy [19]. Governments of today realise the power of the Internet and although it might be argued that the enthusiasm for this medium has waned, the advantages of the Internet still outweigh the disadvantages.

The commercialisation of the Internet and computer-related technology replaced the industrial society with an information-based society and contributed to globalisation. The Internet brought about many advantages but at the same time unlocked several challenges unknown to the physical world. Internet-connected countries battle to control the flow of information across borders for the purpose of national security and law enforcement. The exploitation of the Internet for the commission of serious crimes challenges countries to find methods of controlling cyberspace, whilst at the same time encouraging the continuous growth of the Internet, stimulating technological innovation, enjoying the benefits of the Internet and maintaining a human rights culture.

Governments, therefore, seek solutions to effectively address the disadvantages associated with the Internet, such as the abuse of the Internet for criminal purposes and especially the threat of terrorism. The laws regulating the Internet that aim to address terrorism and other crime, illustrate the problems and challenges governments face when deciding upon the form and feasibility of securing the Internet. Although the aim of these solutions are above reproach, namely the securing of the Internet for purposes of, initially crime prevention, detection, investigation and prosecution and more recently, the preservation of national security, the format of the solution in achieving this aim is still open to debate.

Securing the Internet to combat terrorism and crime bring about many concerns such as the conflict between security and privacy, and in some instances, the right to freedom of expression. How do we reconcile the conflict between security and privacy? Should the conflict be an issue of concern? If affirmative, can we reconcile two such opposite concepts at all, since security falls within the ambit of information security technology and the right to privacy resort within the legal system?

Globally, most legal systems recognise an individual's right to Internet privacy. Yet, statements abound that online privacy is nothing but an illusion that cannot exist in an electronic medium such as the Internet due to the nature and characteristics inherent to the Internet [5]. Are such statements correct? Does privacy protection differ in an electronic medium such as the Internet compared to the physical world?

These issues are of special relevance since the terrorist attack on 11 September 2001 (referred to as 9/11) on the United States served as a catalyst for implementing state control of information in the form of surveillance technology in respect of the Internet [15]. The pressure to secure the Internet has not diminished since 9/11; if anything, it has intensified.

The control governments apply to information on the Internet by means of surveillance, impacts on privacy and security. Monitoring and debating governmental control of the Internet by privacy groups, the Internet user, information security professionals and lawyers are fundamental in preventing abuse of governmental control and in upholding human rights such as the right to privacy. Many questions arise from state control of the Internet, such as: should ISPs carry an obligation in ensuring state control? Does the traffic data retention of all Internet users violate the right to privacy as well as the presumption of innocence until proven guilty? Should search engines reveal user's search terms to a government in conducting research for the possible implementation of legislation or reveal information of an Internet user who criticised a government or filter search terms to comply with a country's laws and regulations, thereby negating the right to privacy and freedom of expression? [10]

The phases of the evolution of Internet legal regulation illustrate the tension between security and privacy: the first phase of self-regulation placed too little emphasis on both security and privacy. As self-regulation evolved into conduct regulation and now state control of information, the emphasis increasingly moved to security resulting in the erosion of Internet privacy.

There exists a famous saying that everyone is better off not seeing how sausages and laws are made [2]. This saying cannot be applicable to the advanced computer technology society we live in as governments the world over try to exert more control over information in the interest of security. Internet security can be undertaken in a non-obvious way, but the infringement of privacy can be very invasive. Scrutiny of the impact of Internet state control on the right to privacy is important as countries have implemented or are in the process of implementing legal and technical measures to exert control of information on the Internet in addressing security issues.

The scrutiny should focus on the conflict between security and privacy in addressing crime and terrorism. The following questions should be addressed in this regard: does the gathering of information on the Internet in the interest of law enforcement and national security violate the right to privacy? If this question is answered in the affirmative, it should be established whether the privacy violation is justifiable in terms of the possible security benefits? When evaluating security and privacy within the context of addressing crime and terrorism the following issues necessitate attention:

- Investigating the existence of a right to privacy with specific reference to the Internet and how Internet privacy differs from privacy protection in the physical world.
- Understanding that governments have increasingly looked for a solution in addressing Internet security to combat crime and now terrorism, and the answer governments have found in the use of surveillance technology on the Internet.
- Evaluating the complexity of ensuring security and protecting the right to privacy on the Internet.
- Before looking at information gathering on the Internet, understanding the approach to Internet privacy protection. Specific reference is made to the legal position of the US and EU regarding the protection of Internet privacy.
- Evaluating the purpose of gathering information on the Internet, namely ensuring law enforcement and national security and the impact information gathering methods have on Internet privacy.

- Understanding that in most instances the information on the Internet is gathered by the Internet Service Provider (indirect surveillance) although it is possible for law enforcement and national security agencies to gather information on the Internet (direct surveillance).

28.2. Recognising a Right to Privacy

The term "privacy" is not a new concept. History shows that early societies had privacy concerns (Privacy and Data Protection Paper, 2005). The protection of privacy as a right is, however, a relatively modern notion. The right to privacy is one of the most important human rights of the modern age and enjoys recognition around the world.

On an international level, countries have reached consensus that the right to privacy should be protected. The United Nations Universal Declaration of Human Rights of 1948 serves as a benchmark for the protection of the right to privacy and specifically territorial and communications privacy. Article 12 of the United Nations Universal Declaration of Human Rights provides:

> "No one shall be subjected to arbitrary interference with his privacy, family, home or correspondence, nor to attacks upon his honour and reputation. Everyone has the right to the protection of the law against such interference or attacks".

The United Nations Universal Declaration of Human Rights recognises limits to the exercise of rights. Certain invasions of privacy may be reasonable or unreasonable. The limits are defined in Article 29 subparagraph 2 as those "determined by law solely for the purpose of securing due to recognition and respect for the rights and freedoms of others and of meeting the just requirements of morality, public order and the general welfare in a democratic society".

Privacy protection should be evaluated within a specific context. When the international community reached an agreement that the right to privacy should be protected in the Universal Declaration, it was reached within the context of a physical world against the background of the Second World War.

Although the right to privacy is recognised as a human right on an international, regional and domestic level, there does not exist a global definition of the right to privacy or approach to privacy protection. This can be explained by understanding that the term, "privacy" is not an abstract term but has only meaning within the context of a national culture, a particular political system and a specific time [16]. Against the background of international and regional benchmarks, each government determines the approach to privacy and the degree and extent of privacy protection within that country.

The interest in privacy increased in the 1960s and 1970s with the advent of computer technology (Privacy and Data Protection Paper, 2005). In the 1990s with the commercialisation of the Internet and the implementation of the worldwide web, Internet privacy became a real concern.

The use of computing technology and the Internet introduced the information age. Information has become a valuable commodity. Computer technology and the Internet provides for the quick and easy collection, storage and exchange of information. Today, personal information (information that identifies a specific person) may be stored on central databases.

In 1890, US Supreme Court Justices Brandeis and Warren gave one of the earliest definitions of privacy as "the right to be let alone." [16, 6, 7]. It is much easier for an individual to exert control over the gathering of information in the physical world than on the Internet (par. 3.3). Computer technology and the Internet challenge the concept of "being left alone". It is possible that the gathering of information can take place out of the public's view and without the public's direct consent [18].

28.3. Understanding Privacy and Security on the Internet

28.3.1. *Introduction*

The Internet was designed as a decentralised "open architecture" with the emphasis on the free flow of information, initially between the military in the United States but after commercialisation, between countries. Neither security nor privacy was the main aim of developing the Internet.

Prior to the historical watershed terrorist attacks on the United States on 9/11, the United States as well as other Internet-connected countries were already grappling with finding solutions to address law enforcement and national security challenges and the difficulties of investigating a crime in an electronic medium such as the Internet.

The 9/11 US terrorist attacks gave the United States and other countries the incentive to implement state surveillance of the Internet or differently phrased, to exercise state control of information generated, send, received and accessed on the Internet.

The transition from conduct regulation to state control of information has been anything but uneventful or above reproach. After 9/11, the following statement was made in respect of the Internet:

> "On 10 September 2001, the Internet was still a place of hopes and dreams that was going to give everyone access to impartial information and undermine dictatorships. A few days later, it had become a lawless place where Al-Qaeda had managed to plan and coordinate its attacks. The Internet began to frighten people. 10 September was the last day of a golden age of free online expression. Since then, Big Brother has loomed ever closer" [3].

The above statement is thought provoking and necessitates further exploration. Various questions come to mind and these questions will be addressed in this chapter. Does there exist any truth in the above statement or is it merely aimed at sensationalism? Did the solutions governments implemented to address crime and terrorism on the Internet bring an end to the right to privacy and to a lesser extent, the right to freedom of expression? Can the limitation of privacy rights be justified in terms of the security benefits derived from state surveillance?

28.3.2. *Understanding the Meaning of Security on the Internet*

As the connection of more countries to the Internet resulted in increased exploitation of the Internet, governments began investigating the use of technological methods in securing the Internet.

Security today is not only relevant from the perspective of a company in the context of e-commerce or an individual user, but also from the perspective of government. Governments face the dilemma of implementing technological solutions aimed at securing the Internet to prevent, detect, investigate and prosecute crime and the increasing threat of the use of the Internet for terrorism. Users also insist on greater governmental online protection against crime.

Crime by means of the use of the Internet is a serious threat to the continuous growth of the Internet and fear of the Internet will neutralise the enormous benefits of the Internet [21].

Governments responded to the threat of crime and terrorism by implementing surveillance technology as a form of securing the Internet. Prior to the implementation of government surveillance, countries mainly, relied on legal regulation of conduct on the Internet by criminalising certain acts. In the case of conduct regulation, technology was not involved and it was purely a legal issue.

"Security" must therefore, be evaluated within the context it is applied. In this chapter, security refers to the use of surveillance technology aimed at securing the Internet against abuse in the form of terrorism and other crime. Surveillance is a form of government (state) control of information. The purpose of Internet surveillance is the gathering of different types of information to assist in the prevention, detection, investigation and prosecution of terrorism and other crime.

28.3.3. *Understanding the Meaning of Privacy on the Internet*

The right to privacy developed within a physical medium and consequently it should be asked whether the privacy right also applies to an electronic medium such as the Internet. If answered in the affirmative, the following question arises, namely how is privacy protected in an electronic medium taking into account that the Internet faces challenges unknown to that of the physical world, such as the fact that there are virtually no online activities or services that guarantee absolute privacy [7]. ISPs and Web sites may, for example, monitor online activities. The ISP can determine which search engine terms were used, which Web sites were visited and the dates, times and duration of online activity.

Securing the Internet by means of surveillance technology grant governments access to information. It is possible for the Internet user to ensure privacy by means of privacy-enhancing technology such as encryption, using anonymous re-mailers. As indicated, the Internet was not designed with security as a priority and the privacy-enhancing tools can also secure the communications of the Internet user. Some may argue that privacy-enhancing technology may also be used to hide criminal activities and that the use thereof should be discouraged or prohibited. Whether the use of privacy-enhancing technology should be prohibited or not, falls outside this discussion.

If the Internet user does not utilise privacy-enhancing technology, does it imply the forfeiture of the legal right to privacy enjoyed in the physical world? Scott McNeally, builder of Sun computers, said in 1999 "you have zero privacy anyway" [5]. Ellison, the chief executive officer of Oracle, supported this viewpoint by stating in 2001 that Internet privacy is "largely an illusion". Some may be of the opinion that the protection and enforcement of privacy is incompatible with an electronic medium such as the Internet due to the characteristics and nature of the Internet. The latter opinion is not supported since the right to privacy is a legal right to which every person including an Internet user is entitled to, irrespective of the communication medium used and/or whether the user makes use of privacy-enhancing technology.

Internet privacy may be divided into two components:

- Data protection or information privacy means the control of an Internet user in respect of who has access to his/her personal information, when and how; and
- Communications privacy means the protection against interference and/or intrusion regarding his/her communication such as Web sites visited, e-mails sent and received and use of search terms.

Much has been written on the data protection or information privacy component of Internet privacy. This component evolves around the protection of consumers where businesses collect, store and trade information about their customers. Countries such as the United States, Canada, Australia and South African use the term "information privacy" whereas the EU member countries use the term "data protection". The two terms are used interchangeably in this chapter.

In this chapter, the emphasis is on the gathering of information to secure the Internet and the effect it has on the right to Internet privacy. Since 2001, many governments have implemented legislation that provide for the gathering of information on the Internet as a form of Internet control.

Internet users have realised that due to the characteristics of the Internet they can lose control over their information if legislation does not secure their control. On the other hand, governments have realised that combating crime and terrorism necessitate control over information. A conflict exists between the control of information that a government seeks in the interest of Internet security and the control the individual user seeks as a right to privacy.

Can the security benefits and the privacy infringement be reconciled? Privacy is not an absolute right. The justifiability of the privacy limitation must be evaluated against the background of a country's approach to privacy protection, the surveillance methods used to gather the information on the Internet and the purpose for gathering the information. Although the information may be gathered for security purposes, the data protection or information protection of the consumer (Internet user) is very relevant.

28.4. Approach to Privacy Protection on the Internet

28.4.1. *Introduction*

As indicated, governments differ in their approach to Internet privacy protection. Since the EU and the United States are identified as the two major "western powers" regarding their influence on the development and implementation of Internet legal regulation, it is relevant to look at their approach to privacy protection [12]. It is also interesting to note that the events of 9/11 provided an incentive to the EU to have a more unified approach to international terrorism and for the EU and the United States to overcome the obstacles regarding international criminal law experienced prior to 9/11 [27].

This does not imply that legal developments in countries outside the EU and the United States are of lesser importance. The US and EU approach to control of the information on the Internet for the purpose of securing the Internet against terrorism and other crime can be used as yardstick when comparing the legal developments in other countries. The global nature of the Internet necessitates countries to harmonise their laws.

The western world has always jealously guarded the protection of human rights, yet new technological and political developments often challenge the human rights culture. As crime on the Internet escalated, countries increasingly looked at methods and approaches in securing the Internet. The US terrorist attack of 9/11 has been one of many terrorist attacks worldwide, but this attack on the United States was of such magnitude that it can be referred to as a globalising event that affected all countries on a global level.

After this attack, the question arose whether it could have been prevented in whole or in part [9] (pp. 1–8). The information had been available but the problem lay with the gathering, analysis and dissemination of information and sharing of information, not only within the territorial borders of the United States between different agencies but also across US borders [9] (pp. 1–8). The United States was the first country to introduce surveillance of the Internet with many other countries following suit (par. 5).

The methods and approaches in addressing law enforcement and national security on the Internet affect the information and communications privacy components of Internet privacy and necessitate attention. Before looking at the information gathering methods, the approach to Internet privacy must be established.

28.4.2. *US*

Contrary to the EU where the right to privacy is a fundamental right protected in the European Convention on Human Rights and Protection of Human Rights of 1950, the constitution of the United States does not explicitly mention a right to privacy [16, 13, 1]. The US Supreme Court has stated that the constitution implies a right to privacy in certain circumstances [8, 1]. Privacy rights have developed as a mixture of state common law, federal and state legislation and constitutional law [1].

The United States does not have encompassing privacy legislation dealing with the handling, collection and management of personal information as in Europe (see par. 4.3 for reference to consumer data protection legislation). The emphasis in the United States has been on self-regulation in respect of commercial entities, with the implementation of sector-specific legislation to deal with privacy violation such as the Gramm-Leach-Bliley Act of 1999 that imposes privacy requirements on financial institutions [7]. The US provides for the limitation of the rights of the government in accordance with federal legislation, the Privacy Act of 1974 that protect the personal information against governmental intrusion.

One of the main laws that regulates online privacy at federal level is the Electronic Communications Privacy Act of 1986 as amended by the Uniting and Strengthening of America by Providing Appropriate Tools required to Intercept and Obstruct Terrorism Act of 2001 (USA PATRIOT ACT). The Electronic Communications Privacy Act was adopted to address the legal privacy issues that were evolving with the growing use of computers and other innovations in electronic communications and updated legislation passed in 1968 that had been designed to clarify what constitutes invasion of privacy when electronic surveillance is involved [4]. The Electronic Communications Privacy Act makes the disclosure of communication content unlawful. The ISP can only release e-mail to a third party in exceptional cases. The courts have held that an individual has no reasonable expectation of privacy in information revealed to third parties, for example, an Internet user cannot enjoy a reasonable expectation of privacy in non-content information sent to an ISP because the user has disclosed the information to the ISP. The US courts have held that in general the Internet user has a reasonable expectation of privacy in content information that is sealed away from the network provider, but does not retain such protection in information disclosed or openly visible to the provider [14].

There have been calls for federal Internet privacy legislation as a means of addressing, for example, the escalating problem of "identity theft" [23]. The federal privacy law will then pre-empt the states from legislating in this area and will be consistent. At present the states can pass privacy legislation. Such general privacy legislation should also provide for an oversight agency to ensure privacy compliance. At present, the United States has the Federal Trade Commission (FTC), a federal agency tasked with the responsibility of protecting online consumers but the FTC has been criticised as not having the teeth to ensure compliance with privacy protection due in main to the lack of comprehensive data protection or information privacy legislation [21].

28.4.3. *EU*

Article 8(1) of the Council of Europe Convention of the Protection of Human Rights and Fundamental Freedoms of 1950 provides that everyone has the right to respect for his private and family life, his home and his correspondence. The right to privacy is not absolute, but may be restricted in terms of Article 8 paragraph 2. Article 8 paragraph 2 states that:

> "there shall be no interference by a public authority with the exercise of this right except such as is
>
> - in accordance with the law;
> - necessary in a democratic society and
> - in the interests of national security, or public safety, or the economic well-being of the country, or for the prevention of disorder or crime, or for the protection of the rights and freedoms of others".

As society evolved, individuals feared they were losing control over their personal information. Advances in computing technology and the Internet unlocked customer data that was traditionally not accessible and the loss of control over personal information became a warranted fear. The EU addressed consumer data protection and implemented two data (information) protection directives, namely, a general directive, Directive 95/46/EC (Data Protection Directive) on the protection of individuals with regard to the processing of personal information and on the free flow of information and a specific directive, 2002/58/EC (the Directive on Privacy and Electronic Communications) in respect of processing of data specifically on an electronic medium such as the Internet. "Processing" is defined in Directive 95/46/EC as the collection, recording, storage, retrieval, use, disclosure by transmission or making available, blocking or destruction. Both directives provide for obtaining consent before processing personal information of the consumer, the Internet user. Directive 95/46/EC was enacted because European citizens felt that they were losing control over their personal information. It also stipulates that personal information can be transmitted outside the EU borders only to a country that guarantees an adequate level of protection (Privacy and Data Protection Paper, 2005). Directive 2002/98/EC limits the ability of member states to intercept electronic communications by prohibiting any form of wide-scale, general or exploratory electronic surveillance other than where necessary to safeguard national security, defence, public security or the prevention, investigation, detection and prosecution of criminal offences.

Difficulties experienced with the prevention, detection and investigation of crime and terrorism led in 2006 to the EU implementing a mandatory Data Retention Directive 2006/24/EC and thereby amending Article 15 of Directive 2002/58/EC that provided for voluntary traffic data retention (not content data retention). By 15 March 2009, all EU member states must have national legislation that provides for traffic data retention of all Internet users. The EU data retention legislation represents proactive policing. Traditionally, laws in the physical world are re-active: a perpetrator is apprehended after a crime is committed. Proactive legislation is pre-emptive with every citizen being a target for suspicion and observation. In other words, the crime is prevented before it is perpetrated or if a crime is committed, the evidence gained before the commission of the crime can be reconstructed. The Data Retention Directive is only applicable to serious crimes due to the implications such collection has on privacy rights of users.

Although data retention is a privacy-destroying tool, it serves as a security mechanism to protect all Internet users against crimes such as "identity theft" and where committed, it assists in the investigation and possible prosecution of the criminals. The ultimate aim of the general retention of traffic data is to ensure harmonised European data retention laws that can assist in tracing the source of the communication, namely the perpetrator of a crime in across border investigations of terrorism and serious crimes [11]. From the perspective of the law enforcement authorities, erasing the "electronic footprints" (traffic data) would have the same effect as wiping off the fingerprints at a crime scene in the physical world. It is for these reasons that the Internet user's privacy is restricted or "curtailed" (not abolished). The law enforcement benefits derived from the surveillance method, data retention, have not escaped criticism [11]. Some have perceived the retention of the electronic "fingerprints" (traffic data) of Internet users as excessive as all Internet users are now treated as potential criminal suspects [8]. Doubts were raised concerning the effectiveness of data retention in addressing crime and national security abuse. Criminals may, for example, find backdoors to keep communications free from state surveillance, such as the use of stenography and peer-to-peer systems.

The argument that traffic data does not deserve privacy protection since the ISP needs this information to ensure communication, is questioned. It cannot be argued that traffic data does not resort under "personal information". Personal information may include traffic data such as the IP number as it may reveal personal information and therefore, fall under the protection of the data protection directive [11]. In some instances, traffic data retention may be more privacy invasive than content data retention. The traffic data can reveal, for example, how many times a certain person sends e-mails to a specific person. The application of data protection rules to the traffic data warehouses might be problematic. Issues such as the security of data storage, usefulness of the data and the evidential value of retained traffic data must be considered [11] (par. 5.2).

Regarding the justifiability of traffic data retention as an exception to Internet privacy in terms of Article 8 of paragraph 2 of the European Convention on Human Rights [11] refer to case law of the European Court of Human Rights. The case law found traffic data retention not justifiable as an exception because it was not necessary, appropriate and proportionate. The authors state as follow:

> "... it seems quite doubtful that general regulations on mandatory retention of traffic data would succeed the challenge of the proportionality test taking into account the immense and undifferentiated scope of intrusion into the private lives of individuals, irrespective of whether or not they are under suspicion".

One should bear in mind that the case law referred to, were all decided prior to 9/11 and other subsequent terrorist attacks in Europe. Furthermore, as stated earlier, the meaning of "privacy" is not static but must reflect the circumstances at a specific time period (par. 2). It is quite probable that in future an EU member's national court as well as the European Court of Human Rights may debate a constitutional challenge to the justifiability of traffic data retention as an exception to Internet privacy. The outcome of such a challenge may not be so obvious as the authors assume. The court will have to balance the security benefits derived from Internet state control of information against the infringement of data protection (information privacy) and communications privacy and determine which one outweighs the other and whether the infringement of Internet privacy for national security and law enforcement purposes, qualify as an exception to Internet privacy in terms of Article 8 paragraph 2 of the European Convention on Human Rights. The court may find that the blanket traffic data retention is in the interest of law enforcement and national security and qualifies as a justifiable limitation as it is only applicable to serious crimes, such as organised crime, money laundering and terrorism to name but a few. Directive 2006/24/EC does not prescribe the serious crimes, but leaves it to each EU member state to determine the serious crimes that will be subject to compulsory traffic data retention.

The argument that traffic data retention comes from the security branch of the EU and that the privacy infringement is automatically justifiable, is not supported. The right to privacy may be limited but the limitation must be justifiable. Justifiability will be established by ensuring that the collection, storage and disclosure of traffic data comply with the Convention on the Protection of Human Rights and the EU data protection regime as outlined in the two directives and is also subject to oversight by the EU data protection commission to ensure checks and balances. The EU Data Protection Directives are not only applicable to the market place but to all entities collecting and storing personal information of the consumer (Internet user) to ensure the safety of the information stored.

28.4.4. *Effect of US and EU Privacy Protection on Other Countries*

As illustrated, the EU and the United States approach the protection of the right to privacy very differently. The EU has comprehensive data protection legislation in place that guarantees privacy protection against governmental and commercial intrusion, referred to as the so-called "hard" data protection laws as opposed to the "soft" laws of the United States [7].

Other countries are well-advised to apply comprehensive protection to privacy on the Internet as implemented by the EU, since the Internet itself is not a privacy-protecting medium [7]. The piecemeal approach of the United States cannot be supported.

Comprehensive privacy legislation may safeguard the Internet user against the commission of crimes such as "identity theft". Data protection or information privacy legislation should provide safeguards and guidelines in respect of the processing of personal information, such as that consumers should know which companies are collecting information from them, what companies are doing with the information, know who may have access to the information and know that the information is held securely by all third parties [21].

It is also important to have a data protection agency that scrutinises the justifiability of privacy violations. For example, where a search engine indicated that they would store the search terms requested by an EU Internet user, the EU data protection commission indicated that the data retention directive was applicable to the providers of a public communications service and not to an information service. Not only must legislation have built-in safeguard mechanisms, but an oversight mechanism must ensure enforcement of the safeguards.

Countries that battle crime and terrorism on the Internet, tread a fine line when it comes to Internet security. O'Harrow wrote a thought-provoking book, *No place to hide* [18] where he refers to a speech by a previous US president, Dwight Eisenhower, who in 1961 said:

> "The potential for the disastrous rise of misplaced power exists and will persist. We must never let the weight of this combination endanger our liberties or democratic processes. We should take nothing for granted. Only an alert and knowledgeable citizenry can compel the proper meshing of the huge industrial and military machinery of defense with our peaceful methods and goals, so that security and liberty may prosper together".

Even today, post 9/11, countries should be mindful of this warning.

28.5. Securing the Internet by Means of Information Gathering

28.5.1. *Introduction*

As indicated here and elsewhere, the 9/11 US terrorist attacks catapulted western countries into implementing Internet security. A few days after the US terrorist attack the UN Security Council adopted Resolution 1,373 on 28 September 2001. This resolution obliges States to take all kinds of measures aimed at the prevention of terrorist acts and at the bringing to justice of those who participated in the financing, planning, preparation or perpetration of such acts or in supporting them [27].

The 9/11 US terrorist attacks showed that terrorists outflanked law enforcement in technology, information and communication [18]. Organised crime and money laundering may be linked to terrorism. The commission of "identity theft" is on the increase. The investigation after 9/11 showed that it is a favourite technique of the terrorists [18]. It is not only terrorism but all serious crimes that impact negatively on the use of the Internet and it can overshadow the many advantages the Internet bring. Addressing crime (which includes terrorism) is non-negotiable, but the tool used to address crime is contentious.

Surveillance was the chosen form for governmental control of information in securing the Internet. Whether this form of Internet state control should have been implemented as a security mechanism, is not part of the discussion. Under discussion is the impact of surveillance as a state control mechanism to ensure Internet security bearing in mind that:

> "The Internet is a tool — a potent tool worthy of safeguards. The safeguards, however, must not destroy the tool's potential for benevolence in an effort to prevent malevolence." [20]

28.5.2. *Purpose of Gathering Information*

Western governments realise that the gathering of information is the key in fighting crime and terrorism. Since 2001, western governments have increasingly implemented legislation providing for the use of surveillance technology to collect information. The information is collected in the interest of law enforcement and national security.

Surveillance allows for the gathering of the following two types of information:

- Content information that includes information concerning the substance, purport or meaning of that communications; and
- Traffic data that is available in the records of transactions of an ISP when a user engages in online activity [17]. Some governments today compel ISPs to retain the following traffic data for a prescribed retention period. Other governments may provide for traffic data preservation.

Traffic data is defined in the EU Mandatory Data Retention Directive 2006/24/EC as follows:

(a) Data necessary to trace and identify the source of a communication;
(b) Data necessary to identify the trace and identify the destination of a communication;
(c) Data necessary to identify the date, time and duration of a communication;
(d) Data necessary to identify the type of communication concerning Internet e-mail and Internet telephony: the Internet service used;
(e) Data necessary to identify the communication device concerning Internet access, Internet e-mail and Internet telephony:

  (i) the calling telephone number for dial-up access and
  (ii) the digital subscriber line (DSL) or other end point of the originator of the communication.

When a person goes online, the ISP automatically records the traffic data as the ISP needs this information for the conveyance of the communication or the billing thereof. Against the latter background, can it be argued that the Internet user does not have a right to privacy in respect of traffic data since it is exchanged and accessed on a public communication system such as the Internet? The Internet user has a right to privacy irrespective of the type of information gathered. The ISP only needs the information for conveyance of the communication and billing purposes but once the ISP has fulfilled this function, the necessity of storage of traffic data must be evaluated within the context of privacy protection and the purpose of continued traffic data retention.

28.5.3. *Information Gathering Methods*

Content and traffic data can be obtained by employing different technological surveillance methods which has been comprehensively discussed elsewhere. The different methods employed affect the right to privacy in varying degrees. Data preservation affects the right of privacy of an individual Internet user who is already earmarked as a suspect. It is justifiable if data preservation complies with legal pre-requisites, for example, prior judicial oversight unless circumstances warrant deviation. The justifiability of the EU compulsory traffic data retention of all users, irrespective of whether they are suspects or not, may be questioned (par. 4.3). Various arguments in support of and against the blanket data retention obligation have been tendered. Data retention as a security method must be weighed against the infringement of privacy. Consent of the data subject is not obtained prior to the interference with the communication of the Internet user. Traffic data may reveal personal information. It should be noted that law enforcement and national security agencies will only request traffic data in respect of the investigation of serious crimes.

Securing the traffic data storage must comply with privacy protection to prevent unauthorised access to the retained data, especially today with the prevalence of "identity theft". In respect of interception and monitoring, the suspect is identified prior to the collection of the content data. The collection of content data is more of a privacy violation than traffic data collection, but in respect of interception it is only applicable to an identified suspect. Interception must also comply with legal pre-requisites. The information gathering methods cannot be applied without legislation providing checks and balances to prevent abuse of governmental powers.

28.5.4. *The Role of the ISP in Securing the Internet*

Regulating the Internet is not easy. In the physical medium enforceability normally do not depend on the assistance of third parties. In respect of the Internet, ISPs administer parts of the networks within the borders of a country within the legal framework of that country. A country must therefore implement legislation that provides for ISP assistance in respect of the control of information. The role of the ISP emphasises the major shift from regulating conduct on the Internet or differently put, criminalising certain conduct on the Internet to control of the information. In respect of regulating control of information on the Internet the active involvement of the ISP, as a third party, is now crucial for the successful implementation of legislation.

The earlier role of the ISP as a mere conduit of information and provider of access to information, has changed to that of a third party involved in the collection of information on behalf of the law enforcement and national security agencies. ISPs have objected to the retention of the traffic data of all users. ISPs have criticised the financial and practical burden in storing data, citing that the problem is not only retaining the data, but maintaining and securing the data warehouses [11]. The effectiveness of retaining data in the prevention and detection of crime and terrorism has been questioned. It has been alleged that where the law enforcement agency or the intelligence agency requests traffic data, such request would generally be of an urgent nature but it may not be so easy for the ISP to quickly comply with the request. The counter argument is that law enforcement and security agencies need information to detect and prevent crime and terrorism. The ultimate purpose of the general retention of traffic data (not content data) is to be able, in the case of a serious crime, to trace and to locate geographically and chronologically the end-user device that was used to transmit the initial information [11]. Many crimes are committed across borders and therefore it is important that countries adhere to harmonised laws and in the case of the EU, harmonised data retention laws.

Can the ISP be held liable when collecting information and disclosing information about its customers to law enforcement and national security agencies? The ISP should be exempt from liability where it acts on the instructions of a law enforcement and national security agency. The compulsory traffic data retention that some governments now require has placed additional burdens on the ISP. The

ISP must ensure the security of the traffic data stored. Comprehensive privacy legislation can serve as a safeguard to abuse of information.

28.5.5. *Consequence of Internet Security*

Edwards and Howells [7] stated "(w)hat the Internet gives with one hand it often takes away with the other".

Securing the Internet by means of governmental control of information and more specifically the use of surveillance technology, undeniably affects the core values of western democracies, namely the right to privacy that consist of data and communications privacy. Differently put, the right to privacy protects the Internet user's right to freedom from processing of personal information without consent and/or the right to freedom against interference of communications without consent. Gathering information on the Internet infringe the right to privacy. The right to data and communications privacy is not an absolute right that may not be restricted. The degree and extent of privacy infringement depends on the form of government control, namely surveillance or censorship (includes surveillance), the surveillance methods employed and the purpose for Internet state control of information.

It should be borne in mind that it is not only governmental control but also the threat of online crime and terrorism that affects the Internet user's right to privacy. The Internet user wants — in some instances — demands, government to grant protection against terrorism and other crime on the Internet as it infringes the right to privacy. The question is how far government protection of the Internet may go before it becomes an unjustifiable invasion of an Internet user's privacy.

The challenges of the Internet in combating, detecting and investigating crime and terrorism resulting in the prosecution of the perpetrator necessitate new law enforcement and national security approaches. Reliance on technology alone will not protect the Internet user's human rights effectively. Legislation must be implemented to safeguard Internet users against governments moving towards a police state or applying surveillance technology excessively without any justification.

As indicated earlier, the supposition that state surveillance of the Internet falls within the security branch of the government and are automatically exempt from the right to privacy protection, is not supported. The privacy infringement must still be necessary, appropriate and proportionate to the security benefits derived. It is essential that in a democratic environment privacy infringement should be open to continuous scrutiny. Legislation must ensure checks and balances to prevent privacy violation without justification and/or excessive privacy infringement. Comprehensive privacy legislation will also safeguard against the commission of crimes. It is therefore important that a country that implements surveillance legislation have comprehensive privacy legislation in place to prevent the abuse of governmental powers. There should be as far as possible, a balance between national security and law enforcement needs with Internet privacy [15, 13]. A government that collects information for security purposes on the Internet must still comply with statutory obligations to prevent the law enforcement and security agencies of abusing their powers.

One of the biggest criticisms levelled against interpol, an international police organisation, is that it does not report to a government. Simon Davies, director of Privacy Internal, a London-based advocacy group that tracks surveillance of civilians, stated in 2008 that Interpol may lack the principle of openness and accountability as it is not subjected to some kind of oversight or scrutiny [25]. It is important that surveillance legislation provide transparency and oversight and that strong privacy legislation protects the individual's right to privacy on the Internet.

28.6. Conclusion

We do not live in an ideal world. Had it been the case, law enforcement and national security on the Internet would not have been a problem with all Internet users being law-abiding citizens and respecting the privacy rights of users.

Crime is without a doubt a serious threat to maximising the full potential of the Internet. Governments must safeguard the Internet. It is not only terrorism but also "cyberwar" and "iWar" that should enjoy attention. "iWar" is defined as "attacks carried over the Internet that target the consumer Internet infrastructure, such as the Web sites that provide access to online banking [22]. Ryan distinguishes "iWar" from what the United States calls "cyberwar" or what China calls "informationalised war". "Cyberwar" is engaged by nation states whereas "iWar" can be wagged by individuals, corporations and communities [22]. In the past, "cyberwar" may have been made off as futuristic and far-fetched, but after 9/11 threats against a communication medium such as the Internet should be taken seriously. Governments should be applauded for taking the threat seriously, but this does not mean that the consumer should not scrutinise the control over information that governments wish to implement. Internet users should not be naïve in thinking that the pressure from government to limit privacy rights will decline in the near future, if anything, it will increase due to globalisation and defined by the Internet, the continuous threat of terrorism, possible cyberwar and the abuse of the Internet for the commission of serious crimes.

We are now at crossroads in respect of Internet security. State control of the Internet can be compared to that of the *djinn* of the legend: once the genie is out of the bottle, its power is unleashed for both good and evil [20]. As long as crime which includes terrorism, cyberwar and iWar threaten the Internet, governmental power over information on the Internet cannot be reversed, but the consumer can determine how the governmental power should be implemented. At the crossroads we have to ask, how much security is necessary to ensure a safe Internet? The erosion of Internet privacy is the price we pay for Internet security. The Internet user must decide whether the privacy erosion is too steep a prize to pay for Internet security. Obviously, governments do not think so.

References

1. M. Agranoff, E-mail: Balancing corporate assets and employee privacy; the privacy papers: Policies for secure personal data, in *The Privacy Papers Managing Technology, Consumer, Employee, and Legislative Actions*, R. Herold (ed.) (CRC Press LLC, USA, 2002), pp. 5 and 41–42.
2. D. Anastasijevic, Birth of a Nation, *Time Magazine* (2008) 48.
3. K. Bowrey, *Law and Internet Culture* (Cambridge University Press, 2005), pp. 8–9, 194–197.
4. Carey and Ustaran, *E-Privacy and Online Data Protection* (Butterworths, Durban, 2002), pp. 1, 29.
5. N. Daves, A Right to Privacy? Get Over It, 2006, available at: http://www.mg.co.za/printPage.aspx?area=/insight/insight_comment_and_analysis/&art.
6. J. Deighton, Market solutions to privacy problems?, in *Digital Anonymity and the Law*, C. Nicoll, J. E. J. Prins and M. J. M. van Dellen (eds.) (T M C Asser Press, The Hague, 2003), pp. 137, 140.
7. L. Edwards and G. Howells, Digital anonymity and the law, in *Digital Anonymity and the Law*, C. Nicoll, J. E. J. Prins and M. J. M. van Dellen (eds.) (T M C Asser Press, The Hague, 2003), pp. 208–209, 214, 216–217, 228–239, 245.
8. D. Eleftheriou, M. Berliri and G. Coraggio, Data protection and e-commerce in the United States and the European Union, *The International Lawyer* **40**(2) (2006) 398–400.
9. C. Fijnaut, J. Wouters and F. Naert, *Legal Instruments in the Fight Against International Terrorism a Transatlantic Dialogue* (Martinus Nijhoff Publishers, Leiden, 2004), pp. 1–8.
10. D. Flint, Don't be evil, *Business Law Review* (2006) 102–104.
11. C. Goemans and J. Dumortier, Enforcement issues — Mandatory retention of traffic data in the EU: Possible impact on privacy and on-line anonymity, *Digital Anonymity and the Law*, C. Nicoll, J. E. J. Prins and M. J. M. van Dellen (eds.) (T M C Asser Press, The Hague, 2003), pp. 161, 164, 167–169, 172–183.
12. J. Goldsmith and T. Wu, *Who Controls the Internet?* (Oxford University Press, 2006), pp. 176–177.
13. J. S. Hiller and R. Cohen, *Internet Law and Policy* (Pearson Education, Inc., USA, 2002), pp. 75–76, 95, 98–100, 169, 170–171.
14. O. S. Kerr, Internet surveillance law after the USA Patriot Act: The big brother that isn't, *Northwestern University Law Review* (2003) 627–630.
15. D. Lyon, *Surveillance After September 11* (Polity Press, Cambridge, UK, 2003), pp. 15, 29, 89, 109–112.
16. G. S. McClellan, *The Right to Privacy* (HW Wilson Company, New York, 1976) pp. 3, 14, 25.
17. C. Nicoll, Digital anonymity and the law, *Digital Anonymity and the Law*, C. Nicoll, J. E. J. Prins and M. J. M. van Dellen (eds.) (T M C Asser Press, The Hague, 2003), pp. 116–119.
18. R. O'Harrow, *No Place to Hide* (Free Press, USA, 2006), pp. 10, 13, 31–32, 53, 78, 300–307.
19. J. Pain, Let's Not Forget 10 September 2001, available at June 12, 2006, http://www.rsf.org/article.php3?id_article=10760.
20. R. S. Poore, Computer forensics and privacy: At what price do we police the Internet? *The Privacy Papers Managing Technology, Consumer, Employee, and Legislative Actions*, R. Herold (ed.) (CRC Press LLC, USA, 2002), pp. 33–34.

21. R. Rushing, A. Schwartz and P. Bruening, Protecting consumers online: Key issues in preventing internet privacy intrusions, fraud and abuse, *Centre for Democracy and Technology* (2006) 1–8.
22. J. Ryan, Outbreak of iWar imminent, *Commercial Crime International* **25**(10) (2008) 10–11.
23. E. Schulz, Personal information comprises: It is time for the U.S. Government to wake up, *Computers and Security* **24** (2005) 261–262.
24. South African Law Reform Commission *Privacy and Data Protection* Discussion Paper 109, Project 124, available at October 2005, www.doj.gov.za/salr/index.htm.
25. V. Walt, Global alert, *Time Magazine* (2008) 30–31.
26. M. M. Watney, The legal conflict between security and privacy in addressing crime and terrorism on the internet, *ISSE/Secure 2007 Securing Electronic Business Processes*, N. Pohlmann, H. Reiner and W. Schneider (eds.) (Vieweg, Wiesbaden, 2007), pp. 26–37.
27. J. Wouters and F. Naert, Police and Judicial cooperation in the European Union and Counterterrorism: An overview, *Legal Instruments in the Fight Against International Terrorism a Transatlantic Dialogue*, C. Fijnaut, J. Wouters and F. Naert (eds.) (Martinus Nijhoff Publishers, Leiden, 2004), p. 138.

Chapter 29

CYBERCRIME

HAMID JAHANKHANI and AMEER AL-NEMRAT
School of Computing, IT and Engineering
University of East London, UK

Today our commonwealth is protected by firewalls rather than firepower. This is an issue of global importance as new technology has provided a world of opportunity for criminals. As a consequence, law enforcement agencies all over the world are struggling to cope. Therefore, today's top priority is to use computer technology to fight computer crime.

29.1. Introduction

Since time immemorial criminal activity has by its very nature drawn together many potential perpetrators of crime. Historically this activity led to an underclass, which in the United Kingdom was countered in Sir Robert Peel's principles of early policing. The original London "peeler's" communicated by means of a whistle. Later, the telephone provided the police of the day with a distributed network of communications posts. These are evidenced by the police boxes which were installed throughout the major cities in the United Kingdom during Edwardian times.

This network provided a means by which the embryonic command and control perspectives were developed for the early metropolitan police forces. In the United States this concept was replicated, most notably in the crime ridden cities of Chicago and Boston. During prohibition these two cities had witnessed the growth of organised crime against a background of criminal focused families, some of which were of Italian extraction and for whom the "Mafia", label became a badge. Such criminal activities are, however, not the sole domain of any particular nation or ethnic group. For example, the activity of East London gangs, Chinese "triads", eastern European and Asian criminal groups having been particularly significant within the United Kingdom during the last 50 years.

The means to communicate provides both the law enforcer and the criminal with the ability to direct resources and share information within their communities, to maximise their operational efficiency, flexibility and speed of response. While the means are identical the ends are clearly not. Indeed, it is, therefore, no surprise that the criminal elements have used communications to further their aims since the 1920s. In particular with the development of early telephone systems into the new telecommunications systems, including the Internet and mobile technologies, have created opportunities for such criminal groups to disseminate information of value in a timely manner.

In many ways, cybercrime is no different to more traditional crime — both involve identifying targets, using surveillance and psychological profiling. The major difference is that the perpetrators of cybercrime are increasingly remote to the scene of the crime. The traditional idea of a criminal gang loses its meaning as members can now reside on different continents without ever having to actually meet.

29.2. Cybercrime

Cybercrime is the world's biggest growth industry and is now costing an estimated €180 billion loss to organisations and individuals, every year. The creation of "virtual identities" gives a greater anonymity to the activities of organised criminals.

E-security is an issue of global importance and the methods cybercriminals use are far-reaching, cunning and technologically advanced. Criminals search out the services of thrill-seeking hackers and "script kiddies" to provide the expertise they need, which can be seen as a modern form of child labour.

The concern about cybercrime prevention has increased significantly among politicians, security specialist, academic institutions and legal professionals, with a staggering array of methods in an attempt to reduce the level of cybercriminal activities in the society.

In the EU and United States, the decision to focus on implications of technical methods for fighting cybercriminal is not arbitrary but comes from the need to do so after the 9/11 attacks. According to Suleyman Anil, who is in charge of protecting NATO against computer attacks, "cyber defence is now mentioned at the highest level along with missile defence and energy security. We have seen more of these attacks and we do not think this problem will disappear soon".

International money laundering is a particular concern in the arena of cybercrime as it can be used to finance and support criminal activities. Internet banking and digital cash are the most common ways of washing dirty money. Criminals try to hide and cover the sources from which their money comes by creating complex layers involving "social engineering" — tricking innocent parties into divulging sensitive information. Phishing, pharming, spyware, bin raiding and public records access are just some of the common techniques used by the criminals. Also, money launderers are moving to exploit other poorly defended message transmission systems and emerging technologies, such as Voice over Internet Protocols (VoIP).

Phishing is an effective means of gathering valuable personal and organisation information. Phishing is one end of a two-ended criminal enterprise in that it is the gathering of the information and the second part is the utilisation of that information for criminal purposes. Phishing is mainly conducted through the medium of e-mail given the ease of use and the relative anonymity of e-mail use. Criminals are able to "mass e-mail" potential victims masquerading as their bank or other party who might have a legitimate interest in contacting them about financial matters. The e-mails are crafted in such a way as to appear as though the request for the information (e.g. names, dates of birth, mother's maiden name, account details, etc.) is being legitimately made. In reality it is not and many willingly provide this information only later to discover that they have unwittingly passed them to an identity thief. Once in possession of this information, the thief can use it to steal an individual's or organisation's identity and divert money away from them.

Bin raiding is a method of obtaining information applies equally to individuals and organisations. As the name suggests, a criminal can rout through dustbins to obtain valuable information about an organisation. This includes obtaining bank account and credit card details, computer passwords, letterheads, signatures and other information which either on their own or added to other information allow a criminal to gain access to an organisation's accounts or those of its clients, trading partners, or suppliers. In a survey commissioned by the security company, Fellowes, it was estimated that 97% of the households, approximately 21 million homes, disposed of information that could be exploited by identity thieves by throwing it in their household refuse [10].

In the United Kingdom it has long been the case that certain personal and business records are freely accessible to the public through public records access. Personal details may be obtained by the payment of a small fee to the General Register Office for documents such as certificates of birth, death, marriages and civil partnerships and there is no scrutiny of the identity of the applicant. Business details and records are obtained in a similar way from Companies House. These resources represent a rich source of information for identity thieves who are then able to utilise them for the purposes of duplicating an identity. Information from the latter source has been used to take over a company's identity entirely by altering the details to the thieves' advantage.

Cross-border cybercrime poses a real threat to global security. Many countries do not have laws in place to combat it, and the international legal framework is patchy. By creating complex and difficult-to-trace Internet layers, which cut across many national borders or, by tricking individuals into releasing their personal data; organised crime is often able to operate virtually undetected.

Internet has all the ingredients needed by organised crime to pursue its damaging business: it's global, it's fast and it's virtual. In the wrong hands, this adds up to the potential to make vast sums of money illegally. In the early days of computers, "computer crime" meant breaking into, or stealing, a single PC. Today, the term spans a wide range of fast-evolving offences. Broadly speaking, cybercrime can be divided into two categories:

- new crimes that are a result of Internet and can only be committed online and
- old-style crimes that use hi-tech and going online.

Organised criminals have the resources to acquire the services of the necessary people. The menace of organised crime and terrorist activity grows ever more sophisticated as the ability to enter, control and destroy our electronic and security systems grows at an equivalent rate. Today, certainly, e-mail and the Internet are the most commonly used forms of communication and information sharing. Just over 1 billion people use the Internet every day. Criminal gangs "buying" thrill-seeking hackers and "script kiddies" to provide the expertise and tools, this is called cyber child labour.

29.3. Cybercrime Profiling

Researchers from different disciplines have attempted to explore different dimensions of the issues that surround cybercrime behaviour. To date, however, despite numerous attempts, there is a lack of agreement on frameworks for either understanding and/or responding to these issues, their impacts and their interrelationships [3]. The lack of classification system is a significant handicap and may be due to the considerable confusion that occurs around the very notion of what constitutes "cybercrime" or computer-related crime and indeed whether it is new or old crime in a new bottle [2].

To develop useful profiles of different offender categories, a large amount of data is required and to improve the reporting of cybercrime, there are need to increase the trust between the public and private sectors, which will result in reporting of cybercrimes when they occur. This will allow researchers to more precisely identify whether or not any unique patterns and characteristics actually exist.

To understand the new trends of the cybercrime and also establishing the appropriate framework that will be the foundation stone to investigate and prosecute the cybercriminals, there is an urgent need for cooperation and harmonisation of public and private sectors to encourage cybercrime reporting. Understanding the steps in the process of committing crime, and understanding the conditions that facilitate its commission, helps us to see how we can intervene to frustrate crime" [16].

Criminal profiling is the process of investigating and examining criminal behaviour to help identify the type of person responsible [15]. The FBI's Hayelwood and Douglas 1980, cited in Johnson 2005 [8], defined profiling as — *An educated attempt to provide ... specific information as to the type of individual who committed a certain crime .... A profile based on characteristics patterns or factors of uniqueness that distinguishes certain individuals from the general population.*

To date, all the national security organisations depend on data and text mining techniques to detect and predict criminal activities, while data mining refers to the exploration and analysis of large quantities of data to discover meaningful patterns and rules [9]. Text mining, sometimes refers to as text data mining, is the process of analysing naturally occurring text for the purposes of extracting and either trivial patterns or knowledge from unstructured text [9]. The objective of many intelligence data analysis projects is to use data mining to find association and/or discover relationships among suspect entities based in historical data, while data mining analysis data from structured database, there is a large volume textual data (e.g. e-mail, telephone conversation and text messages), which crime investigators have to examine which are unstructured.

Data mining is a powerful tool that enables criminal investigator who may lack extensive training as data analysts to explore large database quickly and efficiently. The following are some of the very common techniques:

(a) *Entity extraction*: the process of identifying names, places, dates and other words and phrases that establish the meaning of a body of text — is critical to software systems that process large amounts of unstructured data coming from sources such as e-mail, document files and the Web. By locating certain types of phrases and associating them with a category, applications such as text analysis software can perform functions such as concept extraction.
(b) *Clustering technique*: group data items into classes with similar characteristics to maximise or minimise interclass similarity — for example, to identify suspects who conduct crimes in similar ways or distinguish among groups belonging to different gangs [4].
(c) *Deviation detection*: researcher deploy this technique to detect fraud, network intrusion detection and other crime analysis that involve tracing some activities which can be appear sometimes to be abnormal.
(d) *Classification*: finds common properties among different crime entities and organises them into pre-defined classes. This technique has been used to identify the source of e-mail spamming based on the sender's linguistic patterns and structural features.
(e) *Social network analysis*: describes the roles of and interaction among nodes in a conceptual network. Investigator can use the technique to construct a network that illustrates criminal's roles, the flow of tangible and intangible goods and information [4].

In 1990, the profiler Brent Turvey met with and interviewed an incarcerated serial killer after extensively reviewing crime report, court transcripts and court records. Nothing matched, Turvey could not comprehend how the prisoner's statement could be so contradictory to the information in the crime reports until he realised that the perpetrator was purposely misconstruing the facts to redirect the responsibility of the crime. Turvey felt it was more reliable to look at the forensic evidence and then using the criminal event, to reconstruct the behaviour [12].

The question of why some people commit crime is a subject which has presented criminologists and sociologists a challenge for many years, and, like many such questions, is one to which there is no simple answer [3, 8]. Johnson [8], have done some fine preliminary work developing typologies, classifying the serial computer criminal as crucial in determine the underlying motivating causal factors. Classification is made by assessing and analysing the written, physical and digital behaviour that exist in each attack. Any understanding of crime patterns and offenders motivation should start perhaps with the question of why people commit crime in the first place. Profiling has traditionally been thought of as attempting to reduce the number of possible offenders to the point where traditional methods of investigation can be introduced to solve the case [1]. Interpreting the intrusion from the criminal point of view will greatly assist the investigator in understanding what motivates an offender [8].

In the case of cybercrime as there is a rapid change of the technology, therefore, cybercriminal's behaviour may become dynamic. This change in behaviour will require a reclassification of the typology being currently used. Essentially, cybercriminal's behaviour is evolving and changing overtime with experience where they learn from their actions, or from their friend's experience, which will enhance their skills. The offender signature which is a repetitive ritualistic behaviour that the offender usually displays at every crime scene provides police an appropriate profiling tool [8]. This will give the investigator the opportunity to understand the motivations that drives the offender to perpetrate such crime. This finding will result in assisting the researcher in the classifying of the type of perpetrator that is being sought.

It is important that we consider some of the more prominent theories of criminal behaviour if we are to understand trends and patterns in criminal activity and if we are to understand the behaviour of those sought by profilers. It would be naïve to presume that the reason why most people commit crime can be found in just one theory [1]. However, the choice of which profiling method to use is controversial. On the surface, it appears that deductive profiling is more suited to computer-related cases than using inductive profiling methods such as the FBI or IP method, which are culturally biased and does not make practical sense [12].

Deductive profiling methods (e.g. Behaviour Evidence Analysis (BEA)), unlike the FBI or IB methods, do not rely on a large offender database or on statistical analysis of previously convicted offenders and should be less culturally biased [12]. The BEA was developed by the profiler Brent Turvey, to overcome some of the FBI and Investigative Psychology (IP) models [11, 12, 15]. The BEA method has four steps and two primary phases which includes victimology and computer scene characteristics.

29.4. Corporate Identity Theft

In the United Kingdom, there is mounting evidence to suggest that the financial impact of identity theft is far greater than had previously been appreciated. A recent UK government department study estimated that the losses sustained by the UK economy now stand at approximately £1.75 billion per annum [5]. As is often the case with crime figures, it is difficult to accurately assess the extent of losses sustained as assessment can only be made on the basis of crimes that are reported and recorded. It is also dependent upon the government having a proper understanding of this particular species of fraud, in its various manifestations and the importance of official recording of this crime. Further, the government and others need to recognise the importance of this crime by creating a statutory offence of identity theft as is the case in the United States where identity theft is a specific offence (Identity Theft And Assumption Deterrence Act, 1998).

In the light of the above, a proper understanding of this fraud is essential as the opportunities for criminals to steal the identities of individuals and businesses has increased and become easier due to the pace at which organisations need to interact with each other and due to e-commerce/Internet usage. To raise awareness of this danger trade organisations, law enforcement agencies and government have sought to educate those who would be most affected. This has been effective in that some businesses have become more aware of the dangers posed by this form of criminal activity. However, the speed and ever changing methods of this type of theft makes it difficult for individuals and businesses to be successful in combating this criminal activity. An example of changing methods is that of the taking of identification information from rubbish bins (known as as "bin raiding"). Once upon a time this was sufficient to yield all the necessary details to effect the identity theft. However, as individuals and businesses become more vigilant in the security/disposal of such information, thieves now need to use more sophisticated and anonymous methods to obtain that information [7].

In addition to greater vigilance by some, it is clear that in the last 3–4 years there has been an increasing awareness of the existence, mechanics and possible consequences of individual identity theft. The awareness that has taken place has been achieved by a combination of news reporting, advertising, information from banks and credit card companies, police and government campaigns and the like. The awareness has been such that the issue of individual protection and individual identity theft resulted in a Parliamentary Committee being established to consider the matter [6]. Whilst the awareness of the theft of an individual's identity has been heightened, the same cannot be said of business identity theft.

Estimates of the losses sustained as a result of corporate identity theft vary. One source puts the current losses at just under £100 million with an indication that they are likely to increase to £700 million by 2020 [13, 14].

Corporate identity theft is having a profound effect upon business relationships and perhaps more so than ever before commercial interactions are dependent upon trust between the parties. This trust relates not only to the relationship between the parties but it often extends to the systems and processes established by them; often by the dominant party. When, for whatever reason, these systems and processes have been found wanting there has been a detrimental effect upon the trust that has existed or would have existed between the parties. This was exemplified in recent reports of data theft that took place over an 18-month period from the retailer T K Maxx. The theft resulted in sales dropping dramatically over an extended period. Whilst it was not the company's identity that was stolen, nevertheless this serves to illustrate the speed of impact and how widely confidence in an organisation can be damaged. In this case, T K Maxx, through its market position and financial resources, was able to show that the "breach" in its processes related only to one category of customer, namely credit card users, and that the information itself was somewhat dated. That said, there will be some who will not be sufficiently convinced as to patronise the company in future. A much smaller company would be unable to mount such an effective damage limitation exercise; still less if it had its identity improperly utilised.

The above example and others show that when an incidence of identity theft occurs there can be both micro and macro implications such as the following:

29.4.1. *Micro Implications*

These includes following:

- loss of brand/business confidence
- damage to reputation
- credit rating damaged
- possible breach of legal obligation.

29.4.2. *Macro Implications*

These includes following:

- loss of confidence in a particular sector/industry
- loss of confidence in a particular way of doing business (e.g. fewer people using the Internet)
- insurance claims resulting in increased premiums for all
- losses recouped by increasing prices for goods and services.

Whilst it is important to make a distinction between individual and organisation identity theft it would be a mistake to think of them as always being separate from each other in all cases.

29.5. Money Laundering

A particular concern in the arena of cybercrime is its use in international money laundering, which in turn can be used to finance and support illegal arms sales, smuggling, drug trafficking, prostitution rings, embezzlement, insider trading, bribery and computer fraud. Internet banking and digital cash are the most common ways of washing dirty money. Criminals try to hide and cover the sources from which their money comes by creating complex layers involving "social engineering" — tricking innocent parties into divulging sensitive information.

There are two main categories under which all social engineering attempts can be classified: computer- or technology-based deception and human-based deception. The technology-based approach is to deceive the user into believing that they are interacting with the "real" computer system (such as a popup window, informing the user that the computer application has had a problem), which gets the user to provide confidential information such as personal and network passwords. The human approach is done through deception, by taking advantage of the victim's ignorance and the natural human inclination to be helpful. These subjects in particular, and how to guard against them, are covered in-depth in the security briefings issued by the CPNI.

Within the technology-based approach, the most common method is "phishing", gaining personal information by using fraudulent e-mail messages that appear to come from a legitimate business, such as the victim's own bank. Most businesses and individuals are already aware that criminals utilise such means to strike at potential victims, but it is a growing concern that money launderers are moving to exploit other poorly defended message transmission systems and emerging technologies, such as VoIP.

29.6. Spam

Spam on the Internet started with services like Usenet but has migrated to e-mail. A highly organised criminal industry is utilising such services and will move to exploit other poorly defended message transmission systems including VoIP and other emerging technologies. The Internet e-mail system was designed for an environment where all users could be trusted implicitly, or in the worst case traced and removed from the network. As a result the SMTP mail protocol, used in conjunction with DNS, is very difficult to strengthen against possible abuse.

Spamming and malware used for other criminal activities have formed a symbiotic relationship; each technology now relies on the other to circumvent the current range of anti-spam and anti-virus countermeasures. Seemingly no amount of legislation seems capable of preventing criminals from utilising such technologies for their illegal goals. Even if such individuals could be tracked down, they still have the option of simply moving to a part of the world where the authorities will tolerate them. Technical measures to counter spamming are largely ineffective, as the protocols used over the Internet are inherently open rather than secure. As new countermeasures make it possible to block spammer's hosts, spammers simply move to hijacking innocent computers. IP Telephony and other messaging services are just as vulnerable and there is no reason to believe they will not be attacked and rendered unusable in the very near future.

Criminals have great incentives to write viruses. If they can get their software running on a PC containing sensitive information it can steal it and send it back to its creator, leading to identity fraud, unauthorised access to bank accounts, industrial espionage and a list of other things only limited by a twisted imagination. A major aim for users of computer viruses is the ability to hijack a host and control it remotely, thereby using it as a stepping stone to attack other machines or relay spam. Such machines are referred to as zombies.

29.7. Conclusions

Researchers, academically or commercially, are continually creating filtering and search engines to find and sort documents from multiple resources. Criminals use zero day vulnerabilities to get what they want and anti-forensics techniques to cover their tracks.

Despite a plethora of Internet-related legislation, cybercrime is still a growing stigma for the e-society. It is evident that Internet usage requires laws and regulatory authorities, which should span across national boundaries and legal systems.

One of the consequences of the 11 September 2001 terrorists attack on the United States was the signing on 23 November 2001 of the International Convention on Cybercrime by the United States and 29 other countries. This international treaty aims at enforcing the ability of these nations to combat cybercrime.

References

1. P. B. Ainsworth, *Offender Crime Profiling and Crime Analysis* (Willan Publishing, USA and Canada, 2001).
2. R. Broadhurst and N. Chantler, United Nations Office on Drugs and Crime: Cybercrime Update: Trends and Development Online, (2006), http://www.eprints.qwt.edu.au/archive/00004690.
3. V. Broucek and P. Turner, Winning the battles, losing the war? Rethinking Methodology for Forensic Computing Research, *Journal in Computer Virology* **2**(1) (2006) 3–12.
4. M. Chau, J. Xu and H. Chen, Extracting meaningful entities from police narrative reports, *Proceeding of National Conference for Digital Government Research*, Los Angeles, California, USA, (2000).
5. Home Department Great Britain. Home Office Identity Fraud Steering Committee, *Updated Estimate of the Cost of Identity Fraud to the UK Economy* (The Stationery Office, 2006).
6. House of Lords, Great Britain Parliament, Science and Technology Committee, *Personal Internet Security*, 5th Report of Session 2006–07 (The Stationary Office, London, 2007) (HL Paper 165-I).
7. M. James, Aping to Defraud — Corporate Identities at Stake (Infosecurity Magazine May/June 2006).
8. T. A. Johnson, *Forensic Crime Investigation.* (CRC Press, USA, 2005).
9. P. Kanellis, E. Kiountouzis, N. Kolokotronis and D. Martakos, *Digital Crime and Forensic Science in Cyberspace* (Idea Group Inc. (IGI), USA, 2006).
10. J. Leyden, Brits in their Identity as ID Thieves Prosper. The Register Magazine, 16 October 2006, http://www.theregister.co.uk/2006/10/16/id_fraud_prevention_week/, Accessed on August 2007.
11. N. Nykodym, R. Taylor and J. Vilela, Criminal profiling and insider cybercrime, *Computer Law and Security Report* **21**(5) (2005) 408–414.
12. M. Rogers, The role of criminal profiling in the computer forensics process, *Computer and Security* **22**(4) (2003) 292–298.
13. Royal and Sun Alliance. *Risk uncovered. A Study into the Future Risks Faced by British Businesses*, (2006), http://ww7.investorrelations.co.uk/royalsun/uploads/media/Risky_Business_ReportFINAL.pdf Accessed on 4.6.2007.

14. Royal and Sun Alliance. A Guide to Corporate Identity Theft. Royal and Sun Alliance, Accessed on 4.6.2007. http://ww7.investorrelations.co.uk/royalsun/assets/pdf/CorporateIDTheftGuide.pdf.
15. B. Turvey, *Criminal Profiling, an Introduction to Behavioural Evidence* (Elsevier, UK, 2002).
16. R. Wilson, Understanding the Perpetration of Employee Computer Crime in the Organisational Context. Working paper no.04-2006.

Chapter 30

CYBERCRIME: INNOCENT VICTIMS AND THEIR PLIGHT FOR JUSTICE

HAMID JAHANKHANI and KEVIN ORME*
University of East London, UK
*and *Metropolitan Police, UK*

Victims of cybercrime in the United Kingdom and around the world often find themselves with no means of recourse at a local level. Police authorities have little or no means of investigating e-crime and budgetary controls ensures that only the most serious offences (usually terrorist related or those of a sexual connotation) will ever be investigated. This chapter aims to determine whether the Internet has brought a new set of crimes to the global marketplace and if so tries to determine the response by the relevant governments and investigating authorities to deal with this potential transference of "crime scenes". In addition, it attempts to investigate whether cybercrime is just a mere variation of the existing crimes already catered for by statute. This chapter also attempts to provide examples of specific Internet crimes (cybercrime) by focussing on specific crime types and explaining the associated problems by analysing the law, investigatory procedure, loopholes, remedies and the effect on innocent victims both in the business community and individually.

30.1. Introduction

The onslaught of cybercrime has been difficult to measure with the ever-increasing global capacity of the Internet and free access to it. The anonymity of offenders, hiding behind international barriers, utilising weaknesses in cross boarder jurisdictions and inadequate "mutual assistance" agreements between countries around the World all play an important part in assisting would-be offenders.

Victims in the United Kingdom and around the world often find themselves with no means of recourse at a local level. Police authorities have little or no means of investigating e-crime and budgetary controls ensures that only the most serious offences (usually terrorist related or of a sexual connotation) will ever be investigated.

The impact and influence of the Internet over the past 12 years has been immense. During that time, access to the Internet has grown enormously. In 1996, 3.4 million UK adults were online; by 2006 this had expanded to 28.5 million, "UK Cybercrime Report" [7].

The prevalence of the Internet, its sheer enormity and exponential growth has revolutionised global communications at an unprecedented scale. The manner in which we conduct our business and private affairs have changed drastically with the advent of technological advances — particularly with the open accessibility to e-mail.

Evolvement of effective communication within organisations using networks, internal platforms allowing external communication and un-patched operating systems containing security flaws allowing unauthorised access may potentially leave the back door open and invite a plethora of opportunity seekers, hardened criminals, and organised crime members into their domains.

This phenomenal rise in Internet usage coupled with a better understanding of "computer literacy" generally has meant that "global doors" have been opened to many that ordinarily would have used more traditional means of communication, such as "snail mail". It means that individuals from all walks of life and experience can set themselves up with high-speed Internet access at home or in their business and inadvertently make themselves targets for organised crime syndicates or self-satisfying cyber-villains. Predictably, this has led to exploitation by organised crime gangs.

The rise of this networked society has expanded the range of information available to individuals and changed the way in which we relate to one another in the virtual world as well as in the physical world. However, it also has a dark side: the Internet has proven to be an influence on criminal, as well as legitimate, activity.

The potential of the Internet to facilitate crime is increasingly a matter for public concern. This has given rise to a need to understand and measure cybercrime. However, attempting to quantify the amount of cybercrime is not straightforward according to the "UK Cybercrime Report" [7]. The main reasons for this is that cybercrime has to be measured and is generally examined via a three stage process.

- Firstly, criminal conduct has to be observed.
- That conduct needs to be established as criminal conduct.
- And that conduct has to be recognised and brought to the attention of the authorities in order to be recorded.

It suggests therefore, that if any "one" part of the three stage process fails to be identified, a crime will not be recorded officially as a statistic.

Cybercrime therefore has unique characteristics. They involve factors that may not easily fall into one the categories outlined above. For example, a banking institution that has very small amounts removed from numerous accounts fraudulently may not be recognised by their victims due to the very minor discrepancies.

There will also be circumstances whereby observed conduct goes unnoticed due to the fact the victim does not actually realise that the conduct is criminal such as virus attacks. And of course, there will be instances whereby the victim simply does not report the matter to the authorities.

Current United Kingdom legislation falls somewhat short of being capable of dealing with this exploitation and has been subject to radical proposals for change by the All Party Parliamentary Internet Group (APIG). At present, those proposals have been "shelved" but the groups report "*Revision of the Computer Misuse Act*," [1] will provide a starting point for this analysis as it contains a review of relevant legislation both current and proposed.

30.2. Definition of Cybercrime

The term "cybercrime" has been used to describe a number of different concepts of varying levels of specificity. At its absolute broadest, the term has occasionally been used to refer to any type of illegal activity that results in a pecuniary loss. This would include violent crimes against the person or property such as armed robbery vandalism, or blackmail.

At its next broadest, the term has often been used to refer only to non-violent crimes that result in a pecuniary loss. This would include crimes where a financial loss was an unintended consequence of the perpetrator's actions, or where there was no intent by the perpetrator to realize a financial gain for himself or a related party. (e.g. when a perpetrator hacks into a bank's computer and either accidentally or intentionally deletes an unrelated depositor's account records.)

Although the term "cybercrime" is now in everyday use, the first problem encountered in *measuring* cybercrime is that there is no commonly agreed definition of the term.

Despite the fact that the word "cybercrime" has entered into common usage, many people would find it hard to define the term precisely. Furthermore, there is no catchall term for the tools and software, which are used in the commission of certain online crimes [8].

Cybercrime has become a common term and its usage tends to generalise just about any illegal activity within the Internet environment. Despite an apparent acceptance of and familiarity with the term, there exist dramatically varied views of what Cybercrime *is*. This lack of definitional clarity is problematic as it impacts every facet of e-crime investigation and the reporting process.

Some of the definitions of cybercrime that do loosely exist include:

Cybercrime is described as criminal activity in which the computer or network is an essential part of the crime, this term is also used to include traditional crimes in which computers or networks are used to enable the illicit activity.

Other examples of cybercrime refer to where the computer or network is used as a tool of the criminal activity includes spamming and criminal copyright crimes, particularly those facilitated through peer-to-peer networks, or where the computer or network is a target of criminal activity include unauthorised access (i.e. defeating access controls), malicious code and denial-of-service (DoS) attacks.

Cybercrime could also encompass where the computer or network is a place of criminal activity, which involves the theft of a service (in particular, telecom fraud) and certain financial frauds.

Finally, examples of traditional crimes facilitated through the use of computers or networks include Nigerian (419) Frauds or other gullibility or social engineering frauds (e.g. hacking "phishing", identity theft, child pornography, online gambling, securities fraud, etc.).

Cybercrime in the context of national security may involve hacktivism (online activity intended to influence policy), traditional espionage, or information warfare and related activities.

One of the recent researches showed that a new cybercrime is being perpetrated every 10 s in Britain [10]. During 2006, the computer crooks were able to strike 3.24 million times. Some crimes performed online even surpassed their equivalents in real world. In addition, experts believe that about 90% of cybercrimes stay unreported [10].

According to a study performed by McGuire [13], a specialist in psychology, the University of San Francisco, the majority of teenagers who hack computer systems are doing it for more fun than aiming to cause harm. McGuire [13] also reports that "*often parents cannot understand the motivation of the teenage hackers*". McGuire conducted an anonymous experiment in the area of San Diego by questioning more than 4,800 students.

The results of the McGuire Survey [13] were presented at the American Psychological Association conference as follows;

- 18% of all youngsters confessed of entering and using the information stored on other personal computer or Web site.
- 13% of all the participants mentioned they performed changes in computer systems or computer files.
- The study revealed that only 1 out of 10 hackers were interested in causing certain harm or earns money. Most teenagers performed illegal computer actions of curiosity, to experience excitement.
- 38% of teenagers were involved in software piracy.

30.3. Cybercrime — Offline Crimes Transposed Online

30.3.1. *A Study of S21 Theft Act 1968*

In an effort to demonstrate how predominantly serious offline crime can and has been transposed into cyberspace, we have completed an analysis of S21 Theft Act 1968, blackmail. This offence traditionally is an extremely serious one which would require intense commitment of manpower on behalf of any Police Criminal Investigation Department.

Although this crime is still a serious offence and not traditionally a day-to-day occurrence, the introduction of the Internet has enabled the "modus operandi" of any criminal to perpetrate this offence with considerable ease. The fear of being detected has been reduced considerably as explained here.

The evolvement of the Internet coupled with its phenomenal rise over the past decade in relation to Internet usage has inevitably attracted exploitation by organised crime groups, cyber-villains, and the like. As technology has developed so has the sophistication of organised gangs. The rewards were and still are lucrative with little risk of capture.

Law enforcement agencies were slow to react and the trend was to wait for legislation to change before enforcing new policy to combat Hi-Tech crime issues. The problem is, of course, that as the offences became more technical in nature, the legislation largely stagnated. The appeals process in the United Kingdom can be unbearably long. It may take many years before a trial becomes a "case stated" thus resulting in old laws being outpaced by technological progress.

The Computer Misuse Act 1990 (CMA 1990) was primarily introduced to provide suitable legislation and tackle the growing problems encountered with unlawful access and modification of computer systems. Most of which was previously being prosecuted under the Criminal Damage Act 1971 (CDA 1971). Effectively, this was a "classic statutory" reaction in "loophole plugging" by the Government.

Current United Kingdom legislation falls somewhat short of being capable of dealing with this exploitation and has been subject to radical proposals for change by the APIG. At present, those proposals have been "shelved" but the group's report "*Revision of the Computer Misuse Act*" [1], will provide a starting point for this analysis as it contains a review of relevant legislation both current and proposed.

30.3.2. *Blackmail — the Legislation*

The traditional interpretation conjures up images of the armed offender threatening serious violence against his victim whereas in reality the definition allows a much more subtle approach and can incorporate a more diverse set of circumstances in an effort to accommodate e-crime.

The CMA 1990 in its present form leaves the door slightly open for a wider interpretation of how far we can make the crime fit the legislation.

Here, we essentially look at how this crime is catered for online and offline. The Theft Act definition provides substantial scope to support a conviction under both circumstances, whereas, The CMA 1990 potentially plugs the loopholes for offences online, and may provide an excellent opportunity to for the Crown Prosecution Service to offer alternative charges when supporting a prosecution.

30.3.3. *Blackmail as Defined by the Theft Act 1968*

Blackmail is ensconced in the annals of time and is an offence that had its legislation most recently defined in S21 (1) Theft Act 1968: ... "*A Person is guilty of blackmail if with a view to gain for himself or another or with intent to cause loss to another he makes any unwarranted demand with menaces*".

S21 (2): "*The nature of the act or omission demanded is immaterial and it is also immaterial whether the menaces relate to action to be taken by the person making the demand*".

The essential elements of Blackmail are as follows

- It is not confined to actually obtaining physical possession of the property by menaces.
- It does not include sexual favours.
- Dishonesty is not an essential element but could be present.
- The offence can be tried in England if ANY part of the offence is committed here (Treacy v DPP 1970).
- The posting, making or receipt of the threat MUST take place in England and Wales.

This offence carries a maximum penalty of 14 years imprisonment is an arrestable offence, which can only be tried on indictment at the Crown Court. This section of the Theft Act is a powerful piece of legislation and could be supported with alternative charges under s3 of the CMA 1990.

30.3.4. *The CMA 1990 and Blackmail*

Computers can also be used to commit the offence of Blackmail, for example where a computer virus is introduced to a system (for example a time bomb, whose purpose is to corrupt or delete stored information after the lapse of a period of time), accompanied later by threats that some or all the files on the system will be corrupted unless a sum of money is paid into a particular account. Such a virus may be introduced directly by a hacker, or simply distributed as part of a software package.

The CMA 1990 attempted to plug loopholes in the legislation creating specific offences of hacking and the placing of viruses. S3, CMA 1990 deals with these offences. An offence is committed if: ... "*he does any act, which causes an authorised modification of the contents of any computer; and at the time he does any act, he has requisite intent and the requisite knowledge*".

The S3 offence is aimed at those who introduce worms and viruses to computer systems. It is committed by a person who does an act that causes unauthorised modification of the contents of any computer. If the physical condition of the computer is impaired, whether intentionally or recklessly, an offence under the CDA 1971 may also be committed. Section 3 covers non-tangible damage, which is now excluded from the CDA. Intent is an essential element of the offence and recklessness alone is not sufficient [2].

30.3.5. *Review of Relevant Legislation*

This section will deal with specific types of Blackmail-related attacks and the problems encountered with current legislation, specifically the CMA 1990.

Blackmail by definition requires a demand with menaces and has been extensively tested by the UK courts. With the conception of the Internet, this definition has had to be interpreted more creatively and adapted to suit the cybercrime environment. S3 Computer Misuse Act provides the legislation that can encompass Blackmail offences such as "time bombs". Modification of data, programs or systems is a key issue here and is often associated with Blackmail.

One of the most common forms of attack associated with Blackmail is the DoS attack. This type of attack appears to fall into two categories — the DoS attack and the Distributed Denial of service attack (DDoS). The DoS can be defined as — preventing the normal operation of a computer by bombarding it with spurious traffic. The DDoS — A DDoS attack that is being made from many different locations simultaneously. DoS and DDoS attacks are extremely common on today's Internet with academic studies measuring over 4,000 a week APIG.

An inquiry by the APIG highlighted the potential difficulties being experienced by the courts and their interpretation of the law. On the one hand, the inquiry found a widespread belief that the CMA is not adequate to deal with DoS and DDoS attacks — possibly stemming from Crown Prosecution Service advice in 2002, which said that the section might not stretch to include all such activity.

On the other hand, the Government and the National High Tech Unit (now Serious Organised Crime Agency SOCA) reported beliefs that Section 3 is sufficiently broad to cover DoS attacks. In April 2003, the Internet Crime Forum Legal Subgroup pointed out that the section did *not* require unauthorised access, merely unauthorised "modification of the contents of any computer", expressing the opinion that "the test applied would be whether the attack had rendered unreliable the data stored on a computer or impaired its operation" [9].

The APIG [1] report states that "The reason for this wide disparity of legal opinion and distrust of the efficacy of the current law is that when DoS and DDoS attacks occur on the Internet then it is the particular circumstances of each attack that makes it obvious whether the CMA wording applies ...".

In general, where a DDoS attack takes place then an offence will have been committed because many machines will have been taken over by the attacker and special software installed to implement the attack. Even when a single machine attacks a system, an offence will sometime be committed because the contents of the system will be altered.

However, when the sole effect of an attack is to fill a nearby link with useless traffic, then it may be hard to show the elements of a CMA offence are present, although a DoS attack has certainly occurred.

The report concludes:

> "It is clearly undesirable to have the illegality of an attack that depends upon the exact mechanism used so we are minded to recommend the creation of a new offence of "impairing access to data".

The APIG report recommends that the legislation should be changed to deal with this and other confusions. However, the APIG leaves it open to the government to decide how to do this in terms of the technicalities of statutory updating: whether to amend the 1990 Act; or whether to draft new legislation.

More worrying than nuisance DDoS attacks are those criminals that use the threat of a DDoS attack to Blackmail companies. In most instances, criminal hackers appear to be targeting high-volume, low-value transactional sites such as online bookmakers. Typically, the hackers threaten to disrupt and paralyse bookmakers' Web sites.

The National High Tech Unit (now SOCA) now part of the SOCA investigated six cases of British Internet bookmakers allegedly being Blackmailed by hackers threatening to disrupt online betting ahead of major sports events such as the Grand National, FA Cup final and the European Championship. Analysts estimate that the online betting market is worth over £3bn per year. For online bookmakers, several hours of downtime would be extremely expensive especially if it coincided with an event such as the Grand National, which attracts more than £100 m in bets.

Sites are frequently Blackmailed into paying between £20,000 and £30,000 a year in return for respite from further attacks. However, £30,000 a year may be a relatively small amount compared to the potential losses so some organizations appear to be paying up. For instance, targeted bookmakers say the money the criminals ask for is about 1-h worth of business [4].

The mechanism of the scam has been known for a number of years but the scale of its use has grown dramatically in the last 12 months or so. Unfortunately, law enforcement agencies do not have time to become involved in every DDoS service attack. They are also extremely difficult to anticipate. Most of them are nuisance attacks and are not performed by organised criminals.

However, if accompanied by Blackmail threats, it is a clear indication that a criminal attack is in progress.

Thankfully, in July 2004, the National Hi-Tech Crime Unit (SOCA), in a joint investigation with the Russian Federation, arrested three key members of the Russian gang involved in extortion. The success of the operation was built on the foundation of international partnerships between law enforcement and business. In Russia, the National Hi-Tech Crime Unit (SOCA) worked closely over many months with the Investigation Department of the Investigative Committee attached to the Ministry of Internal Affairs (MVD) and the MVD's computer crimes specialist department [15].

The cash was being transferred by a number of money transfer agencies that helped the National Hi-Tech Crime Unit (SOCA) track the money and thereby identified members of the gang. However, SOCA have stressed that while bookmakers might be under the spotlight now, Web-based extortion is a generic high-tech crime that has been a problem for some time. Therefore, it is clear that the problem of corporate Blackmail is growing and that law enforcement agencies will have to work together on a global level if it is ever going to be beaten.

Clearly, the CMA 1990 is now some 15 years old. Compare this same period in technological years and is easy to understand why there are potential pitfalls when trying to prosecute cybercrime. DoS and DDos attacks were not around when this legislation was being composed and finally implemented. It has to be argued that a simple Dos attack sending e-mails on mass to a company may or may not cause an unauthorised modification or unauthorised access to computer material. Each case will clearly have to be taken on its merits. Unfortunately, the prosecuting authorities in the United Kingdom will generally not prosecute these offences as it will do not justify the expense and demand on resources. The offence will be judged on its likelihood of conviction and what may be clawed back from it.

DDos attacks on the other hand pose a slightly different problem with a host of new considerations. Invariably, they consist of attacks that control many machines that have been corrupted and had special software installed on them. The software is likely to enable unauthorised access and may have made unauthorised modifications to a computer system.

There is therefore more justification to pursue offences of this nature as corroboration which could be sought from the "hijacked" computers. This type of offence will have a much better chance of being considered by specialists such as the National High Tech Unit (SOCA) for prosecution.

This has become a global problem with organised crime gangs increasing the intensity of the attacks. Part of the problem is effective communication and execution of similar laws outside the jurisdiction of the United Kingdom. With a Global capacity for cybercrime, the law around extradition issues may have to be examined. The basic s1 Hacking offence under the CMA 1990 carries a penalty of 6 month imprisonment or a £5,000 fine.

The extradition rules in the United Kingdom are such that the minimum penalty should carry 1-year imprisonment. It should be noted that many Countries use extradition proceedings far more frequently than we do.

If new or amended legislation is to be implemented, serious consideration must be given to increasing penalties or creating an aggravated offence of "hacking" in order to assist law enforcement agents with such extradition difficulties. Its further recommendation is that private prosecutions should be allowed, which would help private companies who have fallen victim to serious attacks.

Recently the Metropolitan Police in London held a meeting at Lords "Met police hooks up with commonwealth to fight cybercrime" [12], whereby the Economic Crime Working Group had a high powered meeting to discuss and share their desire to increase awareness in the business community of cybercrime risks and provide advice to policy makers. It is also designed to provide a forum for the sharing of best practices in fighting cybercrime.

The underlying problem here is that this type of cybercrime is usually not taken too seriously until it really needs attention. What is needed is an amendment to the CMA addressing new offences. The parameters are far too wide presently. It is not sufficient to judge the illegality of an attack. There must be a more proactive approach to protect the business owner by creating new offences of "impairing access to data" as recommended by the APIG report.

In fact, it has never been a more appropriate time. The most recent decision by the courts at a hearing at Wimbledon Magistrates on the 2nd November 05 resulted in an unsuccessful prosecution on a teenage boy that had sent 5 million e-mails to an ex-employer bringing down the company's e-mail server. The judge discussed whether a flood of unsolicited e-mails would not cause unauthorised access or modification as the e-mail server was set up for the purposes of receiving e-mails.

In a written ruling Judge Grant said "In this case the individual e-mails caused to be sent each caused a modification which was in each case an "authorised" modification. Although they were sent in bulk resulting in the overwhelming of the server, the effect on the server is not a modification addressed by Section 3 of the CMA, "Teenager cleared of e-mail attack charge" [8].

There are loopholes within current legislation that need to plugged and the APIG have identified and submitted the appropriate proposals to achieve those goals. We just need the government to wake up and realise the anomalies that exist and submit the draft for legislative confirmation.

30.3.5.1. *CMA 1990 S3A*

The Police and Justice Bill 2006 made some amendments to the UK's CMA. Specifically the new section 3A essentially criminalised the development, ownership or distribution of hacking tools and are detailed as follows:

30.3.5.2. *Making, supplying or obtaining articles for use in computer misuse offences*

After Section 3 of the 1990 Act, there is inserted —

3A Making, supplying or obtaining articles for use in offence under Section 1 or 3:

(1) A person is guilty of an offence if he makes, adapts, supplies or offers to supply any article —
    (a) Intending it to be used to commit, or to assist in the commission of, an offence under Section 1 or 3.
    (b) Believing that it is likely to be so used.

(2) A person is guilty of an offence if he obtains any article with a view to it being supplied for use to commit, or to assist in the commission of, an offence under Section 1 or 3.

(3) In this section, "article" includes any program or data held in electronic form.
(4) A person guilty of an offence under this section shall be liable —
  (a) On summary conviction in England and Wales, to imprisonment for a term not exceeding 12 months or to a fine not exceeding the statutory maximum or to both;
  (b) On summary conviction in Scotland, to imprisonment for a term not exceeding 6 months or to a fine not exceeding the statutory maximum or to both;
  (c) On conviction on indictment, to imprisonment for a term not exceeding 2 years or to a fine or to both.

This amendment does nothing to assist victims of cybercrime and merely potentially incriminates against the legal use of such tools in detecting and preventing offences.

There may be advantage, however. If the article was supplied in the course or connection with fraud, then prosecutors may consider if their case is also an offence contrary to Section 6 and/or Section 7 of the Fraud Act 2006.

An offence of making or supplying articles for use in frauds contrary to Section 7 is punishable by a maximum of 10-year imprisonment and an offence of possession of articles for use in fraud contrary to Section 6 is punishable by a maximum of 5-year imprisonment.

The new Fraud Act is yet to be tested but does offer much higher penalties and may therefore bring the perpetrator into the realms of an extradition treaty should that be necessary.

30.4. Crime Control Strategy

It becomes pretty obvious that cybercrime does not need a physical presence between two parties. The perpetrator and victim can be in two different cities, different counties and different countries. All that is needed is an Internet connection and an attack is viable. Within minutes a victim's computer can be infiltrated, have their identity stolen and commit identity fraud on a massive scale.

Cybercrime unlike real world crime typically does not need one-to-one confrontation. In effect this means that perpetrators can commit numerous crimes quickly in succession with little or no effort due to its sheer automation.

A traditional approach to investigating real world crimes would be to react, investigate and apprehend the offender. This type of investigation will normally focus in on and follow a chain of events. The crime gets reported, then allocated and investigated. Invariably, the officer will have numerous crimes to investigate at any one time and quickly becomes overwhelmed with the tasks in hand, particularly if an extremely serious offence takes precedence over his workload.

An element of prioritisation therefore becomes inevitable, resulting in those more historically reported crimes getting pushed to the "bottom of the pile".

Cybercrime on the other hand tends to go against the traditional process of investigation in that it can be committed by a very small percentage of the population (throughout the world!). This relatively small group can commit crime on a scale that far surpasses what may be achieved in the real world. As a result, the number of cybercrimes will exponentially exceed real world crimes.

The problem is of course that the law enforcement agencies still have to deal with the traditional offline crimes, with no increase in resources, manpower, or training the net result will be a total inadequacy to deal with and apprehend the perpetrators of cybercrime.

Cybercriminals avoid the physical constraints that govern real world crime; funds can be extracted from a Bank and moved into offshore accounts with little effort and less visibility [3]. The typical reactive strategy is far less effective against online crime because the reaction begins well after the crime has been committed.

Funds are often moved swiftly through different banks and often through different jurisdictions. The investigator is then left with making formal requests through governmental bodies; this coupled with the usual bureaucracy impedes investigations considerably.

Obtaining the physical evidence from banking institutions in foreign jurisdictions is a mammoth task even with the appropriate authorities in place.

This culminates in a lack of ability to identify offender profiles and patterns of offence. It makes it difficult to compare facts and figures to real world crimes and patterns. The reason partly for this is that online crimes tend to get grouped with real world crimes. Police authorities tend to "screen out" crimes that do not have enough points to qualify for allocation and in so doing falsely reflect the crime figures and potential clear up rates.

Police authorities also have trouble with identifying cybercrime and consequently do not understand how to categorise offences into a respective group. The "Love Bug" virus is a typical example of these causing billions of dollars of damage in over 20 countries.

The simple fact of the matter is that we do not have accurate statistics because many cybercrimes go undetected and many detected cybercrimes go unreported [13].

30.4.1. *A New Approach*

The current crime control strategy is blatantly not effective against cybercrime. We have to examine the options open to determine new methodology and techniques for dealing with this problem. Current Law enforcement practices particularly in the United Kingdom have shown that there are severe shortfalls in dealing with those crimes that are not considered to be the most serious. Paedophiles, Terrorism and some Sexual Offences do appear to have been resourced by SOCA and discussion around that department is beyond the intended scope of this paper.

It appears that we have three options:

- Modify, adapt and resource our current practices and reactive methods.
- Develop a new non-reactive strategy for cybercrime.
- Develop a combination of the above.

There appears to be four proposed alternatives to improve on this strategy:

- The Council of Europe's Convention on cybercrime.
- Law enforcement pro-activity.
- Civilian pro-activity techniques.
- More officers.

30.4.2. *Convention on Cybercrime*

The convention on cybercrime is intended to improve the law enforcement's ability to react to cybercrime (Council of Europe, 2001) [5]. It seeks to achieve this by

- Harmonising the domestic criminal substantive law (in the area of cybercrime).
- Providing the domestic criminal procedural law powers necessary for the investigation and prosecution of such offences.
- Setting up a fast and effective regime of international co-operation [3].

The premise of the convention — that an international network of consistent laws will improve national law enforcement's ability to react across borders and thereby restore the effectiveness of the current crime control strategy — is unobjectionable. The implementation is where the difficulty lies and to be truly effective this implantation has to be enacted buy every country. It contains 48 articles, 33 of which require parties to adopt legislation or take other implementing measures (Council of Europe, 2001) [5].

This will be a relatively straightforward process for some countries such as the United States, which already has cybercrime legislation in place. It will not be so simple for those countries that do not and potentially by the time they do have the appropriate legislation may well find themselves "out of date" in terms of the developing technology at the current pace.

The bottom line is therefore that although the reaction to transnational crime will improve Law enforcements ability to react we are unlikely to see any improvement in the near future.

30.4.3. *Law Enforcement Pro-Activity*

It has been suggested that law enforcement uses pro-active methods to hit back at the cybercriminal by allowing officers to react and disseminate viruses, malware, DoS attacks, against those that perpetrate online fraud [14]. The potential problems here however are such acts but law enforcement may themselves be deemed criminal offences and illegal in certain jurisdictions.

30.4.4. *Civilian Pro-Active Techniques*

The suggestion here is that civilians react to cybercrime similarly to police by reacting and supplementing the capabilities of the police. Again this raises serious issues around legality [11]. The practical risks involved in authorising victim self-help are too numerous, particularly when the individuals' computer skills may be limited and as a result retaliate against the wrong computer [11]. The risks are severe as the victim may shut down and innocent computer with essential services such as a doctor's surgery, hospitals, or governmental agency.

The other matters to consider are elements of vigilantism which generally is not an accepted practice amongst civilised societies.

30.4.5. *More Police Officers*

This in itself appears to be an obvious solution. However, with the pace at which cybercrime can be perpetrated there is no evidence that by just raising numbers it will improve the efficacy of their investigations. Additionally, resources are already stretched to capacity and it is unlikely that additional resources will become available to recruit, train and equip enough officers to make a reactive strategy a viable approach.

The Metropolitan Police Service (MPS) approach to dealing with e-crime currently is being reviewed. The implementation of an e-crime strategy by the MPS can be broken down into five distinct areas with a project head for each. An overview of that policy are as follows.

30.4.5.1. *Intelligence*

Strategic and tactical analysis, joint intelligence sharing protocols, Distinct Crime recording codes to quantify e-crime, coordinate and publish revised on (Covert Internet Investigations) and Covert Intelligence Source (CHIS) protocols.

30.4.5.2. *Prevention*

Fraud alert Web site, media campaign, crime prevention officers training, Msc projects, awareness seminars, first responders training, olympic threat assessment, mapping advice sites and collation of best practice.

30.4.5.3. *Enforcement*

Review: refresh ACPO manual, training for test purchase officers re Covert Intelligence Investigation, central database for skills and establish international evidential protocols.

30.4.5.4. *Together*

Practitioners forum to capture best practice, conduct a digital forensic best value review. Identify partnership and sponsorship opportunities.

30.4.5.5. *Communication*

To raise awareness of e-crime through publicity both internally and externally, establish change management plan, conduct capability assessment for all MPS high-tech assets, e-crime conference and presentations in both internal and external environments.

The approach by the MPS is all encompassing and has wide implications over time. However, it does little to assist a pro-active preventive policy against e-crime. It is in essence a non-reactive strategy. Community policing for example emphasises cooperation between police and civilians to create a climate in which crime is not tolerated, but is this a viable option for cybercrime? Do we assign officers to patrol cyberspace? The resources required would be immense and simply not available and never will be under this MPS strategy.

However, a variation on community policing could be applied to cyberspace. Community policing relies on active police presence and secondarily on civilian efforts. How about reversing the roles and create a model that relies on active civilian efforts and secondary policing.

30.5. Conclusion

The worldwide growth of personal and business use of the Internet over the past 16 years has been almost exponential. Restrictions on the use of the Internet or how it is used is limited to how you access it, that is, how easily is the Internet accessible to an individual or business. The ability to be able to gain access 24/7 has presented untold opportunities for potential scammer, hacker, and identity thief or just about anybody in the World that has an unlawful intention to commit cybercrime.

There is no omnipotent regulatory body with adequate power to enforce the required regulations. In fact, the regulations do not really exist, other than those written on the statute books of individual Countries. It seems, therefore, although certain jurisdictions are concentrating using resources in building an infrastructure around formalising what constitutes cybercrime, little or no effort or funding is being put into supporting law enforcement to tackle the proliferation of offences being committed.

The opportunity to generate income streams that were not previously possible and this combined with anonymity has increased the opportunity for crime.

Reported cybercrime data in the United Kingdom have been investigated, and qualitative data are difficult to obtain. There are also issues with the statistical validity and robustness of the results and the issues of “under reporting”. The United States appears to have more quantitative data than the United Kingdom and this will have to be relied upon for an overall analysis.

An examination has been undertaken in an attempt to consider whether the Internet has brought about a new breed of criminal, crime type and also if the two together can have formed a sense of "hopelessness" for their victims through lack of reporting conditions and formality in investigative procedures.

There has been no qualitative or quantitative research done in this subject that has been published that could be found, so this is the first detailed investigation into the subject.

The conclusions reached are that there are potentially "new" crimes on the Internet, but only in as far as their complexity. The end result is primarily the same, i.e. some considerable loss or gain, whether it is financial or something that appears intangible.

Crimes that can go undetected because of the lack of specialist knowledge by investigators and front line portals set up for reporting such events, for example, police stations.

Cybercrime offences are crimes that have an underlying element of dishonesty that have been (or were) in existence before the Internet existed. What has changed is the

- Increased targeting of victims
- Cost reduction in targeting those victims
- Increased global capacity and the ability to cross international boundaries with ease
- Complete lack of legal process to deal with the issues
- Increased anonymity
- Extremely small chance of the crime being reported
- Lack of suitable international legislation to bring the suspect to justice
- Total lack of understanding of the gravity of such offences as interpreted by the Courts of Justice.

In real terms, some of the most serious crimes under UK Law have most certainly transposed online. Potentially, life-threatening situations may go undetected, purely to a lack of reporting policy and inadequate training for police officers in this field. In addition, where a reporting policy does exist it merely amounts to a referral to an outside agency to make a decision as to whether the crime should or should not be reported!

The other apprehension held by police forces around the country is that opening the proper channels to cybervictims will be tantamount to opening the "flood gates" to increased crime figures and therefore underperformance issues in detecting crimes for that force area.

Legislation is a major issue, but the attitude towards cybercrime and its victims needs to be addressed if we are to see an effectual clamp down on cyber-villains and the like. Investment is essential, accurate recording of cybercrime does not exist, international legislation and bureaucracy are major barriers for investigators and yet cybercrime is spiralling out of control.

Not a good outlook for the cyber-victim. The "deck" is most certainly "stacked in favour" of the cyber-villain.

References

1. APIG (All Party Parliamentary Internet Group), June 2004, Revision of the, Computer Misuse Act, Report of an Inquiry by the All Party Internet Group, http://www.apcomms.org.uk/apig/, Accessed on 20-07-08.
2. D. Bainbridge, *Introduction to Computer Law.* 5th edition. (Longman, p. 404, 2004).
3. S. W. Brenner and L. L. Clarke, Distributed security: A new model of law enforcement, *John Marshall Journal of Computer & Information Law*, 2005.
4. V. Chaudhary and G. Wood, Web bookies held to ransom, *The Guardian*, February 24, p. 33.
5. Council of Europe, 2001, Convention on Cybercrime (ETC no 185), Budapest, 23.11.2001, http://conventions.coe.int/Treaty/EN/Treaties/Html/185.htm, Accessed on 18-05-08.
6. T. Espiner, Teenager Cleared of Email Attack Charge, ZDNET, http://news.zdnet.co.uk/security/0,1000000189,39235359,00.htm, 02 November 2005, Accessed on 20-03-08.
7. S. Fafinski, UK Cybercrime Report, https://www.garlik.com/press/Garlik_UK_Cybercrime_Report.pdf, Accessed on 20-07-08.
8. S. Gordon and R. Ford, On the definition and classification of cybercrime, *Journal in Computer Virology* (Springer Paris) **2**(1), August 2006.
9. ICF, *Reform of the Computer Misuse Act 1990, ICF Legal Subgroup, 30th April 2003*, http://www.internetcrimeforum.org.uk/cma-icf.pdf, Accessed on 20-07-08.
10. Infoniac, British Cybercrooks Moving at One-Crime Per 10 Seconds Speed, 2007. http://www.infoniac.com/hi-tech/british-cybercrooks-moving-crime-per-seconds-speed.html, Accessed on 03-05-08.
11. C. Karnow, Strike and Counter Strike; The Law on automated Intrusions and Striking back, Blackhat Windows Security, http://www.blackhat.com/presentations/win-usa-03/bh-win-03-karnow-notes.pdf, Accessed on 12-06-08.
12. J. Leyden, East European gangs in online protection racket, *The Register*, November 12, 2003, http://www.theregister.co.uk/2005/10/24/cybercrime_forum/, Accessed on 20-07-08.
13. S. McGuire, Reported by Marilyn Elias, USA TODAY, Most Teen Computer Hackers more Curious than Criminal, 2007. http://www.usatoday.com/news/health/2007-08-19-teen-hackers_N.htm, Accessed on 3-05-08.
14. J. Reidenberg, States and internet enforcement, *University of Ottawa Law & Technology Journal* **1**(213), 2004.
15. World Online Gambling Law Report, 2004, Global Protection Racket Smashed in Joint Operation between UK's National Hi-Tech Crime Unit and Russian Police, July 21, 2004. www.e-comlaw.com/woglr/index.asp, Accessed on 20-07-08.

Chapter 31

INTELLIGENT DECISION SUPPORT SYSTEMS FOR ASSISTANCE IN FORENSIC INVESTIGATION PROCESSES

DALE DZEMYDIENE
Mykolas Romeris University
Ateities str. 20, LT-08303 Vilnius, Lithuania

The purpose of this chapter is to look into the trends of development of intellectual information systems and technologies as well as future possibilities of their implementation in the legal application domain. The question of how to integrate the different pieces of data available into a coherent framework for intelligence purposes has become an important topic. Increased mobility and new communication channels give criminals the capacity to better plan and organise their activities over large geographical areas. Patterns reflecting fraudulent activities are therefore, always more difficult to reveal from the huge quantity of data scattered in files across independent police and legal structures. The main tasks for development of advice, consultative information systems and the methods of implementation are being considered.

31.1. Introduction to Artificial Intelligence Methods in Crime Analysis Domain

The artificial intelligence methods are very important in the development of judicial and legal consultative expert systems [23, 20]. A decision support system (DSS) is a computerised information and knowledge-based system for helping make decisions [5]. A decision is a choice between alternatives based on estimates of the values of those alternatives. Supporting the choice making process involves supporting the estimation, the evaluation and/or the comparison of alternatives, using methods of knowledge representation, recognition of situations, reasoning and modelling.

A key part of law enforcement is to understand those activities, through the development and use of methods, models and tools for collecting and then interpreting the large volume of data available in real-time. Consequently, intelligence programmes have been developed, leading to recognition of a field of activity called crime analysis, which has been described as "the identification of and the provision of insight into the relationship between crime data and other potentially relevant data with a view to police and judicial practice" [38].

Knowledge base can assist, but not replace, the decision-making process by providing access to more timely and accurate information on individual cases, for example, in scheduling cases and assigning them to investigators and in preparing decisions on bail and sentencing. Computerisation can also provide investigators with rapid access to reference material, such statutes and case law. We have good examples of applications of DSS:

- advisory systems in law enforcement [39] and
- software packages that use artificial intelligence techniques to draw conclusions from facts presented by the user [29, 14].

Such systems offer additional advantages that facilitate the decision-making process in individual cases.

Although advisory systems do not and should not provide definitive solutions, they may be helpful in command and control functions in law enforcement, crime-solving (the creation of suspect profiles), computer-aided instruction and programme planning and design.

The use of advisory systems may also improve legal training, for example, providing police with advice on the type of information required by prosecutors to reduce the rate of case attrition.

Legal and organisational barriers, the proliferation of methods for collecting and interpreting data, and the consequent variety of incompatible recording systems make comparisons difficult, when performed. Poor analysis of similar offences crossing different areas is recognised as one of the major weaknesses pertaining to information systems; this has been called "linkage blindness".

The problem is addressed to developing of multi-component knowledge management system for crime investigation domain which will integrate many types of decision support methods: ontology-based system, dynamic modelling for crime scenario analysis, data mining for information retrieval and dissemination [9, 10].

The methods of artificial intelligence helpful in the management of repository with impartial crime information and situation recognition are discussed. We would like to introduce you into the area of different data analysis strategies for crime investigation which can support computer-based system. The integration knowledge components are used for retrieval of relevant information from different database (DB) systems.

The aiding advisory processes are helpful in relevant patterns recognition and crime investigation. The architecture of decision support repository includes a model of the problem solution strategies that allows choosing an appropriate multi-dimensional statistical method, a case-based reasoning (CBR) mechanism. This chapter outlines the different components of decision support and illustrate the integration possibilities with data mining technologies of crime profile data warehouses (DWHs).

31.2. Related Works of Developing Intelligent DSSs in Crime Investigation Domain

The criminality depends on social, political and economic factors and it is one of the most significant indicators, reflecting a social and moral situation of society. Some questions, such as "what influences criminality?" become problematic to answer directly from the data space. To this end the integrated methods of data mining, data structuring and factors evaluation are required.

Our universe of discourse deals with knowledge representation for decision support processes of developing the assistance environment for evaluation of social-economic situation of the different regions. The question of how to integrate the different pieces of data available into a coherent framework for intelligence purposes has become an important topic.

A large volume of data related to crimes is being collected [35]. An intelligent workflow systems for impression evidence that proactively fuses, and learns to fuse, descriptions and images, will require an active cooperation between the end-users (e.g. the police forces), the researchers in information extraction, in text and image mining, in grid technologies and software vendors. The force linking systems, such as LOCARD (Locard Evidence Tracking System, 2005), NPLC [6], IMPRESS [21], combine forensic and physical evidential "hits" to display links between criminals and the evidence, and produces a profile of offenders and of crimes committed. Such systems and imaging workflow systems can all be used as systems for building the profile of a habitual criminal [25]. There are evidence tracking systems that deal with the movement of crime-related exhibits from the crime scene through to the court; this tracking is again performed through a class description, much like that used by freight handling organisations [29, 23, 20].

The component-based knowledge management system is proposed by developing two different parts for recognition of criminality: one part focus on many aspects of how situation of criminality can be recognised and handled; another part — for recognition of crime aspects which are useful in crime investigation processes. We look for the rate of change and revise the patterns of influence space of the state using data mining, and including actions of analysis, risk and influencing factors assessment, profitability, etc. Distinct tasks require different data structures and various data mining exercises [28, 31].

A numerous quantity of multi-dimensional data in DWHs revealed some problems in the organisation of data as well as in supporting data mining techniques for making precise decisions. Nowadays, data mining methods make progress from a simple discovery of frequent patterns and regularities toward knowledge-based and interactive decision support processes in subject-oriented and integrated

DWHs [19, 30]. But relevant patterns are not so easy to be extracted from large and impartial DWHs.

Recent research work has proven the benefits of using repositories based on ontology of the domain and real data of DWHs integrated with data mining techniques, based on statistical multi-dimensional methods [19]. The development of ontological system in this particular field requires the application of additional intelligent methods, their task being the description of intricate decision situations, informal ways of arriving at a decision, etc. The DSS under construction ensures ease of use and clarity of interpretation in the presentation of analysis results from the DWH of crime and social investigation.

This chapter describes an approach of using structured and semi-structured analysis sessions of forensic analysis. A structured analysis session commonly has three forms: a discovery, prediction and forensic analysis activity where we perform specific tasks of crime situation analysis. The knowledge representation based on the ontological view of crime domain using unified modelling language (UML) [4] is useful in such an analysis session for the recognition of the structure of crime information. The time series analysis methods are introduced in a semi-structured analysis session of data mining and allow us to exercise statistical control, forecast the main crime tendencies and support a decision for crime prevention means [12].

Computer systems in forensic intelligence can be classified into groups. Investigative methods are essentially based on analogy. The second class of system pertains to collections that help classify evidence by the type of object that could have transferred a trace.

31.3. Ontological Layer of Crime Analysis Information System

An ontological reasoning service is used to represent a sophisticated conceptual model of document terms, concepts and their relationships; a Web-based open hypermedia link service that can offer a range of different link-providing facilities in a scalable and non-intrusive fashion and integrated to form a conceptual hypermedia system to enable documents to be linked via metadata describing their contents and to improve the consistency and breadth of linking of hypertext documents at retrieval time [1, 26].

Ontology refers to engineering artefacts, constituted by a specific vocabulary used to describe a certain reality, plus a set of explicit assumptions regarding the intended meaning of the worlds [1, 15]. The ontology is a formal model of the kinds of concepts and objects that appear in the real world, together with the relationships between them. Ontology takes a variety of forms, from hierarchical classification schemes to logic-based models. The methodologies proposed for ontology development are: M. Uschold's, M. Grüninger's and M. S. Fox's (TOVE); On-To-Knowledge (OTK) methodology, etc. [33, 27].

We must consider the language L with vocabulary V, and $I{:}V\rightarrow D\cup R$ as a function assigning elements of D to constant symbols of V, and elements of R to predicate symbols of V. The role of ontology can be considered as a set of logical axioms designed to account for the intended meaning of vocabulary [15]. In general, it is not easy to find the right set of axioms.

The creation and implementation of ontology enable us to systematise the collected facts, to evaluate structures of knowledge. Creation of ontology is an interactive process.

Ontology is an emerging instrument for knowledge representation, share, reuse and interoperability. Due to varying understanding of the ontology concept, it is often used by different meanings.

Some of the definitions, used in computer science field, are presented in [16]:

(i) ontology is "a representation of a conceptual system that is characterised by specific logical properties"
(ii) ontology is "a synonym of conceptualisation"
(iii) ontology is "a special kind of knowledge bases"

The first definition accentuates the collection of statements or other semantic definitions for a domain; the second emphasises that we deal with an abstract, simplified view of the part of the world; while the third — that ontology is engineering artefact.

We assume the following definition of ontology: "Ontology is a conceptual specification that describes knowledge about a domain in a manner that is independent of epistemic states and state of affairs" [17]. This definition emphasises that ontologies are universal models of domains or models of known knowledge in a domain.

Ontology can be defined formally, which allows for knowledge sharing and automatic reasoning. Ehrig [13] formally defines ontology as a structure:

$O = (C, T, \leq_{\mathrm{C}}, \leq_{\mathrm{T}}, R, A, \sigma_{\mathrm{R}}, \sigma_{\mathrm{A}}, \leq_{\mathrm{R}}, \leq_{\mathrm{A}}, I, V, \iota_{\mathrm{C}}, \iota_{\mathrm{T}}, \iota_{\mathrm{R}}, \iota_{\mathrm{A}})$, where: C, T, R, A, I and V are disjoint sets for concepts, data types, relations, attributes, instances and data values; partial orders $\leq_{\mathrm{C}}$ on C are *concept hierarchy*, which defines taxonomic structure, $\leq_{\mathrm{T}}$ defines *type hierarchy*, $\leq_{\mathrm{R}}$ defines *relation hierarchy*, $\leq_{\mathrm{A}}$ defines *attribute hierarchy*; functions σ_{R}: $R \to C^2$ and σ_{A}: $R \to C^2$ are called *relation signature* and *attribute signature*; a function ι_{C} is called *concept instantiation*, and functions: ι_{T}, ι_{R}, ι_{A} are data type, relation and attribute instantiations, respectively.

Therefore, we adopt a formal definition of ontology from [16]. Ontology is a 4-tuple $< C, R, I, A >$, where C is a set of classes (concepts), R is a set of relations, I is a set of instances and A is a set of axioms.

Classes (other synonymous terms: concepts, categories and types) represent important concepts of the domain. Classes in the ontology are usually organised in taxonomies, where generalisation-specification mechanisms are applied.

Relations (properties, slots, attributes and roles) represent associations between the concepts of a domain. Most often the "is-a" (hierarchy), "consist-of" (aggregation) and association relationships are used. However, the taxonomical structure is not the only one possible.

Ontology usually contains binary relations, where the first argument is called the domain and the second is called the range. The attributes are sometimes distinguished from relations. The difference between attributes and relations is that the range of an attribute is a data type, not a class.

Instances (individuals) represent individuals in ontology. Instances can be defined in ontology or in DB of factual data.

Formal axioms are used for expressing propositions that are always true, for example, in the e-learning course the same person can not be in the lecturer and in the student role at the same time. Formal axioms are used to infer new knowledge (formal axioms are not analysed in this chapter).

The crime investigation ontology is developing by using the UML according to object-oriented design methodology. We try to illustrate by examples how classes and relations between classes are revealed for description crime ontology. Such ontology is developed for repository which manage different components of the information system.

The ontology of crime investigation should describe the circumstances of crime event, the mechanism of crime and particular regularities of *modus operandi*. The facts about criminals, about traces and evidences, about *modus operandi* and versions are the base for creating the crime analysis information system. The example of ontological description of concepts is represented by object class diagram in Fig. 31.1 and illustrates the character of forensic evidence of crime.

Ontology must satisfy two main requirements:

- it must be formalised in order a computer could process it and
- it must be shared in order a community of experts of some field could agree with it and use it in applications.

Metadata describes other data to enhance its usefulness. The library catalogue or DB schema is canonical examples.

Based on the crime classification in the penal code, it would be: crimes against the humanity (genocide, etc.), crimes against person's life (murder, etc.), crimes against property (thefts, material injury, etc.), economic crimes (burglary, etc.), — in total all about 200 different types of crime (Fig. 31.2).

In addition, the model itself is visually more lucid as opposed to multiple enumerations of instances. The model may be used to automatically generate the instances of various course of action by the application of thorough qualitative imitational modelling.

In its own turn, the creation of information on different types of crime should be based on criminality crime character and its elements (Fig. 31.3).

Crime analysis information should be dynamic, complex, operative, true, trustful, available, safe, ergonomic and scientific.

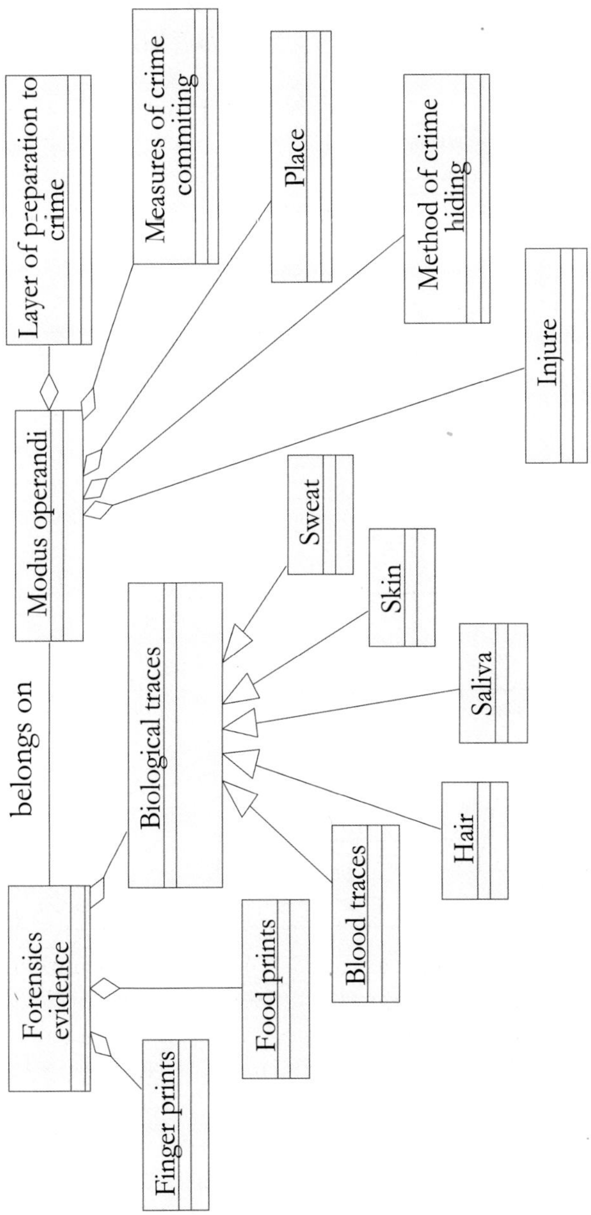

Fig. 31.1. An example of representation of forensic evidence of crime.

Fig. 31.2. The example of crime classification.

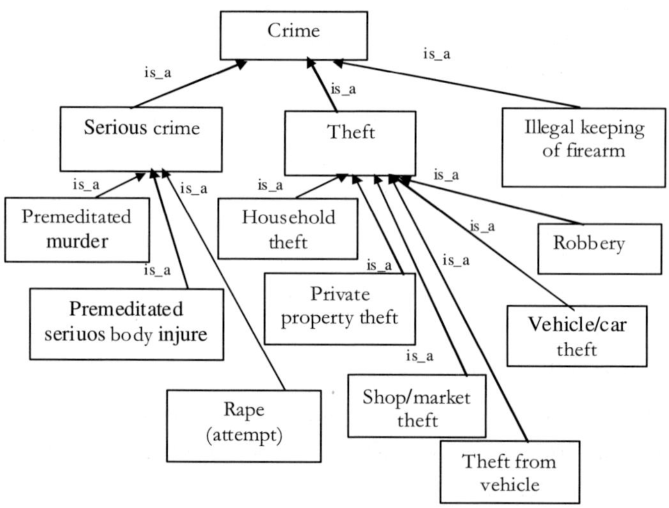

Fig. 31.3. An example of representing crime hierarchy in the ontological layer.

The sources of investigation information could be material and immaterial objects that are connected with the circumstances of the crime event and contain the information on these circumstances.

In addition, the model itself is visually more lucid as opposed to multiple enumerations of instances. The model may be used to automatically generate the instances of various course of action by the application of thorough qualitative imitational modelling.

31.4. Description of Forensic Intelligence Processes for Computer-aided Investigation

Traditionally, each scene of crime is seen as an independent "complete" set of data from which the intelligent investigator can find the signature of the offender. Links between cases are often inferred through police investigation; in a separate process, cases are also compared through the use of physical evidence collected at the scene of crime [18].

Roughly, crime analysis aims at:

- Deciphering the knowledge used by experienced investigators to identify and formalise concepts and notions, and the methods used by them to manage their information.
- Conceiving new methods of treating data, and adapting existing ones, given that computer tools can improve the analysis process in a way that was not previously possible.
- That criminality itself is evolving over time, and that weaknesses observed in the course of an analysis can necessitate upgrades.
- Normalising language and symbols used in the course of the analysis to facilitate teamwork on complex problems, and to disseminate the results of the analysis in a way that is easy to interpret.

Using those methods the advisory system under development must produce relevant conclusions or hypotheses that can help towards concrete actions.

A broad variety of analysis forms could be defined. For instance, crime pattern analysis, profile analysis, case analysis (course of events immediately before, during and after a serious offence), comparative case analysis, etc.

The forensic intelligence process starts with the collection of data and ends with the integration of results into the analysis of crimes under investigation (Fig. 31.4).

In this process, important intermediate steps may be described as follows:

- the acquisition of new data to transform it into a digital or symbolic form suitable for searching for similar items in the DB, and to memorise it in a way that will allow retrieval when needed;
- the search and matching process and
- the interpretation/verification of the returned result.

Introducing model of reasoning from past cases subsumes knowledge-intensive CBR. Such a model can be viewed as containing three phases:

(1) *Activation phase*: retrieve a set of cases that matches the current problem according to some criteria.

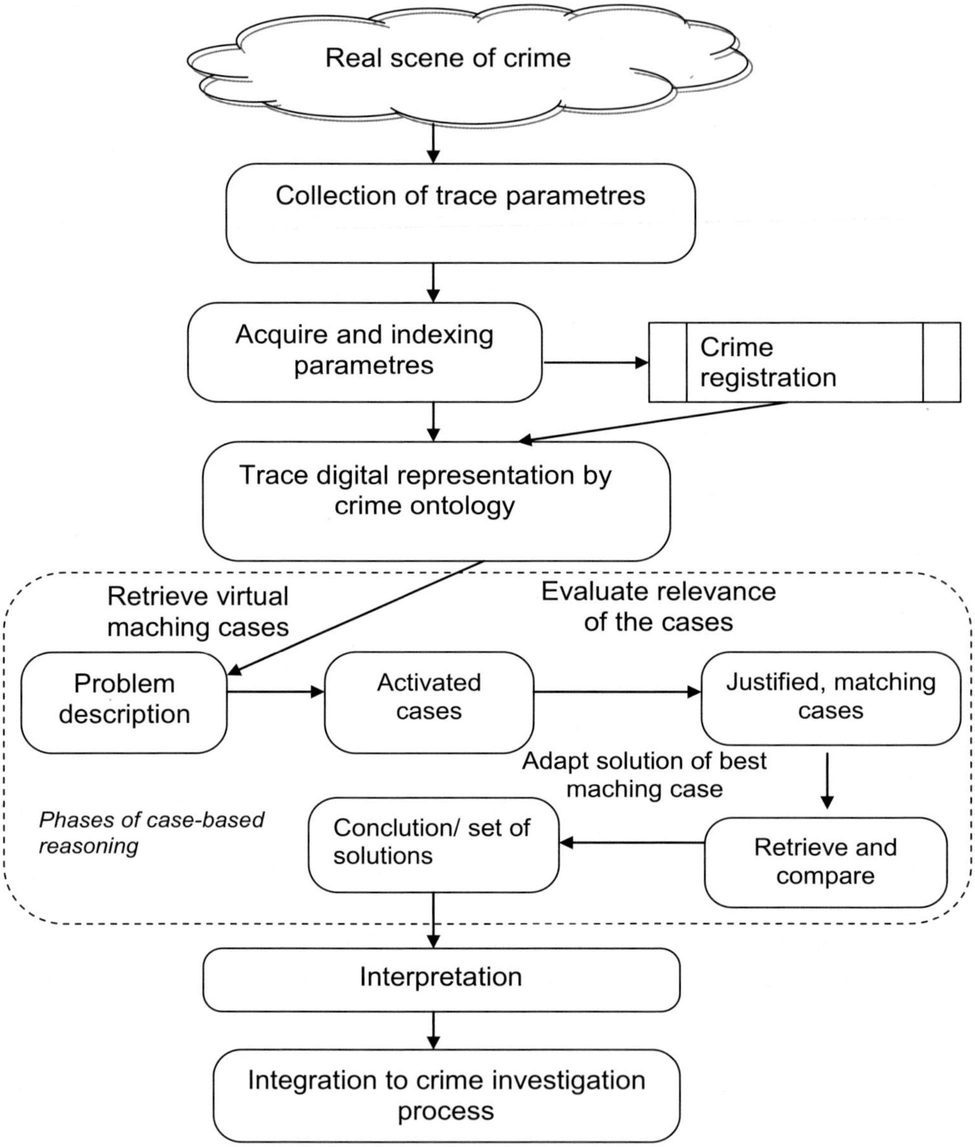

Fig. 31.4. Stages of forensic intelligence process from the collection of data to the dissemination of results.

(2) *Explanation phase*: evaluate the applicability of retrieved cases to the current problem. This involves looking for contradictions, checking constraints, generating and testing expectations set up by the retrieved case.
(3) *Adaptation phase*: adapt the solution of the case to the new problem. A modification process would typically involve a generalisation of the past solution followed by a specialisation satisfying constraints on the current problem.

Similarities found between accessible information can lead to inferences on the profile of offenders, their mode of action and the means they have used. The identification of recidivists is the well-known form of this basic scheme.

Knowledge base can assist, but not replace, the decision-making process by providing access to more timely and accurate information on individual cases, for example, in scheduling cases and assigning them to investigators and in preparing decisions on bail and sentencing.

Computerisation can also provide investigators with rapid access to reference material, such statutes and case law.

From a new situation or a new case, the purpose is to identify known offenders from their antecedents. The search for links between cases is another elementary activity that falls within the same framework; it allows groupings that delineate criminal events. Finally, when objects are found or perpetrators identified all the cases in which there participation or presence can be supposed should be extracted. This process is traditionally incomplete because links may not systematically be searched before the offender has been arrested.

Computer systems in forensic intelligence can be classified into two groups. Investigative methods are essentially based on analogy. Consequently, it is not surprising that most existing computer systems recognised as forensic intelligence systems can be viewed as an aid to performing this type of reasoning process. The second class of system pertains to collections that help classify evidence by the type of object that could have transferred a trace.

The crime analysis information system should secure the component integrity and could operate as an integral system.

The subject area considered by us could be limited by up to the consideration of crime investigation as an ontological point of view based on building of the exterior world model.

The important knowledge in crime investigation is knowledge on crime investigation methods, criminal procedure, other knowledge on special scientific methods and tactics for investigation. Investigators and judges take part in the knowledge process, the essence of which is a reconstruction of the crime event by evidences and traces left in surroundings. These traces are fixed and recognised that for unknown reasons become evidences.

The creation, removal and transformation of evidence information take place step-by-step and reflect multi-stage and structured causative connections that describe the role of actors in different criminal activities (Fig. 31.5).

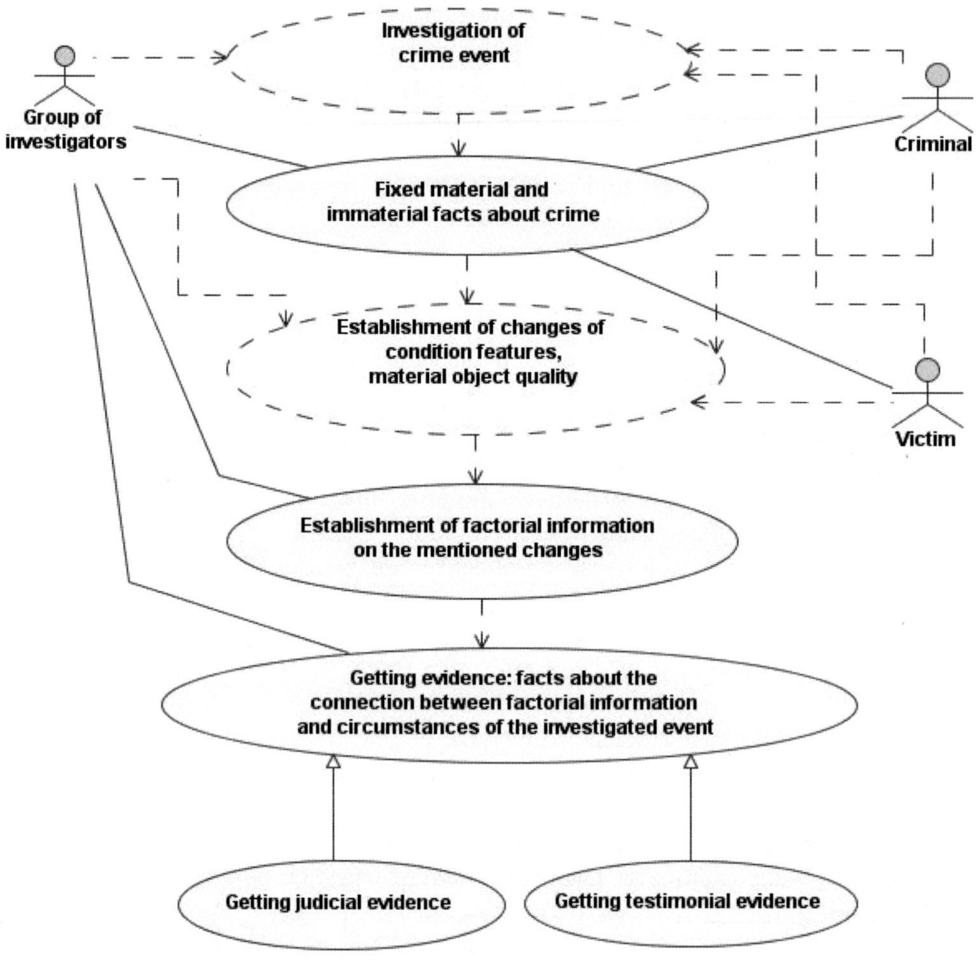

Fig. 31.5. Use case diagram of investigation stages of crime evidence information.

Crime investigation information should describe the circumstances of crime event, the mechanism of crime and particular regularities of *modus operandi*.

The facts about criminals, about traces and evidences, about *modus operandi* and versions are the base for creating the crime analysis information system.

It could be that identifying an offender for a crime is sufficient to put him or her behind bars, without being charged with other offences that could be difficult to prove. This is something that real-time crime analysis avoids. These steps have been partially reproduced in computer applications; the well-known illustrations are automatic fingerprint identification systems and DNA DBs. For instance, from a DNA point of view, it includes the application of the analytical techniques used to identify markers of a sample, together with encoding into the system the obtained values in the form of numbers.

Most information is gathered through a device that is interfaced with a computer, like a scanner or various types of camera. The image obtained is then treated to extract relevant features for encoding. This operation is computerised to different degrees, depending on the trace and its quality. Potentially, all the other marks transferred, or the information left by the offender (from clothing or accessories), could be analysed in a similar way.

Major data flows are considered by separating a criminal event as a process, models of data flows as well as of the processes transforming them are constructed (Fig. 31.6).

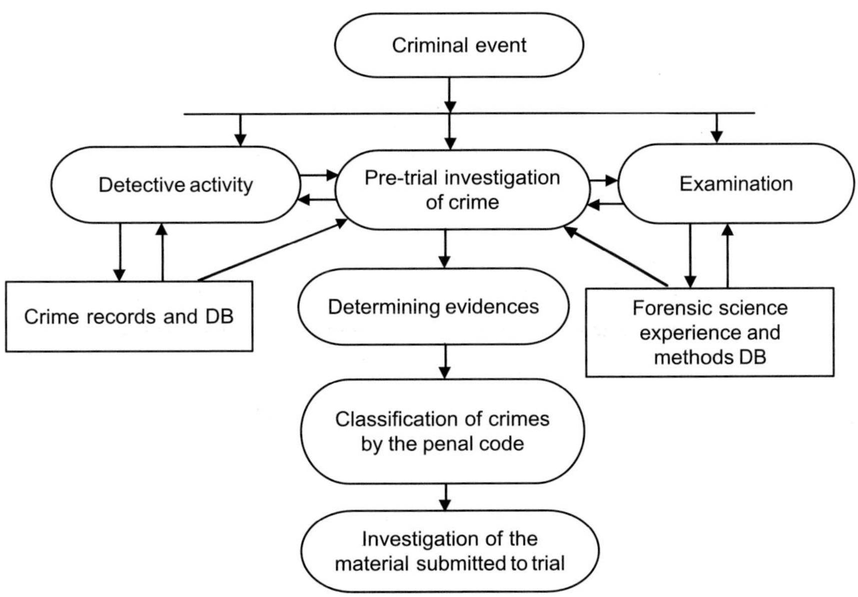

Fig. 31.6. The data flow diagram of the main processes of pre-trial crime investigation.

The investigation of criminal event could be detailed as follows:

- establishment of the sources of the investigated case
- retrieval extraction of information on the investigated event from sources
- implementation of particular circumstances of the investigated event
- creation of a complete information system and determination of actual structure of the investigated event

Thus, all the activities connected with the investigation of crime are aimed at the final result — to define the structure of a criminal event.

Pre-trial crime investigation is based on crime science knowledge about the crime investigation, planning, investigation situations, versions, etc. Here, information about archives of penal crimes as well as statistical information on criminology DBs is of great importance.

The ontology of crime analysis information — first is the reflection of a crime and its investigation; information which describes the circumstances of a crime, its mechanism and particular regularities of crime committing, information on criminals, traces as well as modus operandi, versions and knowledge about the methods of crime investigation, penal procedure and other methodological and tactical recommendations of crime and forensic science.

The observation of the scene of crime is performed without touching anything. Whole surrounding is inspected by fixing all the possible traces from the outlying areas to the centre (inspection of fingerprints, cigarette butts, blood, etc.) — this is primary observing.

A corpse in the centre and its examination (a doctor examines the body, diagnoses the course of death by the signs of violence, he proves it to be not a suicide) and the time of murder by the death marks, the site (the corpse could be transferred); the investigator examines the body and clothes to identify the person, sex, fingerprints, blood, saliva, sperm or other traces on the body or clothes. After all-round observing, everything is fixed by taking pictures and drawing up the report.

If deep knowledge is revealed, the causal structure of the objects in the domain is examined. Qualitatively expressed knowledge is the predominant type of deep knowledge.

There are certain advantages of qualitative deep knowledge representation:

- as a rule, the qualitative perspective is narrower than general considerations about physical or psychological processes, which are to be modelled
- the values of numerous parameters are not necessary for the completion of a model
- computer-based implementation of qualitative modelling is less intricate than multiple modelling of instances
- qualitative modelling can be used to explain various mechanisms of the system

The relations between processes are described by sequence and collaboration diagrams (Fig. 31.7).

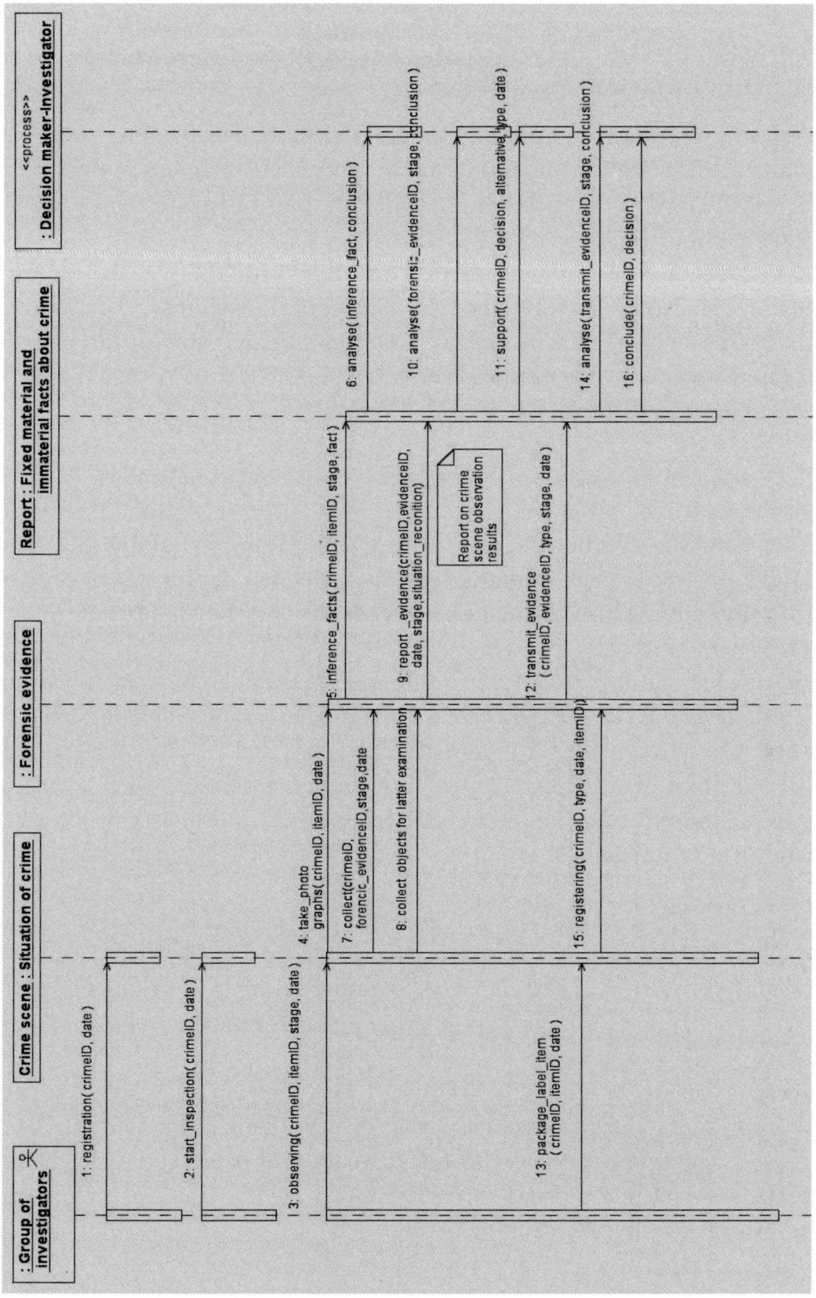

Fig. 31.7. An example of description of crime investigation scenarios by means of sequence diagram.

According to the primary information, some versions are raised and further plan of investigation is drawn up. This is the first important decision for further actions.

31.5. Integration of Ontology into the Scenario Generation Algorithm of Crime Investigation

The integration of ontology, meta-modelling and crime pattern recognition is very important in the DSS. We use the model-based and CBR techniques, derived from the existing technology of compositional modelling and integrate reasoning about evidences based on ontology.

Automatic selection and management of the rules, suitable for the assessment of a particular situation, is performed in the expert systems. Core and shell mechanisms of the expert system ensure the presentation, sorting and selection of the rules [3]. As a result, the user is given a simple interface to enter the primary rules in a suitable natural format, e.g.:

$$IF < \text{situation}_i^V \text{conditions}_\text{k} > THEN < \text{situation}_j^V \text{conclusions}_\text{n} >$$

Image enhancement by computer has emerged as a forensic examination method in its own right. In latent print examination, quantitative digital image processing relative to automated fingerprint identification systems has been in practice for over 20 years [14, 25].

Histogram equalisation: radiometric enhancement wherein concentrations of radiometric intensities in the histogram of an image are reassigned values, effectively, spreading out those intensities to increase contrast and clarity. The values of the required variables are scanned in the information module, which is the main component of the monitoring (i.e. real-time recording of data) system.

For instance, the image of fingerprints stored in the DB is defined as:

$$F_{i,j}^k = (\{x_{i,j}\}, \{y_{i,j}\}, \Delta t_{k,m}, \{d_{i,j}, \ldots, d_{i,n}\}),$$

where i is the identification of person, j is the finger number in the position of two hands $j = \{1, \ldots, 10\}$, k is the corresponds to the time moment then the image is fixed.

Histogram: typically a graph of frequency distribution in which the x axis distribution represents the radiometric levels of gray from dark (left, 0) to light (right, 255, etc.); the number of pixels of each gray level is indicated by the y axis (height) of the vertical line at that intensity. $\Delta t_{k,m}$ is the time interval then the fingerprint is valid for identification of person. The set of detected convolutions is defined as $\{d_{i,j}, \ldots, d_{i,n}\}$.

The image of fingerprint found in the real scene of crime may be defined as:

$$F_{i,j}^{m*}(\{x_{i,j}^*\}, \{y_{i,j}^*\}, t_m, \{d_{i,j}^*, \ldots, d_{i,n}^*\}).$$

The rule for revealing the correspondence of such fingerprints may be constructed as follows:

IF $<\textit{selected method}_w =: \textit{Histogram equalization}>$;
$AND <\{x_{i,j}\} = \{x^*_{i,j}\}>$;
$AND <\{y_{i,j}\} = \{y^*_{i,j}\}>$;
$AND <\Delta t_{k,m},> = t_m>$;
$AND <\{d_{i,j},\ldots,d_{i,n}\} = \{d^*_{i,j},\ldots,d^*_{i,n}\}>$
THEN $< F^k_{i,j} = F^{m*}_{i,j} >$

Thus, if a fingermark, even fragmentary, found at the scene of a crime is compared with the content of a DB; a restricted number of similar fingerprints will be given to the experts who interpret them.

Approximately, we can reveal the conclusion:

$$<\text{fingerprint } F^{m*}_{i,j} \text{ corresponds to finger print } F^k_{i,j}>.$$

The histogram equalisation is radiometric enhancement wherein concentrations of radiometric intensities in the histogram of an image are reassigned values.

Moreover, technological means provide the opportunity to solve problems involving strategic data recording and integrating it with the automatic situation detection and management system. The system offers several ways of making decisions, based on the situation evaluation. The knowledge acquisition methods have the properties of representing deep knowledge [32, 39].

If deep knowledge is revealed, the causal structure of the objects in the domain is examined. Qualitatively, expressed knowledge is the predominant type of deep knowledge. There are certain advantages of qualitative deep knowledge representation:

- as a rule, the qualitative perspective is narrower than general considerations about physical or psychological processes, which are to be modelled
- the values of numerous parameters are not necessary for the completion of a model
- computer-based implementation of qualitative modelling is less intricate than multiple modelling of instances
- qualitative modelling can be used to explain various mechanisms of the system.

In addition, the model itself is visually more lucid as opposed to multiple enumerations of instances. The model may be used to automatically generate the instances of various course of action by the application of thorough qualitative imitational modelling.

Long-term use of the rule-based knowledge representation method in the expert systems has unveiled some of their disadvantages and certain limitations of expert practice expression. The exploitation of the rule-based system is aggravated when the knowledge base is large. The expression of the rules, generalising the use of rules, that is, meta-rules, is quite complicated. There are certain limitations of displaying the deep knowledge. It is difficult to supplement the knowledge base with experience and exceptional cases [7]. Most systems have relatively simple explanation capabilities, by displaying rule sequences based on their use to draw a particular conclusion. It is rather difficult to encode knowledge by means of strictly defined rules in some applied fields.

The aforementioned reasons create the necessity to seek for the methods of display that would allow avoid these disadvantages: case-based learning strategy is integrated with the experts' experience in the MODELER system [37]. Some systems employ case-based learning for the design of knowledge base optimising system [2]. The opportunities of conceptual clusters of machine learning are presented, by use of case-based method for testing and assessing a real-time complex system knowledge base [24, 30].

The CBR is a computer-based method, which analyses the solution of formal solved problems based on analogy or associations with the solution of a current problem. The CBR has several advantages over productive systems: CBR is closer in nature to the factual processes of human-made solutions; the expert, having been presented with a problem, initially compares this problem with formerly solved problems, determines its separate similar parts. Should the case-based system fail to arrive at a desired conclusion, new rules are to be made and added to the knowledge base.

Scenarios are modelled as the causes of evidences and they are inferred based on the evidences they may have produced (Fig. 31.6).

The goal of such part of the DSS — to find a set of possible alternatives of crime causes following from scenarios that support the entire set of available evidence.

This set of possible alternatives of crime *modus operandi* can be defined as:

$$A_C = \{a \in A | \exists s \in S, (\forall c \in C, (S \Rightarrow c)) \wedge (S \Rightarrow a)\},$$

where A_C is the set of all possible alternatives of crime *modus operandi* (e.g. suicide, accident or murder);

S is the set of all consistent crime scenarios,

C is the set of all collected pieces of crime evidence.

The available evidence supports every alternative of crime *modus operandi* that follows from a plausible scenario. For example, the rules are constructed by the ontology presented in the previous examples, and the scenario space is revealed:

$$\begin{aligned} C_1 = \{ & \textit{violence\_injuries}(p), \\ & \textit{ballistic-report\_of\_shoot-mechanism}(p), \\ & \textit{bullets\_cartridges}(p), \\ & \textit{blood\_traces\_analyses}(p) \\ \Rightarrow & \textit{diagnosis}(\textit{murder}(p))\}. \end{aligned}$$

$$C_2 = \{medical\_report\_of\_occasional\_death(p), \\ lack\_of\_collagen(p), \\ accidental\_blood\_vessel\_rupture(p) \\ \Rightarrow diagnosis(accidental\_death(p))\}.$$

In a possible world described by crime event environment, if the situation detection data (conditions) are realised, crime evidences C_1 corresponds to the facts, the alternative decision — conclusion *murder*(p) become true. The conclusion *accidental_death*(p) become true in another situation detection conditions.

The rules are specified for main types of crimes under investigation and maintained in knowledge base of the DSS by aforementioned style.

31.6. Architectural View of the Integration Components of DSS for Crime Investigation Processes

The refinement of knowledge, trends and patterns within the data is the essential condition for a qualified crime detection, social control and crime prevention strategy. Crime is a complex phenomenon of social, political and economical impacts. A crime rate is defined as the number of crimes per unit of population. A crime is considered registered if it is included into the centralised register. The regional offices of crime statistics and research collect a great volume of data from different sources that reflect criminal events and their relations. The data are stored in warehouses with multiplex variables and temporal dimensions.

The principal mission of the criminal information system has been to assist all officials and agencies of the criminal justice system in the fulfillment of their varied responsibilities on a state-wide basis by providing round-the-clock access to needed information.

The unified national criminal information system could be created based on crime characteristics (*corpus delicti*), which include the object of attempt (victim), the crime subject (criminal), the crime situation and way of crime commitment [11]. The criminal information system has to provide the information about similar crimes, which were committed before, have to connect the new crime with already committed. The criminal characteristics consist of information about the object of attempt (victims); information about the crime subject; information about the way of crime commitment; information about the crime situation. The types of situation can be provocative, conflict and accidental, etc. [22].

The advisory system under development must produce relevant conclusions or hypotheses that can help towards concrete actions.

A broad variety of analysis forms could be defined. For instance, crime pattern analysis, profile analysis, case analysis (course of events immediately before, during and after a serious offence), comparative case analysis, etc.

The architecture, shown in Fig. 31.8 is used in DSS with the CBR mechanism for retrieving a new problem solving class and server-based discovery to automatically

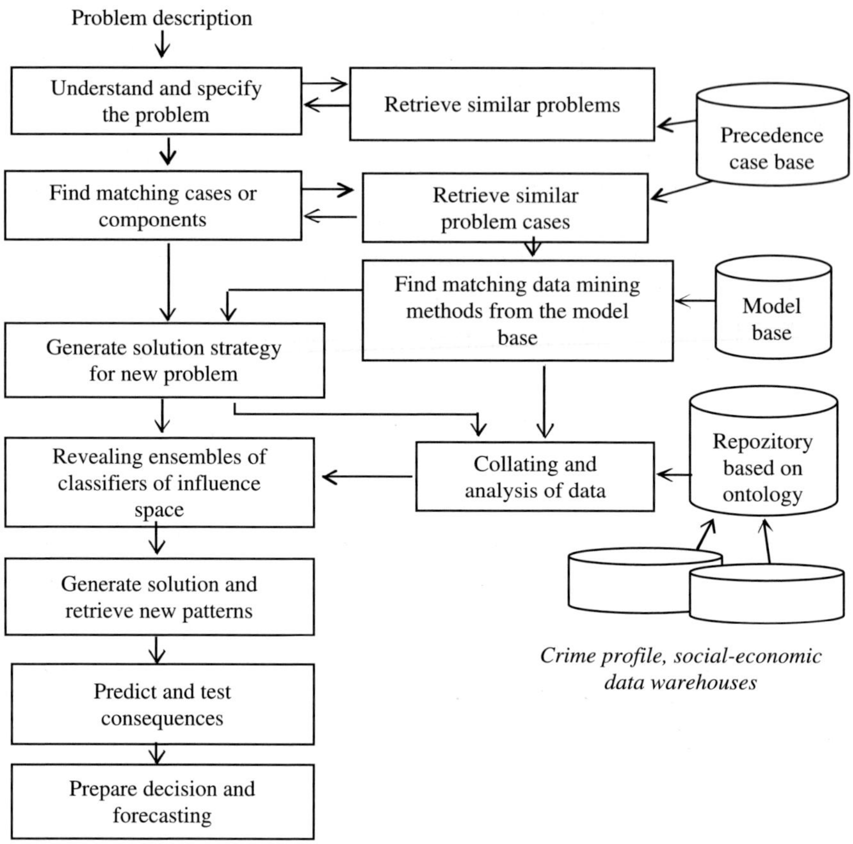

Fig. 31.8. The components of CBR and discovery activities integrated in the architecture of DSS.

find unexpected patterns within the influence space. The discovery modules help in decision as to which patterns should be looked for and submit queries and statistical tests to the search engine.

We deal with two components of recognition: recognition system and predictive modelling system. These two forms of activities are distinct. We look for the rate of change and revise the patterns of influence space of the state in the discovery system. The predictive modelling systems may have the ability to generate hypotheses and test consequences.

The process of revealing ensembles of classifiers of influence space consists of interactions of the retrieval process of factors that have the greatest impact on the problem under consideration. The selection mechanism of the appropriate statistical method is related with the testing of the results and consequences. The CBR approach can help us in such an analysis.

Summarisation of the same data set by two sampling or summarisation methods may result in the same result, and summarisation of the same data set by two methods may produce two different results [28]. Another way could be using a pattern recognition process in a DWH [31]. In this context, research patterns are characterised as generic structures won by experience of modelling in past, they can be abstract, describing the structure of a model or concrete, describing one particular model.

The considered problem is compared with similar problems solved. This is the process of integration a new problem-solving mode into the existing knowledge base. It involves selection of which information from the case to retain, in what form to do that and how to organise the case for the later retrieval from similar problems.

CBR is dependent on the structure and content of its case memory organisation. The case search and matching processes have to be effective.

The ontological view of domain and metadata organisation of aggregation space of DWH can help solving important issues such as: how the cases should be described and represented, how they should be interrelated and how they should relate to the entities such as input features, solutions and problem-solving states.

31.7. Factor Analysis for Recognition of Situation of Criminality

We may analyse the rate of change during some historical period in the conventional way, conceive the reasons of the trends and perform forecasting for various items of influence space. The knowledge representation is based on the ontological view of main social — economic indicators [8, 12].

The various sets of social-economic indicators usable for international and in-country comparisons and measuring economic progress can be get into several meaningful groups:

- Common indicators that characterise specific regions — the total area, population, density, etc.
- Social indicators — structure of population, natural increase, migration, numbers of unemployment, criminality, etc.
- Economic indicators — municipal budget revenue and expenditure, sales of industrial production, foreign investments, etc.

Observation objects of interest (regions, districts, towns, etc.) are selected, i.e., a sample: $D = (d_1, d_2, \ldots, d_N)$. The object of a data set is a unit of data whose features are to be investigated. The objects have respective features — indicators that describe their attributes $X = (x_1, x_2, \ldots, x_n)$. These features are measured within particular time intervals $\Delta t = (\Delta t_1, \Delta t_2, \ldots, \Delta t_k)$. Compose an $(N \times n \times k)$ dimensional matrix that consists of object features in the time intervals considered $Q_{ij\Delta t}$, where i is the object considered, j denotes measured features and Δt is a time interval.

Relative severity of crime problems compared to other areas is social-economic indicators of state and other similar areas: crime rate; structure of population; gross domestic product (GDP) per capita; unemployment; average disposable income; poverty rate and number of population living under poverty rate.

Preparing data for a further analysis, we determine homogeneity of the objects observed by investigating their properties. To this end, the methods of cluster or variance analysis should be applied.

Cluster analysis belongs to classification algorithms and solves an issue how to organise the observed data into meaningful structures. The general categories of the cluster analysis methods are: joining or tree clustering, two-way joining or block clustering and k-means clustering.

If the clusters are clear heuristically, the methods of variance analysis are usually used. This classification problem can be solved in other ways, too: using heuristics or extreme way.

Clusters of objects N are defined by choosing a fixed time interval Δt, and soundness of the clusters formed is verified in other time intervals.

When clusters of objects are formed, the structure of features characterising the clusters is under determination. For this reason, the factor analysis methods are selected for the problem solution. The factor analysis is applied for reducing the number of variables and for detecting a structure in the relationships between the variables.

Having verified the data adequacy/suitability to the factor analysis, variables that are not suitable for the analysis are found and eliminated.

For making the exploratory data analysis, it is recommended first to analyse principal components. A multiple regression analysis determines the relationship between several independent variables and a dependent variable. The regression function can be estimated, using the least squares estimation or any other loss function (non-linear estimation). After the regression equation has been estimated, the prediction can be computed for a set of independent variables.

The significance of the equation is verified by the criterion F, while the influence of the variables selected is analysed by reciprocally comparing standardised regression coefficients.

The target of the research was to explore, estimate and apply the use of multi-variate statistical models in the analysis and prediction of the state situation and tendencies for even distribution of the quality of life, education opportunity, public healthcare, personal career abilities, self-expression, community, culture, recreation — all these things are treated as a part of the quality of life. To estimate the situation and take decisions it is expedient to evaluate and select the main factors that influence the social security.

The aim of factor analysis is to explain the outcome of p variables in the data matrix X using fewer variables, the so-called factors. These factors are interpreted as latent (unobserved) common characteristics of the observed $x \subset \boldsymbol{R}^n$. In the factor analysis every observed $x = (x_1, \ldots, x_n)^T$ can be written as:

$$x_j = \sum_{l=1}^{k} a_{jl} f_l + \varepsilon_j, \quad j = 1, \ldots, n;\ k \leq n,$$

here f_l for $l = 1, \ldots, k$ denotes the factors, ε_j is the residual of x_j on the factors. Given the assumption that the residuals are uncorrelated across the observed variables, the correlations among the observed variables are accounted for the factors.

Multi-dimensional statistics, image identification, CBR and other methods compose the base for data mining. Frequently, these methods are interrelated.

The discovery process is complex, not single-valued and has some different analysis activities. The structured analysis tasks of discovery system are more formal activities in which the influence space of the problem under consideration may be analysed.

31.8. Information Infrastructure of Crime Investigation Context

Computer systems in forensic intelligence can be classified into two groups. Investigative methods are essentially based on analogy. Consequently, it is not surprising that most existing computer systems recognised as forensic intelligence systems can be viewed as an aid to performing this type of reasoning process. The second class of system pertains to collections that help classify evidence by the type of object that could have transferred a trace.

Unified criminal information system (Fig. 31.9) consists of separate DBs, which are ruled of different departments.

As a consequence:

- new structures could be created to exchange of information across countries
- new intelligence structures within the organisations could be created, with an important part dedicated to crime analysis
- structured and normalised methods of analysis of data could be developed

For these purposes a broad variety of computerised tools have been introduced (geographic information systems, meta-modelling of repositories, statistical analysis, qualitative management of DWHs, etc.).

The context of crime investigation includes separate DBs ruled by different departments. The main DBs are: DB of population (information about citizens — all data of identification card and/or passport); DB of transport (information about vehicles registration); DB of firearms (information about registered firearms and owners); DB of wanted persons (information about wanted criminals and missing persons); DB of stolen vehicles (information about stolen vehicles); DB of stolen firearms (information about stolen firearms); DB of crimes and criminals (information about committed crimes).

The people participating in this process for each DB may range from the scene of crime officer to the expert responsible for the interpretation step. To ensure that the process is reliable, and to diminish time and resources needed, three general rules may be applied:

- check existing data, their coherence and their "usefulness"
- write computer programs that implement part of the process, where possible
- where it is not possible, define methods and control their application

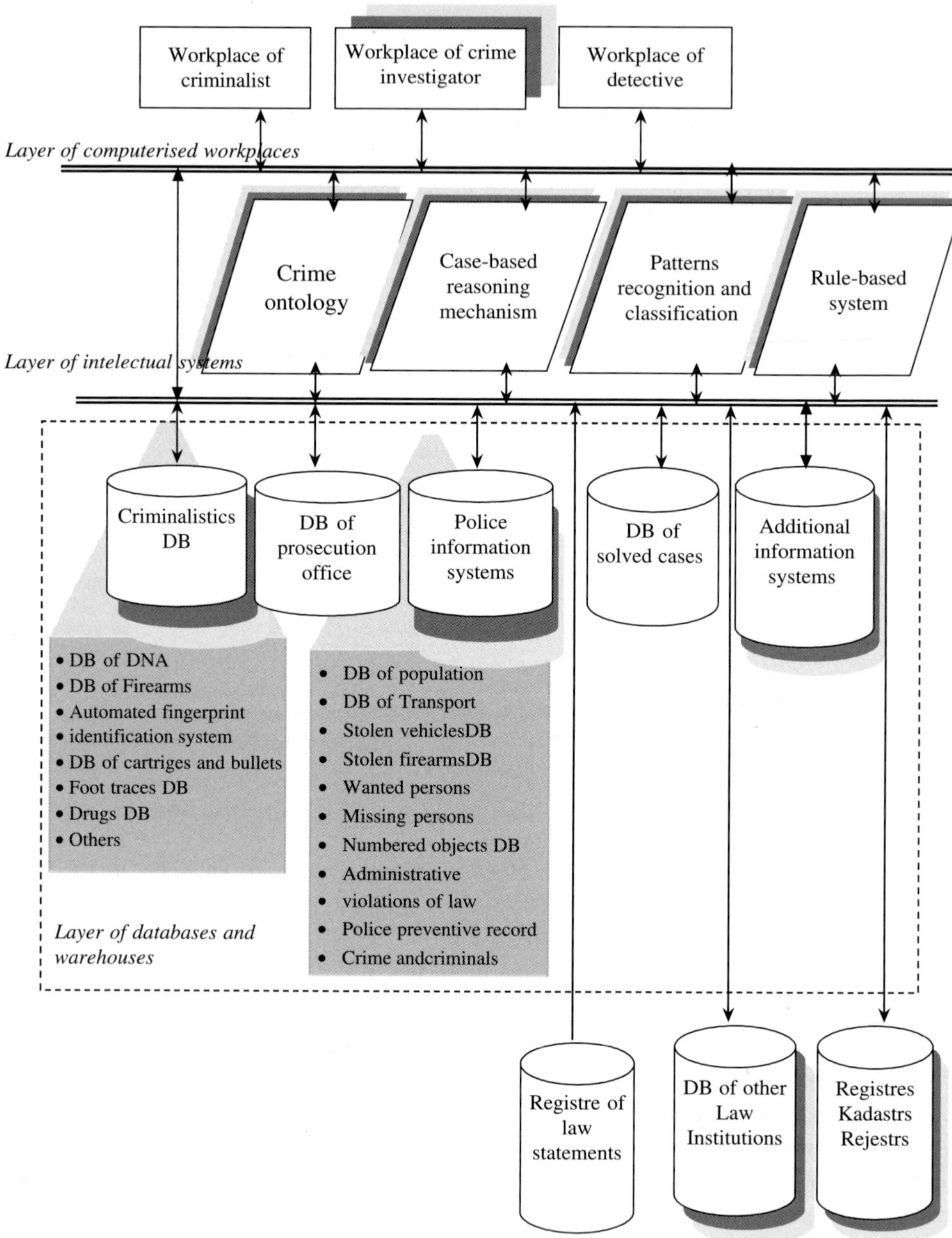

Fig. 31.9. Integration of components of intellectual systems with DBs.

Retrieval and matching collection of data at the scene of crime is always incomplete and imprecise, and collected marks are often fragmentary, even if the investigation is careful and thorough. An object that caused a trace can evolve, and marks or prints can be distorted. The match between recorded data and collected evidence (in its digital form) is therefore generally only partial. A human operator must always interpret a limited set of possible solutions at the end of the chain.

A broad range of retrieval and matching algorithms using various techniques are actually implemented to compare the data automatically. From a practical point of view, DNA and fingerprints provide extreme examples; the problem of matching two DNA markers is very simple from a computing perspective, as the match is generally determined by an "exact" comparison. It is not claimed that the implementation of the whole process, from the collection of data to its storage in the system, is easy, but the "retrieval and matching problems" themselves are not difficult. A retrieval algorithm can be implemented directly with simple computer development tools.

31.9. Conclusion

The knowledge representation methods play an important role in solving decision-making problems for the development of an advisory system in crime investigation processes. The application of artificial intelligence methods in development of advising, expert systems plays an important role in solving decision-making problems of judicial and law enforcement systems.

The DSS allows combinations of crime events and states that produce a given set of evidence from the knowledge base and ontology, helpful in generating scenario fragments from inconsistent situations. The proper crime investigation depends on the quality of the advisory crime analysis information system based on ontology, converting the first outside and inside facts in to criminality analysis processes, rendering opportunities of optimal decision making for the investigator. The main purpose of such a DSS is supplying of information on the crime investigation activity in the crime analysis.

The unified approach of integrating different DBs for aiding advisory processes in relevant patterns recognition and crime investigation is proposed. A key part of this approach enforcement is to understand those activities, through the development and use of methods, models and tools for collecting and then interpreting the large volume of data available in real-time for crime investigation. Consequently, intelligence programs have been developed, leading to recognition of a field of activity called crime analysis, which has been described as "the identification of and the provision of insight into the relationship between crime data and other potentially relevant data with a view to police and judicial practice".

References

1. S. Aitken and J. Curtis, Design of process ontology: Vocabulary, semantics and usage, *LNCS*, ed. V. R. Benjamins (Springer, Spain, 2002), pp. 108–113.
2. R. Barletta, An introduction to case-based reasoning, *AI Expert* (1991) 43–49.
3. I. Bratko and I. Kononenko, Learning diagnostic rules from incomplete and noisy data, *AI Methods in Statistics*, ed. B. Phelps, Gower Technical Press, London, 1987.
4. G. Booch, J. Rumbaugh and I. Jacobson, *Unified Modeling Language User Guide* (Addison-Wesley, 1999).
5. A. Caplinskas, General introduction to artificial intelligence, *Lecturer Notes of the Nordic-Baltic Summer School on Applications of AI to Production Engineering*, eds. K. Wang and H. Pranevicius (KTU Press, Technologija, Kaunas, 1997), pp. 1–38.
6. Centre for Crime Prevention in Lithuania, 2006, http://www.nplc.lt/english/index.html.
7. A. R. Chaturvedi, Acquiring implicit knowledge in a complex domain, *Expert Systems with Applications* **6**(1) (1994) 23–36.
8. J. W. Duncan and A. C. Gross, *Statistics for the 21st Century* (Irwin, USA, 1995).
9. D. Dzemydienė, Representation of decision making processes for the ecological evaluation system. *International Journal "Annals of Operation Research*" (Baltzer Science Publishers, Netherland, 1994) **51**, 349–366.
10. D. Dzemydienė, Temporal information management and decision support for predictive control of environment contamination *processes, Proceedings the 5th East-European Conference 'Advances in Databases and Information Systems' ADBIS'2001*, eds. A. Capliskas and J. Eder (Technika, Vilnius, 2001) **1**, 157–172.
11. D. Dzemydiene and E. Kazemikaitiene, Ontology-based DSS for crime investigation processes, *Information Systems Development: Advances in Theory, Practice, and Education*, eds. O. Vasilecas *et al.* (Springer, 2005), pp. 427–438.
12. D. Dzemydiene and V. Rudzkiene, Multiple regression analysis of crime pattern warehouse for decision support, *LNCS* (Springer, 2002), Vol. 2453, pp. 249–258.
13. M. Ehrig, P. Haase and N. Stojanovic, Similarity for ontologies — a comprehensive framework, *Workshop Enterprise Modelling and Ontology: Ingredients for Interoperability*, at PAKM, 2004.
14. E. R. German, Computer image enhancement of latent print and hard copy output devices, *Proceedings of International Symposium on Latent Print Examination*, U.S. Government Printing Office (Washington, D.C, 1987), pp. 151–152.
15. N. Guarino, Formal ontology and information systems, *Formal Ontology in Information Systems. Proceedings of FOIS'98*, ed. N. Guarino (IOS Press, Italy, Amsterdam, 1998), pp. 3–15.
16. G. Guizzardi, Ontological foundations for structural conceptual models, PhD with Cum Laude. Telematica Institute Fundamental Research Series, Vol. 015, Enschede, The Netherlands, 2005.
17. G. Guizzardi, *On Ontology, Ontologies, Conceptualizations, Modeling Languages, and (Meta)Models, Frontiers in Artificial Intelligence and Applications, Databases and Information Systems IV*, eds. O. Vasilecas, J. Edler and A. Caplinskas (IOS Press, Amsterdam, 2007).
18. B. Hebenton and T. Terry, *Criminal Records* (Brookfield, USA, 1993).
19. H. Hinrichs, Statistical quality control of warehouse data, Databases and Information Systems, eds. J. Barsdinš and A. Caplinskas (Kluwer Academic Publishers, 2001), pp. 69–84.

20. H. J. Holzen, S. Raphael and M. A. Stoll, *Perceived Criminality, Criminality Background Checks and the Racial Hiring Practices of Employers*, 2002.
21. IMPRESS — IMPRession Evidence and Serial Crime Profiling System, 2005, http://www. computing.surrey.ac.uk/ai/impress/full.html.
22. V. Justickis and J. Peckaitis, Imprisonment Today and Tomorrow. International Perspectives on Prisoner's Rights, and Prison Conditions, 2nd edn. (The Hague/London/Boston, Kluwer Law International, 2001), pp. 467–477.
23. G. L. Kovacich and W. Boni, *High-Technology-Crime Investigator's Handbook: Working in the Global Information Environment* (Butterworth-Heinemann, 2000).
24. B. Kuipers, Qualitative simulation of causal explanation, *IEEE Transactions on Systems, Man and Cybernetics*. SMC-**17**(3) (1987) 432–444.
25. H. C. Lee and R. E. Gaensslen, *Advances in Fingerprint Technology* (Elsevier, New York, 1991).
26. B. Leuf, *The Semantic Web: Crafting Infrastructure for Agency* (John Wiley & Sons, Ltd., Chichester, England, 2006).
27. S. Maskeliunas, Ontological engineering: Common approaches and visualization capabilities, *Informatica* **11**(1) (2000) 41–48.
28. K. Parsaye, Rotational Schemas: Multi-Focus Data Structures for Data Mining Information Discovery Inc. 1996.
29. K. Pastra, H. Saggion and Y. Wilks, Intelligent indexing of crime-scene photographs, *IEEE Intelligent Systems, Special Issue on "Advances in Natural Language Processing"* **18**(1) (2003) 55–61.
30. M. Pechenizkiy, S. Puuronen and A. Tsymbal, On the use of information systems research methods in data mining, *Information Systems Development: Advances in Theory, Practice, and Education*, eds. O. Vasilecas *et al.* (Springer, 2005), pp. 487–499.
31. W. Pree, *Design Patterns for Object-Oriented Software Development* (Addison Wesley, 1995).
32. J. R. Quinlan, Induction of decision trees, *Machine Learning* **1** (1986) 81–106.
33. M. Sabou, C. Wroe, C. Goble and G. Mishne, Learning domain ontologies for web service descriptions: An experiment in Bioinformatics, *Proceedings of the 14th International Conference on World Wide Web, Session: Semantic Web*, 2005, pp. 190–198.
34. R. Saferstein, *Criminalistics: An Introduction to Forensic Science*, 4th edition, (USA: Prentice Hall Career & Technology Englewood Cliffs, New Jersey 07632, 1990).
35. United Nations Crime and Justice Information Network, 2006, URL: http://www. uncjin.org/.
36. M. Uschold and M. Grüninger, Ontologies: principles, methods and applications, *Knowledge Engineering Review* **11**(2) (1996) 93–155.
37. R. C. Vellore, A. S. Vinze and A. Sen, MODELER: Incorporating experience to support model formulation — a case-based planning approach, *Expert Systems with Applications* **6**(1) (1994) 37–56.
38. P. B. Weston and K. M. Wells, *Criminal Investigation: Basic Perspectives*, 2 edn. (Pretence Hall, Englewood Cliffs, NJ, 1990).
39. J. Zeleznikow, Knowledge discovery and machine learning in the legal domain, Applied Computing Research Institute, La Trobe University, Australia, 1999.

Glossary

Decision support system (DSS): Decision Support Systems (DSS) are a specific class of computerised information system including knowledge-based systems that supports business and organisational decision-making activities. A properly designed DSS is an interactive software-based system intended to help decision makers compile useful information from raw data, documents, personal knowledge and/or business models to identify and solve problems and make decisions.

Data warehouse (DWH): computer-based information technology for organising complex DBs with functionality of DB management systems working in multi-dimensional data storage and retrieval regime.

UML: unified modelling language.

Chapter 32

BIOINFORMATICS AND BIOMETRICS

KENNETH REVETT
University of Westminster, London, UK

32.1. Introduction to Bioinformatics

Molecular biology is the study of biology at a molecular level — focusing principally on DNA and the processes that transform DNA into biological molecules such as proteins. After more than five decades of extensive and far-reaching research, a central theme has emerged that applies across the phylogenetic scale — and is termed as the central dogma of molecular biology. This dogma can be summarised succinctly in the following manner (see Fig. 32.1):

32.2. DNA Begets RNA and RNA Begets Protein!

As enunciated by the Nobel laureate and co-discoverer of the structure of DNA, Francis Crick has defined the central dogma as:

> "The central dogma of molecular biology deals with the detailed residue-by-residue transfer of sequential information. It states that such information cannot be transferred from protein to either protein or nucleic acid".

How this dogma is manifested is as amazing as it is complex. Proteins are complex molecules that are involved in all life-supporting activities — from respiration to thinking. The structure of a protein is usually represented schematically as a sequence of units that are joined together through chemical bonding. The units are termed amino acids (AAs) and in humans — there are approximately 20 naturally occurring AAs. It is the specific combination of AAs — which is specified through the sequence of nucleotides found in DNA that impart specificity in terms of functionality to proteins. All organisms need to respire and perform similar functions — which implies that there are certain genes in common across

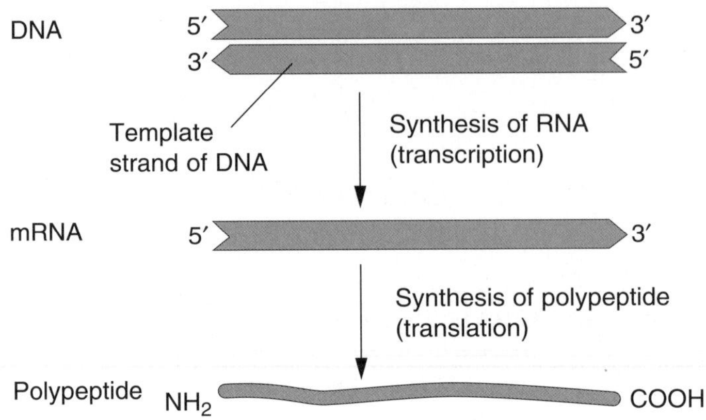

Fig. 32.1. The central dogma in molecular biology, illustrating the sequential (and unidirectional transfer of information from DNA via the RNA intermediate into the ultimate endpoint, polypeptides (proteins)). (Image taken from http://www.cbs.dtu.dk/staff/dave/DNA_CenDog.html, accessed on March 2008).

most organisms. Similarity between the genome of organisms is a major theme within the bioinformatics community.

The genome of an organism consists of the sequence of nucleotides that are contained within the DNA. DNA consists of four constituent elements — which are repeated throughout the length of the genome. For humans, there are approximately 3×10^9 nucleotides contained within 23 pairs of chromosomes. Biological machinery exists which interprets, through the recursive execution of the genetic code, the information content of the genetic code into the formation of a variety of molecules termed proteins. Figure 32.2 depicts the basic genetic codebook used by virtually all organisms. With the discovery of the genetic code — a hybrid science termed bioinformatics has emerged.

Bioinformatics is a computational approach to the elucidation of the information content of genomes. It sits at the border between typical biological science and computer science. The biologist is confronted with a bewildering array of difficulties in the pursuit of elucidating how the genetic code actually operates. How does one perform a typical controlled scientific experiment on a living organism — how can we control particular variables — when we are not sure which are critical for the survival of the organism up to the time when a measurement is required to determine the outcome? This is where the computational approach of bioinformatics has proven to be extremely helpful. By utilising an information-theoretic approach to examining the information content of bio-molecules, bioinformatics has provided a wealth of useful scientific information. The basic strategy of the bioinformatics approach is to focus on the information content — without the need to perform direct biological experiments.

The information content of bio-molecules was originally made possible by elucidation of the genetic code: that is, particular sequences of nucleotides (known

First letter	Second letter: U	C	A	G	Third letter
U	UUU, UUC: Phe UUA, UUG: Leu	UCU, UCC, UCA, UCG: Ser	UAU, UAC: Tyr UAA Stop UAG Stop	UGU, UGC: Cys UGA Stop UGG Trp	U C A G
C	CUU, CUC, CUA, CUG: Leu	CCU, CCC, CCA, CCG: Pro	CAU, CAC: His CAA, CAG: Gln	CGU, CGC, CGA, CGG: Arg	U C A G
A	AUU, AUC, AUA: Ile AUG Met	ACU, ACC, ACA, ACG: Thr	AAU, AAC: Asn AAA, AAG: Lys	AGU, AGC: Ser AGA, AGG: Arg	U C A G
G	GUU, GUC, GUA, GUG: Val	GCU, GCC, GCA, GCG: Ala	GAU, GAC: Asp GAA, GAG: Glu	GGU, GGC, GGA, GGG: Gly	U C A G

Fig. 32.2. The genetic code which map codons (sets of three nucleotides A, C, G, U) into specific AAs (indicated by their three letter abbreviations).

as a triplet or codon) yield, through an intermediary (see mRNA in Fig. 32.1), a particular AA. Strings of AAs form proteins, which are essential elements for all living organisms. The mapping from codons to AAs was the cornerstone in molecular biology (the basis of the central dogma of molecular biology). Now, sequences of nucleotides were translated into strings of AAs (see Fig. 32.2 for details). This is the foundation of modern bioinformatics.

Proteins consist of one or more strings of AAs (sometimes 100s) — of which there are 20 that are naturally occurring. The arrangement of the AAs is termed the primary structure (as depicted in Fig. 32.3). Through complex physical processes involving charge interactions, the primary structure ultimately yields a 3-dimensional structure termed the conformation. It is the 3-dimensional structure (conformation) of proteins that imparts their particular functions. Since most organisms perform similar functions such as respiration, etc., there might

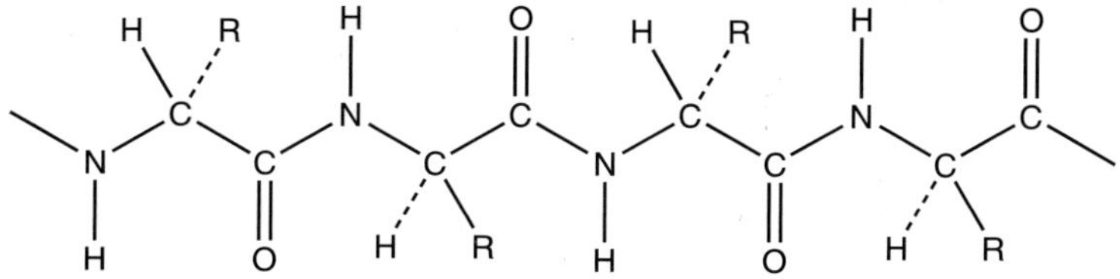

Fig. 32.3. The primary sequence of a polypeptide chain, consisting of two AAs linked up chemically through the peptide bond.

be similarity in the structure of the proteins responsible for these functions. This structural similarity can be assessed by examining the primary structure — the sequence of AAs — as the physics involved in transforming the primary structure to 3D structure is assumed to be universal. So, if we wish to determine the structure of an unknown protein — the typical approach is to compare the primary structure against known protein structures (contained within protein structure databases such as NCBI). The question then becomes how do we compare the primary sequences of proteins?

Needleman and Wunch proposed the notion of a global alignment, which provides a means of examining how similar two sequences are to one another by determining the extent to which two sequence of characters overlap one another [11]. For instance, Fig. 32.4 depicts how two protein sequences may overlap with respect to one another — and forms the basis of a scoring mechanism, which quantifies the similarity between two protein sequences. The score then is a measure of the similarity between two sequences. By comparing an unknown sequence against a database of sequences and ranking the scores in descending order, one can determine which database sequence most closely matches the unknown sample. The function(s) of the matched sample is then used to infer the function of the unknown protein sequence.

The global alignment scoring mechanism essentially compares two sequences component by component (i.e. AA by AA). It requires that the two sequences are the same length — which may limit its utility in some situations. When two

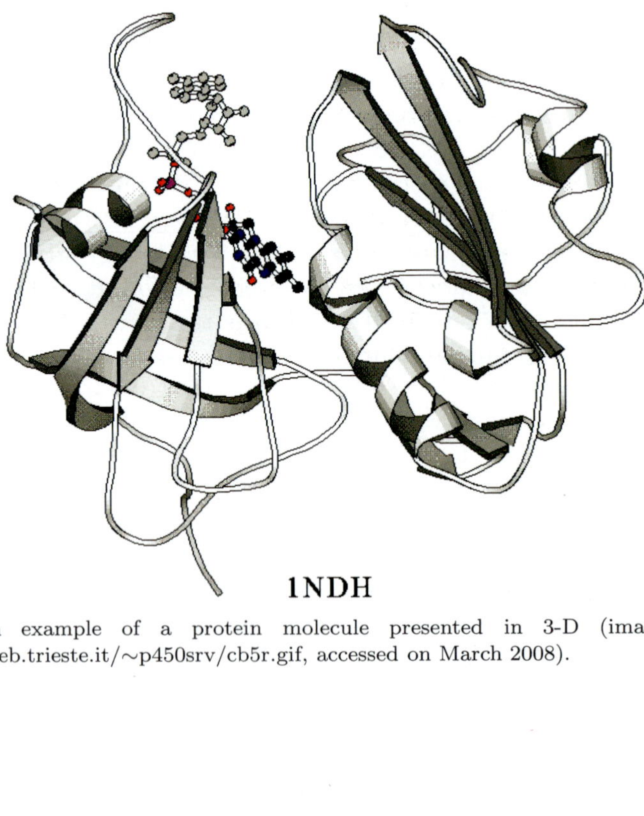

Fig. 32.4. An example of a protein molecule presented in 3-D (image taken from: http://www.icgeb.trieste.it/~p450srv/cb5r.gif, accessed on March 2008).

components match exactly, a positive score is added to the cumulative score. If two components do not match, a negative score is added. To enhance the alignment, gaps can be added — essentially blanks — to either sequence. The gaps are also penalised by adding a negative score, but the magnitude is lower than that for a mismatch (see Fig. 32.4 for details). Though very useful, this approach may fail to find elements within sequences (termed motifs) that are highly conserved from an evolutionary standpoint. There may be regions within the sequence that are highly conserved across organisms because they serve a very useful purpose, though the rest of the protein may have undergone significant levels of mutation. These protein sequences would produce a low global score. To take motifs into account when comparing sequences, Smith and Waterman modified the global alignment algorithm to account for the existence of embedded motifs by a slight modification of the scoring mechanism [16]. Essentially, the Smith and Waterman approach is analogous to finding the longest common sequence (LCS) between two strings, also referred to as the Levenshtein distance between two strings [9].

These alignment tools form the basis of the bioinformatics approach deployed in the biometrics literature. To utilise these tools, one has to convert the biometric data into a sequence of strings from a finite alphabet. Then one can simply compare the strings and score them, much the same way protein or DNA sequences are scored. In what follows are case studies in which this approach has been implemented in two contexts: keystroke dynamics and intruder-detection-based biometrics.

32.3. A Bioinformatics-Based Approach to Keystroke Dynamics

To place these ideas into the current context — imagine that a login attempt can be converted into a string of characters. To do this, the attributes extracted from the login attempt must be mapped onto a string within a given alphabet. The principle attributes extracted from keystroke dynamics include digraphs, trigraphs and keypress duration (dwell time). These attributes are then mapped onto an alphabet — in this work, the AA alphabet, which contains 20 elements. This mapped version of the input attributes becomes a query sequence and the task then is to determine the closest match to a database of sequences derived in exactly the same manner from each user of the system. We then have a similar situation to that of a bioinformaticist — can we determine which sequence in the database is a closest match to the query sequence? This type of question has been addressed repeatedly within the bioinformatics community for over three decades now — with quite satisfactory results in most cases. The principle tool of the bioinformaticist is the sequence alignment problem, which is illustrated in Fig. 32.5.

The strings of characters in Fig. 32.3 represent a shorthand notation for AAs — which can be viewed as a string of these symbols. Of course there is biological meaning to them — but for the most part — bioinformaticists can treat them symbolically simply as a string of (Smith & Waterman, 1975). In this work, the digraph times, dwell time and trigraph times are discretised into an

Fly:	GAKKVIISAPSAD-APM–F
Human:	GAKRVIISAPSAD-APM–F
Yeast:	GAKKVVSTAPSS-TPM–F

Fig. 32.5. Multiple sequence alignment of a portion of the glyceraldehyde3-phosphate dehydrogenase (GADPH) protein from three different animal species. Note the dashes indicate gaps — insertions/deletions in the alignment. Also note that there is a considerable amount of sequence identity — that is, symbols that are the same across all three sequences.

AA alphabet — which yields a string of characters similar to that found in a protein — but considerably shorter (42 residues in length). We can then apply the huge quantity of practical and theoretical research that has been successfully developed in bioinformatics to the task of authentication.

To determine the relative alignment between a given query sequence and a database of 1,000 of sequences, one must have a metric — called the alignment score. For instance, if two elements match at a given position (in vertical register) — then a score of +1 is recorded. If the characters at a particular position do not match — a score of −1 is recorded. Gaps can be introduced — in global alignments — that serve to increase the overall score. If, for instance, two sequences were similar at some point but one had an insertion/deletion mutation, then these mutations would cause a misalignment between the sequences from this point forward — reducing the overall alignment score. Gaps can be placed in either the query or the target sequence and as many as required can be added. They also serve to ensure that the two sequences are of the same length. When gaps are added — they are penalised and reduce the score — but if they are effective at re-aligning the sequence further downstream — then the penalty may be cancelled out by a higher match score. Otherwise, the effect is to reduce the overall score — which is required because any two sequences can be aligned given enough gaps placed within them.

In this work, the sequence alignment algorithm employed is against a set of sequences stored as motifs (obtained via a position-specific scoring matrix (PSSM)). This process entails aligning each column in the series of sequences and calculating the frequency of each AA within each column. A matrix is generated that scores the relative frequency of each of the 20 AAs at each column position. This data can be used to generate a motif — where high scoring positions (corresponding to frequently occurring residues) are important features of the data. There are two dimensions that can be employed in PSSM: the magnitude of the frequency within a column and the number of columns (and whether they are consecutive or not). The process of generating the motif from the training data was extremely fast — in the order of a few milliseconds — which was a constraint placed on this system — it must be able to operate in a real-time environment. One can vary the mapping alphabet — to vary the resolution of the resulting mapped string. For instance, a mapping onto the digits 0–9 would produce a much coarser map that would generate longer motifs. The number of keys on a piano (typically 88) would provide a much more refined mapping — yielding shorter motifs. The choice of the AA alphabet was based on a compromise between motif length between the two extremes mentioned and also to map directly into the bioinformatics literature. The actual extraction of these motifs is described next, followed by a brief description of the dataset.

There were 20 participants in this study, all are computer science undergraduate students from a Polish University. The users were provided with eight-character login IDs and passwords, generated randomly by a computer program. The enrolment process required users to enter their login ID/password 10 times successfully. Each participant enrolled onto a single machine located on campus — for both phases of this study. After successfully enrolling (10 trials), the participants were asked to perform 100 self-logins (for FRR) and 100 attacks on other accounts (for FAR data). The following regime was used for non-enrolment logins: each participant was asked to self-login 100 times over a 7-day period. Therefore, each participant logged into their own account approximately 15 times/day. In addition, students were instructed to login at three different periods of the day: morning (09:00–10:00), noon (12:00–13:00) and early evening (17:00–18:00). At each period, students were asked to either perform self-login or non-self login five times. This simulates the way users would normally access their computer systems, logging in at various periods during the course of a workday.

With the enrolment data collected, the enrolment trials were discretised using the 20-letter AA alphabet. To do this, the largest time (with a resolution of 1 mS) for each of digraphs, dwell times and trigraphs were identified (column-wise) from the enrolment process and used to normalise each attribute. This process was repeated for each of the three sets of attributes collected in this study. The normalised data yields values from 0 to 1. The range is then partitioned equally into 20 bins corresponding to each of the 20 AAs available from the alphabet. Next, each bin was assigned to one of the 20 elements from the AA alphabet (which were arranged in ascending alphabetical order). The corresponding collection of mapped attributes (now contained within a string of AAs) was ordered in the following fashion: digraphs, dwell time and trigraph times, and this order was used consistently throughout this work. The next stage is the development of a motif for each of the enrolment entries — which will be stored in the query database for verification/identification. A position specific scoring matrix (PSSM) was generated from the enrolment trials (10 for each user). Briefly, the frequency of each AA in each column was calculated and stored in a separate table (the PSSM). The data was not converted into a log-likelihood value as the expected frequencies were not available from a limited set of data. Instead, the values can be interpreted as probabilities of a given residue appearing at each position. There are two parameters that were examined with respect to motif formation: the frequency of each AA residue at each position — and the number of elements within the motif (whether they are consecutive or not). These are tunable parameters that can automatically be set based on the data from the enrolment process. The stringency of the motif-based signature is based on these parameters: for high-level security application, positions with a very high frequency — i.e. greater than 80% and for a minimum of 50% of the residues can be deployed. Likewise, reduced stringency is accomplished by relaxing these values. If during the enrolment process, the frequency within a column and the number of consistent columns was below some minimal threshold (50%), then the user would either be requested to re-enrol, or the normalisation time could be increased iteratively until the minimal threshold was achieved.

The normalisation time (and hence the bin times) was stored with each database entry (all three stored separately) to allow for a re-mapping of the attributes to sequence values if it became necessary. The mapped entries from the enrolment process and the resulting motifs for all 20 users formed the query database which was used for both verification and identification. The run-time for the motif extraction phase was fairly constant at approximately 2–8 mS. The efficiency in this particular case is related to the short length of the sequences (many proteins contain 100s of residues) and the large degree of similarity between the sequences — these were generated from the enrolment sequence. The generation of the motif is performed immediately after enrolment — and for all intensive purposes is so short that the user can then login to the system straight away and be authenticated by the resulting model.

The authentication algorithm works as follows: a user enters his/her login credentials. It is discretised into the AA alphabet based on the normalisation time associated with the login details associated with a given login ID/password combination. The authentication sequence is then compared with the stored motif for the login ID and given a score based on the algorithm specified in Eq. (1) using a simple global alignment algorithm. Generally, a match is scored +1 and a mismatch is scored −1, and "−" in either the probe of the query sequence has a score of 0. The score is computed and if it above a specified threshold q then the entry is authentication, else it is rejected. Note that the score "q" can be based on both the frequency threshold at a specific column and the number of columns (as well as whether the columns are contiguous or not). The user has three attempts before system lockout. The value of q was set equal to the motif length (number of non-blank "−" entries) in this particular study — although in principle it could be varied to control the level of security required. The identification algorithm works as follows: the normalisation factor was extracted from the user during the authentication attempt. Then we proceed exactly as in the authentication phase, except we compare the resulting motif against all motifs stored in the database. Duplicate motifs are expected to be a rare event considering the number of expected entries in the motifs and the range of values at each. In this pilot study — the average cardinality of the motifs was 42 (14 digraphs, 16 dwell times and 12 trigraphs). Each motif could contain up to 20 different values — yielding an average of $42^{20}/1$ where "l" is the motif length possible motifs. In actuality, the full range of the sequence alphabet may not be fully covered. If there was a tie between two or more entries from the database, then this login attempt is rejected and the user is asked to re-enter his/her login details. Lastly, please note that the data stored in the DB is updated every time a user successfully logs into the system. The oldest entry is removed, replaced by the latest and the PSSM value is updated and a new motif is generated for each user ID. The next section describes the results obtained from this study, and fill in experimental details as required.

All 20 users were requested to enrol and authenticate during a 1-week period of this study. In Table 32.1, we present a typical enrolment dataset that consists solely of the digraph times for the username/password extracted during the enrolment period. The first stage in our algorithm is to discretise the enrolment data — since the values obtained are essentially continuous (to a resolution of 1 mS). This stage requires obtaining the largest digraph (which in this was 1.37 — see Table 32.1 for an example) and normalising all digraph values. Then binning was performed, where each digraph was assigned its ordinal position within the discretisation alphabet: A = "ACDEFGHIKLMNPQRSTVWY" — the single letter code for AAs in ascending lexical order. This resulted in the following dataset displayed in Table 32.2. In our previous work, the maximal frequency of each element within each column — this corresponds to the range of values that were obtained for each

Table 32.1. Sample of an enrolment of 10 consecutive entries of the username/password for a randomly selected participant. The numbers represent time in mS — and each column represents a digraph (only eight are displayed).

#1	#2	#3	#4	#5	#6	#7	#8
25	34	33	21	31	58	66	63
19	28	25	29	28	69	64	75
23	34	29	23	24	48	78	68
28	31	25	25	38	47	56	78
21	35	33	32	30	44	54	82
19	42	22	24	28	68	78	75
28	34	27	25	26	47	38	59
25	32	23	26	29	78	46	64
30	28	35	41	24	44	65	61
26	24	42	33	26	57	58	69

Table 32.2. Discretisation of an enrolment entry (corresponding to the raw data given in Table 32.1) using the AA alphabet. Note that the maximal normalisation value was 0.063 s.

F	G	E	E	G	F	N	M
E	F	F	E	F	M	S	P
Y	G	E	E	F	H	Q	Q
F	G	F	F	H	I	L	Q
E	G	F	E	G	I	H	P
E	H	E	E	F	Q	Q	P
E	G	E	F	F	I	H	L
E	F	F	F	F	Q	E	K
E	E	F	E	F	I	F	G
E	E	E	E	F	N	F	L

digraph (Revett, 2007). This same approach was taken in this study — as it proved to be quite effective with the previous data collected from a new participant cohort. The consensus sequence for the example in Fig. 32.5 is EGEEFI-, where the "–" symbol indicates that there is no unique dominant symbol within the column(s) — although the exact frequency of each AA is maintained across all columns. The sequence above is the consensus sequence for this enrolment instance. It represents the average behaviour — in that it captures the most frequently entered digraph values. This amounts to an unweighted voting scheme, with a threshold for inclusion (set to 0.5 in this study).

The entries within the consensus sequence (motif) also possess information regarding the typing speed and the consistency possessed by the person entering it. This is the value of the normalisation factor — and influences the spread of the digraphs (and AA symbols) within the generated sequence. If an imposter attempted to input this particular username/password, the maximal digraph value would more than likely differ from this particular one, and the normalisation process may yield a different set of elements within the consensus sequence. Please note that the maximal digraph value for each enrolment is stored with the user's data — and it is this value that is used to discretise all subsequent authentic successful login attempts. Also, note that for each subsequent successful login attempt, the data for this account is updated — such that the oldest entry is removed and the consensus sequence is updated, along with the longest digraph value (the normalisation factor). This allows the system to evolve with the user as his/her typing of their login details changes over time — which invariably happens as the user becomes more familiar with their login details.

For the authentication process (verification and identification), the username and password entered is discretised using the stored maximal value for digraph time. Equation (1) provides the values used when performing the motif matching algorithm. It is a very simple algorithm — which was the intention — as there may be 1,000s of entries in the database. The results from our preliminary study indicate that it took on average 19 mS to compare the motifs over the 20 entries — yielding a value of 106 matches/s. In (Eq. (32.1)), the match indicates the same symbol in each sequence at the same site, mismatch means that a non-match- and non-blank and the "–" — blank indicates that this position was not part of the motif.

$$\begin{array}{rl} +1\} & \text{match} \\ -1\} & \text{mismatch} \\ 0\} & \text{“–” in either sequence} \end{array} \tag{32.1}$$

The algorithm just described may present complications if the enrolment was only marginally successful. That is, in the example provided in Eq. (1) above, there were clear-cut dominant entries within each column (except for column 7). If there were less then a threshold level of unique entries in the consensus sequence (a system defined parameter set at a default value of 0.5), then the system automatically re-calculates the normalisation factor. This re-scaling occurs iteratively until there is at least a threshold level of entries in the consensus sequence. For higher-level security installations, the threshold value can be raised — requiring more consistency during the enrolment. This enhanced stringency though has a negative impact on the false rejection rate as it imparts greater stringency on the user during subsequent login attempts. Also note that the AA alphabet is just one of the many choices — the greater the cardinality, the more refined the bin size for a constant normalisation value. This too results in a reduction in the number of elements contained within the consensus sequence. Our results with a 10-(decimal) and 16-(hexadecimal) lettered alphabet yielded results that reduced the discriminatory capacity of the system. The value for the length of the discretisation alphabet can be examined by looking at the false acceptance rate — which was as high as 13% with the decimal and 6% with the hexadecimal-based alphabets. We, therefore, decided to stick with the AA alphabet and adjust the normalisation factor when necessary. The system as just described resulted in an overall FAR of 0.6% and an FRR of 0.4% — a total error rate of 1.0% (for a consistency threshold of 0.5). This result is comparable to that found in other keystroke dynamic-based systems (XXXXXX). Please note that this result is measured for individual login attempts — that is, 8 attempts were unsuccessful on an individual login attempt (FRR) — but no user was locked out because they failed to login within three attempts.

When the consistency threshold was raised to 0.80 (rounded upwards when necessary), the FAR was 0.0% but the FRR increased to 0.5% — yielding a total error rate of 0.5%. In Table 32.3, the results for FAR/FRR are summarised with respect to the threshold for dominant consensus elements. Note that, in this experiment, the same 20 users were used and each logged into their own accounts 100 times (FRR measurement) and each participant logged into the other 19 accounts 100 times each (a total of 1,900 FRR attempts/account and 100 FAR attempts/account).

In addition to the frequency threshold data that was presented in Table 32.4, another parameter, based on the total number of elements in the consensus sequence was also investigated in this study. The two parameters are somewhat related — in

Table 32.3. Summary of FAR/FRR results as a function of the consistency threshold for elements within the consensus sequence (motif). These results are randomly selected values from a series of 100 experiments.

Threshold	FAR (in %)	FRR (in %)
1.0	0.0	0.8
0.80	0.0	0.5
0.60	0.1	0.2
0.40	0.4	0.0

Table 32.4. Sample results from an experiment in which the length (total number) of consensus sites matched the stored record for a set of randomly selected login attempts.

Threshold	FAR (in %)	FRR (in %)
30	0.0	0.6
25	0.1	0.5
20	0.1	0.1
15	0.3	0.0

Table 32.5. Results indicating the identification capacity of the system with respect to fractional occupancy at each site and the total length of sites consistent with the stored sequence (not contiguous).

Threshold	FAR (in %)	FRR (in %)
0.5 (0.50)	0.0	0.1
0.5 (0.75)	0.0	0.0
0.6 (0.50)	0.0	0.0
0.6 (0.75)	0.0	0.0
0.7 (0.50)	0.0	0.0
0.7 (0.75)	0.0	0.0
0.8 (0.50)	0.0	0.0
0.8 (0.75)	0.0	0.0
0.9 (0.50)	0.0	0.0
0.9 (0.75)	0.0	0.0

that for a given position to be significant, it must have a specific frequency value. In this part of this study, a frequency value of 0.5 was used as a first approximation, and the focus was on the total number of matching entries within the motifs stored with a particular use ID/login sequence. The data in Table 32.3 present a summary of the data as a function of the fractional percentage of the motif cardinality.

Lastly, a critical issue in this chapter was the ability of the system to be able to perform user identification — an issue that has not been addressed within the keystroke dynamics literature at all in any significant degree. The data in Table 32.5 presents the results of the identification, where the 20 users were to be identified based on a single login transaction as a function of column consistency threshold and column length. Note that the threshold for all consistent attributes was approximately 50% (20) or 75% (32).

32.4. Edit Distance Approach to Biometrics

Several papers have been published recently, employing the use of some fundamental algorithms commonly employed in the bioinformatics literature. In 2002, Bergadano published a paper on keystroke dynamics that employed the use of the edit distance between n-graphs generated from keystroke latencies [2]. In this study, a cohort of 44 faculty members from a computing department were asked to enter a long text string containing 683 characters, repeated five times (yielding a total of 220 samples). A further 110 subjects were asked to provide a single sample of the same text (produces the imposter samples). It should be noted that the text string entered contained a mixture of Italian and English words. All participants were native Italian speakers, whom were also familiar with English language. In addition, all participants were used to typing and using computer keyboards, though there was a considerable range in typing speeds and typing skills generally. Also note that all participants used the same keyboard on the same computer in similar circumstances (lighting, office location, etc. — see [2] pp. 372). The introduction of errors was not considered problematic in this study. When a typing mistake was made, users were allowed to correct them — but in essence the actual number of n-graphs was reduced accordingly. The difference in this regard with respect to other studies is that the typing sample containing one or more errors were not discarded. There were a total of 350 unique trigraphs contained in the long text string, but on average users shared approximately 272. By using trigraphs only, this work specifically relies on typing speed. It would be interesting to perform a detailed analysis to determine the effect of and the information content of typing errors.

The task investigated in this study was to be able to assign a typing sample to one of the subjects' in this study (via a distance metric and reference profiles). This study design employed the five samples of the long text string provided by each of the 44 subjects. These samples were designated as the legal authentic users of the system. The system was attacked 110 times by the 110 volunteers whom each provided one sample of the same text (which was used as the imposter samples). In addition, the system was also attacked by the five instances of all the other legal users of the system (a total of 215). When a legal user attacked the system, their entries were removed to ensure that they did not match one of their existing typing samples, as would normally be expected to occur. The authors state that there were a total of 71,500 attacks on the system, yet the total number of attacks indicated in this chapter indicates a smaller figure (325×153) of approximately 49,725. This number reflects 325 attacks ($110 + 215$) entered by 153 ($110 + 43$) possible intruders.

It is very interesting to note that this study does not make use of a single averaged referenced profile like most previous studies. Instead, each of a subject's typing sample is used individually. This has the immediate effect of increasing the numbers with respect to calculating FAR and FRR. Whether using each individual sample versus an averaged reference sample influences FAR/FRR measurements was not directly addressed by this study, nor any other study as far as this author is aware of. This should serve as an interesting area of research to pursue. What the authors of this chapter do present in this regard is the authentication/identification as a function of the number of samples of each user. The data indicates that the classification accuracy varied little when the number of samples was reduced from 4 to 1 (100–96.9%). The classification error was calculated as the number of misclassified cases out of the total number of cases (possible comparisons). The total number of cases was calculated from the sample permutations. For instance, if there were three samples used for a person's typing sample (their model), 10 different models are available for the user (5 choose 3); with 44 users would generate 440 samples. These can be tested against the remaining two samples, yielding a total of 880 test cases.

The authors employ the edit distance as an automated method for calculating the difference between two typing samples (collection of n-graphs). The edit distance is a measure of the entropy between two strings. In the bioinformatics literature, the classic strings are DNA and protein sequences. DNA sequences are simply a string of nucleotides, where each character is contained within the set of (a, c, g, t). A more diverse string is obtainable from protein sequences, which contain 20 elements, usually represented as single-character abbreviations ("acdefghiklmnpqrstvwy"). A common task in bioinformatics is to determine the phylogenetic relationship between two and more protein sequences. One simple method is to directly compare the aligned sequences (by align, they must be coregistered at their start of the sequences) and assigning a score to those elements that are exact matches. The total number of matches normalised by the length of the shortest sequence provides a measure of the similarity between the two strings. The value of this match score varies between 0 and 1 — the later indicating a perfect match. This is rarely the case for full length sequences, but quite often there will be segments within the sequence that do produce a high match score. Usually, these segments (subsequences) may provide valuable insight into functionally important elements within the protein sequence. This issue will be discussed further later in this chapter. For the present study, the authors employ a similarity metric that employs the edit distance, a measure of the order (entropy) of the sequences.

The basic idea behind the edit distance (also termed the Levenshtein distance) is to determine how many moves of the elements in both strings must be made in order to become perfectly aligned. It is a measure of the entropy difference between two systems, which is proportional to the number of shifts that must be made for perfect or best alignment. The strings that must be aligned in this study were based on trigraph latencies, which were arranged in ascending temporal order (always and matched to the particular trigraph). This operation was performed every time a sample was used for authentication against all collected samples in the database. If there were duplicate trigraphs, the average was recorded. If two or more trigraphs yielded the same exact latencies, they were arranged in lexical order (ascending based on the starting character). Note that the temporal resolution of the trigraph latencies was low, but variable — certainly in the order of 10 mS. The reason for this was that the users logged into the system via a terminal connected via a relatively slow modem in a network with variable traffic levels. The degree of disorder (the entropy) was calculated, yielding a maximal disorder equal to the half the length of the string if the two strings were completely reversed. This value was normalised by dividing the entropy by the maximal possible entropy for a string of a given length (both strings must be equal in length) N, yielding a value between 0 and 1 inclusive (see Eq. (32.2)).

$$N = \frac{|\mathrm{V}|^2}{2} \text{(if V was even)}; \ \frac{|\mathrm{V}|^2 - 1}{2} \text{(if V was odd)} \tag{32.2}$$

An example of the algorithm is illustrated in Fig. 32.6(b). The entries in the left panel are the trigraph times sorted in ascending order, with the corresponding trigraphs in the right panels. The numbers above the arrows indicate the number of moves required to match all the trigraph characters.

The entropy (edit distance measure) was used to determine how closely two typing samples were to one another. When comparing a typing sample U against all other typing samples — the basic assumption is that the sample U, should be closer to other samples of the same user then it is to other samples. This notion still requires the use of a similarity metric. The authors use the empirical result that entropy tends to be non-linear function of the number of elements. This results was computed based on randomly generated arrays. They state that in 10 million different randomly sorted arrays of 100 elements, only 573 yielded a disorder measure (edit distance) in the interval [0.44–0.5], the remaining arrays were in the interval of [0.5–0.9] ([2] p. 370). This is a rather unusual result and may need further systematic clarification. They use these results to build a distance metric that can be used to unequivocally discriminate users through their keystroke dynamics (based on the edit distance from trigraphs). Using all typing samples for each user (total of five in this study), they build a distance metric termed the mean difference (md), which is a weighted sum of the differences between a typing sample to be identified and all typing samples for all users in the system. The classification assigns the typing sample with the user in the system that yields the smallest md. This method yielded, with 0% FRR, an FAR (they use IPR) of 2.3%, a fairly significant result.

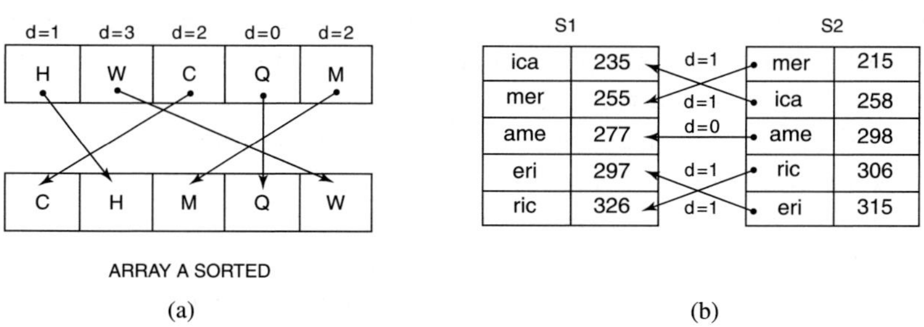

Fig. 32.6. A sample of the algorithm for matching up two sets of digraphs generated from typing samples. On the left is an array version, demonstrating alignment of individual array elements. The same process appears on the right, where the ordering of trigraphs is based on trigraph timings. From [2].

The authors further claim that this classification accuracy can be improved upon based on the following observation. They argue that the absolute mds alone may not be sufficient to correctly classify a sample data to the appropriate owner. If a sample X is to be classified as belonging to user A, then the following constraints must be satisfied: (i) the md of X to A (md(A, X)) is smallest and (ii) md(A, X) is closer to the average value of A (indicated as m(A) in their paper) than to any other md for all other users in the system (md(B, X)). These constraints are implemented as a statistical measure of closeness used by Gaines *et al.*, 1980 and Leggett and Williams, 1993. Their formulation is presented in Eq. (32.3):

$$\mathrm{md}(A, X) < \mathrm{m}(A) + |k(\mathrm{md}(B, X) - \mathrm{m}(A))| \tag{32.3}$$

The previous results used an implicit value of 1.0 for k, and the authors examine the classification accuracy as a function of k. The best results were obtained when $k = 0.66$, resulting in an FAR of 0.1371% and an FRR of 0%. Essentially, k is a stringency factor — which will have an impact on the FAR|FAR values. A small value for k (which is allowed to vary from 1 to 0) makes the system more stringent, and will tend to increase the FRR value, while decreasing FAR. This is indeed born out by their results, with FAR of 0% and FRR of 7.27% with a value of $k = 0.3$.

Lastly, the authors claim that they can produce a system that can yield perfect classification if the following additional constraint is added: the sample must not only be closer to a particular user's model than to any other model (the previous constraint), but it must also be sufficiently close to all the samples in the claimed model. This closeness with respect to the samples in a particular model is implemented using the standard deviation (sd) of the samples in the model. This constraint can be implemented in the following manner (Eq. (32.4)):

$$\mathrm{Md}(A, X) < \mathrm{m}(A) + \alpha^* \ \mathrm{MAX}(A) + \beta^* \mathrm{sd} \ d(A) \tag{32.4}$$

Where α and β are real-valued constants that must be empirically derived for each user (or set of samples). The purpose of these two terms is to strike the required balance between FAR and FRR. The values for the parameters are also affected by the value of k, so for a complete system, all three factors must be determined. The authors claim by a judicious selection of values for the three parameters, an FAR/FRR value of 0% can be obtained. This required tuning the values of these three terms for each user in the system.

The authors claim that a key advance in this work is the notion of using relative rather than absolute keystroke dynamics. This may help to nullify changes in typing speed — provided any change in speed is consistent across all characters. Using absolute values for trigraphs in the decision model may produce misleading results, which will tend to increase the FAR/FRR values. The edit distance is a relative measure of typing dynamics — and will yield the same result independent of typing speed, provided it is uniform across the characters typed. The users used a long text string (683 characters), which was entered five times. This is a rather burdensome enrolment requirement, requiring 3,415 characters to be typed. The authors investigated whether this number could be reduced — not directly but by using a fraction of the trigraphs. As the number of trigraphs used to build the model of the user decreased, the error rates (measured as FAR/FRR) increased considerably, especially with lower values of the parameter k. The FRR increased from 0.9 to 5.4% when half the digraphs were used, and the value of k changed from 1.0 to 0.5. These results were obtained using the full 683 character inputs — but taking either the first half or second half of the trigraphs (or which ever fraction they choose to test with). One might expect that the later trigraphs might be more consistent than the first — based on a minor practise effect. This result implies that reducing the size of the input may significantly reduce the classification accuracy of this method. This effect will have to be determined experimentally.

An extension of the 2002 work by Bergadano was published by two of the co-authors in 2005 [8]. This study cohort was considerably larger, with 205 subjects. The text entry system was also different, in that they used a continuous free text data acquisition protocol. The authors believe that short text entries will not supply the required data to generate an accurate model sufficient for classification purposes. Asking users to type in long text within a single trial may not be feasible — or at least not attractive to the users. Their alternative is to extract keystroke dynamic-related data while subjects are interacting with a computer. This approach is termed free text by the authors — and is also a continuous mode of data collection — in that user data is extracted during and may be used to authenticate after a user has access to the computer system. The authors also employ a different similarity/scoring metric than their 2002 work. Lastly, digraphs were used instead of trigraphs as in their previous work.

Gunetti and Picardi state that there may be a difficulty when using a relative keystroke dynamics scheme for user authentication — as in the Bergadano study [2]. The difficulty can be explained by a simple example: if two users enter the same text with exactly the same dynamics, but at two different typing speeds — they will not be discriminated as different users by the Bergadano scheme. This is a consequence of using a relative scheme. Gunetti proposes in their 2003 paper to extend the classification scheme to include both the relative and absolute measures (referred to as "R" and "A" measures respectively, in their paper). The "R" measure is identical to that employed by Bergadano, based on statistical measures of edit distance values. The "A" measure reflects actual typing speed of each pair of identical n-graphs. More formally, they define a similarity measure metric based on the duration of the same n-graphs occurring is two typing samples $S1$ and $S2$. This metric, which they term $G_{Si,di}$ are treated as *similar* if $1 < \max(d1, d2)/\min(d1, d2) \leq t$, where t is a constant greater than 1. Note that d1 and d2 refer to the duration of the digraphs from typing samples $S1$ and $S2$. They then define the "A" distance between $S1$ and $S2$. Over the common n-graphs for a given value of t is provided in Eq. 32.5.

$$A(S1, S2) = \frac{1 - (\text{number of similar } n\text{-graphs for } S1 \text{ and } S2)}{(\sum n\text{-graphs shared by } S1 \text{ and } S2)}. \tag{32.5}$$

The authors then combine the "R" and "A" metrics for the process of user authentication and/or identification, using essentially an additive formulation. The classification accuracy of this system, in terms of FAR/FRR was similar to the values obtained in the Bergadano study [2]. The authors experimented with various combinations of digraphs, trigraphs and tetragraphs) with "R" and "A" values and the data suggest unequivocally that the "R" values outperformed the "A" value in terms of classification error. The authors then combined "R" and "A" measures to see if this could enhance the classification accuracy of the system. The authors tested various combinations of the two measures and find the classification error was significantly reduced (to less then 1%), similar to the results from using "R" alone in the 2002 Bergadano paper. Note that the error is simply a misclassification error. The authors do report FAR/FRR data — and these results indicate that for a very low value for FAR (less than 1%), the FRR ranged from 5 to 14%.

The addition of absolute typing dynamics (i.e. typing speed) did not appreciably enhance the accuracy of this system (on the contrary it generally reduced the overall FAR/FRR compared to their previous work). Part of the reason for the variability in their results may be due to the long time frame over which data was collected — subjects were given 6 months to enter the required data. The input devices varied between office computers and personal laptops, which might have affected the consistency of the typing dynamics of the study participants. In addition, there are several other issues with this study that need further elaboration. For one, the similarity measure does not allow for duplicate digraph entries, as the quotient in the similarity metric must be larger than 1. Secondly, the "A" measure is really 1-Hamming distance. Though the authors argue against this issue (Gunetti and Picardi, 2003, pp. 321 and 322) for their reply to this issue. The amount of text provided by the users was quite substantial in many cases, with an average character length of 700–900 characters, repeated 14 times. This may be a prohibitive amount of data to collect other than in a continuous free text format. Lastly, the essential difficulties with the 2002 work are generally applicable to this study, as both studies employ essentially the same classification processes. Lastly, one may wish to examine how well edit distance classifies text when employed on much shorter strings.

Choras and Mroczkowski have published a few papers on keystroke dynamics based on a similar strategy [3,4]. In this work, the use of the edit distance for digraphs/trigraphs as defined by Bergadano and Gunetti was employed on a short text string (login ID and password). The 2007a study employed 18 subjects whom were required to enter their login ID and passwords (which was their full name in forename/surname format) 10 times to establish a reference for each of the subjects. To test the system, 20 login attempts by each participant was used for FRR calculations and 10 attempts for FAR calculations. The authors employed a series of thresholds (0.25, 0.3, 0.35 and 0.4) to decide whether or not to accept an authentication attempt. The FRR results yielded an average value of approximately, 22% (range 0–55%). The FAR results were more remarkable, with a value of 0% for all but two users (1.9% and 8.1%, respectively). Their 2007b paper [4] performed a very similar analysis, examining the use of digraphs and trigraphs, via an edit distance measure to estimate values for FAR and FRR. The results indicated that digraphs provided a significantly reduced FAR/FRR compared to the use of trigraphs. The authors indicate that the thresholds should be larger for digraphs than trigraphs, indicating that digraphs were more reliable/discriminating than trigraphs.

The results from the Choras and Mroczkowski studies indicate that using edit distance for short text strings results in a much higher FAR/FRR value than for long text strings. In their studies, they required users to enter their authentication string 10 times — yielding at most 10 samples of each digraph. Although not stated explicitly, the average length of the authentication string should be in the order of 10–15 characters (the average number of characters in a person's name), yielding 13 and 14 digraphs/trigraphs, respectively. This is a relatively small sample size compared to the Bergadano and Gunetti studies. Their use of a hard threshold value for discriminating users may not be a suitable approach as well. One could generally question whether the edit distance per se is a sufficiently robust similarity measure.

32.5. Sequence Alignment and Intruder Detection Based Biometrics

What happens if someone acquires login credentials of another user (legitimate) of a trusted computer system? Well, they have full access to the hijacked user's repertoire of capacities on the system. Hijacking can occur by simply taking over someone's computer while they are away from it — this typically assumes that the lockdown facility is inadequate. In addition, a person can hack into an account, acquiring the username and password. Regardless of the method employed, this hijacker is engaging in a *masquerade* scenario. How can one develop a system to detect a masquerader? There are two basic approaches: *anomaly detection* and *penetration identification* (for further details, see [5, 15]. In anomaly detection, one possesses a representation of a "typical user" — and any deviation from typicality issues a warning signal that a possible intruder may be logged in. Penetration identification relies on possession of intruder signatures — and then any match to one of these signatures issues an intruder alert. In the work by Coull *et al.* [5, 6], a bioinformatics approach to intruder detection has been proposed for [5, 6]). The basis of these works is the use of audit trail information from masquerade attacks and the use of a modified local alignment algorithm for detection.

In [5], the authors introduced their bioinformatics approach to intrusion detection [5]. In this study, the authors employed the SEA data which was introduced into the literature by Schonlau *et al.* [15], which provides data derived from the implementation of the UNIX *acct* auditing facility. The SEA dataset consists of 50 blocks of 100 commands each (5,000 commands) for each user. The data is clean from intrusion detection attempts — and serves as a natural basis for a control group. Data from 70 users were employed in this study (both Coull studies), and split into two groups. One group (50 users) was used as the test subjects and the other (20 users) had their commands interspersed in the test group data. The interspersed group then served as the masqueraders who were to be detected by the ID system. The task then was to determine if a masquerade attempt could be found in the masquerader group based on data from the test group. The sequence of elements, therefore, consisted of commands that users could enter while interacting with the computer system. The test user's audit trails were considerably longer then the masquerade sequences, so a global alignment would consistently produce a low score simply because of the mismatches (which would have to be filled in with gaps) that occur when two sequences are of equal length. A local alignment would be more suitable in this scenario — as it looks for highly conserved (identical if possible) elements within two sequences [17]. The authors claim that the standard Smith and Waterman local alignment would not be the most suitable algorithm — as it allows for blocks of unregistered elements to occur as prefixes and suffixes. That is, the local alignment allows for extended prefix and suffix regions to be excluded during scoring. This facilitates the alignment of local regions within the sequence, without requiring that the entire sequence is aligned. Using a standard dynamic programming approach, scores were computed based on local matching of sequence elements. Essentially, if the elements at a particular position in the sequences matched, a positive score (+1) was placed in the appropriate position within the scoring matrix (see Lesk, 2006 for details on scoring matrices). If a gap had to be included, then a negative score would be incurred, the magnitude of which depended on whether it occurred in the test or masquerade sequence. If the gap was placed in the test sequence, a score of -2 resulted, and -3 if it occurred in the masquerade signature. If there was a mismatch between the two elements, a score of $+0$ was recorded. Lastly, if the score was negative for a match or gap insertion into the test elements, then the score was reset to zero. The purpose of this scoring approach, which varies from the Smith and Waterman approach, was to ensure that gaps would be allowed, without compromising classification accuracy. Resetting the score back to zero allows an arbitrarily long prefix sequence to occur without compromising the over sequencing score. In addition, suffixes can be ignored, at the expense of reducing the score somewhat. In short, their modified scoring sequence is designed to facilitate the alignment between test and masquerade sequences with a minimal number of gaps or mismatches.

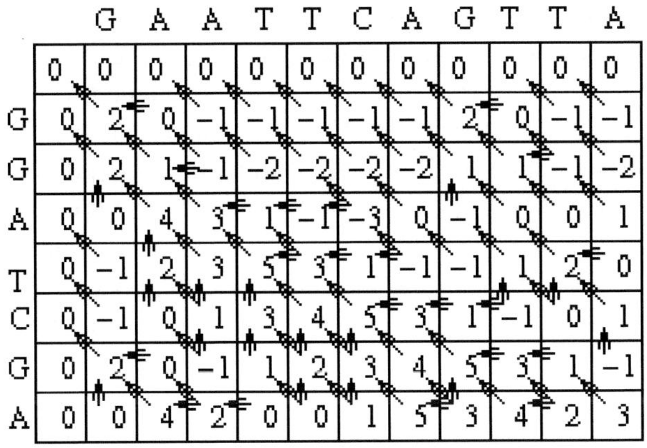

Fig. 32.7. An example of a scoring matrix and the backtracking steps required to compute the overall alignment score.

Once the scoring matrix has been filled (which has the dimensions of the two sequences being assessed), an overall score is assigned to the two sequences, starting from the bottom right and working up to the top left (see Fig. 32.7 for an example of a scoring matrix). The score then reflects the similarity between two sequences. The score can range from 0 to 2*length of the shortest sequence (assumes a perfect match across all sequence elements). If a masquerade is to be detected by this approach, the score should be less then the maximal value — indicating that there is a mismatch between the two sequences. How much can two sequences vary to qualify as a masquerade attempt? This question can be addressed in the current context by determining a threshold that distinguishes normal from masquerade attempts. If the threshold is set too leniently, then the false negative rate will increase, and likewise, the false positive rate will increase if set to stringently. Using these metrics as a guideline, the threshold was determined based on a cross-validation the user's signature against itself. The results from the 2003 work revealed hit rates (true positives) of approximately, 76% with a false positive rate of approximately, 8% [5]. This result was significantly better than all reported results and probably reflects a lower bound on the actual error rate.

In [6], the authors extended the original 2003 work by incorporating a *dynamic signature*. In their previous work, the signature for a user was static — and if the computer system changed in some way, for example, by the addition of new software, which would result in new commands being entering, then their signature would change [5]. This change in signature could be classified as a masquerade attempt, increasing the false negative rate. The key to implementing a dynamic signature is to identify motifs within the signature — that is, regions with high matching rates. If a new command is entered by the user (assumed due to the installation of new software) occurs within one of these conserved motifs, then one can assume that the command is legitimate (what the authors term a "bad" mismatch). To implement a "good" mismatch requires that a given position within a command sequence can be filled with more than one value. That this "good" mismatch occurs within a highly conserved region indicates that it should not reduce the score appreciably — that it reflects natural variation for this particular signature. Using the same SEA dataset and protocol, and deploying signature updating, the results yielded a slightly lower hit rate (69%), but significantly reduced the false positive rate to just under 2%.

32.6. DNA-Based Biometrics

DNA is a polymer consisting of four bases (as nucleotides) that are chemically linked to form long strands, much as proteins consist of polymers of AAs. Our DNA exists naturally as a pair of strands which run anti-parallel to one another as depicted in Fig. 32.8 (see [11] for a comprehensive treatment of this topic). In humans, DNA is packaged into a set of 23 homologous chromosomes — forming our genotype. DNA of course sits at the top of the central dogma, and serves as the information driver for the unfolding of biological development. It is interesting to note that of the 3×10^9 nucleotides contained in our genome, the overwhelming majority of this DNA is not used for the coding of proteins. Generally speaking,

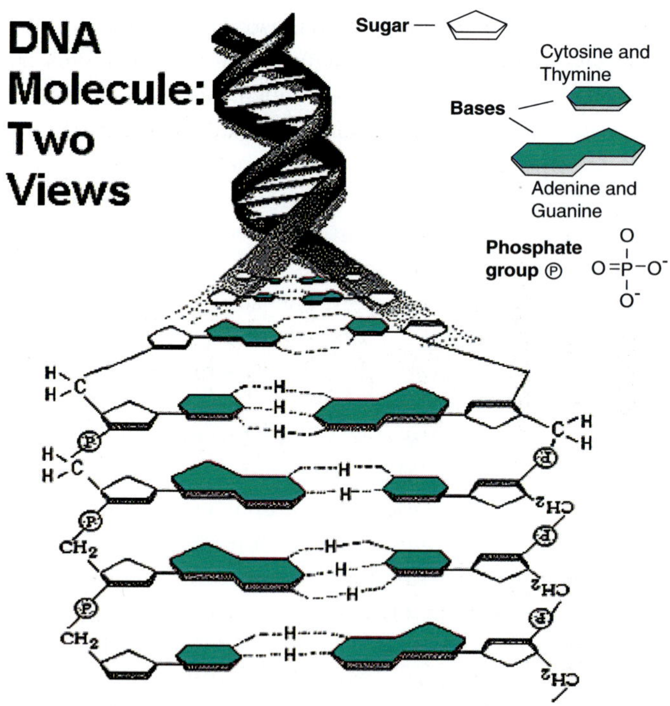

Fig. 32.8. A graphical representation of the double helix as it exists within each human cell (that contains a nucleus).

our genome can be partitioned into two types: coding and non-coding regions. The coding region accounts for approximately 5% our of genome (that is, encodes for proteins)! The function of the balance is not completely understood, but may reflect evolutionary baggage. Another interesting point to note is that our DNA differs from our congeners by approximately 0.1% (3×10^6 nucleotides) found within the non-coding regions. To use DNA as a means of person identification, one must seek uniqueness within this 0.10% of the genome. Though most of the non-coding DNA is composed of randomly arranged nucleotides, there are repetitive patterns that appear a variable number of times — termed variable number tandem repeats (VNTRs). These are short sequences of DNA that appear within specific regions in the non-coding DNA, and appear to be unique to an individual. The location of these VNTRs is constant across individuals, but the number of repeats is unique, and forms the basis of the technique termed DNA fingerprinting. Fig. 32.9 depicts the distribution of the regions where VNTRs are searched for during the process of DNA fingerprinting. The number of VNTRs at these locations are unique to each individual, and depending on the number of sites examined, one can obtain signature of an individual that is unique (chance of a duplicate match is less than 1 in 1×10^9). Even identical twins have different genomes.

DNA fingerprinting is a very effective biometric tool, as it is unique and is resilient to change during the lifespan of the person. The difficulty with this approach is that it requires sophisticated equipment and time to obtain the fingerprint. These factors make it difficult to use in real-time — and provide at

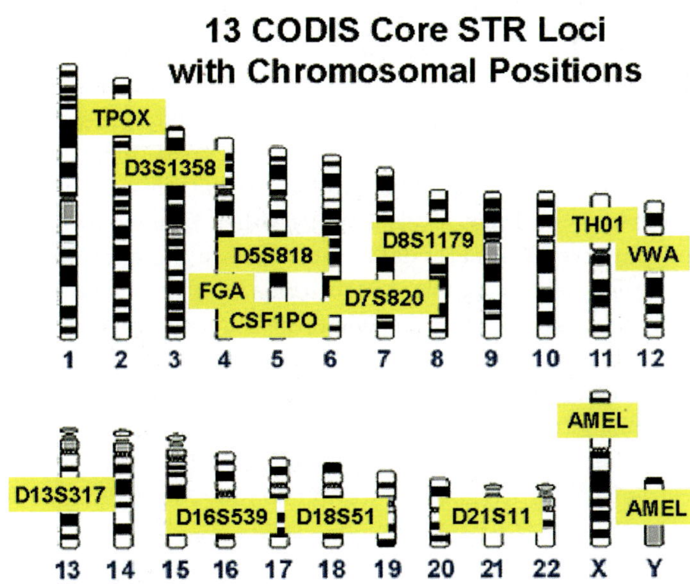

Fig. 32.9. The chromosomal locations of regions known to contain VNTRs which are searched during DNA fingerprinting. Taken from: http://en.wikipedia.org/wiki/Combined_DNA_Index_System.

present at least, an insurmountable barrier to routine deployment. If these issues can be solved, this approach is by far the most reliable and most difficult to forge.

The predominant use of DNA fingerprinting is of course in forensics — the scientific study of crime evidence. There is a large FBI-based data repository called CODIS (*CO*mbined *D*NA *I*ndex *S*ystem), as well as several other databases (notably in the United Kingdom) that house the vast majority of DNA fingerprinting data. As of May 2007, 177,870 forensic profiles and 4,582,516 offender profiles have been accumulated (http://www.fbi.gov/hq/lab/codis/national.htm), making it the largest DNA databank in the world, surpassing the United Kindom's National DNA Datatbase, which consisted of an estimated 3,976,090 profiles as of June 2007 (http://www.publications.parliament.uk). As of the same date, CODIS has produced over 49,400 matches to requests, assisting in more than 50,343 investigations (http://www.fbi.gov/hq/lab/codis/aidedmap.htm). The public perception of DNA databases has generally been positive (at least in the United States), and the number of entries is continuing to rise. Privacy issues have been addressed in the development of DNA databases, and all personal information associated with a DNA sample is removed before the sample is stored. The last step is to improve the technological barriers to real-time sample acquisition. One can only assume that this is simply a matter of time.

32.7. Conclusion

The deployment of a molecular biological approach to biometrics provides a novel capacity to implement biometric solutions in both a behavioural as well as a physiological manner. Using the computational tools developed to analyse biological molecules such as DNA and proteins, novel tools such as the edit distance and variants of the sequence alignment algorithm have been successfully adopted to work within the biometrics domain. The case studies presented in this chapter with respect to keystroke dynamics provides evidence that this approach is not only feasible, but provides accuracy levels rivaling other behavioural biometrics, and even lower end physiological methods as well. We are pursuing further work that will seek to apply sequence alignment algorithms to other behavioural biometrics such as mouse dynamics. Generally speaking, the sequence approach can be applied to any data domain that can be delineated as a sequence of events with a finite set of labels (alphabet). The advantages of the merger between molecular biology and behavioural biometrics are many: the algorithms have been thoroughly studied and are relatively computationally efficient and the non-invasiveness of behavioural biometrics ensures that real-time, accurate and non-invasive techniques will continue to flourish.

With respect to the physiological side of molecular biology, that is, DNA fingerprinting, the technique is the most accurate of all, with an error rate of less than 1 in 1×10^9. The remaining issue with respect to DNA fingerprinting is the ability to carry out the analysis in real-time. Waiting 30 min is not realistic if it is to be routinely deployed. The deployment time is a technological issue, for which there does not appear to be any theoretical limit to — and one would assume that this issue is surmountable. The other issue is the public perception of this biometric. In the United States, public perception of DNA fingerprinting has moved towards a growing acceptance of this technique. Whether this is true on an international scale remains to be determined through proper research methods. It will certainly be influenced by the frequency of database and related breeches of security. Herein lies the catch-22: to enhance security — we require more stringent access control mechanisms — but public perception influences the rate at which research developments in the field reach the public sector. How do we solve this critical issue?

References

1. A. A. E. Ahmed and I Traore. Anomaly intrusion detection based on biometrics, *Proceedings of the 2005 IEEE Workshop on Information Assurance and Security*, United States Military Academy, West Point, NY, 2005, pp. 452–453.
2. F. Bergadano, D. Gunetti and C. Picardi, User authentication through keystroke dynamics, *ACM Transactions on Information and System Security* **5**(4) (2002) 367–397.
3. M. Choras and P. Mroczkowski, Keystroke dynamics for biometrics identification, *ICANNGA 2007*, eds. B. Beliczynski *et al.* (LNCS 4432, 2007a), 424–431.
4. M. Choras and P. Mroczkowski, Recognizing individual typing patterns, *IbPRIA 2007, Part II*, eds. J. Marti *et al.* (LNCS 4478, 2007b), 323–330.
5. S. E. Coull, J. W. Branch, B. K. Szymanski and E. A. Breimer, Intrusion detection: A bioinformatics approach, *Proceedings of the 19th Annual Computer Security Applications Conference*, Las Vegas, USA, 2003, pp. 24–33.
6. S. E. Coull and B. K. Szymanski, Sequence alignment for masquerade detection, *Computational Statistics and Data Analysis* **52**(8) (2008) 4116–4131.
7. F. Crick, Central dogma of molecular biology, *Nature* **227** (1970) 561–563.
8. D. Gunetti and C. Picardi, Keystroke analysis of free text, *ACM Trans. Inf. Syst. Security* **8**(3) (2005) 312–347.
9. Jones and Pevzner, *An Introduction to Bioinformatics Algorithms* (MIT Press, Cambridge, MA, USA, 2004).
10. A. M. Lesk, *Introduction to Bioinformatics*, 2nd ed. (Oxford University Press, Oxford, UK, 2005).
11. B. Lewin, *Genese VII* (Oxford University Press, Oxford, UK, 2000).
12. S. B. Needleman and C. D. Wunsch, A general method applicable to the search for similarities in the amino acid sequence of two proteins, *J. Mol. Biol.* **48** (1970) 443–453.
13. K. Revett, On the use of multiple sequence alignment for user authentication via keystroke dynamics, *International Conference on Global eSecurity 2007 (ICGeS)*, University of East London, 2007a, pp. 112–120.
14. K. Revett, A bioinformatics based approach to behavioral biometrics, *Proceedings in the Frontiers in the Convergence of Bioscience and Information Technologies*, Cheju Island, Korea, 2007b, pp. 665–670.
15. M. Schonlau, W. DuMouchel, W. Ju, A. F. Karr, M. Theus and Y. Vardi, Computer intrusion: Detecting Masquerades, *Statistical Science* **16**(1) (2001) 58–74.

16. T. F. Smith and M. S. Waterman, Identification of common molecular subsequences, *J. Mol. Biol.* **147** (1981) 95–197.
17. G. Tandon, P. Chan and D. Mitra, MORPHEUS: Motif oriented representations to purge hostile events from unlabeled sequences, *Proceedings of VizSEC/DMSEC'04*, Washington, DC, USA 2004.

Chapter 33

CRIMINAL DATA MINING

VIKTORAS JUSTICKIS
Professor of Criminology and Forensic Psychology,
Mykolas Romeris University, Vilnius, Lithuania

33.1. Data Mining. Definition and Related Concepts

33.1.1. *Definition*

Data mining (DM) is the exploration of sets of data (usually large ones) in order to discover their features, patterns, regularities that can be useful in solution of practical and theoretical problems.

The use of the term "data analysis" (DA) in science and applications is rather amorphous. Most often it is associated with processing of large amounts of information, discovering of "hidden information" in these data, use of rather sophisticated, "non-trivial" statistical procedures and interactive data processing [1–4].

The above-mentioned definition of DM includes such concepts as "data" and their sets, their "exploration", "new features, patterns, regularities" and "practical and theoretical problems" whose solutions are to be promoted using results of DM.

These key concepts are discussed in turn.

Data. The data in the DM definition above can be data sets from different areas (finance, mass media, Internet, medicine, industry, radio astronomy and many others). DM is usually associated with large quantities of data.

Exploration usually is seen in DM in a very broad way. It includes processing of data by quite different, usually statistical, methods. Lists of statistical procedures used in DM usually include data classification, measuring of connections between variables, prediction, affinity grouping, clustering and description [1]. In fact, lists of statistical procedures used in DM can include all methods, from the most elementary to the most sophisticated. There is hardly any statistical procedure that could not be used for DM.

New features, patterns, regularities supposed to be revealed by DM also are seen in a very broad way. They include both basic ones (e.g. averages, modes, medians, frequencies, percents, etc.) and outcomes of the sophisticated ones (e.g. of cluster analysis, regression, structural equation and neural network methods).

Knowledge to be discovered by DM also is seen in the broadest way. They can be new peculiarities of these data, of their interconnection (correlations, concordances, associations, etc.), their general structure (clusters, groups, latent factors), etc.

Practical and theoretical problems whose solutions are to be promoted using results of DM. There are no limits in the range of problems that can be studied and/or solved using DM. Most often DM contributes to their solution by providing information that is necessary for solution of the problem. That can be done by revealing latent (unobservable) factors causing this problem events (criminality, economical and social crises, sudden changes in economical and social situation, etc.), by specifying shapes of the problem (e.g. detecting space and time pattern of some dangerous processes).

33.1.2. *DM and Associated Concepts*

There are several other concepts that are of key importance in data exploring. These are in some respects associated with DM, in other contrasted. Boundaries and interrelations between them and DM have to be explored. Two such concepts are of special importance:

Experimental method (EM)
Date Analysis (DA).

DM and EM. DM is usually contrasted with the EM. Relations between DM and its "opposite" EM are especially important for understanding of DM.

First of all, EM is "hypothesis guided". Its motto is: "First hypothesis — then data". The first step in EM is formulation of the hypothesis. This is done based on current theory. This hypothesis determines what will be tested, which variables will be studied and how these variables will be operated. This operation produces outcome data that are explored to test the hypothesis. So, in an experiment data that are not available, they are specially "produced" to verify the hypothesis. An experiment usually tests relations between variables. No wonder that amount of data in experiment is usually small.

The EM follows strict rules. It provides conclusions on causal interactions between verified variables.

DM is opposite to the EM literally in all respects.

1. As opposed to the EM, DM is not hypothesis guided. Instead it is "data guided". Its motto is: "First data — then hypotheses".
   In DM data are not produced. Instead some available data, usually "convenience" or "opportunity" sets of data are used. This means that these data are not specially gathered for DM, they have been collected for some other use.

Such data are explored with the aim of finding "something interesting" — any feature or regularity of interest. When the latter are found, suggestions (hypothesis) on their nature and reasons behind them are made.

2. DM (again contrary to EM) is not any definite line of successive actions. Instead DM is a cyclic, "shuttle" process. The analyst examines their data, searching for their regularities. These regularities are studied, and then the analysis is continued. This cycle can occur many times. It is continued till the analyst decides that no further interesting regularities can be found and that they already "understands the data" and they have got all information they had been searching for.
3. DM (contrary to EM) does not provide any direct evidence of causal interaction between variables. It can only establish interconnections between them.

DM and DA. DA is the second concept associated with DM. Sometimes both are seen as allied or even identical. However, their interrelation should be specified.

Both DM and DA are similar to each other and contrasted to EM in several important respects.

Both start with data and are "data guided". Both try to discover some regularities in their data.

Similar to DM, DA does not provide direct evidence on causal relations between variables.

As for differences, the line of demarcation between DM and DA is very unclear.

The criteria of triviality or non-triviality of statistical procedures (and, accordingly, of received results) are most often used to distinguish them [5, 6]. DA is associated with the use of rather simple, usual, "trivial" statistical procedures and DM with sophisticated, new, uncommon, "non-trivial" ones. However, there are no different lists of procedures used for DA and those for DM. The great majority of "non-trivial" procedures that today are considered as typical DM procedures (cluster, factor, regression analyses, pattern recognition, neuron nets methods) were widely used long before the DM appeared. On the other side, the most simple and "trivial" statistical procedures, like calculation of basic statistical characteristics (averages, dispersions, frequencies, etc.) are widely used in the modern DM. Actually, manuals for DA and for DM include identical sets of statistical procedures [7, 2].

This chapter is based upon the idea that the criteria of triviality cannot be used to distinguish DM and DA. The point is that the sophisticated, "non-trivial" statistical methods do not consist of any separate group of methods. Their use is not independent on the use of simple "trivial" ones. They are not two different ways in exploring data: DM and DA. In fact, examination of data using rather simple "trivial" tools (usually associated with the DA) and their exploration with sophisticated "non-trivial" ones (usually seen as more special for DM) are parts of the same process. They are interchanging and actively interacting links of the single data examination. Thus, the DM *is not* the name of some group of statistical procedures (e.g. sophisticated, "non-trivial" ones).

The distinctive feature of DM is that its core is *a meta-procedure* guiding data examination, its *strategy and tactics*, including the use of both simple and sophisticated statistical procedures.

The prefix *meta-* usually means "higher level". For example, "meta-language" means a language designed to describe different languages, "meta-galaxy" means a galaxy including many usual galaxies. In the case of DM, meta-procedure means a *procedure showing how different statistical procedures should be used when exploring some kind of data.*

This meta-procedure can be presented as a special *DM algorithm*, Fig. 33.1.

In this respect, the very use of sophisticated statistical procedures (even when they are used to process large amounts of information and provides "non-trivial" outcomes) cannot be called DM. Only when this use is integrated with, and guided by, a strategic meta-procedure, determining the selection and use of these sophisticated procedures and indicating ways in which they are combined with simple ones, is DM.

33.2. Crime DM — Integration of Knowledge on Crime and Statistics

Every strategy is a very general rule or set of rules of action. It shows its general direction and/or succession.

An efficient strategy has to be based upon generalisation of knowledge on the area under consideration.

In the case of the data examination such a strategy has to consider both:

The knowledge of the nature of data under examination;
The statistical procedures that used for their processing.

The knowledge of the nature of data under examination. As mentioned above, data can originate from a variety of sources. However, when coded as figures or symbols, they become deceptively similar. Examining astronomic data on stars, biological ones on genes, or criminological ones on committed offences, one can see the same figures.

In reality though, all these data are very different both in events they represent and in the way they do it. The same figure representing a number of light years between stars and willingness of a criminal to commit an offence are different not only in objects they show, but also in interrelation between the figure and this object. Exploration of data cannot be blind to the special nature. Their nature is the first to be considered when both selecting the way of their statistical analysis and interpreting its results.

The statistical procedures that used for their processing. Knowledge of statistical procedures is the second point to be regarded in designing the strategy of data examination. Modern statistics includes many hundreds of statistical procedures. However, none of them is universal. No one can be applied for any purpose and every one has limitations. Depending on the purpose of data examination and the nature of data some of these limitations can be important, other ones not.

The strategy of data examination must find the best correspondence between specifics of data and particularities of statistical procedures. It is just DM that suppose to ensure this *correspondence.* DM suggests which procedures and in what sequence it is to be used when processing the data. In this sense, DM is an integration of two sets of demands to exploration of data: those of data and those of statistic.

DM is a relatively new concept. No special strategy or meta-procedures were necessary in the over 20 years ago. At that time, statistical procedures were not very numerous and therefore were quite comprehensive. In that situation, the analyst's experience was quite sufficient to select an appropriate tool for data exploration.

Contrary to this, today there are a large variety of different statistical tools and methods available. Each of them has its own prevalence and restrictions, its one destination and particular features.

The data itself has also changed. Especially important is the great increase of very special social data, with their great degree of vagueness and uncertainty, highly intuitive nature, ambiguity about their interpretations, etc.

In this new situation, the matching of data specific problems with suitable analytical tools becomes complicated. Therefore, general meta-procedures facilitating such matching are highly necessary. Therefore, the appearance of DM is a response to this challenge and attempt to find ways to solve these difficulties.

The next parts of this chapter focus upon special nature of crime data and specifically DM in this area.

First, the specific of knowledge on crime will be discussed.

This includes introductory concepts of criminal behaviour and criminal data, and the requirements for its processing. Then specifics of interpretations of crime data and use of criminological theory for it will be discussed.

This will provide the basis to outline the strategy for exploring crime data and for designing of the meta-procedure showing the choice of statistical procedures to use, their consistency and the way in which they have to be used.

33.3. Criminality — Its Nature, Crime Data and their Processing

33.3.1. *Introductory Concepts*

Criminality is one of the most important threats for individual and general safety. Thefts, murders, bodily injuries, rapes, corruption, offences against environment, financial system, etc. endanger personal and public safety. Their cost comes to many billions dollars, millions of human lives, moral decline, distrust in law, disordered relations in society [8–11]. Its consequences affect millions of people:

offenders, victims, their families. Criminality and its control attract great social and economical resources, which otherwise could be used to heighten living standards, improve public health, education, etc. Criminality endangers the basic social values like human life, health, dignity, property, etc.

In this way, criminality endangers the very foundation of the society.

Criminality is a *mass phenomenon*, consisting of many thousands individual offences committed in the world, in a country, a region, city, village.

A *Criminal event* is a single manifestation of criminality. This term denotes both an individual manifestation (a single offence) and mass event. The latter includes all changes in criminality, its manifestations in phenomena closely related to criminality-victimisation, damage, psychological consequences, like fear of crime.

An increase in the number of some offences, change in the ratio of repeated crimes, intensification of crime causing processes, like crime caused by consumption of drugs or alcohol, are all referred to as criminal events.

33.3.2. *Sources and Use of Crime Data*

Criminal events are represented by crime data. So, a single offence is represented in the related police record, mass events, for example, disturbances in public peace on streets in some city can be represented by crime statistic in this city.

33.3.2.1. *Sources of crime data*

There are very many sources of data on crimes.

Official statistics. Every offence that is known to the police must be registered, which produces criminal statistic data. Every step in the further processing of this offence is to be recorded. This produces a great deal of further data on it. There are data produced by police, courts, correction institutions and other institutions reliable for crime control. No wonder that criminality brings huge amounts of data. There is a long history of their study and interpretation [12].

Official crime statistics are gathered and reported by many countries and are of interest to several international organisations, including *Interpol* and the *United Nations.* A part of them is published as a national or regional crime statistic [13–15]. The most impressive source of crime data is a World UNO data bank representing criminal situation and activities of criminal justice in the most countries in the world [16, 17].

The official crime statistics is not the only official source of crime data. The others are public health, finances, government and ecology statistics. Their data also comprise information on violation of regulations in all these areas [18–20].

Empirical studies. The second source of criminality data are empirical studies, especially sociological mass pulls. Many of them are carried out in a typical DM manner. No hypotheses are checked. Such studies are carried out to determine and assess the current criminal situation in a country, region, community.

From the point of view of security, the most important of these studies are victimological ones. Their respondents are asked whether they were victims of crime.

The prevalence of such investigation is, first of all, that they are free of many problems met by official registration. There are many offences that do not get known to police because their victims had not informed police [21]. Victimological polls are much more able to show these offences than official statistics. It is so, first of all, because people are asked directly. Therefore, they provide information even on offences that were not reported to police and escaped registration.

Second, victimological pulls usually are not carried out by police or other criminal justice institutions. The point is that police and other criminal justice institutions are responsible for criminal situation. Therefore, they are interested in embellishment of criminal statistics. This often causes them to resist registration, to hide some offences, especially if organisation of the registration provides opportunities to do so. Opposite to this, victimological studies are most often carried out by impartial scientific institutions and public organisations.

Last but not least is international comparability of victimological data. Because laws vary between jurisdictions, comparing of crime statistics between, and even within countries can be difficult.

As opposed to official crime statistics, based upon national criminal law, which is different in different countries, the same victimological questions can be asked in different countries. This provided opportunity for wide international studies of criminal victimisation, to get much better idea on individual safety in different countries [22, 23]. Research using a series of victim surveys in 18 countries of the European Union funded by the European Commission has reported that the level of crime in Europe has fallen back to the levels of 1990 [24]. Similar trends are observed in the USA, Canada, Australia and other industrialised countries.

In all these cases, the aim is to collect as many as possible data and this way to provide an opportunity to get insight into criminal situation.

Another source of crime data are so called self-reports. They are anonymous pools in which respondents answer whether they committed single offences. The base for this research tool was a striking finding that people willingly provide information about their committed offences if they can provide this information anonymously [25].

To summarise, large amounts of different crime data are available.

Before the era of DM the use of these criminal data sets was rather restricted. They were "an elephant that gave birth to a mouse". Billions of pieces of national crime data were collected to provide few figures on crime. Most often it was the number of crimes committed and data on most important offences, like those against human life, property, public peace. Impressive international victimological projects interrogated many thousands people in different countries, only to get a few figures.

DM revolutionised the use of all these huge treasuries of crime information. Its methods proved to be able to extract great amounts of highly important additional information.

Interconnections between different kinds of criminality, clusters of criminality, models of interconnections between crime variables, latent structures of crime data, all of this provided quite new opportunities for crime prognosis and crime control.

Now exploration of crime data and especially crime DM is intensively developing in several directions.

33.3.2.2. *Use of crime data. New opportunities provided by DM*

The main directions of use of crime DM will be outlined and DM opens new horizons in each of them.

Informational. It is a processing of crime data aimed to characterise a current criminal situation. Informational processing provides different indicators showing general number of offences and offenders, rates of different offences, also data showing reaction of criminal justice to offences: numbers of people arrested, of cases in courts, persons convicted, released, etc.

Informational use of crime data is their most early and traditional use [26]. Since the very beginnings of crime statistics in 1805, the most simple statistical techniques have been used in processing of crime data. They were percents, frequencies, averages. Their use became a tradition in characterising criminality. These techniques are basis of the great majority of national crime statistics. Still today reviewing thick volumes of national crime statistic, one usually finds only most primitive data. They are numbers of crimes and offenders in the country and its single regions, their ratio per 10,000 population, their changes during last years, numbers of people convicted and incarcerated.

The overwhelming information inherent in all these stores of information remains hidden.

Today, DM is revolutionising informational use of crime data.

Traditional scarce description of crime situations now can be supplemented by plenty of additional data on interconnections between different sides of criminality, between criminality and associated variables, clusters of different crimes can be discovered, latent variables affecting criminality can be revealed.

Like all areas of crime DM, this area is in continuing development. Most national crime statistic agencies still traditionally publish only most primary characteristics of criminality, without any in-depth research of their deeper regularities. Traditionally, the latter is supposed to be the remit of the researcher — the criminologists. Such view is usually explained in traditional terms of "objectivity". According to it, "simple" indicators, like percents, averages, etc. are believed to be more "objective" and accurate than more sophisticated ones, like "clusters", latent factors and other results of DM. Therefore, only they are calculated and presented to the public. Actually, this belief has never been substantiated and belongs rather to superstitions. Despite of it, this view is very enduring and is one of the strongest handicaps in further developing crime data exploration.

Secondly, criminologists and others who are supposed to use more sophisticated statistical methods usually do not belong to these agencies and have a rather restricted approach to data bank. Therefore, the traditional "minimalist" approach to processing of crime data still dominates. Overcoming of this barrier, seems to bring a real break-through both in characterisation of criminal situations and in its understanding by public. The latter is especially important, considering the prevailing simplified public view on criminality and ways in which it can be affected. Demonstration of factors, regularities and structures behind the elementary characteristics of crime situation is necessary for understanding the highly complicated nature of criminality.

Specification of a criminal policy. A crime policy is supposed to discover the most important crime problems and to direct crime control actions to their solution. A prerequisite of an efficient criminal policy is reliable knowledge of the distribution of offences in time and/or space. This application of DM is often called — "crime mapping".

In a narrow sense it means the study of territorial distribution of crimes. Different territorial units can be extremely different in intensity of crime. This means necessity of quite different approaches, especially in terms of distribution of resources and effort [27–31].

In a broader sense the "mapping" also includes "time mapping" — seasonal, day and night distribution of the general criminality and single offences [32–38].

Also in this area, the use of more sophisticated statistical tools introduces new opportunities. There are opportunities to discover and explore multi-dimensional patterns of criminality.

Discovery of such patterns is of crucial importance for the modern crime control. It is so, first of all, for the huge diversity of criminality. In the most national legislations, 100 and more acts are defined as offences. Their interconnections can be highly complicated. All this is even more complicated by many regional and time variations in criminality. All this produces a highly confusing picture. In this situation, reduction of this variety to few general patterns in criminality is the only way to make a criminal situation visible and comprehensible.

The knowledge of these patterns and their distribution is also highly important for efficient crime control. It shows ways to concentrate and coordinate crime control activities. The search for patterns of criminality — their clusters, latent factors, decision — tries is a golden way for specifying such strategic directions [39, 32].

Revealing of hidden criminality. Any crime tends to be hidden. Offences escape detection by camouflaging, by being tightly interconnected with highly appreciated activities: forced labor can be masked as working education, a guise of a parent love can cover the sexual exploitation of a child. An especially good shelter for criminal activity is excess of information. For example, the best cover for a criminal bank transaction are the billions of legal ones occurring around. In this situation detecting of the illegal transaction is a much more difficult task than searching for a needle in a haystack.

DM revolutionises this area too.

It provides new opportunities for detection of unusual and criminally suspicious multi-dimensional crime patterns. Many criminal activities that cannot be detected using an one-dimensional approach become visible when searched with the use of multi-dimensional techniques. The list of such criminal activities include financial, insurance and credit card frauds, computer crimes. This includes, first of all, detection of suspicious financial actions [32, 40–42] (Babcock, McGee and Kolbasuk, 2004).

The second direction is detecting criminal organisations and their activities, especially money laundering [43–45].

Crime profiling. Research of typical clusters in committing crimes.

A detective investigating an offence examines its traces, tools, victims, way in which the offence was committed, history of similar offences, crime scenario, all this is important to get an idea on the offender and the offence. Usually a detective tries to recognise patterns of signs telling on the offender, its motives, providing useful information for solution of the crime. The Detective's ability to recognise such "patterns of signs" is highly dependent on their intuition, experience and personality.

DM revolutionises this area too. DM of many similar crimes allows discovery some stable and informative patterns in their traces. Based upon primary data on an offence, such a pattern predicts its still covered distinctive feature, e.g. the most probable personal traits of the criminal.

An important direction in use of DM is analysis of crime scenes and ways in which the criminal acted in it (modus operandi). This is used to facilitate detection, forecasting and prevention of other similar crimes [46–53].

33.3.3. *Interpretation of Crime Data — Criminological Theory and Crime DM*

In this part the nature of criminality is discussed. This will help to outline specifics in processing its data.

33.3.3.1. *Criminal behaviour*

Criminal behaviour is a concept describing one's behaviour immediately before, during and after committing an offence, and the total of inner processes (motivational, cognitive, etc.) that caused, accompanied and followed it.

The concept of criminal behaviour is fundamental for explaining criminality.

It is so, first of all, because any criminal event consists of individual criminal actions. Each of them is a behaviour of a single person, caused by their individual motivations, attitudes, situations they met, perception of this situation, reactions to it, interactions with the victim, other participating people, etc. [54].

Second, the criminal behaviour takes a central place among the total of factors causing a crime. Any changes in one's motivation, in the crime situation, the perception of this situation and in the behaviour of other people can determine whether and how an offence will be committed. The latter, in turn, determines manifestations of this offence and the traces it leaves. Specifics of one's motivation, perception and situation are of the greatest importance for the prevention of crime. They determine how it can be done.

Criminal behaviour is also the most important factor mediating interconnection between global social, economical, political, environment factors and criminality. All general crime control measures (including the law, crime prevention, police activities, etc.) and all social, cultural, moral, political factors (socialisation, education, moral, etc.) can bring any changes in criminality only through criminal behaviour, acting factors immediately preceding and following a crime [55].

Third, this central position of criminal behaviour is of key importance for exploration of crime data. These data are only more or less accurate manifestations of this behaviour. Therefore, just changes in criminal behaviour and in factors most immediately causing and shaping them should be seen behind crime data, when interpreting them. Local patterns of criminality studied by *crime mapping* also are, first of all, manifestations of local crime situations and peculiarities of individual motivations in them.

This is true also for designing actions supposed to control criminality and the general criminal policy. Changes in the *national criminal policy* can affect the national criminality only if it changes individual situations and/or motivation. For all these reasons the criminal behaviour is a central topic of criminology — a science studying crime [55]. Criminology is responsible for description of criminal behaviour, discovering its reasons and the way for explaining it. Therefore, criminological knowledge of criminal behaviour is important at all stages of every study of criminality. In every crime-related problem, it is the criminological theory that suggests what should be explored, what kind of information should be collected, what hypotheses should be set up and verified, how the results should be interpreted and what actions can be taken to solve the problem.

Criminology describes and explains criminal behaviour using *criminological theories*. Thus, the essence and some distinctive features of criminological theories should be discussed.

33.3.3.2. *Criminological theory and their role in processing of crime data*

The aim of any science is to explain events. The basis of such explanation is the proved scientific theory. The theory provides a scientific model for these events. It is scientifically proven information on the factor or sets of factors which caused this event. The situation of theories in criminology (like in many other behaviour sciences) is rather special [56] and very different from one in exact sciences. In the latter only one theory is supposed to be correct. For example, once the theory of gravity was substantiated, it is unnecessary to re-check it each time when having to do with manifestations of gravitation.

The situation in criminology is quite different.

First, it has numerous criminological theories. Second, each one of them can provide its own explanation of a criminal event. Third, none of these theories has any predetermined power.

For example, every modern criminological theory can provide its own explanation of increase or decrease of criminality in some location. However, none of these explanations has any advantage over other ones. Contrary to a theory in exact sciences, any explanation proposed by criminological theory has to be verified. There are always many other theories which have similar scientific status but provides different explanations.

Old wisdom illustrates this situation in criminology (and many other social sciences).

It tells about three blind people who had never seen any elephant and tried to understand what it was like. The elephant was nearby. One blind groped for its leg and said "The elephant is a pale". The second one fumbled its stomach and rejoined "No! The elephant is a barrel". The third one touched its trunk and said "both you are wrong! The elephant is a rope!". This story is a good metaphor of the modern criminology and its theories. Instead of one generally admitted theory of criminal behaviour, there are hundreds of theories. Each one of them is different because it "touches" the "elephant" (criminal behaviour) from a different perspective. Every one pretends to explain the whole "elephant" — to indicate the main factor or the set of factors causing this behaviour. In doing this (also, like these blind people), every theory is both correct and wrong. It is correct, because shows its special side of the "elephant" — criminal behaviour. It is wrong because is not able to explain the whole elephant.

The situation is complicated even more because there is no current way to integrate different theories. Our elephant is not any combination of a pole, barrel and rope. There are many things that need to be discovered to make the integration possible. Also integration of the current criminological theories of criminal behaviour is still not possible because the criminal behaviour, like the elephant, is much more than the combination of all current theories could show.

Therefore, the scientific status of modern criminological theories is different than in exact sciences. When dealing with a criminal behaviour (for example, when searching for reasons of street criminality) *it is not possible to say in advance which of many modern criminological theories will be able to explain it.* Instead, every criminal event proposes its own (hypothetical) explanation. Every one indicates a *different set of factors* supposed to cause the criminal behaviour in question.

It means that it is up to a researcher to review all possible explanations and to find which one is appropriate.

The situation that every criminological theory can provide its own explanation of any criminal event is highly important for exploring (and processing) crime data.

This situation raises a very special challenge: When trying to explain a criminal event, all criminal theories have to be considered. In other words, exploration of crime data has to be done with due regard of all existing criminological theories. If some theory is missed, there is a risk that this missed theory was the correct one.

Most social sciences also have many competing theories, however, criminology is extreme in this respect. The point is that it studies criminality by integrating the theories of many other sciences. Thus, criminology exceeds many other social sciences in the number of theories. Representative books on criminological theories expound hundreds of them [57].

Therefore, analysing crime data can be a titanic task. One has to consider these data from the point of view of all possible criminological theories. It is not all. Exploring crime data, one has to have a good idea on probable ways in which every theory can manifest itself in these data.

The challenge of this extremely multi-theory nature of criminology can be met in several ways. The EM deals with this in a most simple and consistent way. Its approach is to verify all possible explanations in sequence, one after another, so that no possible explanation will be lost. Each one of them will be verified and the outcome of every verification will be a clear statement: of whether this explanation is correct or not.

However, in case of crime exploration, this prevalence of experimental approach is outweighed by its labor-intensiveness. Considering the great number of criminological theories to be verified it means an endless job.

DM tries to solve this problem in the most direct way. When exploring data, it tries to discover manifestations of different theories. Doing this it tries to suggest which of many criminological theories is the most promising in explaining the criminal behaviour in question.

The situation met in exploration of crime data could be illustrated by a broader example.

33.3.4. *Extended Example of Crime DM*

The government of a city "B" faced the persistent increase in violation of the public peace on its streets. Offenders bother or attack passer-bys, damage public property, engage in hooliganism, etc.

The authorities of the city have to develop a set of efficient measures to control these crimes. To do this, a well-established knowledge of factors causing it is necessary.

The first way to acquire it is the experimental approach.

In this case, possible explanations of the event have to be set up. Then every probable explanation should be tested using the experimental approach: organising experimental and control groups, manipulating variables, comparing outcomes. Every criminological theory can propose its own explanation and it is necessary to check each of them. Considering large number of criminological theories this work can take a considerable time.

In the same situation, DM proposes a different approach. It proposes to try to recognise the proper explanation right away by exploring available data on street criminality in the city. This should be done by examining data and trying to recognise in them manifestations of some criminological theory.

Usually a city has plenty of data that could be examined. They can be data acquired by registration of street offences, public surveys, interrogation of crime victims and witnesses, police officers' observations, psychological and other examinations of offenders, etc. All these data can build a data bank of the street criminality in city B. In DM, an analyst explores these data and tries to discern manifestations of hidden factors causing street criminality in this city.

There are several possible cases found whilst exploring this data

The first case. Reviewing the data the analyst has noticed that the overwhelming part of these crimes are committed by low-educated, aggressive men, acting in gangs and belonging to local region populated with similar people. This makes quite probable that of all criminological theories the most proper in this case is the theory of "lower class sub-culture". This theory was developed by Miller [58]. It states that the most important source of criminal behaviour is the lower class criminal sub-culture. It is the set of values and views of groups of people, belonging to the educational and financial bottom of society and sharing a value system which is rather opposite to the rest of society. The core of this value systems are "masculinity", aggressive self-affirmation, rejection of long-time life perspective in favor of the current moment, hostility against the values of the middle class (education, career, rationality, property, etc.).

This value system directly or indirectly causes signs which were noticed when observing the data: low education of offenders, their high aggressiveness, etc.

All of this provides a good reason to believe that the criminal behaviour in question was manifestation of this sub-culture.

The second case. Surveying the same data, the analyst has seen a quite different picture. It was noticed that the majority of street offences had been committed by young people with a long history of unsuccessful attempts to achieve any higher economical and educational status. The data can show that they are rebellious and their offences have signs of protesting behaviour. All this may suggest that this criminal behaviour should be explained in terms of Robert Merton "norms-aims" and "norms-frame" theory [59].

This theory explains criminal behaviour in terms of discrepancy between the individual "norms-goals" which establish the aims that a person has to achieve in their life, and the "norms-means", which define (and therefore restrict) ways in which these targets can be researched. The pattern of signs discovered when exploring the data (young offenders' age, their history of unsuccessful attempts to achieve any higher economical and educational status, prevalence of disappointment and protesting actions) give a different explanation than the previous one. They provide a good reason to believe that it is the set of factors described by R. K. Merton's theory that is responsible for street offences in the city.

The third case. When exploring the data, the analyst found a different picture. The analyst discovered a strong trend for street offences to be committed in places with insufficient lighting, also in remote, rarely visited streets. This may suggest that criminal events in the city should be explained in terms of control theory [60].

This theory explains criminal behaviour as a result of weakened social control over one's behaviour. This suggestion can be supported by data on the personality of offenders showing that they have characteristically impaired social links.

The fourth case. Observations of police officers show association of street offences with outward appearance of buildings on different streets. Presence of graffiti, broken windows, dilapidated buildings, pile of garbage is associated with increased level of criminality. This suggests an explanation in terms of the "broken" window theory [61]. This theory explains offences as a result of interpretation by possible offenders of "signs".

The list of further explanations and theories can be very long.

As mentioned above, all criminological theories could be used to explain every criminality-related phenomenon, including street offences in the city B.

What the analyst finds when reviewing their data is *a promising explanation.* It is a situation when examining the data, a set of manifestations characteristic for some criminological theory is found. It does not mean that just this theory is the correct one in the given case. It only means that the in-depth research of this explanation is promising. This includes the study of variables indicated by these explanations, of their values and interconnections. This study should show whether these values and connections confirm those predicted by the promising theory. Also, additional information supposed to support or reject this explanation can be gathered. Statistical procedures that for different reasons could not be applied to all data but only to smaller part of them can be used and provide further information on this promising set.

This additional information increases or reduces certainty that the proper explanation is found. For example, in our example, the primary general examination of data supposed that W. Miller's "lower class" theory is the most promising. However, a following study showed that no further manifestations of this theory are found. The majority of offenders do not belong to the lowest class of society. Also interconnections predicted by this theory are not found. This has at least two consequences.

First, for a given moment, it refutes the belief that this explanation is promising.

Second, it provides new information on *all* data. For example, the finding that offenders do not belong to the lowest social group enlarges probability of several other explanations.

Thus, the in-depth research of promising explanations enhances the total amount of available information. This way it also increases the probability of the proper explanation to be recognised.

Criminology is able to provide numerous explanations of every criminal event that it would be impossible to try to test experimentally each of hundreds possible explanation. DM seems to be the only choice in this situation. It provides a good chance for discovering the proper explanation much quicker.

However, as it is usual in this life, DM solves one great problem but causes a new, also great one. The ratio of its prevalence and shortcomings is important.

Crime DM — Prevalence and shortcomings

Prevalence. Considering the huge amount of possible data and great number of possible explanations to be checked, checking in line of all possible explanations is likened to the search for a needle in a haystack. DM can make the identification of a correct explanation much shorter, in success, it can be found right away. It is in a striking contrast with the labor intensive and tardy EM. The great prevalence over the experimental approach and its possible benefits are quite obvious.

However, free cheese can be found only in a mouse trap. A list of shortcomings of DM follows:

1. *Dependence of DM success on available data.* DM usually uses "convenience" or "opportunity" data. In our example it was data acquired by registration of street offences, public surveys, interrogation of crime victims and witnesses, police officers' observations, data received in psychological and other examinations of offenders. It is true that in examining these data, it is possible to notice manifestations of the criminological theories that are able to explain criminal data. However, there is no guarantee of it. It is quite possible that for different reasons the proper theory was not able to manifest itself in our data. In this case the proper explanation will not be found.
2. *Indirect manifestations of the proper theory.* When examining data, it is not always possible to see factors causing a criminal behaviour directly. They manifest themselves indirectly, which means through the chain of mediating variables. These mediating factors can vary and therefore reduce or distort the manifestation of theory. The effect of these mediating factors can cause the real factors causing criminal event remain unnoticed, their manifestation can be reduced or distorted. For this reason, again, the proper explanation can be missed.
3. *Dependence of DM on subjective skills and other traits of the analyst.* An analyst is supposed to "notice" or to "recognise" indirect manifestations of the proper explanation. Therefore, the success is dependent on the analyst's expedience, skill, attentiveness, effort, also on their prejudices and biases. Many psychological factors affecting the analyst's perception and recognition are involved in this process. In psychology, the analyst's task is described as "figure-ground perception". Many psychologically caused illusions and deceptions of this type have been described [62]. An analyst can be under influence of each one of them.

   An analyst has to take many intuitive and therefore poorly controlled decisions. For example, when noticing probable manifestations, the analyst has to answer the question of whether these manifestations are strong enough to assume the action of the hidden causing factors, described by some theory. The answer to this question strongly depends on the analyst's ideas on this theory. It is important how much the analyst believes that this theory is correct and how probable its manifestation are in the given situation.

4. *The DM does not reveal causal relationships between variables.* It is only the interconnection (correlations, associations, etc.) between variables that DM discovers. However, it cannot tell which of them is causal and which not. This shortcoming is especially important for crime DM. To be able to control criminal behaviour it is necessary to know its reason and its moving causes.

These moving causes can be discovered only by special methods, able to reveal causal interactions. DM is not able to do it. The most that can be achieved using DM is an *indirect* indication on probable causal relation between them. It is possible to identify a strong interconnection between variables such as an indirect indication on a probable causal relation between them. But it still has to be verified using EMs.

33.3.5. *Conclusion — Basic Demands to General Algorithm of Crime DM*

The review of specific feature of criminological data leads to several conclusions important for their exploration by DM.

Any crime data are manifestations of the criminal behaviour behind them.

Any criminal behaviour can be explained from the perspective of all modern criminological theories.

Therefore, review of crime data has to be done from perspective of all modern criminological theories. Any of them can propose its own explanation of the criminal event under consideration.

Explanation of any of the criminal event is done by indicating the factor or the set of factors that may cause a criminal behaviour under consideration.

A causing factor (or their set) manifests itself in data indirectly by affecting observable variables, their values, interconnections and the structure of the data. Manifestations produced by a causing factor (or the set of factors) present itself in the given data by showing its own characteristic pattern of observable signs.

This means that in exploration of data, factors causing the criminal event should be searched for by looking for its characteristic pattern in the manifestation of the data.

Though any criminological theory can propose a probable explanation of data, it is not known in advance, which theory will be correct. Therefore, manifestations of any one should be sought. This means that it is not possible to single out any group of criminological theories in advance as the most promising for some specific criminal data. In the extended example, there is no special criminological theory of street offences. All criminological theories explaining criminal behaviour generally have to be involved in an explanation of the single criminal behaviour on streets. Therefore, the available data on these criminal events should be reviewed from the perspective of all criminological theories.

The main result of the overview of data is discovery of the patterns of manifestations characteristic for some criminological theory. This theory provides a promising explanation of the criminal event.

This promising explanation should be subjected to further in-depth study. It should provide additional information that will increase or reduce probability that it is the proper explanation.

This additional information also can intensify manifestations of other patterns, representing other explanations.

These intensified patterns can also be those that already were examined and qualified as unpromising.

It means that in course of crime data exploration no reduction of the list of possible explanations occurs. Crime data are analysed in the perspective of all possible criminological theories, during the whole exploration.

This suggests a specific schema of crime data exploration. It is a "serial checking and returning of explanations proposed by all criminological theories". An earlier rejected explanation is included into number of those that can be recognised as promising later. This schema consists of two blocks representing key stages of DM: General and special ones.

For the general one, all data should be overviewed in the perspective of explanations proposed by all criminological theories and the most promising explanation should be found. On the special one, the in-depth exploration of the promising explanation should be done.

This cyclic interchange of general and specific exploration of data should be continued until the correct (*proper*) explanation is found. The term "proper" can have two different meanings. (1) For different reasons (they will be discussed later), it can be admitted to be the true one, really determining the criminal event. In this case, the term "proper explanation" just means that no further search is necessary. (2) An explanation can be admitted as not the final one, but promising enough to be worth further exploration by stronger, though more laborious, EM. Both possibilities will be discussed in greater details later.

All of this supports the scheme of the interactive cyclic algorithm. Its basis is a continuous cyclic interchange of general review of all, and specifically the in-depth, exploration of a part of the data.

This scheme meets the most important specifics of crime data, the opportunity of their exploration in the perspective of all criminological theories. An analyst (criminologist), armed with knowledge of all modern criminological theories examines their data. The essence of this examination is the continuous alternation of two perspectives: (1) General (global) when all data is examined from the point of view of all criminological theories and (2) Special, the in-depth view of only one theory. It means that the analyst continually returns to the general criminological perspective of their data. Armed with their knowledge of criminological theories and considering particularities of their data, the analyst considers probable manifestations of different theories in their data and checks, whether any of these manifestations can be observed. The following in-depth examination of the noticed manifestation also supports this general criminological perspective by mining of additional information and its addition to that which is already available. This facilitates the following general examination, enhancing the chances of the manifestation of the proper theory.

It can be supposed that such a scheme of data exploration gradually increases the probability for the true explanation to be detected.

It can also be supposed that such an interactive process reduces subjectivity of crime data exploration. As shown before, DM relies to the great extent on the analyst's experience and intuition. This means that the proper explanation can go unnoticed for numerous psychological effects are able to mask its manifestations. This is especially true for crime data considering the variety of their possible interpretations.

The described interactive scheme of crime data processing can prevent these effects to a significant degree. It can do it, first, increasing amount of information that can be used for recognition of the proper explanation. Second, it does it many times, iteratively modifying *manifestations* of the proper theory in the given data. Being transformed and supplemented, the pattern of manifestation that has not been noticed on the previous cycle can get more distinctive and be noticed on the following one.

33.3.6. *Interactive Crime DM Algorithm*

33.3.6.1. *Structure of the crime DM algorithm*

All of these considerations on the specifics of a crime, on its explanation with criminological theories and exploration of crime data, provide the basis for designing the algorithm for crime DA.

The algorithm consists of two general parts: interactive and completion ones (Fig. 33.1). The first one includes two interacting blocks: the general and the special one. The second includes operations and decisions to be taken when the proper explanation is considered to be found.

33.3.6.2. *Part one. Interactive exploring of data*

The first part includes interaction of the two main stages of DM:

1. The general overview of all data from the perspective of all possible explanations aimed at discerning the most promising one.
2. The in-depth exploration of the promising explanation. As opposed to the general overview, in this stage only a small part of data is examined. The aim of this stage is to receive further information on this part of data and to assess the most promising explanation.

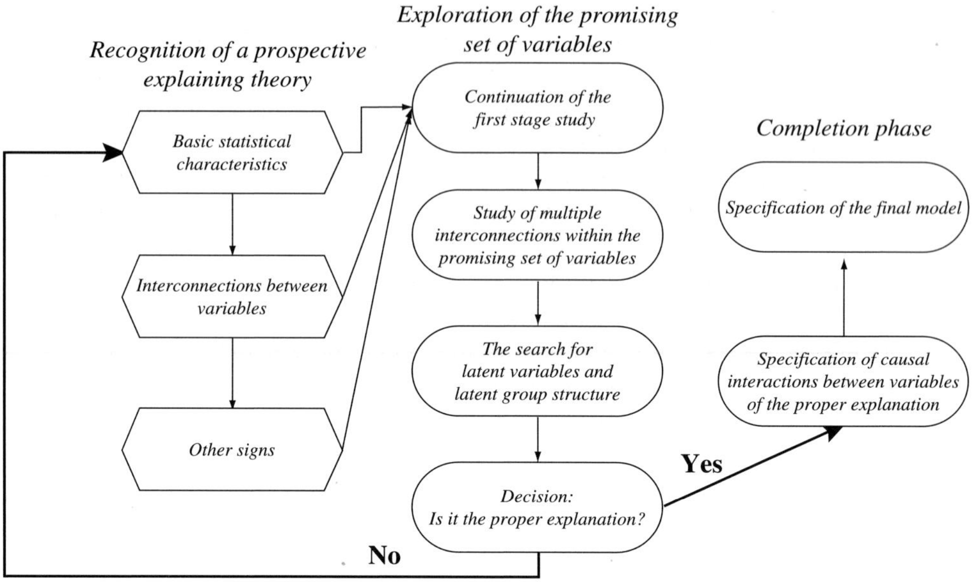

Fig. 33.1. The crime data-mining algorithm.

33.3.6.2.1. General stage. Recognition of the promising explaining theory

In this stage all parameters of the data are examined. The first task in DM is to identify and determine the data possible manifestations of any theory.

Statistical and other procedures are used in this stage to assist in doing this and provide signs and patterns of manifestation.

The main classes of such signs are: (See Fig. 33.1).

1. *Basic statistical characteristics.* These are basic statistical characteristics of all variables. Such characteristics can be averages, modes, medians, frequencies, percents, etc. (Hand and Smyth, 2001) [63].
   They provide the first opportunity to notice manifestations of an explaining theory. So, in the example of street offences, it was high averages predicted by theory variables that enabled the analyst to recognise its manifestation.
2. *Interconnections between variables.* These characteristics can be coefficients of correlation, contingences, association, etc. (Hand and Smyth, 2001). Their calculation and examination is usually the next step of the general exploration. The usual way to present these coefficients of association is the matrix of interconnections between all variables. Its survey provides the next opportunity to recognise pattern of interconnections predicted by some theory, after the examination of basic statistical characteristics.
   So, in the case of street offences determined by criminal sub-culture, not only averages of this sub-culture parameters (aggressiveness, masculinity orientations,

belonging to gangs, rejection of middle class values, etc.), but also significant correlations between them can be characteristic and be noticed by the analyst. Thus, if manifestations of some theory were not recognised surveying basic statistics of variables, they can be noticed exploring their interconnections. This is the second opportunity to recognise this promising explanation.

3. *Other signs.* Beside basic statistics and interconnections many other features of data can be manifestations of causal factors predicted by the relevant theory. Many of them can appear when processing data for quite different aims. Such signs can be found by investigating the structure of data, reducing their dimensionality, searching for clusters, etc. So, in the example of street offences, the cluster analysis can reveal a group of population strikingly different from the rest of the population, with characteristic traits of criminal sub-culture and active involvement in street offences.

33.3.6.2.2. Specific stage of DM. Exploration of a promising explanation

The aim of the first stage was to identify a promising set of variables. As shown above, an outcome of this stage is highly dependent on personal factors, the analyst's power of observation, attentiveness and expedience. Therefore, the aim of the second stage of DM is to get further evidences showing that the choice was correct (or to state that it was wrong).

Contrary to the first stage, on which the total of data was explored, the second stage is concentrated on manifestations of the promising explanation.

The special stage consists of two sub-stages.

First sub-stage. Continuation of the first stage study.

There is an assumption that a promising explanation was noticed on the beginning of the first stage. For example, this can occur immediately after examining the basic statistical characteristics. Then the first step on the specific stage can be the completing of the following general stage actions, calculations of interconnections and multiple connections for variables involved in the promising explanation.

The aim of this is to ascertain whether observed coefficients confirm the analyst's predictions of the promising theory.

In the example on street offences, suppose that a promising explanation was identified on the general stage by noticing characteristic pattern of averages. Then a specific stage can be started with calculation and examination of correlations between the related variables.

Thus, the outcome of the first sub-stage is the further increase or decrease of certainty that the set under consideration is the true one.

The second sub-stage is a further study of the promising explanation using the study of multiple interconnections within the promising set of variables.

The next step can be calculation of multiple interconnections between variables of the promising set. The aim is to explore how much they confirm the prediction of the theory under consideration.

This can be done by using different measures of joint interconnections between three and more variables [64, 1]. Canonical, multiple, partial correlations are some of these measures [65].

This provides an opportunity to reveal an interconnection between two sets of variables. The first set can be characteristics of a criminality under study, the second — variables representing the set of explaining factors. In our example (street criminality), it can be a set of variables representing criminal behaviour (its frequency, heaviness, caused damage, etc.), on the one side, and variables representing a possible explanation, on the other.

The coefficient of multiple correlations is a measure of interconnection between one variable and the set of others. The first one can be the data on number of crimes, the second variables can represent a possible explanation. In the street offences example, an increase in every variable of criminal sub-culture enhances one's total affiliation in this sub-culture, which, in turn, causes an increase in all related variables. Thus, a high value of related multiple correlation (or any other indicator of multiple interconnection) can be one more manifestation of the explaining theory. Again, it is a new opportunity to recognise this manifestation or (if it has been already recognised) a new reason to believe that its explanation is valid.

Criminological exploration of data extensively uses procedures measuring multiple interconnections. Many criminological variables can be measured only by dichotomy (yes-no, like sexes), nominal (qualitative categories, like one's nation), ordinate (ranges of intensity) scales, etc.

This makes popular multiple measures and mathematical models able to operate with such variables.

The most popular in crime data exploration is logistic regression. For theory see Ref. 66, for examples of use for criminological DA see Refs. 67, 68.

The third sub-stage is the search for latent variables and latent group structure.

The factor or the set of factors causing a criminality most often are latent. However, they cause visible interconnections between observable variables.

The third sub-stage procedures provide opportunities to see whether manifestations of these latent variables confirm predictions of the theory.

Most criminological theories are based upon rather simple set of variables. Often it is one (e.g. one's involvement into criminal sub-culture in sub-culture theory) or few variables (e.g. one's adherence to norms-goals and to norms-frames in R. Merton theory).

The structure of their action is also rather simple. A latent factor directly causes its observable manifestations: criminal behaviour, personality of perpetrator, actions of victims, etc. (See Fig. 33.2)

Statistical procedures most often used to discover such latent variables are so called "common factor analysis". It is an approach providing an opportunity to explain the variance of a set of variables as effect of some latent factors [69].

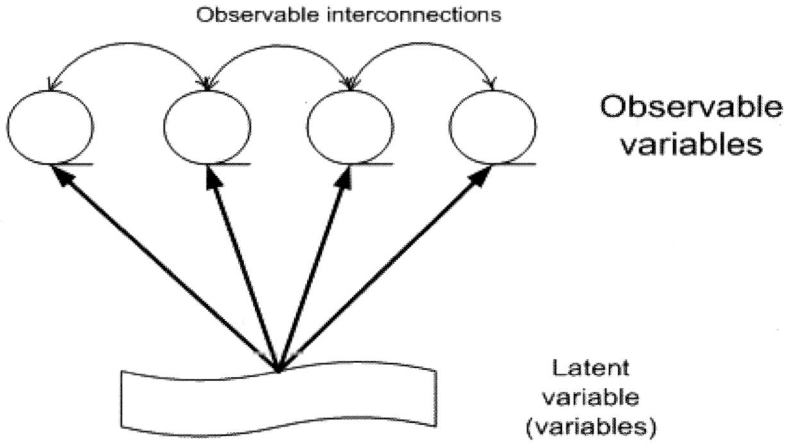

Fig. 33.2. One latent variable model.

Use of the factor analysis provides an opportunity to verify:

1. whether interconnection between observable variables can be accounted for and the effect of one or several latent factors;
2. are these latent factors similar to those predicted by theory;
3. what part of general variation of observed variables can be explained by the action of latent variables.

If they are able to explain a considerable part of this variation, it is strong evidence that real interaction confirms the theory.

Very often the interconnection between observable and latent factors predicted by a criminological theory is much more complicated. For example, the "broken window" theory supposes a chain interaction between observable and latent factors. In this chain, "broken windows" (outward appearance of a street and its buildings) causes two latent factors: it stimulates one's "destructive feelings" (person sees street as a place in which destruction is a proper behaviour) and it weakens the social control (a person perceives a street as a place where antisocial, aggressive actions are permitted). These two latent factors together both stimulate further destruction (new "broken windows") and stimulate offences (Fig. 33.3).

An appropriate solution for such a complex interaction is a structural equation model. These procedures provide an opportunity to verify:

1. whether interconnection between observable variables can be accounted for by a complex model which is suggested by some criminological theory;
2. whether the latent factors and the scheme of their interaction with each other and with observable variables are similar to those predicted by the criminological theory under consideration;

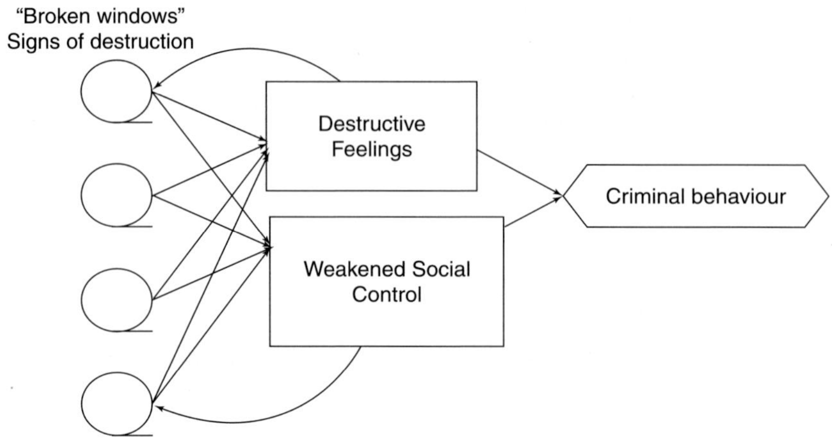

Fig. 33.3. Complicated model including latent variables.

3. what part of general variation of observed variables can be explained by the supposed model of interaction.

If they are able to explain a considerable part of this variation, it is strong evidence that the real interaction confirms the theory. For a theory and experience in its use in crime analysis see Refs. 70, 71.

K-means clustering as a method for discovering a latent variable.

A great part of criminological theories propose models based upon some group or classes. It can be typologies of offenders, settlements, situations, etc. A good example is R. Merton's theory. According to this theory, a person can select four different solutions for tension between the norms-goals and norms-frames. Every selection causes quite different development of one's personality, life situation, and even offences committed. This means that empirical data on personal traits, one's situation and violations of law should reveal four types of offenders, very different in many respects. K-means clustering can be used to test whether such types really exist. (See Ref. [72] for theory, Refs. [73–75] for example of use). It is especially suitable because it presuppose (states in advance) the number of classes that supposed to be revealed. Discovery of four classes with traits similar to those predicted by R. Merton's theory could mean that this theory could be used to explain the criminality under consideration.

This provides a decision whether the set under examination is really promising.

By the end of any special stage a lot of new information on the set of variables under examination is gathered. This information changes the primary probability that this set is the proper one and can explain the criminal event under consideration (for example, can explain the crucial part of its variation). This probability can either increase or decrease.

This probability provides a basis for crucial (and, in fact, highly intuitive) "either-or" decisions and whether this probability is high enough to stop the search for new sets of variables. This decision can be taken on any sub-stage of the special stage. It means that any moment the analyst can decide that they can stop both examination of the set and the search for new sets of variables. It also means that they can proceed to final stage of DM.

33.3.6.2.3. Interaction between general and specific stages of DM

Nature and problems of the interaction. The crime DM is highly interactive. From the general stage, the exploration of all data and recognition of a promising set of variables. Exploration of these few variables either enhances the certainty that they are just proper variables or reduces it (See Fig. 33.1). In the first case ("Yes"), DM passes to final verification and specification of the model. The interaction between general and special stages of DM stops.

In the second one ("No"), it returns to the general stage for new exploration of all data and search of a new set of promising variables. The interaction between general and special stages of DM continues.

The most frequent problem situation is when no set of variables can explain the criminal events under consideration.

The most common explanation is a *poor quality of data.* As mentioned, the data used by DM has often been gathered for some other purpose. So, official crime statistics are a by-product of processing of crimes by institutions of criminal justice. Therefore, defects in functioning of these institutions cause defects in these data. Bureaucratic carelessness causes gaps and distortions, bureaucratic formalities causes "blindness" to new, unusual or different data, the selfish craving for best prestige brings embellishments, and the institution's policy stimulates tendentiousness of a "registration policy".

All of these cause a "noise" and distortions in data. These noises and distortions are able to conceal manifestations of the proper theory.

Therefore, the revision of data is the first reaction to a situation when no set of variables can explain the criminal events.

The second reason is the lack of a proper explanation among suggested ones. As mentioned above, modern criminology has great number of criminological theories and each one can provide its own explanation of a criminal event. Therefore, it is possible that the proper one was not considered when examining data, especially during the general stage of DM.

One more situation is that reasons behind the criminal event are not described by known criminological theories. As mentioned, the "elephant" — criminal behaviour is much more than only a combination of a pole, barrel and rope.

In this situation an analyst has to turn into a criminologist-researcher and to propose, to develop, and to substantiate a new criminological theory.

The fourth possible reason is the "human factor" involved in DM. As shown above, the core of DM — the general examination of all data with the aim of recognising the most promising set of factors by its manifestations — is fraught with subjectivity. This subjectivity can be caused by psychological factors (perception psychology), the analyst's bias, their theoretical preferences, specific life experiences, thoroughness, conscientiousness and diligence.

As usual, remedies are the analyst's commitment, the most possible transparency of their decisions, and an independent re-verification of their results by other analysts. DM proposes many additional tools to facilitate the finding of the proper set of data. The most powerful of them are those providing visualisation of data and results of their processing.

The reasons for which the search for the proper explanation is unsuccessful can be very different.

One of the many possible reasons can be a "locality effect". It is a situation when different parts of data should be explained by different theories. A theory which is generally correct for the total of data can be not correct for their specific part.

In the street violence example, a promising set of variables can be true for the whole city but wrong for one of its parts. For example, R. Merton's theory can be adequate when explaining criminal events in some city in terms of tensions between norms-goals and norms-frames. However, slums of the city can be populated with a quite different population (migrants, socially maladjusted, etc.), flooded with criminal gangs and, therefore, quite different regularities can be present. If this part of the city is large enough, it can blur the effects predicted by R. Merton's theory. After this part of city is excluded from the data, the theory will start working.

This means that a quite different explanation has to be found for this part of the city. It can be, for example, Miller's criminal sub-culture theory, mentioned above.

"Locality effect" is a dangerous and insidious enemy of an analyst. Often enough no signs warn of it. Even if suspected, it is difficult to find which part of data can cause it. How a locality phenomenon in data can be searched for?

Of course, the practical knowledge of the object and data is crucial. One of the statistical procedures that could be useful is cluster analysis, especially nearest k-neighbour and quickly developing "self-organising maps" procedures (For theory and application see Refs. [76, 77, 47] (Schroeder *et al.*, 2003). Both are able to find unevenness in distribution of data and ways to discover a highly distinctive part of the data within them.

33.3.6.2.4. Visualisation of DM

Visualisation includes procedures to support the recognition of a promising set of variables.

DM usually processes large amounts of data. Many hundreds of variables are quite usual in DM, especially in its general stage. Processing of these amounts of data often produces many hundreds indicators or coefficients. A good example is the most popular tool for exploring of interconnections between all variables and their correlation (or other coefficients) matrix. The number of coefficients in such a matrix increases in a quadratic mode with the growth of the matrix size. For example, a matrix reflecting interconnections between 20 variables (which is quite modest viewing over all data) contains 400 coefficients, a 30 variable matrix contains 900 of them.

There is no doubt that this amount of coefficients makes the risk of the manifestations of the proper set of variable going unnoticed very high. There is a great need of tools able to make these manifestations more visible. Therefore, visualisation is of great importance for efficient DM.

DM proposes several groups of tools facilitating visualisation. They are "secondary procedures" which process results of previous processing. These results can be basic statistical indicators, interconnection coefficients or any other data; and any others.

1. *Filtering.* These are methods to "remove noise". Of all results they eliminate those which certainly are of no interest and just complicate identification of the promising set. The most common example of these techniques is the "exploratory matrix" in which all insignificant correlations of all correlations with value below some level are eliminated. This technique greatly reduces the number of coefficients to explore. In this way, recognition of strong connections becomes more possible even in situation of large-size matrixes, in which many variables are included.
2. *Analytical grouping.* This is the selection and grouping of results supposed to be the direct or indirect manifestation of a common phenomenon behind them. For example, when exploring crime data it can be a group of indicators manifesting intensity of criminal activity in a region. This group could include number of offences, the proportion of serious ones, the damage caused by offences, etc. Grouping and joint observing can facilitate discovery of patterns that include crime intensity. Another example could be group of averages or other indexes reflecting aggressiveness. It can include such variables which are supposed to be manifestations of offender's aggressive feelings: number of violent offences, number of other violent actions, senseless and unmotivated demolition of property, etc. The analytical grouping means joint observation of such interrelated groups of variables.
   The next step in this direction is to substitute these groups of related variables with their "representative" — the scale. This can be done using tools proposed by the modern representational measurement theory [78, 79] (Canady and Babcock, 2007).

3. *Change of modality.* Both psychology and everyday experience show that changes in the way in which some data are presented can facilitate their exploration and helps to identify important details. Changes of data modality include processing of data from digital form into graphical form (and vice versa), the use of different graphic or digital presentation forms. Such transformations are able to "de-conjure" the perception of analyst, to make them able to notice details which used to escape their notice [80].
4. *Initial general reduction of the dimensionality of data.* As seen above, multi-dimensionality is one of the greater hindrances in identification of manifestations of a theory in data. The pattern of values and connections between the variables and the characteristics for manifestation of a specific theory gets lost among many dozens or hundreds other variables, their values, coefficients, etc.

The reduction of dimensionality means transfer from multi-dimensional to a one or few dimensional description of data.

The general reduction is a reduction of dimensionality *of all* variables of data.

Such a crucial reduction of all data on the very beginning of their exploration can look very tempting to the analyst. Seemingly, it can spare them from the abundance of excessive dimensionality and visualise all data using only a two or three dimension model. This seems be able, like a magic wand, to solve all problems caused by great dimensionality: data can be visualised directly, "localities" can be revealed, clusters of variables will be discovered, etc.

However, dimensionality reduction is also one of the most problematic and dangerous moves.

Its weak point is the dependence of its accuracy from dimensionality of data. Usually every increase in dimensionality is followed by accelerated demand on increase of number of observations (Hand, 2001) [81]. Popular analytics' wisdom says that the number of observation should be at least ten times greater than dimensionality. In fact, this "rule of thumb" quickly loses its accuracy with the further increase of the number of dimensions, typically exceeding 10–15 (Hand, 2001) [81]. The further increase of dimensionality brings rapid growth in inaccuracy of the reduction results.

33.3.6.3. *Part two. Completion Phase*

33.3.6.3.5. Proper explanation — Specification of causal interactions between variables

After some set of variables was identified as determining the criminal events, the causal interactions have to be specified.

DM can show interconnections between variables, however, it is not able to reveal their causal nature, i.e. to identify which of the connected variables is the reason and which one is its consequence.

However, discovery of causal relations between variables is often highly desirable. It is necessary every time the outcomes of DM are used to affect the criminal event. In example of street offences, even the strongest correlation between lighting of street and criminality does not prove causal relation between them. This connection does not mean that it is just poor lighting which causes increase in criminality. It is quite possible that both the lighting is the worse and criminality is higher in the poorest regions of the city. Therefore, there is no basis to believe that improving of lighting can reduce criminality. Paradoxically, it is quite probable that in these regions improved lighting can facilitate criminality making easier offences.

Therefore, after stating the most proper explanation, DM has to pass on the baton to methods able to determine the directions of causality.

Interactions supposed to be causal have to be verified in an experiment. The latter has to be arranged according the methodology of this method, including the organisation of an experimental and control groups, manipulating with the independent variable supposed to be a cause, measuring consequences of these manipulations for dependent variable.

Therefore, in many cases the experiment is the inevitable continuation of DM.

Such an experiment is the natural successor of DM. It focuses upon the most important outcome of DM — the proper explanation and examines whether relations between corresponding variables are causal ones.

The succession of DM by experimental examination of the proper explanation seems to be the golden rule in DM. It should be done always when outcomes of DM are supposed to be used as a basis for any practical actions.

However, there are also several situations in which following experimental research is advisable but not indispensable.

1. *Detection of the hidden criminality. Crime profiling. Research of typical clusters in crime committing.* In all these applications, the accent is set on revealing some patterns of variables. They can be signs of illegal bank interactions, of events preceding bank robbery, or of criminal organisation activities. In all these cases much less emphasis is placed on the reasons causing them.
   This is so for several reasons.
   In many cases, it is presupposed that these reasons, and therefore the ways in which this events should be controlled, are known. So, bank analytics search to detect illegal interactions to eliminate them. They are not so much interested on reasons of this criminal behaviour. They know what to do after such interactions are discovered.
   Police departments fighting against organised criminality has tools and legal power to fight a criminal organisation after it is detected. The problem is to identify it. It is quite natural that they are interested in detection of hidden activities and not in revealing their reasons. Such an approach is characteristic for so called "symptomatic approach". In the examples above, both the bank and the police department focus on suppression of criminality and not on clarification

of its reasons. Prevalence of such an approach is its simplicity. Its weakness is its short-sightedness (it removes only the given criminal event without preventing such events in a long-term perspective).

Its second weakness is the lack of feed-back and efficiency control. Without exact knowledge on causes of an event it is impossible to say whether any actions taken have been really successful. For example, if these actions were followed by a reduction in criminality, it could happen on its own, spontaneously. It is even possible that spontaneous reduction in criminality occurred despite our actions. It could happen if actions supposed to reduce criminality actually stimulate it. Only causal knowledge could show the real effect of these actions.

2. *Situations in which the possibility to use experimental approach is restricted.* An integral part of experimental approach is manipulation with people, their behaviour, living conditions, etc. When dealing with countries, regions, cities, districts this faces great legal, moral, political, psychological, etc. problems. For example, to examine the efficiency of incarceration some offenders (an experimental group) should and other ones (similar to the first one but the control group) should not be incarcerated. On the one hand, such a procedure strongly contradicts the law and morality. However, on the other one, there is no other way to get any certain knowledge on causal effect of incarceration. In such a situation, if possible, manifold partial substitutions for the full ("controlled") experiment are used [82, 83].

   Another choice is just to presume causal relations. It goes without saying that efficiency of this strategy depends on accuracy of these presumptions.
3. *Distribution of offences in time and/or space* ("*crime mapping*"). The DM can be the final stage of DA when simple "geographic" knowledge on different sides of criminal behaviour is needed. All that is required to be known in this situation is what, how often, how intense and where it happens. In such a case, it is supposed that reasons of criminality on some space (country, region, city, etc) are well known and are similar in all its parts. It is presupposed that the only difference between these parts is intensity of criminality.

Such an approach is also tightly connected with symptomatic view of criminal behaviour. It is a situation in which crime control agencies do not try to address the real reasons of local differences in criminality. All what is tried to do is to discover most affected areas and to concentrate efforts on them.

Again, actions addressing only "symptoms" — local differences in manifestations, without considering their reasons, are quite likely to be inefficient and/or their effects to be instable [82].

33.3.6.3.6. Specification of the final model

This is the final stage. By this stage the searched set of variables has been found, the connections between them have been revealed and their nature and direction has been specified.

In the final stage, all this has to be incorporated into the final statistical model that, on the one side, generalises outcomes of data exploring and, on the other one, is to be a basis for practical use. Practical uses include, correcting crime policy, to design crime control measures and crime prevention programs. Usually, this model is presented in shape of equation or algorithm that formalises relations between independent factors (those causing changes in criminal behaviour) and dependent ones (parameters of this behaviour). It should be a model both precise and convenient enough to be successfully used for practical or scientific needs.

Often enough such a model has been developed on previous stages on DM, especially on the special stage when examining multiple interconnections between variables.

Two interconnected problems are crucial in the course of the final tweaking of the model, and its adjustment for the further use. They are exactness and simplicity of the model.

Exactness of a model is its ability to forecast exact values of dependent variables (parameters of criminal behaviour). The real meaning of the dependent variables is compared with those forecasted by the model.

Modern statistical tools provide the opportunity to reach every level of precision of the model. This is especially true for most powerful neural network methods. The latter brought significant progress in modelling complicated curvilinear interconnections between many variables [84]. Using these methods the analyst can reach any degree of adjustment of their model to their data.

On the other side, there are some restrictions to such adjustment.

1. *Precision of the model. Adjustment of the model to data.* There exists a curvilinear correspondence between local precision of a model (its adaptation to data on which it was built) and its general precision. It can be seen that with increasing local precision the general precision increases only to a certain point. After this point, increased local precision is followed by a decreased general one [81]. This means that it is not a good idea to try to reach a maximum possible degree of preciseness of a model. Paradoxically, a precise model can be better than a very precise one. Therefore, an analyst should "stop half-way" through their model. However, most often the point to stop can be chosen only intuitively.
2. *Complexity of the model.* Most often an increase in precision of a model can be achieved by its complication and sophistication. It can be done using sophisticated functions, additional variables, etc. Again, statistical research showed that often an increase in complexity also brings a progressive increase in

demands on data, especially to the number of observations. Therefore, also here "simplicity wins" and some "optimal" complexity of a model should be found [81], pp. 18–22.

3. *Practical considerations.* Greater complexity of a model makes it less "foolproof". The more complex a model is the more opportunity for mistakes it provides. The more simple it is, the less demanding it is for qualification, expedience and diligence of people applying it, the easier is to control its use.

   The proper complexity level, like many other decisions in DM, is found amongst these and many other considerations.

References

1. G. S. Berry and G. Linoff. *Data Mining Techniques: For Marketing, Sales, and Customer Relationship Management* (John Wiley & Sons, New York, Chichester *et al.*, 2004).
2. D. Hand, H. Mannila and P. Smyth. *Principles of Data Mining* (Bradford Book — The MIT Press, London, 2001).
3. P. Giudici. *Applied Data Mining: Statistical Methods for Business and Industry* (John Wiley, 2003).
4. W. Kloesgen and J. Zytkow (eds). *Handbook of Data Mining and Knowledge Discovery* (Oxford University Press, 2002).
5. U. Fayyad, G. Piatetsky-Shapiro and P. Smyth. The KDD process for extracting useful knowledge from volumes of data, *Commun. ACM* **39**(11) (1996) 27–41.
6. G. C. Oatley, B. W. Ewart and J. Zeleznikow. Decision support systems for police: Lessons from the application of data mining techniques to 'Soft' forensic evidence, *Artif. Intel. Law* **14**(1) (2006) 35–100.
7. A. F. Siegel. *Statistics and DA: An Introduction* (Wiley & Sons, New York, 1988).
8. D. T. Lykken. The causes and costs of crime and a controversial cure, *J. Pers.* **68**(3) (2000) 559–605.
9. M. T. French, K. E. McCollister, P. Alexandre, D. D. Chitwood and C. B. McCoy. Revolving roles in drug-related crime: The cost of chronic drug users as victims and perpetrators, *J. Quant. Criminol.* **20**(3) (2006) 217–241.
10. B. L. Benson, I. S. Leburn and D. W. Rasmussin. The impact of drug enforcement on crime: An investigation of the opportunity cost of police resources, *J. Drug Issues* **31**(4) (2001) 989–1006.
11. J. J. M. Van Dijk, R. Manchin, J. Van Kesteren, S. Nevala and G. Hideg. *The Burden of Crime in the EU. Research Report: A Comparative Analysis of the European Crime and Safety Survey* (*EU ICS*) (2005). Accessed at [1] January 28th http://www.vartotojai.eu/magazine/6/5.pdf.
12. M. Maguire. Crime Data and Statistics, *Oxford Handbook of Criminology*, 4th ed. eds. Maguire, Morgan and Reiner (Oxford University Press, 2007), pp. 241–301.
13. FBI, Uniform Crime Reporting (UCR). Program (2006) In http://www.fbi.gov/ucr/cius2006/offenses/index.html.
14. Bundeskriminalamt. *Police Crime Statistics 2006 — Federal Republic of Germany.* http://www.bka.de/pks/pks2006ev/pcs_2006.pdf.
15. M. F. Aebi, K. Aromaa, B. A. Cavarlay, Gordon Barclay, G. Beata, H. von Hofer, V. Hysi, J.-M. Jehle, M. Killias, P. Smit and C. Tavares. *European Sourcebook of Crime and Criminal Justice Statistics* (2006) http://www.europeansourcebook.org/esb3_Full.pdf.

16. United Nations, *National Accounts, Statistics: Analysis of main aggregates and detailed tables, 2000* (United Nations Publ., New York, 2002).
17. K. Aromaa, S. Leppä, S.i Nevala and N. Ollus (eds.), *Sixth United Nations Survey on Crime Trends and Criminal Justice Systems*, Helsinki: European Institute for Crime Prevention and Control, affiliated with the United Nations (HEUNI) (2003).
18. M. Watson. Environmental offences: The reality of environmental crime, *Environ. Law Rev.* **7**(3) (2005) 190–200.
19. L. E. Korsell. Big stick, little stick: Strategies for controlling and combating environmental crime. *J. Scand. Stud. Criminol. Crime Prev.* **2**(2) (2001) 127–148.
20. T. L. Leap. *Dishonest Dollars: The Dynamics of White-Collar Crime* (Cornell University Press, Ithaca, 2007).
21. J. T. Kaariainen. Trust in the police in 16 European Countries: A multilevel analysis, *Eur. J. Criminol.* **4**(4) (2007) 409–435.
22. H. J. Schneider. Victimological developments in the world during the past three decades (I&II): A study of comparative victimology, *Int. J. Offender Ther. Comp. Criminol.* **45**(4) (2001) 449–468, 539–556.
23. J. van Wilsem. Criminal victimization in cross-national perspective. An analysis of rates of theft, violence and vandalism across 27 countries, *Eur. J. Criminol.* **1**(1) (2004) 89–109.
24. J. N. Kesteren, P. van Mayhew and P. Nieuwbeerta. Criminal victimisation in seventeen industrialised countries: Key-findings from the 2000 international Crime Victims Survey, The Hague, Ministry of Justice, WODC.
25. F. Xitao, B. C. Miller, P. Kyung-Eun, B. W. Winward, M. Christensen, H. D. Grotevant, R. H. Tai. An exploratory study about inaccuracy and invalidity in adolescent, *Self-Report Surveys, Field Meth.* **18**(3) (2006) 223–244.
26. C. Emsley. The history of crime and crime control institutions. *Oxford Handbook of Criminology* (Oxford University press, 2002), pp. 203–230.
27. K. Pease. What to do about it? Lets turn off our minds and GIS, *Mapping and Analysing Crime Data — Lessons from Research and Practice*, eds. A. Hirschfield and K. Bowers (Taylor and Francis, London and New York, 2001), pp. 225–237.
28. G. C. Oatley and B. W. Ewart. Constructing a Bayesian belief network to determine the likelihood of burglary, *Proceedings of the Fifth International Conference on Forensic Statistics* (*ICFS5*), Isola di San Servolo, Venice, Italy, 2002.
29. R. M. Leary. Home office policing & reducing crime unit, *Evaluation of the Impact of the FLINTS Software System in West Midlands and Elsewhere*, Home Office, London.
30. A. Hirschfield. Decision support in crime prevention: DA, policy evaluation and GIS, *Mapping and Analysing Crime Data — Lessons from Research and Practice* (Taylor and Francis, London and New York, 2001), pp. 237–269.
31. R. Adderley and P. B. Musgrove. Data mining at the West Midlands Police: A study of bogus official burglaries, *BCS Special Group Expert Systems*, ES99 (Springer-Verlag, London, 1999), pp. 191–203.
32. C. McCue. *Using Data Mining to Predict and Prevent Violent Crimes* (2004). Available at http://www.spss.com/dirvideo/richmond.htm?source=dmp age&zone=rtsidebar.
33. I. D. Wilson, J. Corcoran and J. A. Ware. Predicting the geo-temporal variations of crime and disorder, *Proceedings of the Sixth Annual International Crime Mapping Research Conference: Bridging the Gap between Research and Practice*, Denver, Colorado, 2002.
34. D. Williamson, S. McLafferty, P. McGuire, T. Ross, J. Mollenkopf, V. Goldsmith and S. Quinn. Tools in the spatial analysis of crime, *Mapping and Analysing Crime Data —*

Lessons from Research and Practice, eds., A. Hirschfield and K. Bowers (Taylor and Francis, London and New York, 2001) pp. 187–203.

35. D. K. Rossmo. Geographic Profiling, *Offender Profiling — Theory, Research and Practice.* eds., J. L. Jackson and D. A. Bekerian (John Wiley and Sons, 1997), pp. 159–177.
36. D. K. Rossmo. Overview: Multivariate spatial profiles as a tool in crime investigation, *Crime Analysis Through Computer Mapping*, eds., C. R. Block, M. Dabdoub and S. Fregly (Police Executive Research Forum: Washington, DC, 1995), pp. 65–97.
37. D. K. Rossmo. Multivariate spatial profiles as a tool in crime investigation, eds., C. R. Block and M. Dabdoub, *Proceedings of the Workshop on Crime Analysis Through Computer Mapping*, Illinois Criminal Justice Information Authority and Loyola University Sociology Department, Chicago.
38. N. Levine. *CrimeStat: A Spatial Statistics Program for the Analysis of Crime Incident Locations (v 2.0)*, Ned Levine & Associates, Houston, TX, and the National Institute of Justice, Washington, DC, May 2002.
39. S. V. Nath. Crime pattern detection using data mining, *Proceedings — 2006 IEEE/WIC/ACM International Conference on Web Intelligence and Intelligent Agent Technology* (*WI-IAT 2006 Workshops Proceedings*), Article number 4053200, (2007) pp. 41–44.
40. T. Cho, K. Wendy and B. J. Gaines. Breaking the (Benford). law: Statistical fraud detection in campaign finance, *Am. Statistician* **61**(3) (2007) 218–223.
41. N. J. Morley, L. J. Ball and T. C. Ormerod. How the detection of insurance fraud succeeds and fails, *Psychol. Crime Law* **12**(2) (2006) 163–180.
42. M. Agyemang, K. Barker and R. Alhajj. Web outlier mining: Discovering outliers from web datasets, *Intell. DA* **9**(5) (2005) 473–486.
43. V. E. Krebs. Mapping networks of terrorist cells, *Connections* **24**(3) (2001) 43–52.
44. J. Xu and H. Chen. Criminal network analysis and visualization. *Commun. ACM* **48**(6) (2005) 100–107.
45. J. S. Zdanowicz. Detecting money laundering and terrorist financing via data mining, *Commun. ACM* **47**(5) (2004) 53–55.
46. M. Strano. A neural network applied to criminal psychological profiling: An Italian initiative, *Int. J. Offender Ther. Comp. Criminol.* **48**(4) (2004) 495–503.
47. R. Adderley and P. Musgrove. Modus operandi modelling of group offending: A data-mining case study, *Int. J. Police Sci. Manage.* Winter, **5**(4) (2003) 265–276.
48. K. Yokota and S. Watanabe. Computer based retrieval of suspects using similarity of modus operandi, *Int. J. Police Sci. Manage.* **4**(1) (2002) 5–15.
49. O. Ribaux and P. Margot. Inference structures for crime analysis and intelligence: The example of burglary using forensic science data, *Forensic Sci. Int.* **100** (1999) 193–210.
50. J. H. A. Ratcliffe and M. J. McCullagh. Crime, repeat victimisation and GIS, *Mapping and Analysing Crime Data — Lessons from Research and Practice*, eds., A. Hirschfield and K. Bowers (Taylor and Francis, London and New York, 2001), pp. 61–93.
51. E. J. Green, C. E. Booth and M. D. Biderman. Cluster analysis of burglary M/O's, *J. Police Sci. Admin.* **4** (1976) 382–388.
52. J. Corcoran, I. D. Wilson, O. M. Lewis and J. A. Ware. Data clustering and rule abduction to facilitate crime hot spot prediction, *Lecturer Notes Comput. Sci.* **2206** (2001) 807–822.
53. R. Adderley and P. B. Musgrove. Data mining case study: Modelling the behaviour of offenders who commit serious sexual assaults, *Proceedings of the Seventh ACM*

SIGKDD International Conference on Knowledge Discovery and Data Mining, August 26–29, San Francisco, CA, USA. ACM, 2001, pp. 215–220.
54. F. Schmalleger. *Criminology Today: An Integrative Introduction*, 3rd ed. update (Prentice Hall, Pearson, 2004).
55. J. Muncie. *Criminology*, Vol. 1–3, (Sage Publications, London, 2006).
56. R. H. Burke. *An Introduction to Criminological Theory* (Willan Publishing, Portland, 2003).
57. R. Lilly, F. T. Cullen and R. A. Ball. *Criminological Theory: Context and Consequences* (Saga Publications, Thousand Oakes, California, 2007).
58. W. B. Miller. Lower class culture as generating milieu of gang delinquency, *J. Soc. Issues* **14** (1958) 5–19.
59. R. K. Merton. *Social Theory and Social Structure*, Enlarged ed., (The Free Press, New York, 1968 [1949]).
60. R. M. Gottfredson and T. Hirschi. *A General Theory of Crime* (Stanford University press, Chicago, 1990).
61. G. L. Kelling and J. Q. Wilson. Broken windows: The police and neighborhood safety, *The Atlantic Monthly* (1982) March.
62. M. Bach. *Optical Illusions & Visual Phenomena* (2007) //http://www.michaelbach.de/ot/.
63. K. C. Basile, J. Chen, M. C. Black and L. E. Saltzman. Prevalence and characteristics of sexual violence victimization among U.S. adults, 2001–2003, Violence Victims, **22**(4) (2007) 437–450.
64. H. Chen, W. Chung, J. J. Xu, G. Wang and Y. Qin and M. Chau. Crime data mining: A general framework and some examples, *IEEE Comput.* **37**(4) (2004).
65. C. T. Taft, P. P. Resick, J. Panuzio, D. S. Vogt and M. B. Mechanic. Coping among victims of relationship abuse: A longitudinal examination, *Violence Victims* **22**(4) (2007) 408–420.
66. D. W. Hosmer and S. Lemeshow. *Applied Logistic Regression*, 2nd ed. (Wiley, New York: Chichester, 2000).
67. A. I. Cass. Routine activities and sexual assault: an analysis of individual- and school-level factors, *Violence Victims* **22**(3) (2007) 350–367.
68. T. I. Herrenkohl and R. Kosterman. Youth violence trajectories and proximal characteristics of intimate partner violence, *Violence Victims* **22**(3) (2007) 259–275.
69. L. Tucker and R. MacCallum. *Exploratory Factor Analysis* (1993) http://www.unc.edu/~rcm/book/factornew.htm.
70. M. A. Kernic and A. E. Bonomi. Female victims of domestic violence: Which victims do police refer to crisis intervention? *Violence Victims* **22**(4) (2007) 463–475.
71. N. Godbout, Y. Lussier and S. Sabourin. Early abuse experiences and subsequent gender differences in couple adjustment, *Violence Victims* **21**(6) (2006) 744–762.
72. J. A. Hartigan. *Clustering Algorithms* (John Wiley & Sons, Inc, New York, 1975).
73. T. Grubesic. On the application of fuzzy clustering for crime hot spot detection, *J. Quant. Criminol.* **22**(1) (2006) 77–105.
74. A. Rasmussen, M. S. Aber and A. Bhana. Adolescent coping and neighborhood violence: Perceptions, exposure, and urban youths' efforts to deal with danger, *Am. J. Community Psychol.* **33**(1/2) (2004) 61–75.
75. E. D. Felix, S. D. McMahon. Gender and multiple forms of peer victimization: How do they influence adolescent psychosocial adjustment? *Violence Victims* **21**(6) (2006) 707–726.
76. T. Kohonen. Self-organizing Maps, 2nd ed. (Springer, 1997).

77. G. Shakhnarovish, T. Darrell and P. Indyk, (eds). *Nearest-Neighbor Methods in Learning and Vision* (The MIT Press, 2005).
78. D. H. Krantz, R. D. Luce, P. Suppes and A. Tversky. *Foundations of Measurement*, Vol. 1–3 (Academic Press, New York, 1971).
79. B. H. Spitzberg and A. E. Veksler. Stalkers: The personality of pursuit: Personality attributions of unwanted pursuers and stalkers, *Violence Victims* **22**(3) (2007) 275–291.
80. L. Wilkinson. *The Grammar of Graphics* (Springer Verlag, New York, 1999).
81. S. Raudys. *Statistical and Neural Classifiers: an Integrated Approach to Design.* (Springer, London, 2001).
82. J. Pearl. *Causality: Models, Reasoning and Inference* (Cambridge University Press, 2000).
83. D. C. Montgomery. *Design and Analysis of Experiments* (John Wiley & Sons, 2005).
84. S. Haykin. *Neural Networks: A Comprehensive Foundation*, 2nd ed. (Prentice Hall, New Jersey, 1998).
85. R. Adderley, J. W. Bond and M. Townsley, Predicting crime scene attendance, *Int. J. Police Sci. Manage.* **9**(4) (2007) 312–323.
86. R. Adderley, J. W. Bond and M. Townsley. Use of data mining techniques to model crime scene investigator performance, *Knowledge-Based Syst.* **20**(2) (2007) 170–176.
87. C. Babcock and M. Kolbasuk. Filter out the frauds, *Inform Week* (995) (2007) 45–50.
88. P. K. Chan, F. Wei, A. L. Prodomidis and S. J. Stolfo. Distributed data mining in credit card fraud detection, *IEEE Expert Intell. Syst. Appl.* **14**(6) (1999) 67.
89. H. Chen, H. Atabakhsh, T. Petersen, J. Schroeder, T. Buetow, L. Chaboya, C. O'Toole, M. Chau, T. Cushna, D. Casey and Z. Huang. COPLINK: visualization for crime analysis, *Proceedings of the Annual National Conference on Digital Government Research*, May 18–21 Boston, MA (2003), pp. 1–6.
90. D. M. Costa, B. Canady and J. C. Babcock. Preliminary report on the accountability scale: A change and outcome measure for intimate partner violence research, *Violence Victims* **22**(5) (2007) 515–533; Crime and Criminal Justice in Europe and North America 1995–1997. Report on the. Criminal Victimization in eleven Industrialized Countries. Key findings from the 1996 International Crime Victims Survey. The Hague: Ministry of Justice, WODC.
91. H. Etorf and H. Spengler. *Crime in Europe. Causes and Consequences* (Springer Verlag, New York, 2002).
92. J. Y. Halpern. Reasoning About Uncertainty, (MIT Press, Cambridge, 2003).
93. G. C. Oatley and B. W. Ewart. Crimes analysis software: 'Pins in Maps', clustering and bayes net prediction, *Expert Syst. Appl.* **25**(4) (2003) 569–588.
94. S. E. Palmer. *Vision Science: Photons to Phenomenology* (MIT Press, Cambridge, MA, 1999) (See Chapter 6).
95. M. Rand. The national crime victimization survey: 34 years of measuring crime in the United States, *Stat. J. UN Econ. Comm. Eur.* **23**(4) (2006) 289–301.
96. J. H. Ratcliffe. Aoristic signatures and the spatio-temporal analysis of high volume crime patterns, *J. Quant. Criminol.* **18**(1) (March 2002).
97. J. P. Stephen. *Criminology*, 3rd ed. (Oxford University Press, 2006).
98. R. Watkins, K. M. Reynolds, R. Demara, M. Georgiopoulos, A. Gonzalez and R. Eaglin. Tracking dirty proceeds: Exploring data mining technologies as tools to investigate money laundering, *Police Practice Res.* **4**(2) (2003) 163.
99. C. S. Widom, A. M. Schuck and H. R. White. An examination of pathways from childhood victimization to violence: The role of early aggression and problematic alcohol use, *Violence Victims* **21**(6) (2006) 675–692.

Index